GEOGRAPHY
THE WORLD AND ITS PEOPLE

GEOGRAPHY
THE WORLD AND ITS PEOPLE

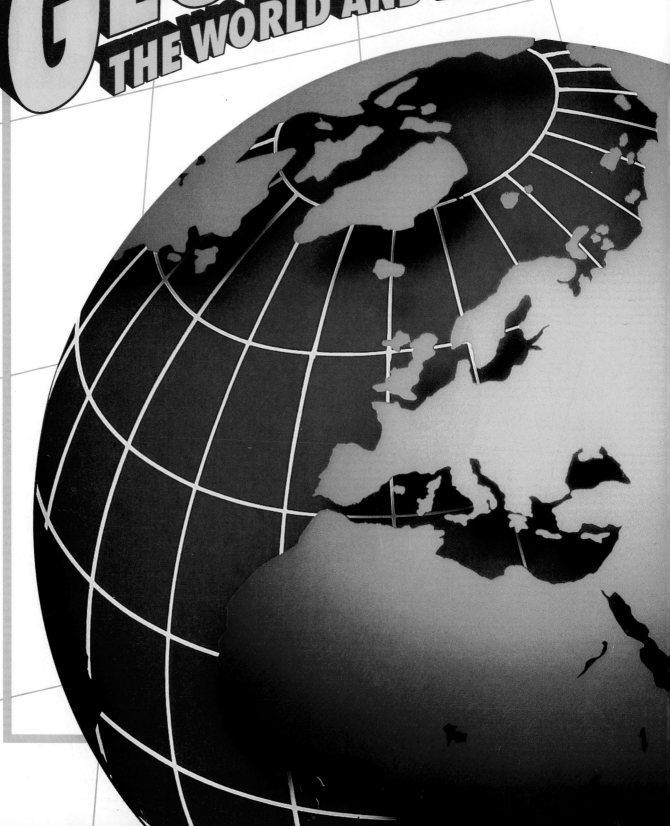

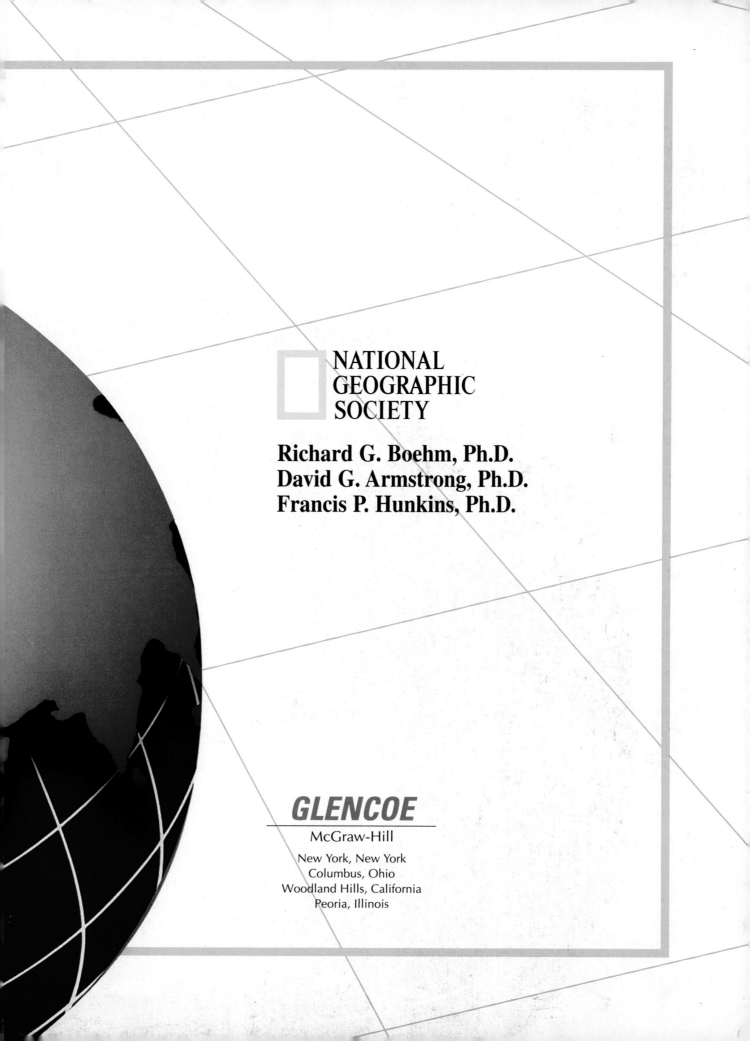

NATIONAL GEOGRAPHIC SOCIETY

Richard G. Boehm, Ph.D.
David G. Armstrong, Ph.D.
Francis P. Hunkins, Ph.D.

GLENCOE

McGraw-Hill

New York, New York
Columbus, Ohio
Woodland Hills, California
Peoria, Illinois

About the Authors

National Geographic Society

The National Geographic Society, founded in 1888 for the increase and diffusion of geographic knowledge, is the world's largest nonprofit scientific and educational organization. Since its earliest days, the Society has used sophisticated communication technologies, from color photography to holography, to convey geographic knowledge to a worldwide membership. The Educational Media Division supports the Society's mission by developing innovative educational programs—ranging from traditional print materials to multimedia programs including CD-ROMS, videodiscs, and software.

Richard G. Boehm

Richard G. Boehm, Ph.D., was one of seven authors of *Geography for Life,* national standards in geography, prepared under Goals 2000: Educate America Act. He was an author of the *Guidelines for Geographic Education,* in which the five themes of geography were first articulated. In 1990 Dr. Boehm was designated "Distinguished Geography Educator" by the National Geographic Society. In 1991 he received the George J. Miller award from the National Council for Geographic Education (NCGE) for distinguished service to geographic education. He was President of the NCGE and has twice won the *Journal of Geography* award for best article. He has received the NCGE's "Distinguished Teaching Achievement" award and currently occupies the Jesse H. Jones Distinguished Chair in Geographic Education at Southwest Texas State University.

David G. Armstrong

David G. Armstrong, Ph.D., is Dean of the School of Education at the University of North Carolina at Greensboro. A social studies education specialist with additional advanced training in geography, Dr. Armstrong was educated at Stanford University, University of Montana, and University of Washington. He taught at the secondary level in the state of Washington before beginning a career in higher education. Dr. Armstrong has written books for students at the secondary and university levels, as well as for teachers and university professors. He maintains an active interest in travel, teaching, and social studies education.

Francis P. Hunkins

Francis P. Hunkins, Ph.D., is Professor of Education at the University of Washington. He began his professional career as a teacher in Massachusetts. He received his masters degree in education from Boston University and his doctorate from Kent State University with a major in general curriculum and a minor in geography. Dr. Hunkins has written numerous books and articles dealing with general curriculum, social studies, and questioning and thinking for students and educators at elementary, middle school, high school, and university levels. As a student of geography, he has traveled extensively to all of the continents, with the exception of Antarctica.

Glencoe/McGraw-Hill

A Division of The McGraw-Hill Companies

Printed in the United States of America.

Send all inquiries to:
Glencoe/McGraw-Hill, 8787 Orion Place, Columbus, Ohio 43240-4027

ISBN 0-02-821485-4 (Student Edition) 0-02-821486-2 (Teacher's Wraparound Edition)

8 9 10 11   027/043   05 04 03 02 01

Consultants

General Content Consultant
NATIONAL GEOGRAPHIC SOCIETY

Multicultural Consultants
Ricardo L. García, Ed.D.
Professor
University of Wisconsin
Stevens Point, Wisconsin

Sayyid M. Syeed, Ph.D.
Secretary General, Islamic Society of North America
Plainfield, Indiana

Introduction to Geography
Theodore H. Schmudde, Ph.D.
Professor of Geography
University of Tennessee
Knoxville, Tennessee

United States and Canada
Frederick Isele, Ph.D.
Former Assistant Professor
School of Education
Indiana State University
Terre Haute, Indiana

Latin America
Frank de Varona
Region Superintendent
Dade County Public Schools
Hialeah, Florida

Europe
Sarah W. Bednarz, Ph.D.
Assistant Professor
Department of Geography
Texas A&M University
College Station, Texas

Russia and the Independent Republics
A. Naklowycz
President, Ukrainian-American Academic Association of California
Carmichael, California

Alfred Bell, Ed.S.
Social Studies Supervisor
Knox County Schools
Knoxville, Tennessee

North Africa and Southwest Asia
Mounir Farah, Ph.D.
Associate Director,
Middle East Studies Program
University of Arkansas
Fayetteville, Arkansas

Africa South of the Sahara
Mary Jane Fraser, Ph.D.
Teacher and Social Studies Curriculum Project Leader
Seattle High School
Seattle, Washington

Asia
Kenji K. Oshiro, Ph.D.
Professor of Geography
Wright State University
Dayton, Ohio

Australia, Oceania, and Antarctica
Marianne Kenney
Social Studies Specialist
Colorado Department of Education
Denver, Colorado

Teacher Reviewers

Charles M. Bateman
Social Studies Teacher and Department Chair
Halls Middle School
Knox County District
Knoxville, Tennessee

Rebecca A. Corley
Social Studies Teacher
Evans Junior High School
Lubbock Independent School District
Lubbock, Texas

Carole Mayrose
Social Studies Teacher
Northview High School
Clay Community School District
Brazil, Indiana

Don Mendenhall
Social Studies Teacher and Department Chair
Coleman Junior High School
Van Buren, Arkansas

Rebecca M. Revis
Social Studies Teacher
Sylvan Hills Junior High School
Pulaski County Special School District
Sherwood, Arkansas

Marita E. Sesler
Social Studies Specialist
Knoxville County School District
Knoxville, Tennessee

Cathy Walter
World Geography Teacher
Colony Middle School
Palmer, Alaska

Contents

CONTENTS

Features

BUILDING GEOGRAPHY SKILLS

Using Technology

MAKING CONNECTIONS

| MATH | SCIENCE | HISTORY | LITERATURE | TECHNOLOGY |

Koala Mom and Joey

Teen Scene

What in the World?

NATIONAL GEOGRAPHIC SOCIETY **Picture Atlas of the World**

Surfing the "Net"

Maps

CONTENTS

Graphs, Tables, and Diagrams

REFERENCE

ATLAS

Contents

ATLAS KEY

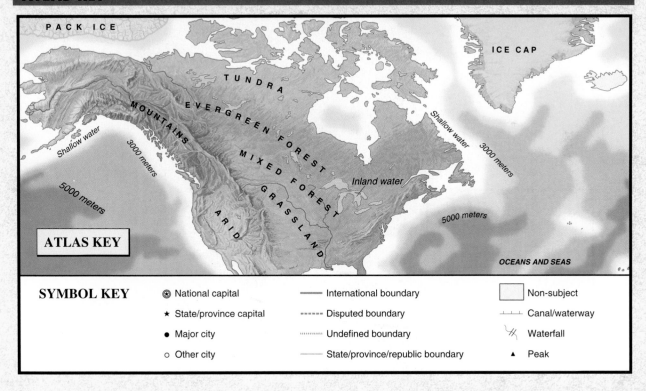

PACK ICE

ICE CAP

TUNDRA

EVERGREEN FOREST

MOUNTAINS

Shallow water

Shallow water

3000 meters

3000 meters

5000 meters

MIXED FOREST

Inland water

GRASSLAND

ARID

5000 meters

ATLAS KEY

OCEANS AND SEAS

SYMBOL KEY			
⊛ National capital	—— International boundary	▭ Non-subject
★ State/province capital	----- Disputed boundary	⊣⊢ Canal/waterway
● Major city	.... Undefined boundary	⤨ Waterfall
○ Other city	—— State/province/republic boundary	▲ Peak

REFERENCE ATLAS

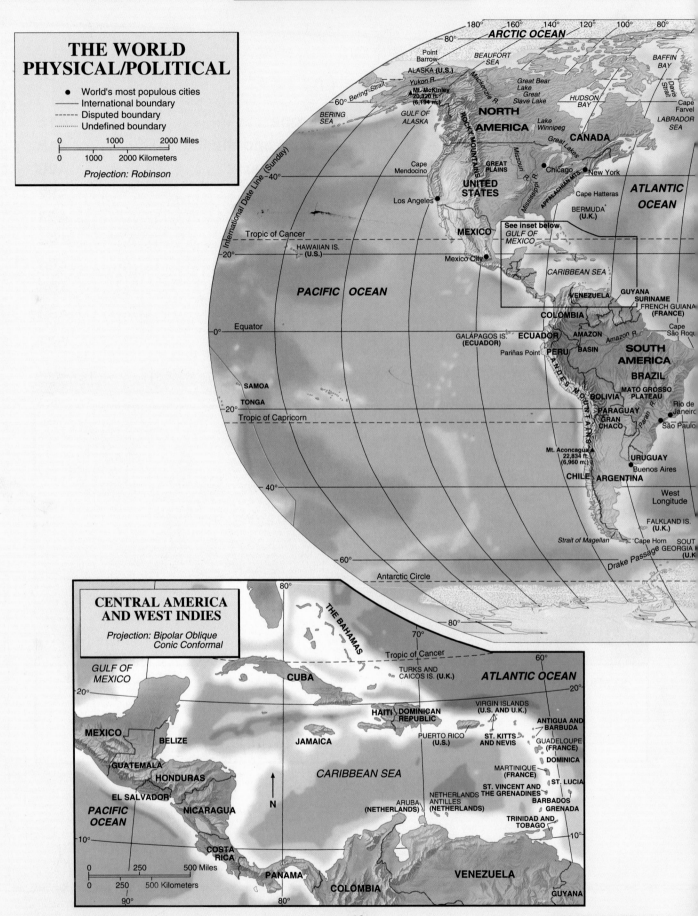

THE WORLD PHYSICAL/POLITICAL

- • World's most populous cities
- —— International boundary
- ----- Disputed boundary
- ········· Undefined boundary

| 0 | 1000 | 2000 Miles |
| 0 | 1000 | 2000 Kilometers |

Projection: Robinson

CENTRAL AMERICA AND WEST INDIES

Projection: Bipolar Oblique Conic Conformal

| 0 | 250 | 500 Miles |
| 0 | 250 | 500 Kilometers |

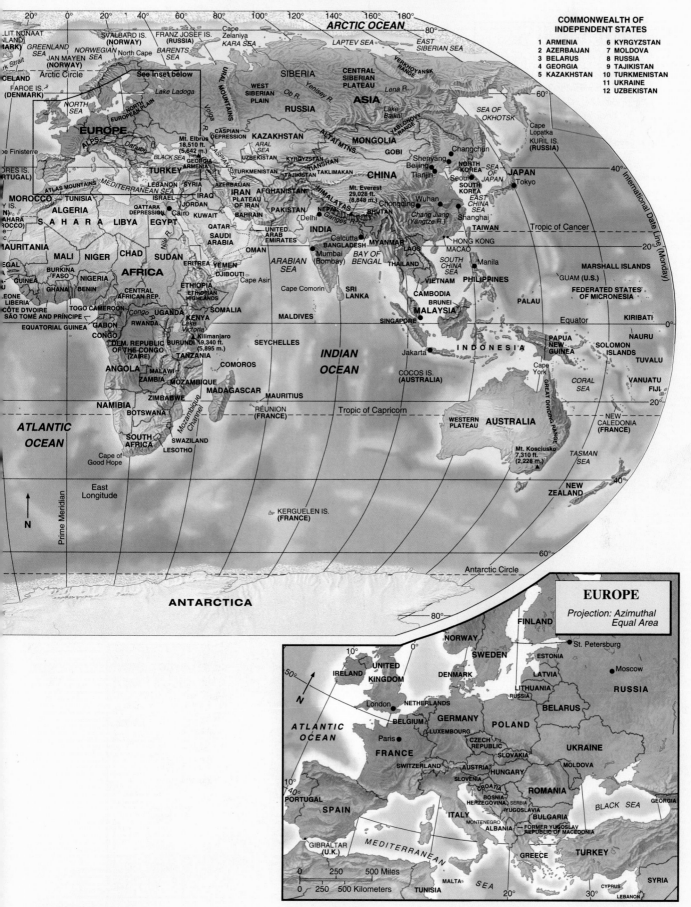

COMMONWEALTH OF INDEPENDENT STATES

1 ARMENIA
2 AZERBAIJAN
3 BELARUS
4 GEORGIA
5 KAZAKHSTAN
6 KYRGYZSTAN
7 MOLDOVA
8 RUSSIA
9 TAJIKISTAN
10 TURKMENISTAN
11 UKRAINE
12 UZBEKISTAN

EUROPE

Projection: Azimuthal Equal Area

REFERENCE ATLAS

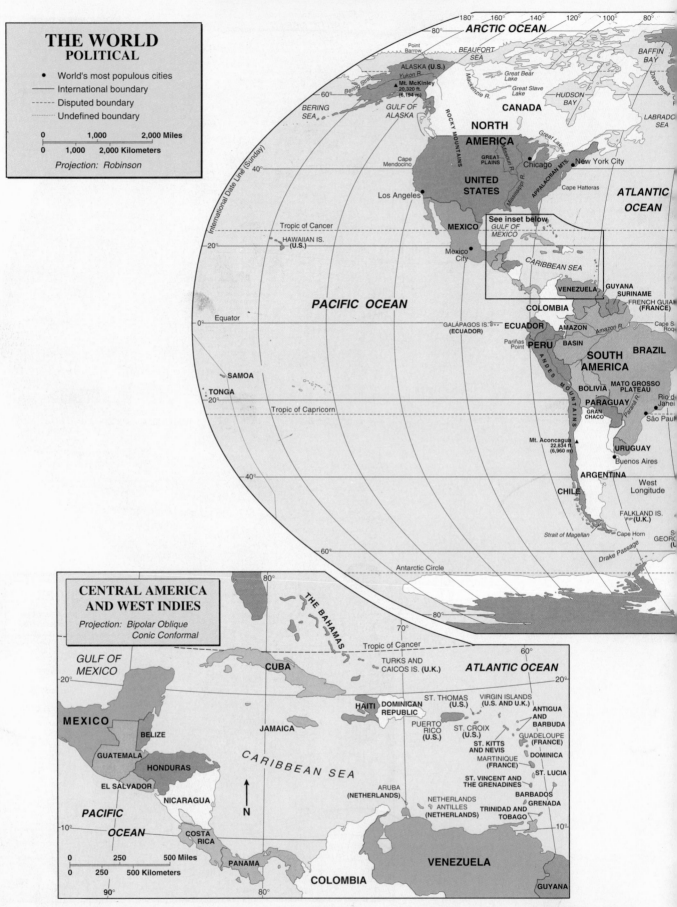

THE WORLD
POLITICAL

- World's most populous cities
— International boundary
--- Disputed boundary
······ Undefined boundary

| 0 | 1,000 | 2,000 Miles |
| 0 | 1,000 | 2,000 Kilometers |

Projection: Robinson

ARCTIC OCEAN

BEAUFORT SEA

Point Barrow

BAFFIN BAY

ALASKA (U.S.)
Yukon R.
▲ Mt. McKinley
20,320 ft.
(6,194 m)

Great Bear Lake

Davis Strait

Mackenzie R.

Great Slave Lake

CANADA

HUDSON BAY

LABRADOR SEA

BERING SEA

Bering Strait

GULF OF ALASKA

ROCKY MOUNTAINS

NORTH AMERICA

Great Lakes

Cape Mendocino

GREAT PLAINS

Chicago

New York City

APPALACHIAN MTS.

UNITED STATES

Missouri R.

Mississippi R.

ATLANTIC OCEAN

Cape Hatteras

Los Angeles

Tropic of Cancer

MEXICO

See inset below
GULF OF MEXICO

HAWAIIAN IS. (U.S.)

Mexico City

CARIBBEAN SEA

VENEZUELA

GUYANA

SURINAME

FRENCH GUIANA (FRANCE)

COLOMBIA

PACIFIC OCEAN

Equator

GALÁPAGOS IS. (ECUADOR)

ECUADOR

AMAZON

Amazon R.

Cape S. Roq

Pariñas Point

PERU

BASIN

SOUTH AMERICA

BRAZIL

SAMOA

TONGA

Tropic of Capricorn

ANDES MOUNTAINS

BOLIVIA

MATO GROSSO PLATEAU

PARAGUAY

GRAN CHACO

Paraná R.

Rio d Janei

São Pau

Mt. Aconcagua
22,834 ft.
(6,960 m) ▲

URUGUAY

Buenos Aires

ARGENTINA

West Longitude

CHILE

FALKLAND IS. (U.K.)

Strait of Magellan

Cape Horn

GEOR (U

Drake Passage

Antarctic Circle

International Date Line (Sunday)

CENTRAL AMERICA
AND WEST INDIES

*Projection: Bipolar Oblique
Conic Conformal*

THE BAHAMAS

80°

70°

80°

Tropic of Cancer

60°

GULF OF MEXICO

CUBA

TURKS AND CAICOS IS. (U.K.)

ATLANTIC OCEAN

20°

20°

ST. THOMAS (U.S.)

VIRGIN ISLANDS (U.S. AND U.K.)

ANTIGUA AND BARBUDA

HAITI

DOMINICAN REPUBLIC

MEXICO

BELIZE

JAMAICA

PUERTO RICO (U.S.)

ST. CROIX (U.S.)

GUADELOUPE (FRANCE)

ST. KITTS AND NEVIS

DOMINICA

GUATEMALA

CARIBBEAN SEA

MARTINIQUE (FRANCE)

ST. LUCIA

HONDURAS

ST. VINCENT AND THE GRENADINES

EL SALVADOR

BARBADOS

GRENADA

NICARAGUA

ARUBA (NETHERLANDS)

PACIFIC

NETHERLANDS ANTILLES (NETHERLANDS)

TRINIDAD AND TOBAGO

10°

OCEAN

COSTA RICA

N

10°

| 0 | 250 | 500 Miles |
| 0 | 250 | 500 Kilometers |

PANAMA

VENEZUELA

COLOMBIA

90°

80°

GUYANA

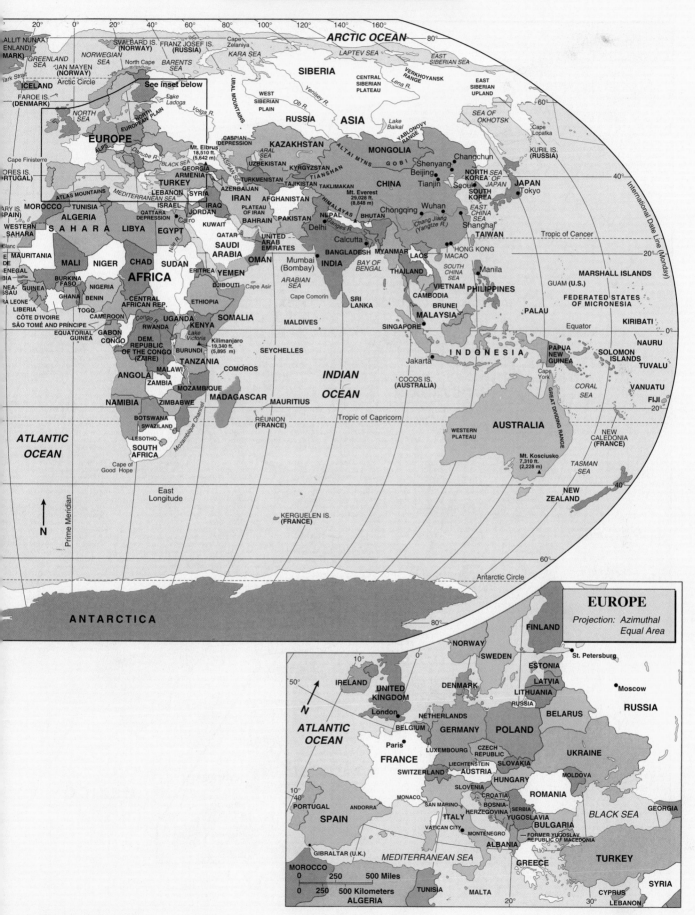

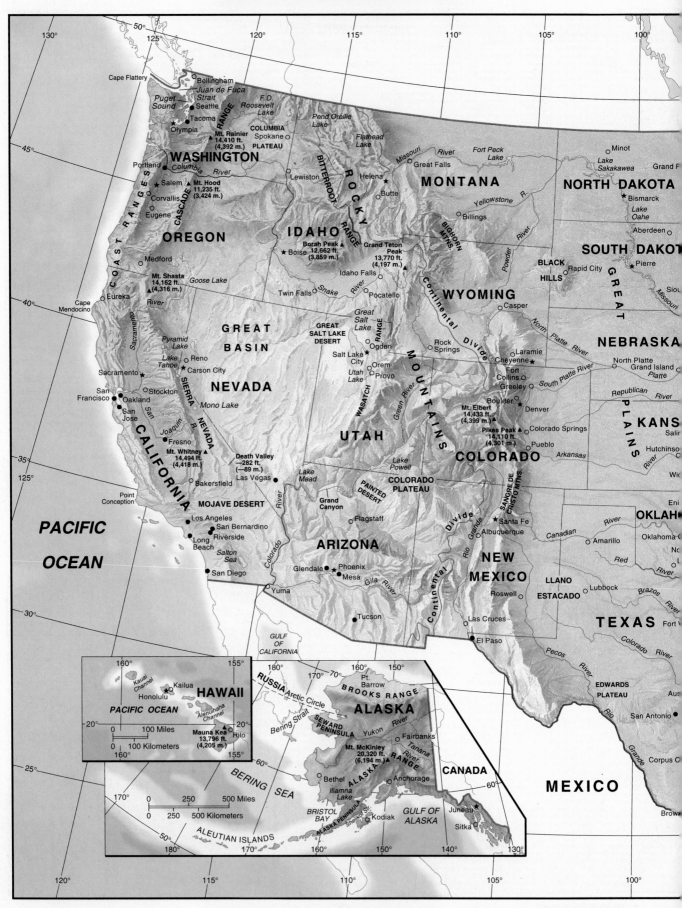

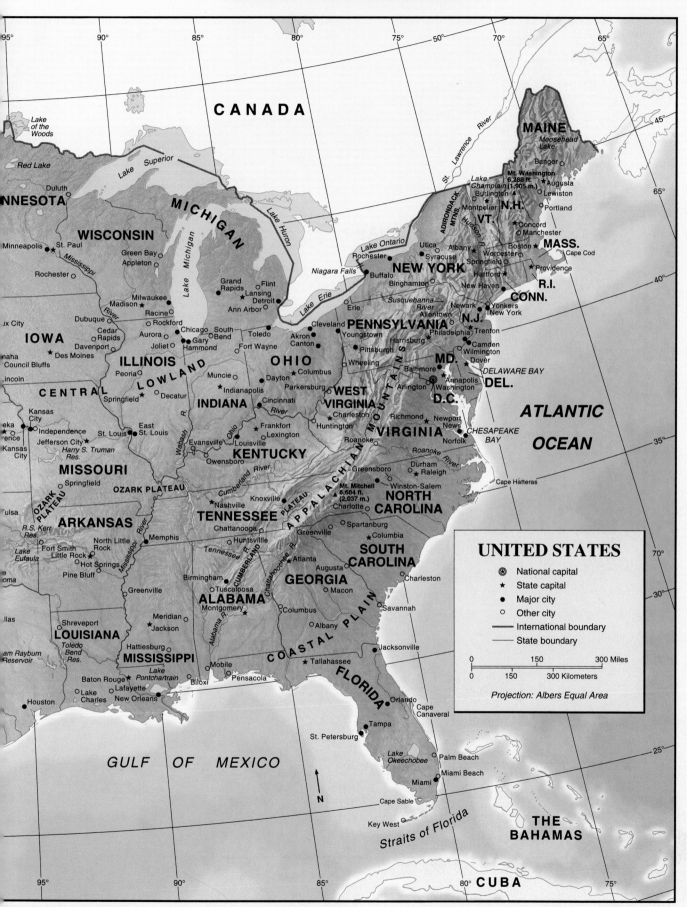

UNITED STATES

- National capital
- ★ State capital
- ● Major city
- ○ Other city
- — International boundary
- — State boundary

| 0 | 150 | 300 Miles |
| 0 | 150 | 300 Kilometers |

Projection: Albers Equal Area

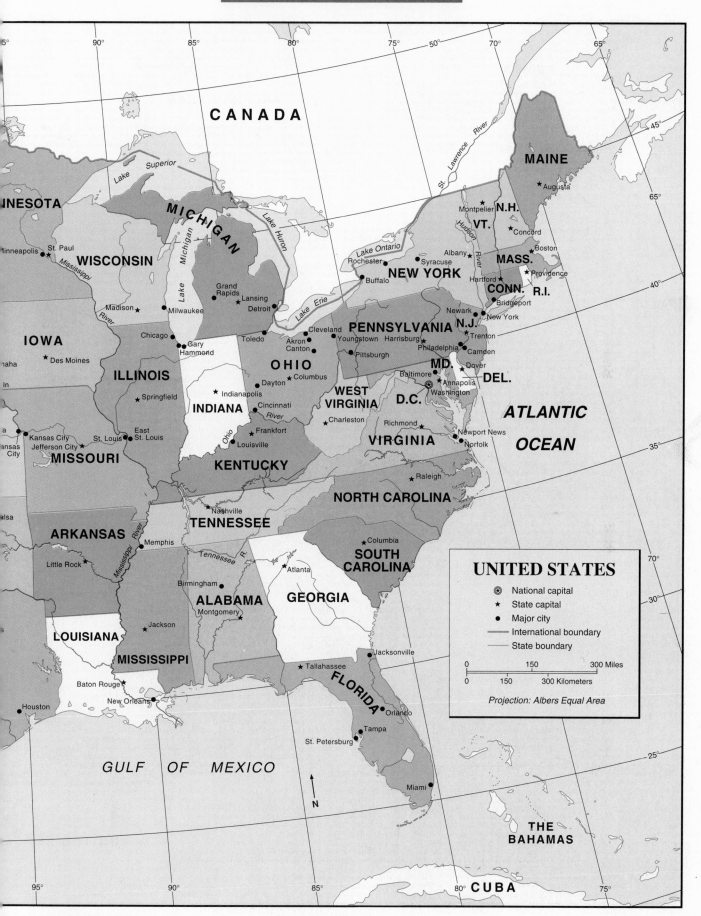

UNITED STATES

⊚ National capital
★ State capital
● Major city
— International boundary
— State boundary

| 0 | | 150 | | 300 Miles |
| 0 | 150 | | 300 Kilometers | |

Projection: Albers Equal Area

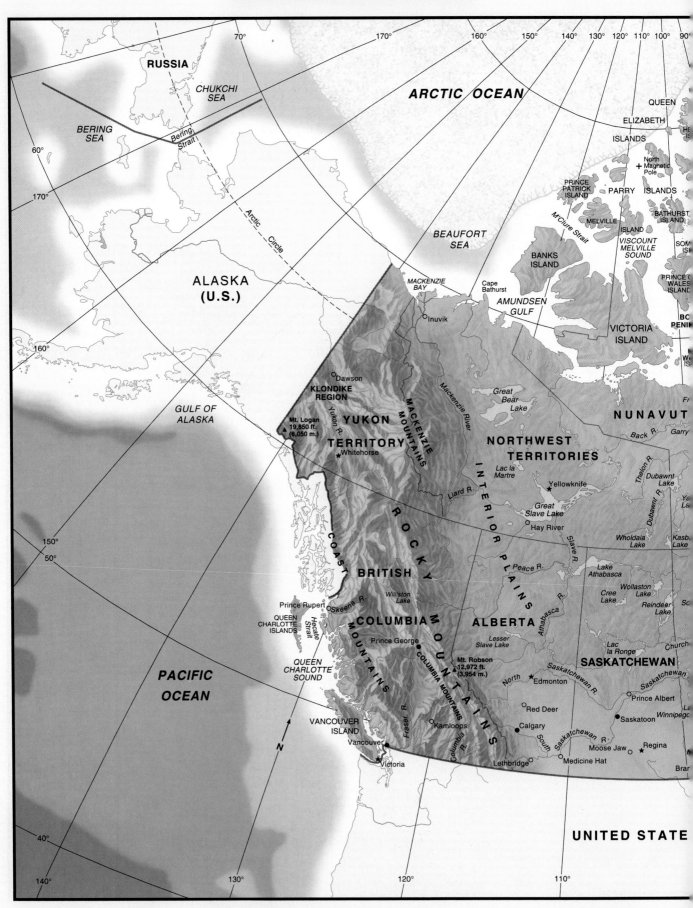

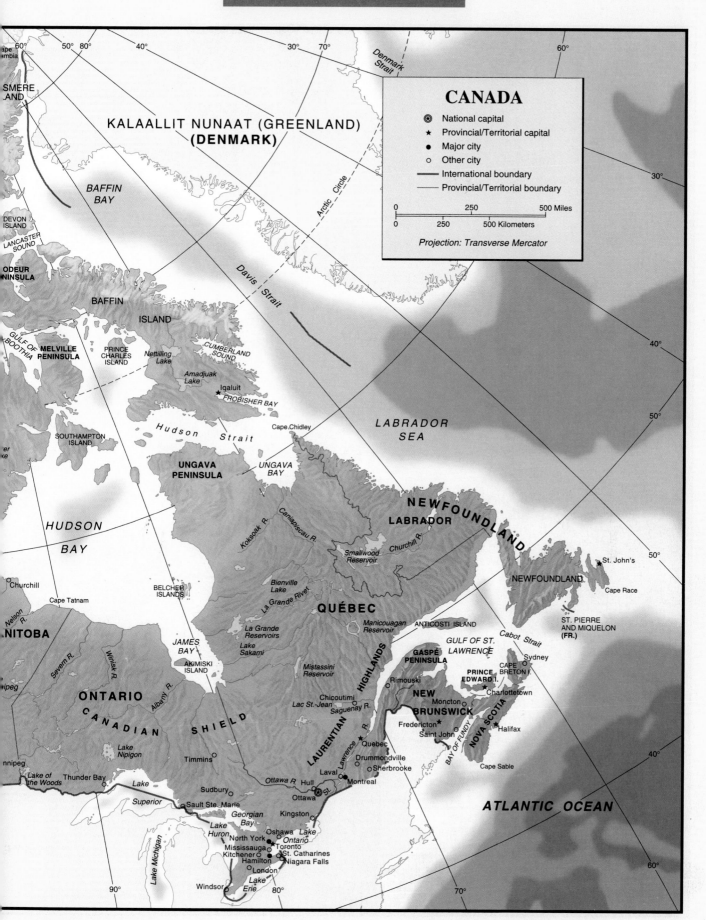

CANADA

- ⊛ National capital
- ★ Provincial/Territorial capital
- ● Major city
- ○ Other city
- —— International boundary
- —— Provincial/Territorial boundary

0		250		500 Miles
0	250		500 Kilometers	

Projection: Transverse Mercator

KALAALLIT NUNAAT (GREENLAND)
(DENMARK)

BAFFIN
BAY

DEVON
ISLAND

LANCASTER
SOUND

ODEUR
NINSULA

BAFFIN

ISLAND

GULF OF
BOOTHIA

MELVILLE
PENINSULA

PRINCE
CHARLES
ISLAND

Nettilling
Lake

CUMBERLAND
SOUND

Amadjuak
Lake

★ Iqaluit
FROBISHER BAY

Hudson   Strait

Cape Chidley

SOUTHAMPTON
ISLAND

UNGAVA
PENINSULA

UNGAVA
BAY

*LABRADOR
SEA*

*HUDSON

BAY*

Churchill

Cape Tatnam

Nelson
R.

NITOBA

Koksoak R.

Caniapiscau R.

NEWFOUNDLAND

LABRADOR

Smallwood
Reservoir

Churchill R.

Bienville
Lake

La Grande Rivet

BELCHER
ISLANDS

QUÉBEC

St. John's

NEWFOUNDLAND

Cape Race

ST. PIERRE
AND MIQUELON
(FR.)

La Grande
Reservoirs

Lake
Sakami

*JAMES

BAY*

AKIMISKI
ISLAND

Manicouagan
Reservoir

ANTICOSTI ISLAND

Cabot Strait

Mistassini
Reservoir

GASPÉ
PENINSULA

GULF OF ST.
LAWRENCE

Sydney

CAPE
BRETON I.

Severn R.

Winisk R.

ONTARIO

C A N A D I A N   S H I E L D

Chicoutimi

Lac St.-Jean

Saguenay R.

LAURENTIAN

HIGHLANDS

Rimouski

NEW

BRUNSWICK

PRINCE
EDWARD I.

Charlottetown

Moncton

NOVA SCOTIA

Albany R.

Lake
Nipigon

Timmins

Fredericton ★

Saint John

BAY OF FUNDY

Halifax ★

nnipeg

Lake of
the Woods

Thunder Bay

Lake

Superior

Sudbury

Sault Ste. Marie

*Georgian
Bay*

*Lake
Huron*

Lake Michigan

St. Lawrence R.

Quebec ★

Drummondville

Sherbrooke

Laval

Hull

Montreal

Ottawa R.

Ottawa ⊛

Kingston

Oshawa

North York

Mississauga

Kitchener

Hamilton

Toronto

*Lake
Ontario*

St. Catharines

Niagara Falls

London

Windsor

*Lake
Erie*

Cape Sable

ATLANTIC OCEAN

SMERE
AND

60°

50° 80°

40°

30° 70°

*Denmark
Strait*

60°

Arctic Circle

30°

Davis   Strait

40°

50°

50°

40°

60°

90°

80°

70°

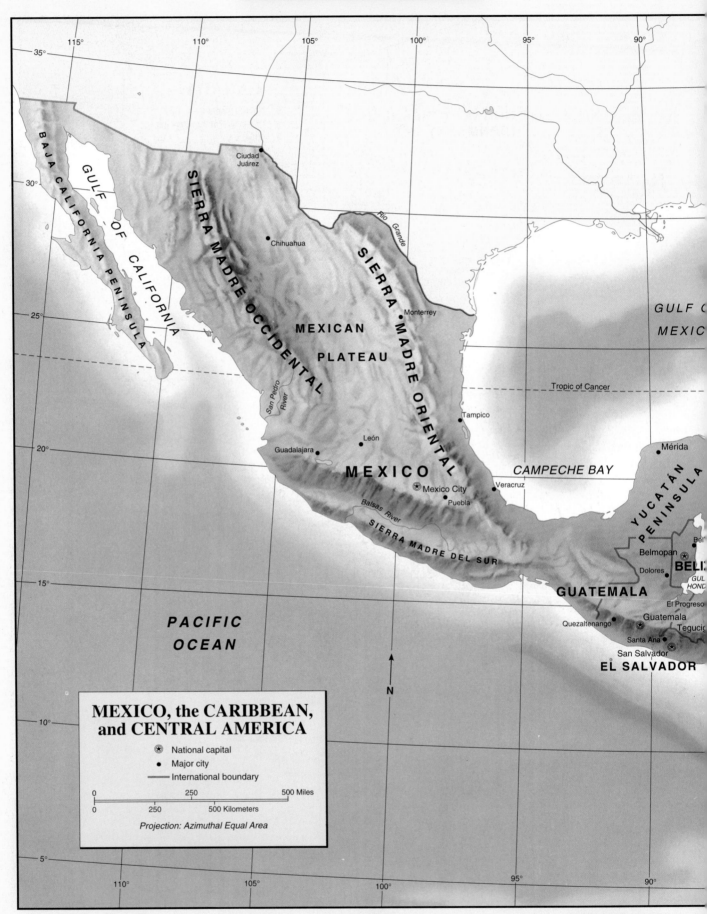

MEXICO, the CARIBBEAN, and CENTRAL AMERICA

⊛ National capital
• Major city
— International boundary

0   250   500 Miles
0   250   500 Kilometers

Projection: Azimuthal Equal Area

BAJA CALIFORNIA PENINSULA
GULF OF CALIFORNIA
SIERRA MADRE OCCIDENTAL
Ciudad Juárez
Chihuahua
SIERRA MADRE ORIENTAL
MEXICAN PLATEAU
Monterrey
Rio Grande
San Pedro River
Tampico
Tropic of Cancer
León
Guadalajara
MEXICO
Mexico City
Puebla
Veracruz
Balsas River
CAMPECHE BAY
Mérida
YUCATÁN PENINSULA
SIERRA MADRE DEL SUR
GULF OF MEXICO
GUATEMALA
Belmopan
Dolores
BELIZE
El Progreso
Guatemala
Quezaltenango
Santa Ana
Tegucig
San Salvador
EL SALVADOR
PACIFIC OCEAN
N

A12

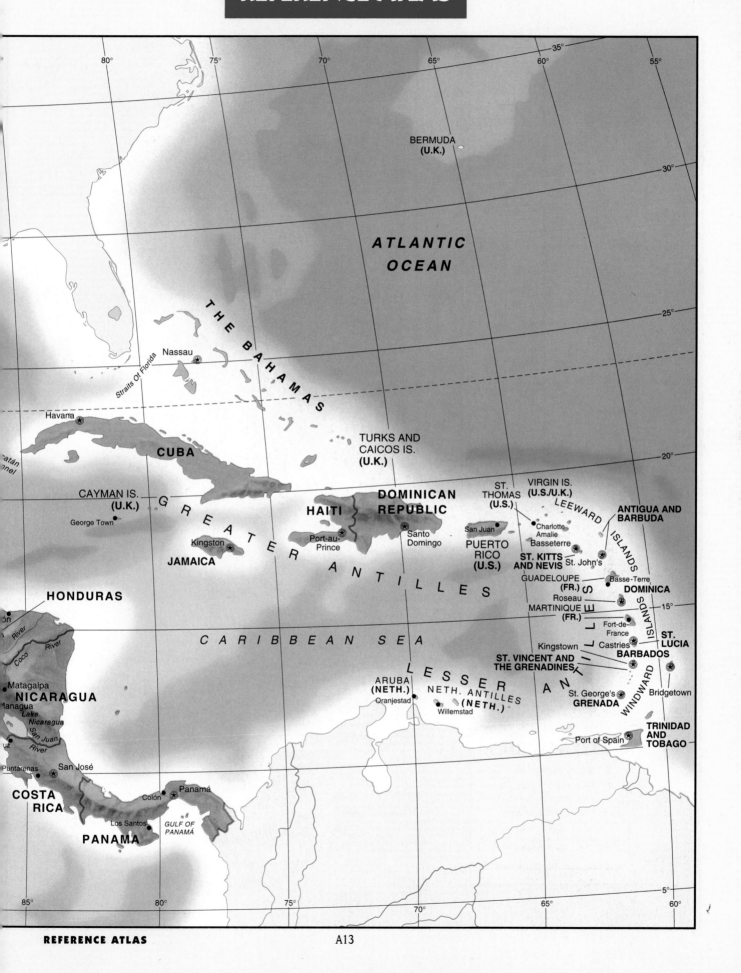

ATLANTIC OCEAN

BERMUDA (U.K.)

THE BAHAMAS

Nassau

Straits Of Florida

Havana

CUBA

Yucatán Channel

CAYMAN IS. (U.K.)

George Town

GREATER ANTILLES

Kingston

JAMAICA

HAITI

Port-au-Prince

DOMINICAN REPUBLIC

Santo Domingo

TURKS AND CAICOS IS. (U.K.)

ST. THOMAS (U.S.)

VIRGIN IS. (U.S./U.K.)

San Juan

PUERTO RICO (U.S.)

Charlotte Amalie

Basseterre

ST. KITTS AND NEVIS

St. John's

LEEWARD ISLANDS

ANTIGUA AND BARBUDA

GUADELOUPE (FR.)

Basse-Terre

DOMINICA

Roseau

MARTINIQUE (FR.)

Fort-de-France

HONDURAS

CARIBBEAN SEA

River

Coco River

Matagalpa

NICARAGUA

Managua

Lake Nicaragua

San Juan River

Puntarenas

San José

COSTA RICA

Colón

Panamá

Los Santos

GULF OF PANAMÁ

PANAMA

LESSER ANTILLES

ARUBA (NETH.)

Oranjestad

NETH. ANTILLES (NETH.)

Willemstad

St. George's

GRENADA

Kingstown

ST. VINCENT AND THE GRENADINES

Castries

ST. LUCIA

BARBADOS

Bridgetown

WINDWARD ISLANDS

Port of Spain

TRINIDAD AND TOBAGO

80° 75° 70° 65° 60° 55°

35° 30° 25° 20° 15° 10° 5°

85° 80° 75° 70° 65° 60°

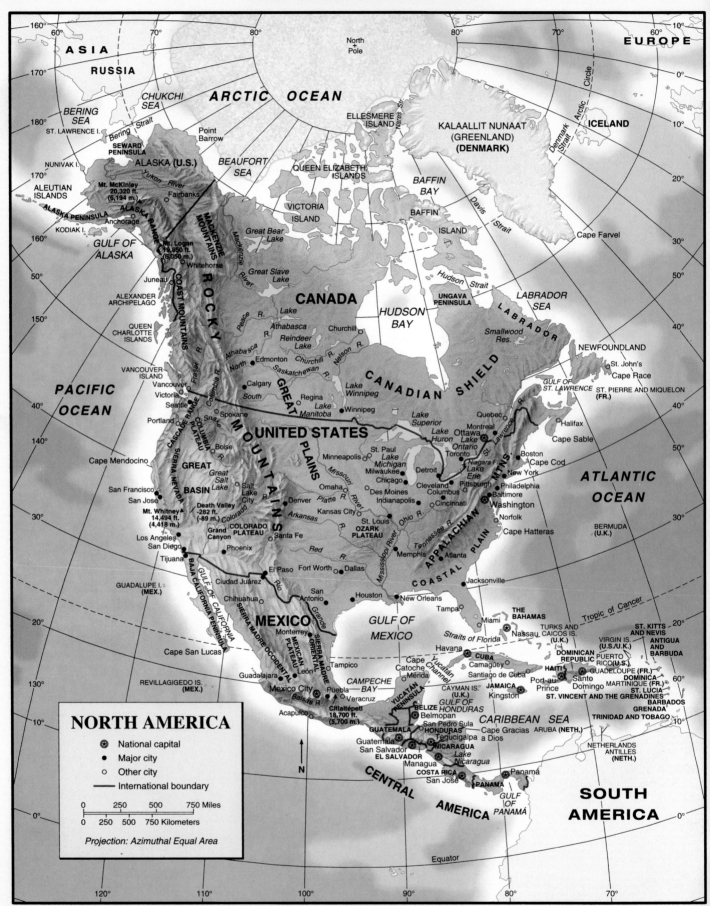

NORTH AMERICA

- ⊛ National capital
- ● Major city
- ○ Other city
- — International boundary

0   250   500   750 Miles
0   250   500   750 Kilometers

Projection: Azimuthal Equal Area

SOUTH AMERICA

- ⊛ National capital
- • Major city
- ○ Other city
- — International boundary

0   250   500 Miles
0 250 500 Kilometers

Projection: Azimuthal Equal Area

EUROPE

⊛ National capital
● Major city
○ Other city
— International boundary
— Republic boundary
┤├ Canal

0   100   200   300 Miles
0   100   200   300 Kilometers

Projection: Azimuthal Equal Area

ICELAND
Reykjavik

Arctic Circle

NORWEGIAN SEA

FAROE IS. (DEN.)

SHETLAND IS. (U.K.)

Trondheim

NORWAY

SCANDINAVIAN HIGHLANDS

Gaidhöpiggen 8,097 ft. (2,468 m.)

Bergen
Oslo

GULF OF BO

SWEDEN

Uppsala

Lake Vänern

Stockholm

ALA

HIIU
SAAREM
GOTLAN

Lake Vättern

OLAND I.

OUTER HEBRIDES IS.
Cape Wrath

ORKNEY ISLANDS

SCOTLAND

Skagerrak

Kattegat

Göteborg

JUTLAND

NORTH SEA

NORTHERN IRELAND (U.K.)
Glasgow
Edinburgh

PENNINE RANGE

UNITED KINGDOM

Copenhagen
DENMARK
Odense

Malmö
BORNHOLM I.

BALTIC SEA

Belfast

IRISH SEA

ISLE OF MAN

Dublin
IRELAND
Cork

Manchester
Liverpool
Leeds
Sheffield
ENGLAND
Birmingham

Kiel
Rostock

Hamburg
Elbe R.

Bremen

Szczecin

Gdańsk

NOR

POLAN

Wars

Cape Clear

St. George's Channel

WALES
Cardiff

London

Bristol

NETHERLANDS
Amsterdam
The Hague
Rotterdam
Antwerp

Mittelland Canal

Hannover

Berlin

Magdeburg

Oder R.

Poznań

Wrocław

POLAN

ATLANTIC OCEAN

English Channel

Strait of Dover

GUERNSEY I. (U.K.)
JERSEY I. (U.K.)

Le Havre

Seine River

BELGIUM
Brussels
Liège
LUXEMBOURG
Luxembourg

Essen
Cologne
Bonn
Frankfurt

Dortmund

Leipzig
Dresden
Chemnitz

GERMANY

Prague
CZECH REPUBLIC

Ostrava
Brno

Kat

SLOV
Bratislav

BRETON PEN.

Paris

Marne R.

Marne-Rhine Canal

Rhine R.

Nantes

Loire

Loire River

FRANCE

Strasbourg

Stuttgart

Danube R.

Munich

Bodensee

Zürich
LIECHTENSTEIN
Vaduz

Linz

Salzburg

Vienna

AUSTRIA
Graz

Budapest

HUNGAR

Pécs

Cape Finisterre

BAY OF BISCAY

Bordeaux

Garonne R.

CENTRAL MASSIF

Lyon

Rhône R.

Lausanne
Geneva
L. Geneva
SWITZERLAND
Bern

Mt. Blanc 15,771 ft. (4,807 m.)

Innsbruck

ALPS

Mt. Rosa 12,203 ft. (4,634 m.)

Turin

PO VALLEY

Milan
Po R.

Venice

Ljubljana
SLOVENIA

L. Balaton

Zagreb

CROATIA

DINARIC ALPS

Sava R.

Novi

Belgra

Bilbao

CANTABRIAN MTNS.

Porto

Valladolid

Duero River

PYRENEES

Toulouse
Midi Canal

Aneto Peak 11,168 ft. (3,404 m.)

ANDORRA
Andorra la Vella

Montpellier

Marseille

GULF OF LION

Nice

Monaco
MONACO

Genoa

Bologna

Florence

SAN MARINO
San Marino

APENNINES

BOSNIA-
HERZEGOVIN

Split

FORMER YUGOSL
REPU
MACE

MONTEN

PORTUGAL

IBERIAN

Zaragoza
Madrid

SPAIN

Tagus River

Lisbon
Setúbal

Guadiana River

Duero River

PENINSULA

SIERRA MORENA

Valencia

Barcelona

CORSICA (FR.)

VATICAN CITY
Rome

ITALY

ADRIATIC SEA

Bari

Tirané

ALBA

Cape St. Vincent

Seville

Granada

Murcia

Palma

BALEARIC IS. (SP.)

SARDINIA (IT.)

Naples

Málaga

Strait of Gibraltar

GIBRALTAR (U.K.)

Cagliari

TYRRHENIAN SEA

G. OF TARANTO

MEDITERRANEAN

Strait of Sicily

Palermo

SICILY
Catania

IONIAN SEA

KEFALLIN

AFRICA

PANTELLERIA (IT.)

MALTA
Valletta

SEA

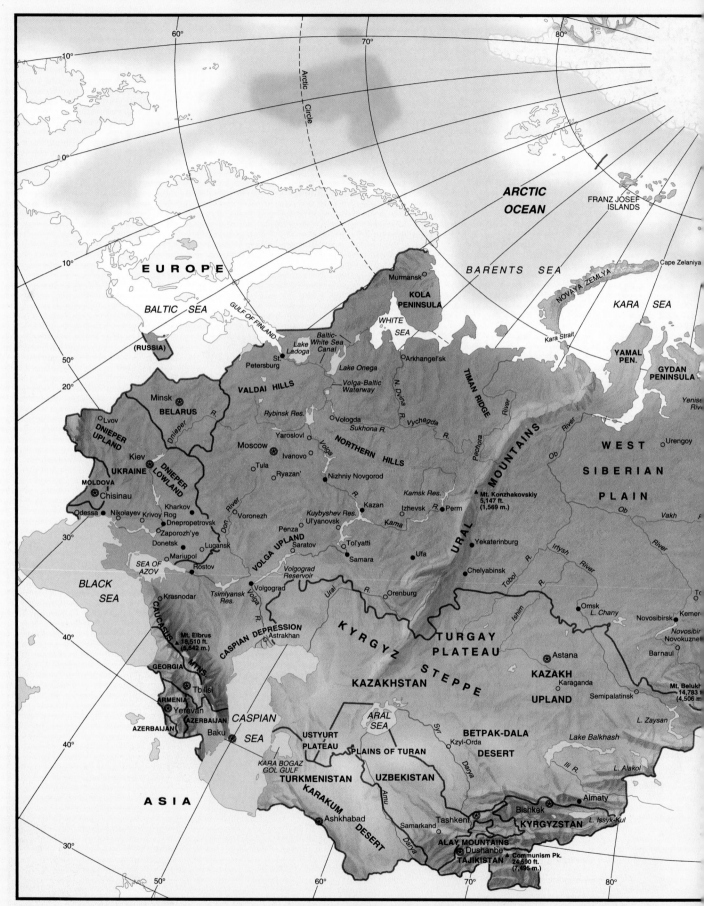

ARCTIC OCEAN

FRANZ JOSEF ISLANDS

Cape Zelaniya

BARENTS SEA

NOVAYA ZEMLYA

KARA SEA

EUROPE

BALTIC SEA

GULF OF FINLAND

(RUSSIA)

Murmansk

KOLA PENINSULA

WHITE SEA

Arkhangel'sk

Kara Strait

YAMAL PEN.

GYDAN PENINSULA

Yenisei

Baltic-White Sea Canal

Lake Ladoga

St. Petersburg

Lake Onega

Volga-Baltic Waterway

N. Dvina R.

Vychegda

TIMAN RIDGE

Pechora River

WEST SIBERIAN PLAIN

Urengoy

VALDAI HILLS

Minsk

BELARUS

Lvov

DNIEPER UPLAND

Dnieper R.

Vologda

Sukhona R.

Ob River

Kiev

DNIEPER LOWLAND

UKRAINE

MOLDOVA

Chisinau

Odessa

Nikolayev

Krivoy Rog

Kharkov

Dnepropetrovsk

Zaporozh'ye

Donetsk

Lugansk

Mariupol

Rostov

SEA OF AZOV

Moscow

Yaroslovl

Ivanovo

Tula

Ryazan'

Don River

Volga R.

NORTHERN HILLS

Nizhniy Novgorod

Kazan

Kamsk Res.

Izhevsk

Perm

Kuybyshev Res.

Ul'yanovsk

Kama R.

Mt. Konzhakovskiy 5,147 ft. (1,569 m.)

URAL MOUNTAINS

Yekaterinburg

Ob

Vakh

Voronezh

Penza

Saratov

VOLGA UPLAND

Tol'yatti

Samara

Ufa

Chelyabinsk

Irtysh River

BLACK SEA

Krasnodar

Tsimlyansk Res.

Volgograd

Volgograd Reservoir

Ural R.

Orenburg

Tobol R.

Ishim

Omsk

L. Chany

Novosibirsk

Kemer

CAUCASUS MTS.

Mt. Elbrus 18,510 ft. (5,642 m.)

CASPIAN DEPRESSION

Astrakhan

KYRGYZ

TURGAY PLATEAU

Astana

Novosibir Novokuzne

GEORGIA

Tbilisi

STEPPE

KAZAKH UPLAND

Barnaul

ARMENIA

Yerevan

AZERBAIJAN

AZERBAIJAN

Baku

CASPIAN SEA

KAZAKHSTAN

Karaganda

Semipalatinsk

Mt. Belukh 14,783 (4,506 m

L. Zaysan

USTYURT PLATEAU

ARAL SEA

Syr

BETPAK-DALA DESERT

Kzyl-Orda

Lake Balkhash

KARA BOGAZ GOL GULF

PLAINS OF TURAN

Darya

Ili R.

L. Alakol

TURKMENISTAN

UZBEKISTAN

ASIA

KARAKUM

DESERT

Amu

Darya

Ashkhabad

Samarkand

Tashkent

Bishkek

KYRGYZSTAN

Almaty

L. Issyk-Kul

ALAY MOUNTAINS

Dushanbe

TAJIKISTAN

Communism Pk. 24,590 ft. (7,495 m.)

RUSSIA AND
THE EURASIAN REPUBLICS

⊛ National capital
● Major city
○ Other city
— International boundary

0   250   500 Miles
0   250   500 Kilometers

Projection: Two-Point Equidistant

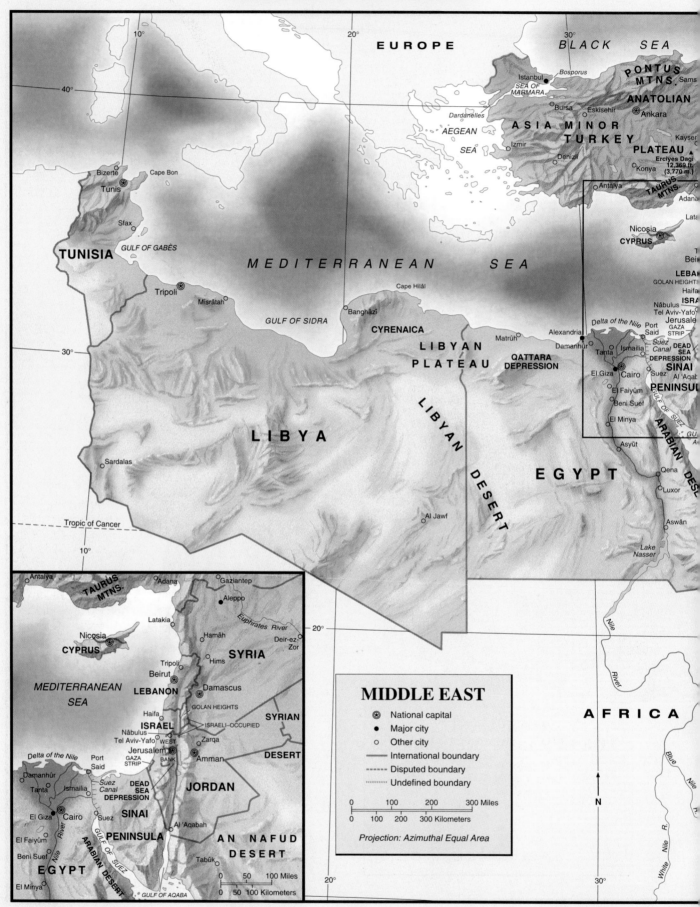

EUROPE

BLACK SEA

10°

20°

30°

Istanbul · Bosporus

PONTUS MTNS. · Sams

SEA OF MARMARA

40°

ANATOLIAN

Dardanelles

Bursa · Eskişehir

ASIA MINOR

Ankara

AEGEAN SEA

TURKEY

İzmir

PLATEAU

Denizli

Konya

Erciyes Dagi 12,369 ft. (3,770 m.)

Kayser

Bizerte

Cape Bon

Antalya

TAURUS MTNS.

Adana

Tunis

Lata

Nicosia

Nicosia

Sfax

CYPRUS

Bei

TUNISIA

GULF OF GABÈS

MEDITERRANEAN SEA

LEBAN

GOLAN HEIGHTS

Haifa

Tripoli

Cape Hilâl

Nâbulus

ISRA

Misrâtah

Tel Aviv-Yafo

Jerusale

GAZA STRIP

Banghāzī

Delta of the Nile

Port Said

DEAD SEA DEPRESSION

GULF OF SIDRA

CYRENAICA

Matrûh

Alexandria

Suez Canal

SINAI

Al 'Aqab

30°

LIBYAN PLATEAU

QATTARA DEPRESSION

Damanhûr

Tanta · Ismailia

PENINSUL

El Giza · Cairo

Suez

ARABIAN DESL

GU

El Faiyûm

LIBYA

Beni Suef

GULF OF SUEZ

LIBYAN DESERT

El Minya

EGYPT

Sardalas

Asyût

Qena

Luxor

Tropic of Cancer

Aswân

10°

Al Jawf

Lake Nasser

Nile

Nile River

20°

AFRICA

Nile R.

Blue Nile

White Nile R.

Inset map

Antalya

TAURUS MTNS.

Adana

Gaziantep

Nicosia

Aleppo

Latakia

Euphrates River

CYPRUS

Hamâh

Deir-ez-Zor

MEDITERRANEAN SEA

Tripoli

Hims

SYRIA

Beirut

Damascus

LEBANON

GOLAN HEIGHTS

SYRIAN

Haifa

ISRAELI-OCCUPIED

Nâbulus

ISRAEL

Zarqa

DESERT

Tel Aviv-Yafo

WEST

Delta of the Nile

Port Said

Jerusalem

Amman

Damanhûr

Ismailia

Suez Canal

GAZA STRIP

BANK

Tanta

DEAD SEA DEPRESSION

JORDAN

El Giza · Cairo

SINAI

El Faiyûm

Suez

PENINSULA

Al 'Aqabah

Beni Suef

ARABIAN DESERT

AN NAFUD

EGYPT

GULF OF SUEZ

DESERT

El Minya

Tabûk

GULF OF AQABA

MIDDLE EAST

⊛ National capital

● Major city

○ Other city

— International boundary

---- Disputed boundary

······ Undefined boundary

0	100	200	300 Miles
0	100	200	300 Kilometers

Projection: Azimuthal Equal Area

N

0	50	100 Miles
0	50	100 Kilometers

20°

20°

30°

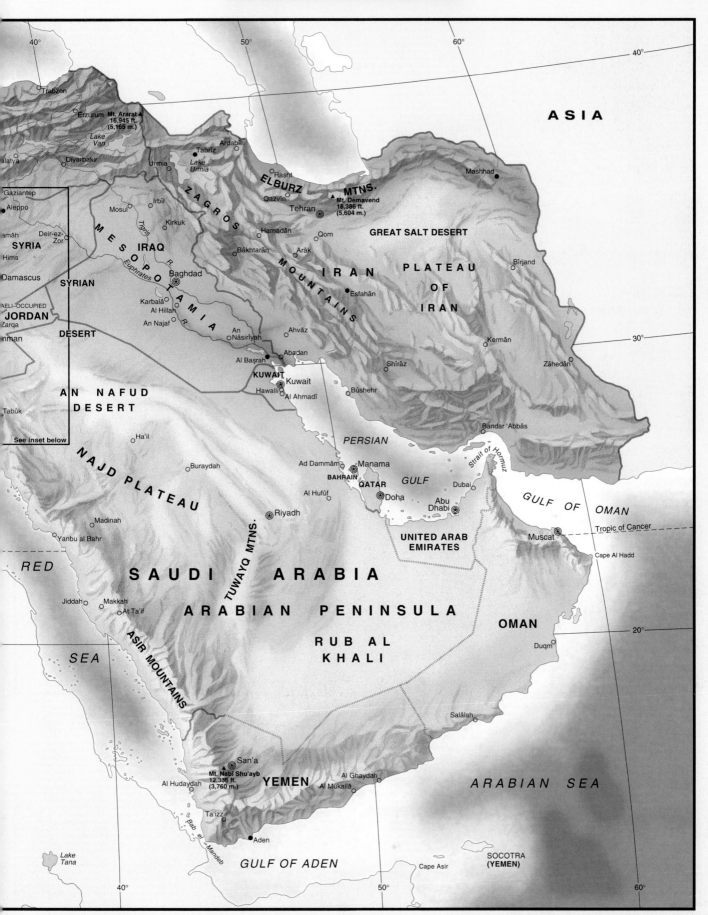

ASIA

Trabzon

Erzurum · Mt. Ararat ▲ 16,945 ft. (5,165 m.)

Lake Van

Malatya · Diyarbakir ○ · Ardabil

Urmia · Tabriz ● · *Lake Urmia*

Gaziantep ○

Aleppo ○ · Mosul ○ · Irbil ○ · Kirkuk ○

Rasht ● · Mashhad ●

ELBURZ MTNS.

Qazvin ○ · ▲ Mt. Demavend 18,386 ft. (5,604 m.)

Tehran ◎

amah ○ · Deir-ez-Zor ○

SYRIA

Hims ○

MESOPOTAMIA · IRAQ

Tigris R.

Hamadan ○

GREAT SALT DESERT

Qom ○

Damascus ○

SYRIAN · Euphrates R. · Baghdad ◎

Bakhtaran ○ · Arak ○

IRAN

PLATEAU OF IRAN

Birjand ○

AELI-OCCUPIED · Karbala ○

JORDAN · Al Hillah ○

Zarqa ○ · An Najaf ○

DESERT

Esfahan ●

nman ○

An Nasiriyah ○

Ahvaz ○

Kerman ○

Zahedan ○

Tabuk ○

AN NAFUD DESERT

Al Basrah ◎ · Abadan ○

KUWAIT

Hawalli ● · Kuwait ◎

Al Ahmadi ○

Bushehr ○

Shiraz ●

Bandar 'Abbas ○

See inset below

Ha'il ○

PERSIAN

Strait of Hormuz

NAJD PLATEAU

Buraydah ○

Ad Dammam ○ · Manama ◎

BAHRAIN

GULF

Dubai ○

GULF OF OMAN

Al Hufuf ○ · QATAR · Doha ◎

Abu Dhabi ◎

Madinah ○

Riyadh ◎

UNITED ARAB EMIRATES

Tropic of Cancer

Muscat ◎

Cape Al Hadd

Yanbu al Bahr ○

TUWAYQ MTNS.

RED

SAUDI ARABIA

OMAN

Jiddah ○ · Makkah ○

At Ta'if ○

ARABIAN PENINSULA

SEA

ASIR MOUNTAINS

RUB AL KHALI

Duqm ○

Salalah ○

San'a ◎ · ▲ Mt. Nabi Shu'ayb 12,336 ft. (3,760 m.)

Al Hudaydah ○

YEMEN

Al Ghaydah ○

Al Mukalla ○

ARABIAN SEA

Ta'izz ○

Bab el Mandeb

Aden ●

Lake Tana

GULF OF ADEN

Cape Asir

SOCOTRA (YEMEN)

40° · 50° · 60°

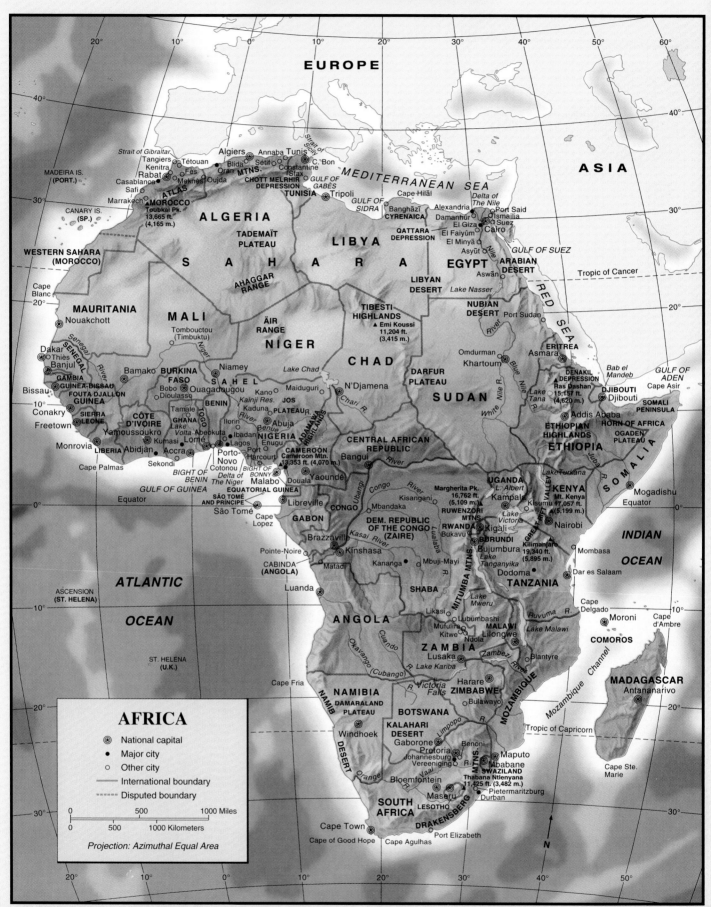

AFRICA

⊚ National capital
● Major city
○ Other city
— International boundary
--- Disputed boundary

0   500   1000 Miles
0   500   1000 Kilometers

Projection: Azimuthal Equal Area

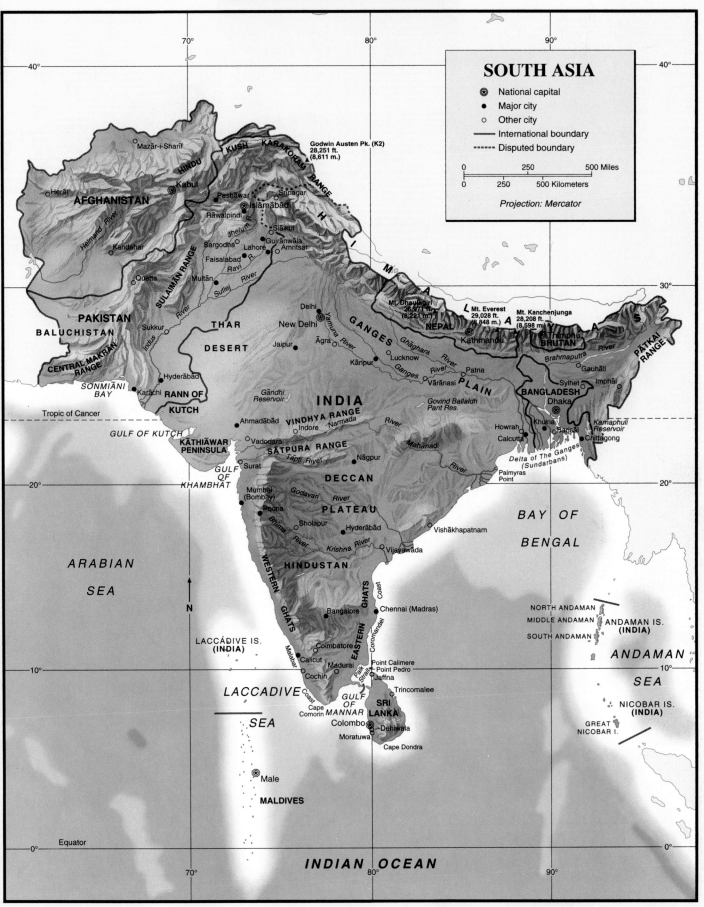

SOUTH ASIA

⊛ National capital
● Major city
○ Other city
— International boundary
---- Disputed boundary

| 0 | 250 | 500 Miles |

| 0 | 250 | 500 Kilometers |

Projection: Mercator

40°   70°   80°   90°   40°

Mazār-i-Sharīf

Godwin Austen Pk. (K2)
28,251 ft.
(8,611 m.)

HINDU KUSH   KARAKORAM RANGE

Herāt
AFGHANISTAN   Kabul ⊛
Peshāwar   Srīnagar
Rāwalpindi   Islāmābād ⊛
Kandahar   Siālkot
Helmand River   Jhelum   Sargodha   Gujrānwāla
Quetta   Faisalābad   Lahore   Amritsar
SULAIMAN RANGE   Ravi   R.
Multan   Sutlej   River

30°

PAKISTAN   Delhi   Yamuna   GANGES   Mt. Dhaulāgiri   Mt. Everest   Mt. Kanchenjunga
26,971 ft.   29,028 ft.   28,208 ft.
BALUCHISTAN   THAR   New Delhi   (8,221 m.)   (8,848 m.)   (8,598 m.)
Sukkur   Jaipur   Āgra   Ghāghara   River   NEPAL   Thimphu
Indus   DESERT   Lucknow   Kathmandu   BHUTAN
Hyderābād   Kānpur   Ganges   River   Patna   Gauhāti   Patkai Range
CENTRAL MAKRAN   Sonmiāni   Vārānasi   PLAIN   Brahmaputra   River
RANGE   BAY   Karāchi   RANN OF   Sylhet   Imphāl
KUTCH   Gāndhi   INDIA   Govind Ballalah   BANGLADESH
Tropic of Cancer   Reservoir   Pant Res.   Dhaka
Ahmadābād   VINDHYA RANGE   River   Howrah   Khulna   Karnaphuli
GULF OF KUTCH   Indore   Narmada   Calcutta   Barisal   Reservoir
KĀTHIĀWAR   Vadodara   SĀTPURA RANGE   Mahānadi   Chittagong
PENINSULA   Surat   Tāpti   River   Delta of The Ganges
GULF   DECCAN   River   (Sundarbans)
OF   Mumbai   Godāvari   River   Palmyras
KHAMBHĀT   (Bombay)   PLATEAU   Point

20°

ARABIAN   Poona   Bhima   Hyderābād   Vishākhapatnam   BAY   OF
SEA   Sholapur   BENGAL
Krishna   River
HINDUSTAN   Vijayawāda

N

WESTERN   Bangalore   Chennai (Madras)   NORTH ANDAMAN   ANDAMAN IS.
GHATS   EASTERN   MIDDLE ANDAMAN   (INDIA)
GHATS   Coromandel   SOUTH ANDAMAN
LACCĀDIVE IS.   Coimbatore   Coast   ANDAMAN
(INDIA)   Calicut   Madurai   Point Calimere
Malabar   Cochin   Point Pedro   SEA
10°   Paik   Jaffna   NICOBAR IS.
Strait   Trincomalee   (INDIA)
LACCADIVE   Coast   GULF   SRI
Cape   OF   LANKA   GREAT
SEA   Comorin   MANNAR   Colombo ⊛   NICOBAR I.
Dehiwala
Moratuwa
Cape Dondra

⊛ Male

MALDIVES

0°   Equator   0°

INDIAN   OCEAN

70°   80°   90°

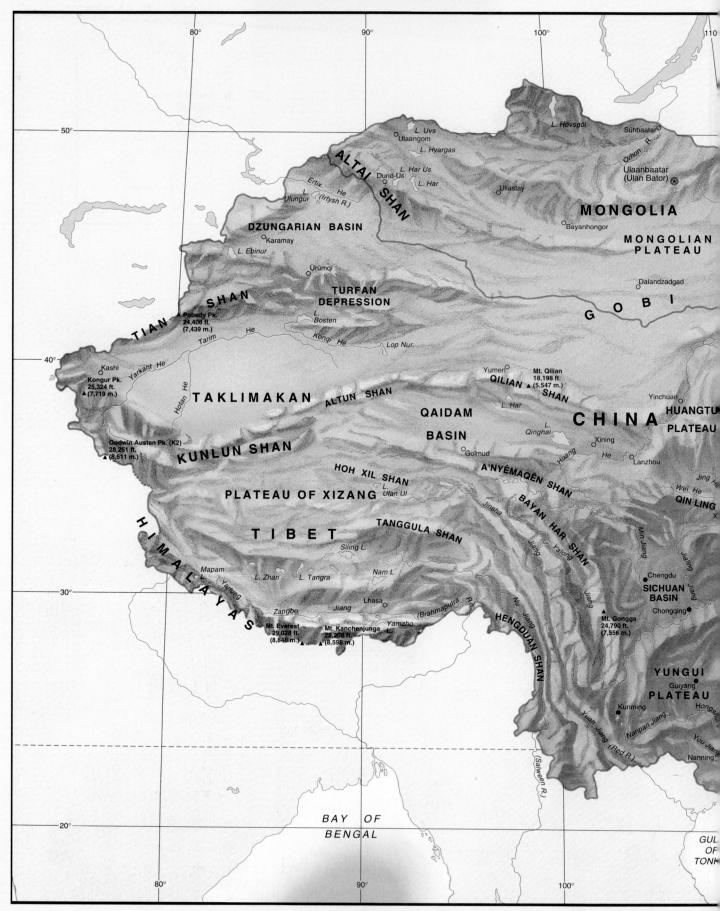

MONGOLIA

MONGOLIAN
PLATEAU

ALTAI SHAN

DZUNGARIAN BASIN

L. Uvs
Ulaangom
L. Hyargas
Dund-Us
L. Har Us
L. Har

L. Hövsgöl
Sühbaatar
Orhon R.
Ulaanbaatar
(Ulan Bator)

Uliastay

Bayanhongor

Dalandzadgad

G O B I

Karamay
L. Ebinur

Ürümqi

Ertix He
(Irtysh R.)
L. Ulungur

TIAN SHAN

TURFAN
DEPRESSION

Pobedy Pk.
24,406 ft.
(7,439 m.)

He
Tarim
Kongi He
L. Bosten
Lop Nur

Kashi
Kongur Pk.
25,324 ft.
▲ (7,719 m.)

Yarkant He
Hotan He

TAKLIMAKAN

ALTUN SHAN

Godwin Austen Pk. (K2)
28,251 ft.
▲ (8,611 m.)

KUNLUN SHAN

HOH XIL SHAN

L.
Ulan Ul

PLATEAU OF XIZANG

TIBET

TANGGULA SHAN

Siling L.

Yumen
QILIAN

Mt. Qilian
18,198 ft.
▲ (5,547 m.)

SHAN

L. Har

QAIDAM

BASIN

Golmud

L.
Qinghai

Xining

Huang

He

Yinchuan

CHINA

HUANGTU
PLATEAU

Lanzhou

A'NYÊMAQÊN SHAN

BAYAN HAR SHAN

Jing He
Wei He
QIN LING

Min Jiang
Jialing

Jiang
Jinsha
Yalong

Jiang
Chengdu

SICHUAN
BASIN

Chongqing

H
I
M
A
L
A
Y
A
S

Mapam
L.

L. Zhari
L. Tangra

Nam L.

Lhasa

Mt. Everest
29,028 ft.
(8,848 m.) ▲

Yarlung
Zangbo

Jiang

Yamzho
L.

Mt. Kanchenjunga
28,208 ft.
▲ (8,598 m.)

(Brahmaputra R.)

Nu Jiang

HENGDUAN SHAN

Mt. Gongga
24,790 ft.
(7,556 m.)

YUNGUI
PLATEAU

Guiyang

Kunming

Xuan Jiang (Red R.)
Nanpan Jiang

Hongshui

You Jiang

Nanning

BAY OF
BENGAL

(Salween R.)

GULF
OF
TONKI

120° 130° 140° 150°

Amur
Heilong Jiang
R.

Choybalsan
yant-Uhaa
INNER MONGOLIA
L. Hulun
L. Buyr
Hailar He
XIAO HINGGAN LING
Qiqihar
Nen Jiang
NORTHEAST (MANCHURIAN) PLAIN
Songhua Jiang
(Sungari R.)
Amur R.
50°

SHAN
otou
Datong
Sanggan He
Beijing
Tangshan
Tianjin
Xar Moron He
Liao He
Changchun
NORTHEAST (MANCHURIA)
Harbin
Jilin
Songhua Res.
L. Khanka
Ussuri R.
Asahikawa
HOKKAIDŌ
Sapporo
Kushiro
Cape Erimo

Fushun
Shenyang
Anshan
Fengcheng
Dandong
Sinuiju
Yalu Jiang
Tumen Jiang
Ch'ongjin
NORTH KOREA
Kimch'aek
Hakodate
SEA OF JAPAN
40°
Akita
HONSHŪ

LULIANG SHAN
TAIHANG SHAN
Taiyuan
Shijiazhuang
NORTH CHINA
Quzhou
Jinan
Zibo
Tai'an
Zaozhuang
Zhengzhou
PLAIN
LIAODONG PENINSULA
Dalian
KOREA BAY
Namp'o
Hamhung
Wonsan
Pyongyang
Kaesong
Seoul
Ansong
Inchon
SOUTH KOREA
Taegu
Niigata
Sendai
Toyama
Kanazawa
Utsunomiya
J A P A N
Mt. Fuji 12,388 ft. (3,776 m.) ▲
Kyōto
Tokyo
Cape Inubō
Kawasaki
Yokohama
Nagoya

BO HAI
Tianjin
YELLOW SEA
SHANDONG PENINSULA
Cape Chengshan
Qingdao
Quzhou

Fen He
Wei Res.
Luoyang
Huang He
Kuai He
Hongze Res.
Grand Canal
Huai He
Huainan
Gaoyou Res.
Kunsan
Kwangju
Mokp'o
Pusan
Korea Strait
Kobe
Osaka
Hiroshima
Kitakyūshū
SHIKOKU
Kochi

ABA SHAN
Han Shui
Chang Jiang
Wuhan
L. Chao
L. Tai
Huzhou
Nanjing
Hangzhou
Shaoxing
Shanghai
CHEJU IS. (S. KOR.)
Fukuoka
Nagasaki
KYŪSHŪ
Kagoshima
Cape Sata
PACIFIC OCEAN
30°

Dongting L.
Yueyang
Jiang
Changsha
L. Poyang
Nanchang
EAST CHINA SEA

WUYI SHAN
Gan Jiang
Xiang
Fuzhou
Dongshan
Formosa Strait
RYUKYU IS. (JAP.)
OKINAWA
N

Taipei
Chilung
Hsinchu
T'aichung
TAIWAN
Yü Shan 13,113 ft. ▲ (3,997 m.)
Chiai
Fengshan
Kaohsiung
Tropic of Cancer

Xi Jiang
Guangzhou
Kowloon
Victoria
Macao **HONG KONG**
MACAO
LEIZHOU PENINSULA
Cape Tungku
yang
INAN
Bashi Channel
Luzon Strait
20°

SOUTH CHINA SEA
PHILIPPINE SEA

EAST ASIA
⊛ National capital
● Major city
○ Other city
── International boundary
---- Disputed boundary

0   150   300 Miles
0   150   300 Kilometers

Projection: Robinson

110° 120° 130° 140°

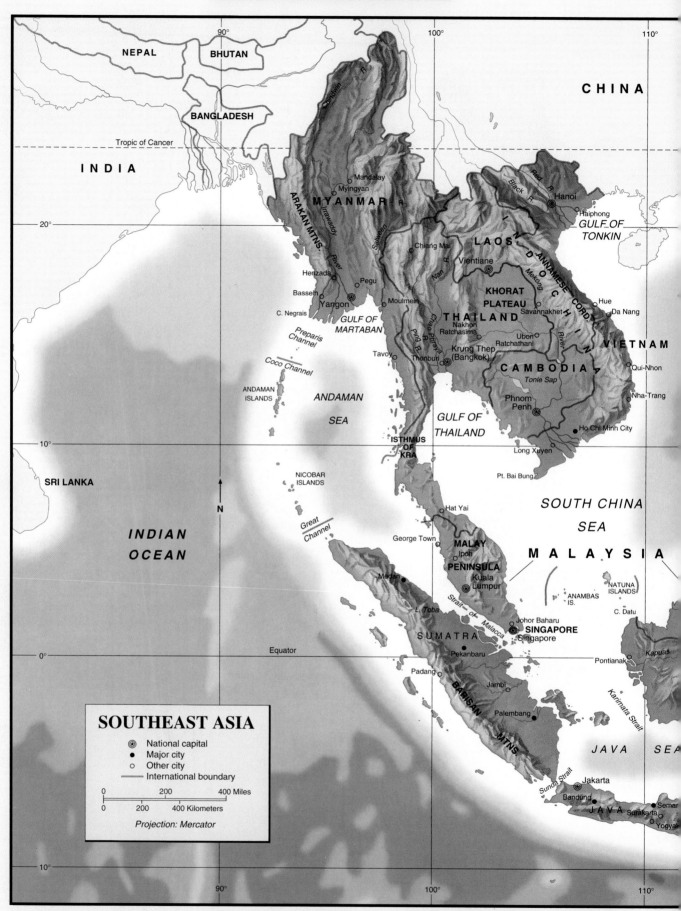

SOUTHEAST ASIA

⊛ National capital
● Major city
○ Other city
— International boundary

0   200   400 Miles
0   200   400 Kilometers

Projection: Mercator

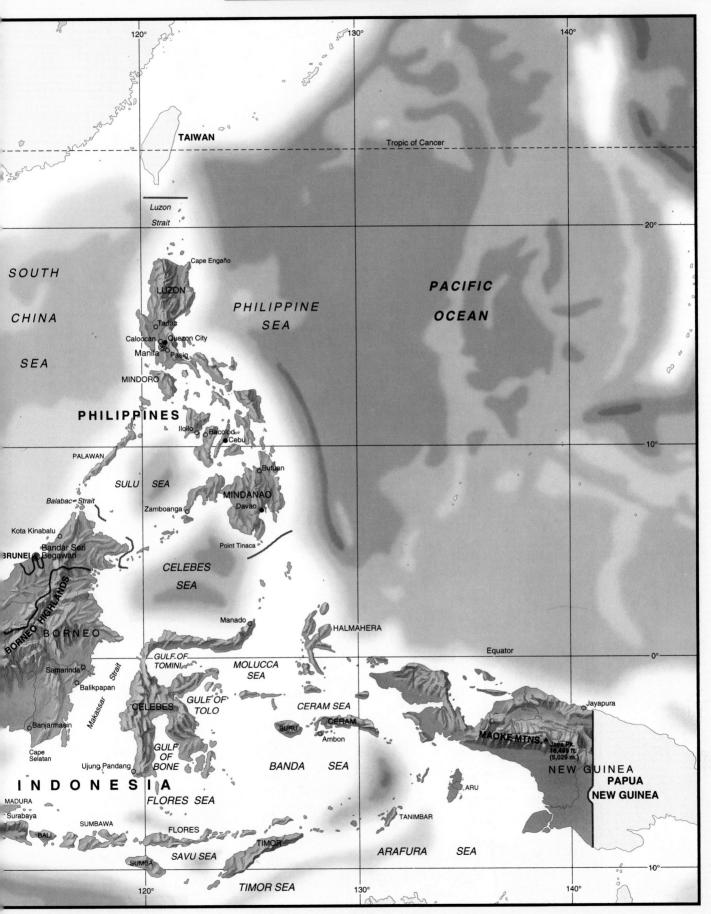

TAIWAN

Tropic of Cancer

Luzon
Strait

20°

SOUTH

CHINA

SEA

Cape Engaño

LUZON

PHILIPPINE

SEA

PACIFIC

OCEAN

Tarlac
Caloocan   Quezon City
Manila   Pasig

MINDORO

PHILIPPINES

Iloilo   Bacolod
Cebu

10°

PALAWAN

Butuan

SULU   SEA

MINDANAO
Davao

Balabac Strait

Zamboanga

Kota Kinabalu

Point Tinaca

Bandar Seri
BRUNEI   Begawan

CELEBES

SEA

BORNEO HIGHLANDS

Manado

HALMAHERA

BORNEO

Equator

0°

GULF OF
TOMINI

MOLUCCA

SEA

Samarinda

Jayapura

Makassar Strait

GULF OF
TOLO

CERAM SEA

Balikpapan

CELEBES

CERAM

BURU

Banjarmasin

Ambon

MAOKE MTNS.

Cape
Selatan

GULF
OF
BONE

BANDA   SEA

Jaya Pk.
16,499 ft.
(5,029 m.)

Ujung Pandang

NEW GUINEA

PAPUA
NEW GUINEA

I N D O N E S I A

MADURA

FLORES SEA

ARU

Surabaya

SUMBAWA

BALI

FLORES

TANIMBAR

TIMOR

ARAFURA   SEA

10°

SUMBA

SAVU SEA

TIMOR SEA

120°

130°

140°

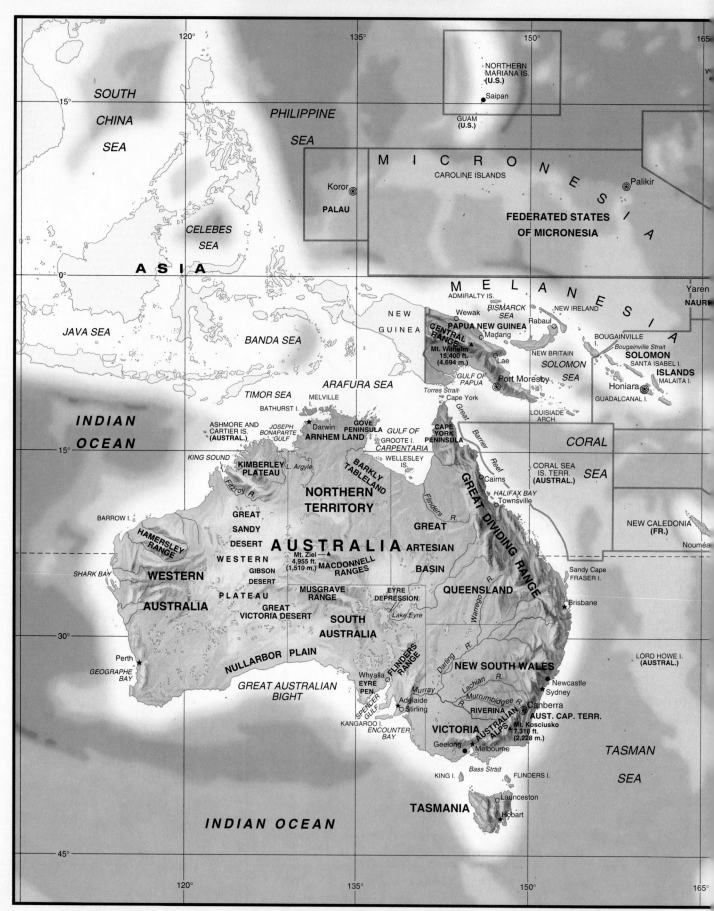

120°   135°   150°   165°

NORTHERN
MARIANA IS.
(U.S.)
● Saipan

GUAM
(U.S.)

SOUTH
CHINA
SEA

PHILIPPINE
SEA

M I C R O N E S I A

CAROLINE ISLANDS

Koror ⊛
PALAU

● Palikir

FEDERATED STATES
OF MICRONESIA

CELEBES
SEA

A S I A

0°

JAVA SEA

M E L A N E S I A

Yaren
NAURU

ADMIRALTY IS.

NEW
GUINEA

Wewak ●

BISMARCK
SEA

.NEW IRELAND

PAPUA NEW GUINEA

CENTRAL
RANGES
Mt. Wilhelm
15,400 ft.
(4,694 m.)

● Madang

Rabaul ●

BOUGAINVILLE
I.
Bougainville Strait

BANDA SEA

Lae ●

NEW BRITAIN

SOLOMON
SEA

SOLOMON
SANTA ISABEL I.
ISLANDS
MALAITA I.

ARAFURA SEA

GULF OF
PAPUA

Port Moresby ●

Honiara ●
GUADALCANAL I.

TIMOR SEA

Torres Strait

MELVILLE
I.

Cape York

INDIAN
OCEAN

BATHURST I.

LOUISIADE
ARCH.

CORAL

ASHMORE AND
CARTIER IS.
(AUSTRAL.)

JOSEPH
BONAPARTE
GULF

★ Darwin
ARNHEM LAND

GOVE
PENINSULA

GULF OF
CARPENTARIA

GROOTE I.

CAPE
YORK
PENINSULA

CORAL SEA
IS. TERR.
(AUSTRAL.)

SEA

15°

KING SOUND

KIMBERLEY
PLATEAU

L. Argyle

WELLESLEY
IS.

Great

NEW CALEDONIA
(FR.)

Fitzroy R.

NORTHERN
TERRITORY

BARKLY
TABLELAND

Flinders
R.

GREAT

Reef

Barrier

Cairns ●

Dividing

HALIFAX BAY
● Townsville

Nouméa ●

BARROW I.

HAMERSLEY
RANGE

GREAT
SANDY
DESERT

AUSTRALIA

Mt. Ziel ▲
4,955 ft.
(1,510 m.)

MACDONNELL
RANGES

GREAT
ARTESIAN

Range

Sandy Cape
FRASER I.

SHARK BAY

WESTERN

GIBSON
DESERT

MUSGRAVE
RANGE

EYRE
DEPRESSION

BASIN

QUEENSLAND

R.

Brisbane ★

AUSTRALIA

PLATEAU

GREAT
VICTORIA DESERT

Lake Eyre

SOUTH
AUSTRALIA

Warrego

30°

Perth ★

GEOGRAPHE
BAY

NULLARBOR   PLAIN

Whyalla ●

FLINDERS
RANGE

Darling

R.

NEW SOUTH WALES

Newcastle ●

LORD HOWE I.
(AUSTRAL.)

GREAT AUSTRALIAN
BIGHT

EYRE
PEN.

SPENCER
GULF

Murray

R.

Lachlan

Murrumbidgee R.

R.

● Sydney

KANGAROO I.

Adelaide ★
Stirling ●

ENCOUNTER
BAY

VICTORIA

RIVERINA

● Canberra
AUSTRALIAN
ALPS

AUST. CAP. TERR.

Mt. Kosciusko
7,310 ft.
(2,228 m.)

TASMAN

Geelong ●
★ Melbourne

SEA

KING I.

Bass Strait

FLINDERS I.

TASMANIA

Launceston ●

INDIAN OCEAN

45°

Hobart ●

120°   135°   150°   165°

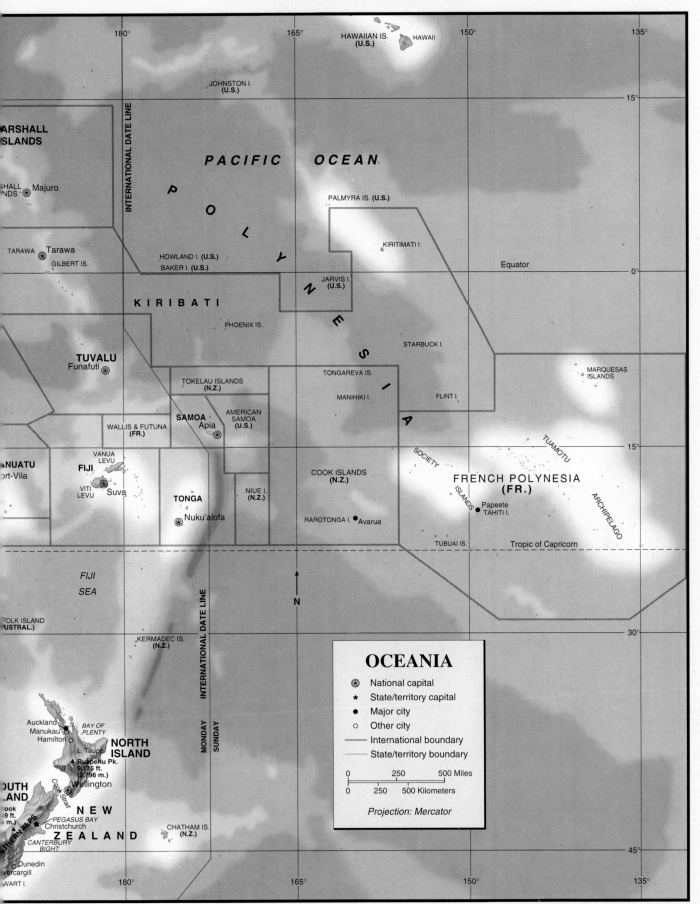

180°   165°   HAWAIIAN IS. (U.S.)   HAWAII   150°   135°

MARSHALL ISLANDS

JOHNSTON I. (U.S.)

15°

PACIFIC OCEAN

SHALL NDS ⊛ Majuro

PALMYRA IS. (U.S.)

P O L Y N E S I A

KIRITIMATI I.

TARAWA ⊛ Tarawa
GILBERT IS.

HOWLAND I. (U.S.)
BAKER I. (U.S.)

Equator   0°

JARVIS I. (U.S.)

KIRIBATI

PHOENIX IS.

STARBUCK I.

MARQUESAS ISLANDS

TUVALU
Funafuti ⊛

TOKELAU ISLANDS (N.Z.)

TONGAREVA IS.

MANIHIKI I.

FLINT I.

WALLIS & FUTUNA (FR.)

SAMOA
Apia ⊛

AMERICAN SAMOA (U.S.)

SOCIETY

TUAMOTU

15°

VANUATU
ort-Vila

VANUA LEVU

FIJI

VITI LEVU   Suva

COOK ISLANDS (N.Z.)

TONGA
⊛ Nuku'alofa

NIUE I. (N.Z.)

ISLANDS

FRENCH POLYNESIA (FR.)

Papeete
TAHITI I.

ARCHIPELAGO

RAROTONGA I. ● Avarua

TUBUAI IS.

Tropic of Capricorn

FIJI SEA

FOLK ISLAND USTRAL.)

30°

KERMADEC IS. (N.Z.)

INTERNATIONAL DATE LINE

↑ N

MONDAY | SUNDAY

OCEANIA

⊛ National capital
★ State/territory capital
● Major city
○ Other city
— International boundary
— State/territory boundary

| 0 | 250 | 500 Miles |
| 0 | 250 | 500 Kilometers |

Projection: Mercator

Auckland
Manukau ○
Hamilton ○

BAY OF PLENTY

NORTH ISLAND

L. Taupo
▲ Ruapehu Pk.
9,175 ft.
(2,796 m.)
○ Wellington

Cook Strait

OUTH AND

ook 9 ft. m.)

PEGASUS BAY
Christchurch

CANTERBURY BIGHT

N E W

CHATHAM IS. (N.Z.)

Z E A L A N D

45°

○ Dunedin
vercargill

WART I.

180°   165°   150°   135°

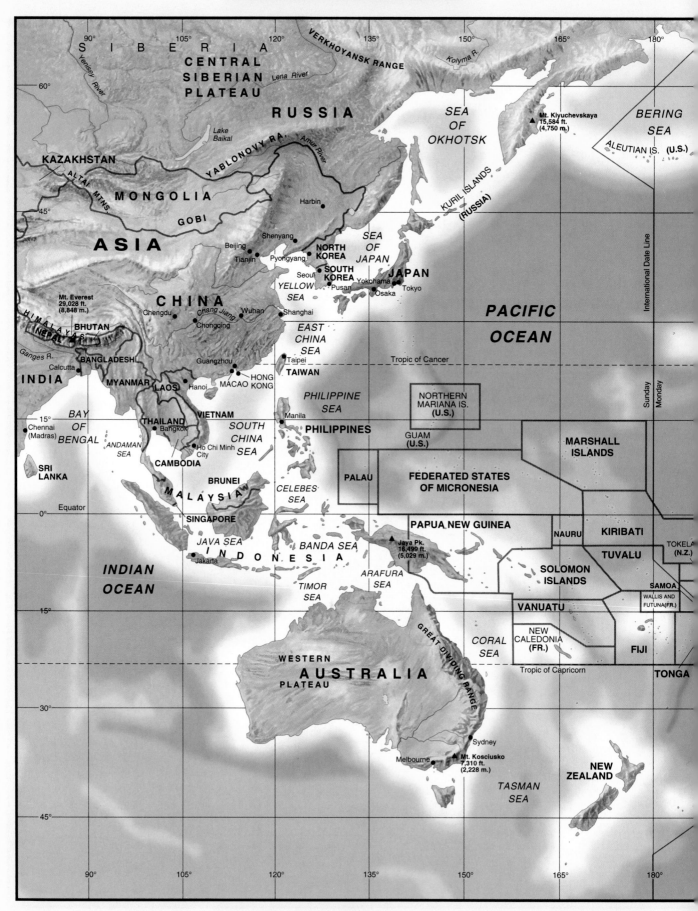

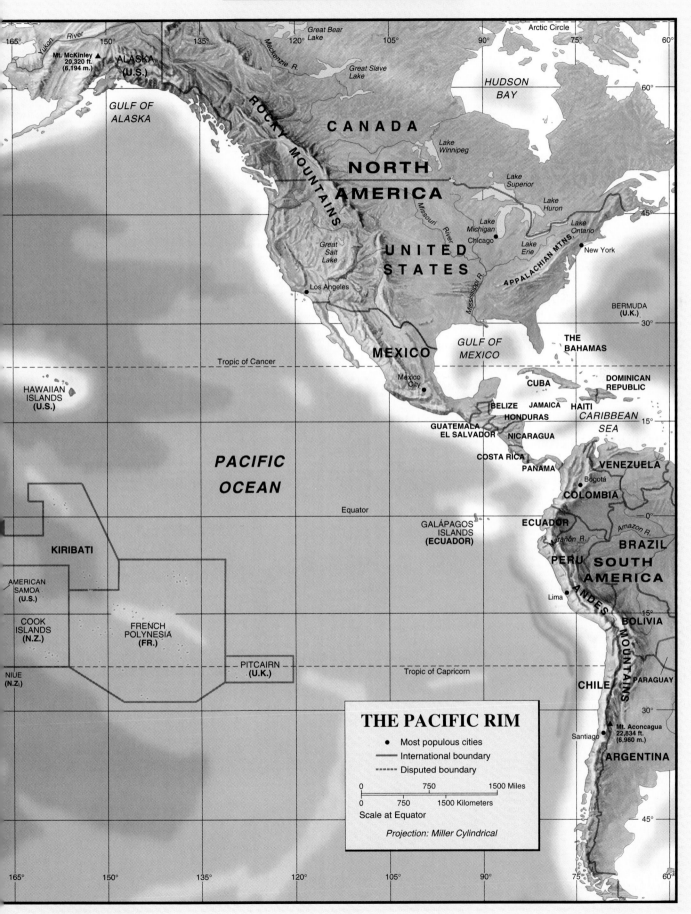

THE PACIFIC RIM

- • Most populous cities
- —— International boundary
- ······ Disputed boundary

0   750   1500 Miles

0   750   1500 Kilometers

Scale at Equator

Projection: Miller Cylindrical

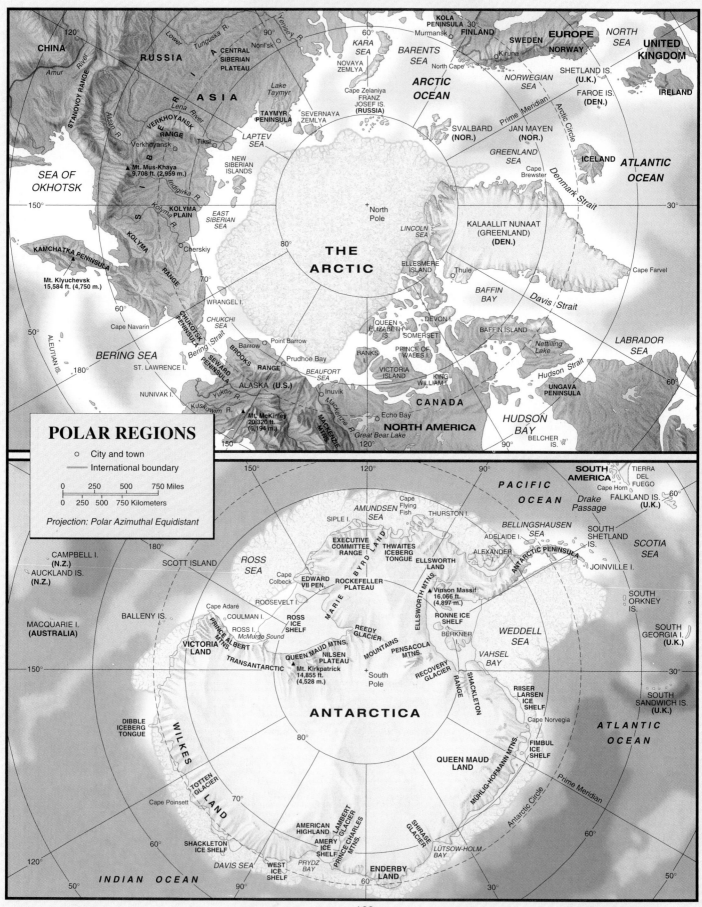

POLAR REGIONS

○ City and town
—— International boundary

| 0 | 250 | 500 | 750 Miles |
| 0 | 250 | 500 | 750 Kilometers |

Projection: Polar Azimuthal Equidistant

THE ARCTIC

CHINA · RUSSIA · CENTRAL SIBERIAN PLATEAU · ASIA · Amur · River · Lower · Tunguska R. · Yenisey R. · Noril'sk · Lake Taymyr · Lena River · KARA SEA · NOVAYA ZEMLYA · Cape Zelaniya · FRANZ JOSEF IS. (RUSSIA) · SEVERNAYA ZEMLYA · KOLA PENINSULA · Murmansk · FINLAND · SWEDEN · NORWAY · Kiruna · North Cape · BARENTS SEA · EUROPE · NORTH SEA · UNITED KINGDOM · IRELAND · ARCTIC OCEAN · NORWEGIAN SEA · SHETLAND IS. (U.K.) · FAROE IS. (DEN.) · Prime Meridian · Arctic Circle · SVALBARD (NOR.) · JAN MAYEN (NOR.) · GREENLAND SEA · Cape Brewster · ICELAND · ATLANTIC OCEAN · Denmark Strait · STANOVOY RANGE · Aldan R. · VERKHOYANSK RANGE · Verkhoyansk · Tiksi · LAPTEV SEA · ▲ Mt. Mus-Khaya 9,708 ft. (2,959 m.) · NEW SIBERIAN ISLANDS · Indigirka R. · SEA OF OKHOTSK · KOLYMA R. · KOLYMA PLAIN · EAST SIBERIAN SEA · North Pole · LINCOLN SEA · KALAALLIT NUNAAT (GREENLAND) (DEN.) · 80° · 70° · Cherskiy · KAMCHATKA PENINSULA · ▲ Mt. Klyuchevsk 15,584 ft. (4,750 m.) · KOLYMA RANGE · WRANGEL I. · CHUKCHI SEA · ELLESMERE ISLAND · Thule · BAFFIN BAY · Davis Strait · Cape Farvel · ALEUTIAN IS. · BERING SEA · CHUKOTSK PENINSULA · Cape Navarin · Bering Strait · ST. LAWRENCE I. · BROOKS RANGE · Barrow · Point Barrow · BEAUFORT SEA · Prudhoe Bay · QUEEN ELIZABETH IS. · DEVON I. · SOMERSET · BANKS I. · PRINCE OF WALES I. · BAFFIN ISLAND · Nettilling Lake · LABRADOR SEA · NUNIVAK I. · SEWARD PENINSULA · ALASKA (U.S.) · Yukon R. · Inuvik · VICTORIA ISLAND · KING WILLIAM I. · Hudson Strait · UNGAVA PENINSULA · Kuskokwim R. · ▲ Mt. McKinley 20,320 ft. (6,194 m.) · MACKENZIE MTNS. · Echo Bay · Mackenzie R. · Great Bear Lake · CANADA · NORTH AMERICA · HUDSON BAY · BELCHER IS.

ANTARCTICA

SOUTH AMERICA · TIERRA DEL FUEGO · Cape Horn · FALKLAND IS. (U.K.) · PACIFIC OCEAN · Drake Passage · AMUNDSEN SEA · Cape Flying Fish · SIPLE I. · THURSTON I. · BELLINGSHAUSEN SEA · ADELAIDE I. · ALEXANDER I. · ANTARCTIC PENINSULA · SOUTH SHETLAND IS. · SCOTIA SEA · JOINVILLE I. · EXECUTIVE COMMITTEE RANGE · BYRD LAND · THWAITES ICEBERG TONGUE · ELLSWORTH LAND · ROSS SEA · Cape Colbeck · EDWARD VII PEN. · ROCKEFELLER PLATEAU · MARIE · ELLSWORTH MTNS. · ▲ Vinson Massif 16,066 ft. (4,897 m.) · RONNE ICE SHELF · WEDDELL SEA · SOUTH ORKNEY IS. · SOUTH GEORGIA I. (U.K.) · ROOSEVELT I. · ROSS ICE SHELF · REEDY GLACIER · BERKNER I. · Cape Adaré · COULMAN I. · PRINCE ALBERT MTNS. · ROSS I. · McMurdo Sound · QUEEN MAUD MTNS. · NILSEN PLATEAU · MOUNTAINS · PENSACOLA MTNS. · VAHSEL BAY · SOUTH SANDWICH IS. (U.K.) · VICTORIA LAND · TRANSANTARCTIC · ▲ Mt. Kirkpatrick 14,855 ft. (4,528 m.) · South Pole · RECOVERY GLACIER · SHACKLETON RANGE · RIISER-LARSEN ICE SHELF · Cape Norvegia · ATLANTIC OCEAN · CAMPBELL I. (N.Z.) · AUCKLAND IS. (N.Z.) · SCOTT ISLAND · BALLENY IS. · MACQUARIE I. (AUSTRALIA) · WILKES · DIBBLE ICEBERG TONGUE · LAND · TOTTEN GLACIER · Cape Poinsett · FIMBUL ICE SHELF · MÜHLIG-HOFMANN MTNS. · QUEEN MAUD LAND · Prime Meridian · Antarctic Circle · AMERICAN HIGHLAND · LAMBERT GLACIER · PRINCE CHARLES MTNS. · SHIRASE GLACIER · LÜTZOW-HOLM BAY · SHACKLETON ICE SHELF · AMERY ICE SHELF · WEST ICE SHELF · PRYDZ BAY · ENDERBY LAND · DAVIS SEA · INDIAN OCEAN

Geography and Map Skills Handbook

CONTENTS

Globes and Map Projections

Building Map Skills

Reading Graphs and Charts

Globes and Map Projections

PREVIEW

Words to Know
- projection
- hemisphere
- latitude
- longitude
- grid system
- absolute location
- great circle route

Read to Learn . . .
1. how mapmakers represent a round globe on a flat map.
2. what projections are useful for different purposes.
3. how to find an exact location.

Globes and the Earth

Although a globe is the most accurate way to represent the earth, using one presents some serious disadvantages. First, it is hard to carry a globe around in your pocket. Then try to imagine a globe large enough to show your community in detail. For these reasons, geographers use maps.

A globe—like the earth—is a sphere. A map, however, is a flat piece of paper. These facts explain why a flat map always distorts the surface of the earth it is showing.

Drawing Maps

Imagine taking the whole peel from an orange and trying to flatten it on a table. You would either have to cut it or stretch parts of it. Mapmakers face a similar problem in showing the surface of the round earth on a flat map. The different ways they have found to do this are called **projections**. Mapmakers call these different ways projections because a source of light, such as a flashlight, is placed inside of a globe and the light projects, or throws, the image onto a flat piece of paper.

Goode's Interrupted Projection

Goode's projection is called an equal-area projection. This means the projection quite accurately presents the size and shape of the continents. Distances—especially in the oceans—are less accurate. Researchers might use this type of projection to compare continent statistics according to area.

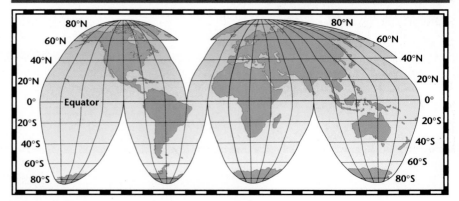

Goode's Interrupted Projection

Mercator Projection

A Mercator projection was one of the earliest types of maps drawn. Mapmaker Gerardus Mercator first created this projection in 1569. This kind of projection shows land shapes fairly accurately, but not size or distance. Areas that are distant from the Equator are quite distorted on this projection. Alaska, for example, appears much larger on a Mercator map than it does on a globe. This map projection does show true directions, however, making it very useful for sea travel.

Mercator Projection

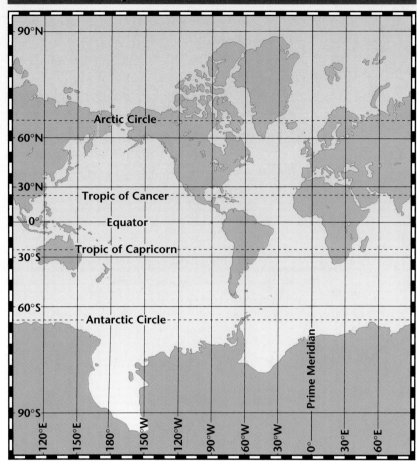

Robinson Projection

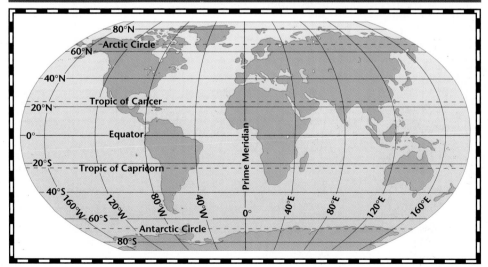

The Robinson projection shows both the size and shape of oceans and continents quite accurately. The shapes of the continents also appear much as they do on a globe. The areas most distorted on this projection are near the North and South poles. Textbook and atlas maps are often Robinson projections.

Hemispheres

Remember that a globe is the most accurate way to represent the earth. To locate places on the earth, geographers set up a system of imaginary lines that crisscross the globe. One of these lines, the Equator, circles the middle of the earth like a belt. It divides the earth into "half spheres," or **hemispheres**. Everything north of the Equator is in the Northern Hemisphere. Everything south of the Equator is in the Southern Hemisphere.

Another imaginary line running from north to south divides the earth into half spheres in the other direction. Find this line—the Prime Meridian—on a globe. It is located at 0° longitude. Everything east of the Prime Meridian is in the Eastern Hemisphere. Everything west of the Prime Meridian is in the Western Hemisphere. North America lies in the Northern Hemisphere and the Western Hemisphere.

Hemispheres

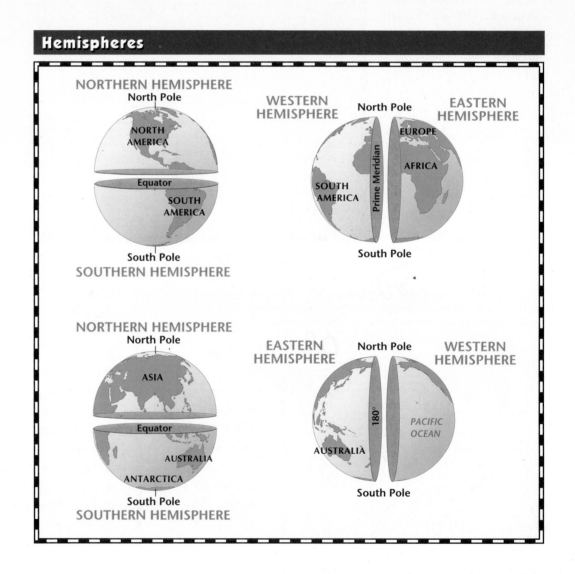

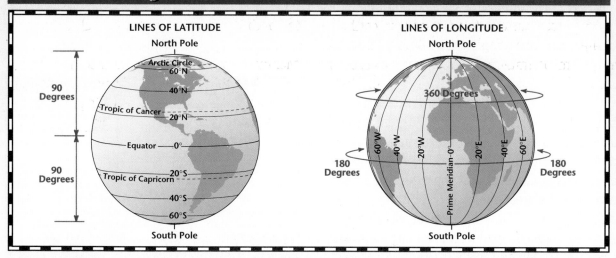

LINES OF LATITUDE

North Pole
Arctic Circle
60°N
40°N
Tropic of Cancer
20°N
Equator — 0°
Tropic of Capricorn
20°S
40°S
60°S
South Pole

90 Degrees
90 Degrees

LINES OF LONGITUDE

North Pole
360 Degrees
60°W 40°W 20°W Prime Meridian–0° 20°E 40°E 60°E
180 Degrees
180 Degrees
South Pole

Latitude and Longitude

Parallels The Equator and the Prime Meridian are the starting points for two sets of lines used to find any location. Parallels circle the earth and show **latitude**, which is distance measured in degrees north and south of the Equator at 0° latitude. The letter *N* or *S* following the degree symbol tells you if the location is north or south of the Equator. The North Pole is at 90° North *(N)* latitude, and the South Pole is at 90° South *(S)* latitude.

Two important parallels in between the poles are the Tropic of Cancer at 23½°N latitude and the Tropic of Capricorn at 23½°S latitude. You can also find the Arctic Circle at 66½°N latitude and the Antarctic Circle at 66½°S latitude.

Meridians Meridians run north to south from pole to pole. These lines signify **longitude**, which is distance measured in degrees east *(E)* or west *(W)* of the Prime Meridian at 0° longitude. On the opposite side of the earth is the International Date Line, or the 180° meridian.

Lines of latitude and longitude cross each other in the form of a **grid system**. Knowing a place's latitude and longitude allows you to locate it exactly on a map or globe. You can name an **absolute location** by naming the latitude and longitude lines that cross nearest to that location. For example, Tokyo is located at about 36°N latitude and 140°E longitude. Where those two lines cross is the absolute location of the city.

Japan

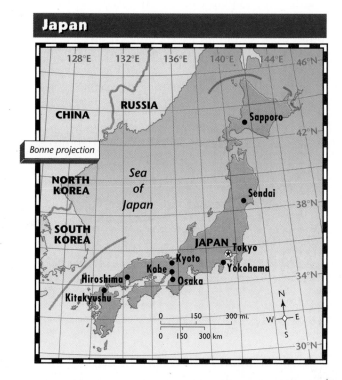

128°E 132°E 136°E 140°E 144°E 46°N

CHINA
RUSSIA
Sapporo
42°N

Bonne projection

NORTH KOREA
Sea of Japan
Sendai
38°N

SOUTH KOREA
JAPAN Tokyo
Kyoto Yokohama
Kobe
Hiroshima Osaka
34°N

Kitakyushu

0 150 300 mi.
0 150 300 km

N
W   E
S

30°N

Great Circles

The idea of a great circle illustrates one important difference between using a map and using a globe. Because both the earth and a globe are round, they accurately show great circles.

A great circle is the shortest distance between two points on earth. A great circle is any circle you can draw on the earth that divides it into two equal parts. A line drawn along the Equator around the entire earth is an example of a great circle. Traveling along a great circle is called following a **great circle route**. Airplane pilots and ship captains often use great circle routes to shorten their trips and cut down on the amount of fuel needed. These routes often take pilots over the North Pole. A trip between Moscow, Russia, and New York, New York, over the North Pole is an example of a great circle route. Other great circles lie along the routes between Hong Kong, China, and Washington, D.C., and between Cape Town, South Africa, and Washington, D.C. Find these routes on a globe.

The great circle route between two points may not appear to be the shortest distance on a flat map. On this map, the great circle route between Tokyo, Japan, and Los Angeles,

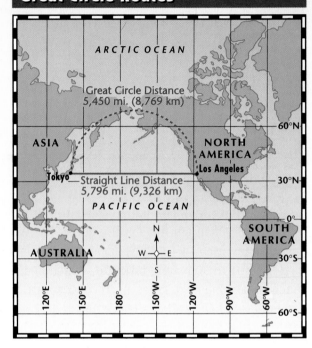

Great Circle Routes

ARCTIC OCEAN

Great Circle Distance 5,450 mi. (8,769 km)

ASIA

NORTH AMERICA
Los Angeles

Tokyo — Straight Line Distance 5,796 mi. (9,326 km)

PACIFIC OCEAN

SOUTH AMERICA

AUSTRALIA

60°N
30°N
0°
30°S
60°S

120°E 150°E 180° 150°W 120°W 90°W 60°W

California, appears to be far longer than the straight line route. Actually, the great circle is more than 345 miles (550 km) shorter!

SECTION 1 ASSESSMENT

REVIEWING GLOBES AND MAP PROJECTIONS

1. **Define the following:** projection, hemisphere, latitude, longitude, grid system, absolute location, great circle route.
2. Why does every map projection distort some parts of the earth?
3. What imaginary line divides the earth into the Northern Hemisphere and Southern Hemisphere? Where is it located?
4. **ACTIVITY** Look at a globe or world map to discover more of earth's great circles. Make a list of airplane flights that could follow great circle routes between New York City and other major world cities.

SECTION 2

Building Map Skills

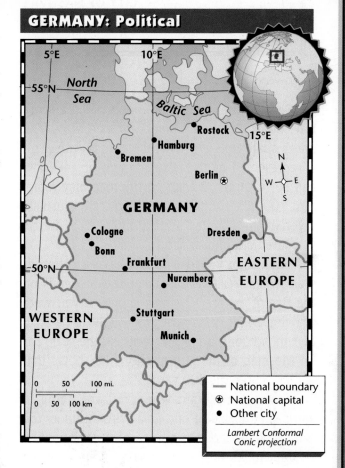

PREVIEW

Words to Know

- key
- cardinal direction
- compass rose
- intermediate direction
- scale bar
- scale
- relief
- elevation
- contour line
- elevation profile
- population density
- climate region

Read to Learn . . .

1. how you can find a place's exact location on a map.
2. how to recognize the special parts of a map.
3. how to identify different kinds of maps.

Parts of Maps

Maps can direct you down the street, across the country, or around the world. There are as many different kinds of maps as there are uses for them. Being able to read a map begins with learning about its parts.

Map Key The map **key** unlocks the information presented on the map. The key explains the symbols used on the map. On this map of Germany, for example, dots mark cities and towns.

On a road map, the key tells what map lines stand for paved roads, dirt roads, and interstate highways. A pine tree symbol may represent a park, while an airplane is often the symbol for an airport.

Compass Rose An important first step in reading any map is to find the direction marker. A map has a symbol that tells you where the **cardinal directions**—north, south, east, and west—are positioned. Sometimes all of these directions are shown with a **compass rose**. An intermediate direction, such as southeast, may also be on the compass rose. **Intermediate directions** fall between the cardinal directions.

GERMANY: Political

North Sea · Baltic Sea · Rostock · Hamburg · Bremen · Berlin ✪ · GERMANY · Cologne · Bonn · Dresden · Frankfurt · Nuremberg · EASTERN EUROPE · WESTERN EUROPE · Stuttgart · Munich

5°E · 10°E · 55°N · 15°E · 50°N

0 50 100 mi.
0 50 100 km

— National boundary
✪ National capital
• Other city

Lambert Conformal Conic projection

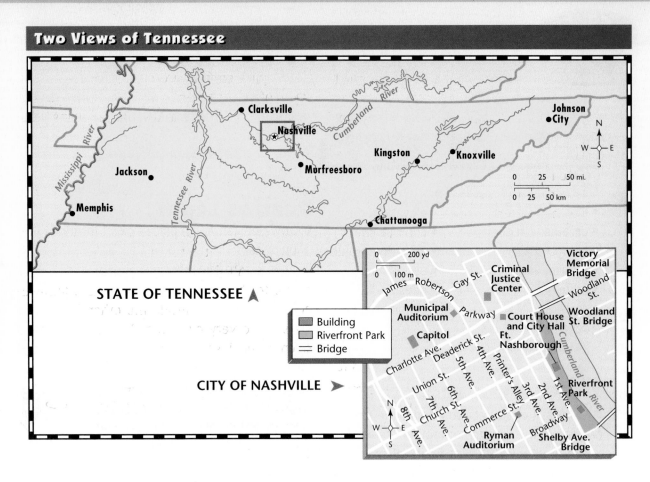

STATE OF TENNESSEE ▲

CITY OF NASHVILLE ➤

Building
Riverfront Park
Bridge

Scale A measuring line, often called a **scale bar**, helps you find distance on the map. The map's **scale** tells you what distance on the earth is represented by the measurement on the scale bar. For example, 1 inch on the map may represent 100 miles on the earth. Map scale is usually given in both miles and kilometers, a metric measurement of distance.

Each map you use might have a different scale. What scale a mapmaker uses depends on the size of the area the map shows. If you were drawing a map of a house, you might use a scale of 1 inch equals 5 feet. In contrast, the scale bar on the map of the city of Nashville shows that ⅜-inch represents 200 yards.

Scale is important when you are trying to compare the size of one area with another. You can't make a true comparison of the continental United States and Africa unless you are using maps drawn to the same scale. You will find similar maps comparing different countries to the continental United States (not including Alaska and Hawaii) throughout your book.

AFRICA AND THE UNITED STATES: Land Comparison

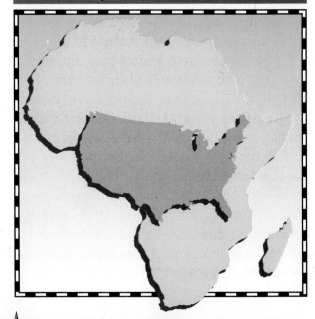

Different Kinds of Maps

Maps that show a wide range of general information about an area are called general purpose maps. Two of the most common general purpose maps are political and physical maps.

Political Maps Political maps generally show political, or human-made, divisions of countries or regions. The political map of Germany on page 7, for example, shows boundaries between Germany and other countries. It also shows cities within Germany and bodies of water surrounding Germany.

Physical Maps A physical map shows the physical features of an area, such as its mountains and rivers. The colors used for some features on physical maps are standard—brown or green for land, blue for water. Physical maps use colors and shadings to show **relief**—how flat or rugged the land surface is. Colors also may be used to show **elevation**—the height of an area above sea level.

Mapmakers also use color to show rainfall, types of soil, or plant life. Physical maps, like political maps, have a key that explains what each color and symbol stands for.

Contour Maps One kind of physical map scientists use may not look like a map at all to some people. This map—called a contour map—shows elevation. A contour map has **contour lines**—one for each major level of elevation. All the land at the same elevation is connected by a line. These lines usually form circles or ovals—one inside the other. If contour lines come very close together, the surface is steep. If the lines are spread far apart, the land is flat or rises very gradually.

Another way to show relief is to look at the landscape from the side, or profile. This **elevation profile** is a cutaway diagram. It clearly shows level land, hills, and steeper mountains.

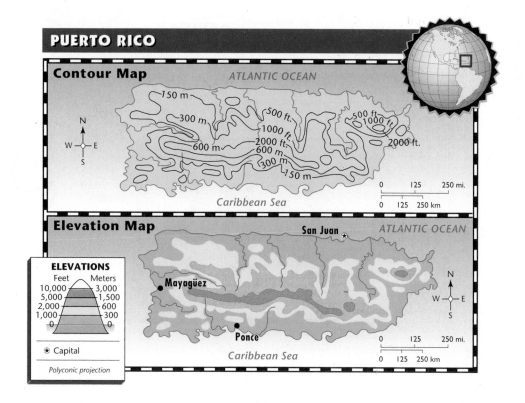

Special Purpose Maps

Besides showing political or physical features, some maps have a special purpose. They may show cultural features, historical changes, unique physical features, population, or climates. The map's title tells what kind of specialized information it shows. Colors and symbols in the map key are especially important on this type of map.

When looking at a special purpose map, it is important to remember that you are seeing only the information that relates to the map's special purpose. For example, on a population map you will see indications of where and how many people live in a country. The map will probably not show you anything about what crops the people there raise.

Land Use and Resources Maps Look at this map of land use and resources in South Africa. Colors identify parts of the area where people follow certain ways of life, such as farming or herding. Places where manufacturing takes place are also marked with color. The symbols stand for the different kinds of natural resources found in or on the land. The number and type of symbols can vary from map to map depending upon the area shown. Diamonds, for example, are not found in all parts of the world as they are in South Africa. So diamonds might not always be in a resource map key. Resources can include minerals, timber, and important crops.

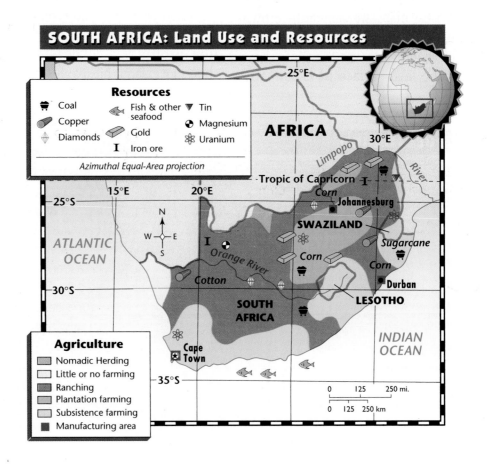

SOUTH AFRICA: Land Use and Resources

Resources
- Coal
- Copper
- Diamonds
- Fish & other seafood
- Gold
- Iron ore
- Tin
- Magnesium
- Uranium

Azimuthal Equal-Area projection

Agriculture
- Nomadic Herding
- Little or no farming
- Ranching
- Plantation farming
- Subsistence farming
- Manufacturing area

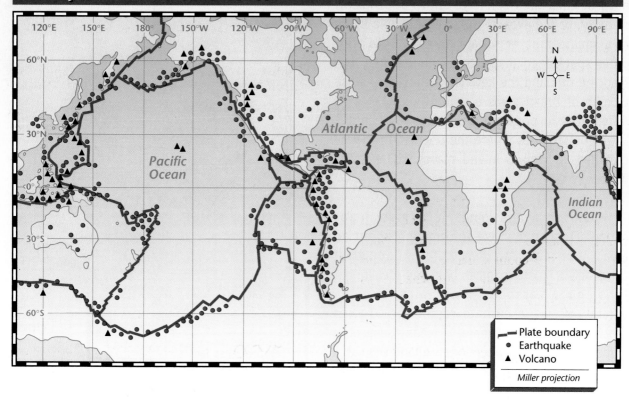

Earthquake and Volcano Zones

- Plate boundary
- Earthquake
- ▲ Volcano

Miller projection

Some special purpose maps have a very limited use. You may have seen a map in your school, for example, that shows the boundaries of your school district. A map such as that would probably only be useful to people in your community and state. Other special purpose maps that show a wide area, such as the entire world, hold information that affects many more people.

Geological Maps Special purpose maps can show information about something that you cannot readily see, such as geological information. Geological maps explain the structure of the earth and how it is believed to have been formed. In addition to giving you new information, such maps might also help you make

connections between events and their causes. It is often helpful to compare a special purpose map to another map to receive a complete picture of how the earth's elements work together.

Look at the Earthquake and Volcano Zones map above. It gives you a view of three things: the location of major plates that make up the earth's crust, where volcanoes have erupted, and where earthquakes have occurred most frequently. You will read in Chapter 2 about the plates that make up the earth's crust and how they move. On this map you can see—and make the connection—that earthquakes and volcanoes seem to occur where these plates meet. Geologists call the area of earthquake and volcano activity near the Pacific Ocean the Ring of Fire.

Population Density Maps Another special purpose map uses colors to show **population density**, or the average number of people living in a square mile or square kilometer. As with other specialized maps, it is important to first read the title and the key. The population density map of Brazil gives a striking picture of its differences in population density. You can see large coastal cities where the population is very crowded, as well as inland spaces where fewer people live.

Climate Maps What do climate maps show? They summarize information about an area's rain, snow, and temperatures throughout the year. A climate map can help you see similarities and differences in climates between different places. A **climate region**, broad areas that have the same climate, is readily seen on a climate map. Geographers often divide the earth into four major climate regions—tropical, dry, mid-latitude, and high latitude. In which climate region do you live?

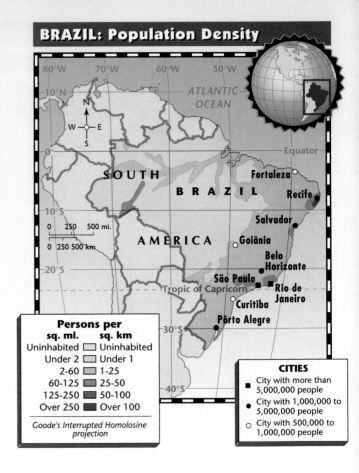

BRAZIL: Population Density

Persons per

sq. mi.	sq. km
Uninhabited	Uninhabited
Under 2	Under 1
2-60	1-25
60-125	25-50
125-250	50-100
Over 250	Over 100

Goode's Interrupted Homolosine projection

CITIES
- City with more than 5,000,000 people
- City with 1,000,000 to 5,000,000 people
- City with 500,000 to 1,000,000 people

SECTION 2 ASSESSMENT

Reviewing Map Skills

1. **Define the following:** key, cardinal direction, compass rose, intermediate direction, scale bar, scale, relief, elevation, contour line, elevation profile, population density, climate region.
2. What kinds of features does a physical map include?
3. What special purpose do geological maps serve?
4. To draw a map showing your route to school, would you use a large scale or a small scale? Why?
5. **ACTIVITY** Use the following grid system "addresses" to find the three cities located at these coordinates: 55°N, 35°E; 23°S, 43°W; 1°N, 103°E.

SECTION 3 Reading Graphs and Charts

PREVIEW

Words to Know

- axis
- bar graph
- line graph
- circle graph
- pictograph
- climograph
- climate
- diagram
- flow chart
- chart
- table

Read to Learn . . .

1. how different types of graphs present information.
2. how charts, tables, and diagrams make data easier to understand.
3. how to read a flow chart.

Graphs

Think of the different places you see graphs and diagrams. They appear in newspapers, magazines, and even on television.

Graphs summarize and present information visually. Each part of a graph gives useful information. First read the graph's title to find out its subject. Then read the labels along the graph's **axes**—the vertical and horizontal lines along the bottom and sides of the graph. One axis will tell you what is being measured. The other axis tells what units of measurement are being used.

Bar Graphs Look carefully at this **bar graph** of the world's five most heavily populated countries. The vertical axis lists the countries. The horizontal axis gives the units of population measurement. By comparing the lengths of the bars you can quickly tell which country has the largest population. Bar graphs are especially useful for comparing quantities, and they may show the bars running across the graph or rising up from the bottom.

Comparing Population

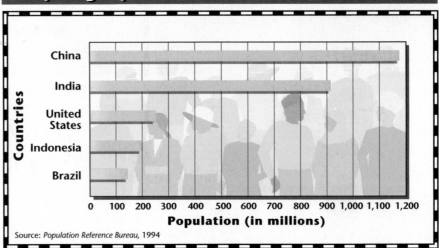

Source: *Population Reference Bureau, 1994*

Line Graphs A tool that is especially good for plotting changes in something over a period of time is a **line graph**. The amounts being plotted on this graph are shown by dots. These dots are connected by a line. Line graphs sometimes have two or more lines plotted on them.

This line graph shows how the population density of India has changed since 1978. The vertical axis lists people per square mile. The horizontal axis shows the passage of time in four-year periods from 1978 to 1994.

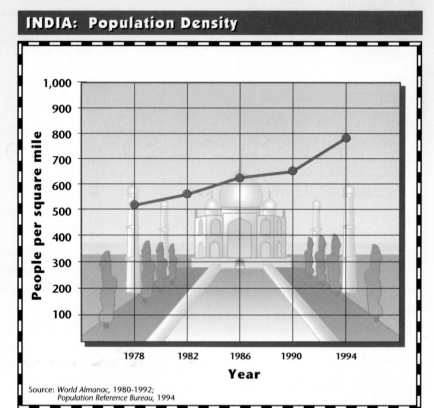

INDIA: Population Density

Source: *World Almanac, 1980-1992; Population Reference Bureau, 1994*

Earth's Water and Land Area

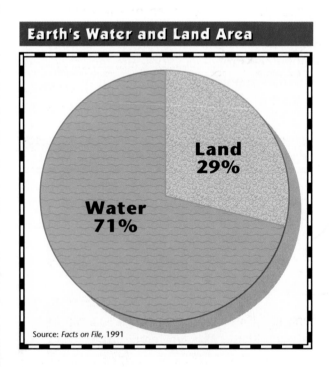

Land 29%

Water 71%

Source: *Facts on File, 1991*

Circle Graphs When you want to show how the *whole* of something is divided into its *parts*, you should use a **circle graph**. Because of their shape, circle graphs are often called pie graphs. Each "slice" represents a part or percentage of the whole "pie." Circle graphs are also good for making comparisons between two or more whole things that are broken into parts.

On this graph, the circle represents the whole surface area of the earth. Land and water each occupy a certain percentage of that whole, as shown by the way the circle is divided. You could easily compare this to another circle graph showing the total surface of another planet with land and water indicated.

Pictographs Like bar graphs, pictographs are good for making comparisons. **Pictographs** use rows of small symbols or pictures, each standing for an amount. This pictograph shows oil production in Venezuela. The key tells you that each oil rig stands for 20 million metric tons of oil. Pictographs are read like a bar graph. The total number of oil rigs in a row adds up to the production for that year.

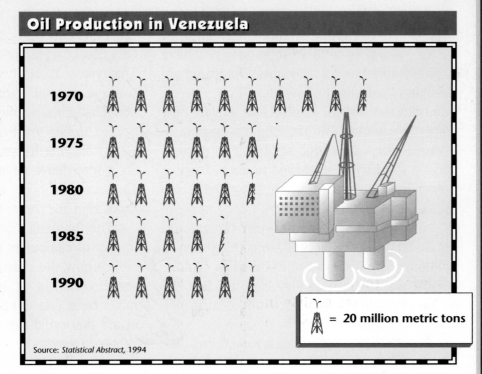

Oil Production in Venezuela

1970

1975

1980

1985

1990

= 20 million metric tons

Source: *Statistical Abstract, 1994*

Climographs A **climograph**, or climate graph, combines a line graph and a bar graph. It gives an overall picture of the **climate**—the long-term weather patterns—in a specific place. Because climographs include several kinds of information, you need to read them carefully.

Note that the vertical bars on the climograph represent average amounts of precipitation in each month of the year. To measure these bars, use the scale on the right-hand axis of the graphs.

The line above the bars, like those on other line graphs, represents changes in the average temperature based on the scale on the left. The months are abbreviated on the bottom axis of the graph.

Climograph: New York City

°F	°C		In.	Cm
100	37.8		20	50.8
90	32.2		18	45.7
80	26.7		16	40.6
70	21.1		14	35.6
60	15.6		12	30.5
50	10.0		10	25.4
40	4.4		8	20.3
30	-1.1		6	15.2
20	-6.7		4	10.1
10	-12.2		2	5.1
0	-17.8		0	0

AVERAGE MONTHLY TEMPERATURE

AVERAGE MONTHLY PRECIPITATION

J F M A M J J A S O N D

Source: *World Weather Guide, 1994*

Diagrams

If you enjoy science, mechanics, or working with your hands, you have probably come across diagrams. **Diagrams** are drawings that either show steps in a process, point out the parts of an object, or explain how something works. You can follow a diagram to connect parts of a stereo system or to repair a bicycle. If you have ever tried to use a diagram, you know that the simpler it is, the easier it is to follow.

Another kind of diagram can show you a cross section—or profile—of something such as landforms. Profiles can be helpful when comparing the elevations of an area. You will find elevation profile diagrams in each unit of this book.

Flow Chart A **flow chart** combines elements of a diagram and a chart. It can show the cause-and-effect process of how things happen. Or a flow chart can present the steps in a process. This type of chart often uses arrows to help you see how one step leads to another. For example, you might draw a flow chart to describe the steps needed to film a video. Arrows would link each step, from first to last. A flow chart could also show how things are interconnected. A chart of a governmental body would show how an idea flows from one part to another.

The flow chart on this page illustrates the impact of computers on different areas of life. Notice how the chart uses the arrows to show the direction in which the effects move. You might be able to think of other areas of computer use that could have been added to the chart. What are they?

The Impact of the Computer

COMPUTER

HOME
- Write letters
- Make budgets
- Play games
- Subscribe to on-line services

SCHOOL
- Write reports
- Use programs
  - For learning
  - For review

BUSINESS
- Keep records
  - Number of products sold
  - Customer information
  - Bills to mail out

Charts and Tables

Charts and **tables** present organized facts and statistics so they are easier to read. These graphics can also show how things are organized or how things are related. To understand a chart or table, first read the title. It tells you what information the chart or table contains. Next, read the labels at the top of each column and the left-hand side of the table or chart. They will tell you what data you are working with.

Tables are especially useful in organizing statistics in an easy-to-read way. This table shows data about the world's leading wheat-producing countries. Others might show many more columns and rows of numbers that would be difficult to understand if they were not organized in a table.

World's Leading Wheat Producers

COUNTRY	METRIC TONS PRODUCED
China	101,003,000
United States	66,920,000
India	55,087,000
Russia	46,000,000
France	32,600,000
Canada	29,871,000
Ukraine	19,473,000
Turkey	19,318,000
Kazakhstan	18,500,000
Pakistan	15,684,000

Source: *The Statesman's Yearbook, 1994-1995*

SECTION 3 ASSESSMENT

Reviewing Graphs and Charts

1. **Define the following:** axis, bar graph, line graph, circle graph, pictograph, climograph, climate, diagram, flow chart, chart, table.
2. If you wanted to show how the population of Mexico City changed from 1850 to 1990, what kind of graph would you draw?
3. What two features of climate does a climograph show?
4. **ACTIVITY** Draw a flow chart showing and labeling the steps in some simple process—for example, making a sandwich or using a pay telephone. Give your flow chart to a classmate to follow it and complete the task described. Write a sentence or two about how easy your chart was to follow. See if your classmate agrees or has suggestions for improvement.

Unit 1

Geography of the World

Christian Science Mapparium, United States

EXPLORING THE INTERNET

To learn more about the geography and people of the world, visit the Glencoe Social Studies Web site at **www.glencoe.com** for information, activities, and links to other sites.

GeoJournal Activity

You are about to journey to remote rain forests, bleak deserts, bustling marketplaces, and mountainous villages. You are entering the world of geography—the study of the earth and its people. Imagine that you could visit any place in the world. Where would you go? What would you want to see? Jot down your thoughts in your journal.

Chapter 1 Looking at the Earth

MAP STUDY ACTIVITY

In this chapter you will learn how geographers look at the earth and its people.

1. What are the earth's seven continents?
2. What are the earth's four major oceans?

SECTION 1
Using the Geography Themes

PREVIEW

Words to Know
- geography
- absolute location
- hemisphere
- latitude
- longitude
- grid system
- relative location
- place
- environment
- movement
- region

Places to Locate
- Equator
- Prime Meridian

Read to Learn . . .
1. how geographers study the earth.
2. how place and location can mean different things.
3. how people relate to their environment and to each other.

Our earth is a fascinating place! Look at this photo of Tokyo, Japan, with its modern buildings. Akio Mansai lives and goes to school in this city. He and his classmates learn about the earth and its geography—the same things you will learn as you read this book.

Geography is the study of the earth in all of its variety. When you study geography, you learn about the earth's land, water, and plant and animal life. You analyze where people are, how they live, and what they do and believe. You especially look at places people have created and try to understand *how* and *why* they are different.

Geographers study the earth as the home of people. Five geographic themes—location, place, human/environment interaction, movement, and region—are used here to help you think like a geographer.

Location

In geography, *location* means knowing where you are. Every place on the earth can be given an exact position on the globe, which is called its **absolute location**. To help geographers mark the absolute location of a place, a network or grid of imaginary lines is placed on the earth.

Equator The Equator is an imaginary line that circles the earth midway between the North Pole and the South Pole. It divides the earth into two hemispheres, or halves. The Northern Hemisphere includes all of the land and water between the Equator and the North Pole. The Southern Hemisphere includes all of the land and water between the Equator and the South Pole.

Latitude and Longitude Other imaginary lines called lines of **latitude**, or parallels, circle the earth parallel to the Equator. They measure distance north or south of the Equator in degrees. The Equator is designated as 0°, and the poles are at 90° North and 90° South.

Lines of **longitude**, or meridians, run from the North Pole to the South Pole. They are numbered in degrees east or west of a starting line called the Prime Meridian, which is at 0° longitude. On the opposite side of the earth from the Prime Meridian is the International Date Line, or 180° longitude.

Lines of latitude and longitude cross one another in the form of a **grid system**. If you know a place's exact latitude north or south of the Equator, and its exact longitude east or west of the Prime Meridian, you can easily mark the absolute location of that place on a map or globe.

Relative Location You also can locate a place by finding out how far and in what direction it is from somewhere else. This is called **relative location** because you are learning where a place is *in relation* to another place.

Place

Place has a special meaning in geography. It means more than where a place *is*. It also describes what a place is *like*. That is, what features make this location similar to or different from another place?

These features may be physical characteristics, such as land shape, plants, animal life, or climate. They also may be characteristics of people and the things they have created, including their language, clothing, buildings, music, or ways of making a living.

The geography theme of *place* describes physical characteristics of land, such as these jagged peaks in the High Tatra mountains of Slovakia *(left)*. Place, however, also describes characteristics of cultural groups, such as the lifestyle of nomads in Tibet *(right)*.
PLACE: What can you learn about the lifestyle of the Tibetan nomads from this photo?

Characteristics of Place

Human/Environment Interaction

Wherever humans have lived or traveled, they have changed their **environment**, or natural surroundings. People have blasted through mountains to build roads, cut down forests, built houses, and used grasslands to graze herds. Some human actions have damaged the natural environment, and some have not.

The environment influences the way people live. People adapt their lives to some environmental conditions. To live in a cold climate, for example, people must invent ways to protect themselves and make a living in the cold. To live in a place that is dry, people have to develop ways to provide water.

Movement

The theme of **movement** helps geographers understand the relationship among places. Movement describes how people in one place make contact with people from another place. People, ideas, information, and products are constantly moving around the world. They travel instantly by telephone, computer, or satellite, or go more slowly by car, train, or ship.

When people in one place want something that is not found in their area, they trade with people in areas that have what they want. People in Japan, for example, may listen to music from the United States, while people in Canada eat bananas shipped from Central America.

Region

Geographers often think about the world in **regions**, or areas that share some common characteristics. Regions can be quite small—your county, city, or neighborhood can be a region. They also can be huge—the western United States is a region.

An area can be called a region because of its physical features such as landscape or climate. A region also can be determined by human traits such as language, political boundaries, religion, or the kinds of work the people do.

High-speed Movement

This French high-speed train is one example of how people, products, and ideas move from one place to another.
MOVEMENT: Why might people board this train?

SECTION 1 ASSESSMENT

REVIEWING TERMS AND FACTS

1. **Define the following:** geography, absolute location, hemisphere, latitude, longitude, grid system, relative location, place, environment, movement, region.
2. **LOCATION** What is the starting line for determining longitude?
3. **PLACE** What two kinds of features are used to describe place?

MAP STUDY ACTIVITIES

4. Turn to the world map on page 24. What large land area is both west of the Prime Meridian and entirely north of the Equator?

BUILDING GEOGRAPHY SKILLS

Using a Map Key

To understand what a map is showing, you must read the **map key,** or legend. The map key explains the meaning of special colors, symbols, and lines on the map.

On some maps, for example, colors represent different heights of land. On other maps, colors might stand for climate areas or population. Lines may stand for rivers, railroads, streets, or boundaries. The map key also explains the meaning of other symbols found on the map. To use a map key, follow these steps:

- Read the map title.
- Study the map key to find out what special information it gives.
- Find examples of each map key color, line, or symbol on the map.

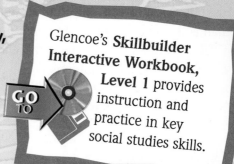

Glencoe's **Skillbuilder Interactive Workbook, Level 1** provides instruction and practice in key social studies skills.

The World

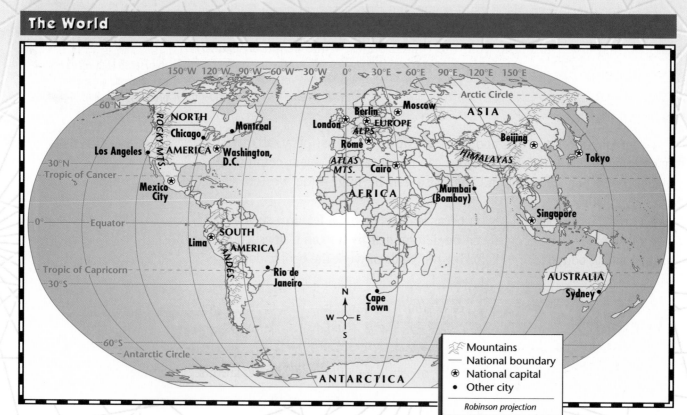

Mountains
National boundary
National capital
Other city

Robinson projection

Geography Skills Practice

1. What are the Himalayas and the Andes? How can you tell?

2. How are national boundaries shown?

3. What does the symbol ✹ represent?

Planet Earth

PREVIEW

Words to Know
- galaxy
- solar system
- orbit
- revolution
- leap year
- axis
- solstice
- equinox

Places to Locate
- Earth
- sun
- moon

Read to Learn . . .
1. how Earth moves in space.
2. why Earth's seasons change.

The first humans to travel in space were amazed by this view of Earth. The beautiful blue oceans and swirling clouds made the planet look like a marble against a dark sky.

Our planet Earth is one of a group of planets revolving around the sun. The sun is just one of hundreds of millions of stars in a **galaxy**, or a huge system of stars. Earth's most important companion in space is the sun. Earth is a member of the **solar system**—planets and other bodies that revolve around our sun.

Location

The Earth in Space

The sun, Earth, eight other planets, their moons, and some smaller asteroids make up our solar system. The planets travel in **orbits**, or elliptical paths, around the sun. Look at the diagram of the solar system on page 26 to see Earth's position in the solar system.

Planets are sometimes classified into two types—those like Earth and those like Jupiter. The Earth-like planets are Mercury, Venus, Mars, and Pluto, which is so far away that little is known about it. These small, solid planets rotate slowly and have few or no moons. The other outer planets—Jupiter, Saturn, Neptune, and Uranus—are huge, rapidly spinning, and surrounded by gassy atmospheres and many moons.

Sun, Earth, and Moon Life on Earth could not exist without heat and light from the sun. About 93 million miles (150 million km) from Earth, the sun is made up mostly of intensely hot gases. Its great mass creates a strong pull of gravity—enough to keep the planets revolving around it.

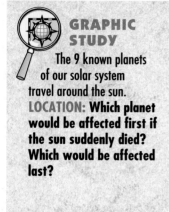

The 9 known planets of our solar system travel around the sun.
LOCATION: Which planet would be affected first if the sun suddenly died? Which would be affected last?

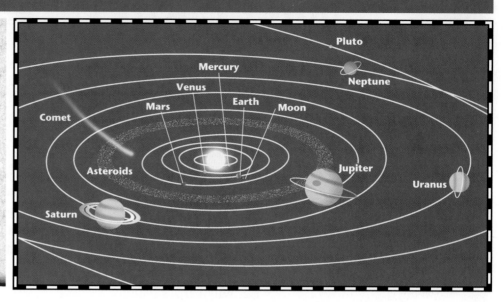

In the same way, the moon revolves around Earth—about once every 30 days. A cold, rocky sphere with no water and no atmosphere, the moon gives off no light of its own. When you see the moon shining, it is actually reflecting light from the sun. The relative positions of the sun, moon, and Earth determine whether you see a "new" or "full" moon.

Earth makes one **revolution,** or complete trip around the sun, in 365¼ days. This is what we define as one year. The extra one-fourth of a day is the reason there is a **leap year,** an extra day in the calendar every four years.

As Earth makes a revolution, it also rotates, or spins, on its axis. The **axis** is an imaginary line that runs through the earth's center between the North and South poles. As Earth turns toward and away from the sun every 24 hours, different areas are in sunlight and in darkness, causing day and night.

Place
The Seasons

Do you live in a place where the seasons change? Or does your weather stay similar year-round? Because Earth is always tilted on its axis, seasons change as Earth travels its year-long orbit around the sun. To see why this happens, look at the four globes in the diagram on page 27. Notice how sunlight falls directly on different parts of Earth at different times during the year.

On about June 21 the North Pole is tilted toward the sun. The sun appears directly overhead at the line of latitude called the Tropic of Cancer. This day is the summer **solstice,** or beginning of summer, in the Northern Hemisphere. It is the day there with the most hours of sunlight.

Six months later—about December 22—the North Pole is tilted away. The sun's direct rays strike the line of latitude known as the Tropic of Capricorn. This is the winter solstice—a time of winter in the Northern Hemisphere, but the beginning of summer in the Southern Hemisphere.

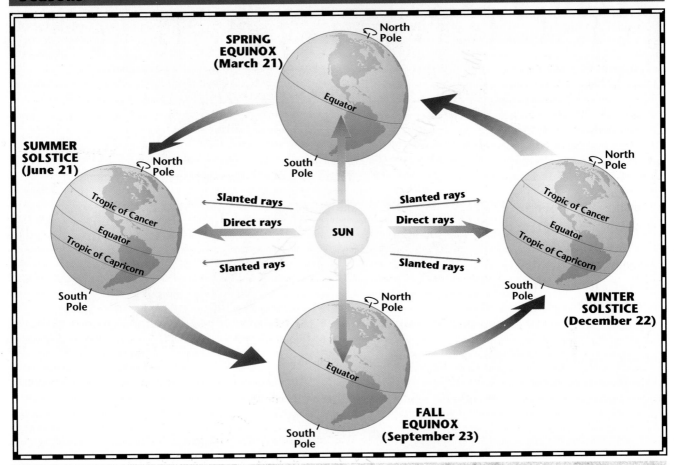

SPRING EQUINOX (March 21)

North Pole

Equator

South Pole

SUMMER SOLSTICE (June 21)

North Pole

Tropic of Cancer

Equator

Tropic of Capricorn

South Pole

Slanted rays

Direct rays

Slanted rays

SUN

Slanted rays

Direct rays

Slanted rays

WINTER SOLSTICE (December 22)

North Pole

Tropic of Cancer

Equator

Tropic of Capricorn

South Pole

FALL EQUINOX (September 23)

North Pole

Equator

South Pole

GRAPHIC STUDY

The tilt of the earth as it revolves around the sun causes the seasons to change.
LOCATION: When it is summer in the Northern Hemisphere, what season is it in the Southern Hemisphere?

Midway between the two solstices, about September 23 and March 21, the sun's rays are directly overhead at the Equator. These are **equinoxes,** when day and night in both hemispheres are of equal length.

SECTION 2 ASSESSMENT

REVIEWING TERMS AND FACTS

1. Define the following: galaxy, solar system, orbit, revolution, leap year, axis, solstice, equinox.
2. PLACE What makes the moon shine?
3. MOVEMENT How do Earth's movements affect the length of the year?

GRAPHIC STUDY ACTIVITIES

4. Looking at the diagram on page 26, describe Earth's location relative to other planets.
5. Look at the diagram above. In June, where do the sun's direct rays hit Earth?

MAKING CONNECTIONS

MEASURING EARTHQUAKES ▽ △

MATH   **SCIENCE**   **HISTORY**   **LITERATURE**   **TECHNOLOGY**

Imagine! The ground shook so hard that the Mississippi River actually changed its course! It was the early 1800s when one of the worst earthquakes in United States history hit New Madrid, Missouri.

Earthquakes happen when great cracks, or faults, in the earth's crust slip. To measure the energy an earthquake releases, scientists use two tools: the seismograph and the Richter scale.

Earthquake Damage, Los Angeles, California

SEISMOGRAPH A seismograph records the earth's tremors, or movements, during an earthquake. A sensitive writing tip draws a line on a turning drum. A tiny tremor thousands of miles away will show up as a wiggly line on the record sheet. The larger the earthquake, the larger the wiggle. The seismograph's recordings measure a quake's movements as well as help scientists locate the earthquake and determine its strength.

THE RICHTER SCALE Developed by Charles Richter in 1935, the Richter scale is a mathematical tool that measures how much energy an earthquake releases. An increase of one number on the scale means a release of 32 times more energy.

You would probably not feel a quake that measures 2.0 on the Richter scale, for example. A quake that measures 5.0, however, releases much more energy and would be felt by everyone in the area. Any earthquake that measures 6.0 or more is considered a major quake. The giant forces unleashed by earthquakes can topple buildings and bridges.

Making the Connection ▽ △

1. What does a seismograph do?
2. What is the Richter scale?

Landforms

Words to Know

- landform
- core
- mantle
- crust
- magma
- continent
- plate tectonics
- fault
- earthquake
- tsunami
- weathering
- erosion
- plateau
- isthmus
- peninsula
- strait
- atmosphere

Places to Locate

- Mount Fuji
- Hawaii

Read to Learn . . .

1. how the earth's structure is layered.
2. how forces change landforms.
3. about the earth's major landforms.

Millions of years ago, these steep valleys in Norway were enlarged by moving ice called glaciers. Norway's landscape is a spectacular example of the forces that constantly change the earth's surface.

Humans occupy only a small percentage of the earth's surface. That surface, however, is very different from place to place. Landscapes and their **landforms,** or individual features, influence where people live and how they relate to their environment.

Place

The Changing Planet

Earth's different landscapes formed over millions of years and are still changing today. A number of forces have shaped past and present landforms. No one has been able to dig deeper than a few kilometers below the earth's crust. Still, geologists have developed a picture of the layers in the earth's structure.

Inside the Earth The inside of the earth is made up of three layers—the **core,** the **mantle,** and the **crust.** In the center of the planet is a dense core of hot metal, probably iron mixed with nickel. This core is divided into a solid inner core and an outer core of melted, liquid metal.

The landforms in Death Valley, California *(left)*, are quite different from those found on this Pacific atoll *(right)*. **HUMAN/ENVIRONMENT INTERACTION: Why do you think it would be difficult for people to live in either of these places?**

Around the core is the mantle, a rock layer about 1,800 miles (2,900 km) thick. Mantle rock is mostly solid, but in it are pockets of **magma**, or melted rock, that flow to the surface when a volcano erupts.

The outer layer, or the earth's crust, is the only layer scientists have studied firsthand. The crust is relatively thin—about 5 to 30 miles (8 to 50 km). It includes both the ocean floors and large land areas known as **continents**. The crust is thinnest on the ocean floor. It is thicker below the continents.

Plate Tectonics Scientists have developed a theory about the earth's structure called **plate tectonics**. This theory states that the crust is not an unbroken shell but consists of moving plates, or huge slabs of rock. These plates fit together like a jigsaw puzzle. They float—sometimes slowly, sometimes suddenly—atop soft rock in the mantle. The plates move in different directions. Some push against each other, others pull apart, and still others slide past one another. Oceans and continents ride on top of the gigantic plates. Scientists claim that, for billions of years, continents have been moving away or toward each other in a process known as continental drift.

Forces of Change **Tectonic activity** and other forces inside the earth have caused changes in the earth's surface. Over time, they have shifted rock layers and folded the crust as easily as if it were paper. They also have built mountains, split continents, and caused **faults**, or cracks, in the earth's crust. Shifts along a fault can cause **earthquakes**, or violent jolts. In coastal areas undersea earthquakes can cause huge waves, known as **tsunamis**, which are sometimes as much as 50 feet (15 m) high!

Another force inside the earth that shapes landforms is volcanic activity. A volcano forms when magma breaks through the crust as lava. How does this shape the land? Volcanoes may explode in a fiery burst of ashes and rock. Then cone-shaped mountains such as Japan's Mount Fuji result. Sometimes lava flows slowly, building up flat mountains like those in Hawaii.

After landforms have been created on the surface, other forces work to change them. A process called **weathering** breaks surface rocks into gravel, sand, or soil. Water, chemicals, and frost cause weathering.

Water, wind, and ice also work on landforms through **erosion**, or the wearing away of the surface. Riverbanks, for example, erode as the river water washes away soil.

Place
Types of Landforms

The earth's land surface consists of seven continents: North America, South America, Europe, Africa, Asia, Australia, and Antarctica. All have a variety of landforms—even icy Antarctica. The ocean floor also has many landforms.

Landforms and Water Bodies

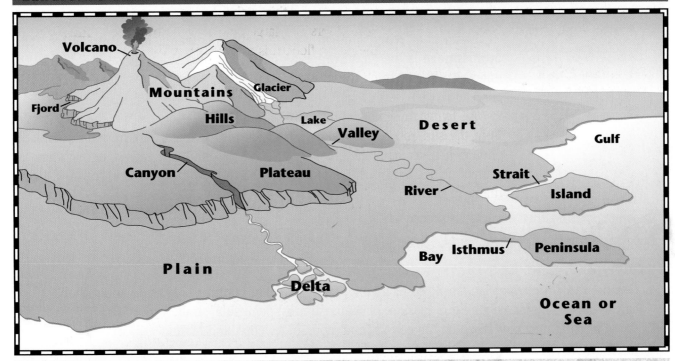

Volcano · Mountains · Glacier · Fjord · Hills · Lake · Valley · Desert · Gulf · Canyon · Plateau · Strait · River · Island · Plain · Bay · Isthmus · Peninsula · Delta · Ocean or Sea

GRAPHIC STUDY

The earth has many different types of landforms and bodies of water.
PLACE: How are plains and plateaus similar? How are they different?

Major Landforms Major landforms include mountains, hills, plateaus, and plains. Look at the illustration on page 31 to see how these are different from one another. Mountains, which can be higher than 20,000 feet (6,100 m), have high peaks, as well as steep or rugged slopes. Hills are lower and more rounded, though they are still higher than the country around them.

Both plateaus and plains are flat, but a **plateau** rises above the land around it. A steep cliff forms at least one side of a plateau. Plains are flat or gently rolling. Smaller landforms include valleys and canyons.

Geographers describe some landforms by their relationship to larger land areas or to bodies of water. An **isthmus** is a narrow piece of land that connects two larger pieces of land. A **peninsula** is a piece of land that is surrounded by water on three sides. A body of land smaller than a continent and surrounded by water is called an island.

Water Bodies About 70 percent of the earth's surface is water. Oceans are the earth's largest bodies of water and can be more than 35,000 feet (11,000 m) deep. Smaller bodies of salt water, which are at least partly enclosed by land, are called seas, gulfs, or bays. A **strait** is a narrow body of water between two pieces of land. Lakes, streams, and rivers are freshwater landforms. You will learn more about the earth's water in Chapter 2.

Place
The Atmosphere

Earth is a living, changing planet of land, water, and atmosphere. The **atmosphere** is the air surrounding Earth. It is a cushion of gases about 1,000 miles (1,600 km) thick. About 99 percent of the atmosphere is made up of nitrogen and oxygen, with 1 percent being other gases.

All living things depend on the atmosphere. It screens out dangerous rays from the sun and reflects some heat back into space. The atmosphere holds in enough heat to make life possible, just as a greenhouse keeps in enough heat to protect plants. Without this protection, Earth would be too cold for most living things to survive.

SECTION 3 ASSESSMENT

REVIEWING TERMS AND FACTS

1. Define the following: landform, core, mantle, crust, magma, continent, plate tectonics, fault, earthquake, tsunami, weathering, erosion, plateau, isthmus, peninsula, strait, atmosphere.

2. PLACE What materials make up the earth's inner core, outer core, and mantle?

3. MOVEMENT How do wind, water, and ice affect landforms on the earth's surface?

GRAPHIC STUDY ACTIVITIES

4. Look at the diagram on page 31. How are a lake and a bay similar? How are they different?

5. On the same diagram, locate a peninsula. How would you describe this landform?

Chapter 1 Highlights

Important Things to Know About Looking at the Earth

SECTION 1 USING THE GEOGRAPHY THEMES

- Location tells you where a place is found.
- Place tells you about the physical and human characteristics of a place.
- Human actions have changed the environment, and the environment has forced humans to adapt.
- People, ideas, information, and products move from place to place.
- Geographers divide the earth into regions based on common physical or human features.

SECTION 2 PLANET EARTH

- Earth, its moon, and other planets are part of our solar system.
- The tilt of Earth on its axis and its revolution around the sun cause the changes in seasons.
- The rotation of Earth on its axis causes areas to have day and night.

SECTION 3 LANDFORMS

- The earth has an inner and outer core, a mantle, and an outer crust.
- Forces inside the earth, such as volcanic activity, create landforms.
- Wind, water, and ice are surface forces that change landforms.
- Major types of landforms include mountains, hills, plateaus, and plains.
- The atmosphere is a cushion of gases that surrounds the earth.

LANDSAT image of Tokyo ▶

REVIEWING KEY TERMS

Match the numbered terms in Set A with their lettered definitions in Set B.

A

1. plate
2. latitude
3. relative location
4. geography
5. revolution
6. solar system
7. weathering

B

A. process by which water and chemicals break down rocks
B. huge slab of the earth
C. imaginary lines that measure distance north and south of the Equator
D. study of the earth
E. planets and other objects that revolve around the sun
F. where a place is in relation to another place
G. complete trip around the sun by a planet

Mental Mapping Activity

Draw a sketch that shows Earth in relation to the sun at the times of the solstices and the equinoxes. Draw four small globes in their correct positions for June 21, December 22, March 21, and September 23. Label the following terms:

- Equator
- North Pole
- South Pole
- Tropic of Cancer
- Tropic of Capricorn

REVIEWING THE MAIN IDEAS

Section 1
1. PLACE What are some of the human characteristics that describe the geography of a place?
2. REGION What factors do geographers use to define a region?

Section 2
3. MOVEMENT How does Earth's orbit and rotation influence the length of the year and the change from day to night?
4. LOCATION When it is summer in the Northern Hemisphere, why is it winter in the Southern Hemisphere?

Section 3
5. PLACE What landforms are formed by volcanic activity?
6. MOVEMENT What forces cause weathering and erosion?

CRITICAL THINKING ACTIVITIES

1. **Drawing Conclusions** Why is the theme of movement particularly important in geography today?
2. **Analyzing Information** What are two physical features found in the place where you live? What are two human features?

CONNECTIONS TO WORLD HISTORY

Cooperative Learning Activity Work in seven groups to research when human beings first appeared on each of the continents. Each group should choose a continent and investigate when and where the earliest human beings have been reported on that continent. As a class, create an illustrated time line that shows when humans first appeared on each continent. Display your time line for other classes to share.

GeoJournal Writing Activity

How often do you come in contact with something or someone from another part of the world? For two or three days, use your journal to keep track of all your contacts that involve the geographic theme of movement. Start by reading the labels in your clothes. Where were your clothes made? Is your favorite TV show broadcast from another country? What about the food you had for dinner? Finish by sketching a world map that locates the sources of all your contacts.

TECHNOLOGY ACTIVITY

Using the Internet Search the Internet for recent geographical news. Narrow your search by using words such as *geography* and *National Geographic Society.* Summarize your findings into a visual display on a poster, and display the poster in class.

PLACE LOCATION ACTIVITY: THE WORLD

Match the letters on the map with the places and physical features of the world. Write your answers on a separate sheet of paper.

1. Asia
2. Australia
3. South America
4. Antarctica
5. Himalayas
6. Andes
7. Africa
8. Rocky Mountains

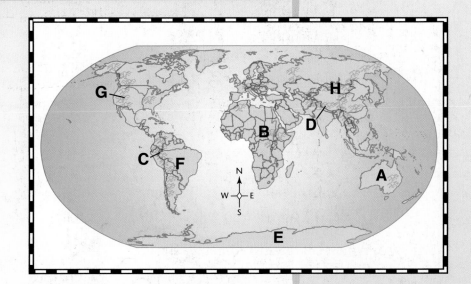

Chapter 2 Water, Climate, and Vegetation

World Ocean Currents

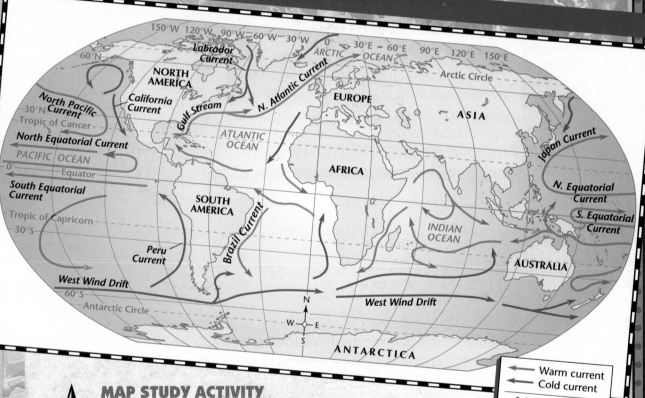

Warm current
Cold current

Robinson projection

MAP STUDY ACTIVITY

In this chapter you will learn about the ways the world's water is used.

1. What ocean current runs along North America's southeastern coast?

2. Which continent's climate is most likely to be influenced by the Peru Current?

Earth's Water

Words to Know
- vegetation
- water vapor
- water cycle
- evaporation
- condensation
- precipitation
- groundwater
- aquifer

Places to Locate
- Pacific Ocean
- Atlantic Ocean
- Indian Ocean
- Arctic Ocean
- Antarctica

Read to Learn . . .
1. how the earth's water moves in a cycle.
2. where people get freshwater.

This fisher in Asia catches snapper for his fish market. People all over the world depend on the oceans. To them the water is a source of food.

Water, however, is vital in many other ways to life on the earth.

Water covers about 70 percent of the earth's surface. Almost all of it is saltwater found in the oceans. Overall, the earth has plenty of water. Some areas, however, never have enough water to support life while other places get too much.

Why do some places get more water than others? Climate determines the amount of water a place receives, which in turn determines its **vegetation,** or plant life. Together water, climate, and vegetation influence how people in a given area live.

Movement
The Water Cycle

Water exists all around you in different forms. Rivers, lakes, and oceans contain water in liquid form. The atmosphere holds **water vapor,** or water in the form of gas. Glaciers and ice sheets are large masses of water in a frozen form. The total amount of water on the earth does not change. It is just constantly moving—from the oceans to the air to the ground and finally back to the oceans. This process is called the **water cycle.**

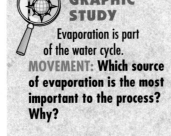

GRAPHIC STUDY

Evaporation is part of the water cycle.

MOVEMENT: Which source of evaporation is the most important to the process? Why?

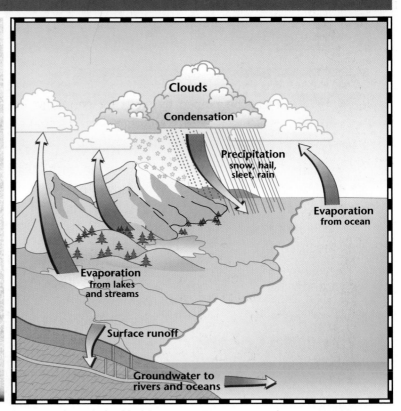

Look at the diagram above and follow the water cycle. The sun drives the cycle by evaporating water from the surface of oceans, lakes, and streams. In **evaporation** the sun's heat turns water into vapor. The amount of water vapor the air holds depends on its temperature. Warm air holds more vapor than cool air.

Warm air tends to rise and cool. As this happens, water vapor changes into a liquid—a process called **condensation**. Tiny droplets of water come together to form clouds. Eventually, the water falls as **precipitation**—rain, snow, or sleet, depending on the air temperature. This precipitation falls to the surface, where it soaks into the ground and collects in streams and lakes to return to the oceans. Soon most of it evaporates, and the cycle begins again.

Human/Environment Interaction

Types of Water

It's a hot day and you rush home for a glass of water. What if you turned on the faucet and nothing came out? All living things need water to survive. Think about the many ways you use water in just a single day—to bathe, to brush your teeth, to cook your food, to quench your thirst. The waters of the earth do more than just meet human needs, however. They also are home to millions of kinds of plants and animals. Water at the earth's surface can be freshwater or salt water. Freshwater is the most important to humans.

Freshwater Only about 3 percent of the water on the earth is freshwater. Of that, almost 2 percent is frozen in glaciers and ice sheets. Lakes and rivers are the important sources of the remaining usable freshwater.

Another source of freshwater is **groundwater**, or water that fills tiny cracks and holes in the rock layers below the surface of the earth. Groundwater can be tapped by wells for people to use. An underground rock layer so rich in water that water actually flows through it is called an **aquifer**. In regions with little rainfall, both farmers and city dwellers sometimes have to depend on aquifers and other groundwater for most of their water supply.

Ocean Water Did you know that you could sail from one ocean to another without touching land? All the oceans on the earth are part of a huge, continuous body of salt water—97 percent of the planet's water. The four major oceans are the Pacific Ocean, the Atlantic Ocean, the Indian Ocean, and the Arctic Ocean. Look at the map on page 42 to see how the four oceans are really one huge body of water. Which ocean do you live closest to?

The Pacific Ocean is both the largest and the deepest ocean. It covers 64,000,000 square miles (165,760,000 sq. km)—more than all the land areas of the earth combined. If you set Mount Everest, the earth's tallest mountain, in the deepest part of the Pacific, it would still be more than one mile below the surface!

As you learned in Chapter 1, smaller bodies of salt water are called seas, gulfs, bays, or straits. Look back at page 31 to see these features again.

This surfer demonstrates one exciting use for the world's water.
PLACE: What are the four major oceans?

SECTION 1 ASSESSMENT

REVIEWING TERMS AND FACTS

1. Define the following: vegetation, water vapor, water cycle, evaporation, condensation, precipitation, groundwater, aquifer.

2. MOVEMENT How does water enter the air during the water cycle?

3. HUMAN/ENVIRONMENT INTERACTION Where do people get freshwater for drinking and growing crops?

4. PLACE What do the oceans of the earth have in common?

GRAPHIC STUDY ACTIVITIES

5. Turn to the diagram on page 38. What happens to water that falls on the earth's surface?

6. Using the same diagram, describe what part mountains play in the water cycle.

BUILDING GEOGRAPHY SKILLS

Using Directions

To describe locations, we use the **cardinal directions** of north, south, east, and west. North and south are the directions of the North and South poles. If you stand facing north, east is the direction to your right—toward the rising sun. West is the direction on your left.

On most maps, you will find a **compass rose** showing the position of these directions. You might also see **intermediate directions**—those that fall between the cardinal directions. For example, the direction northeast falls between north and east. To use directions on a map, do the following:

- Use the compass rose to identify the four directions.
- Choose two features on the map.
- Determine whether one feature is north, south, east, or west of the other feature.

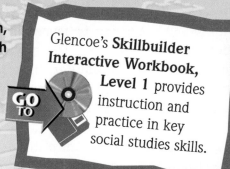

Glencoe's **Skillbuilder Interactive Workbook, Level 1** provides instruction and practice in key social studies skills.

Berlin, Massachusetts

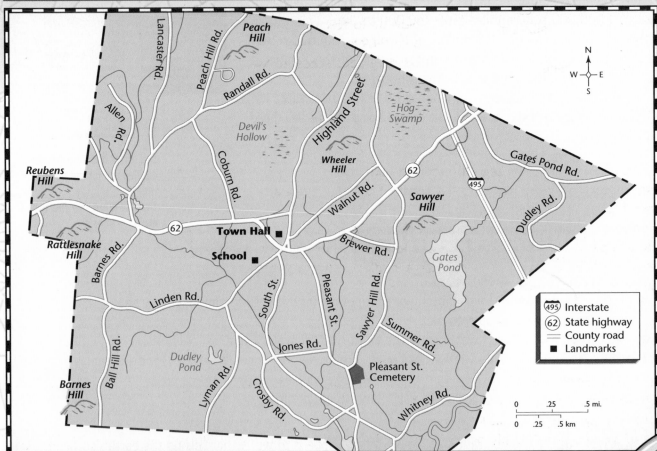

Geography Skills Practice

1. Does Hog Swamp lie north or south of Route 62?

2. From Gates Pond, what direction is Reubens Hill?

Influences on Climate

Words to Know
- weather
- climate
- tropics
- current
- hurricane
- typhoon
- monsoon
- rain shadow

Places to Locate
- Tropic of Cancer
- Tropic of Capricorn
- Equator
- Gulf Stream

Read to Learn . . .
1. what creates a particular climate.
2. how moving wind and water circulate the sun's heat.
3. what causes a rain shadow.

The power of nature's storms is awesome. The winds of a hurricane blow up to 130 miles per hour (209 km per hour), bending trees almost to the ground!

If you call a friend and say, "It's hot and rainy today," you are talking about the weather. **Weather** refers to the constant changes in the air during a short period of time. But if you say, "Summers are usually hot and rainy in my part of the country," then you have described your climate. **Climate** is the usual pattern of weather events in an area over a long period of time. Climate is determined by latitude, by location near large bodies of water, and sometimes by position near mountains.

Location
Latitude and Climate

As you learned in Chapter 1, the earth's tilt and rotation cause different regions to receive different amounts of heat and light from the sun. The sun's direct rays fall year-round at low latitudes near the Equator. This area—known as the **tropics**—lies mainly between the Tropic of Cancer and the Tropic of Capricorn. If you lived in the tropics, you would almost always enjoy a hot climate, unless you lived at a high elevation where temperatures are cooler.

Outside the tropics the sun is never directly overhead. In the high-latitude regions around the North and South poles, darkness descends for six months each year. There the sun's rays hit the earth indirectly at a slant. Thus, climates in these regions are always cool or cold even in midsummer.

Movement
Wind, Water, and Currents

In addition to latitude, the movement of air and water helps create Earth's climates. Moving air and water help circulate the sun's heat around the globe. In the ocean, the moving streams of water are called **currents**.

Near the Equator, air and water are most intensely heated. Warm winds and water currents move from the tropics toward the poles, bringing warmth with them. A large, warm-water ocean current known as the Gulf Stream flows from the Gulf of Mexico through the cool Atlantic Ocean along the east coast of North America. Find the Gulf Stream on the map below. Notice how the Gulf Stream then crosses the Atlantic, bringing warm water and mild air to western Europe.

Large bodies of water influence climates in another way. Water temperatures do not change as much or as fast as land temperatures do. Thus, air over

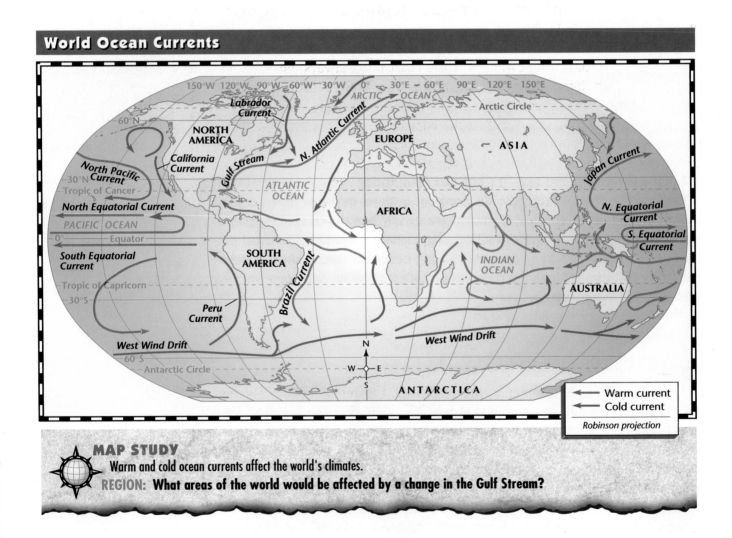

World Ocean Currents

MAP STUDY
Warm and cold ocean currents affect the world's climates.
REGION: What areas of the world would be affected by a change in the Gulf Stream?

42

UNIT 1

large bodies of water is warmer in winter and cooler in summer. This keeps coastal air temperatures moderate.

Storms Moving wind and water may make climates milder, but they also cause storms. **Hurricanes**, or violent tropical storm systems, form over the warm Atlantic Ocean at certain times of the year, usually in the summer and fall. Hurricanes bring high winds and drenching rain to the islands in the Caribbean Sea and to North America. They also rip through East Asia, although hurricanes that form in the Pacific Ocean are called **typhoons**.

Storms and heavy rains are not always destructive. In fact, during various seasons of the year, storms are an important part of some climate patterns. People in South Asia depend on **monsoons**, or seasonal winds, for the rain that waters their crops.

Rain Shadow Breezes blowing over the coast don't always move inland. Why? Natural landforms sometimes get in the way. When moist winds blow from the ocean toward a coastal mountain range, for example, the winds are forced upward over the mountains. As the winds rise, air cools and loses moisture as rain or snow. The climate on this windward, or coastal, side of mountains is moist and often foggy. Trees and vegetation are thick and green.

By the time the air reaches the inland side of the mountains, it is cool and dry. This creates a **rain shadow**, the dry area on the inland side of the mountains. The dry air of a rain shadow warms up again as it moves down mountainsides, giving the region a dry or desert climate.

The heavily forested Cascade Range in Washington state faces the ocean *(left)*. The inland side of these mountains is a partly dry wheat-growing area *(right)*.
LOCATION: On which side of these mountains is the rain shadow?

SECTION 2 ASSESSMENT

REVIEWING TERMS AND FACTS
1. Define: weather, climate, tropics, current, hurricane, typhoon, monsoon, rain shadow.
2. PLACE Why are climates warm or hot year-round in places near the Equator?
3. HUMAN/ENVIRONMENT INTERACTION Why are monsoons both helpful and harmful?

MAP STUDY ACTIVITIES

4. Turn to the map on page 42. Locate the Tropic of Cancer and Tropic of Capricorn. At what latitudes are they located? What type of currents—warm or cold— is between them?

MAKING CONNECTIONS

MATH **SCIENCE** **HISTORY** **LITERATURE** **TECHNOLOGY**

Have you heard recently about unusual weather in various parts of the world? There have been powerful typhoons along California's coast, paralyzing ice storms in Canada, flash floods in South America and Africa, and fierce drought in Southeast Asia. What's causing these extremes of weather around the globe? No one knows for sure, but many scientists say one of the major causes is El Niño.

FORMATION El Niño is a climatic change that starts in the tropical Pacific Ocean and sets off changes in the atmosphere. In El Niño, winds blowing east to west across the Pacific near the Equator weaken and sometimes reverse direction. This change in direction allows a large mass of warm water near Australia to move east toward South America. Along South America's Pacific coast, the warm water mass displaces the cold Peru Current flowing north. Warmer water brings more evaporation, and heavy rains fall over South America.

EFFECTS El Niño also affects other parts of the world. It can cause severe storms in some areas and drought in others. It also interferes

with the flow of jet streams, or high-altitude air currents that shape the world's weather. The map below shows how El Niño affects winter weather in North America. The polar jet stream moves far north of its usual flow.

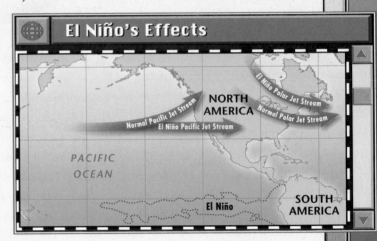

El Niño's Effects

Cold, dry Arctic air is kept out of the eastern United States, which experiences a warmer winter. Meanwhile, the Pacific jet stream moves farther south than usual. It brings fierce storms to the California coast, which normally enjoys mild, sunny weather. El Niño used to occur every four to five years but now happens more often. Recent El Niños have caused thousands of deaths and brought billions of dollars in damages around the world.

Making the Connection

1. Under what conditions does El Niño form?
2. What are the global effects of El Niño?

SECTION 3

Climate and Vegetation

PREVIEW

Words to Know
- rain forest
- savanna
- marine west coast climate
- Mediterranean climate
- humid continental climate
- humid subtropical climate
- tundra
- permafrost
- elevation
- timberline
- steppe

Places to Locate
- Amazon River basin
- Mediterranean Sea

- Arctic Circle
- Antarctica
- Greenland

Read to Learn . . .
1. what major world climate regions are like.
2. where each world climate region is located.
3. what kinds of vegetation grow in each world climate region.

R eds, pinks, and purples blanket a usually brown

desert. For a short time—when a desert is showered with rain—the landscape may bloom. During most of the year, however, deserts are an unfriendly place for most living things.

T he world's climates can be organized into four major regions: tropical, mid-latitude, high latitude, and dry. Some of these regions are determined by their latitude, or distance from the Equator. Others are based on their altitude, or height above sea level.

Region
Tropical Climates

Tropical climate regions get their name from the tropics, the areas along the Equator reaching to about 30°N and 30°S. Temperatures here change little from season to season. The warm tropical climate region can be separated into two types. The tropical rain forest climate is wet in most months, with up to 100 inches (254 cm) of rain a year. The tropical savanna climate has two distinct seasons—one wet and one dry.

Tropical Rain Forest Climate In some parts of the tropics, the growing season lasts all year. In these areas, rain and heat produce lush vegetation and dense forests called **rain forests**. These forests are home to millions of kinds of

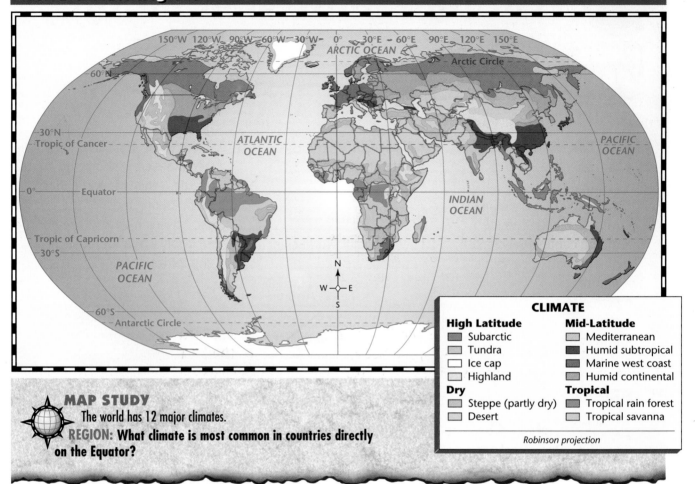

World Climate Regions

CLIMATE

High Latitude
- Subarctic
- Tundra
- Ice cap
- Highland

Dry
- Steppe (partly dry)
- Desert

Mid-Latitude
- Mediterranean
- Humid subtropical
- Marine west coast
- Humid continental

Tropical
- Tropical rain forest
- Tropical savanna

Robinson projection

MAP STUDY

The world has 12 major climates.

REGION: What climate is most common in countries directly on the Equator?

plant and animal life. Tall hardwood trees such as mahogany, teak, and ebony form the canopy, or top layer, of the forest. The vegetation at the canopy layer is so thick that little sunlight reaches the forest floor. The Amazon River basin in South America is the world's largest rain forest area.

Tropical Savanna Climate In other parts of the tropics, such as southern India and eastern Africa, most of the year's rain falls during the wet season. The rest of the year is hot and dry. **Savannas**, or broad grasslands with few trees, occur in this climate region. Find the tropical savanna climate areas on the map above.

Region
Mid-Latitude Climates

Mid-latitude, or moderate, climates are found in the middle latitudes of the Northern and Southern hemispheres. They extend from about 30°N to 60°N and from 30°S to 60°S of the Equator. Most of the world's people—and probably you, too—live in this region.

The mid-latitude region has a greater variety of climates than other regions. This variety results from a mix of air masses—warm air coming from the tropics and cool air coming from the polar regions. In most places, temperatures change with the seasons.

Marine West Coast Climate Coastal areas, where winds blow from the ocean, usually have a mild **marine west coast climate**. If you lived in one of these areas, your winters would be rainy and mild, and your summers cool.

Mediterranean Climate Another mid-latitude coastal climate is called a **Mediterranean climate** because it is similar to the climate found around the Mediterranean Sea. This climate has mild, rainy winters and hot, dry summers.

Humid Continental Climate If you live far from the oceans in inland areas of North America, Europe, or Asia, you face a harsher **humid continental climate**. In these areas, winters can be long, cold, and snowy. Summers are short but may be very hot.

Humid Subtropical Climate Mid-latitude regions close to the tropics have a **humid subtropical climate**. Rain falls throughout the year but is heaviest during the hot and humid summer months. Humid subtropical winters are generally short and mild.

Region
High Latitude Climates

High latitude climate regions lie mostly in the high latitudes of each hemisphere, from 60°N to the North Pole and 60°S to the South Pole. Climates are cold everywhere in the high latitude regions, some more severe than others.

Subarctic Climate Just below the Arctic Circle lie the subarctic areas. The few people living here face severely cold and bitter winters, but temperatures do rise above freezing during summer months. Huge evergreen forests grow in the subarctic region, especially in northern Russia.

Tundra Climate The climate of the Arctic tundra area is harsh and dry. The **tundra** is a vast rolling plain without trees. The top few inches of the ground thaw during summer months, however. This allows sturdy grasses, small berry bushes, and wildflowers to sprout. In parts of the tundra and subarctic regions, the lower layers of soil are known as **permafrost** because they stay permanently frozen.

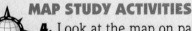

This village in the Swiss Alps has cooler temperatures than areas at lower elevations. **LOCATION: What is a timberline?**

Ice Cap Climate On the polar ice caps and the great ice sheets of Antarctica and Greenland, the climate is bitterly cold. Monthly temperatures average below freezing. Temperatures in Antarctica have been measured at –128°F (–103°C)! Although no vegetation grows here, some fungus-like plants can live on rocks.

Highland Climate A highland, or mountainous, climate has cool or cold temperatures year-round. The **elevation,** or height above sea level, of a place changes its climate dramatically. Higher into the mountains, the air becomes thinner. It cannot hold the heat from the sun, so the temperature drops. Even in the tropics, snow covers the peaks of high mountains.

On those peaks, you come to the **timberline**, the elevation above which no trees grow. Beyond the timberline, only small shrubs and wildflowers grow in meadows.

Region
Dry Climates

Dry climate refers to dry or partially dry areas that receive little or no rainfall. Temperatures can be extremely hot during the day and cold at night. Dry climates can also have severely cold winters.

Desert Climate You can find dry climate regions at any latitude. The driest, with less than 10 inches (25 cm) of rainfall a year, are called deserts. Only scattered plants such as cacti can survive the dry desert climate. With roots close to the surface, cacti can collect any rainfall.

Steppe Climate Many deserts are surrounded by partly dry grasslands known as **steppes**. They get more rain than deserts, averaging 10 to 20 inches (25 to 51 cm) a year. Bushes and short grasses cover the steppe landscape.

SECTION 3 ASSESSMENT

REVIEWING TERMS AND FACTS
1. Define: rain forest, savanna, marine west coast climate, Mediterranean climate, humid continental climate, humid subtropical climate, tundra, permafrost, elevation, timberline, steppe.
2. LOCATION Where are the moderate climate regions located?
3. PLACE How are climates of inland areas different from those nearer coasts? Why?

MAP STUDY ACTIVITIES
4. Look at the map on page 46. In which climate region do you live?
5. Using the same map, name two countries that have high latitude climates.

Chapter 2 Highlights

Important Things to Know About Water, Climate, and Vegetation

SECTION 1 EARTH'S WATER

- Water constantly moves in a cycle from oceans, to air, to land, and back to oceans.
- Rivers and lakes provide important sources of freshwater.
- Freshwater also is stored in the earth's underground layers of rock.
- The world's four oceans are the Atlantic Ocean, Pacific Ocean, Indian Ocean, and Arctic Ocean.

SECTION 2 INFLUENCES ON CLIMATE

- The kind of climate a place has depends mainly on its latitude.
- Landforms and nearby bodies of water also influence climate.
- Air and water move around the earth as warm or cold winds and currents.
- A large body of water keeps nearby land temperatures moderate.
- Areas in a rain shadow behind coastal mountains often have dry climates.

SECTION 3 CLIMATE AND VEGETATION

- Four major climate regions are tropical, mid-latitude, high latitude, and dry.
- Tropical climates can be wet, with rain forests, or dry, with savannas.
- In the mid-latitude region, climate is affected by distance from the ocean.
- Places with high latitude climates do not have much vegetation.

Ocean Life ▶

REVIEWING KEY TERMS

Match the numbered terms in Set A with their lettered definitions in Set B.

A
1. aquifer
2. climate
3. hurricane
4. marine west coast
5. permafrost
6. rain shadow
7. tundra
8. vegetation

B
A. long-term patterns of weather
B. layer of soil that stays frozen
C. violent storm over a tropical ocean
D. underground rock layer storing water
E. plants that grow naturally in a location
F. dry region inland from coastal mountains
G. mild climate with rainy winters and cool summers
H. high latitude climate region lacking trees

REVIEWING THE MAIN IDEAS

Mental Mapping Activity

Sketch a map of the world showing the continents. Then shade and label the following:
- Equator
- Atlantic Ocean
- Pacific Ocean
- Indian Ocean
- mid-latitude climate regions

Section 1
1. MOVEMENT What are the major steps in the water cycle?
2. HUMAN/ENVIRONMENT INTERACTION From what sources do people get usable freshwater?

Section 2
3. PLACE Why are climates near the poles always cool or cold?
4. PLACE What effect does a large body of water have on climates in nearby places? Why?
5. REGION Why are desert areas often found on the inland slopes of coastal mountains?

Section 3
6. LOCATION At what latitudes are rain forests found?
7. PLACE What creates a marine west coast climate?
8. PLACE What kinds of plants grow in the tundra?

CRITICAL THINKING ACTIVITIES

1. **Determining Cause and Effect** Many ancient civilizations developed near rivers. Why do you think people settled there?
2. **Synthesizing Information** Mount Kilimanjaro, the tallest mountain in Africa, is located near the Equator. What kind of climate would you find at its base? At its peak?

CONNECTIONS TO ECONOMICS

Cooperative Learning Activity Climate determines the types of economic activities that develop in various parts of the world. Work in groups to compare economic activities in the Eastern and Western hemispheres. Each group should choose one country from each hemisphere to compare. List the different economic activities that could take place in both countries, and those that could not because of climate. Report your findings to the other groups.

GeoJournal Writing Activity

Water is essential for all living things. In your journal, describe a memorable experience you had that involved water. You may want to describe an adventure you had swimming or fishing, for example. Put your ideas in the form of a poem or song.

TECHNOLOGY ACTIVITY

Building a Database Make a chart comparing the influences of latitude, wind, and water currents on climate. Put this information into a database, with separate fields for each type of information. Print and distribute your database to the rest of the class.

PLACE LOCATION ACTIVITY: WORLD OCEANS AND CURRENTS

Match the letters on the map with the bodies of water and ocean currents. Write your answers on a separate sheet of paper.

1. Gulf Stream
2. Atlantic Ocean
3. Indian Ocean
4. California Current
5. Japan Current
6. Arctic Ocean

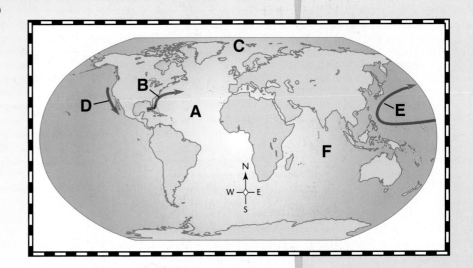

GeoLab ACTIVITY

From the classroom of
**Eleanor Bloom,
Bloom-Carroll Schools,
Lithopolis, Ohio**

WATER: It's not all wet!

Background

Water—the world's most precious resource—is vital to all life. It is used for drinking, irrigation, industry, transportation, and energy. But is all water the same? Let's take a look at the various kinds of water to find out what's really in this precious resource.

Moose Pond in the White Mountains, New Hampshire

Believe it or NOT!

About 97 percent of the earth's water has salt in it and is not suitable for drinking. About 2 percent of the world's water is in the form of glaciers or ice caps. Only about 1 percent of the earth's water is available for drinking.

Your view of pond water may look like this magnified droplet.

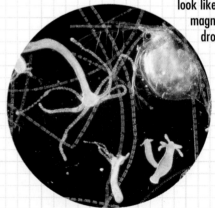

Materials

- 8 empty baby-food-size jars
- microscope
- blank microscope slides and slide covers
- eyedropper
- 8 different types of water: distilled water, spring water, salt water, pond water, soapy water, faucet water, well water, filtered water

What To Do

A. Fill each of the jars with a different type of water. Label the jars.

B. Allow the jars to sit for 24 hours.

C. Observe and write down what you see in each jar the next day.

D. Shake each jar, then use the eyedropper to take a sample from each jar. Drip each sample onto a slide. You will end up with 8 different slides.

E. Put a slide cover over each sample and label each slide.

F. Observe each slide under the microscope.

G. Make drawings of what you see and label each drawing.

Lab Activity Report

1. After 24 hours, which two water samples differ the most?

2. Describe the differences between the water samples.

3. What did you learn about water?

4. Drawing Conclusions What type of water would make the best drinking water?

Go A Step Further!

Activity
The process of desalination removes salt from water. Find out more about desalination at the library. Report to your class about this process. Use a world map to show what countries rely on desalination for the majority of their drinking water.

3 The World's People

World Culture Regions

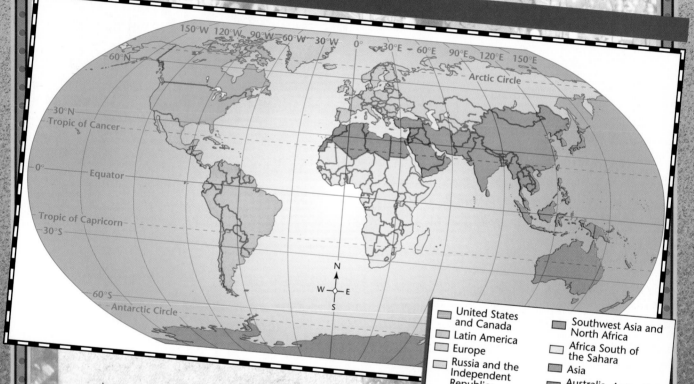

150°W 120°W 90°W 60°W 30°W 0° 30°E 60°E 90°E 120°E 150°E

60°N

Arctic Circle

30°N
Tropic of Cancer

0° — Equator

Tropic of Capricorn
30°S

60°S — Antarctic Circle

N
W ✛ E
S

☐ United States and Canada	☐ Southwest Asia and North Africa
☐ Latin America	☐ Africa South of the Sahara
☐ Europe	☐ Asia
☐ Russia and the Independent Republics	☐ Australia, Antarctica, and Oceania

Robinson projection

MAP STUDY ACTIVITY

What is important to you? Do you think people in other countries believe as you do? In Chapter 3 you will learn about the world's people.

1. How many culture regions are shown on the map above?
2. In what culture region do you live?

SECTION 1 Culture

If you wake up to rock music, put on denim jeans, drink orange juice for breakfast, and speak English, those things are part of your culture. If you eat flat bread for breakfast, speak Arabic, and wear a long cotton robe to protect you from the hot sun, *those* things are part of your culture.

Place
What Is Culture?

When some people hear the word *culture*, they think of priceless paintings and classical symphonies. **Culture**, as used in geography, is the way of life of a group of people who share similar beliefs and customs. These people may speak the same language, follow the same religion, and dress in a certain way. The culture of a people also includes their government, their music and literature, and the ways they make a living. What things are important in your culture?

Early Cultures Some 4,000 to 5,000 years ago, at least four cultures arose in Asia and Africa. One developed in China along a river called the Huang He. Another developed near the Indus River in South Asia, a third between the

Tigris and Euphrates rivers in Southwest Asia, and a fourth along the Nile River in North Africa.

All four river-valley cultures developed agriculture and ways of irrigating, or bringing water to the land. Why was irrigation important? Farming produced more food than hunting and gathering, which meant that larger populations could develop. People then learned trades, built cities, and made laws.

The river-valley cultures eventually became **civilizations,** or highly developed cultures. These cultures spread their knowledge and skills from one area to another, a process known as **cultural diffusion**.

Region
Culture Regions Today

Geographers today often divide the world into areas called culture regions. These regions may be based on the kind of government, social groups, economic systems, languages, or religions found in each.

Governments The kind of government, or political system, a society has reflects its culture. Until a few hundred years ago, most countries had authoritarian systems in which one person ruled with unlimited power.

World Culture Regions

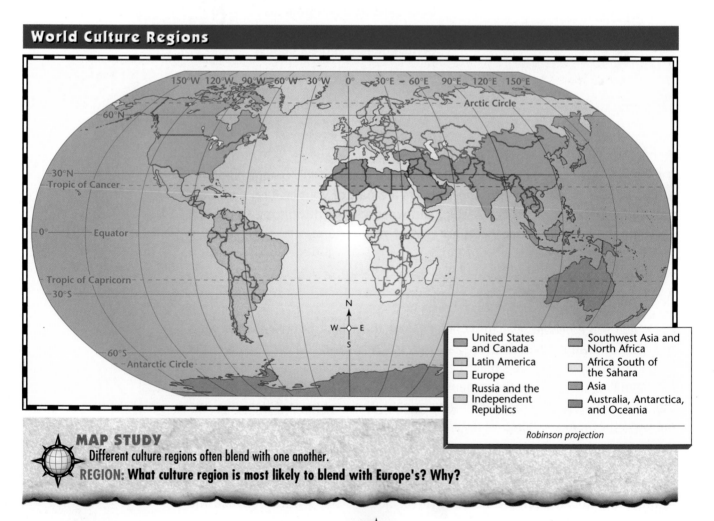

United States and Canada	Southwest Asia and North Africa
Latin America	Africa South of the Sahara
Europe	Asia
Russia and the Independent Republics	Australia, Antarctica, and Oceania

Robinson projection

MAP STUDY
Different culture regions often blend with one another.
REGION: What culture region is most likely to blend with Europe's? Why?

UNIT 1

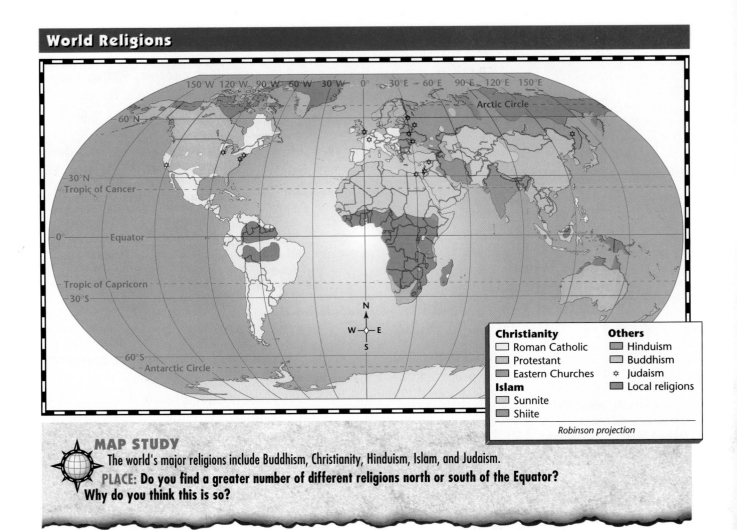

MAP STUDY

The world's major religions include Buddhism, Christianity, Hinduism, Islam, and Judaism.

PLACE: Do you find a greater number of different religions north or south of the Equator? Why do you think this is so?

When the people of a country hold the powers of government, we think of that government as a democracy. Citizens choose their leaders by voting. Once in power, leaders in a democracy are expected to obey a constitution or other long-standing traditions that require them to respect individual freedoms.

Language Language is a powerful tool, offering a way for people to share information. Sharing a language is one of the strongest unifying forces for a culture. Languages spoken in a culture region often belong to the same **language family**, or group of languages having similar beginnings. Romance languages, for example, come from Latin, the language of ancient Rome. Spanish, Portuguese, French, Italian, and Romanian are in the Romance language family.

Religion Another important part of culture is religion. Some of the major world religions are Buddhism, Christianity, Hinduism, Islam, and Judaism. The map above shows you the areas where people practice these religions.

Social Groups Another way of organizing culture regions is by social groups. A country may have many different groups of people living within its

In rural Germany, grapes are harvested to make wine *(above)*. In the city of Stuttgart, workers assemble Mercedes-Benz autos *(right)*.
PLACE: What are two important products produced in your state?

borders. These differences are sometimes based on income, or how much money people earn.

Sociologists, or scientists who study social groups, measure how well a society meets the needs of its people. The **standard of living** is a measure of people's quality of life based on their income and the material goods they have. Another measure of a society's well-being is its **literacy rate**, or the percentage of people who can read and write.

Economies All societies have economic systems. An economic system tells how people produce goods, what goods they produce, and how those goods are then bought and sold. In most democratic societies, a market economy prevails. This is an economic system based on **free enterprise**, where people start and run businesses to make a profit with little interference from government.

Another economic system is **socialism**. In a socialist system, the government sets goals, may run some businesses, and tends to have a central role in the economy.

SECTION 1 ASSESSMENT

REVIEWING TERMS AND FACTS

1. Define the following: culture, civilization, cultural diffusion, language family, standard of living, literacy rate, free enterprise, socialism.

2. LOCATION Along what rivers were four ancient cultures located?

3. REGION What three factors help identify a culture region?

MAP STUDY ACTIVITIES

4. Turn to the map on page 56. What is the name of the culture region that includes Mexico and South America?

BUILDING GEOGRAPHY SKILLS

Using Latitude and Longitude

To find an exact location, geographers use a set of imaginary lines. One set of lines—latitude lines—circles the earth's surface east to west. Each line of latitude—also called a parallel—is numbered from 1° to 90° and followed by an N or S to show it is north or south of the Equator.

A second set of longitude lines runs vertically from the North Pole to the South Pole. Each of these lines is called a meridian. The starting point for meridians, or 0° longitude, is called the Prime Meridian. Longitude lines are numbered from 1° to 180° followed by an E or W—east or west of the Prime Meridian. To find latitude and longitude, follow these steps:

- Choose a location on a map or globe.
- Identify the number of the nearest parallel, or line of latitude.
- Identify the number of the nearest meridian, or line of longitude, that crosses it.

Glencoe's **Skillbuilder Interactive Workbook, Level 1** provides instruction and practice in key social studies skills.

The World

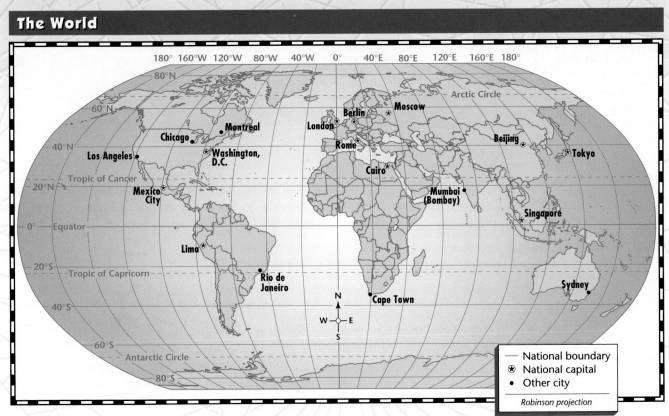

Geography Skills Practice

1. What is the exact location of Washington, D.C.?

2. What cities on the map lie south of 0° latitude?

PREVIEW

Words to Know
- emigrate
- refugee
- population density
- urbanization
- developed country
- developing country
- demographer
- birthrate
- death rate
- famine

Places to Locate
- China
- Vietnam
- Egypt

Read to Learn . . .
1. where most people in the world live.
2. how scientists measure population.
3. how the earth's population is changing.

Tokyo is one of the most crowded places in the world. At rush hour, thousands of Japanese workers cram the subways for the hectic ride to work or home.

Although Tokyo is crowded, some parts of the world have few people. What makes one area crowded and another empty? Climate, culture, and jobs are some of the things that help determine where people live.

Movement
Population Patterns

Some families live in the same town or on the same land for generations. Other people move frequently from place to place. In some cases people choose to leave the country in which they were born. They **emigrate**, or move, to another country. You may know of people who are forced to flee their country because of wars, food shortages, or other problems. They become **refugees**, or people who flee to another country for refuge from persecution or disaster.

Population Distribution People live on only a small part of the earth. As you learned earlier, land covers only about 30 percent of the earth's surface, and half of this land is not useful to humans. People cannot make homes, grow crops, or graze animals on land covered with ice, deserts, or high mountains.

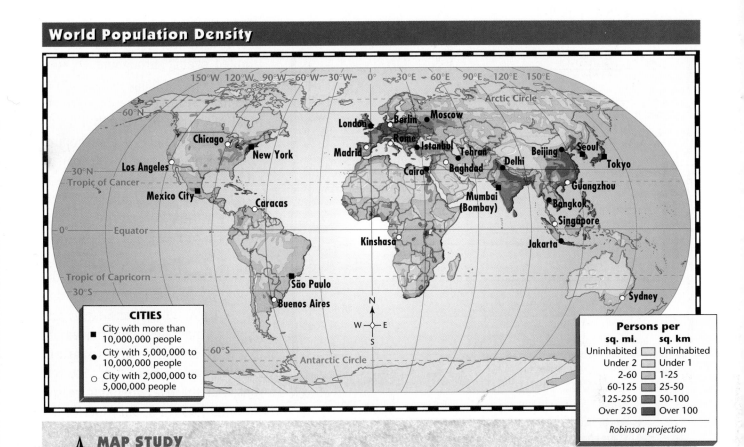

The world's total population is about 6 billion.

PLACE: How would you describe the population density of Kinshasa, Democratic Republic of the Congo, in Africa?

Even on the usable land, population is not distributed, or spread, evenly. One reason is that people naturally choose to live in places with plentiful water, good land, and a favorable climate. Other reasons may lie in a people's history and culture.

Look at the map above to see how the world's population is distributed. Notice that people are most concentrated in western Europe and eastern and southern Asia. The United States also has areas of dense population.

Asia is a huge continent in land area. It also has the largest population—nearly 3.6 billion. Africa is the second-largest continent in area and in overall population, with 770 million people. Europe is much smaller in land area, but its 730 million people make it third in population size.

Population Density One way to look at population is by measuring **population density**—the average number of people living in a square mile or square kilometer. Population density gives a general idea of how crowded a country or region is. For example, the countries of Vietnam and Congo have about the same land area. The population density in Vietnam is 625 people per square mile (241 people per sq. km). Congo has an average of only 20 people per square mile (8 people per sq. km).

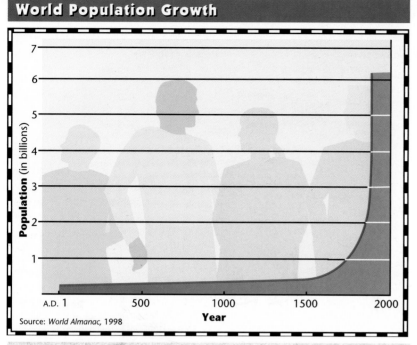

World Population Growth

Source: *World Almanac*, 1998

GRAPHIC STUDY
Between A.D. 1 and 1500, the world population grew by only 300 million people.
PLACE: By about how much has the world population grown in just the last 500 years?

Notice that density is an *average*. It assumes that people are distributed evenly throughout an area. But this seldom happens. A country may have several large cities where most of its people actually live. In Egypt, for example, overall population density is 171 people per square mile (61 people per sq. km). In reality, about 99 percent of Egypt's people live within 20 miles of the Nile River. The rest of Egypt is desert.

Urbanization Throughout the world, populations are changing as people leave villages and farms and move to the cities. This movement to cities is called **urbanization**. People move to cities for many reasons, but the overwhelming reason is to find jobs. In South America, for example, economic hardships have caused millions of people to move to cities such as São Paulo in Brazil and Buenos Aires, Argentina.

Urban areas are usually centers of industrialization. Countries that are industrialized are called **developed countries**. Those countries that are working toward industrialization are called **developing countries**. The economies of these countries depend mainly on agriculture and developing modern industries.

Human/Environment Interaction
Population Growth

How fast has the earth's population grown? In 1800 the number of people in the world totaled about 857 million. During the next hundred years the population doubled to nearly 1.7 billion. By the late 1990s, this figure had risen to about 6 billion people. **Demographers**, or scientists who study population, believe that the earth will have more than 7 billion people by the year 2010. Look at the graph above to see how population has grown since 1800.

Measuring Growth Populations are growing at different rates in different places. To find the growth rate in a specific area, demographers compare the birthrate with the death rate. The **birthrate** is the number of children born each year for every 1,000 people. The **death rate** is the number of deaths for every 1,000 people.

62

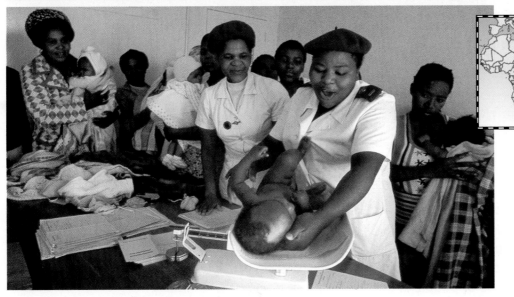

As more children receive medical care in this South African clinic, the country's death rate will drop.
PLACE: What other factor can reduce a country's death rate?

Growth rates tend to be high in developing countries. Better health care and living conditions have cut the death rate, and people are living longer. Birthrates also may remain high because some cultures favor large families.

Population Challenges Rapid population growth presents many challenges. A growing population requires more food. Since 1950 food production fortunately has increased faster than population on all continents except Africa. Millions of people, however, still suffer **famine**, or lack of food. Another challenge is that expanding populations use up resources more rapidly than stable populations. Some developing countries face shortages of water, housing, and jobs. Others face the threat of AIDS, a worldwide disease that has claimed hundreds of thousands of lives.

SECTION 2 ASSESSMENT

REVIEWING TERMS AND FACTS

1. Define the following: emigrate, refugee, population density, urbanization, developed country, developing country, demographer, birthrate, death rate, famine.

2. MOVEMENT What are some reasons that people leave their home countries?

3. HUMAN/ENVIRONMENT INTERACTION Why is population distributed unevenly on the earth's land surfaces?

4. PLACE What data does a country's birthrate give?

MAP STUDY ACTIVITIES

5. Look at the graph on page 62. What major changes occurred between 1400 and 1900?

6. Turn to the map on page 61. What are the two largest cities shown south of the Equator?

MAKING CONNECTIONS

COUNTING HEADS

MATH | **SCIENCE** | **HISTORY** | **LITERATURE** | **TECHNOLOGY**

Have you ever counted the number of people in a room? Geographers often rely on similar head counts of people in an area to understand the number and distribution of the many different kinds of people in our world.

WHAT DEMOGRAPHERS STUDY Demographers are the scientists who take the head counts. They collect and examine information about people—births and deaths, marriages, divorces, ages, and national backgrounds. Demographers figure out a population's *density,* or the number of people living in a given area. They also analyze *distribution,* or where people live. They even look at the movement of people from place to place.

DEMOGRAPHER'S TOOLS Two principal tools that a demographer uses are a census and a record of vital statistics. A *census* is a government's formal count of its population. Such a count keeps track of the number of people and obtains information about age, employment, income, and other characteristics. You are counted in the United States Census once every 10 years.

Vital statistics are records of basic human events—births, diseases, marriages, divorces,

Planning for Growth

and deaths. These statistics reveal what is happening or has happened to an area's population.

WHY DEMOGRAPHERS? Demographers analyze information to discover how different groups of people react to events and other people. Their findings often explain how changes in a society have affected people. For example, a rising cost of living in the United States has resulted in smaller families.

Just as often, demographic findings can predict likely events, giving people time to prepare and plan for changes. If many births are recorded, for example, cities may plan to build more schools.

Making the Connection

1. What do demographers do?
2. How can demographic findings be used?

Resources

PREVIEW

Words to Know
- natural resource
- renewable resource
- nonrenewable resource
- fossil fuel
- subsistence farming
- service industry
- pollution
- pesticide
- acid rain
- hydroelectric power
- solar energy

Places to Locate
- Pennsylvania
- Arabian Peninsula

Read to Learn . . .
1. what renewable and non-renewable resources are.
2. how people use resources to make a living.
3. how overusing resources may threaten the environment.

Green fields of corn have dotted North America since ancient times. This rich harvest of corn reflects the abundance of natural resources found in the region. What are natural resources, and why are they so important?

A **natural resource** is anything from the natural environment that people use to meet their needs. Natural resources include fertile soil, clean water, minerals, trees, and energy sources. Human skills and labor are also valuable natural resources. People use all these resources to improve their lives.

Human/Environment Interaction

Types of Resources

The value of resources changes as people discover new uses and new technology. Trees, for example, have been a valuable resource throughout history. For thousands of years people used wood to stay warm, cook food, and build homes, vehicles, and furniture.

Oil, on the other hand, was a gooey nuisance to the people of central Pennsylvania until the oil industry developed. Oil-based products then began to power cars and heat homes. Great underground pools of oil were discovered in Southwest Asia in the 1930s. Today the countries surrounding the Persian Gulf are among the richest on the earth.

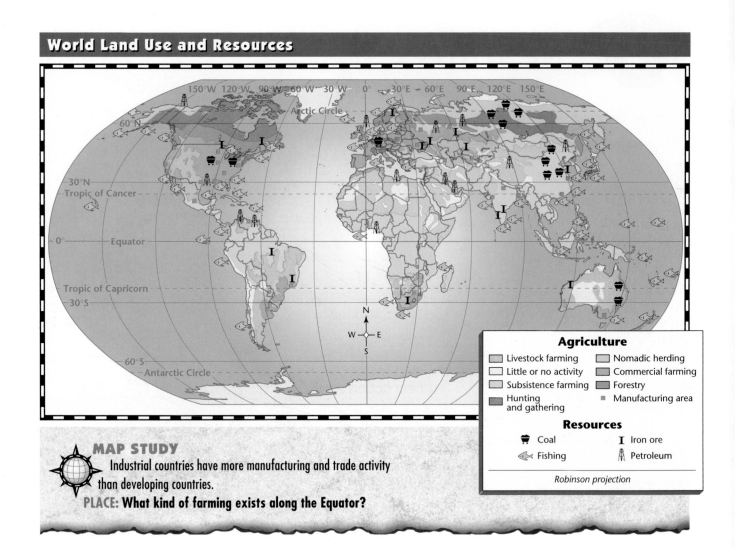

MAP STUDY
Industrial countries have more manufacturing and trade activity than developing countries.
PLACE: What kind of farming exists along the Equator?

Renewable Resources Some natural resources can be replaced as they are used up. These **renewable resources** can be replaced naturally or grown fairly quickly. Forests, grasslands, plant and animal life, and rich soil all can be renewable resources if people manage them carefully. A lumber company concerned about future growth can replant as many trees as it cuts. Fishing and whaling fleets can limit the number of fish and whales they catch in certain parts of the ocean.

Nonrenewable Resources Metals and other minerals found in the earth's crust are **nonrenewable resources**. They cannot be replaced because they were formed over millions of years by geologic forces within the earth.

One important group of nonrenewable resources is **fossil fuels**—coal, oil, and natural gas. Industries and people depend on these fuels for energy and as raw materials for plastics and other goods. We also use up large amounts of other metals and minerals, such as iron, aluminum, and phosphates. Some of these can be reused, but they cannot be replaced.

Human/Environment Interaction
Making a Living

People use natural resources to make the things they need and want. People also use natural resources to earn a living for themselves and their families.

Farming One of our basic needs is food. Although cities are growing through urbanization, about half the people in the world still live in rural areas. Most farmers live by **subsistence farming**, or growing only enough to feed themselves. They work in small fields with hand tools and use animals for heavy work. Sometimes they sell part of a harvest.

Commercial agriculture, on the other hand, usually operates on a large scale, with farmers producing food crops and livestock to sell. Huge plots of land may be farmed using machinery.

Industry Other working people, particularly in cities, have jobs in industry. Some industries produce goods, ranging from nails to shoes to basketballs to military jets. Other industries supply services. People in **service industries** do many different kinds of jobs. They may be computer operators, car mechanics, cooks, doctors, musicians, or taxi drivers. How many service industries have you depended on today?

WHAT IN THE WORLD?

All Wet

Do you waste water? The average American uses about 60 gallons (227 l) every day—maybe more. If you love long showers, remember that 6 or 7 gallons (23 to 27 l) of water are going down the drain every minute. Filling up a bathtub is not good either. That uses about 36 gallons (about 137 l).

Human/Environment Interaction
Environmental Challenges

The use of natural resources affects the environment. Many human activities can cause **pollution**—putting impure or poisonous substances into the land, water, and air.

Land and Water Only about 11 percent of the earth's surface has land good enough for farming. Chemicals that farmers use may improve their crops, but some also may damage the land. **Pesticides**, or chemicals that kill insects, can pollute rivers and groundwater. Other human activities also pollute soil and water. These include oil spills from tanker ships and illegal dumping of waste products, such as untreated sewage.

Air Industries and vehicles that burn fossil fuels are the main sources of air pollution. Throughout the world, fumes from cars and other vehicles pollute the air. The chemicals in air pollution can seriously damage people's health.

These chemicals combined with precipitation may fall as **acid rain**, or rain carrying large amounts of sulfuric acid. Acid rain eats away the surfaces of buildings, kills fish, and can destroy entire forests.

Careless use of resources has caused oil spills in Alaska *(right)* and acid rain damage to forests in eastern Europe *(above).*
HUMAN/ENVIRONMENT INTERACTION: How could the damage shown here have been prevented?

Some scientists believe that increasing amounts of certain chemicals in the atmosphere are gradually warming the earth. This global warming, they claim, will raise global temperatures a few degrees during the next century. Glaciers and parts of ice caps may melt, increasing sea levels throughout the world.

Energy Developed nations and developing nations both need safe, dependable sources of energy. Fossil fuels are most often used to generate electricity, heat buildings, run machinery, and power vehicles. Fossil fuels, however, are nonrenewable resources. In addition, they contribute to air pollution. So today many countries are trying to discover new ways of using renewable energy sources. Two of these ways are **hydroelectric power**, the energy generated by falling water, and **solar energy**, or energy produced by the heat of the sun.

SECTION 3 ASSESSMENT

REVIEWING TERMS AND FACTS

1. Define the following: natural resource, renewable resource, nonrenewable resource, fossil fuel, subsistence farming, service industry, pollution, pesticide, acid rain, hydroelectric power, solar energy.

2. HUMAN/ENVIRONMENT INTERACTION Why are metals a nonrenewable resource?

3. HUMAN/ENVIRONMENT INTERACTION How are service industries different from other industries?

4. HUMAN/ENVIRONMENT INTERACTION How might farmers pollute the environment?

MAP STUDY ACTIVITIES

5. Look at the map on page 66. What areas of the world have large forests?

6. Using the same map, find two areas where people carry on subsistence farming.

Chapter 3 Highlights

Important Things to Know About the World's People

SECTION 1 CULTURE

- In geography, culture means a group of people who share similar beliefs and customs.
- A people's culture includes their government, economy, language, religion, and social organization.
- Four ancient cultures developed along river valleys.
- Geographers study the world in terms of culture regions.

SECTION 2 POPULATION

- People live on only about 15 percent of the world's land.
- Population is distributed very unevenly over the earth's surface.
- Population can be measured in terms of density, or the average number of people living in a square mile or square kilometer.
- Developed countries are industrialized. Developing countries are working toward that goal.
- By the late 1990s, the world's population had risen to about 6 billion.
- Rapid population growth threatens the world's supply of food and resources.

SECTION 3 RESOURCES

- Natural resources are renewable or nonrenewable.
- Renewable resources such as forests, grasslands, and animal life can be replaced—if managed carefully.
- Nonrenewable resources such as metals and fossil fuels cannot be replaced once they are used.
- About half of the world's people live by farming—most by subsistence farming.
- People who work in industry are employed in either manufacturing or service jobs.

Rio de Janeiro, Brazil ▶

REVIEWING KEY TERMS

Match the numbered terms in Set A with their lettered definitions in Set B.

A

1. birthrate
2. subsistence farming
3. emigrate
4. fossil fuel
5. standard of living
6. renewable resource
7. urbanization
8. free enterprise
9. socialism

B

A. a measure of the quality of life

B. the number of children born each year for every 1,000 people

C. the movement of people to the cities

D. to move to another country

E. resource that can be replaced

F. people run businesses to make a profit

G. farmers grow only enough for themselves

H. economic system in which government runs some businesses

I. resource that can't be replaced

Mental Mapping Activity

Sketch a map of your state or community. Shade in and label the following resources or uses of land in your area. Add other resources if you need to, and don't forget to provide a map key.

- Forestry
- Commercial farming
- Manufacturing area
- Coal
- Petroleum

REVIEWING THE MAIN IDEAS

Section 1

1. PLACE What makes up the culture of a region?
2. REGION What are some measures used to judge the quality of life in a culture?

Section 2

3. MOVEMENT Why is world population distributed unevenly?
4. PLACE What does population density measure?

Section 3

5. HUMAN/ENVIRONMENT INTERACTION What is the difference between renewable and nonrenewable resources?
6. MOVEMENT What are three kinds of jobs people hold in service industries?

CRITICAL THINKING ACTIVITIES

1. **Making Comparisons** Why do population growth rates differ in developed countries and developing countries?
2. **Analyzing Information** Why are human skills and labor considered natural resources?

CONNECTIONS TO WORLD CULTURES

Cooperative Learning Activity Working in groups, refer to the culture regions map on page 56. Each group should find information about the population of more than one country in a culture region. Include information about literacy rate, birthrate, death rate, percentage of urban/rural dwellers, and standard of living. Brainstorm ways to present your information and make a presentation to the class.

GeoJournal Writing Activity

Families and individuals often have personal cultural traditions. For example, your family may have holiday customs or favorite foods that come from another culture region. In your journal, describe some elements of your home culture, including influences from other culture regions.

TECHNOLOGY ACTIVITY

Using a Spreadsheet Select three countries and find their population data in 10-year increments. Use your data to create a spreadsheet. In cells A2, A3, and A4, type in the names of your selected countries. In cells B2 through F2, type in the years *1950–2000*. In cells B2 through G2, type in the data for your first country. Do the same for the other two countries. Then create a bar graph showing population changes for the countries over a 50-year period.

PLACE LOCATION ACTIVITY: WORLD CULTURE REGIONS

Match the letters on the map with the culture regions of the world. Write your answers on a separate sheet of paper.

1. Africa South of the Sahara
2. United States and Canada
3. Australia, Oceania, and Antarctica
4. Latin America
5. Europe
6. Asia
7. Southwest Asia and North Africa

8. Russia and the Independent Republics

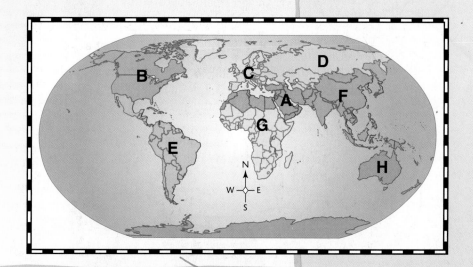

Danger: Ozone Loss

PROBLEM

Ozone: At ground level, it's a harmful gas and a major ingredient in smog. In the upper atmosphere, it's a lifesaver. Ozone acts as a sunscreen, blocking harmful ultraviolet rays that could kill all life on Earth. But manufactured chemicals, such as chlorofluorocarbons (CFCs), have thinned the ozone layer around the globe. This created an ozone hole over Antarctica—increasing the ultraviolet radiation reaching Earth. Higher levels of radiation lead to skin cancers and other health problems, and damage food chains and crops.

SOLUTIONS

● CFCs must go. An agreement approved by 74 nations calls for a ban on CFCs and other ozone-depleting chemicals by the year 2000.
● Ozone-friendly materials are replacing CFCs: Ammonia and helium are being used as coolants in refrigerators and air conditioners; propane and butane, in aerosol sprays; and water and terpenes, in cleaners.
● CFCs in existing air conditioners, refrigerators, and other products are being recycled.

Fun in the sun is not what it used to be. Sunbathers in Australia—before heading for the beach—tune in to television reports on daily ultraviolet levels. Australia has one of the world's highest rates of skin cancers.

OZONE FACT BANK

🍃 About 90 percent of all ozone in the atmosphere occurs in the stratosphere, approximately 8–30 miles (15–48 km) above Earth.

🍃 If compressed, scattered ozone molecules in the ozone layer above Earth would be about as thick as a pane of glass.

🍃 At certain times of the year, the ozone hole over Antarctica is about the size of the continental United States.

🍃 A polystyrene cup, when broken down, can add a billion CFC molecules to the atmosphere—and destroy a hundred trillion molecules of ozone. If production of CFCs stopped today, scientists estimate that it would take 50–100 years for the ozone layer to return to normal.

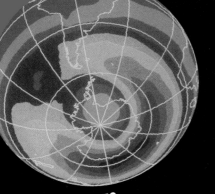

Loss of the protective ozone blanket is greatest over the South Pole. These images taken by the *Nimbus 7* satellite show the growing ozone hole, the black areas appearing in October 1983 and October 1989.

October 1979

October 1983

October 1989

environmental activities

TAKE A CLOSER LOOK

1 Why is the ozone layer important to all life on Earth?

2 What manufactured chemicals are damaging the ozone layer? Where are these chemicals found?

good planets are hard to find

WHAT CAN YOU DO?

🍃 Stop and think...before you use any can of aerosol spray containing CFCs.

🍃 Use popcorn or newspaper for packing material instead of polystyrene pieces.

🍃 If your family owns an air conditioner, have it checked for leaks. They cause 20 percent of CFCs leaked yearly in the United States.

🍃 Organize a "Sun Alert" campaign to warn students of the danger of skin cancer from overexposure to the sun.

TEEN TRIBUTE

Students at the George C. Soule School in Freeport, Maine, were concerned about the use of fast-food packages made with polystyrene, a plastic that contains CFCs. The students asked the town council to ban polystyrene packaging. A major fast-food chain argued against the ban, but the council sided with the students. Polystyrene packaging was banished from Freeport!

A Chinese family bikes home—carrying a new refrigerator charged with ozone-depleting CFCs.

Unit 2
The United States and Canada

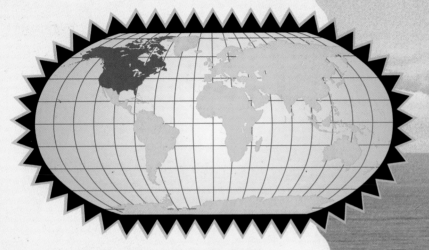

What Makes the United States and Canada a Region?

The people share . . .

- a varied physical environment.

- mid-latitude climates and fertile soils.

- a shift from a manufacturing economy to service industries.

- a population of mixed cultures.

To find out more about the United States and Canada, see the Unit Atlas on pages 76–87.

Heceta Head Lighthouse, United States

EXPLORING THE INTERNET

To learn more about the United States and Canada, visit the Glencoe Social Studies Web site at **www.glencoe.com** for information, activities, and links to other sites.

GeoJournal Activity

The United States and Canada are a cultural kaleidoscope—a mixture of the many ethnic groups who have settled in North America. Find out what groups of immigrants settled in your local area. In your journal, write an essay about the cultural influences these people have had on your town or city.

UNIT 2 ATLAS

76

NATIONAL GEOGRAPHIC SOCIETY

IMAGES *of the* WORLD

1. Native American, New Mexico
2. Lighthouse on the Atlantic Ocean, Nova Scotia
3. Boy facing a logging truck, Oregon
4. Outfitter on a camping trip, Alberta
5. Egret in a salt marsh, Louisiana
6. Fireworks over the *Statue of Liberty,* New York City
7. Red gabled buildings, Nova Scotia

All photos viewed against bison grazing, Wyoming.

Regional Focus

The United States and Canada are huge countries. Together they reach from the Arctic Ocean in the north to the country of Mexico in the south. West to east, they stretch from the Pacific Ocean to the Atlantic Ocean. The United States and Canada—with Mexico—form the continent of North America.

Region
The Land

The United States and Canada cover more than 7 million square miles (18 million sq. km). They share many of the same landscapes. Mountain ranges, basins, and plateaus run through the western part of both countries. The largest range is the massive Rocky Mountains, stretching more than 3,000 miles (4,800 km) from Alaska to New Mexico.

East of the Rockies spread the broad, rolling Great Plains. This flat area runs through the central part of the United States and Canada. The plains reach to the Canadian Shield—a huge, mineral-rich area of ancient rock—in Canada's far north. Find the Rockies and the Great Plains on the map on page 82.

Low mountains and coastal plains cover the eastern part of the United States and Canada. Here you will discover North America's second-longest mountain range, the Appalachians. East and south of these low mountains lie coastal lowlands that fan out westward to Texas.

Waterways The Mississippi River is the largest river system in North America. It flows from near the United States-Canadian border in the north to the Gulf of Mexico in the south. The Missouri River, the Ohio River, and their tributaries merge into the Mississippi River system. The largest lake system is the Great Lakes—Superior, Huron, Michigan, Erie, and Ontario. The waters of these five lakes are interconnected and eventually meet with the St. Lawrence River, which flows into the Atlantic Ocean.

Place
Climate and Vegetation

The vast size of landforms affects climate, soils, and vegetation in the United States and Canada. The forests and flatlands in far northern Canada and Alaska have short, cool summers and long, cold winters. The Pacific coast from southern Alaska to northern California has a mild, humid climate. Mist-covered forests and lush

Southwest Vegetation

The huge saguaro cactus grows only in the desert climates of the southwestern United States and Mexico.
PLACE: Where do you find the only deserts in the United States and Canada?

green plants cover this area. Pacific mountain ranges keep ocean rain clouds from reaching inland areas. Just east of the mountains lie the only deserts in the region.

The Great Plains The Great Plains has a humid continental climate of cold winters and hot summers. This region receives rain from the Gulf of Mexico and the far north. Before settlers arrived in the 1800s, grasses covered much of the Great Plains. Now the land is covered with farms and cities.

Other Areas The Northeast region of the United States and Canada has the same climate as the Great Plains. Most of the southern states, however, are in a humid subtropical region that experiences mild winters. Only the tip of Florida is far enough south to have a tropical climate. The Hawaiian Islands—the only other area of the United States with a tropical climate—lie 2,400 miles (3,862 km) from the United States mainland in the Pacific Ocean.

Region
Economy

The United States and Canada base their economies on the free enterprise system. This means that individuals and groups—rather than the government—run businesses. Skilled workers, many natural resources, and the use of advanced technology have brought prosperity to the United States and Canada.

Agriculture United States and Canadian farmers raise many meats, grains, vegetables, and fruits. Up-to-date farming methods have increased the size of farms and the amount of food produced. At the same time, they have decreased the number of farms and farm workers. Fertile soil and many waterways make the Great Plains, the Pacific coast, and the Northeast major agricultural areas.

The Pacific Coast

The city of Vancouver on Canada's Pacific coast is the country's busiest port.
PLACE: On what does Canada base its economy?

Industry The United States and Canada are rich in such mineral resources as copper, iron ore, nickel, silver, oil, natural gas, and coal. Both countries are world leaders in manufactured goods. Their service industries—stores, banking, insurance, and entertainment—provide jobs for the largest number of people.

Movement
People

Throughout their histories, the United States and Canada have become home to settlers from almost every part of the world. Some of these people came to find a better way of life. Still others were forced to come as enslaved laborers.

Ethnic Groups Today the people of this region come from many different backgrounds. Most Americans and Canadians are European in origin. Others are African Americans, Hispanics, Asian Americans, and Native Americans. English is the major language of the United States and Canada, but French is the dominant language of Canada's Quebec Province. Other languages are also spoken in both countries.

Population About 300 million people live in the United States and Canada. In land area, Canada is the second-largest country in the world. Yet most of its 31 million people live in urban areas close to its border with the United States. The United States is the world's third-largest country in population. Most of its 270 million people live in urban areas along the coasts or in the Great Lakes region. Among the largest cities are New York City, Los Angeles, Chicago, and Dallas. The area from Washington, D.C., to Boston is almost continuously urban and is called a megalopolis.

Place
Culture

At least 12,500 years ago, the ancestors of Native Americans settled North America. Europeans arrived between the 1500s and 1700s. During the late 1700s, colonists along the Atlantic coast successfully fought the British for their freedom. They founded the United States of America. In the 1800s the new Americans moved into new lands, fought a great Civil War, and then reunited. Immigrants arriving by the thousands transformed the economy from an agricultural one to an industrial one. During the 1900s, the United States fought two world wars and became a major world power.

Canada, on the other hand, stayed apart from the United States. It gradually won its independence from the British during the 1800s and early 1900s. As in the United States, many immigrants made their homes in Canada. People moved westward and developed the economy, although differences between English-speaking and French-speaking Canadians threatened national unity. In the 1900s, Canada began to play an active role in world affairs.

Nature's Light Show

The colorful glow of the northern lights flickers in Alaska and northern Canada.
REGION: What do you think causes the northern lights?

Picture Atlas of the World

Economic Geography

Using the CD-ROM The physical features of a region influence its type of economic activity. As a result, the economies of the United States and Canada differ from region to region. Assemble a file of economic activity information. (See the User's Guide for information on how to use the Collector button.) Include essays on the United States and Canada and these photographs: New York City, the Hawaiian pineapple industry, Newfoundland fishermen, and Saskatchewan wheat fields. Then use the file to answer the following questions:

1. Many of the largest cities in the United States are located east of the Mississippi River. Why were cities able to grow and prosper on the eastern coast of the United States?
2. What makes Hawaii one of the world's top producers of pineapples?
3. How has economic activity in Newfoundland contributed to the decline of the physical environment?
4. Which region of Canada is known as the country's granary?

Surfing the "Net"

Regional Trade

In the early 1990s the United States and Canada, along with Mexico, established an agreement called the North American Free Trade Agreement, or NAFTA. These countries wanted to create an economic trading bloc. However, there were some groups in the United States and Canada that were opposed to NAFTA. To find out more about the reasons why NAFTA was opposed, look on the Internet.

Getting There

Follow these steps to gather information on why certain groups were opposed to NAFTA.
1. Use a search engine. Type in the name *North American Free Trade Agreement*.
2. After typing *North American Free Trade*

Agreement, enter words like the following to focus your search: *NAFTA, economics, government, United States,* and *Canada*.
3. The search engine will provide you with a number of links to follow.

What To Do When You Are There

Organize the class into two groups. One group will represent the people who were opposed to NAFTA, and one group will represent the people who supported NAFTA. Click on the links to navigate through the pages of information and gather your findings. Each group should use a word processor to create a report about why they oppose or support NAFTA. Share the reports with the class.

Physical Geography

THE UNITED STATES AND CANADA: Physical

ELEVATIONS

Feet	Meters
10,000	3,000
5,000	1,500
2,000	600
1,000	300
0	0

⊛ National capital
▲ Mountain peak

Lambert Equal-Area projection

Hawaii (U.S.)

Map Study

1. **REGION** What physical region covers much of the central part of the United States and Canada?

2. **PLACE** What mountain range runs along the eastern coast of the United States and Canada?

UNIT 2

ELEVATION PROFILES

UNITED STATES

14,000 ft.   4,267 m

Rocky Mountains

Sierra Nevada

Atlantic Ocean

10,000 ft.   3,048 m

Pacific Ocean

Appalachian Mountains

Atlantic Coastal Plain

Great Plains

5,000 ft.   1,524 m

Central Lowlands

2,000 ft.
1,000 ft.   610 m
Sea level   305 m
Sea level

←———— West to East at 40°N latitude ————→

CANADA

10,000 ft.   3,048 m

Coast Mountains

Rocky Mountains

Laurentian Highlands

Atlantic Ocean

Canadian Shield

Great Plains

Gulf of St. Lawrence

5,000 ft.   1,524 m

Pacific Ocean

2,000 ft.
1,000 ft.   610 m
Sea level   305 m
   Sea level

←———— West to East at 50°N latitude ————→

Source: *Goode's World Atlas,* 19th edition

THE UNITED STATES AND CANADA: Land Comparison

GeoFacts

Highest point: Mount McKinley (Alaska) 20,320 ft. (6,194 m) high

Lowest point: Death Valley (California) 282 ft. (86 m) below sea level

Longest river: Mackenzie River (Canada) 2,635 mi. (4,240 km) long

Largest lake: Lake Superior 31,700 sq. mi. (82,103 sq. km)

Highest waterfall: Yosemite Falls (California) 2,425 ft. (739 m) high

Graphic Study

1. **LOCATION** In what general area—east or west—are the highest elevations in the United States and Canada found?
2. **PLACE** How would you describe the size of the United States compared with the size of Canada?

Cultural Geography

THE UNITED STATES AND CANADA: Political

RUSSIA

GREENLAND

Ellesmere Island

ARCTIC OCEAN

- National boundary
- State/provincial boundary
- ⊛ National capital

Lambert Equal-Area projection

Alaska (U.S.)

Yukon Territory

Arctic Circle

Northwest Territories

Nunavut

British Columbia

Alberta

C A N A D A

Hudson Bay

Newfoundland

Hawaii (U.S.)

160°W   20°N
0   100 mi.
0   100 km

Manitoba

Saskatchewan

Quebec

Ontario

New Brunswick

Prince Edward Is.

Washington

Montana

North Dakota

Minn.

Ottawa ⊛

Vt.   Me.

Nova Scotia

Oregon

Idaho

Wyoming

South Dakota

Wis.

Mich.

N.H.

N.Y.

Mass.
R.I.

Nebraska

Iowa

Ill.

Ind.

Ohio

Pa.

Conn.
N.J.

ATLANTIC OCEAN

Nevada

Utah

Colorado

Kansas

Missouri

Ky.

W.Va.

Va.

Md.
Del.

Washington, D.C.

PACIFIC OCEAN

California

U N I T E D

Arizona

New Mexico

Oklahoma

Ark.

Tenn.

N.C.

S.C.

S T A T E S

Miss.

Ala.

Ga.

0   250   500 mi.
0   250   500 km

Texas

La.

Fla.

N
W   E
S

Gulf of Mexico

MEXICO

Tropic of Cancer

Map Study

1. **LOCATION** What U.S. state lies farthest east?
2. **PLACE** What is the capital of Canada?

THE UNITED STATES AND CANADA: Ethnic Groups

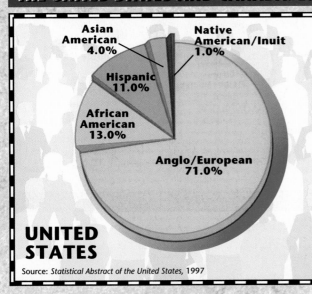

Asian American 4.0%

Native American/Inuit 1.0%

Hispanic 11.0%

African American 13.0%

Anglo/European 71.0%

UNITED STATES

Source: *Statistical Abstract of the United States, 1997*

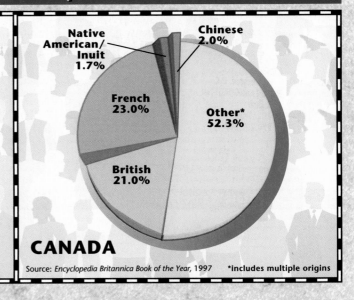

Native American/Inuit 1.7%

Chinese 2.0%

French 23.0%

Other* 52.3%

British 21.0%

CANADA

Source: *Encyclopedia Britannica Book of the Year, 1997*   *includes multiple origins

GeoFacts

Biggest country (land area): Canada 3,849,674 sq. mi. (9,970,610 sq. km)

Smallest country (land area): United States 3,536,340 sq. mi. (9,159,121 sq. km)

Largest city (population): New York City (1997) 16,332,000; (2015 projected) 17,602,000

Highest population density: United States 76 people per sq. mi. (29 people per sq. km)

Lowest population density: Canada 9 people per sq. mi. (4 people per sq. km)

COMPARING POPULATION: Canada and the United States

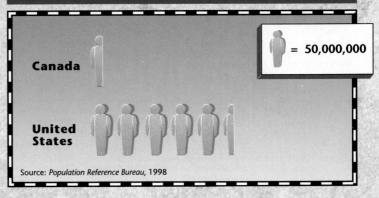

Canada

United States

= 50,000,000

Source: *Population Reference Bureau, 1998*

Graphic Study

1. **PLACE** What is the population of Canada?
2. **MOVEMENT** Which country has a larger Native American population?

UNIT 2 ATLAS

Countries at a Glance

United States

Washington, D.C. ★

CAPITAL:
Washington, D.C.
MAJOR LANGUAGE(S):
English
POPULATION:
270,200,000

LANDMASS:
3,536,340 sq. mi./
9,159,121 sq. km
MONEY:
U.S. Dollar
MAJOR EXPORT:
Machinery
MAJOR IMPORT:
Machinery

Canada

Ottawa ★

CAPITAL:
Ottawa
MAJOR LANGUAGE(S):
English, French
POPULATION:
31,000,000

LANDMASS:
3,849,674 sq. mi./
9,970,610 sq. km
MONEY:
Canadian Dollar
MAJOR EXPORT:
Motor Vehicles
MAJOR IMPORT:
Motor Vehicles

U.S. State Name Meaning and Origin

ALABAMA
Montgomery ★
"thicket clearers"
(Choctaw)

ALASKA
Juneau ★
"the great land" (Aleut)

ARIZONA
★ Phoenix
"little spring" (Papago),
or "dry land" (Spanish)

ARKANSAS
★ Little Rock
"downstream people"
(Quapaw)

Sacramento ★
CALIFORNIA
unknown meaning (Spanish)

Denver ★
COLORADO
"red" (Spanish)

CONNECTICUT
Hartford ★
"beside the long tidal river"
(Native American)

Dover ★
DELAWARE
named for Virginia's
colonial governor,
Baron De La Warr

★ Tallahassee
FLORIDA
"feast of flowers"
(Spanish)

★ Atlanta
GEORGIA
named for England's
King George II

Honolulu ★
HAWAII
unknown meaning
(native Hawaiian)

IDAHO
★ Boise
unknown meaning
(Native American)

ILLINOIS
★ Springfield
"tribe of superior men"
(Native American)

INDIANA
Indianapolis ★
"land of Indians"
(European American)

IOWA
★
Des Moines
unknown meaning
(Native American)

Topeka ★
KANSAS
"people of the
south wind" (Sioux)

★ Frankfort
KENTUCKY
"land of tomorrow"
(Iroquoian)

LOUISIANA
Baton ★
Rouge
named for France's
King Louis XIV

MAINE
Augusta
★
named for an ancient
French province

MARYLAND
Annapolis ★
named in honor of the wife
of England's King Charles II

Boston ★
MASSACHUSETTS
"great mountain place"
(Native American)

MICHIGAN
Lansing ★
"great lake" (Ojibway)

MINNESOTA
Saint ★
Paul
"sky-tinted water"
(Sioux)

MISSISSIPPI
★ Jackson
"father of the waters"
(Native American)

MISSOURI
Jefferson ★
City
"town of the large canoes"
(Native American)

MONTANA
★ Helena
"mountainous" (Spanish)

NEBRASKA
Lincoln ★
"flat water"
(Native American)

NEVADA
★ Carson City
"snowcapped"
(Spanish)

NEW HAMPSHIRE
Concord ★
named for Hampshire,
a county in England

NEW JERSEY
★ Trenton
named for Isle of Jersey,
a British territory

Countries/States/Provinces not drawn to scale.

U.S. State Name Meaning and Origin

NEW MEXICO
★ Santa Fe

named for the state's former colonial ruler, Mexico

NEW YORK
Albany ★

named in honor of the English Duke of York

NORTH CAROLINA Raleigh ★

named in honor of England's King Charles I

NORTH DAKOTA
★ Bismarck

named for the Dakota, a Native American group

OHIO
Columbus ★

"great river" (Native American)

OKLAHOMA
Oklahoma City ★

"red people" (Choctaw)

OREGON
★ Salem

unknown meaning and origin

PENNSYLVANIA
Harrisburg ★

"Penn's woodland" named for the father of Pennsylvania's founder, William Penn

Providence ★
RHODE ISLAND

unknown meaning and origin

SOUTH CAROLINA
Columbia ★

named for England's King Charles I

SOUTH DAKOTA
★ Pierre

named for the Dakota, a Native American group

Nashville ★ **TENNESSEE**

named for Tanasi, "Cherokee villages" (Cherokee)

TEXAS
★ Austin

"friends" (Tejas)

Salt Lake City ★
UTAH

"people of the mountains" (Ute)

★ Montpelier
VERMONT

"green mountain" (French)

Richmond ★
VIRGINIA

named for the unmarried Queen Elizabeth I of England, known as "the Virgin Queen"

★ Olympia
WASHINGTON

named in honor of George Washington

Charleston ★
WEST VIRGINIA

began as the western part of Virginia before becoming a state in 1863

WISCONSIN
Madison ★

"grassy place" (Chippewa)

WYOMING
Cheyenne ★

"upon the great plain" (Delaware)

Canadian Province Name Meaning and Origin

ALBERTA
Edmonton ★

named for the daughter of England's Queen Victoria

BRITISH COLUMBIA

Victoria ★

named for Christopher Columbus and the province's British heritage

MANITOBA
Winnipeg ★

"strait of the great spirit" (Algonquian)

NEW BRUNSWICK
Fredericton ★

named for English royal family of Brunswick-Luneburg

NEWFOUNDLAND
St. John's ★

"new found land," named by European explorer John Cabot in 1497

NOVA SCOTIA
Halifax ★

Latin term for "New Scotland," based on province's Scottish heritage

ONTARIO
Toronto ★

meaning unknown (Iroquoian)

PRINCE EDWARD ISLAND
Charlottetown ★

named for the son of England's King George III

QUEBEC
Quebec ★

"place where the river narrows" (Algonquian)

SASKATCHEWAN
Regina ★

"fast flowing river" (Cree)

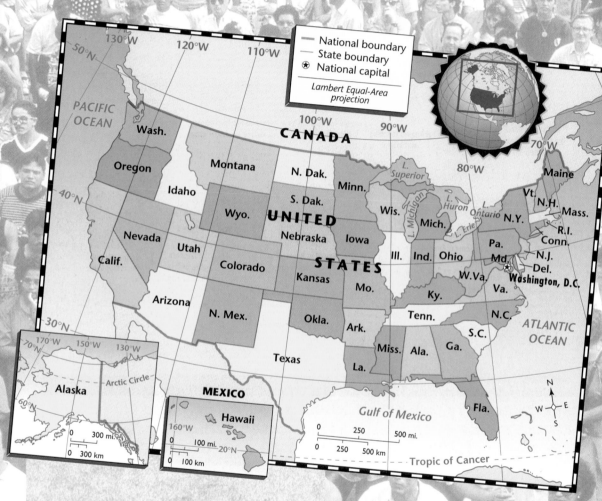

National boundary
State boundary
⊛ National capital

Lambert Equal-Area projection

130°W 120°W 110°W

50°N

PACIFIC
OCEAN

100°W CANADA 90°W

70°W

Wash.

Oregon

Montana N. Dak. Minn.

L. Superior

Maine

40°N

Idaho

S. Dak.

Wis.

L. Michigan

L. Huron

L. Ontario

Vt. N.H.

Mass.

Wyo. UNITED

Mich.

L. Erie

N.Y.

R.I.
Conn.

Nevada Utah

Nebraska Iowa

Pa.

N.J.

Calif.

Colorado STATES

Ill. Ind. Ohio

Md.

Del.

Kansas

Mo.

W.Va. Va. ⊛ Washington, D.C.

Arizona

Ky.

N. Mex.

Okla.

Ark.

Tenn.

N.C.

30°N

170°W 150°W 130°W

70°N

Texas

Miss. Ala. Ga.

S.C.

ATLANTIC
OCEAN

La.

Alaska Arctic Circle

60°N

MEXICO

160°W Hawaii

20°N

Fla.

Gulf of Mexico

N

W   E

S

0   300 mi.
0   300 km

0   100 mi.
0   100 km

0   250   500 mi.
0   250   500 km

Tropic of Cancer

MAP STUDY ACTIVITY

You may think you already know everything about the United States.

1. Do you know what state has the largest population?
2. What is the distance from the country's east coast to west coast?
3. What is the longest river in the United States?

SECTION 1

The Land

PREVIEW

Words to Know
- contiguous
- urban
- megalopolis
- rural
- coral reef

Places to Locate
- Appalachian Mountains
- Mississippi River
- Great Lakes
- Great Plains
- Rocky Mountains
- Sierra Nevada Mountains
- Hawaiian Islands

Read to Learn . . .
1. what landforms are found in the United States.
2. what climates occur in the United States.

If you like to hike and camp, the Great Smoky Mountains are for you! Each year, thousands of vacationers visit the Smokies, which lie on the border between North Carolina and Tennessee. Why were they given this name? A smoky blue haze covers these high and rugged peaks in the Appalachian Mountain range.

The Great Smoky Mountains are only one of many exciting places in the United States. Look at the map on page 88 and imagine traveling through some of the 50 states. Start in the northeast. Hike along the Maine coast, and get soaked with ocean spray. Head south and clap your hands to country music in Nashville, Tennessee. Go west and take a mule ride a mile down into the Grand Canyon. End your journey on the west coast in California. All of the United States—from coast to coast—offers many sights, sounds, and adventures!

Location

A Huge Country

Most of the United States—48 of the 50 states—stretches 2,807 miles (4,517 km) across the entire middle part of North America. These 48 states are **contiguous**, or joined together inside a common boundary. The map on page 88 shows you that they touch three major bodies of water—the Atlantic Ocean, the Gulf of Mexico, and the Pacific Ocean.

Two states lie apart from the 48 contiguous states. Alaska, the largest state, spreads over the northwestern corner of North America. Hawaii, the newest state, lies far out in the Pacific Ocean. It is about 2,400 miles (3,862 km) southwest of the California coast.

The United States—the world's fourth-largest country in size—has a total land area of 3,536,340 square miles (9,159,121 sq. km). Only Russia, Canada, and China are larger.

Region
From Sea to Shining Sea

Like a patchwork quilt, the United States consists of regional patches of different landscapes. Look at the map below to find the five physical regions of the United States.

The Coastal Plains A broad lowland runs like a quilt's border along the Atlantic Ocean and the Gulf of Mexico. Geographers divide this lowland into two parts: the Atlantic Coastal Plain and the Gulf Coastal Plain.

The Atlantic Coastal Plain borders the Atlantic coast from Massachusetts to Florida. Many of the region's deepwater ports provided excellent harbors for the first settlers' ships. Port cities such as Boston and New York City have devel-

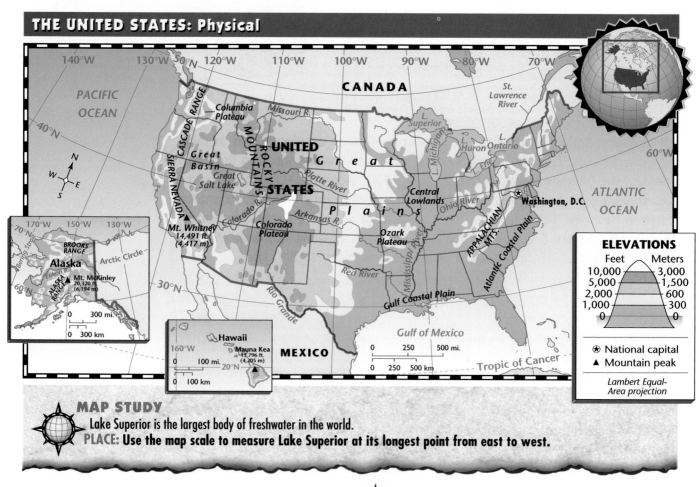

MAP STUDY
Lake Superior is the largest body of freshwater in the world.
PLACE: Use the map scale to measure Lake Superior at its longest point from east to west.

The United States has five physical regions. Florida's wetland Everglades is in the Coastal Plains *(left)*. The Golden Gate Bridge, linking San Francisco to northern California, is in the Pacific Region *(right)*.
REGION: Which area looks more heavily populated?

oped worldwide trade networks. You can drive for hundreds of miles and not always tell where one **urban**, or densely populated, area ends and another begins. Geographers refer to this urban pattern as a **megalopolis**, or super city. Farther south are more **rural**, or sparsely populated, areas. In Florida, you will even discover uninhabited marshes and swamps.

The Gulf Coastal Plain hugs the Gulf of Mexico from Florida to Texas. Look at the map on page 90. You can see that the Gulf Coastal Plain is much wider than the Atlantic Coastal Plain. The Mississippi River—the longest river in the country—drains much of the Gulf Coastal Plain. This river runs 2,340 miles (3,765 km) from the north central United States to the Gulf of Mexico. Barges carry goods to and from cities along the Mississippi—some all the way to the port city of New Orleans.

The Appalachian Mountains Curving west along the Atlantic Coastal Plain are the Appalachian (A•puh•LAY•chuhn) Mountains. The second-longest mountain range in North America, the Appalachians run almost 1,500 miles (2,400 km) from eastern Canada to western Alabama.

The Appalachians are the oldest mountain range on the continent. How do geographers know this? The rounded mountain peaks show their age. Erosion has worn them down over time. The average height of the Appalachian ranges is less than 6,000 feet (2,000 m).

The Interior Plains The central part of the country's landscape quilt is the Interior Plains. The eastern region of these plains is called the Central Lowlands. Traveling through this area, you will see thick forests, broad river valleys, rolling flatlands, and grassy hills.

The Gateway Arch in St. Louis welcomes visitors to this Mississippi River port. **REGION: In what physical region is St. Louis located?**

The largest group of freshwater lakes in the world—the Great Lakes—lie in the northern part of the Central Lowlands. Glaciers formed Lake Superior, Lake Michigan, Lake Huron, Lake Erie, and Lake Ontario millions of years ago. The waters of these connected lakes flow into the St. Lawrence River, which empties into the Atlantic Ocean.

Farther west the landscape, blanketed with fields of grain and pastures, takes on a checkerboard pattern. In the Interior Plains region lie the Great Plains—a broad, high area. The Great Plains rises in elevation from about 2,500 feet (762 m) in the east to about 6,500 feet (981 m) in the west.

The Rocky Mountains The western part of the Great Plains meets the rugged Rocky Mountains. The Rockies are the largest mountain range in North America, stretching all the way from Alaska to Mexico. In the contiguous United States, the Rocky Mountains run more than 1,100 miles (1,770 km) from north to south.

Look at the physical map on page 90. The Rockies—with some peaks rising more than 14,000 feet (4,270 m)—are higher and more rugged than the Appalachians. Movements of plates under the earth's crust formed the Rockies millions of years ago. A ridge of these mountains, the Continental Divide, separates the rivers and streams flowing west to the Pacific Ocean from those flowing east toward the Mississippi River. Several important rivers—including the Colorado, Missouri, Arkansas, and Rio Grande—begin in the Rocky Mountains.

Just west of the Rockies lies an area of largely empty basins and plateaus. A large valley there, called the Great Basin, holds the Great Salt Lake. If you ever try swimming in it, you will discover that the lake's high salt levels make it easy for you to float.

To the southwest—in California—sits the lowest and hottest place in the United States. It is Death Valley, which lies about 282 feet (86 m) *below* sea level. Summer temperatures in Death Valley often climb to 125°F (52°C).

The Pacific The Pacific Ocean forms the western border of the United States's landscape quilt. Near the coast lie two major mountain ranges: the Sierra Nevadas and the Cascade Range. Like the Rockies, these Pacific ranges were formed by plate movements. The melting of these plates created volcanoes and lava flows, which formed the Cascade Range. Earthquakes and volcanic eruptions still shake the region. West of the Pacific mountain ranges are coastal lowlands and fertile valleys.

Alaska and Hawaii are part of the Pacific region. Glaciers, islands, and bays line Alaska's southern coastline. Central and southern mountain ranges are broken up by lowlands and plateaus. Mount McKinley, North America's highest mountain at 20,320 feet (6,194 m), towers over this area. A vast plain stretches along the coast of the Arctic Ocean in Alaska's far north.

The Hawaiian Islands were formed by eruptions of volcanic mountains on the ocean floor. Some of the islands have **coral reefs**. These are submerged or low-lying structures formed over time from the skeletons of small sea animals.

Region
Climate

Tropical climates, mid-latitude climates, high latitude climates—all are found in the United States. Why is there so much variety? The country's huge size, changing elevations, and the flow of its ocean and wind currents create these differences.

Mid-Latitude Climates Most of the United States lies in mid-latitude climate regions. The map on page 94 shows you the parts of the country that have a humid continental climate. Winters in these areas are cold and moist; summers, long and hot. Rain falls throughout the year. Snow often blankets this area in winter, especially around the Great Lakes.

The southeastern United States has a humid subtropical climate. Winters are mild and cool, and summers are hot and humid. Severe thunderstorms, including destructive tornadoes, are common in this region during summer months.

The area along the Pacific coast from northern California to southeastern Alaska has a marine west coast climate. Temperatures here are mild year-round, and Pacific winds bring plenty of rainfall. Southern California has a Mediterranean climate of dry, warm summers and rainy, mild winters.

Water Power Along the Colorado River

The energy of the Colorado River is harnessed by the Glen Canyon Dam in Arizona. **HUMAN/ENVIRONMENT INTERACTION: What type of energy can be created by flowing water?**

Dry Climate A dry climate prevails in the plateaus and basins between the Pacific mountain ranges and the Rockies. Hot, dry air gets trapped here when the Pacific ranges block humid ocean winds. This region is dotted with deserts, including the Great Salt Lake Desert, the Black Rock Desert, and Death Valley. The western part of the Great Plains also has a dry climate. Summers here tend to be hot and dry, and winters very cold.

High Latitude Climates You will find a highland climate if you visit the mountains in the western part of the contiguous United States. To the north,

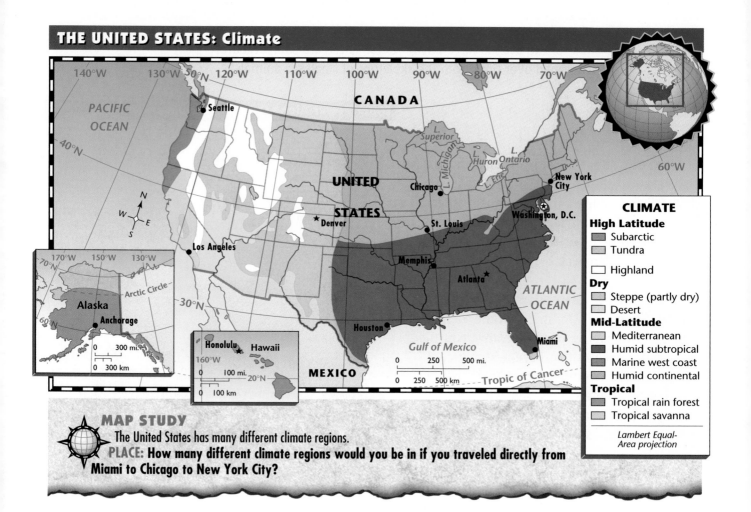

MAP STUDY
The United States has many different climate regions.
PLACE: How many different climate regions would you be in if you traveled directly from Miami to Chicago to New York City?

Alaska is the only part of the country to have subarctic and tundra climates. Winters are bitterly cold and summers are very cool along the Arctic Circle.

Tropical Climates What is the wettest state in the United States? Hawaii, which has a tropical climate, claims that distinction. You can find warm temperatures and plenty of rainfall at the southern tip of Florida, too.

SECTION 1 ASSESSMENT

REVIEWING TERMS AND FACTS

1. Define the following: contiguous, urban, megalopolis, rural, coral reef.

2. LOCATION Where are the Great Lakes located?

3. PLACE What is the largest mountain range in the contiguous United States?

4. REGION Why does the United States have so many different types of climates?

MAP STUDY ACTIVITIES

5. Look at the physical map on page 90. What high, flat area covers much of the central United States?

6. Study the climate map above. In what climate region is Memphis, Tennessee?

UNIT 2

BUILDING GEOGRAPHY SKILLS

Using Scale

Model cars or airplanes are like the larger versions, except for size. Models are made to scale—1 inch stands for a larger measurement in the real vehicles. **Scale** is also used to represent size and distance on maps. For example, 1 inch on a map may represent 100 miles (161 km) on the earth's surface. A map scale is usually shown with a **scale bar**. This bar shows you how much real distance on the earth is shown by a measurement on the map. To use scale to find distance, follow these steps:

- On the scale bar, find the unit of measurement.

- Using this unit, measure the distance between two points.

- Multiply that number by the miles or kilometers each unit stands for.

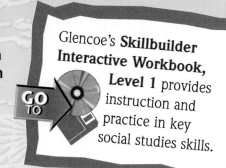

Glencoe's **Skillbuilder Interactive Workbook, Level 1** provides instruction and practice in key social studies skills.

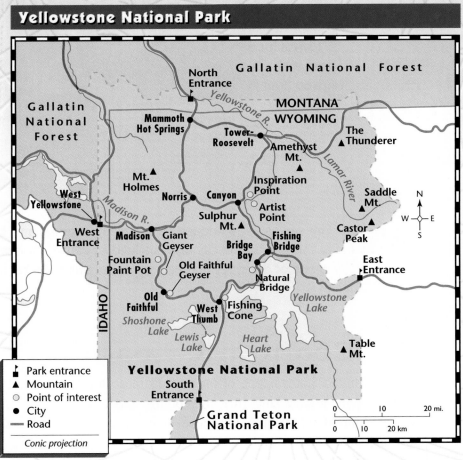

Yellowstone National Park

Gallatin National Forest
North Entrance
Gallatin National Forest
MONTANA
WYOMING
Yellowstone R.
Mammoth Hot Springs
Tower-Roosevelt
The Thunderer
Amethyst Mt.
Mt. Holmes
West Yellowstone
Norris
Canyon
Inspiration Point
Saddle Mt.
Madison R.
Sulphur Mt.
Artist Point
Lamar River
West Entrance
Madison
Giant Geyser
Bridge Bay
Fishing Bridge
Castor Peak
Fountain Paint Pot
Old Faithful Geyser
Natural Bridge
East Entrance
Old Faithful
West Thumb
Fishing Cone
Yellowstone Lake
IDAHO
Shoshone Lake
Lewis Lake
Heart Lake
Table Mt.
Yellowstone National Park
South Entrance
Grand Teton National Park

■ Park entrance
▲ Mountain
○ Point of interest
● City
— Road

Conic projection

0   10   20 mi.
0   10   20 km

Geography Skills Practice

1. In what three states does Yellowstone National Park lie?

2. By road, about how far is the North Entrance from West Thumb?

3. About how much farther by road is it from Old Faithful Geyser to the South Entrance than from the geyser to Bridge Bay?

The Economy

PREVIEW

Words to Know
- free enterprise system
- service industry
- farm belt
- dry farming
- acid rain
- interdependent

Places to Locate
- Northeast
- South
- Midwest
- Interior West
- Pacific

Read to Learn . . .
1. how people in the United States earn their livings.
2. why the United States ranks as a world economic leader.
3. what economic challenges the United States faces today.

Two 110-story towers loom above the Hudson River in the Financial District of New York City. This is the World Trade Center. From the top floor of the Trade Center, you can view the skyscrapers and harbor of New York City.

New York City serves as headquarters for many national and international businesses and banks. The city also is a leading center for overseas trade. New York City is one of the cities that help make the United States a world economic leader.

Place
An Economic Leader

The United States is one of the world's most developed countries. In addition to having a large land area, the country is rich in natural resources and in skilled, hardworking people. All of these benefits provide the United States with a strong, productive economy.

The American economy is based on the **free enterprise system.** Under free enterprise, people own and run businesses with limited government controls. America's economic strength was first built on agriculture, which remains important. The United States also is strong in science, technology, education, and medicine.

Economic Regions

Geographers divide the United States into regions based on economic activity. These regions are the Northeast, the South, the Midwest, the Interior West, and the Pacific. How do the people in your community make their livings?

The Northeast The Northeast is one of the oldest manufacturing areas in the United States. By the early 1800s, New England had launched the nation's Industrial Revolution. Today you can find all types of goods and machinery manufactured here. The map below shows you that coal mining, for example, is an important economic activity in Pennsylvania and West Virginia.

The deepwater harbors and manufacturing economies of Boston, New York City, Philadelphia, and Baltimore make the Northeast a major center of world trade. New York City is also a world center of fashion, entertainment, publishing, and communications.

Many Northeastern cities employ thousands in **service industries** such as business, finance, banking, and insurance. Service industries are businesses that provide services to customers rather than producing farm or industrial products.

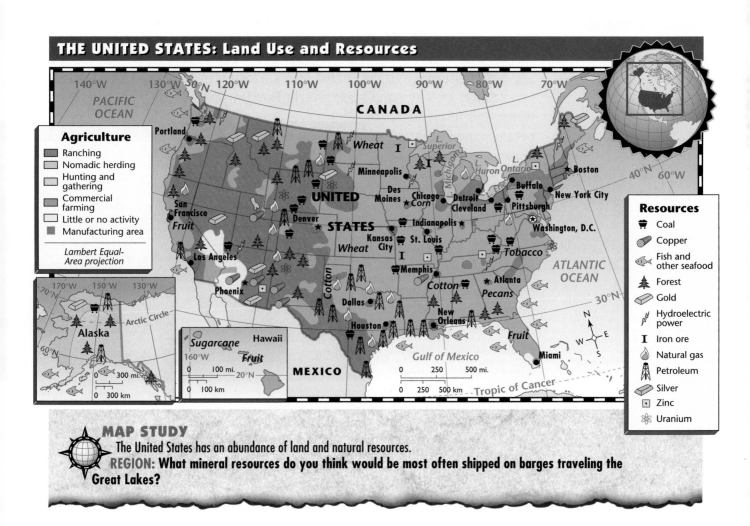

THE UNITED STATES: Land Use and Resources

Agriculture
- Ranching
- Nomadic herding
- Hunting and gathering
- Commercial farming
- Little or no activity
- Manufacturing area

Lambert Equal-Area projection

Resources
- Coal
- Copper
- Fish and other seafood
- Forest
- Gold
- Hydroelectric power
- Iron ore
- Natural gas
- Petroleum
- Silver
- Zinc
- Uranium

MAP STUDY
The United States has an abundance of land and natural resources.
REGION: What mineral resources do you think would be most often shipped on barges traveling the Great Lakes?

Computer research and development is an important service industry in Silicon Valley, California.
HUMAN/ENVIRONMENT INTERACTION: What is a service industry?

The national capital, Washington, D.C., offers service-industry jobs in government and tourism.

Agriculture is also important to the economy of the Northeast. Farmers grow fruits and vegetables and raise dairy cattle and chickens. Fishing is a leading industry along the Atlantic coast.

The South As in the Northeast, the economy in the South is varied. The people here work in manufacturing, farming, and fishing. Industry, however, has become the South's main source of income. Oil-based products, textiles, electrical equipment, and airplane parts emerge from factories and refineries here.

In recent years the South has attracted new businesses and people. Service industries have grown, as have large entertainment centers such as Seaworld and Walt Disney World in Orlando, Florida. Seaside resorts in Florida, South Carolina, and Mississippi attract tourists from all over the world. The map on page 97 shows that inland cities such as Atlanta and Dallas are major manufacturing areas. Large port cities—Houston, Miami, and New Orleans—are busy manufacturing and shipping centers.

Agriculture has always been a major economic activity in the South. Texas—with more farms than any other state—raises livestock and grows crops such as wheat and cotton. The South's warm, wet climate favors crops not usually grown elsewhere in the United States. Farmers in Louisiana and Arkansas, for example, grow rice and sugarcane. In Florida and Texas, they cultivate citrus fruits; and in Georgia, pecans and peanuts. Look at the map on page 97 to see where some agricultural products are grown in the South.

The Midwest In this region lies the American **farm belt**. Flat land and fertile soil cover much of the Midwest. Productive farms supply huge crops of corn, soybeans, and grains such as oats and wheat. In some areas of the Great Plains, farmers use **dry farming** to grow a certain kind of wheat. Dry

Waves of Grain

Contour plowing follows the curves of the land and is used to reduce water runoff.
HUMAN/ENVIRONMENT INTERACTION: Why do you think it is important to keep water from running off of farmland?

farming is a way of plowing land so that it holds rainwater. Dairy products and livestock are also important in the midwestern economy.

When you hear the term "Midwest," you probably don't think of large port cities for shipping. But the Midwest has a number of waterways and ports. The St. Lawrence Seaway links the Great Lakes to the Atlantic Ocean. The Mississippi and Ohio rivers wind through the Midwest into the Gulf Coastal Plain. These waterways are deep enough for ships to transport farm and industrial products to other parts of North America and the world.

In the late 1800s, the Midwest drew many immigrants and became a leading industrial area. Workers today produce iron, steel, heavy machinery, and cars. Industrial cities such as Chicago, Detroit, Milwaukee, and Cleveland dot the coasts of the Great Lakes. St. Louis—near the spot where the Mississippi River meets the Missouri River—is an important shipping and transportation center.

The Interior West A dry climate affects economic activity in the broad Interior West. Ranching is more common than farming in this region. A single cattle or sheep ranch may cover 2,000 acres (809 ha) or more. With limited water sources, farming depends on irrigation. Potatoes, hay, wheat, barley, and sugar beets are the region's major crops.

The Interior West has plenty of rich mineral and energy resources. Look at the map on page 97. You will see that workers in the region mine silver and copper and drill for oil.

Have you ever visited Arizona's Grand Canyon? Tourism is a rapidly growing industry in the Interior West. The cities of Albuquerque, Denver, Phoenix, and Salt Lake City are tourist and transportation centers as well as places of government and business.

The Pacific The Pacific region boasts economic activity from farming, manufacturing, and tourism. Some of the best farmland in the region is in central California. Many of the fruits and vegetables you eat every day come from here. Crops such as sugarcane, pineapples, coffee, and rice flourish in Hawaii's tropical climate and fertile volcanic soil. Fishing is also a major industry along the Pacific region's coastal waterways.

Lumber and mining are important sources of income in many states. California has gold, lead, and copper. Alaska has vast oil reserves. Industries hum in the large cities of the Pacific region—Los Angeles, San Francisco, San Diego, and Seattle. Among the leading manufactured products are airplanes, computers, and computer software. Los Angeles is also the world capital of the movie industry, and San Diego is home to the United States Navy's Pacific fleet.

An estimated 47.5 million people visit the United States each year. Many of them begin their trip on America's Pacific coast, lured by snowcapped mountains, golden beaches, and crystal-clear lakes. Tourism is Hawaii's major source of income. The tropical beauty of the Hawaiian Islands attracts millions of tourist dollars annually.

The Skyline of Los Angeles, California

More than one-third of California's residents live in the sprawling city of Los Angeles. **REGION: What is Los Angeles famous for?**

Human/Environment Interaction
Today's Challenges

Americans use the latest technology to develop the country's natural resources. In some cases, however, people use resources in ways that are harmful to the environment. And some resources are being completely used up!

Use of Resources Factories in the United States use oil, gas, and coal to manufacture goods. The burning of these fuels creates pollution. When air pollution mixes with water vapor and falls to the earth in rain or snow, this is called **acid rain**. Acid rain kills fish, pollutes lakes and rivers, and damages forests in the United States and across the border in Canada.

Overusing resources is another problem. Forests are cut to provide lumber and paper products Americans want. Fishing off the coasts has led to a decline in many species of fish. In recent years, national and state governments have passed laws to protect natural resources.

Growth of Industry High-technology industries help the United States maintain its leadership role in world manufacturing. As computers and robots entered factories, however, many American workers lost their jobs. Some of the country's workforce have had to learn new skills or start new careers.

Service industries are a major part of the United States economy. Service industries such as banking, medical care, education, and government employ more workers than any other industry.

World Trade The United States leads the world in the total value of its imports and exports. Millions of jobs are linked to the exporting and importing of goods. In recent years, the economies of the United States and other countries have become more **interdependent**, or reliant on each other. In 1993 the North American Free Trade Agreement (NAFTA) was approved. It has increased trade among the United States, Canada, and Mexico.

SECTION 2 ASSESSMENT

REVIEWING TERMS AND FACTS

1. Define the following: free enterprise system, service industry, farm belt, dry farming, acid rain, interdependent.

2. REGION Why is the Interior West different from other economic regions in the United States?

3. HUMAN/ENVIRONMENT INTERACTION How has acid rain affected the United States?

4. PLACE What form of economic activity employs the most people in the United States?

MAP STUDY ACTIVITIES

5. Look at the land use map on page 97. What is the leading mineral resource found near the Gulf of Mexico?

MAKING CONNECTIONS

| MATH | SCIENCE | HISTORY | **LITERATURE** | TECHNOLOGY |

American author John Steinbeck and his dog Charley traveled the country in search of America. This excerpt describes his feelings as he drove through the Bad Lands of South Dakota.

I went into a state of flight, running to get away from the unearthly landscape. And then the late afternoon changed everything. As the sun angled, the buttes and coulees [dry streambeds], the cliffs and sculptured hills and ravines lost their burned and dreadful look and glowed with yellow and rich browns and a hundred variations of red and silver gray, all picked out by streaks of coal black. It was so beautiful that I stopped near a thicket of dwarfed and wind-warped cedars and junipers, and once stopped I was caught, trapped in color and dazzled by the clarity of the light. Against the descending sun the battlements were dark and clean-lined, while to the east, where the uninhibited light poured slantwise, the strange landscape shouted with color. And the night, far from being frightful, was lovely beyond thought, for the stars were close, and although there was no moon the starlight made a silver glow in the sky. The air cut the nostrils with dry frost. And for pure pleasure I collected a pile of dry dead cedar branches and

built a small fire just to smell the perfume of the burning wood and to hear the excited crackle of the branches. My fire made a dome of yellow light over me, and nearby I heard a screech owl hunting and a barking of coyotes, not howling but the short chuckling bark of the dark of the moon. This is one of the few places I have ever seen where the night was

The Bad Lands

friendlier than the day. And I can easily see how people are driven back to the Bad Lands.

From *Travels with Charley* by John Steinbeck. Copyright © 1961, 1962 by The Curtis Publishing Co., © 1962 by John Steinbeck, renewed © 1990 by Elaine Steinbeck, Thom Steinbeck, and John Steinbeck IV. Used by permission of Viking Penguin, a division of Penguin Books USA Inc.

Making the Connection

1. How did Steinbeck's feelings toward the Bad Lands change as the day wore on?
2. How does the landscape of the Bad Lands compare with the landscape in your community?

SECTION 3

The People

PREVIEW

Words to Know
- immigrant
- colony
- revolution
- republic
- multicultural
- ethnic group
- mobile
- national park

Places to Locate
- Mississippi River
- Sunbelt
- Chicago

Read to Learn . . .
1. how the United States began.
2. why the United States is a land of many cultures.
3. how the arts have developed in the United States.

Every year, millions of people join in a huge, coast-to-coast birthday party on the Fourth of July. How do you celebrate this national holiday that marks the birth of the United States?

Americans have much in common, such as celebrating the birth of their nation on the Fourth of July. At the same time, they follow many different ways of life. Americans trace their roots to a variety of places around the world. Either they—or their ancestors—were **immigrants**, or people from other lands who have come to live permanently in the United States.

Place

Influences of the Past

The first Americans were nomads who followed their herds. Moving from campsite to campsite, these people from northern Asia worked their way south. They were the first immigrants to North America. These first Americans came to the country we now call the United States thousands of years ago. Over the centuries other people from Europe, Africa, and other parts of Asia and the Americas followed.

102

Early Period The first people, some experts believe, arrived 12,500 or more years ago. They separated into different groups as they moved into almost every region of what is now the United States. The Native Americans developed cultures influenced by the environments in the areas where they settled.

In the 1600s and 1700s, Europeans settled in North America. They came seeking land, riches, and the right to live freely. European groups from Spain, France, Great Britain, and other countries set up **colonies**, or overseas settlements tied to a parent country. In settling the land, however, the Europeans and their descendants often grabbed land and killed Native Americans.

Government of the People By the late 1700s, people living in the British colonies along the Atlantic coast had started to think of themselves as Americans. They fought a war with the British that ended British rule in the American colonies. This **revolution**, or sudden political change, produced the independent country we know as the United States of America.

In 1787 a group of American leaders wrote a plan of government for the country. It became the Constitution of the United States. The Constitution created a **republic**—a form of government in which the people elect their own officials, including the leader of the country. Meanwhile, the American colonies had become states. The national government and the state governments still share the task of ruling the country.

Adobe Village in Taos, New Mexico

Some Pueblo Native Americans still live in traditional adobe villages in New Mexico.
MOVEMENT: From what continent did the first Americans migrate?

Branches of the United States Government

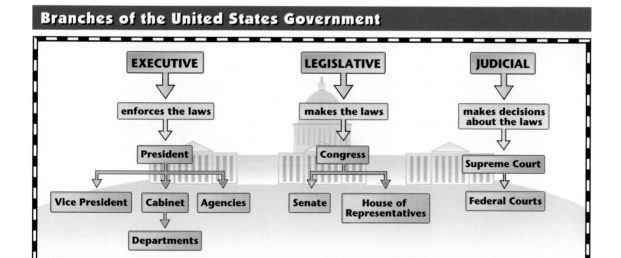

EXECUTIVE	LEGISLATIVE	JUDICIAL
enforces the laws	makes the laws	makes decisions about the laws
President	Congress	Supreme Court
Vice President, Cabinet, Agencies	Senate, House of Representatives	Federal Courts
Departments		

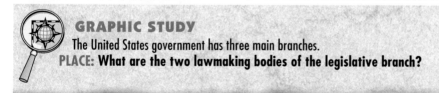

GRAPHIC STUDY
The United States government has three main branches.
PLACE: What are the two lawmaking bodies of the legislative branch?

Teenage Doctor

By age 7, Masoud Karkehabadi had completed high school. By age 13, he had graduated from college. By age 14, Masoud may cure Parkinson's disease. Masoud is an extremely intelligent 13-year-old who lives in California. He has been doing research to find a cure for Parkinson's disease since he was 11. "I believe my intelligence is a gift," said Masoud. "I want to use it to the best of my capacity to help society." Next year, Masoud plans to start his first year of medical school. In his free time, he plays baseball and street hockey and takes care of his pet iguana, Abi.

Industry and Expansion Over the next 200 years, the growth of industry changed the United States into an economic giant. By the mid-1800s, factories were using steam-powered machines—and later electricity—to make goods. Where did workers for these factories come from? Americans moved from farms to the cities, and thousands of immigrants poured into the United States.

Through wars, treaties, and purchases, the United States gained control of the lands west of the Mississippi River. The railroads and low prices for land drew people to this western region. Settlers there supplied minerals and agricultural products to other parts of the country.

A World Power During the 1900s, the United States became a leader in world affairs. As industry grew in the United States, foreign trade became more important. During and after World War I and World War II, the vast resources of the United States aided allies around the world. Along with weapons and manufactured goods, American culture also spread overseas and changed the way of life in many places.

Movement

One Out of Many

About 270.2 million people live in the United States. The variety of its people makes the United States a **multicultural** country—a country that has many different cultures.

A Variety of People

The United States is home to people of many different ethnic groups. An **ethnic group** is a group of people who share a common culture, language, or history. The graph on page 85 shows some of the different ethnic groups that make up the American population.

English is the major language of the United States, but many other languages are spoken. For example, many Spanish-speaking people live in the Southwest, in Florida, and in major cities such as New York City and Chicago.

WHAT IN THE WORLD?

An American Sound

In the late 1800s a type of music known as jazz was born in the United States. It is believed to be the only musical art form to originate in the United States. Jazz's multicultural roots blend African rhythms, African American religious music, and European harmonies.

UNIT 2

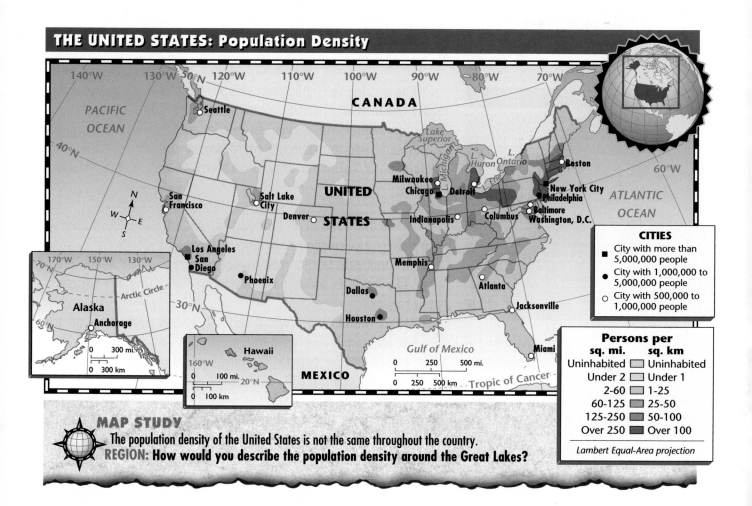

THE UNITED STATES: Population Density

CITIES
- City with more than 5,000,000 people
- City with 1,000,000 to 5,000,000 people
- City with 500,000 to 1,000,000 people

Persons per	
sq. mi.	sq. km
Uninhabited	Uninhabited
Under 2	Under 1
2-60	1-25
60-125	25-50
125-250	50-100
Over 250	Over 100

Lambert Equal-Area projection

MAP STUDY
The population density of the United States is not the same throughout the country.
REGION: How would you describe the population density around the Great Lakes?

Religion has always been an important influence in American life. Most Americans follow some form of Christianity. Judaism, Islam, and Buddhism are also important religions in the United States.

A Nation on the Move Go, go, go! Americans have always been a **mobile** people. This means they move from place to place. Americans today move from city to city to get better jobs, better homes, or better educations.

Which city and state do you call home? You might live in Fairview—the most common city name in the country—or in Uncle Sam, Louisiana. The map above shows you that most people in the United States live in or near large cities. The Northeast, parts of the Midwest, and the Pacific coast have the greatest number of people. Since the 1970s, the fastest-growing areas in the United States have been the South and the Southwest. Industrial growth and a pleasant climate have drawn people to this area known as the Sunbelt.

How People Live Compared to most other countries, the United States enjoys a high standard of living. For most people food is abundant, not too costly, and easily available. Americans, on average, can expect to live about 76 years. Medical advances have helped bring about longer lives.

Place
The Arts and Recreation

Just for Fun

Baseball is one of the most popular sports in the country. **PLACE: What is a popular recreational activity in your area?**

The arts in the United States show the variety of American life. Since colonial days artists, writers, and builders have developed uniquely American styles. In recent times, leisure time for Americans has increased. People across the country play and watch sports, camp and travel, read novels, and attend plays and movies.

Art and Literature The earliest Americans used materials from their environments to create works of art. For example, Native Americans in the West made pottery from clay found in the area. In every age, artists have drawn on the beauty and variety of their region.

Author Mark Twain wrote about boyhood adventures on the Mississippi River. Washington Irving wrote about Dutch settlers in New York state. Willa Cather presented the experiences of pioneers on the Great Plains. Georgia O'Keeffe painted the colorful cliffs and deserts of the Southwest.

American literature is also filled with the stories, folktales, and myths of many ethnic groups. Native American folktales explain the mystery of nature. African American gospel songs tell about the strength of a people in overcoming adversity.

Sports and Recreation Americans are enthusiastic sports fans and players. What sports are most popular? Baseball, football, and basketball are played by both professionals and amateurs. Americans also enjoy physical activities such as biking, skiing, golfing, and jogging. They explore the outdoors and many visit **national parks** set aside for recreation and for protection of wilderness and wildlife.

SECTION 3 ASSESSMENT

REVIEWING TERMS AND FACTS
1. **Define the following:** immigrant, colony, revolution, republic, multicultural, ethnic group, mobile, national park.
2. **PLACE** Why is the United States known as a multicultural country?
3. **MOVEMENT** Why are Americans today so mobile?
4. **PLACE** What areas of the United States have the fastest-growing population?

MAP STUDY ACTIVITIES
5. Turn to the map on page 105. What is the population density of Miami, Florida?
6. Using the same map, name two sparsely populated areas of the United States.

UNIT 2

Chapter 4 Highlights

Important Things to Know About the United States

SECTION 1 THE LAND

- The United States is the world's fourth-largest country in land area.
- The country consists of 48 contiguous states plus Alaska and Hawaii.
- Five physical regions are found in the country—the Coastal Plains, the Appalachian Mountains, the Interior Plains, the Rocky Mountains, and the Pacific region.
- Almost every type of climate region is found in the United States.

SECTION 2 THE ECONOMY

- The economic regions of the United States are the Northeast, South, Midwest, Interior West, and Pacific regions.
- A wealth of natural resources and skilled workers have helped to make the United States a world leader in farming and industry.
- Pollution and overuse of resources are serious environmental issues in many areas of the country.
- Service industries employ more workers than any other industry.

SECTION 3 THE PEOPLE

- Americans trace their roots to every part of the world.
- The United States became independent in the late 1700s. The Constitution established a new republic with a democratic form of government.
- By the 1900s the country had become an industrial power and a leader in world affairs.
- About 270 million people live in the United States, most in urban centers along the coasts.

Festival in Washington, D.C. ▶

Chapter 4 Assessment and Activities

REVIEWING KEY TERMS

Match the numbered terms in Set A with their lettered definitions in Set B.

A

1. free enterprise
2. contiguous
3. national park
4. dry farming
5. mobile
6. farm belt
7. acid rain

B

A. areas joined together inside a common boundary
B. moving from place to place
C. system in which people own and run businesses with little government control
D. plowing land so that it holds rainwater
E. pollution mixed with rain or snow
F. land set aside for protection of wilderness
G. an area of many productive farms

Mental Mapping Activity

Draw a freehand map of the 50 states of the United States. Label the following items:

- Alaska
- New England
- Great Lakes
- Mississippi River
- Rocky Mountains
- Great Plains

REVIEWING THE MAIN IDEAS

Section 1

1. **REGION** What are the five physical regions of the United States?
2. **LOCATION** In what climate region is Los Angeles located?

Section 2

3. **PLACE** Name three industrial cities of the Midwest.
4. **REGION** Why is farming limited in the Interior West?
5. **MOVEMENT** What is the purpose of NAFTA?

Section 3

6. **MOVEMENT** From which three European countries did most early settlers come?
7. **PLACE** What three branches of government does the United States have?

CRITICAL THINKING ACTIVITIES

1. **Evaluating Information** Has the United States made wise use of its resources? Explain.
2. **Analyzing Information** How does the art of Native Americans of the Southwest reflect their environment?

CONNECTIONS TO WORLD HISTORY

Cooperative Learning Activity Work in groups with each group choosing one state of the United States. Research and prepare a report on the history of the first people who settled that state. When research is complete, share your information with your group. As a group, prepare a report with a poster or map that shows the group's findings.

TECHNOLOGY ACTIVITY

Developing Multimedia Presentations

Work with a partner or group to research how your state's climate influences its economy, tourist attractions, types of clothing, and seasonal foods. Use your research and a video camera to develop a commercial promoting your state. Incorporate music, acting, and computer graphics, if possible, in your video. Present your commercial to the rest of the class.

GeoJournal Writing Activity

Find a spot outdoors. Sit and make yourself comfortable. Facing west, draw in your journal what you see on the land and in the sky. Observe any shadows cast by the objects in your view. Then face north and draw what you see. Continue until you have drawn the eastern and southern views as well. What features are in your drawings?

PLACE LOCATION ACTIVITY: THE UNITED STATES

Match the letters on the map with the places and physical features of the United States. Write your answers on a separate sheet of paper.

1. Washington, D.C.
2. Appalachian Mountains
3. Detroit
4. Gulf of Mexico
5. California
6. Lake Superior
7. Great Salt Lake
8. Seattle
9. Mississippi River
10. Ohio River

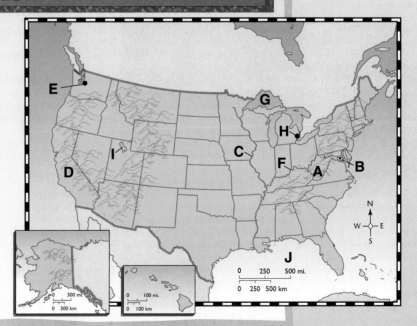

From the classroom of Debora Bittner, San Antonio, Texas

THE CONTINENTAL DIVIDE

Background

In North America, a high ridge of the Rockies known as the Continental Divide separates the waters flowing west to the Pacific Ocean from those flowing east toward the Mississippi River and the Atlantic Ocean. Because rivers flow downhill, whatever landforms are in their paths can direct which way water systems flow. This activity will demonstrate how mountain ranges and elevation can direct the flow of water.

Continental Divide

In Europe, the continental divide is found in the Alpine Mountain System, which includes the Alps. Streams and rivers drain to the Atlantic Ocean and the Arctic Ocean in one direction, and to the Mediterranean Sea and the Black Sea in the other direction.

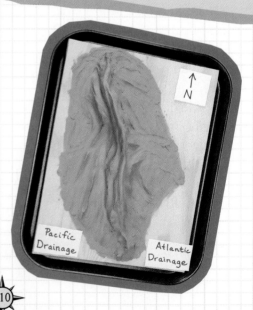

Materials

- 1 piece of plywood—18" × 24"
- a tray large enough for the plywood to fit into
- 1 quart of water
- 2 pounds of modeling clay

What To Do

A. Choose an area in the western United States that contains a portion of the Continental Divide.

B. Build a model of the relief of this area from the modeling clay. Be sure to include different levels of elevation.

C. After the model is completed, identify the north, south, east, and west of your model.

D. Label the area to the west of the mountains "Pacific Drainage."

E. Label the area east of the mountains "Atlantic Drainage."

F. Slowly and carefully pour the water on the peaks of the mountains. Move slowly from north to south as you pour. Take note of the direction and flow of the water.

Lab Activity Report

1. In which direction did most water flow?

2. Why were there different amounts flowing in different directions?

3. How does the elevation of the mountains increase or decrease drainage?

4. Drawing Conclusions How will various flows affect the people living in the drainage areas?

Go A Step Further

Activity

Find out the position of the continental divide in South America. What effect does the drainage of the Amazon River have on the Brazilian rain forest? Summarize your findings in a two-paragraph report.

Cultural HERITAGE:

THE UNITED STATES AND CANADA

MUSIC ▶ ▶ ▶ ▶

All kinds of music are popular in the United States and Canada, but jazz is thought to be the only form of music that originated in the United States. The earliest jazz musicians were African American.

▲ ▲ ▲ ▲
FOLK ART

Since colonial times, people in the United States and Canada have made quilts as decorative covers for their beds. Quilts are made from scraps of colorful fabric sewn together in a patchwork design.

STAINED GLASS ▶ ▶ ▶ ▶

The delicate blossoms and leaves of this stained-glass lamp were created by American artist Louis Tiffany in the 1890s. Tiffany was born in New York City and created many beautiful Tiffany lamps.

Canada's French heritage is reflected in many of its buildings. The Chateau Frontenac, in the city of Quebec, is a large hotel patterned after a French castle.

WOOD CARVING
▶ ▶ ▶ ▶

Totem poles were carved by Native Americans who lived in the northwestern United States and Canada. Many of the figures on totems represent clans or families. A clan's totem might include carvings of birds, fish, plants or other objects found in nature.

PAINTING ▲ ▲ ▲ ▲

Farms and country life were the subjects of many paintings by American artist Winslow Homer. Homer, born in Boston in 1836, was a realist. He tried to paint his subjects as they appeared.

APPRECIATING CULTURE

1. Quilts are called folk art because they were made by everyday people for other people to use or enjoy. They were not created as art to be put into museums. What objects in your home might be considered folk art in the future?

2. Some Native American clans or families were recognized by their totem poles. If you created a totem pole to represent the people who live in your home, what characters or objects would you use?

ARCTIC OCEAN

Alaska (U.S.)

Greenland (DENMARK)

Baffin Bay

Yukon Territory
★ Whitehorse

Northwest Territories
★ Yellowknife

Arctic Circle

Nunavut

Iqaluit ★

National boundary
Provincial boundary
⊛ National capital
★ Provincial capital
● Other city

Lambert Equal-Area projection

British Columbia

Alberta
Edmonton ★

Saskatchewan

Manitoba

C A N A D A

Hudson Bay

Newfoundland

St. John's ●

Calgary ●

Victoria ★ ● Vancouver

● Saskatoon
Regina ★

Winnipeg ★

Ontario

Quebec

Prince Edward Is.

Charlottetown ★
Fredericton ★
Quebec ★

Halifax ●

U N I T E D   S T A T E S

Sudbury ●
Kingston ●
Windsor ●

Montreal ●
Ottawa ⊛
Toronto ★

New Brunswick

Nova Scotia

ATLANTIC OCEAN

PACIFIC OCEAN

0   250   500 mi.
0   250   500 km

N
W   E
S

MAP STUDY ACTIVITY

As you read Chapter 5, you will learn about Canada, the second-largest country in the world.

1. What is the northernmost territory of Canada?
2. What is Canada's national capital?
3. In what province is the national capital located?

SECTION 1

The Land

Words to Know
• prairie
• cordillera

Places to Locate
• Appalachian Highlands
• St. Lawrence River
• Canadian Shield
• Hudson Bay
• Rocky Mountains
• Coast Mountains

Read to Learn . . .
1. how Canada's landscapes differ from region to region.
2. where the oldest rock formations in North America are found.
3. how climate affects where Canadians live.

Imagine yourself in this photograph. Do you hear the waves of the Atlantic crashing against the rocky coast? This lovely place is Cape

Breton Island, one of Canada's many scenic areas.

Vikings first landed their longboats on its eastern coast around A.D. 1000. Niagara Falls thunders in the southeast. Grizzly bears roam its western territories. What country are we describing? It is Canada.

Location
A Northern Land

The map on page 114 shows you that Canada lies north of the contiguous United States. Between the two countries lies the longest undefended border in the world. Like the United States, Canada has the Atlantic Ocean on its eastern coast and the Pacific Ocean on its western coast. The Arctic Ocean lies north of Canada.

The second-largest country in the world in area, Canada covers 3,849,674 square miles (9,970,610 sq. km). Only Russia is larger. Canada contains 10 provinces and 3 territories. Look at the map on page 114 to find the provinces of Newfoundland, Nova Scotia, New Brunswick, Prince Edward Island, Quebec (kwih•BEHK), Ontario, Manitoba, Saskatchewan (suh•SKA•chuh•wuhn), Alberta, and British Columbia.

The wide-open prairies of Canada's Interior Plains are dotted with grain elevators filled with wheat and other grains.
REGION: What physical region in the United States is an extension of Canada's Interior Plains?

Now look for the Northwest Territories and the Yukon Territory. In 1999 a third territory—Nunavut (NOO•nuh•VUHT)—was carved out of part of the Northwest Territories. This area is claimed by the Inuit.

Regions
Landforms

Canada—from an Iroquois word for "village"—can be divided into six major physical regions: the Appalachian Highlands, the St. Lawrence and the Great Lakes Lowlands, the Canadian Shield and the Arctic Islands, the Interior Plains, the Rocky Mountains, and the Pacific Coast. From border to border, Canada has more lakes and inland waterways than any other country in the world.

The Appalachian Highlands Along Canada's southeastern Atlantic coast stretch the Appalachian Highlands. They continue through the eastern United States as the Appalachian Mountains. Traveling around this part of Canada, you will see rolling hills and low mountains. The valleys between are dotted with farms. Forests blanket much of the Appalachian Highlands. Many deepwater harbors nestle along the region's jagged, rocky coast.

The St. Lawrence and Great Lakes Lowlands Sprawling urban areas and slow-moving barges are a common sight in this region. The St. Lawrence River and the Great Lakes form the major waterways leading

WHAT IN THE WORLD?

Nature's Flowerpots

On the shoreline of Ontario's Georgian Bay are strange-looking landforms called flowerpots. Wind and water have worn away the bases of these tall rock formations. Their wide tops are covered with flowers, plants, and small shrubs—just like a flowerpot!

into central Canada. Find the St. Lawrence River valley on the map below. Now find the peninsula in Ontario that is bordered by Lake Erie, Lake Ontario, and Lake Huron. These areas are lowlands with rich soil and vital transportation routes. You can find most of Canada's people, farms, and industries here. The world-famous Niagara Falls—where the waters of Lake Erie flow into Lake Ontario—also impresses visitors to the region.

The Canadian Shield and the Arctic Islands Picture a huge, abandoned parking lot. The concrete slabs have split and buckled, and wildflowers poke through the cracks and puddles. The Canadian Shield looks a little like this, except that ice and snow cover its 2 million square miles (5,180,000 sq. km)—more than half of Canada's entire land area!

Wrapped around Hudson Bay is the huge, horseshoe-shaped Canadian Shield. Hills worn down by erosion and hundreds of lakes carved by glaciers cover much of the Shield. Scientists claim this region holds some of the oldest rock formations—some more than 3 billion years old—in North America.

What plants can thrive in the cold northern areas of the Canadian Shield? Mosses and small shrubs grow here. In the southern part, evergreen forests provide shelter and food to deer, elk, and moose. Although the soil here is not good for farming, it is rich in mineral resources such as iron ore, copper, nickel, and gold.

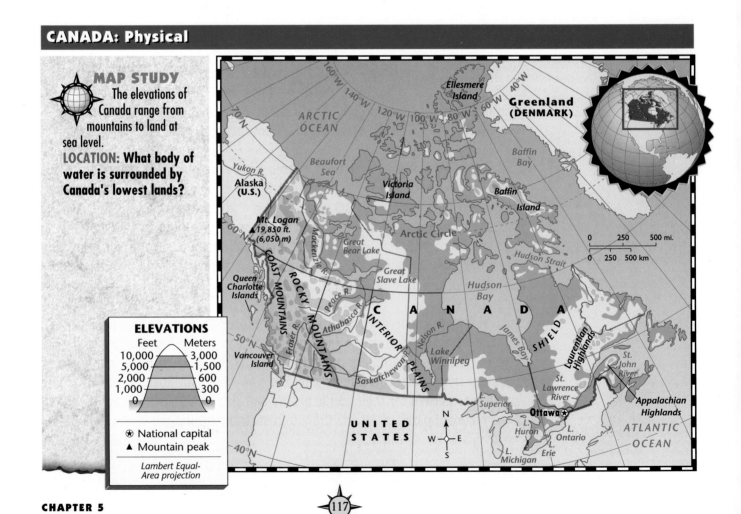

CANADA: Physical

MAP STUDY The elevations of Canada range from mountains to land at sea level.
LOCATION: What body of water is surrounded by Canada's lowest lands?

ELEVATIONS
Feet	Meters
10,000	3,000
5,000	1,500
2,000	600
1,000	300
0	0

⊛ National capital
▲ Mountain peak

Lambert Equal-Area projection

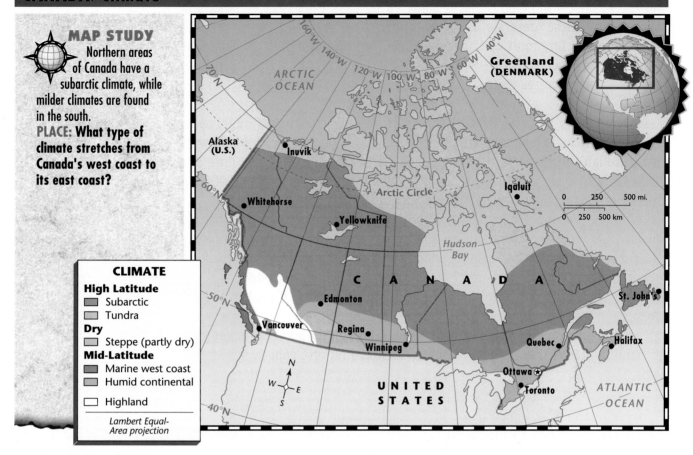

MAP STUDY
Northern areas of Canada have a subarctic climate, while milder climates are found in the south.
PLACE: What type of climate stretches from Canada's west coast to its east coast?

CLIMATE

High Latitude
- Subarctic
- Tundra

Dry
- Steppe (partly dry)

Mid-Latitude
- Marine west coast
- Humid continental

- Highland

Lambert Equal-Area projection

To the north lie Canada's Arctic Islands. Ten large islands and hundreds of small ones lie almost entirely north of the Arctic Circle. Because of their far northern location, the Arctic Islands have no forests. Only tundra plants such as grasses and lichens grow here. Glacial ice covers the northernmost islands.

The Interior Plains Canada shares landforms with its neighbor to the south. The Great Plains of the United States become the Interior Plains in Canada. You see many forests and lakes in the northern part of the Interior Plains. The southern part of these plains is a huge rolling **prairie,** or inland grassland, with very fertile soil. Wheat and other grainfields echo with the rumble of tractors. Rich mineral resources such as petroleum and natural gas are also found in the Interior Plains.

The Rocky Mountains Another landform shared by Canada and the United States lies west of the Interior Plains. The Rocky Mountains are part of an area called the Cordillera (KAW•duhl•YEHR•uh). A **cordillera** is a group of mountain ranges that run side by side. Like the American Rockies, the Canadian Rockies are known for their scenic beauty and rich mineral resources. Banff and Jasper National parks in Alberta draw tourists from around the world.

The Pacific Coast Mount Logan—Canada's highest peak—towers 19,850 feet (6,050 m) over the Pacific coastal region. A major group of the Cordillera—the Coast Mountains—rise above this western part of Canada. Like the Rockies, the Coast Mountains cross into the United States. Near Canada's border with Alaska, several mountains soar more than 15,000 feet (4,572 m).

Place

Climate

Traveling Along the Trans-Canada Highway

Canada's ten provinces are linked by a highway that extends from the Atlantic to the Pacific coasts.
MOVEMENT: What three areas of Canada are not linked by the Trans-Canada Highway?

One thing Canada does not share with the United States is climate. The map on page 118 shows you that Canada generally has a cool or cold climate because it lies in the high latitudes of the Western Hemisphere. As you read in Chapter 2, however, differences in climate can be caused by a place's location or its nearness to oceans.

In northern Canada people shiver in the cold polar climate. Farther south, between 50° and 70° North latitude, you find a subarctic climate with short, cool summers and long, cold winters. The southeastern part of Canada has a humid continental climate. Most Canadians live in this area.

The southwestern Pacific coast is Canada's only area of wet, mild winter climate. There the Coast Mountains cause warm winds from the Pacific Ocean to release moisture. The western, or windward, side of the mountains gets more rain and warmer temperatures than any other part of Canada. On the eastern side of the mountains, however, the climate is dry or partly dry in the rain shadow.

SECTION 1 ASSESSMENT

REVIEWING TERMS AND FACTS

1. Define the following: prairie, cordillera.
2. PLACE What is the landscape of the Appalachian Highlands like?
3. PLACE In what physical region do most Canadians live?
4. REGION What area of Canada enjoys a mild climate and plenty of rainfall?

MAP STUDY ACTIVITIES

5. Turn to the physical map on page 117. Which of the Great Lakes lies farthest north?
6. Look at the climate map on page 118. In which climate region do you find Ottawa?

BUILDING GEOGRAPHY SKILLS

Using the Internet

It is difficult to imagine the shape of land you have not seen. Physical features such as mountains, plains, oceans, and rivers are shown on maps. A map that shows these natural features and creates an impression of the lay of the land is called a *physical map.* Physical maps use colors and shading to show relief—how flat or rugged the land surface is. Colors are also used to show the land's elevation, or height above sea level.

Use the Internet to help you learn to read physical maps.

Go to the National Geographic Web site at www.nationalgeographic.com to find physical maps of several countries around the world. To read a physical map, apply these steps:

• Identify the region shown on the map.

• Use the map key to find the meaning of colors and symbols.

• Find important physical features including elevation, rivers, and coastlines.

• Imagine the actual shape of the land.

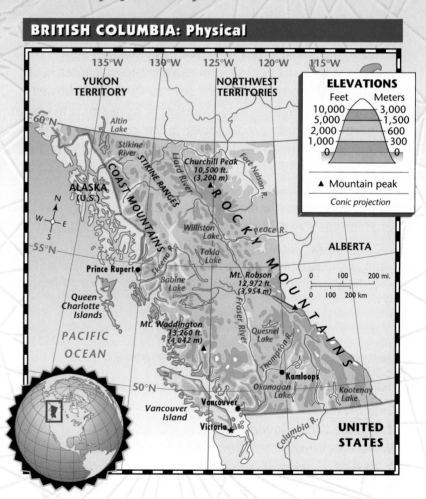

BRITISH COLUMBIA: Physical

ELEVATIONS

Feet	Meters
10,000	3,000
5,000	1,500
2,000	600
1,000	300
0	0

▲ Mountain peak

Conic projection

Technology Skills Practice

1. Compare the physical maps you found on the Internet. What types of physical features are shown on each map?

2. How is relief represented on each map?

3. What is the highest point above sea level on each map? The lowest point?

120

SECTION 2 ·
The Economy

PREVIEW

Words to Know
- fossil fuel
- newsprint

Places to Locate
- Newfoundland
- Maritime Provinces
- Quebec
- Ontario
- Prairie Provinces
- British Columbia

Read to Learn . . .
1. what natural resources Canada has.
2. how the Canadian people earn their livings.
3. what challenges face Canada and its economy today.

Hundreds of bays and fishing villages dot the jagged Atlantic coast of Newfoundland. Not far from the village shown here lie the Grand Banks—one of the great fishing areas of the world. Fleets from Canada and many other countries harvest fish in the Grand Banks.

Fishing is Canada's oldest industry. It is only one of many economic activities that Canadians carry on today. Productive farms, mines, and businesses help make Canada one of the world's most economically developed countries.

Region
Economic Specialties

Canada's economy is very similar to that of the United States. The country is known for its rich farmland, many natural resources, and skilled workers. Service industries, manufacturing, and farming are the country's major economic activities. Canada, too, has an economy based on free enterprise. The Canadian government, however, plays an active part in some economic activities, such as broadcasting, transportation, and health care.

Like the economy of the United States, Canada's economy differs from region to region. Geographers group Canada's provinces and territories into six economic regions: Newfoundland and the Maritime Provinces, Quebec, Ontario, the Prairie Provinces, British Columbia, and the North.

Agriculture
- Ranching
- Nomadic herding
- Hunting and gathering
- Forestry
- Commercial farming
- Subsistence farming
- Little or no activity
- Manufacturing area

Resources
- Coal
- Copper
- Fish and other seafood
- Gold
- I Iron ore
- Petroleum
- Silver
- Uranium
- Zinc

Lambert Equal-Area projection

ARCTIC OCEAN

Greenland (DENMARK)

Alaska (U.S.)

Arctic Circle

0   250   500 mi.
0   250   500 km

Hudson Bay

C A N A D A

Corner Brook
St. John's
Sydney
Edmonton
Calgary
Regina
Victoria   Vancouver   *Wheat*
Winnipeg
Quebec
Halifax
Saint John
Montreal
Ottawa
Hamilton   Toronto
Windsor

UNITED STATES

ATLANTIC OCEAN

N   W   E   S

Newfoundland and the Maritime Provinces Canada's easternmost economic region is formed by Newfoundland and the Maritime Provinces. These provinces include Nova Scotia, New Brunswick, and Prince Edward Island. Fishing and shipbuilding were major economic activities here. Because of overfishing, however, only 3 percent of the region's workers now make a living from the sea.

By looking at the map above, you can see why farming is limited in this region. A short growing season and thin, rocky soil discourage large-scale farming. Small farms throughout the Maritime Provinces, however, grow crops such as potatoes and apples.

Manufacturing and mining provide jobs for most people in this area of Canada. Halifax, the capital of Nova Scotia, is a major shipping center. Its harbor remains open in winter when ice closes most other eastern Canadian ports. Tourism is also important to the economy.

Quebec About 25 percent of Canada's people live in Quebec, the largest province in area. Manufacturing and service industries are major economic activities here. Montreal on the St. Lawrence River is Canada's second-largest city and the center of Quebec's economic and social life. To the north, Quebec's mines provide iron ore, copper, and gold. Other economic activities in Quebec are agriculture and fishing.

Ontario Ontario is Canada's second-largest province, but it has the most people and the greatest wealth. The map on page 122 shows you that Ontario is rich in natural resources. This is one reason Ontario is the leading manufacturing region of Canada. The region produces more than half of Canada's manufactured goods.

Toronto, the capital of Ontario, is Canada's largest city. It is also the country's chief manufacturing, financial, and communications center. As a center of industry, Ontario has faced industrial pollution and other threats to the environment.

Ontario borders the Great Lakes and the St. Lawrence River. To open the Great Lakes to ocean shipping, the United States and Canada built the St. Lawrence Seaway. Look at the map and diagram below. You see how this system of locks and canals allows ships to travel between the Great Lakes and the Atlantic Ocean. The St. Lawrence Seaway has helped make Ontario a major shipping area.

Farmers in southern Ontario grow grains, fruits, and vegetables and raise beef and dairy cattle. Canada's capital city, Ottawa, lies in Ontario near the Quebec border. Many Canadians work in government offices in Ottawa.

The Prairie Provinces The heartland of Canada overflows with resources. The Prairie Provinces are Manitoba, Saskatchewan, and Alberta. The map on page 122 shows you that farming and raising cattle are major economic activities in this region. The Prairie Provinces are important wheat producers. Ranchers in Alberta also raise most of Canada's cattle.

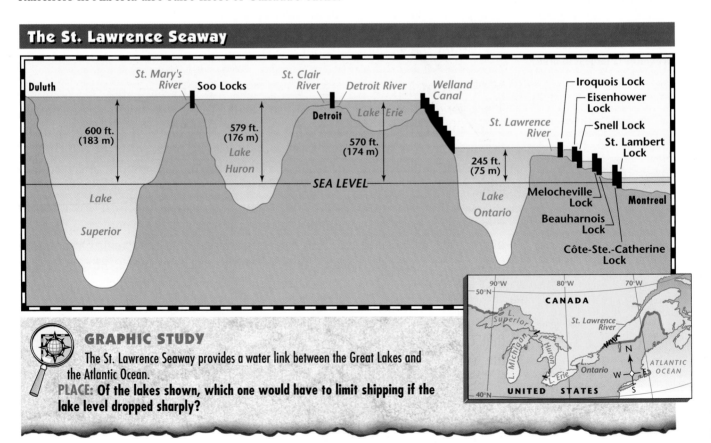

The St. Lawrence Seaway

GRAPHIC STUDY
The St. Lawrence Seaway provides a water link between the Great Lakes and the Atlantic Ocean.
PLACE: Of the lakes shown, which one would have to limit shipping if the lake level dropped sharply?

123

A vast prairie surrounds the rapidly growing city of Calgary, Alberta *(left)*. A skywalk links business and shopping areas in Toronto, Ontario *(right)*.
PLACE: Which of Canada's provinces is largest in area?

Some of the world's largest reserves of fossil fuels—oil and natural gas—were discovered in Alberta and Saskatchewan. A **fossil fuel** is a fuel that is formed from the remains of prehistoric plants and animals.

The Prairie Provinces also have some of the fastest-growing cities in Canada. Calgary and Edmonton, both in Alberta, are leading centers for the oil industry. Winnipeg, in Manitoba, is a major transportation and business center that links the eastern and western parts of Canada.

British Columbia The far west of Canada is a treasure chest of resources. Thick forests blanket much of British Columbia. As you might guess, timber, pulp, and paper industries provide much of British Columbia's income. The province helps to make Canada the world's leading producer of **newsprint,** or paper that is made into newspapers. The mining of coal, copper, and lead also adds to the wealth of British Columbia.

Agriculture and fishing are strong economic activities in British Columbia. In river valleys between mountains, farmers raise cattle and poultry and grow many fruits and vegetables. Fishing fleets sailing out into the Pacific Ocean bring their catch of salmon, halibut, and herring to port. Vancouver, British Columbia's largest city, is a bustling trade center and Canada's major Pacific port.

The North Geese, moose, bear, and reindeer lay claim to northern regions that cover 40 percent of Canada's land. The North economic region includes the Yukon Territory, the Northwest Territories, and Nunavut. The far north is Arctic wilderness, and the south produces forest products and minerals. The Yukon Territory, the Northwest Territories, and Nunavut are still largely undeveloped.

Place
Today's Challenges

The differences that give Canada a rich, diverse economy also provide challenges to the country. One issue concerns two of Canada's cultures. Others involve trade and environmental issues.

Regional Differences Not only physical and economic differences but also cultural differences divide the regions of Canada. About 80 percent of Quebec's population is French-speaking. The rest of Canada, however, is largely English-speaking. Many people in Quebec want freedom for their province. In 1995 voters there narrowly defeated a proposal for independence.

NAFTA Since World War II, Canada has strengthened its economic ties with the United States. Today about 70 percent of Canada's trade is with the United States. In contrast, only about 20 percent of United States trade is with Canada. Some Canadians want even closer economic ties with the United States. Others fear loss of Canadian jobs to the United States.

In 1993 the three major countries of North America—Canada, the United States, and Mexico—created a new trading partnership. They approved the North American Free Trade Agreement (NAFTA). This agreement aims eventually to do away with trade barriers among the three countries.

Pollution The protection of its natural resources is a major challenge facing Canada. The national government works with the provinces to deal with air and water pollution. Environmental agreements between Canada and the United States aimed at reducing pollution have also been made.

SECTION 2 ASSESSMENT

REVIEWING TERMS AND FACTS
1. **Define the following:** fossil fuel, newsprint.
2. **PLACE** Why is Ontario a major shipping area?
3. **REGION** How are the Prairie Provinces like the Great Plains of the United States?
4. **MOVEMENT** How do Canadians view close economic ties between Canada and the United States?

MAP STUDY ACTIVITIES
5. Look at the map on page 122. In which provinces is fishing a major economic activity?
6. Using the same map, name the mineral found north of the Arctic Circle.

MAKING CONNECTIONS

GOLD! GOLD! GOLD!

MATH | **SCIENCE** | **HISTORY** | **LITERATURE** | **TECHNOLOGY**

"Gold! Gold! Gold!" shouted the headline of the *Seattle Post-Intelligencer* on July 17, 1897. Gold seekers from Canada's Klondike region in the Yukon Territory had found gold dust and nuggets. Within 10 days, ships began landing excited gold seekers in the Alaskan ports of Skagway and Dyea. Getting to the goldfields, however, was a difficult problem. Miners had to brave mountains, glaciers, and the Yukon River.

NO SAFE WAY Three routes led to the Klondike's goldfields in Dawson, Canada. One crossed a glacier; the second and third went through mountain passes. None were safe.

The Malspina Glacier, near the Yukon Territory's southern border, was a dangerous, slippery ice field. Huge cracks lay hidden near its surface. Many people died or were injured in falls. Others were blinded by the sunlight on the ice.

Another route, the 45-mile-long (73-km) Skagway Trail, began with an easy climb. The sight facing the miners from the top of the trail, however, was row upon row of boulder-covered hills and deep, muck-filled swamps.

The third route, Dyea Trail, was only 33 miles (54 km) long, but it led to Chilkoot

Pass. Crossing this pass meant making a very steep climb thousands of feet up a snow-covered mountain face. Most miners were not experienced in this type of rugged travel.

Chilkoot Pass, Alaska

THE YUKON RIVER If they made it past the coastal mountains, gold seekers still had to build boats and sail down the Yukon River. They faced rapids, a deadly whirlpool, and huge boulders before reaching the goldfields at Dawson.

Nearly 40,000 people made it to Dawson during the Klondike's three-year gold rush. Only 200 to 300 miners actually found gold. Was it worth the trip?

Making the Connection

1. Why did so many people head for the Klondike in 1897?
2. What geographical barriers did gold seekers face?

The People

PREVIEW

Words to Know
- parliamentary democracy
- prime minister
- bilingual

Places to Locate
- Quebec
- Montreal

Read to Learn . . .
1. how Canada gained its independence.
2. what groups make up the Canadian people.
3. where most Canadians live.

Founded by French explorers in 1608, Quebec is one of the oldest cities in North America. It has many picturesque buildings, churches, and even a fortress overlooking the St. Lawrence River. In Quebec, you can see signs in French and taste mouth-watering French cooking in the restaurants.

Like the United States, Canada is a multicultural country. The circle graph on page 85 shows you the groups that make up Canada's population. Native Americans and Inuit form a small but growing part of the population. Most Canadians, however, share a European background.

Place
Influences of the Past

The great migration of the first Asians to North America brought Canada its earliest people some 12,500 years ago. They later became the Native Americans and the Inuit. Many other groups—most of them European—later settled in Canada.

The French In 1534 French explorer Jacques Cartier sailed into the St. Lawrence River valley and claimed the area for France. French explorers, settlers, and missionaries later founded cities such as Quebec and Montreal. For almost 230 years, France ruled the area around the St. Lawrence and the Great Lakes as New France.

Quebec's annual winter carnival includes ice sculpting, fireworks, and canoe races *(left)*. Canada's national police force is the Royal Canadian Mounted Police *(right)*.
PLACE: What are the two official languages of Canada?

The British In 1497 John Cabot landed on the Atlantic coast near Newfoundland and claimed the area for England. In the 1600s and 1700s, the British and the French fought over land in North America. The British won control of New France. After the American Revolution, many Americans loyal to the British settled in Canada.

The Birth of Canada Several cultures emerged in Canada in the 1700s. Both English-speaking and French-speaking settlers arrived. The western part— Ontario—became a colony for English-speaking settlers. The eastern part with its large French-speaking population became the colony of Quebec. In 1867 the British united the provinces of Ontario, Quebec, Nova Scotia, and New Brunswick into one nation known as the Dominion of Canada.

Prince Edward Island, the western territories, and Newfoundland eventually joined the Dominion as provinces. In 1982 Canadians claimed the right to change their constitution without British approval. The British king or queen reigns, but the British government has no real power in Canada.

Canada's Government Features borrowed from both the United Kingdom and the United States formed Canada's government. The Canadians have a British-style **parliamentary democracy.** Voters elect representatives to a lawmaking body called Parliament. The head of the largest political party in Parliament is named the **prime minister,** or leader. Over the years, power has shifted from the national government to the provinces.

Movement

A Changing Population

Although Canada is large in land area, it has only about 31 million people. The population has grown rapidly since World War II, however, because of

heavy immigration and a high birthrate. Look at the population density map below. It shows you that most Canadians live in cities near the Canadian-United States border. This heavily populated area covers only about one-tenth of the country!

Canada is a **bilingual** country, or a country with two official languages—English and French. But other European languages, as well as Native American languages, are also spoken.

Canada has a long heritage of religious freedom. Although most Canadians are Roman Catholics or Protestants, other Canadians follow Judaism, Buddhism, Hinduism, or Islam.

Place
The Arts and Recreation

Canada's culture is a blend of many heritages. Each province holds on to its ethnic traditions with pride.

Literature, Art, Music From poetry to novels to drama, Canadian authors write in either English or French about many subjects. The works of Canadian painters and sculptors are as varied as Canada's people. In the early 1900s, a group of artists known as the Group of Seven painted the northern Canadian

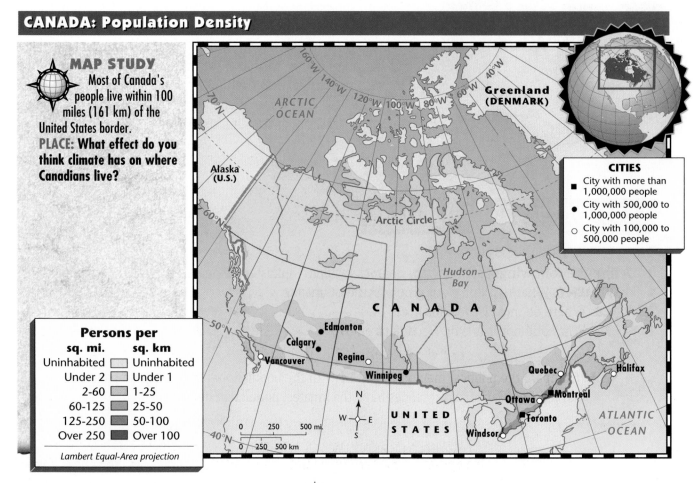

CANADA: Population Density

MAP STUDY
Most of Canada's people live within 100 miles (161 km) of the United States border.
PLACE: What effect do you think climate has on where Canadians live?

CITIES
■ City with more than 1,000,000 people
● City with 500,000 to 1,000,000 people
○ City with 100,000 to 500,000 people

Persons per
sq. mi.	sq. km
Uninhabited	Uninhabited
Under 2	Under 1
2-60	1-25
60-125	25-50
125-250	50-100
Over 250	Over 100

Lambert Equal-Area projection

wilderness in brilliant colors. One of these artists, Emily Carr, painted British Columbia's green forests and colorful Native American totem poles.

Canadians share a love for all kinds of music—from jazz to classical. Music of different cultures—such as Scottish folk music, reggae from the Caribbean area, and African American rap—is also popular. Canadians of Native American, Ukrainian, and African ancestry host many festivals where their ethnic dances are performed.

Food and Recreation Canadians enjoy the same foods as their neighbors in the United States. A variety of meats and seafood—beef, chicken, and fish—are served with vegetables and potatoes. Ethnic foods are popular from coast to coast.

French Canadians create their own special dishes. Thirteen-year-old Marie Villeneuve and her family live in Quebec. At Christmas they enjoy hearty pea soup, a meat pie called a *tourtière,* and a dessert made of maple syrup.

Canadians enjoy a variety of activities, especially outdoor sports. You will find local parks or national parks crowded with sports enthusiasts. One of these sports fans is Michael Evans. He is a 15-year-old Canadian who lives in Toronto. Michael, like many Canadians, enjoys playing and watching ice hockey. He also takes part in winter sports, including skiing, skating, and snowshoeing. During the summer Michael goes sailing on Lake Ontario and plays baseball. He tries to attend all the baseball games played at Toronto's large indoor stadium, the Skydome.

Teen Scene

Let's See the Puck!

Anton Lafonte has been playing hockey since he was four. As soon as the water freezes on the lake near his home, he and his friends choose up teams for the season. Because winters are so long in northern Quebec, Anton has had plenty of ice time to fine-tune his game. This year Anton has recruited two new players for his team— his younger sisters Yvette and Jennifer.

SECTION 3 ASSESSMENT

REVIEWING TERMS AND FACTS
1. **Define the following:** parliamentary democracy, prime minister, bilingual.
2. **MOVEMENT** What two European groups settled Canada?
3. **PLACE** Where do most Canadians live?
4. **PLACE** What outdoor sports do Canadians enjoy?

MAP STUDY ACTIVITIES

5. Compare the political map on page 114 with the population density map on page 129. Which two Canadian provinces have the greatest population densities?

Chapter 5 Highlights

Important Things to Know About Canada

SECTION 1 THE LAND

- Canada covers most of the northern part of North America.
- Canada has a variety of landforms, including mountains, lowlands, prairies, and Arctic wilderness.
- Most of Canada has a cool or cold climate. Milder temperatures are found in the southern part of the country.

SECTION 2 THE ECONOMY

- Canada has one of the world's most developed economies. It is rich in natural resources, farmland, and skilled workers.
- Workers in Canada's cities fill jobs in banking, communication, and other service industries. Rural Canadians mostly grow grain crops and raise livestock.
- Canada faces three major challenges: holding its separate regions together, working out its trade relationship with the United States, and dealing with environmental issues.

SECTION 3 THE PEOPLE

- Native Americans and Inuit were the first Canadians. Europeans, mainly British and French, later settled Canadian lands and founded the modern country of Canada.
- Cultural differences exist between French-speaking Quebec and the rest of Canada, which is largely English-speaking.
- Canada today is a multicultural country with many different peoples from throughout the world.
- Many people in Quebec want freedom for their province. In 1995 voters there narrowly defeated a proposal for independence.

Canadians cheering in Toronto's Skydome ▶

REVIEWING KEY TERMS

Match the numbered terms in Set A with their lettered definitions in Set B.

A

1. cordillera
2. parliamentary democracy
3. prairie
4. prime minister
5. fossil fuel
6. bilingual

A. voters elect representatives to a lawmaking body called Parliament
B. area in which two languages are spoken
C. head of government chosen from the largest political party in Parliament

B

D. group of mountain ranges that run side by side
E. inland area of grasslands
F. product formed from the remains of prehistoric plants and animals

Mental Mapping Activity

Draw a freehand map of Canada. Label the following items:
- Nova Scotia
- British Columbia
- Quebec
- Coast Mountains
- Montreal
- Ontario
- Arctic Ocean

REVIEWING THE MAIN IDEAS

Section 1
1. LOCATION What three oceans surround Canada?
2. REGION Which physical region is mostly rolling hills and low mountains?

Section 2
3. REGION In which economic region do you find wheat and oil?
4. HUMAN/ENVIRONMENT INTERACTION What are two important economic activities in British Columbia?
5. PLACE Which province is mostly French in culture and language?

Section 3
6. MOVEMENT Who were the first people to settle Canada?
7. PLACE What are Canada's two official languages?

CRITICAL THINKING ACTIVITIES

1. **Making Comparisons** Compare the overall climate of Canada with the overall climate of the United States.
2. **Analyzing Information** Why do most Canadians live near the border of the United States?

CONNECTIONS TO ECONOMICS

Cooperative Learning Activity Work in a group of four to learn more about the economic activities of the major cities in Canada. Each group will choose one of the following cities: Montreal, Ottawa, Toronto, Winnipeg, or Vancouver. Each member will research one of the major industries in your city. Share your information with your group. Prepare a commercial using charts, posters, or music that encourages companies to do business in your city.

GeoJournal Writing Activity

Imagine that you and your family are moving to Canada. Describe which province or territory you would like to live in. Why did you choose this area? How would your life change if you moved there?

TECHNOLOGY ACTIVITY

Using E-Mail Use your local library or access the Internet to locate an E-mail address for the Nunavut Planning Commission in Canada. Compose a letter requesting information about the new Canadian territory of Nunavut. Using this information, create a bulletin board showing the location of the new territory, as well as a chronology of its creation. Provide photographs and captions with your display.

PLACE LOCATION ACTIVITY: CANADA

Match the letters on the map with the places and physical features of Canada. Write your answers on a separate sheet of paper.

1. Ontario
2. St. Lawrence River
3. Winnipeg
4. Hudson Bay
5. Vancouver
6. Yukon Territory
7. Newfoundland
8. Rocky Mountains
9. Ottawa

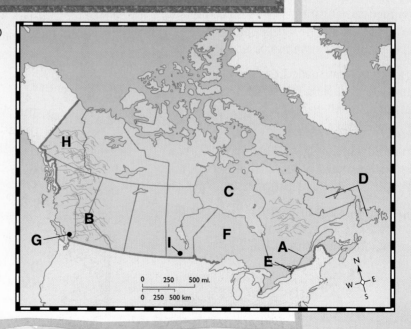

EYE ON THE ENVIRONMENT

Rats! They thrive around trash heaps such as this in New York City's East Harlem.

United States and Canada

TRASH

PROBLEM

The United States and Canada generate a lot of trash—approximately 200 million tons a year. That's about 4 pounds per person per day. Most of this garbage is burned in incinerators or dumped in landfills. But incinerators can pollute the air, and landfills are running out of space. Nearly half of all landfills in the United States will be land*fulls* by the year 2000.

SOLUTIONS

● Environmentalists want paper manufacturers to pay a tax on each ton of new paper produced—which should encourage the use of recycled paper.

● The list of products made from recycled materials includes crushed glass mixed with asphalt, a paving material called glasphalt, and recycled plastic made into park benches and clothing.

● A firm in Eugene, Oregon, makes building panels called Enviroboards out of recycled plastic, newspaper, cardboard, rubber, and polystyrene.

● A hospital in Burlington, Vermont, recycles 600–900 pounds of kitchen waste a day. The food waste is composted, and the compost is used on vegetables grown nearby for the hospital.

Printed on recycled paper.

TRASH FACT BANK

🍃 In the United States, 19 percent of trash is recycled, 17 percent incinerated, and 64 percent dumped in landfills.

🍃 Every $1,000 spent in fast-food stores creates 200 pounds of trash.

🍃 The garbage thrown away each year in the United States could fill garbage trucks lined up bumper-to-bumper on a four-lane highway circling the earth.

🍃 Paper, construction debris, and yard cuttings take up nearly 90 percent of the space in landfills.

🍃 A ton of recycled paper saves more than 4,000 kilowatt-hours of energy, 7,000 gallons of water, and 17 trees.

This landfill near Washington, D.C., is part of the solution for getting rid of trash—and part of the problem. Landfills in the United States are overflowing, and it's difficult to open new ones.

TEEN TRIBUTE

As part of an ecology club project, Taz Guishard tends red worms living in a bin in his classroom at Franko Middle School in Mount Vernon, New York. The worms recycle organic waste in a process called vermicomposting. One pound of worms can chew through one pound of waste and change it into rich soil in one month. "A lot of people talk about saving the earth," says Taz, "but I'm doing something about it."

environmental activities

CANADA

REDUCE • RECYCLE • REUSE • RESTORE •

TAKE A CLOSER LOOK

1 What happens to most garbage in the United States and Canada?

2 What is expected to happen to landfills in the United States by the year 2000?

WHAT CAN YOU DO?

🍃 Stop and think before you throw ANYTHING away!

🍃 Recycle aluminum—you can save 90 percent of the energy needed to make new cans by reusing the old.

🍃 Organize a recycling program for your lunchroom. Could left-over food go to a homeless shelter?

🍃 Boycott fast-food restaurants that do not use recycled materials for packaging.

This boy takes a step to solve the trash problem by picking up litter. Much of the trash taken to disposal areas can be recycled and reused.

Unit 3 Latin America

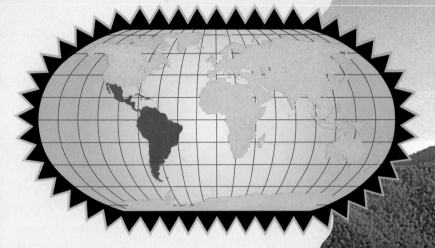

What Makes Latin America a Region?

The people share . . .

- a strong Spanish or Portuguese influence on culture and language.
- an African and Native American heritage.
- a mostly tropical or subtropical climate.
- the world's largest zone of tropical rain forest climate.

To find out more about Latin America, see the Unit Atlas on pages 138–151.

Machu Picchu, Peru

EXPLORING THE INTERNET

To learn more about Latin America, visit the Glencoe Social Studies Web site at **www.glencoe.com** for information, activities, and links to other sites.

GeoJournal Activity
As you read about this region, use each of your senses. Record your ideas about the tastes, smells, feel, look, and sounds that make up Latin America.

UNIT 3 ATLAS

NATIONAL
GEOGRAPHIC
SOCIETY

IMAGES
of the
WORLD

1. **Girl in traditional dress, Guatemala**
2. **Crowds in movie district, Buenos Aires**
3. **Gaucho swinging a rope, Argentina**
4. **Maya potter, Mexico**
5. **Men drilling for oil, Brazil**
6. **Rio de Janeiro at night**
7. **Desert landscape, Mexico**

All photos viewed against a panorama of Iguaçú Falls, Brazil.

138

UNIT 3 ATLAS

Regional Focus

Latin America is the name given to the vast region that lies south of the United States. The map on page 144 shows you that Latin America begins at the Rio Grande, a river dividing the United States and Mexico. It ends at the southern tip of the continent of South America.

Region
The Land

Latin America covers some 7,900,000 square miles (20,400,000 sq. km)— about 16 percent of the earth's surface. This region includes Mexico and the countries of Central America and South America. It also takes in the West Indies island countries in the Caribbean Sea.

Cruising the Magdalena River in Colombia

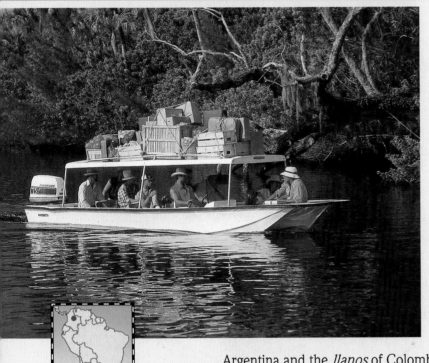

A river ferry carries passengers and their luggage down the Magdalena River.
LOCATION: What other major rivers are found in South America?

Mountains High mountains can be found in much of Latin America. In Mexico, three mountain ranges meet to form what looks like the letter *Y.* Mountains also run south through much of Central America. Most of the West Indies islands are the tops of volcanic mountains rising up from the Caribbean Sea. Along the west coast of South America, you will find the Andes—the world's longest mountain chain. Volcanic activity and earthquakes are common in these mountainous areas of Latin America.

Plains Vast plains cover other areas of Latin America. You will find plains along the coasts of Mexico and Central America. Broad inland plains also can be found in parts of South America. The two major South American plains are the *pampas* of Argentina and the *llanos* of Colombia and Venezuela. These plains areas are made up of rolling grassland with few trees.

Rivers The map on page 144 shows you that South America has five major rivers: the Magdalena, the Orinoco, the Río de la Plata, the São Francisco, and the Amazon. The Amazon is the longest river in the Western Hemisphere. It stretches for about 4,000 miles (6,436 km) across South America from the Andes eastward to the Atlantic Ocean. It carries 20 percent of the world's unfrozen freshwater and drains a basin of more than 3 million square miles (7.8 million sq. km).

Natural Resources The countries of Mexico and Venezuela are among the world's leading producers of oil and natural gas. Latin America in general is rich in natural resources such as copper, iron ore, silver, tin, lead, oil, and natural gas. Rivers and waterfalls provide electric power to many countries. Farmers use the rich soil in many areas of Latin America to grow grains, fruits, and coffee.

Region
Climate and Vegetation

Latin America has a variety of climates and different types of vegetation. Traveling through the region, you can journey through deserts and rain forests and from grassy plains to empty plateaus.

Elevation Most of Latin America lies in the tropics and has some form of tropical climate. Elevation, however, affects climate throughout the region. Land at low elevations is hot and humid with green tropical vegetation. Crops such as sugarcane, bananas, and cacao beans grow well here. At higher elevations, you will notice that the climate becomes milder and temperatures cooler. Crops such as coffee, corn, and wheat grow in this area. In the highest areas, you will see very little plant life and may even spot frost and snow.

Rain Forests Rain forests cover lowland areas of Latin America. They are rainy and hot all the time. Rain forests have more kinds of plants and animals than anywhere else on the earth. The world's largest tropical rain forest is in Brazil in the Amazon basin. Other rain forests grow along the eastern coast of Central America and in some of the Caribbean islands.

Amazon Rain Forest

Plants in the Amazon basin provide sources for one-fourth of the world's medicines. **LOCATION: Where are other rain forests in Latin America?**

UNIT 3 ATLAS

Goods to Market

Workers load cases of bananas onto a waiting ship in the port of Guayaquil *(top)*. A woman from Ecuador weaves hats that she will sell in the city's market *(bottom)*.
PLACE: What percentage of Latin Americans live in cities?

Region
Economy

Latin America's economies are based partly on agriculture. Fertile soil and a warm, wet climate allow farmers to grow tropical crops—coffee, bananas, and sugarcane—in parts of the region. In addition, Latin Americans also make use of grasslands to raise livestock. Argentina, Mexico, and Brazil are among the world's leaders in cattle raising and meat production.

Service industries and manufacturing have grown rapidly in Latin America. In countries such as Venezuela, Mexico, and Brazil, workers produce cars, textiles, cement, chemicals, and electrical goods. Some countries, however, lack the money, skilled labor, and transportation systems needed to build industries. Rugged mountains and thick tropical vegetation provide physical barriers to movement.

Movement
The People

Latin America has about 500 million people, about 9 percent of the world's population. Most Latin Americans live either along the coasts of South America or in a broad strip of land reaching from Mexico into Central America. In recent years, many Latin Americans have moved from the countryside to the cities. About 70 percent of the region's people are now city dwellers. Four Latin American cities rank among the 15 largest urban areas in the world. They are Mexico City, Mexico; São Paulo and Rio de Janeiro, Brazil; and Buenos Aires, Argentina.

Latin Americans include Native Americans, Europeans, Africans, and Asians. All of these groups have left their mark on the culture of the region. Ancient Native American peoples, such as the Maya, Aztec, and Inca, had advanced civilizations long before Europeans arrived in the Americas. During the 1500s, Europeans enslaved and destroyed the Native American civilizations. They later enslaved Africans and brought them to work their plantations.

Spain and Portugal ruled most of Latin America from the 1500s to the early 1800s. The Spanish and Portuguese brought the Roman Catholic faith and their languages to the region. Because these languages are based on Latin, the region became known as Latin America. Most Latin Americans today are Roman Catholics.

Many Latin American countries won their freedom in the early 1800s. Wealthy landowners and military officials, however, controlled governments and often ignored the needs of poor farmers and workers. By the mid-1900s, the rise of cities and industries challenged these leaders. With wide public support, new democratic governments emerged.

NATIONAL GEOGRAPHIC SOCIETY

Picture Atlas of the World

A Regional Study

Using the CD-ROM Create a file about Latin America. (See the User's Guide on how to use the Collector button.) Read the essays and vital statistics, look at the photos, map, and videos; and listen to the musical selections for each country. Make sure your file includes the following information:

1. most heavily and sparsely populated countries

2. major cities in the region

3. major physical features of the region

4. examples of colonial influences in architecture, religions, music, and languages

5. examples of Native American culture and traditions

Use your file to write an essay describing the regional characteristics of Latin America.

Surfing the "Net"

Latin American Culture

Latin American culture has been influenced by its history of colonial rule. Over time groups from Spain and Portugal, as well as Britain, France, and the Netherlands, settled in the region. These groups introduced elements of their cultures to the region. Some of these European influences can still be seen in Latin American culture today. To find out more about the influence of colonial rule on Latin American culture, search the Internet.

Getting There

Follow these steps to gather information on Latin American culture.

1. Use a search engine. Type in the words *Latin America.*

2. Enter words like the following to focus your search: *culture, history, colonization,* and *Europe.*

3. The search engine will provide you with a number of links to follow.

What To Do When You Are There

Click on the links to navigate through the pages of information and gather your findings. Use your findings to create an illustrated time line that identifies when and where each group of Europeans settled. Be sure to describe what each group contributed to Latin American culture. Display your time line for your classmates. Discuss the similarities and differences you found about colonial influences on Latin American culture.

Physical Geography

LATIN AMERICA: Physical

120°W   100°W   **UNITED STATES**   80°W   60°W   40°W

ATLANTIC OCEAN

Rio Grande

Baja California   **Plateau of Mexico**   SIERRA MADRE OCCIDENTAL   SIERRA MADRE ORIENTAL

MEXICO

Gulf of Mexico

BAHAMAS

Tropic of Cancer

20°N

Yucatán Peninsula   SIERRA MADRE DEL SUR

CUBA

W E S T   I N D I E S

Greater Antilles

BELIZE   **JAMAICA**

HAITI   **DOM. REP.**

Puerto Rico (U.S.)

HONDURAS

Lesser Antilles

GUATEMALA
EL SALVADOR
NICARAGUA
COSTA RICA

Caribbean Sea

PANAMA   *Isthmus of Panama*

Lake Maracaibo

Magdalena R.

GUYANA
SURINAME
French Guiana (Fr.)

Orinoco River

LLANOS   **VENEZUELA**

COLOMBIA

Guiana Highlands

0°   Equator

Rio Negro

AMAZON

Galápagos Islands (Ecuador)

ECUADOR

Amazon River

BASIN

PERU

Rio Madeira

B R A Z I L

Brazilian

A N D E S   M O U N T A I N S

Lake Titicaca

BOLIVIA

Altiplano

Rio São Francisco

Highlands

20°S

Atacama Desert

GRAN CHACO

Paraná R.

PARAGUAY

Tropic of Capricorn

PACIFIC OCEAN

Paraguay R.

Aconcagua 22,834 ft. (6,960 m)

ARGENTINA   **URUGUAY**

C H I L E

PAMPAS

Río de la Plata

40°S

Falkland (Malvinas) Is. (U.K.)

PATAGONIA

Tierra del Fuego
Cape Horn

Strait of Magellan

ELEVATIONS

Feet	Meters
10,000	3,000
5,000	1,500
2,000	600
1,000	300
0	0

▲ Mountain peak

Goode's Interrupted Homolosine projection

0   300   600 mi.

0   300   600 km

N   W   E   S

Map Study

1. PLACE What mountain chain runs along the western coast of South America?

2. LOCATION What body of water lies south of the West Indies?

ELEVATION PROFILE: South America

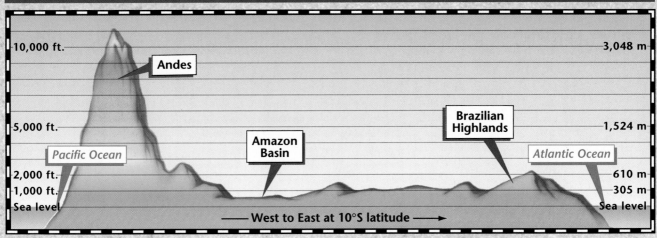

10,000 ft. — 3,048 m

Andes

Pacific Ocean

5,000 ft. — 1,524 m

Amazon Basin

Brazilian Highlands

Atlantic Ocean

2,000 ft. — 610 m
1,000 ft. — 305 m
Sea level — Sea level

← West to East at 10°S latitude →

Source: *Goode's World Atlas,* 19th edition

LATIN AMERICA AND THE UNITED STATES: Land Comparison

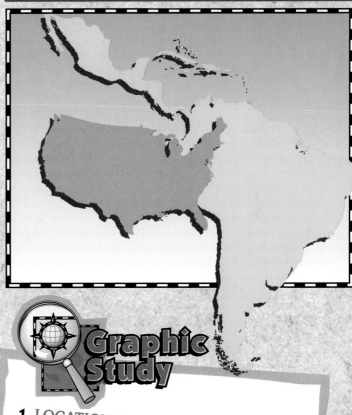

GeoFacts

Highest point: Mt. Aconcagua (Argentina) 22,834 ft. (6,960 m)

Lowest point: Valdes Peninsula (Argentina) 131 ft. (40 m) below sea level

Longest river: Amazon River (Brazil) 4,000 mi. (6,436 km) long

Largest lake: Lake Maracaibo (Venezuela) 5,217 sq. mi. (13,512 sq. km)

Highest waterfall: Angel Falls (Venezuela) 3,212 ft. (979 m)

Largest rain forest: Amazon rain forest (Brazil) 2,700,000 sq. mi. (6,993,000 sq. km)

Highest lake: Lake Titicaca (Peru and Bolivia) 12,500 ft. (3,811 m) in altitude

Graphic Study

1. LOCATION What is the highest point in South America?

2. PLACE Using the land comparison map, compare the land area of the United States to that of Latin America.

Cultural Geography

LATIN AMERICA: Political

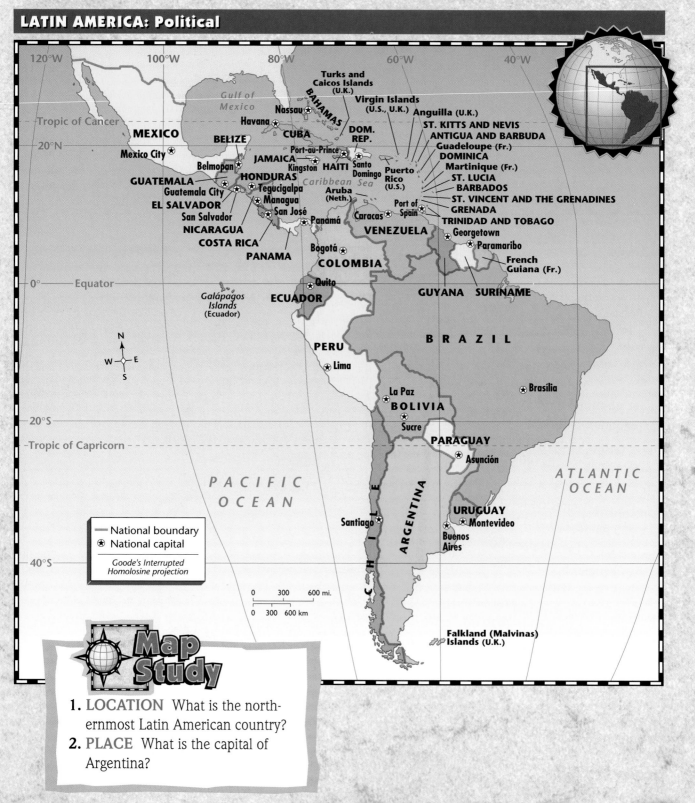

120°W 100°W 80°W 60°W 40°W

Tropic of Cancer

20°N

Gulf of Mexico

Turks and Caicos Islands (U.K.)

BAHAMAS

Nassau

Virgin Islands (U.S., U.K.)

Anguilla (U.K.)

Havana

ST. KITTS AND NEVIS

MEXICO

CUBA

DOM. REP.

ANTIGUA AND BARBUDA

Guadeloupe (Fr.)

BELIZE

DOMINICA

Mexico City

Port-au-Prince

JAMAICA

Santo Domingo

Martinique (Fr.)

Belmopan

Kingston

HAITI

Puerto Rico (U.S.)

ST. LUCIA

GUATEMALA

HONDURAS

Caribbean Sea

BARBADOS

Guatemala City

Tegucigalpa

Aruba (Neth.)

ST. VINCENT AND THE GRENADINES

EL SALVADOR

Managua

Port of Spain

GRENADA

San Salvador

San José

Panamá

Caracas

TRINIDAD AND TOBAGO

NICARAGUA

VENEZUELA

Georgetown

COSTA RICA

Bogotá

Paramaribo

PANAMA

COLOMBIA

French Guiana (Fr.)

0° Equator

Quito

GUYANA

SURINAME

Galápagos Islands (Ecuador)

ECUADOR

PERU

B R A Z I L

Lima

N

W E

S

La Paz

Brasília

BOLIVIA

20°S

Sucre

Tropic of Capricorn

PARAGUAY

Asunción

PACIFIC OCEAN

ATLANTIC OCEAN

C H I L E

A R G E N T I N A

URUGUAY

Santiago

Montevideo

Buenos Aires

— National boundary

✹ National capital

Goode's Interrupted Homolosine projection

40°S

0   300   600 mi.

0   300   600 km

Falkland (Malvinas) Islands (U.K.)

Map Study

1. **LOCATION** What is the northernmost Latin American country?
2. **PLACE** What is the capital of Argentina?

LATIN AMERICA: Largest Urban Areas

Mexico City, Mexico

São Paulo, Brazil

Rio de Janeiro, Brazil

Buenos Aires, Argentina

Lima, Peru

Bogotá, Colombia

Santiago, Chile

Guadalajara, Mexico

Caracas, Venezuela

= 3,000,000

Source: *The World Almanac, 1998*

GeoFacts

Biggest country (land area): Brazil
3,265,060 sq. mi.
(8,456,505 sq. km)

Smallest country (land area):
Grenada 130 sq. mi. (337 sq. km)

Largest city (population): Mexico
City (1997) 24,000,000; (2015
projected) 32,000,000

Highest population density:
Barbados 1,596 people per sq.
mi. (616 people per sq. km)

Lowest population density: French
Guiana 5 people per sq. mi.
(2 people per sq. km)

COMPARING POPULATION: Latin America and the United States

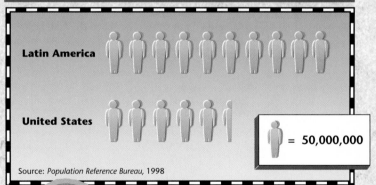

Latin America

United States

= 50,000,000

Source: *Population Reference Bureau, 1998*

 Graphic Study

1. **PLACE** How does the population of Latin America compare to the population of the United States?

2. **PLACE** What are the three most populated urban areas in Latin America?

Countries at a Glance

Antigua and Barbuda

  St. Johns

LANDMASS:
170 sq. mi./
440 sq. km
MONEY:
Eastern Caribbean dollar
MAJOR EXPORT:
Petroleum Products
MAJOR IMPORT:
Food and Live Animals

CAPITAL:
St. Johns
MAJOR LANGUAGE(S):
English
POPULATION:
100,000

Argentina

 Buenos Aires

CAPITAL:
Buenos Aires
MAJOR LANGUAGE(S):
Spanish
POPULATION:
36,100,000

LANDMASS:
1,056,640 sq. mi./
2,736,698 sq. km
MONEY:
Peso
MAJOR EXPORT:
Food and Live Animals
MAJOR IMPORT:
Machinery

Bahamas

 Nassau

CAPITAL:
Nassau
MAJOR LANGUAGE(S):
English
POPULATION:
300,000

LANDMASS:
3,860 sq. mi./
9,997 sq. km
MONEY:
Dollar
MAJOR EXPORT:
Petroleum
MAJOR IMPORT:
Fuels

Barbados

 Bridgetown

CAPITAL:
Bridgetown
MAJOR LANGUAGE(S):
English
POPULATION:
300,000

LANDMASS:
166 sq. mi./
430 sq. km
MONEY:
Dollar
MAJOR EXPORT:
Sugar
MAJOR IMPORT:
Machinery

Belize

 Belmopan

CAPITAL:
Belmopan
MAJOR LANGUAGE(S):
English, Spanish,
native Creole dialects
POPULATION:
200,000

LANDMASS:
8,800 sq. mi./
22,792 sq. km
MONEY:
Belize Dollar
MAJOR EXPORT:
Sugar
MAJOR IMPORT:
Machinery

Bolivia

 La Paz, Sucre

CAPITALS:
La Paz, Sucre
MAJOR LANGUAGE(S):
Spanish, Quechua, Aymara
POPULATION:
8,000,000

LANDMASS:
418,680 sq. mi./
1,084,381 sq. km
MONEY:
Boliviano
MAJOR EXPORT:
Zinc
MAJOR IMPORT:
Raw Materials

Brazil

 Brasilia

CAPITAL:
Brasilia
MAJOR LANGUAGE(S):
Portuguese
POPULATION:
162,100,000

LANDMASS:
3,265,060 sq. mi./
8,456,505 sq. km
MONEY:
Cruzeiro
MAJOR EXPORT:
Machinery
MAJOR IMPORT:
Petroleum

Chile

 Santiago

CAPITAL:
Santiago
MAJOR LANGUAGE(S):
Spanish
POPULATION:
14,800,000

LANDMASS:
289,110 sq. mi./
748,795 sq. km
MONEY:
Peso
MAJOR EXPORT:
Paper Products
MAJOR IMPORT:
Intermediate Goods

Colombia

 Bogotá

CAPITAL:
Bogotá
MAJOR LANGUAGE(S):
Spanish
POPULATION:
38,600,000

LANDMASS:
401,040 sq. mi./
1,038,694 sq. km
MONEY:
Peso
MAJOR EXPORT:
Petroleum
MAJOR IMPORT:
Machinery

Costa Rica

 San José

CAPITAL:
San José
MAJOR LANGUAGE(S):
Spanish
POPULATION:
3,500,000

LANDMASS:
19,710 sq. mi./
51,049 sq. km
MONEY:
Colón
MAJOR EXPORT:
Clothing
MAJOR IMPORT:
Petroleum

Countries not drawn to scale.

Cuba

Havana

LANDMASS:
42,400 sq. mi./
109,816 sq. km
MONEY:
Peso
MAJOR EXPORT:
Sugar
MAJOR IMPORT:
Fuels

CAPITAL:
Havana
MAJOR LANGUAGE(S):
Spanish
POPULATION:
11,100,000

Dominica

Roseau

LANDMASS:
290 sq. mi./
751 sq. km
MONEY:
Eastern Caribbean Dollar
MAJOR EXPORT:
Bananas
MAJOR IMPORT:
Machinery

CAPITAL:
Roseau
MAJOR LANGUAGE(S):
English, French, Creole
POPULATION:
100,000

Dominican Republic

Santo
Domingo

LANDMASS:
18,680 sq. mi./
48,381 sq. km
MONEY:
Peso
MAJOR EXPORT:
Ferronickel
MAJOR IMPORT:
Petroleum

CAPITAL:
Santo Domingo
MAJOR LANGUAGE(S):
Spanish
POPULATION:
8,300,000

Ecuador

Quito

LANDMASS:
106,890 sq. mi./
276,845 sq. km
MONEY:
Sucre
MAJOR EXPORT:
Petroleum
MAJOR IMPORT:
Raw Materials

CAPITAL:
Quito
MAJOR LANGUAGE(S):
Spanish, Quechua,
Jivaroan
POPULATION:
12,200,000

El Salvador

San Salvador

LANDMASS:
8,000 sq. mi./
20,720 sq. km
MONEY:
Colón
MAJOR EXPORT:
Coffee
MAJOR IMPORT:
Chemicals

CAPITAL:
San Salvador
MAJOR LANGUAGE(S):
Spanish
POPULATION:
5,800,000

French Guiana*

Cayenne

LANDMASS:
34,035 sq. mi./
88,150 sq. km
MONEY:
French Franc
MAJOR EXPORT:
Shrimp
MAJOR IMPORT:
Food

CAPITAL:
Cayenne
MAJOR LANGUAGE(S):
French
POPULATION:
200,000

Grenada

St. George's

LANDMASS:
130 sq. mi./
337 sq. km
MONEY:
Eastern Caribbean Dollar
MAJOR EXPORT:
Bananas
MAJOR IMPORT:
Machinery

CAPITAL:
St. George's
MAJOR LANGUAGE(S):
English, French patois
POPULATION:
100,000

Guatemala

Guatemala

LANDMASS:
41,860 sq. mi./
108,417 sq. km
MONEY:
Quetzal
MAJOR EXPORT:
Coffee
MAJOR IMPORT:
Petroleum

CAPITAL:
Guatemala
MAJOR LANGUAGE(S):
Spanish, Mayan languages
POPULATION:
11,600,000

Guyana

Georgetown

LANDMASS:
76,000 sq. mi./
196,840 sq. km
MONEY:
Dollar
MAJOR EXPORT:
Sugar
MAJOR IMPORT:
Fuels

CAPITAL:
Georgetown
MAJOR LANGUAGE(S):
English, Amerindian
dialects
POPULATION:
700,000

Haiti

Port-au-Prince

LANDMASS:
10,640 sq. mi./
27,558 sq. km
MONEY:
Gourde
MAJOR EXPORT:
Textiles
MAJOR IMPORT:
Food and Live Animals

CAPITAL:
Port-au-Prince
MAJOR LANGUAGE(S):
French, Creole
POPULATION:
7,500,000

*Territory of France

UNIT 3 ATLAS

Countries at a Glance

Honduras

Tegucigalpa

LANDMASS:
43,200 sq. mi./
111,888 sq. km

CAPITAL:
Tegucigalpa
MAJOR LANGUAGE(S):
Spanish
POPULATION:
5,900,000

MONEY:
Lempira
MAJOR EXPORT:
Bananas
MAJOR IMPORT:
Machinery

Jamaica

Kingston

LANDMASS:
4,180 sq. mi./
10,826 sq. km

CAPITAL:
Kingston
MAJOR LANGUAGE(S):
English, Jamaican Creole
POPULATION:
2,600,000

MONEY:
Dollar
MAJOR EXPORT:
Alumina
MAJOR IMPORT:
Fuels

Mexico

Mexico City

LANDMASS:
736,950 sq. mi./
1,908,700 sq. km

CAPITAL:
Mexico City
MAJOR LANGUAGE(S):
Spanish, Amerindian
languages
POPULATION:
97,500,000

MONEY:
New Peso
MAJOR EXPORT:
Machinery
MAJOR IMPORT:
Machinery

Nicaragua

Managua

LANDMASS:
46,873 sq. mi./
121,401 sq. km

CAPITAL:
Managua
MAJOR LANGUAGE(S):
Spanish
POPULATION:
4,800,000

MONEY:
Cordoba Oro
MAJOR EXPORT:
Coffee
MAJOR IMPORT:
Petroleum

Panama

Panamá

LANDMASS:
28,737 sq. mi./
74,428 sq. km

CAPITAL:
Panamá
MAJOR LANGUAGE(S):
Spanish, English
POPULATION:
2,800,000

MONEY:
Balboa
MAJOR EXPORT:
Bananas
MAJOR IMPORT:
Fuels

Paraguay

Asunción

LANDMASS:
153,400 sq. mi./
397,306 sq. km

CAPITAL:
Asunción
MAJOR LANGUAGE(S):
Spanish, Guaraní
POPULATION:
5,200,000

MONEY:
Guaraní
MAJOR EXPORT:
Cotton Products
MAJOR IMPORT:
Machinery

Peru

Lima

LANDMASS:
494,210 sq. mi./
1,280,004 sq. km

CAPITAL:
Lima
MAJOR LANGUAGE(S):
Spanish, Quechua,
Aymara
POPULATION:
26,100,000

MONEY:
Sol
MAJOR EXPORT:
Copper
MAJOR IMPORT:
Machinery

Puerto Rico*

San Juan

LANDMASS:
3,420 sq. mi./
8,858 sq. km

CAPITAL:
San Juan
MAJOR LANGUAGE(S):
Spanish
POPULATION:
3,900,000

MONEY:
Dollar
MAJOR EXPORT:
Chemical Products
MAJOR IMPORT:
Chemicals

St. Kitts-Nevis

Basseterre

LANDMASS:
140 sq. mi./
363 sq. km

CAPITAL:
Basseterre
MAJOR LANGUAGE(S):
English
POPULATION:
40,000

MONEY:
Eastern Caribbean Dollar
MAJOR EXPORT:
Sugar and Molasses
MAJOR IMPORT:
Food and Live Animals

St. Lucia

Castries

LANDMASS:
236 sq. mi./
611 sq. km

CAPITAL:
Castries
MAJOR LANGUAGE(S):
English, French patois
POPULATION:
100,000

MONEY:
Eastern Caribbean Dollar
MAJOR EXPORT:
Bananas
MAJOR IMPORT:
Machinery

Countries not drawn to scale.

*U.S. Commonwealth

Ocho Rios, Jamaica

Beautiful beaches and a warm climate make tourism a main source of income for Jamaica and other islands in Latin America.

St. Vincent and the Grenadines

 Kingstown

CAPITAL:
Kingstown
MAJOR LANGUAGE(S):
English
POPULATION:
100,000

LANDMASS:
150 sq. mi./
389 sq. km
MONEY:
Eastern Caribbean Dollar
MAJOR EXPORT:
Bananas
MAJOR IMPORT:
Food Products

Suriname

 Paramaribo

CAPITAL:
Paramaribo
MAJOR LANGUAGE(S):
Dutch, Sranantonga, English
POPULATION:
400,000

LANDMASS:
60,230 sq. mi./
155,996 sq. km
MONEY:
Guilder
MAJOR EXPORT:
Alumina
MAJOR IMPORT:
Machinery

Uruguay

 Montevideo

CAPITAL:
Montevideo
MAJOR LANGUAGE(S):
Spanish
POPULATION:
3,200,000

LANDMASS:
67,490 sq. mi./
174,799 sq. km
MONEY:
New Peso
MAJOR EXPORT:
Textiles
MAJOR IMPORT:
Machinery

Trinidad and Tobago

 Port of Spain

CAPITAL:
Port of Spain
MAJOR LANGUAGE(S):
English
POPULATION:
1,300,000

LANDMASS:
1,980 sq. mi./
5,128 sq. km
MONEY:
Dollar
MAJOR EXPORT:
Petroleum
MAJOR IMPORT:
Food

Venezuela

 Caracas

CAPITAL:
Caracas
MAJOR LANGUAGE(S):
Spanish
POPULATION:
23,300,000

LANDMASS:
340,560 sq. mi./
882,050 sq. km
MONEY:
Bolivar
MAJOR EXPORT:
Petroleum
MAJOR IMPORT:
Machinery

Map labels

UNITED STATES

BAJA CALIFORNIA NORTE

SONORA

CHIHUAHUA

COAHUILA

BAJA CALIFORNIA SUR

Gulf of California

Tropic of Cancer

SINALOA

DURANGO

Monterrey •

NUEVO

LEÓN

TAMAULIPAS

ZACATECAS

SAN LUIS POTOSÍ

NAYARIT

AGUASCALIENTES

GUANAJUATO

QUERÉTARO

Guadalajara •

León •

HIDALGO

JALISCO

TLAXCALA

Mexico City ✪

COLIMA

MICHOACÁN

MÉXICO

PACIFIC OCEAN

MORELOS

GUERRERO

Acapulco •

Puebla • Veracruz •
PUEBLA

VERACRUZ

Bay of Campeche

YUCATÁN

QUINTANA ROO

CAMPECHE

TABASCO

OAXACA

CHIAPAS

Caribbean Sea

CENTRAL AMERICA

FEDERAL DISTRICT

Gulf of Mexico

CUBA

30°N

20°N

110°W   100°W   90°W

N
W   E
S

Legend:
- National boundary
- State boundary
- ✪ National capital
- • Other city

Polyconic projection

0   125   250 mi.
0   125   250 km

MAP STUDY ACTIVITY

As you read Chapter 6, you will learn about Mexico, the most northern country in Latin America.

1. How many states make up the country of Mexico?

2. What is the capital of Mexico?

The Land

PREVIEW

Words to Know
- land bridge
- peninsula
- basin
- latitude
- altitude

Places to Locate
- Isthmus of Tehuantepec
- Baja California
- Yucatán Peninsula
- Sierra Madre Occidental
- Sierra Madre Oriental
- Sierra Madre del Sur
- Plateau of Mexico

- Mexico City
- Rio Grande

Read to Learn . . .
1. where Mexico is located.
2. why Mexico is called the "land of the shaking earth."
3. what climates are found in Mexico.

Clara Ponce is a teenager who lives in Mazatlán, a large port city and tourist resort on the coast of the Pacific Ocean. When Clara's parents take her and her sisters on vacation, they have some spectacular choices— Mexico's sunny beaches, hot deserts, misty rain forests, or snowy mountains.

Clara is proud of the Spanish and Native American heritage that Mexico shares with its Latin American neighbors to the south. Mexico also has strong links with its northern neighbor, the United States.

Location
Bridging Two Continents

Mexico forms part of a **land bridge,** or narrow strip of land that joins two larger landmasses. This land bridge connects North America and South America. Look at the map on page 154. You can see that Mexico sweeps to the southeast like a crooked funnel. The widest part of Mexico borders the United States in the north. Farther south, Mexico reaches its narrowest point at the Isthmus of Tehuantepec (tuh•WAHN•tuh•PEHK). Here, only 140 miles (225.3 km) separates the Pacific Ocean from the Gulf of Mexico.

The Pacific Ocean borders Mexico on the west. Extending south along this west coast is Baja (BAH•hah) California. It is a long, thin **peninsula,** or piece of land surrounded by water on three sides. On the eastern side, the Gulf of

Mexico and the Caribbean Sea wash Mexico's shores. Jutting far out into the Gulf is another large peninsula—the Yucatán Peninsula. Find Baja California and the Yucatán Peninsula on the map below.

Place
Land of the Shaking Earth

From spectacular mountains and volcanoes to deep valleys, Mexico has a rugged landscape. Why? The country sits where some plates of the earth's crust have collided for billions of years. These collisions formed mountains and volcanoes and caused terrifying earthquakes. One famous volcano was named Popocatepetl (POH•puh•KAH•tuh•PEH•tuhl), or "smoky mountain," by ancient Native Americans. The Aztec, one of the early peoples of Mexico, called their country the "land of the shaking earth." The ground is still moving! Mexico City, the capital, has been shaken by many earthquakes in the past 50 years.

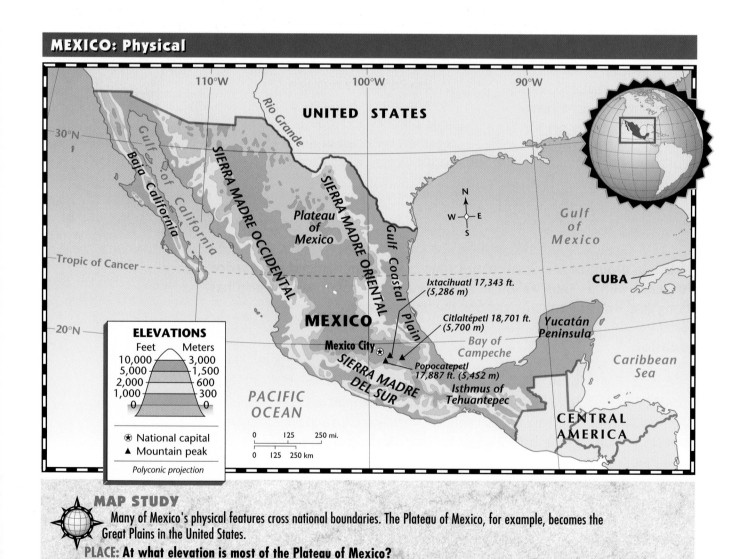

MEXICO: Physical

ELEVATIONS

Feet	Meters
10,000	3,000
5,000	1,500
2,000	600
1,000	300
0	0

⊛ National capital
▲ Mountain peak

Polyconic projection

Ixtacihuatl 17,343 ft. (5,286 m)
Citlaltépetl 18,701 ft. (5,700 m)
Popocatepetl 17,887 ft. (5,452 m)

0   125   250 mi.
0   125   250 km

MAP STUDY
Many of Mexico's physical features cross national boundaries. The Plateau of Mexico, for example, becomes the Great Plains in the United States.

PLACE: At what elevation is most of the Plateau of Mexico?

A one-lane road crisscrosses the jagged peaks of the Sierra Madre Occidental.
LOCATION: Along which coast of Mexico does the Occidental range extend?

Mountains and Plateau Three major mountain ranges tower over Mexico. One range—the Sierra Madre Occidental—runs north and south along western Mexico near the Pacific Ocean. Clara's village lies in this lightly populated area. Another major range, the Sierra Madre Oriental, runs along Mexico's eastern side. The steep ridges and deep canyons of yet another range—the Sierra Madre del Sur—rise in southwestern Mexico.

The Sierra Madres surround the large, flat center of the country—the Plateau of Mexico—which covers 40 percent of the country. You find mostly deserts and grassy plains in the northern part of the Plateau. As you move south, the Plateau steadily rises in elevation with occasional **basins**, or broad, flat valleys. Snowcapped mountains, some of which are volcanoes, rise above the basins. Volcanic eruptions have left fertile soil in the basins. The map on page 154 shows you the location of the mountain ranges and plateau.

Coastal Lowlands Mexico's coastal plains stretch along the Pacific Ocean and the Gulf of Mexico. Many of the country's longest rivers flow through these coastal plains. The Rio Grande, one of Mexico's great rivers, empties into the Gulf of Mexico. The Rio Grande forms about 1,300 miles (2,090 km) of Mexico's border with the United States.

Region
Climate

By looking at the map on page 156, you see that Mexico has many different climates. What causes such differences? As you read in Chapter 2, **latitude**—or location north or south of the Equator—affects temperatures. The Tropic of Cancer runs through the center of Mexico. It marks the northern

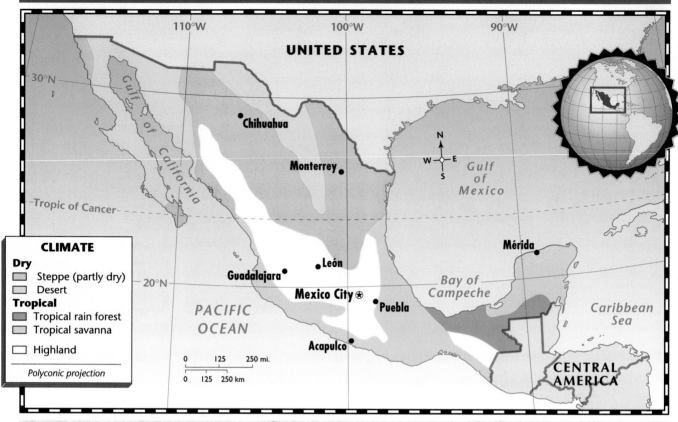

110°W 100°W 90°W

UNITED STATES

30°N

• Chihuahua

Gulf of California

Monterrey •

Gulf of Mexico

Tropic of Cancer

CLIMATE
Dry
Steppe (partly dry)
Desert
Tropical
Tropical rain forest
Tropical savanna
Highland
Polyconic projection

• León

Mérida •

20°N

Guadalajara •

PACIFIC OCEAN

Mexico City ⊛

Bay of Campeche

Puebla

Caribbean Sea

0 125 250 mi.

0 125 250 km

Acapulco •

CENTRAL AMERICA

MAP STUDY

Mexico's many climates are caused by the country's latitude as well as altitude.

LOCATION: Near what line of latitude is Mexico City?

A Rugged Land

In 1521, a Spanish soldier named Hernán Cortés conquered Mexico. When he returned to Spain, the Spanish king asked him to describe Mexico. Legend says that Cortés crumpled a piece of paper and threw it on a table, saying, "This, your Majesty, is the land of Mexico." In this way he showed that Mexico is a very mountainous country.

edge of the tropics, where temperatures are generally warm year-round.

Altitude Zones Mexico's mountains also affect its temperatures. The diagram on page 157 shows that the country's mountains create three **altitude** zones. You could travel through each of these zones in a day's trip across the Sierra Madres.

Hot Land Your trip begins as you stand on the coastal plain near the sea in the *tierra caliente* (tee•AY•rah kah•lee•AYN•teh), which means "hot land" in Spanish. Hot and humid all year, the *tierra caliente* temperatures average 77°F (25°C) to 82°F (28°C). Dense rain forests or tall grasses blanket much of this zone, although bananas, rice, sugarcane, and oranges also grow here.

Mexico's Altitude Zones

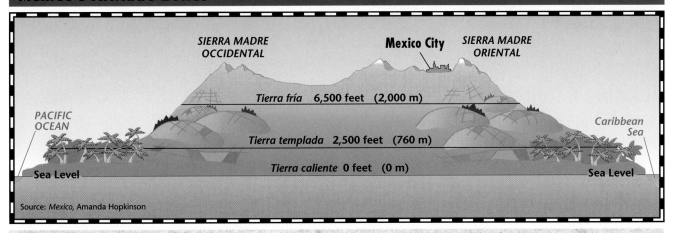

SIERRA MADRE OCCIDENTAL

Mexico City

SIERRA MADRE ORIENTAL

Tierra fría   6,500 feet   (2,000 m)

PACIFIC OCEAN

Caribbean Sea

Tierra templada   2,500 feet   (760 m)

Tierra caliente   0 feet   (0 m)

Sea Level

Sea Level

Source: *Mexico*, Amanda Hopkinson

GRAPHIC STUDY

In the diagram, you see that the *tierra caliente* includes the low coastal plains. Hot, humid climates prevail along the coasts, with temperatures ranging from 60° to 120°F (15.6° to 48.9°C).

PLACE: What kinds of vegetation are shown growing in the *tierra caliente*?

Temperate Land As you climb into the mountains, the temperatures become a little cooler. The trees are larger and have more leaves. You are in the middle elevation—the *tierra templada* (tehm•PLAH•dah), or "temperate land" where temperatures hover around 70°F (21°C). Farmers can grow a wide variety of crops here.

Cold Land As you reach the top of the mountains, the climate becomes very cold. You are traveling through the *tierra fría* (FREE•ah), or "cold land." During the year, temperatures here average below 68°F (20°C). Short, stunted trees and a few wildflowers cling to rocky soil. As you travel higher, plant life becomes very scarce and then disappears under snow-topped peaks.

SECTION 1 ASSESSMENT

REVIEWING TERMS AND FACTS

1. **Define the following:** land bridge, peninsula, basin, latitude, altitude.
2. **LOCATION** What country borders Mexico to the north?
3. **PLACE** How does the landscape of the Plateau of Mexico differ from north to south?
4. **PLACE** What three altitude zones would you go through if you crossed Mexico's mountains?

MAP STUDY ACTIVITIES

5. Turn to the map of Mexico on page 154. What bodies of water surround Mexico?
6. Look at the map on page 156. What climates are found in northern Mexico?

BUILDING GEOGRAPHY SKILLS

Reading a Relief Map

Differences in the elevation, or height, of land areas are called *relief*. In a relief map, colors or shadings show areas of different elevations.

Relief maps are an important source of geographic information. To read a relief map, follow these steps:

- Read the map title to identify the land area shown on the map.
- Use the map key to determine what elevations are shown on the map.
- Identify the areas of highest and lowest elevations on the map.

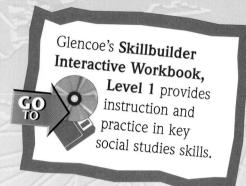

Glencoe's *Skillbuilder Interactive Workbook,* **Level 1** provides instruction and practice in key social studies skills.

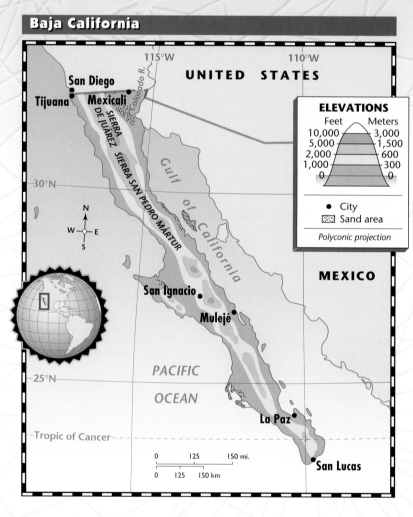

Baja California

ELEVATIONS

Feet	Meters
10,000	3,000
5,000	1,500
2,000	600
1,000	300
0	0

● City
▦ Sand area

Polyconic projection

Geography Skills Practice

1. What area of Mexico is shown on the map?
2. At what elevation is San Ignacio?
3. What is the elevation of most of the area along the west coast of Baja?

SECTION 2

The Economy

PREVIEW

Words to Know
- service industry
- *maquiladora*
- subsistence farm
- plantation
- industrialized
- smog

Places to Locate
- Guadalajara
- León
- Puebla

Read to Learn . . .
1. how Mexicans earn a living.
2. what the three economic regions of Mexico are.
3. what economic challenges face modern Mexico.

Oil rigs stand offshore in the Gulf of Mexico as reminders of Mexico's hopes for a better economy. Every time workers strike oil underneath the Gulf, Mexico improves its position as one of the world's top oil producers.

Oil fuels Mexican businesses and homes. It is also exported, which brings much money into the country. The map on page 160 shows that oil is only one of many minerals found in Mexico.

Region

Economic Regions

Manufacturing and mining are vital to the Mexican economy. Almost one-fifth of the world's silver is mined in Mexico. **Service industries** also strengthen the country's economy. A service industry, you remember, is a business that provides services to people instead of making goods. Banking and tourism are Mexico's main service industries. The geography of Mexico influences its economy. Many geographers divide Mexico into three economic regions: central Mexico, the north, and the south.

Although rich in minerals, Mexico is poor in fertile land. The country's many mountains, deserts, and rain forests limit the land that can be farmed to only 11 percent of the total area. Farmers make good use of this land by growing coffee, corn, cotton, oranges, and sugarcane.

Mexico's resources differ greatly as you move from the north to the south.

PLACE: What resources are found in the north? In the south?

Central Mexico Central Mexico is the economic heart of the country. This area is home to more than half of the population and offers favorable conditions for agriculture. Large industrial cities, such as Mexico City and Guadalajara (GWAH•duhl•uh•HAHR•uh), also flourish here. Other cities in central Mexico include León (lay•OHN) and Puebla (PWEH•bluh). Find these cities on the map on page 161.

Mexico City is the giant hub of central Mexico. About 24 million people live in Mexico City and its suburbs, making it one of the largest urban areas in the world. It is also the center of Mexico's government.

The North The northern economic region includes Baja California and the northern part of the Plateau of Mexico. Much of the land in the north is too dry to farm, but farmers irrigate in some places to grow cotton, fruits, cereals, and vegetables. In hilly areas ranchers raise cattle, sheep, goats, and pigs. Ranching, as we know it today, originated in Mexico. *Vaqueros*, or cowhands, developed the tools to herd, rope, and brand cattle.

The largest city in this region is Monterrey, which leads the country in steel production. Workers in the region mine copper, silver, lead, and zinc. Other workers are employed in *maquiladoras* (mah•KEE•luh•DOH•rahz), factories that assemble parts shipped from other countries. Workers in *maquiladoras* assemble automobiles, stereo systems, computers, and other electronic products. These assembled goods are then sent to foreign countries, especially the United States.

The South The southern region stretches from Mexico City to the Yucatán Peninsula. People have lived here since at least 2000 B.C. Those living in this area are among the poorest in Mexico. Farmers in many mountain villages are subsistence farmers. **Subsistence farms**, small plots where farmers grow only enough food to feed their families, are common in this area. In the valleys, wealthy farmers grow coffee or sugarcane on **plantations**, large farms that raise a single crop for money. Tourism is also important in the south. The region's beautiful beaches and historic monuments draw visitors from all over the world.

Human/Environment Interaction
Economic Challenges

During the past 50 years, Mexico has **industrialized**. What does this mean? It means that Mexico has become less a country of farms and villages and more a country of factories and cities. Many challenges arise with industrial growth, however. They include conserving land, controlling pollution, creating new jobs, and increasing trade with other countries.

Pollution Industrial growth in Mexico affects the surrounding environment. As you read earlier, mountains encircle Mexico City, blocking the flow of air. Mexico City's many factories and cars pollute the air, leaving a thick haze of

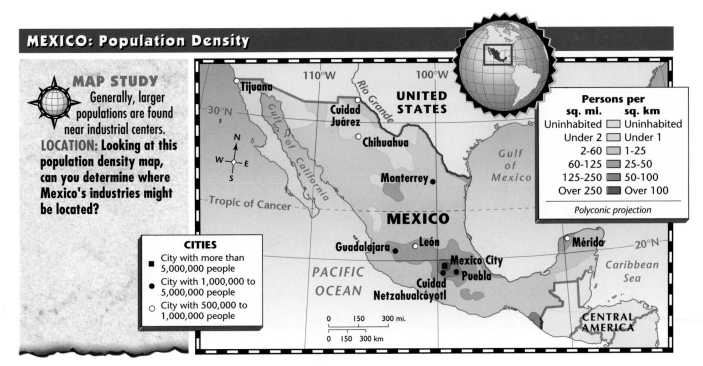

MEXICO: Population Density

MAP STUDY
Generally, larger populations are found near industrial centers.
LOCATION: Looking at this population density map, can you determine where Mexico's industries might be located?

CITIES
- City with more than 5,000,000 people
- City with 1,000,000 to 5,000,000 people
- City with 500,000 to 1,000,000 people

Persons per	
sq. mi.	sq. km
Uninhabited	Uninhabited
Under 2	Under 1
2-60	1-25
60-125	25-50
125-250	50-100
Over 250	Over 100

Polyconic projection

A Floating Village?

On the outskirts of Mexico City lies the floating village of Xochimilco (SOH•chih•MEEL•koh).

When it was settled in the 1200s, crops were planted on mud-covered rafts held together by reeds. Roots from the plants eventually anchored the floating gardens to the bottom of the lake surrounding Mexico City. Xochimilco's farmland is still a sort of raft, and visitors travel on canals instead of roads.

smog, or fog mixed with smoke and chemicals, to settle over the city. Schoolchildren wear masks at recess to filter out the pollution, and sometimes the city completely shuts down because people must stay indoors.

In the countryside, farmers often burn wild vegetation to make room for crops, destroying thousands of acres of forests. Mexico must find ways to use and conserve its resources without spoiling the countryside.

Population Changes Mexico's population is growing twice as fast as the population of the United States. Because health care and diet have improved, Mexico's birthrate is rising and more people are living longer. With this increase in population, Mexico cannot provide jobs for all the people who want to work.

Look at the map on page 161. It shows that most of Mexico's 98 million people live in the southern part of the Plateau of Mexico. The many people living in this one area strain resources. Crowded conditions and unemployment are common. Many Mexican people in search of work move to the United States. Some enter the United States illegally, without proper travel papers or official permission to work.

Free Trade In 1993 Mexico, the United States, and Canada approved the North American Free Trade Agreement, or NAFTA. This agreement eventually will allow goods and money to move freely among these three countries. It has already begun to create many new jobs in Mexico. Those who support NAFTA hope that increased trade will help Mexico's economy grow and improve life for all Mexicans.

SECTION 2 ASSESSMENT

REVIEWING TERMS AND FACTS

1. **Define the following:** service industry, *maquiladora,* subsistence farm, plantation, industrialized, smog.
2. **PLACE** What is Mexico's most important mineral?
3. **REGION** What are Mexico's three economic regions?
4. **HUMAN/ENVIRONMENT INTERACTION** What environmental problems face those who live in Mexico City?

MAP STUDY ACTIVITIES

5. Look at the map on page 160. What minerals, besides oil, are found in Mexico?
6. According to the map on page 160, what kinds of farming take place on the Yucatán Peninsula?

THE ANCIENT MAYA

MATH	SCIENCE	HISTORY	LITERATURE	TECHNOLOGY

The Maya (MY•uh) were Native Americans who lived in the Yucatán Peninsula and Central America. From about A.D. 250 to A.D. 900, their civilization flourished, making advances in astronomy, architecture, and mathematics. They also invented a sophisticated form of writing hundreds of years before Europeans came to the Americas.

RELIGION Religion was the strongest force in Maya society. The Maya worshipped many gods and goddesses, including a corn god, a rain god, and a moon goddess. To keep accurate records for their religious festivals, the Maya became skilled astronomers. They used their knowledge of the sun and stars to predict eclipses and to develop a calendar. The Maya calendar had 360 days and 5 "unlucky" days to equal 365 days.

ARCHITECTURE Ruins of Maya cities lie hidden in the steamy rain forests of the Yucatán Peninsula. One of the greatest Maya cities was Chichén Itzá (CHIH•chehn EET•suh). Most Maya cities had religious temples built as huge pyramids. The main pyramid at Chichén Itzá reveals the organized thinking of Maya architects. The pyramid has 4 sides, each with a stairway of 91 steps running to the top of a platform—a total of 364 steps. The platform at the top brings the total to 365, the number of days in a year.

The Main Pyramid at Chichén Itzá

MATHEMATICS The Maya were also skilled in mathematics. They created a number system based on the number 20 that allowed them to count into the millions. Hundreds of years before the Arabs, the Maya developed the concept of zero.

Maya civilization mysteriously collapsed sometime after A.D. 800. No one knows why the Maya abandoned their cities. Descendants of the Maya, however, today live in southern Mexico and nearby areas of Guatemala.

Making the Connection

1. Where in Mexico did the Maya live?
2. What were some Maya achievements in architecture? In mathematics?

SECTION 3

The People

PREVIEW

Words to Know
- colony
- *mestizo*
- adobe

Places to Locate
- Tenochtitlán

Read to Learn . . .
1. what groups influenced Mexican culture.
2. how city life differs from country life in Mexico.
3. what makes up Mexican culture today.

Thousands of tourists visit Mexico City every year. They often visit the unique Plaza of Three Cultures. Here, in a single place, you can see the ruins of an ancient Native American monument, a stone church built by the Spaniards, and modern glass and steel apartment buildings!

Mexico City displays the mix of cultures—old and new—that make Mexico a fascinating country. Two groups—Native Americans and Spaniards—have made important contributions to Mexico's growth as a nation.

Place
Influences of the Past

Historians usually divide Mexican history into three periods. The first is the age of great Native American civilizations. The second is the time of Spanish rule, and the third is the age of modern Mexico.

Native Americans The first people to live in Mexico came from Asia. Later known as Native Americans, these people traveled south through North America and entered Mexico thousands of years ago. Look at the map on page 165 to locate Mexico's major Native American civilizations.

One group, the Maya, flourished in the Yucatán area between A.D. 250 and 900. They built cities around towering stone temples in thick rain forests. These temples honored Maya gods and rulers. Turn to page 163 to read about the ancient Maya.

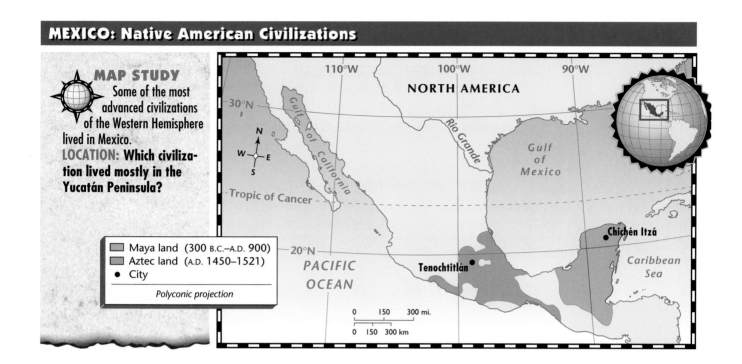

MAP STUDY
Some of the most advanced civilizations of the Western Hemisphere lived in Mexico.
LOCATION: Which civilization lived mostly in the Yucatán Peninsula?

Maya land (300 B.C.–A.D. 900)
Aztec land (A.D. 1450–1521)
• City

Polyconic projection

Around A.D. 1200 the Mexica, whom the Spanish called the Aztec, built a city named Tenochtitlán (tay•NOHCH•teet•LAHN) in central Mexico. Mexico City is located in this area today. The Aztec were fierce warriors as well as builders and traders. Merchants in Tenochtitlán set up marketplaces, which were filled with pottery, woven baskets, cloth, gold, and silver.

The Spanish Heritage In 1519 a Spanish army led by Hernán Cortés arrived on Mexican soil, and, 2 years later, conquered the Aztec. Mexico remained a Spanish **colony,** or overseas territory, for about 300 years. During that time many Spaniards controlled the lives of Native Americans, forcing them to work on farms and in mines. As a result, the Spanish and Native American cultures mixed. The term **mestizo** (meh•STEE•zoh) refers to a person of mixed Native American and European heritage. Today about 60 percent of Mexico's people are *mestizos.* Another 30 percent are Native American.

You can also see the influence of the Spanish on religion in Mexico. Most Mexicans today—about 90 percent—are Roman Catholic. Festivals honor the country's patron saint, the Virgin of Guadalupe (GWA•duhl•OO•pay).

The Zócalo, Mexico City

Each year thousands of people gather in the Zócalo, or Constitution Square, in remembrance of the Mexican revolution.
PLACE: How many years did the Mexican revolution last?

Modern Mexico Mexico gained its freedom from Spain in 1821. During much of the 1800s, a few wealthy families, the army, and church leaders controlled the government. They did little to improve life for Mexico's many poor farmers. In 1910 discontent exploded in a revolution.

The revolution lasted until the 1920s and greatly influenced Mexico's government. Like the United States, Mexico is a federal republic. A national government and 31 state governments share powers. Look at the map on page 152 to see the states that make up Mexico.

Until recently, one political party ruled Mexico. Then, in the 1990s, many Mexicans, especially poor Native American farmers, demanded reforms. Now other political parties are beginning to win elections.

Place

City Life

Bustling cities dot Mexico, providing homes for three out of four Mexicans. These cities have both modern and older sections. Guadalajara, in central Mexico, has broad streets lined with palm trees. The city's tall modern buildings and colorful shops convey a sense of excitement and progress. Guadalajara's citizens also enjoy historic areas with parks, fountains, and splendid old churches.

Areas of beautifully preserved old homes are found in many Mexican cities. These homes, usually made of **adobe** (uh•DOH•bee), or sun-dried clay brick, often are built around courtyards with fountains and pots of blooming plants. Houses in poor urban areas are made of scraps of wood, metal, or whatever materials can be found. Most of these lack electricity and running water.

Guadalajara, Mexico

Nicknamed the "Pearl of the West," Guadalajara has more than three million people. **MOVEMENT: Do most people in Mexico live in rural or urban areas?**

Country Life

Country life is very different from city life in Mexico. Most Mexican villages are very poor. Some village homes are built of cement blocks with a flat, red-tiled roof, and others are made of sheet metal, straw, or clay. Homes line narrow streets that lead to a central plaza, which may include a small church, a few shops, and the local government building.

Almost every village has a marketplace where men, women, and children sell or trade clothes, food, baskets, or pottery. Some show their goods in a large building that resembles a warehouse. Others display their wares outside on brightly colored shawls spread upon the street.

The Arts and Recreation

Mexican art explores and represents the pride of the people in their achievements and their heritage. Early Native Americans gave Mexico beautiful temples and dramatic sculptures. Spaniards and Native Americans later built and decorated many of the country's churches. Modern composers, artists, and writers have also expressed the spirit of Mexico. Craftworkers in Acapulco, for example, combine ancient and modern skills to make silver jewelry, colorful pottery, and handblown glass.

Painters and Writers Mexican artists and writers have created many national treasures. In the early 1900s, Mexican artists produced beautiful murals, or wall paintings. Among the most famous mural painters were José Clemente Orozco and Diego Rivera, who painted scenes from Mexican history. Much of Mexican literature, such as works by modern writers Carlos Fuentes and Octavio Paz, describes daily life.

Music and Dance Mexicans listen to and compose all kinds of music—from modern rock to classical. Traditional Mexican music is played by mariachi (mahr•ee•AH•chee) bands. If you listen to a typical mariachi band, you will hear a singer, two violinists, two guitarists, two horn players, and a bass player. These musicians dress in colorful outfits and wear wide-brimmed hats called sombreros.

Teen Scene

In Touch With the Past

Like most of his friends, Benito Carrea loves to listen to Mexican rock music. On Sundays, however, Benito leaves his radio at home and goes to the park with his family. He takes the flute and practices the traditional music his father and grandfather taught him. Benito's family are descendants of the Maya. More than 1,700 years ago, the Maya played flutes similar to Benito's along with drums, shells, and rattles made from gourds.

Mariachi bands play lively Mexican music.
PLACE: **What musical instruments do you recognize in this photo?**

Celebrations Listen to the excited voices and music in the plaza! What is going on? Throughout the year, Mexicans enjoy special celebrations called fiestas. The word *fiesta* means "feast day" in Spanish. Fiestas are festivals that include parades, fireworks, music, and dancing. Independence Day (September 15-16) and Cinco de Mayo (May 5) are patriotic days of celebration.

Most Mexicans also celebrate Christmas. Actors perform religious plays every evening from mid-December until Christmas Day. After each play, children gather around a piñata (peen•YAH•tuh), a hollow figure of stiff paper filled with toys and candy. Wearing blindfolds, the children try to break the piñata open with a stick.

Foods Traditional Mexican food combines Spanish and Native American cooking. Corn, a crop grown since ancient times, has always been the most important food in Mexico. Mexicans make tortillas, a thin flat cornmeal bread shaped by hand and cooked on a griddle. A folded tortilla filled with vegetables, cheese, beans, or meat becomes a taco. As you probably know, tacos are now as popular in the United States as they are in Mexico.

Sports If you like soccer, visit Mexico and join the crowds. Soccer is the country's most popular sport. Major soccer games are played in Mexico City's Azteca Stadium, which holds about 100,000 people. Baseball, which came to Mexico from the United States more than 50 years ago, also draws large crowds.

Bullfighting, introduced by Spaniards, is a favorite spectator sport for tourists. Many Mexicans admire expert bullfighters for their bravery. Other Mexicans, however, think that bullfighting is cruel because of the way that bulls are killed during the event.

SECTION 3 ASSESSMENT

REVIEWING TERMS AND FACTS
1. **Define the following:** colony, *mestizo,* adobe.
2. **PLACE** Name two groups that influenced modern Mexican culture.
3. **PLACE** How does city life differ from country life in Mexico?
4. **REGION** What celebrations are common throughout Mexico?

MAP STUDY ACTIVITIES
5. Turn to the map on page 165. Name an ancient civilization that lived in central Mexico.
6. Using the same map, name the major city of the Aztec Empire.

Chapter 6 Highlights

Important Things to Know About Mexico

SECTION 1 THE LAND

- Mexico is part of the land bridge connecting North America and South America.
- Major landforms in Mexico are mountains, plateaus, and lowlands.
- Much of Mexico's climate is tropical or desert.

SECTION 2 THE ECONOMY

- Central Mexico has the largest cities and the greatest industrial development of Mexico's three economic regions.
- Oil is Mexico's most important product.
- Coffee, rice, sugarcane, and fruit are grown on large and small farms.
- Challenges facing Mexico include encouraging industry, creating jobs, and protecting the environment.

SECTION 3 THE PEOPLE

- Native Americans were the earliest people to live in Mexico. Spanish explorers and settlers arrived in the 1500s.
- The Roman Catholic religion remains an important influence on the people of Mexico.
- Mexico is a federal republic. Its 31 states share power with the national government.
- The culture of modern Mexico mixes old and new art, music, and literature grown from Native American, Spanish, and Mexican roots.

A market in Guanajuato, Mexico ▶

REVIEWING KEY TERMS

Match the numbered terms in Set A with their lettered definitions in Set B.

A

1. smog
2. subsistence farm
3. plantation
4. industrialized
5. *maquiladora*
6. adobe
7. *mestizo*
8. land bridge

B

A. growth of factories and cities
B. large farm that grows a single crop
C. factory that assembles parts from other countries
D. person of mixed Native American and European ancestry
E. sun-dried clay brick
F. narrow strip of land that connects two larger land masses
G. to grow only enough to feed your family
H. fog mixed with smoke and chemicals

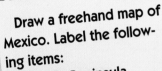

Mental Mapping Activity

Draw a freehand map of Mexico. Label the following items:

- Yucatán Peninsula
- Gulf of Mexico
- Baja California
- Pacific Ocean
- Gulf of California
- Mexico City

REVIEWING THE MAIN IDEAS

Section 1

1. **REGION** What continents does Mexico connect?
2. **PLACE** Name Mexico's three major mountain ranges.
3. **LOCATION** How does Mexico's latitude affect its climate?

Section 2

4. **LOCATION** Where are most of Mexico's oil fields located?
5. **PLACE** What are Mexico's main service industries?
6. **HUMAN/ENVIRONMENT INTERACTION** What economic activities take place in the north?

Section 3

7. **MOVEMENT** Name two Native American civilizations that influenced Mexico's history.
8. **MOVEMENT** What European nation once ruled Mexico?
9. **PLACE** What are Mexico's most popular sports?

CRITICAL THINKING ACTIVITIES

1. **Determining Cause and Effect** Why are temperatures in Mexico City—which is south of the Tropic of Cancer—usually not as hot in the summer as temperatures in northern Mexico?
2. **Evaluating Information** What examples of Mexican culture do you find in your own community? What conclusions can you draw from this knowledge?

CONNECTIONS TO GOVERNMENT

Cooperative Learning Activity Work in groups to learn more about Mexico. Each group will choose one of the following topics to research: (a) what powers the states have; (b) what powers are given to the federal govenment; (c) how government officials are elected; (d) what political parties have power. After your research is complete, share your information with your group. As a group, prepare a written report or poster that illustrates the group's findings.

GeoJournal Writing Activity

Review your notes for the GeoJournal Activity on page 137. Now imagine you are a European immigrant who came to Mexico during the days of Spanish colonial settlement. Write a letter to a friend or relative in Europe describing your new home.

TECHNOLOGY ACTIVITY

Developing Multimedia Presentations

Imagine that you own a company that is about to set up operations in Latin America. Your company may be a clothing manufacturer, a coffee plantation, or a banking and financial center. Select a country and research its business customs. Develop a video for the employees that are new to the country. Demonstrate procedures and key words your employees will need to know in order to conduct business in the country.

PLACE LOCATION ACTIVITY: MEXICO

Match the letters on the map with the places and physical features of Mexico. Write your answers on a separate sheet of paper.

1. Sierra Madre Occidental
2. Sierra Madre Oriental
3. Sierra Madre del Sur
4. Plateau of Mexico
5. Mexico City
6. Isthmus of Tehuantepec
7. Acapulco
8. Guadalajara
9. Gulf of Mexico
10. Rio Grande

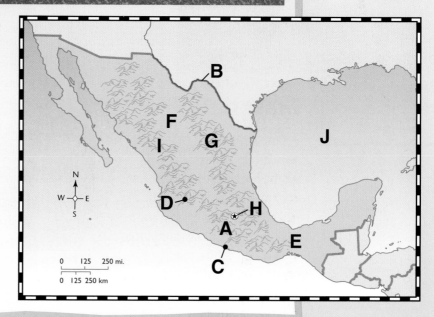

GeoLab ACTIVITY

VOLCANOES: Powerful Giants

Background

Fierce giants named Irazú, Cotopaxi, and Lascar threaten Latin America. Why don't the people try to destroy these giants? They can't—the giants are all volcanoes. In Chapter 1 you learned that volcanoes occur where hot, fluid lava forces its way up to the surface of the earth by extreme pressure. Find out more about erupting volcanoes by making your own volcanic eruption.

Mount Kilauea Erupts!

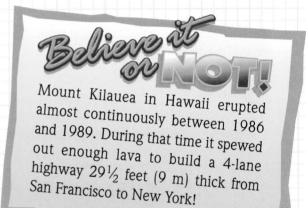

Mount Kilauea in Hawaii erupted almost continuously between 1986 and 1989. During that time it spewed out enough lava to build a 4-lane highway 29½ feet (9 m) thick from San Francisco to New York!

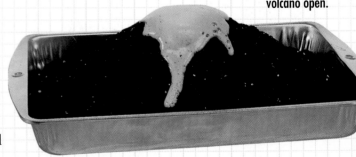

Build your mountain carefully to leave the mouth of the volcano open.

Materials

- sand or dirt, slightly damp
- 1/2 cup water
- small, empty tin can
- empty quart bottle
- 4 tablespoons baking soda
- 1/4 cup dishwashing liquid
- 1/4 cup white vinegar
- red food coloring (optional)
- long-handled spoon
- large baking pan

What To Do

A. Place the baking pan on a level surface.

B. Put the tin can in the center of the pan and build a mountain of sand around it, leaving the top of the can uncovered. (This makes the crater of your volcano.)

C. Pour the baking soda into the can.

D. In the quart bottle mix the water, dishwashing liquid, vinegar, and a few drops of red coloring.

E. Pour only half of the mixture into the tin can. (You can repeat your eruption with the rest later.)

F. Step back and watch your volcano erupt. If the lava does not flow from your volcano right away, carefully stir the can's mixture with the spoon.

Lab Activity Report

1. Did the lava flow more to one side of your volcano, or evenly all around it?

2. Describe how fast the lava flowed at the beginning compared with the end of the eruption.

3. What did you learn about volcanoes from this experiment?

4. Drawing Conclusions How do you think an erupting volcano might affect the wildlife that lives around it?

Go A Step Further!

Activity

Find out what destructive volcano erupted in the West Indies in 1902. What happened to the people and the land around it? Write a newspaper article describing the event as if you had been there.

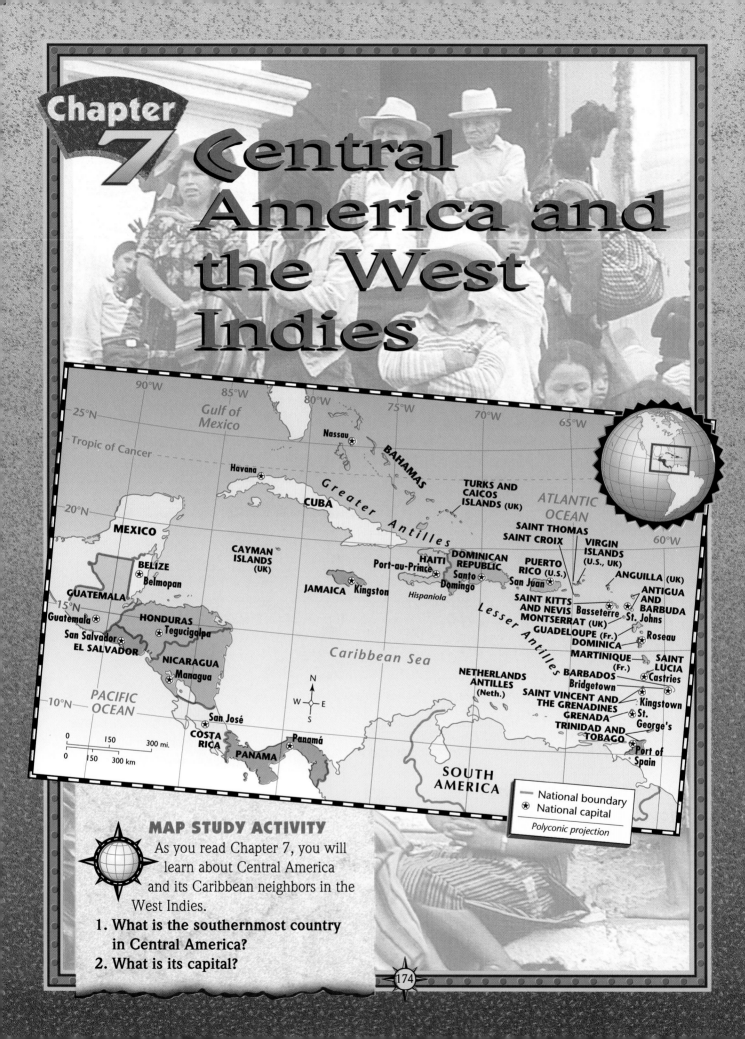

Chapter 7

Central America and the West Indies

National boundary
⊛ **National capital**

Polyconic projection

MAP STUDY ACTIVITY

As you read Chapter 7, you will learn about Central America and its Caribbean neighbors in the West Indies.

1. **What is the southernmost country in Central America?**
2. **What is its capital?**

Central America

Words to Know

- hurricane
- plantation
- chicle
- nutrient
- bauxite
- *ladino*
- literacy rate
- isthmus

Places to Locate

- Guatemala
- Belize
- El Salvador
- Honduras
- Nicaragua
- Costa Rica
- Panama
- Central Highlands
- Pacific Lowlands
- Caribbean Lowlands
- Panama Canal

Read to Learn . . .

1. where Central America is located.
2. how farming supports the economy of Central America.
3. what groups of people settled Central America.

Imagine a fiery red sea of lava rolling toward you. For those who lived in the shadow of Mount Izalco, this was not just a daydream. Today, however, Mount Izalco is an inactive volcano in Central America.

Central America is part of the land bridge that lies between the continents of North America and South America. The map on page 174 shows that Central America includes seven countries: Belize, Guatemala, Honduras, El Salvador, Nicaragua, Costa Rica, and Panama.

These countries all have mountainous inland areas, and most have two coastlines—one on the Caribbean Sea and the other on the Pacific Ocean. All except Belize share a mostly Native American and Spanish cultural heritage. Belize, once ruled by the British, has strong British and African influences.

Place

The Land

Central America extends more than 1,000 miles (1,600 km) north to south and measures 300 miles (483 km) at its widest point. Look at the map on page 174. You see that the region stretches from Mexico southward to South America. If you travel across it from east to west, you will journey from the Caribbean Sea to the Pacific Ocean.

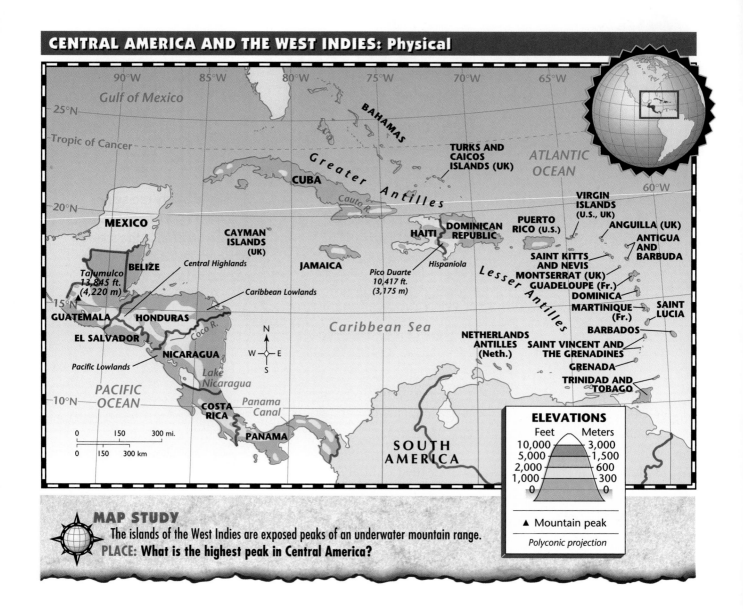

MAP STUDY
The islands of the West Indies are exposed peaks of an underwater mountain range.
PLACE: What is the highest peak in Central America?

Landforms Volcanic eruptions are part of life in Central America. Like Mexico, these countries have a rugged landscape and many active volcanoes. A chain of volcanic mountains called the Central Highlands rises like a backbone along most of the region. Volcanic material that has broken down over the centuries has left rich, fertile soil. Throughout Central America, farmers grow coffee, bananas, sugarcane, and other crops for export.

The map above shows you that on either side of the Central Highlands lie coastal plains. The Pacific Lowlands extend along the Pacific side. The Caribbean Lowlands lie on the Caribbean side.

Climate Central America's climate is mostly tropical, but there are some differences from country to country. What causes these differences? In general, altitude and location on the continent determine climate. Mountains can block the movement of winds and moisture.

176

Look at the map on page 184. The central part of the region has mountains and highland areas that are dry and cool year-round. In the Pacific Lowlands on the west coast, a tropical savanna climate prevails. Temperatures are warm and rain is plentiful from May through November. From December through April, the climate is hot and dry.

If you don't mind a daily rain shower, you would be happy living in the Caribbean Lowlands. These eastern lowlands have a hot, tropical rain forest climate year-round. Here you could expect about 100 inches (254 cm) of rainfall each year and cool breezes from the Caribbean Sea. Those breezes, however, can turn into deadly hurricanes during the summer and autumn months. **Hurricanes** are fierce storms with winds of more than 74 miles (119 km) per hour. They pose a threat for those who live on the coastal plains.

WHAT IN THE WORLD?

Hurricane Hugo

Hurricanes can be deadly, and Hurricane Hugo certainly was. In 1989 its 220-mile-per-hour (about 350-km-per-hour) winds stripped the Caribbean of vegetation and devastated almost every island in its path. Hugo finally blew itself out after smashing into the United States. More than 500 lives were lost, and property damage totaled between $2 billion and $4 billion.

Region
The Economy

Fertile soil and rain forests rank as Central America's most important resources. The economies of different countries depend on the exporting of agricultural products from the farms and wood products from the rain forests to world markets.

Farming Two kinds of farmers form the base of Central America's economy. Owners of **plantations**—large farms that grow produce for sale—raise coffee, bananas, and sugarcane. They export their harvest to the United States, western Europe, and other parts of the world. International trade is not as important to the other type of farmer—the subsistence farmer. Most raise small crops of corn, beans, rice, and livestock to feed their families. At nearby village marketplaces, they sell extra food for other supplies.

Rain Forests Under the green canopy of the Central American rain forests, great treasures can be found. The dense forests offer valuable woods—mahogany, balsa, and teak—which are exported. Workers also tap the sapodilla tree for **chicle** (CHIH•kuhl), a substance used in making chewing gum.

Scientists use trees and plants from the rain forest in medical research or to make new medicines. Unusual animals found nowhere else on earth also roam among the rain forest plants and trees.

In the Caribbean Lowlands, farmers have cleared rain forest areas to raise crops. Because of heavy rains throughout the year, the soil erodes and loses many of its **nutrients,** or minerals that supply food to plants. Farmers in this area now earn most of their living from raising livestock.

Many Central Americans are worried about the rapid clearing of so many acres of rain forest. What are they doing about it? Costa Rica has turned some forest areas into national parks. Other countries control all logging operations and assist workers in replanting cleared areas. In this way, Central American governments hope to save some of the rain forest for future generations. You can read more about the problems facing the rain forests on page 236.

Industry Missing from the skylines of most major Central American cities are the towers and chimneys of industry. Although the region has a number of small industries, not much manufacturing has been developed. Most of the countries are poor in mineral resources, especially fuels.

Guatemala and Costa Rica are exceptions. Guatemala sends crude, or unrefined, oil to overseas markets. Costa Rica exports **bauxite,** a mineral used to make aluminum. The map below shows other resources found in Central America.

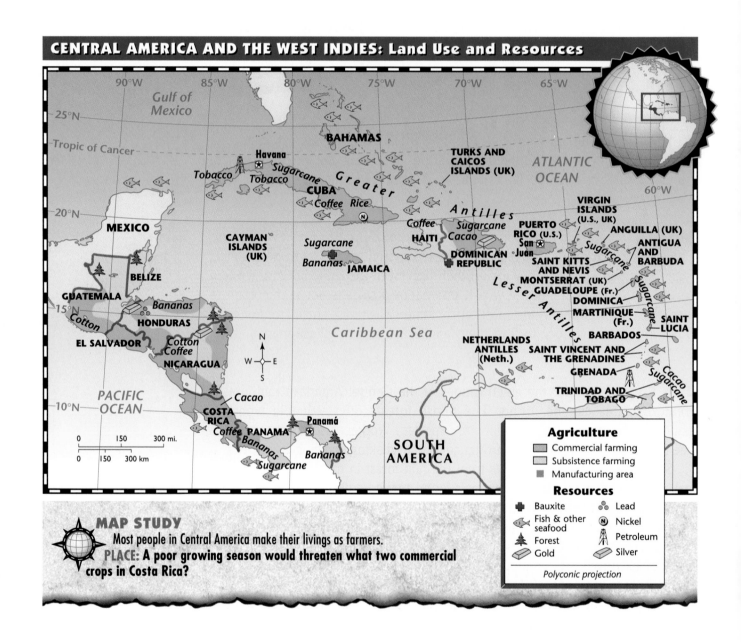

CENTRAL AMERICA AND THE WEST INDIES: Land Use and Resources

MAP STUDY
Most people in Central America make their livings as farmers.
PLACE: A poor growing season would threaten what two commercial crops in Costa Rica?

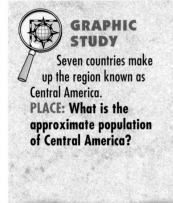

GRAPHIC STUDY

Seven countries make up the region known as Central America.
PLACE: What is the approximate population of Central America?

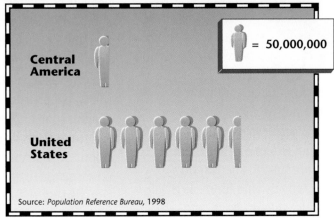

Central America

United States

👤 = 50,000,000

Source: *Population Reference Bureau, 1998*

Movement
The People

In Guatemala the cooking stones of the ancient Maya stand cold—overgrown with creeping vines. Steep stairways lead to empty temples. Water canals lay dry and cracked. The ancient Maya have long since disappeared—but their descendents live on in Central America.

Influences of the Past Like Mexico, Central America is a mix of cultures. The Maya settled throughout Central America about 250 to 400 B.C. As you learned in Chapter 6, the Maya developed an advanced civilization. After A.D. 800 the Maya mysteriously left their cities and scattered. Today some of their descendents live in Guatemala as well as in parts of Mexico.

In the late 1400s, the Spanish settled in the Central American region. In the 1500s they claimed territory along the Caribbean coast. For the next 300 years Spanish landowners forced Native Americans to work on plantations. Eventually, the cultures blended. Native Americans gradually adopted the Spanish language and many adopted the Roman Catholic religion.

Settled by the British in the 1600s, the area that is now the country of Belize eventually came under British rule. The British enslaved Africans and brought them to work in Belize's forests. Africans are the largest ethnic group in Belize.

Independence Most Central American countries gained independence by 1821. With help from the United States, Panama won independence from Colombia in 1903. The last Central American country to gain independence was Belize, which ceased to be a colony of the United Kingdom in 1981.

Revolutions have rocked most of the countries of Central America from the mid-1800s to the present. In many of these countries the people continue to fight for the government that best meets their needs.

Panajachel, Guatemala

Lake Atitlán in the highlands of Guatemala is surrounded by volcanic peaks.
HUMAN/ENVIRONMENT INTERACTION: How does the lake provide a living for this Guatemalan?

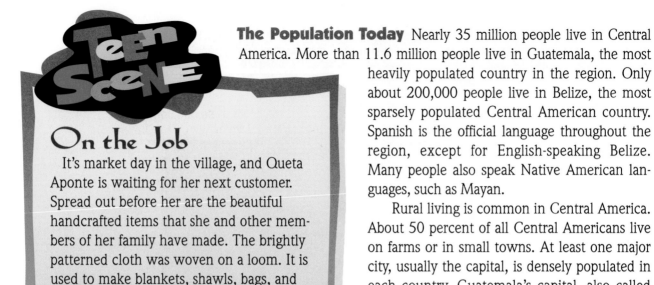
On the Job

It's market day in the village, and Queta Aponte is waiting for her next customer. Spread out before her are the beautiful handcrafted items that she and other members of her family have made. The brightly patterned cloth was woven on a loom. It is used to make blankets, shawls, bags, and small stuffed animals. Queta started selling her family's goods at the weekly market when she was seven years old. The money she earns will help to buy food and other things that her family needs.

The Population Today Nearly 35 million people live in Central America. More than 11.6 million people live in Guatemala, the most heavily populated country in the region. Only about 200,000 people live in Belize, the most sparsely populated Central American country. Spanish is the official language throughout the region, except for English-speaking Belize. Many people also speak Native American languages, such as Mayan.

Rural living is common in Central America. About 50 percent of all Central Americans live on farms or in small towns. At least one major city, usually the capital, is densely populated in each country. Guatemala's capital, also called Guatemala, ranks highest in population among Central American cities with about 1,800,000 people.

People in urban areas hold manufacturing or service industry jobs, or they work on farms outside the cities. Those living in coastal areas harvest shrimp, lobster, and other seafood to sell in city markets or for export.

Celebrations During the steamy month of August, Nicaraguan teenager Juan Ruiz and his friends celebrate the festival of Santo Domingo (Saint Dominic). They join the crowds overflowing the streets of Managua, Nicaragua's capital. Special foods, music, and dancing highlight the festivities. When there are no festivals to attend, Juan and his friends like to play baseball and soccer.

Place
National Profiles

To better understand the variety of people and places in Central America, look closely at Guatemala, Costa Rica, and Panama.

Guatemala The beautiful lakes of Guatemala reflect its majestic volcanoes. Most Guatemalans live in the southern Central Highlands area. The culture of Guatemalans comes from both Native American and Spanish influences. More than 40 percent of the people follow a rural way of life similar to that of their Native American ancestors. These people live in villages and usually do not travel beyond their country's borders. At an area market, you can tell exactly where villagers come from. How? Look at their clothing—each Native American village has its own special colors and styles of dress.

Guatemalans who speak Spanish and practice European ways are called **ladinos.** Most ladinos live in cities in modern houses or apartments and work as laborers or businesspeople. From 1960 to 1996, a civil war was fought between the government and groups living in the highlands. About 150,000 Guatemalans died in this war.

Costa Rica Costa Ricans like to say, *"Darse buena vida,"* which is Spanish for "Enjoy the great life!" Why? Costa Rica offers one of the highest standards of living in the world. It also has one of the highest **literacy rates,** or percentage of people who can read and write. Most Costa Ricans are of Spanish ancestry and only a small number claim Native American heritage.

Wars have rarely occurred in this country. Costa Rica has enjoyed good relations with its neighbors. A well-developed democratic government rules, supported by a police force. Today Costa Ricans boast that their country has more schools than police barracks.

Most of the country's 3.5 million people live in the cool Central Highlands. The capital, San José, is in this area, as are most of Costa Rica's other cities. Near the highlands, farmers grow the country's major export—coffee.

Panama Panama is one of the major crossroads of the world. Across it stretches the Panama Canal. This southernmost Central American country lies on an **isthmus.** Linking Central and South America, the isthmus of Panama also separates the Caribbean Sea and the Pacific Ocean.

In 1903 the United States helped Panama win its independence from Colombia. Then the United States proposed building a canal across Panama to shorten shipping time between the Atlantic and Pacific oceans. On the map on page 182 you can see how this route crosses Panama. Engineers completed the Panama Canal—one of the engineering wonders of the world—in 1914. Since then the United States has held ownership of the canal and the land on either side of it. In the year 2000 Panama will gain control of the canal zone.

Nearly one-half of Panama's 2.8 million people live and work in the area near the canal. Most are of mixed Native American and Spanish ancestry and speak Spanish, the official language. English, however, is also widely spoken. The national capital, Panamá—located on the Pacific side of the isthmus—is the leading center of culture and industry.

SECTION 1 ASSESSMENT

REVIEWING TERMS AND FACTS
1. **Define:** hurricane, plantation, chicle, nutrient, bauxite, *ladino,* literacy rate, isthmus.
2. **LOCATION** What continents does Central America link?
3. **MOVEMENT** What groups of people have influenced the development of Central America?

MAP STUDY ACTIVITIES
4. Turn to the physical map on page 176. Which country in Central America has the highest elevation?
5. Look at the land use map on page 178. What resources are found in the region?

MAKING CONNECTIONS

BUILDING THE PANAMA CANAL ▼ ▲

MATH   SCIENCE   HISTORY   LITERATURE   **TECHNOLOGY**

On September 25, 1513, Spanish explorer Vasco Núñez de Balboa sighted the Pacific Ocean from a hill in what is now Panama. He became the first—but not the last—to dream of a canal through Panama that would connect the Atlantic and Pacific oceans. Not until 400 years later did scientists and engineers begin a 10-year battle with Panama's swampy jungles and soft volcanic soil to make the canal a reality.

JUNGLES AND SWAMPS The thick semitropical jungles and swamps of east Panama provided excellent breeding grounds for disease-carrying mosquitoes and rats. From 1904 to 1906, medical scientist Colonel William C. Gorgas headed teams that cleared brush, drained swamps, and cut down tropical forest. By 1906 Gorgas had wiped out yellow fever and bubonic plague, making Panama safe for workers.

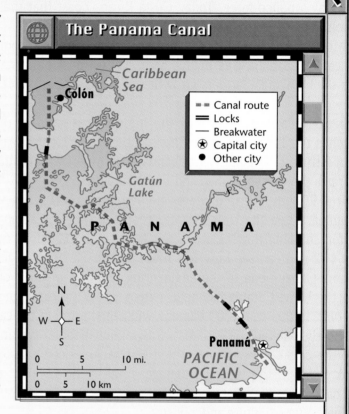

The Panama Canal

Caribbean Sea
Colón
Gatún Lake
P A N A M A
Panamá
PACIFIC OCEAN

- - - Canal route
=== Locks
— Breakwater
⊛ Capital city
● Other city

N W E S

0   5   10 mi.
0 5 10 km

MOUNTAINS OF SOFT SOIL After the canal area was disease-free, the digging began. The canal's course ran through hills of soft volcanic soil. As soon as a shovelful of soil was removed, more slid into its place. Massive landslides occurred as the heavier hilltops lost the support of their subsoil. To keep digging was the only solution: First the hilltops were dug away, then the lower layers of soil. Crews dug more than 211 million cubic yards (161 million cu. m) of dirt before the canal was completed!

Making the Connection ▼ ▲

1. Why was a canal through Panama desirable?
2. What geographic obstacles did Panama present to the building of a canal? How were these challenges met?

SECTION 2

The West Indies

PREVIEW

Words to Know
- archipelago
- colony
- communism
- cooperative
- light industry
- dialect
- commonwealth

- Hispaniola
- Jamaica
- Puerto Rico

Places to Locate
- Caribbean Sea
- Greater Antilles
- Lesser Antilles
- Cuba
- Dominican Republic
- Haiti

Read to Learn . . .
1. what landforms and climates are common to the West Indies.
2. how people in the West Indies earn a living.
3. what cultures are found in the West Indies.

Robinson Crusoe, the fictional character who was shipwrecked, may have seen a tropical view just like this when he washed ashore. This beautiful bay is in Saint Lucia in the Caribbean Sea.

The Caribbean **archipelago,** or group of islands, is called the West Indies. It sweeps, like the tail of a cat, from Florida to the northeast coast of South America.

The islands of the Bahamas lie southeast of Florida. The rest of the northern Caribbean area, or the Greater Antilles, includes the large islands of Jamaica, Cuba, Hispaniola (HIHS•puh•NYOH•luh), and Puerto Rico. The Lesser Antilles includes a number of smaller islands rising from the blue waters of the southern Caribbean.

Place
The Land

If you flew above the clouds over a mountain range, what would you see? Probably only the tops of mountains would be visible. That is what you see when you're looking at the islands of the West Indies—the tops of mountains.

Many islands are really an underwater chain of mountains, some of which are active volcanoes. Others are limestone mountains pushed up from the ocean floor by pressures under the earth's crust.

Landforms The islands of the West Indies cover an area of about 90,699 square miles (234,909 sq. km). Most Caribbean islands have central highlands that slope down to the sea. Strips of fertile coastal plain line these coasts.

Look at the map on page 174 to see how the West Indies islands vary in size. Cuba, the largest island, covering about 42,400 square miles (109,816 sq. km), is the size of the state of Ohio. Anguilla (an•GWIH•luh) is one of the smallest islands, just 34 square miles (88 sq. km). It is about half the size of Salt Lake City, Utah.

Climate The picture-postcard view of coconut palms and sunny beaches in the West Indies can be enjoyed year-round. The islands of the West Indies lie in the tropics, and most enjoy a fairly constant tropical savanna climate. Cool northeast breezes sweep across the Caribbean Sea, keeping temperatures between 70°F (20°C) and 85°F (30°C) and bringing gentle rains.

CENTRAL AMERICA AND THE WEST INDIES: Climate

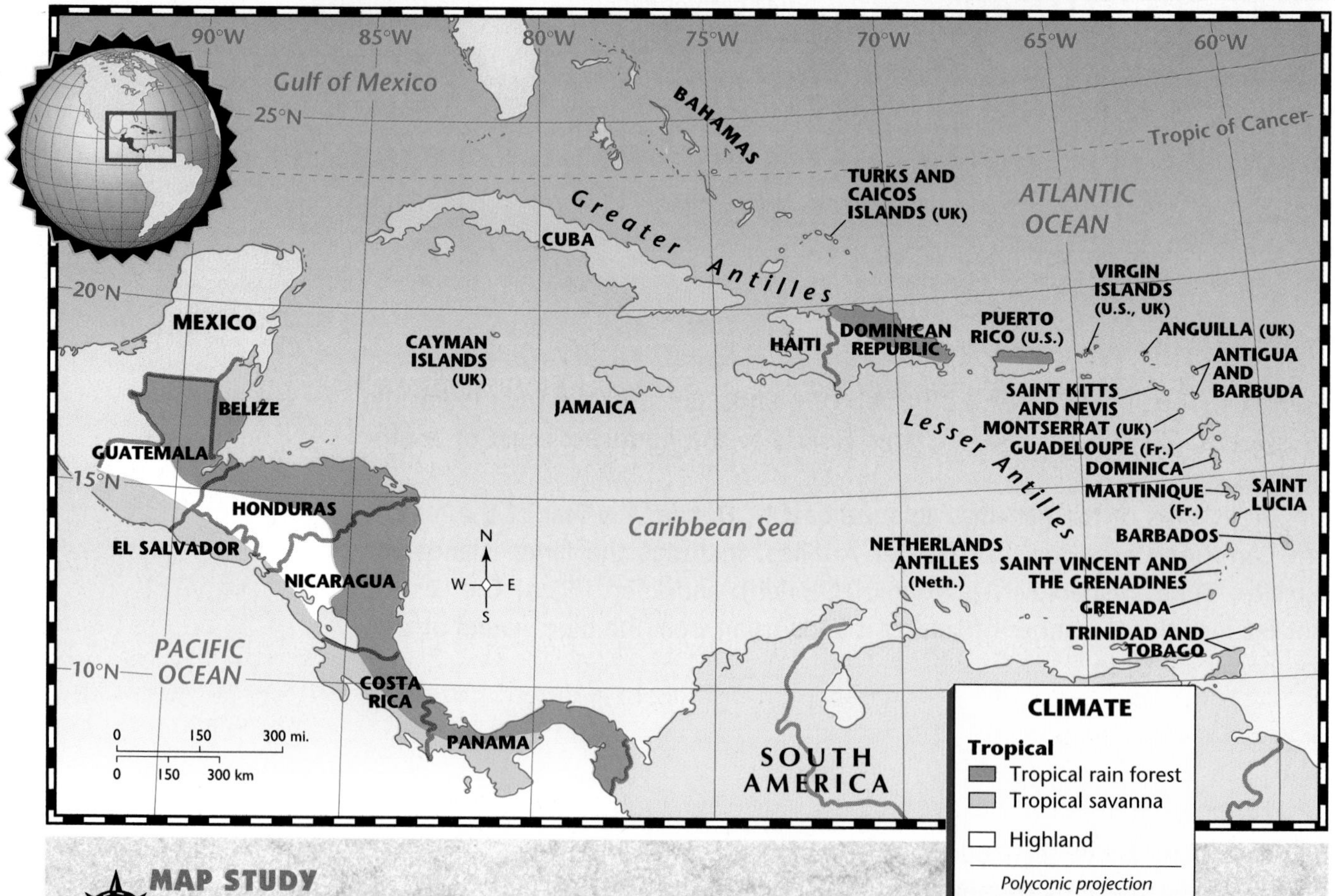

CLIMATE

Tropical
- Tropical rain forest
- Tropical savanna
- Highland

Polyconic projection

MAP STUDY
Most of the islands of the West Indies lie south of the Tropic of Cancer.
PLACE: How is the climate of the West Indies similar to the climate of Central America? How is it different?

Caribbean rains are not *always* gentle, however. For more than half of the year, hurricanes batter the West Indies. The word "hurricane" comes from the Taíno people of Cuba, who worshiped a god of storms named Hurakan.

The Economy

Farming and tourism are the most important economic activities in the West Indies. Many people, however, cannot find jobs or have to work for low pay. Governments in the West Indies are trying to develop new industries that will bring more jobs to the region.

One common economic problem is that many islands have a farm economy based on only one crop. Sugar, for example, makes up 75 percent of the exports of the islands of Saint Kitts and Nevis. If an island's major crop fails, or if too much sugar is produced and prices fall, the economy is in serious trouble.

Island Life

Farming Wealthy West Indian plantation owners grow crops such as sugarcane, bananas, coffee, and tobacco for export. Many full-time laborers are needed to work the plantations. Look at the map on page 178 to see where plantation crops are grown. The map also shows that some West Indians are subsistence farmers who own or rent small plots of land.

Tourism A warm climate, beautiful beaches, and friendly people draw large numbers of tourists to the West Indies. Airlines and cruise ships from the United States and Europe make regular stops in the Caribbean region. The service industries created by tourism provide jobs and a strong base for the economies of many island nations.

Mining, Manufacturing, and Trade Mining and manufacturing generally play a small role in the economies of the West Indies. Some island countries, however, do contain mineral resources. Find these areas on the map on page 178. Jamaica, for example, is a leading world producer of bauxite, used to make aluminum, and Trinidad exports crude oil.

Trading ships from Europe and the rest of Latin America have sailed to the West Indies since the late 1500s. The building of the Panama Canal also brought trade and commerce to the region.

Workers harvest sugarcane in the Dominican Republic *(left).* Trinidad's annual carnival attracts many tourists *(right).* **HUMAN/ENVIRONMENT INTERACTION: What are the two most important economic activities in the West Indies?**

Map

CITIES
- ■ City with more than 5,000,000 people
- ● City with 1,000,000 to 5,000,000 people
- ○ City with 500,000 to 1,000,000 people

Gulf of Mexico

25°N

ATLANTIC OCEAN

BAHAMAS

Tropic of Cancer

Havana

Greater Antilles

CUBA

TURKS AND CAICOS ISLANDS (UK)

20°N

MEXICO

CAYMAN ISLANDS (UK)

VIRGIN ISLANDS (U.S., UK)

DOMINICAN REPUBLIC

PUERTO RICO (U.S.)

ANGUILLA (UK)

HAITI

Port-au-Prince

San Juan

ANTIGUA AND BARBUDA

Santo Domingo

SAINT KITTS AND NEVIS

JAMAICA

MONTSERRAT (UK)

GUADELOUPE (Fr.)

BELIZE

DOMINICA

GUATEMALA

15°N

HONDURAS

Lesser Antilles

MARTINIQUE (Fr.)

SAINT LUCIA

Guatemala

Tegucigalpa

Caribbean Sea

BARBADOS

EL SALVADOR

NICARAGUA

NETHERLANDS ANTILLES (Neth.)

SAINT VINCENT AND THE GRENADINES

Managua

GRENADA

PACIFIC OCEAN

10°N

TRINIDAD AND TOBAGO

San José

SOUTH AMERICA

COSTA RICA

Panamá

PANAMA

N W E S

0 150 300 mi.
0 150 300 km

Persons per

sq. mi.	sq. km
Uninhabited	Uninhabited
Under 2	Under 1
2-60	1-25
60-125	25-50
125-250	50-100
Over 250	Over 100

Polyconic projection

MAP STUDY

In Central America more people live on the Pacific coast than on the Atlantic coast.

PLACE: Which island nation is the most densely populated?

Movement

The People

The cultures of Africans, Native Americans, Europeans, and Asians have mixed and mingled to create the cultures of the West Indies. Why do you find so many groups represented here? Think about the location of the islands. Over the centuries, a variety of peoples have passed through and settled in this region.

Influences of the Past When Christopher Columbus sailed into the harbor of Hispaniola in 1492, who met him? As you might think, it was a Native American group—the Taíno. At their first meal together, a dish made of vegetables and meat called pepper pot stew was probably served.

The first permanent European settlement in the Western Hemisphere was established in 1498 by the Spanish in the present-day Dominican Republic.

186

UNIT 3

During the next 200 years, the Spanish, the English, the French, and the Dutch also founded **colonies,** or overseas settlements, on many of the islands.

Forced Labor The Europeans forced the peoples of the islands to farm the land and to work in the mines. After most Native Americans died from European diseases or harsh treatment, the Europeans turned to Africa for workers. They purchased Africans and shipped them to the islands to work the sugar plantations. Most of the Caribbean population today traces its ancestry to Africa.

An upset in the Caribbean plantation system took place in the mid-1800s. World sugar prices collapsed, and many plantations closed. After that, the ruling European governments passed laws that ended the enslavement of workers.

Plantation owners still in need of workers brought them from Asia, particularly India. The Asians agreed to work a set number of years in return for free travel to the West Indies. Today large Asian populations live in Trinidad, Jamaica, and other islands.

Independence During the twentieth century most of the smaller Caribbean islands won their freedom from colonial rule. Today some countries of the West Indies are republics. Others are constitutional monarchies.

Cuba is the only country in the Western Hemisphere with a government based on the ideas of **communism.** Communism favors strong government control of the society and the economy. Some Caribbean islands still have close links to European nations or to the United States.

A Mix of People About 37 million people live in the West Indies. Find the island of Cuba on the map on page 186. With more than 11.1 million people, Cuba is the most heavily populated island in the West Indies. Five of the Caribbean's smallest countries—Antigua and Barbuda, Dominica, Grenada, Saint Lucia, and Saint Vincent and the Grenadines—have only about 100,000 people each.

Most West Indians have added some aspects of European cultures to their own. Many have adopted European languages and the Christian religion. Some have mixed them with their Native American and African cultures.

Culture Music and dance are important to the cultures of the Caribbean islands. The bell-like tones of the steel drum, a musical instrument developed in Trinidad, are part of the rich musical heritage of the region. Enslaved Africans created a kind of music called calypso. Its strong rhythms have spread around the world.

Daily Life Rajo Hassanali, a young woman of Asian descent, lives in Trinidad. When Rajo is not in school, she works on a nearby cocoa plantation. Many of her friends travel to the cities to work in hotels, restaurants, or shops. During their free time the young people enjoy baseball, basketball, and soccer. Urban areas of the West Indies are home to about 60 percent of the people. The other 40 percent of the population live and work in the countryside.

Place
Caribbean Islands

The islands of the West Indies share many similarities and differences. Some of these differences can be seen in Cuba, Haiti, Jamaica, and Puerto Rico.

Historic Landmark in Haiti

Haitian ruler Henri Christophe had this fortress built in the early 1800s to protect the island from invaders.
PLACE: What country ruled Haiti until 1804?

Cuba One of the world's top sugar producers, Cuba lies about 90 miles (about 140 km) south of Florida. About 50 percent of the Cuban population are black, and another 25 percent are of mixed African and Spanish ancestry. Most Cuban farmers work on **cooperatives,** or farms owned and operated by the government. In addition to sugarcane, they grow coffee, tobacco, rice, and fruit. In Havana, Cuba's capital, workers in **light industries** produce food products, cigars, and household goods.

Currently, Cuba is a communist state led by dictator Fidel Castro. Cuba won its independence from Spain in 1898. A revolution led by Castro in 1959 set up a communist government. Opposed by the United States, Castro turned to the Soviet Union for support. When the Soviet Union broke up in 1989, it withdrew its support from Castro. Since then the Cuban economy has collapsed, and many Cubans live in poverty.

Haiti Haiti shares the island of Hispaniola with the Dominican Republic. More than 90 percent of Haiti's 7.5 million people are of African ancestry. Power struggles and civil war have left Haiti's economy in ruins. Haiti's people are generally poor and live in rural areas. Coffee, the major export crop, is shipped through Port-au-Prince, the country's capital.

Haiti won its independence from France in 1804. It was the second independent republic in the Western Hemisphere (after the United States). During most of the 1900s, dictators ruled Haiti. In 1990 Haitian voters elected Jean-Bertrand Aristide (zhawn behr•trahn ah•rih•STEED) as a democratic president. Military leaders, however, forced Aristide out of the country. With the help of the United States, Aristide finally returned to power in 1994. Two years later, Aristide completed his term, and René Préval was elected president.

Jamaica What languages would you expect to hear if you visited Jamaica? If you said English, you are correct. But other languages such as Creole, a mix of English and French, are also spoken there. Some Jamaicans also speak **dialects,** or local forms of a language, such as patois.

More than 90 percent of Jamaica's 2.6 million people are of African or mixed African and European backgrounds. The other 10 percent are of Asian ancestry.

Once a British colony, Jamaica became independent in 1962. About 52 percent of the Jamaican population live in urban areas. Kingston is the country's capital and largest city.

This island country is known for its misty blue mountains and sunny tropical beaches. Jamaica's economy relies mainly on tourism. One of the few Caribbean islands with mining interests, the country exports bauxite in addition to bananas, sugar, and coffee.

Puerto Rico To be or not to be . . . a state? That is a question Puerto Ricans ask themselves every few years. Should they officially become a state in the United States? Their last vote on the matter was no.

Today about 2.8 million Americans are of Puerto Rican ancestry. Puerto Rico was a Spanish colony from 1508 to 1898. After the Spanish-American War in 1898, Puerto Rico came under the control of the United States. Since 1952 the island has been a **commonwealth**, a partly self-governing territory, under American protection.

Most of Puerto Rico's 3.7 million people are of Spanish ancestry. Many of them, however, have mixed European, African, and Native American backgrounds. About 75 percent of the Puerto Rican population live in towns and cities. San Juan, the capital and largest city, is home to about 1 million people.

Puerto Rico boasts more industry than any other island in the West Indies. Factories there make chemicals, clothing, medicines, and metal products. Puerto Rico is an island of mountains, rain forests, and coastal lowlands. Agriculture and tourism form the basis of Puerto Rico's economy. Sugarcane and coffee rank as its major crops.

WHAT IN THE WORLD?

Reggae Rhythms

One of Jamaica's most well-known exports is reggae music, a style of music that combines American pop music with African rhythms. Bob Marley, the most famous performer of reggae, lived in Jamaica all of his life. He is a national hero today.

SECTION 2 ASSESSMENT

REVIEWING TERMS AND FACTS

1. Define: archipelago, colony, communism, cooperative, light industry, dialect, commonwealth.

2. PLACE What landforms are typical in the West Indies? How do they influence climate?

3. PLACE What is the capital of Puerto Rico?

4. MOVEMENT What different groups have settled in the West Indies in the last 400 years?

MAP STUDY ACTIVITIES

5. Turn to the climate map on page 184. Why do the West Indies have a tropical climate?

6. Look at the population density map on page 186. Name two of the most densely populated islands in the Greater Antilles.

BUILDING GEOGRAPHY SKILLS

Interpreting an Elevation Profile

You've learned that differences in land elevation are often shown on relief maps. Another way to show elevation is on **elevation profiles**. An elevation profile is a diagram that shows a side view of the land-forms in an area.

Suppose you could slice right through a country from top to bottom and could look at the inside, or *cross section*. The *cross section*, or elevation profile, below pictures the island of Jamaica. It shows how far Jamaica landforms extend below or above sea level.

Use these steps to understand an elevation profile:

• Read the profile title.

• Read the labels to identify the different landforms shown.

• Compare the highest and lowest points in the profile.

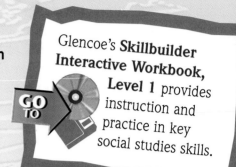

Glencoe's **Skillbuilder Interactive Workbook, Level 1** provides instruction and practice in key social studies skills.

JAMAICA: Elevation Profile

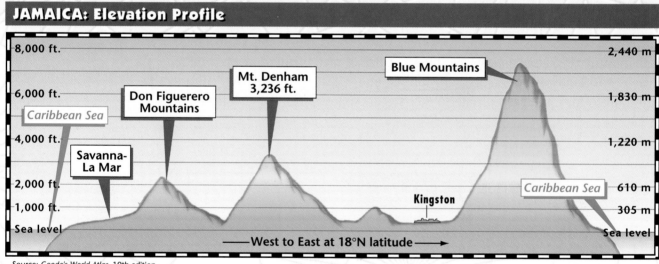

8,000 ft. — 2,440 m

6,000 ft. — 1,830 m

Caribbean Sea

4,000 ft. — 1,220 m

Don Figuerero Mountains

Mt. Denham 3,236 ft.

Blue Mountains

Savanna-La Mar

2,000 ft. — 610 m

Caribbean Sea

1,000 ft. — 305 m

Kingston

Sea level — Sea level

—— West to East at 18°N latitude ——

Source: *Goode's World Atlas*, 19th edition

Geography Skills Practice

1. At about what elevation is Kingston?

2. What are the highest mountains, and where are they located?

3. Where are the lowest regions?

4. Along what line of latitude was this cross section taken?

190

Chapter 7 Highlights

Important Things to Know About Central America and the West Indies

SECTION 1 CENTRAL AMERICA

- Central America extends from Mexico south to South America.
- Central America includes seven countries: Belize, Guatemala, Honduras, El Salvador, Nicaragua, Costa Rica, and Panama.
- The landscape of Central America drops from inland highlands to coastal plains.
- Central America's climate is mostly tropical savanna or tropical rain forest.
- Central America's economy relies on farm products and rain forest resources.
- About 35 million people live in Central America. Guatemala, with 11.6 million people, is the most heavily populated country in the region.

SECTION 2 THE WEST INDIES

- The West Indies consists of two major island groups: the Greater Antilles and the Lesser Antilles.
- The islands of the Bahamas are also part of the West Indies. They lie southeast of Florida.
- Most of the islands of the West Indies have a tropical savanna climate with wet and dry seasons.
- Farming and tourism are the major economic activities of the West Indies.
- The cultures of the West Indies mix African, Native American, and European influences.

Market in Chichicastenango, Guatemala ▶

REVIEWING KEY TERMS

Match the numbered terms in Set A with their lettered definitions in Set B.

A

1. bauxite
2. *ladino*
3. archipelago
4. light industry
5. chicle
6. isthmus
7. dialect

B

A. substance used to make chewing gum
B. local form of a language
C. group of islands
D. mineral used to make aluminum
E. producing household goods
F. Guatemalan who speaks Spanish and follows European ways
G. narrow piece of land connecting two large landmasses

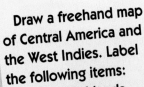

Mental Mapping Activity

Draw a freehand map of Central America and the West Indies. Label the following items:
• Central Highlands
• Pacific Ocean
• Caribbean Sea
• Cuba
• Greater Antilles
• Lesser Antilles
• Isthmus of Panama

REVIEWING THE MAIN IDEAS

Section 1

1. **PLACE** What countries make up Central America?
2. **REGION** How have rain forests boosted the economy of Central America?
3. **MOVEMENT** What European people conquered most of Central America in the 1500s?
4. **HUMAN/ENVIRONMENT INTERACTION** How did the United States change the landscape of Panama?

Section 2

5. **LOCATION** Where are the Bahamas located?
6. **REGION** What is the largest island in the West Indies?
7. **HUMAN/ENVIRONMENT INTERACTION** What two kinds of farms are found in the West Indies?
8. **PLACE** What two islands have large Asian populations?

CRITICAL THINKING ACTIVITIES

1. **Determining Cause and Effect** Why was Panama chosen as a good location to build a canal?
2. **Evaluating Information** Have early groups had a lasting influence on the West Indies? Explain.

CONNECTIONS TO WORLD HISTORY

Cooperative Learning Activity In groups of three, investigate which Caribbean nations were at one time colonies; what European country ruled these colonies; when the colonies gained their independence; and how these countries celebrate their independence days. Each member of the group can research a different group of islands. Present your findings to the rest of the class. Try to include at least one visual element.

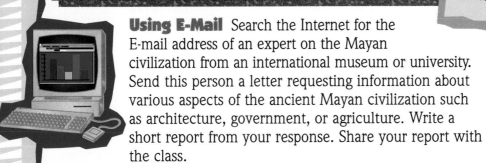

GeoJournal Writing Activity

Imagine you are a Costa Rican teenager with a pen pal in the United States. Write a letter to your pen pal describing what you would most like to know about a teenager's life in the United States.

TECHNOLOGY ACTIVITY

Using E-Mail Search the Internet for the E-mail address of an expert on the Mayan civilization from an international museum or university. Send this person a letter requesting information about various aspects of the ancient Mayan civilization such as architecture, government, or agriculture. Write a short report from your response. Share your report with the class.

PLACE LOCATION ACTIVITY: CENTRAL AMERICA AND THE WEST INDIES

Match the letters on the map with the places of Central America and the West Indies. Write your answers on a separate sheet of paper.

1. Costa Rica
2. Trinidad and Tobago
3. Panama Canal
4. Cuba
5. Atlantic Ocean
6. Caribbean Sea
7. Port-au-Prince
8. Guatemala
9. Puerto Rico

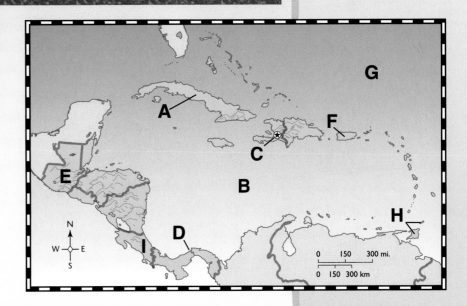

8 Brazil and Its Neighbors

National boundary
National capital
Other city

Goode's Interrupted
Homolosine projection

80°W · 70°W · 60°W · 50°W · 40°W

10°N

Maracaibo

Caracas

VENEZUELA · GUYANA · SURINAME

ATLANTIC OCEAN

Georgetown · Paramaribo

Cayenne

FRENCH GUIANA (Fr.)

0°

SOUTH

Equator

10°S

BRAZIL

Recife

AMERICA

Brasília

20°S

Tropic of Capricorn

PARAGUAY

Rio de Janeiro

PACIFIC OCEAN

Asunción

São Paulo

30°S

URUGUAY

Montevideo

N
W E
S

0   200   400 mi.
0   200   400 km

MAP STUDY ACTIVITY

In this chapter you will read about Brazil, the largest country in South America, and its neighbors.

1. What is the capital of Brazil?
2. What other countries border this huge country?

Brazil

Words to Know

- basin
- *selva*
- escarpment
- sisal
- *favela*

Places to Locate

- Amazon River
- Paraná River
- São Francisco River
- Brazilian Highlands
- Great Escarpment
- São Paulo
- Rio de Janeiro
- Brasília

Read to Learn . . .

1. where Brazil is located.
2. what landforms and climates are found in Brazil.
3. what natural resources Brazil's economy depends on.

Imagine standing inside this photograph of the Amazon rain forest—the largest rain forest in the world. Can you hear this parrot chatter? Do you feel the stickiness of the steamy, tropical air? Lush flowers and trees surround you so thickly that you can't see the sky!

The Amazon rain forest covers the northern half of South America. Flowing through this vast, flat region is the mighty Amazon River. What country lays claim to this tropical beauty? The Amazon River and its surrounding rain forest are located mostly in the country of Brazil.

Region

The Land

The largest country in South America, Brazil has a land area of 3,265,060 square miles (8,456,505 sq. km). Because it covers such a large area, Brazil encompasses many types of landforms. The map on page 196 shows you that Brazil has lowland river valleys as well as highland areas and coastal plains.

Rivers and Lowlands The Amazon is the world's second longest river, winding almost 4,000 miles (6,436 km) from the Andes to the Atlantic Ocean. On its journey to the ocean, the Amazon passes through a wide, flat basin. A **basin** is a low area entirely surrounded by higher land. Most of the Amazon basin is covered with thick tropical forests, which Brazilians call the *selva*.

In addition to the Amazon basin, Brazil has two other lower land areas. They are located along the Paraná River and the São Francisco River. Trace these rivers on the map below. Both rivers begin in the center of Brazil, but they flow in opposite directions. The Paraná flows to the southwest, and the São Francisco flows to the northeast.

Highlands Don't picture Brazil as one big rain forest. Highland areas, in fact, cover more than one-half of the country. The Brazilian Highlands—the largest highland area—begin south of the Amazon River basin and cover much of east central Brazil. Low mountain ranges in the highlands drop sharply to the Atlantic Ocean, a drop called the Great Escarpment. An **escarpment** is a steep cliff between a higher and lower surface.

The Climate Brazil's climates are as varied as its landforms. If you like steam baths, you'd love living in the Amazon basin. With its tropical rain forest climate, the basin has hot, steamy temperatures and torrents of rain year-round. In most of Brazil's highland areas, you find a tropical savanna climate, with wet

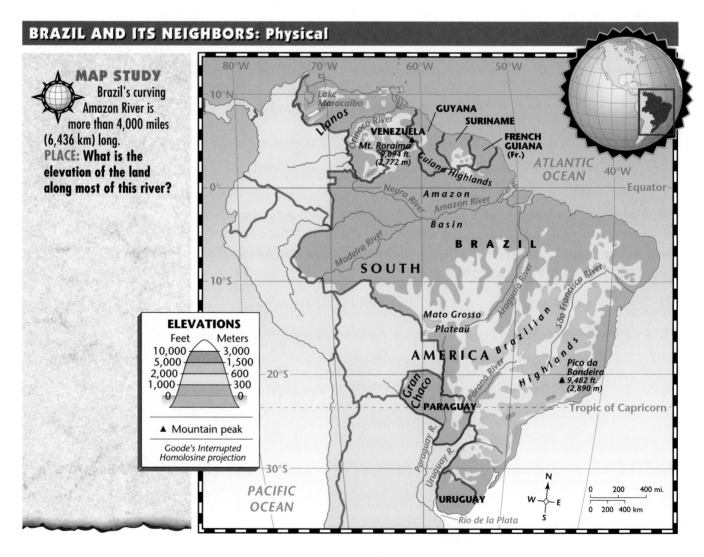

BRAZIL AND ITS NEIGHBORS: Physical

MAP STUDY
Brazil's curving Amazon River is more than 4,000 miles (6,436 km) long.
PLACE: What is the elevation of the land along most of this river?

ELEVATIONS
Feet	Meters
10,000	3,000
5,000	1,500
2,000	600
1,000	300
0	0

▲ Mountain peak

Goode's Interrupted Homolosine projection

and dry seasons. For those who prefer a drier climate, moderate temperatures prevail in the southern part of the country year-round.

Region
Economic Regions

Brazil's diverse economy includes agriculture, manufacturing, and service industries. Agriculture has been important in the country for centuries, and rapid industrial growth during this century has made Brazil the leading manufacturing nation in South America.

Setting the Style

In Brazil traditional clothing styles differ from region to region. In the northeast, women wear colorful skirts, bright blouses, and much jewelry. In the south, cowhands wear baggy pants, felt hats with wide brims, and ponchos, or sleeveless coverings.

The North For many years the Amazon rain forest was a mysterious region whose secrets were guarded by the Native Americans living there. In recent years, however, new roads have allowed more people to settle in the north. The map on page 204 reveals that the north is rich in mahogany, teak, and other forest woods. The area also contains minerals such as bauxite and iron ore. As a result, the Brazilian government has encouraged logging and mining in the north.

Many people worry that the increasing use of forest resources is destroying the rain forest and its wildlife. They also fear that economic development threatens the survival of Native Americans still living in the area. Turn to page 236 to learn more about the destruction of the rain forests.

The Northeast In the northeast region, farmers and ranchers have cleared coastal rain forests to grow crops and raise cattle. Sugarcane, cotton, cacao, and **sisal**—a plant fiber used to make rope—are this region's major farm products. Overgrazing in this dry area has ruined much of the land, however.

The Southeast The southeast is rich in mineral resources and fertile farmland. This region boasts one of the largest iron-ore deposits in the world, and the area also supplies coffee drinkers with their favorite beverage. The graph on page 198 shows that Brazil is the world's leading producer of coffee.

The towering forests of the north are matched by the towering skyscrapers of the southeast. Here stand Brazil's major cities and centers of industry. São Paulo (sown POW•loo) is home to more than 21 million people, making it one of the fastest-growing urban areas in the world. It is also Brazil's leading trade and industrial center. Rio de Janeiro (REE•oh day zhuh•NEHR•oh) with a population of about 13 million, is a favorite tourist spot. Tourists flock to the city for its beautiful coastline, sandy beaches, and colorful festivals.

The South The south's vast plains support huge herds of cattle. Brazilian beef is exported all over the world. Another product of the south is *yerba maté,* a tea-like drink that is popular throughout the southern part of South America. *Yerba maté* is made from the leaves of holly trees that grow in the region.

Maracana Stadium in Rio de Janeiro is the world's largest stadium.
PLACE: What is the Brazilian word for soccer?

West Central Region Inland highlands and plateaus cover the west central region, which is very isolated and has few settlers. The soil here is not fertile, and the land has been overgrazed. The Brazilian government hopes to encourage better farming practices that will boost the economy of this region.

Movement
People

With more than 162 million people, Brazil has the largest and fastest-growing population of any Latin American nation. Unlike most of the countries in South America, Brazil has a culture that is largely Portuguese rather than Spanish. The Portuguese were the first and largest European group to colonize Brazil. Today Brazilians are of European, African, Native American, Japanese, or mixed ancestry. Brazilians speak their own version of the Portuguese language.

Influences of the Past Native Americans were the first people to live in Brazil. In the 1500s the Portuguese took control of the northeast and other coastal areas. They forced Native Americans to work on sugar plantations or in mines. Many Native Americans died from disease or overwork. Needing workers, the Portuguese traders purchased Africans and then shipped them to

Leading Coffee-Producing Countries

GRAPHIC STUDY

Brazil's warm, moist climate and fertile volcanic soil make it the world's leading coffee producer.
PLACE: What other Latin American countries are among the top seven coffee-producing countries?

🍵 = .25 billion pounds

Coffee Beans Produced in One Year

Billions of pounds / Billions of kilograms

Brazil · Colombia · Indonesia · Mexico · Guatemala · Ethiopia · Uganda

Source: *World Book,* 1998

Brazil. Enslavement finally ended in 1888, but many Africans remained. Today about 6 percent of Brazil's population are of purely African ancestry. African traditions and customs have influenced Brazil's religions, music, dance, and foods.

In 1822 Brazil became a monarchy after winning its independence from Portugal. About 70 years later it became a republic. During the twentieth century Brazil's government has seesawed between dictatorship and democracy.

Brazilians Today Look at the map on page 209. You see that most of Brazil's people live along the country's Atlantic coast. The Brazilian government has tried to encourage people to move from these crowded coastal areas to less populated inland areas. In the early 1960s Brazil moved its capital from the coastal city of Rio de Janeiro to the newly built inland city of Brasília. With about 2 million people, Brasília is a modern and rapidly growing city.

Most rural Brazilians live in villages of one- or two-room houses made of stone or brick. Some work on plantations or ranches, and others cultivate subsistence farms. About 76 percent of Brazil's people live in cities. Many city dwellers are very poor and live in *favelas*, or slum areas. Others have good jobs in industry and government and live in city apartments or suburban houses.

Antonio Vargas is a Brazilian teenager who lives in Brasília. He enjoys showing visitors Brasília's modern government buildings and university. But Antonio often visits his cousins in Rio de Janeiro. Rio, as it is usually called, is said to be one of the world's most beautiful cities. Buildings from Brazil's early history stand side by side with modern skyscrapers. Busy downtown streets lead to white, sandy beaches along Guanabara Bay.

Brazilian Sports and Celebrations It is safe to say that *fútbol,* or soccer, is a way of life in Brazil. Every village has some kind of soccer field, and the larger cities have stadiums. In Rio de Janeiro, the stadium can seat 220,000 fans! The most famous soccer player in the world, Pelé, is from Brazil, and the country frequently competes for the World Cup.

Brazilians enjoy many celebrations. The festival known as Carnival is celebrated just before the beginning of Lent, the Christian holy season that comes before Easter. The most spectacular Carnival is held each year in Rio de Janeiro.

SECTION 1 ASSESSMENT

REVIEWING TERMS AND FACTS

1. Define the following: basin, *selva,* escarpment, sisal, *favela.*

2. REGION What kinds of landforms and climates does Brazil have?

3. PLACE How does city life in Brazil differ from rural life?

MAP STUDY ACTIVITIES

4. Look at the political map on page 194. How far from Rio de Janeiro is Brasília?

5. Study the physical map on page 196. How does northern Brazil's landscape differ from southern Brazil's?

BUILDING GEOGRAPHY SKILLS

Reading a Political Map

Think about the last sports event you participated in or watched. Boundaries—the outer limits of the playing area—are set up by the rules of the sport. In the same way, political areas such as cities, states, and countries have boundaries. People living within these borders follow a common set of rules or laws.

Political maps show these boundaries. Lines represent the boundaries between political areas. These areas are often shown in contrasting colors. Political maps also include cities and important physical features. To read a political map, apply these steps:

- Read the map title.
- Identify the political areas on the map.
- Locate the major cities.

GO TO Glencoe's **Skillbuilder Interactive Workbook, Level 1** provides instruction and practice in key social studies skills.

VENEZUELA: Political

- —— National boundary
- ⊛ National capital
- ● Other city

Azimuthal Equal-Area projection

0   75   150 mi.
0   75   150 km

Geography Skills Practice

1. What countries are shown on this map?

2. Name the capital city and three other major cities of Venezuela.

3. Where are most of the cities located?

4. How would you explain why cities are located in these areas?

Caribbean South America

PREVIEW

Words to Know
- *llanos*
- hydroelectric power
- altitude
- *caudillo*

Places to Locate
- Venezuela
- Caribbean Sea
- Andean Highlands
- Caracas
- Lake Maracaibo
- Orinoco River
- Guyana
- Suriname
- French Guiana

Read to Learn . . .
1. what landscapes and climates are found in Caribbean South America.
2. what early groups influenced Caribbean South America.
3. how countries in Caribbean South America use their resources.

A ngel Falls—the highest waterfall in the world— roars down a cliff in a highland area of the country of Venezuela. If you stood at the bottom of the waterfall, you'd strain your neck as you gazed 3,212 feet (979 m) up to the top!

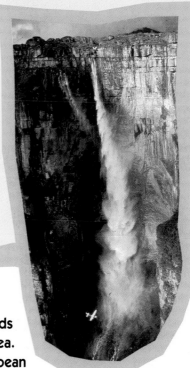

V enezuela is located in Caribbean South America. This region extends across the northern part of South America along the Caribbean Sea. Besides Venezuela, two other countries and a territory border the Caribbean Sea: Guyana, Suriname, and French Guiana.

Place
Venezuela

Venezuela is the largest country in Caribbean South America. Its boundaries touch the countries of Brazil, Colombia, and Guyana. The waters of the Caribbean Sea and the Atlantic Ocean wash its shores.

The Land What types of landforms would you see in Venezuela if you could look down on it from space? Looking northwest you would see the Maracaibo (MAR•uh•KY•boh) basin. This lowland coastal area surrounds Lake Maracaibo, the largest lake in South America. To the east you would see a hilly area known as the Andean Highlands—home to most of Venezuela's people.

In the south you would see the Guiana Highlands with dense rain forests and very few people. Grassy plains known as the ***llanos*** lie between the Andean and Guiana highlands. The *llanos* have many ranches, farms, and oil fields. Venezuela's most important river—the Orinoco—flows across the *llanos*. It is a valuable source of **hydroelectric power**, or water-generated electricity, for Venezuela's cities.

The Climate Because it is close to the Equator, Venezuela has mostly a tropical rain forest climate. Cool, dry winds from the northeast, however, keep temperatures moderate. As in Mexico, temperatures in Venezuela also differ with **altitude**, or height above sea level. The lowland Maracaibo basin and inland river valleys are hot and rainy. As you travel up into the highland areas, you are usually warm in the daytime but cool at night.

The Economy Venezuelans once depended on crops such as coffee and cacao to earn their livings. Since the 1920s, though, petroleum has changed all that. Today Venezuela, a world leader in oil production, is one of the wealthiest countries in South America. Bauxite, coal, diamonds, and gold also are mined there.

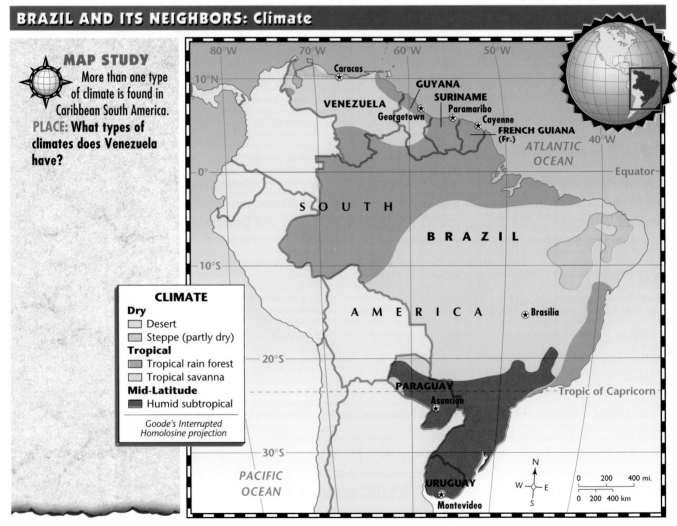

BRAZIL AND ITS NEIGHBORS: Climate

MAP STUDY
More than one type of climate is found in Caribbean South America.
PLACE: What types of climates does Venezuela have?

CLIMATE
Dry
◻ Desert
◻ Steppe (partly dry)
Tropical
◻ Tropical rain forest
◻ Tropical savanna
Mid-Latitude
◼ Humid subtropical

Goode's Interrupted Homolosine projection

202

UNIT 3

About 90 percent of Venezuela's people work in service industries or manufacturing. Agriculture, however, remains an important part of the country's economy. Landowners and ranchers export beef and dairy cattle. Smaller farms grow bananas, coffee, corn, and rice.

The Past Spanish explorers gave Venezuela its name. With its rivers and natural canals, the country reminded them of Venice, Italy. Like many other Latin American nations, Venezuela was a Spanish colony from the early 1500s to the early 1800s. In 1821 South Americans rejoiced as Simón Bolívar (see•MOHN buh•LEE•VAHR) and his soldiers freed the northern part of the continent from Spanish rule. Nine years later Venezuela became an independent republic. During most of the 1800s and 1900s, the country was governed by harsh military rulers known as *caudillos.* But since the 1960s Venezuela has developed into a stable democracy.

The People Most of Venezuela's 23.3 million people are a mix of European, African, and Native American backgrounds. Spanish is the major language of the country; the major religion is Roman Catholicism.

A remarkable 86 percent of the population of Venezuela live in cities. María González, a teenager, is among the 3.2 million people who live in Caracas (kuh•RAH•kuhs), Venezuela's capital and largest city. Traveling on freeways, her school bus passes towering skyscrapers on its daily route. Like many people in Caracas, María and her family live in a high-rise apartment. Because of oil, the people of Venezuela enjoy one of the highest standards of living in South America.

Venezuela's modern capital is nestled in a valley surrounded by mountains.
PLACE: What natural resource gives Venezuelans one of the highest standards of living in South America?

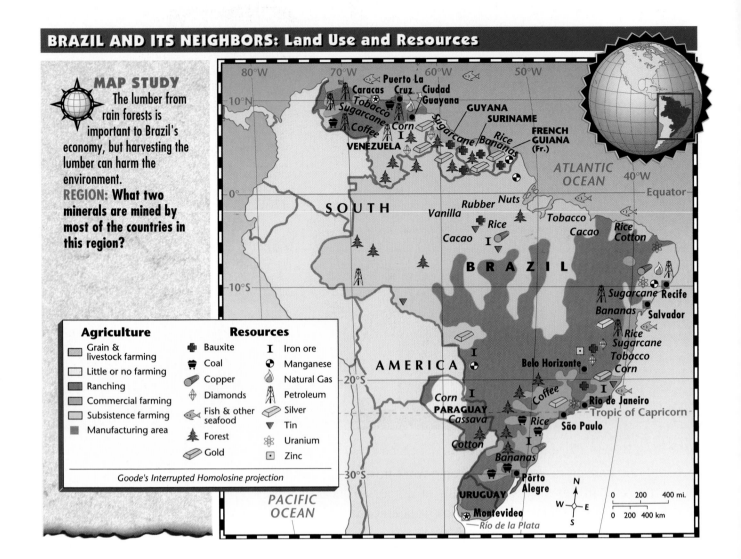

MAP STUDY

The lumber from rain forests is important to Brazil's economy, but harvesting the lumber can harm the environment.

REGION: What two minerals are mined by most of the countries in this region?

Agriculture

- Grain & livestock farming
- Little or no farming
- Ranching
- Commercial farming
- Subsistence farming
- Manufacturing area

Resources

- Bauxite
- Coal
- Copper
- Diamonds
- Fish & other seafood
- Forest
- Gold
- Iron ore
- Manganese
- Natural Gas
- Petroleum
- Silver
- Tin
- Uranium
- Zinc

Goode's Interrupted Homolosine projection

Place
The Guianas

In addition to Venezuela, Caribbean South America includes the countries of Guyana (gy•A•nuh), Suriname (sur•uh•NAH•muh) and the territory of French Guiana (gee•A•nuh).

Guyana Guyana lies just north of the Equator. About the size of the state of South Dakota, Guyana's high inland plateau is covered by thick rain forests. Closer to the coast the land gradually descends to a low flatland. The climate of Guyana is largely tropical rain forest.

When the British arrived in Guyana in the early 1800s, they forced the Native Americans and, later, Africans to work on sugarcane plantations. Asians came and worked voluntarily. Guyana finally won its independence in 1966. Today about one-half of Guyana's population are of Asian ancestry. Another one-third are of African ancestry. The rest of Guyana's people are Native Americans and Europeans. Most live along the Caribbean coast where Georgetown, the capital, is the major city.

Guyana's past still influences its economy. The map on page 204 shows you that one of Guyana's major products is sugar. Other people earn their living mining gold and bauxite, or working in urban service industries.

Suriname Suriname ranks as the smallest independent country in South America in both area and population. Native Americans were the first settlers. In the late 1600s the region came under Dutch rule. The Dutch bought African people—and later brought in Asians—to work on their large plantations. In 1975 Suriname won its freedom from the Dutch. Most of Suriname's people live along the coast. Paramaribo (PAR•uh•MAR•uh•BOH) is the capital and chief port.

Suriname's economy depends on agriculture and mining. Rice is the country's largest crop, followed by bananas, coconuts, and sugar. Suriname also mines bauxite and gold. The ring of buzz saws throughout the thick rain forests signals a thriving lumber industry.

French Guiana To the east lies French Guiana with a low coastal plain and inland rain forests. The climate inland is hot and humid. Cooling ocean winds however, keep coastal areas comfortable at about 80°F (27°C).

The French settled this area during the early 1600s. They, too, founded a colony of sugarcane plantations worked by enslaved Africans. The colony also served as a place for French prisoners from the 1790s to the 1940s.

Today French Guiana is considered part of France. The French government builds hospitals and schools in the capital, Cayenne (ky•EHN), and other cities. It also provides government jobs for many of French Guiana's people. Most are of African or mixed African and European ancestry.

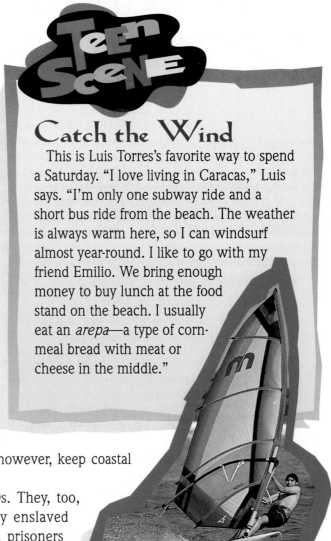

Catch the Wind

This is Luis Torres's favorite way to spend a Saturday. "I love living in Caracas," Luis says. "I'm only one subway ride and a short bus ride from the beach. The weather is always warm here, so I can windsurf almost year-round. I like to go with my friend Emilio. We bring enough money to buy lunch at the food stand on the beach. I usually eat an *arepa*—a type of cornmeal bread with meat or cheese in the middle."

SECTION 2 ASSESSMENT

REVIEWING TERMS AND FACTS

1. Define the following: *llanos,* hydroelectric power, altitude, *caudillo.*

2. LOCATION Why are Brazil's northern neighbors known as Caribbean South America?

3. HUMAN/ENVIRONMENT INTERACTION What resource has changed Venezuela's economy?

4. PLACE What European countries have influenced the Guianas?

MAP STUDY ACTIVITIES

5. Look at the land use map on page 204. What are the major resources of Caribbean South America?

MAKING CONNECTIONS

Latin American author Victor Villaseñor's grandparents grew up in Mexico during the early 1900s. He wrote his family history in a novel called *Rain of Gold*. In this passage, Villaseñor tells about the girlhood journey of his grandmother, Lupe, through the country-side of northern Mexico.

Copper Canyon

Lupe came to the short new pines that had grown after the white pine forest had been burnt by the meteorite. The sound of the waterfalls was devastating. Carefully, Lupe continued through the young white pines and went out through the loose rock break on the north rim of the canyon.

. . . Suddenly, passing through the break, the whole world opened up. She was above the mountain peaks and flat mesas piled up for hundreds of miles in every direction. There was beauty to the left, beauty to the right. Beauty surrounded Lupe. This was the high country of northwest Mexico, unmapped and uncharted. Some fingers of the canyon of La Barranca del Cobre [Copper Canyon] ran deeper than the Grand Canyon of Arizona.

Following one of the little waterways, Lupe crossed the meadows of tightly woven wild-flowers of blue and yellow and red and pink.

It was quiet up here without the waterfall's terrible roar. Her brother's little dog flushed out deer and mountain quail as they went along.

Crossing a tiny creek, Lupe saw the fresh tracks in the cold mud of the terrible jaguar. The little dog's hair came up on his back. Lupe petted him, looking around carefully, but saw nothing. Jaguars, after all, were fairly common, and so people were more respectful than afraid of them, just as they were of any natural force.

Excerpt from *Rain of Gold* by Victor Villaseñor is reprinted with permission from the publisher (Houston: Arte Publico Press—University of Houston, 1991)

Making the Connection

1. Describe the countryside that Lupe sees on her journey.
2. The jaguar was an animal that early Native Americans prized highly. Why do you think that was so?

SECTION 3 ·
Uruguay and Paraguay

PREVIEW

Words to Know
- landlocked
- buffer state
- welfare state
- *gaucho*
- cassava

Places to Locate
- Uruguay
- Montevideo
- Río de la Plata
- Paraguay
- Gran Chaco
- Asunción

Read to Learn . . .
1. where Uruguay and Paraguay are located.
2. how the people of Uruguay and Paraguay earn a living.
3. what cultures have influenced Uruguay and Paraguay.

Imagine strolling through this elegant park in Montevideo, the capital of Uruguay. Through the trees, you see modern office buildings. Seagulls from the nearby beaches swoop by.

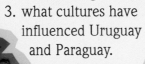

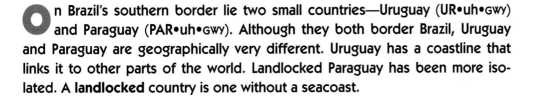

On Brazil's southern border lie two small countries—Uruguay (UR•uh•GWY) and Paraguay (PAR•uh•GWY). Although they both border Brazil, Uruguay and Paraguay are geographically very different. Uruguay has a coastline that links it to other parts of the world. Landlocked Paraguay has been more isolated. A **landlocked** country is one without a seacoast.

Place
Uruguay

Find Uruguay on the map on page 194. See how Brazil borders it on the north? The rest of Uruguay faces water: the Atlantic Ocean, the Río de la Plata, and the Uruguay River.

Uruguay's motto should be "Moderation in All Things." The country's landscape is undramatic, with low-lying interior grasslands and narrow, fertile coastal plains. Its climate is also generally moderate. Most of the country experiences a humid subtropical climate. Uruguay's grasslands, fertile soil, and mild climate help farmers to grow wheat and ranchers to raise livestock.

The Economy The map on page 204 shows you that Uruguay's major economic activities are raising sheep and cattle. The country's leading exports are primarily animal products—wool, beef, and hides. Uruguay also has factories that make tires, textiles, and leather goods. Industry, however, is limited because of Uruguay's lack of mineral resources.

The beautiful beaches of Punta del Este make it a popular resort city.
PLACE: What three bodies of water border southern Uruguay?

Influences of the Past Uruguay won its independence from Spain in 1811. Moderation guides its politics. For most of its history, the country has been a buffer state between Brazil and Argentina. A **buffer state** is a small country located between larger, often hostile, neighbors. Uruguay maintains good relations with Argentina and Brazil.

Today the government of Uruguay spends more on education than it does on defense. It has turned Uruguay into a **welfare state**, a country that uses tax money to provide its people with help if they are sick, needy, or out of work.

The People Uruguay's 3.2 million people are in many ways more European than Latin American. Most of them are European in ancestry. They are descended from Spanish and Italian settlers who came to Uruguay during the 1800s and early 1900s. Only a small number of people have Native American or African ancestry. Spanish is the official language, and the Roman Catholic faith is the major religion. About 1.2 million Uruguayans live in the coastal city of Montevideo (MAHN•tuh•vuh•DAY•OH), the country's capital.

Most of Uruguay's people enjoy a comfortable standard of living. City dwellers hold jobs in government, business, and industry. They live in apartments or single-family houses. Unskilled workers, however, often live in tiny shacks edging the cities.

Only 10 percent of Uruguay's population live in rural areas. Many of them own or rent small farms. Others work as *gauchos*, or cowhands, on huge ranches. Legends about the *gauchos* have inspired Uruguay's folk music and literature.

If you lived in Uruguay you would eat a lot of meat, especially beef. You would enjoy sipping *yerba maté*, a tealike drink, through a straw from a gourd or the dried shell of a fruit. In your spare time, you and other Uruguayans might enjoy soccer and *gaucho* rodeos.

Place
Paraguay

Landlocked Paraguay is located near the center of South America. Two contrasting geographic areas make up Paraguay. The Paraguay River divides Paraguay into east and west regions. Rolling hills, forests, streams, rivers, and a humid subtropical climate make the eastern region a pleasant place to live. Most of Paraguay's people live here.

West of the Paraguay River is a plains region called the Gran Chaco. It covers three-fifths of Paraguay but holds less than 5 percent of its people. Winter droughts and summer floods often affect this area. Grasses, palm trees, and quebracho (kay•BRAH•choh) trees thrive in a tropical savanna climate. Quebracho trees are a source of tannin, a chemical used to process leather.

The Economy Forestry and farming are the major economic activities in Paraguay. Large cattle ranches cover much of the country. Most farmers, however, own or settle on small, fertile plots of land and grow corn, cotton, or cassava.

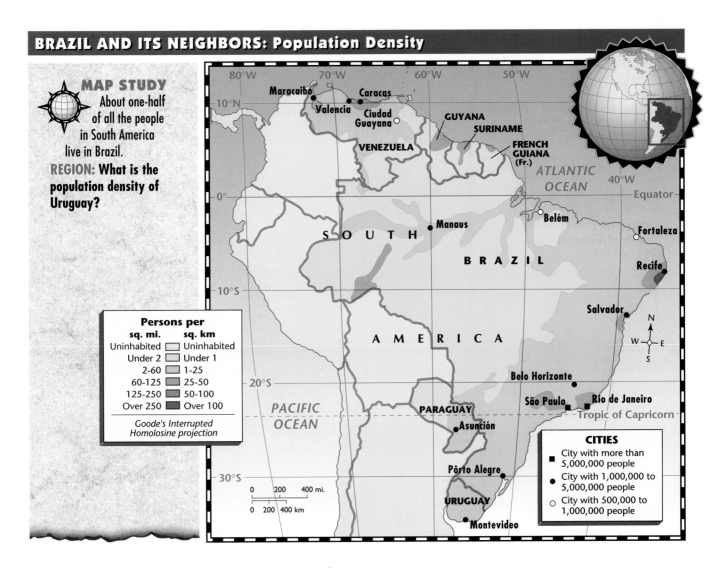

BRAZIL AND ITS NEIGHBORS: Population Density

MAP STUDY
About one-half of all the people in South America live in Brazil.

REGION: What is the population density of Uruguay?

Persons per

sq. mi.	sq. km
Uninhabited	Uninhabited
Under 2	Under 1
2-60	1-25
60-125	25-50
125-250	50-100
Over 250	Over 100

Goode's Interrupted Homolosine projection

CITIES
- ■ City with more than 5,000,000 people
- ● City with 1,000,000 to 5,000,000 people
- ○ City with 500,000 to 1,000,000 people

The ruins of a mission mark the Roman Catholic influence on Paraguay *(left)*. A modern-day gaucho drinks his *yerba maté (right)*.
MOVEMENT: What country sent Roman Catholic missionaries to Paraguay?

Cassava is a plant with roots that can be ground up and eaten or used to make pudding such as tapioca. These roots are very nutritious. You may have eaten pudding made from Paraguay's cassava.

The People A Native American group called the Guaraní were the first people to live in Paraguay. By the 1500s settlers and Roman Catholic missionaries from Spain arrived in the country. Paraguayans today are mostly of mixed Guaraní and Spanish ancestry. They speak two languages—Spanish and Guaraní—and practice the Roman Catholic faith. About one-half of the population live in cities. Asunción (uh•SOONT•see•OHN), with about 1 million people, is the capital and largest city.

Paraguayan arts are influenced by Guaraní culture. Guaraní lace is Paraguay's most famous handicraft. Like their neighbors, the people of Paraguay enjoy meat dishes and sip *yerba maté,* a tealike drink. Although soccer is Paraguay's most popular sport, people also enjoy basketball, horse racing, and swimming.

SECTION 3 ASSESSMENT

REVIEWING TERMS AND FACTS
1. Define the following: landlocked, buffer state, welfare state, *gaucho,* cassava.
2. MOVEMENT Where did most of Uruguay's people come from?
3. PLACE What two languages are spoken in Paraguay?

MAP STUDY ACTIVITIES
4. Look at the population density map on page 209. Where do most of Paraguay's people live?

Chapter 8 Highlights

Important Things to Know About Brazil and Its Neighbors

SECTION 1 BRAZIL

- Brazil is South America's largest country in size and population.
- The Amazon River basin holds the world's largest rain forest.
- Rapid industrial growth during the 1900s has made Brazil the leading manufacturing nation in South America.
- Many people fear that economic development in the Amazon rain forest threatens the forest, its wildlife, and the Native Americans living there.
- The people of Brazil are of European, African, Native American, or mixed backgrounds.

SECTION 2 CARIBBEAN SOUTH AMERICA

- Caribbean South America includes Venezuela, Guyana, Suriname, and French Guiana.
- Oil has made Venezuela one of the wealthiest countries in South America.
- The people of the Guianas trace their heritages to Asian, African, European, and early Native American cultures.

SECTION 3 URUGUAY AND PARAGUAY

- Most of Uruguay's people are of European descent. They live along the coast.
- Paraguay is a landlocked country. Its people speak two languages: Spanish and Guaraní.
- Uruguay's economy depends on livestock raising. Paraguay relies on forestry and farming.

Carnival in Rio de Janeiro, Brazil ▶

REVIEWING KEY TERMS

Match the numbered terms in Set A with their lettered definitions in Set B.

A

1. sisal
2. *favela*
3. landlocked
4. *yerba maté*
5. *gaucho*
6. *selva*
7. *llanos*
8. buffer state
9. c*audillo*

B

A. having no seacoast
B. plains area in Caribbean South America
C. Amazon rain forest
D. small country between two larger, hostile countries
E. military leader
F. slum area in Brazil
G. South American cowhand
H. plant fiber used to make rope
I. tealike drink

Mental Mapping Activity

Draw a freehand map of Brazil, Caribbean South America, Uruguay, and Paraguay. Label the following items:
• Caribbean Sea
• Atlantic Ocean
• Amazon River
• Brasília
• Caracas
• Asunción
• Montevideo

REVIEWING THE MAIN IDEAS

Section 1
1. PLACE About how much of Brazil is highlands?
2. REGION What kind of economy does Brazil have?
3. HUMAN/ENVIRONMENT INTERACTION How has the Brazilian government tried to develop the country's inland areas?

Section 2
4. REGION Where do most of Venezuela's people live?
5. HUMAN/ENVIRONMENT INTERACTION What two resources are mined in the Guianas?

Section 3
6. REGION What two landscapes are found in Paraguay?
7. PLACE In what major way do Paraguay and Uruguay differ in their geography?

CRITICAL THINKING ACTIVITIES

1. **Analyzing Information** Why are some people concerned about Brazil's rain forests?
2. **Drawing Conclusions** Why do you think the Paraguay River is important to the people of landlocked Paraguay?

212

CONNECTIONS TO WORLD CULTURES

Cooperative Learning Activity Working in groups of three, learn more about daily life in Brazil, Caribbean South America, Uruguay, and Paraguay. Each group will choose one country, then each person will select one of the following topics to research: (a) holidays and festivals; (b) the arts; or (c) sports and leisure activities. As a group, prepare a travel brochure to encourage tourists to visit your country.

GeoJournal Writing Activity

Imagine what it would be like to be a rural Brazilian moving to the city of São Paulo. Write a letter home about your experiences in the city.

TECHNOLOGY ACTIVITY

Using the Internet Search the Internet for photographs that provide additional information about the Amazon rain forest. Narrow your search by using words such as *tropical rain forest*, *wildlife*, or *Native Americans*. Print the photographs and write captions describing the characteristics of the rain forest shown in the picture. Display the photographs and their captions on the classroom bulletin board.

PLACE LOCATION ACTIVITY: BRAZIL AND ITS NEIGHBORS

Match the letters on the map with the places of Brazil and its neighbors. Write your answers on a separate sheet of paper.

1. Río de la Plata
2. Caracas
3. Suriname
4. Brasília
5. Uruguay
6. Rio de Janeiro
7. Orinoco River
8. Brazilian Highlands
9. Amazon River

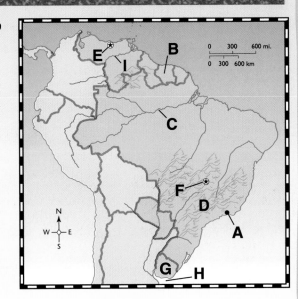

Cultural HERITAGE:
LATIN AMERICA

FABRIC ART ▶ ▶ ▶ ▶

This colorful cloth animal is a *mola*, a type of folk art made by the Cuna Indians of Panama. Designs often include abstract patterns and images of native plants and animals.

◀ ◀ ◀ ◀ MUSIC

Native American and mestizo musicians play many instruments, including the panpipes pictured here. These are similar to instruments used by their ancestors long ago.

METALWORKING ▶ ▶ ▶ ▶

Native American artisans who lived in what is present-day Colombia crafted this gold piece centuries ago. The Native Americans used gold objects for decoration and religious purposes. Today museums house artifacts such as these.

Diego Rivera was a Mexican artist famous for bold murals with large, simplified figures. His works illustrate the culture and history of Mexico. Some of his murals are in the National Palace in Mexico City.

DANCE ► ► ► ►

Many Latin American countries sponsor national dance companies that preserve traditional dances. Dancers such as this with the Ballet Folklórico of Mexico perform colorful productions of Native American and Spanish traditional dances.

APPRECIATING CULTURE

1. Which type of art shown here appeals most to you? Why?

2. Latin America has a rich and varied cultural history. In what ways have Latin Americans tried to preserve this heritage?

ARCHITECTURE ▲ ▲ ▲ ▲

In the late 1600s Europeans built many cathedrals in the baroque style in Latin America. The baroque style features carved columns, ornamental sculptures, colored tile, gold, and silver.

The Andean Countries

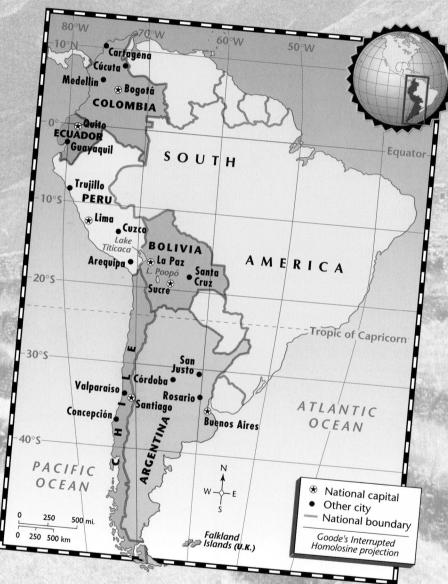

National capital ✪
Other city ●
National boundary ━━━

Goode's Interrupted
Homolosine projection

MAP STUDY ACTIVITY

In Chapter 9 you will learn about the countries that the Andes run through.

1. What is the capital of Argentina?
2. What is unusual about Bolivia's capital?

Colombia

PREVIEW

Words to Know
- cordillera
- *llanos*
- cash crop
- *campesino*

Places to Locate
- Colombia
- Andes
- Magdalena River
- Bogotá

Read to Learn . . .
1. where Colombia is located.
2. what products Colombia exports.
3. how Colombia became independent.

Towering peaks of the Andes stand guard over the city of Bogotá, the capital of Colombia. The 6 million people of Bogotá are nestled in a high basin of the Andes—the world's longest mountain range. It stretches more than 4,500 miles (7,240 km) along the western length of South America!

Colombia shares the majestic Andes with the countries of Ecuador, Peru, Bolivia, Chile, and Argentina. Together these six South American lands are known as the Andean countries.

Region
The Land

Colombia is the only country in South America that borders both the Caribbean Sea and the Pacific Ocean. Its land area totals 401,040 square miles (1,038,694 sq. km)—about three-fourths the size of Alaska.

Colombia's Landforms Narrow ribbons of coastal lowlands stretch along the Caribbean Sea and the Pacific Ocean. Bananas, cotton, and sugarcane are grown on large plantations along the Caribbean. Busy coastal ports handle Colombia's trade. In the west, thick forests spread over the Pacific lowlands. Few people live here.

The Andes spread outward through the center of Colombia. The Andes form a **cordillera,** or a group of mountain chains that run side by side. Colombia's

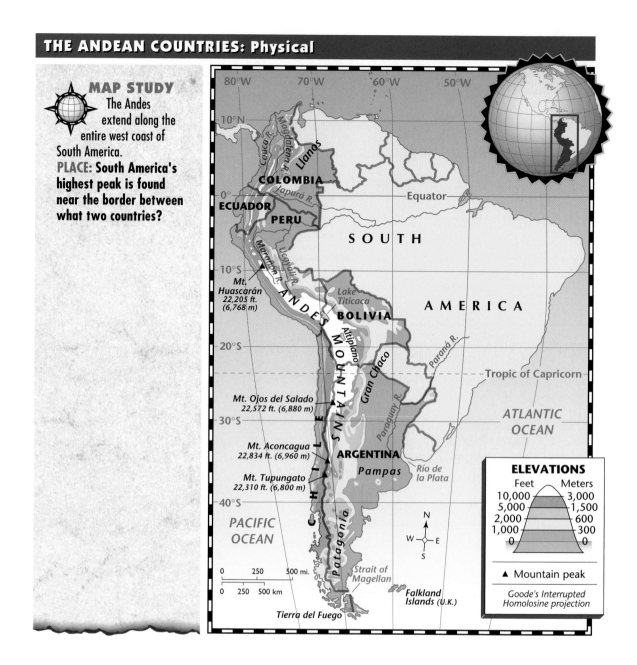

MAP STUDY The Andes extend along the entire west coast of South America.
PLACE: South America's highest peak is found near the border between what two countries?

major river, the Magdalena, winds between the central and eastern Andean chains to the Caribbean Sea. Most of Colombia's cities—including its capital, Bogotá (BOH•guh•TAH)—lie in the valleys of this mountainous area.

Vast plains cover the remaining 60 percent of Colombia. Tropical rain forests spread across the southeast plain into the Amazon River basin in neighboring Brazil. Colorful birds such as toucans, parrots, and parakeets screech from the trees of the tropical forests. Only a few Native American groups live in this hot, steamy region.

In the northeast you find hot grasslands called the **llanos.** Here on large estates ranchers drive their cattle across the rolling plains.

The Climate Look at the climate map on page 223. You can see that Colombia lies entirely within the tropics. Yet Colombia does not have a totally tropical climate.

In the high altitudes of the Andes, temperatures are very cool for a tropical area. Bogotá, for example, lies at 8,355 feet (2,547 m) above sea level. High temperatures average only 67°F (20°C). Low temperatures average about 50°F (10°C).

Human/Environment Interaction
The Economy

Did you know that the average American eats 27.3 pounds (12.4 kg) of bananas a year? Colombia ranks as the second-largest Latin American supplier of this fruit. Most Colombians earn a living as farmers, factory workers, or miners.

Colombian factories produce automobiles, machinery, clothing, and food products. These industries are supported by the rich minerals found in the country, including coal, iron ore, petroleum, and natural gas. A leading supplier of gold, Colombia also produces 90 percent of the world's emeralds.

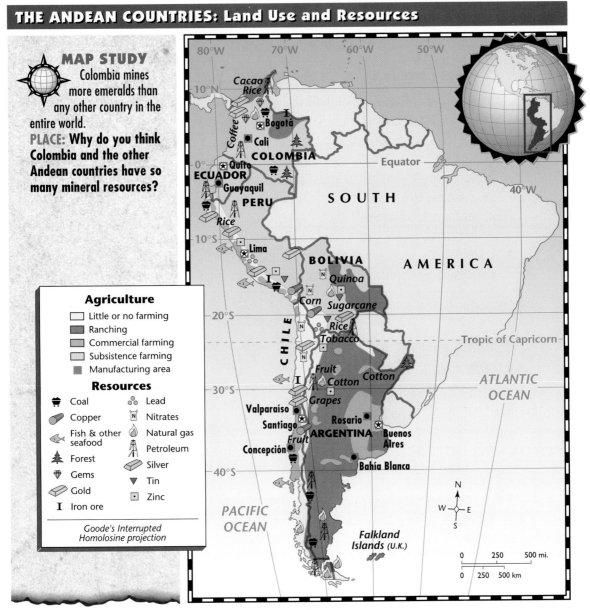

THE ANDEAN COUNTRIES: Land Use and Resources

MAP STUDY
Colombia mines more emeralds than any other country in the entire world.
PLACE: Why do you think Colombia and the other Andean countries have so many mineral resources?

Agriculture
- Little or no farming
- Ranching
- Commercial farming
- Subsistence farming
- Manufacturing area

Resources
- Coal
- Copper
- Fish & other seafood
- Forest
- Gems
- Gold
- I Iron ore
- Lead
- N Nitrates
- Natural gas
- Petroleum
- Silver
- Tin
- Zinc

Goode's Interrupted Homolosine projection

Agriculture Differences in land elevation allow Colombians to grow a variety of crops. Coffee is the country's major **cash crop**—a crop that is usually sold for export. Colombian coffee—grown on large plantations—is known all over the world for its rich flavor. Colombia also exports cotton, tobacco, and cut flowers. Huge herds of cattle roam large ranches in the *llanos.* The rain forests of the eastern plains also supply a valuable resource—lumber.

Movement
The People

The population of Colombia totals about 38.6 million. Most Colombians live in the valleys of the Andes. Nearly all Colombians speak Spanish, but they are of mixed European, African, and Native American backgrounds. The Roman Catholic faith is Colombia's major religion.

Independence Colombia, Venezuela, Ecuador, and Panama became independent as one country in 1819. Símon Bolívar, whom you read about in Chapter 8, led this struggle for independence. Over time Colombia's neighbors broke away from Colombia and became separate countries.

During the 1800s and 1900s, political disagreements divided Colombia. In the 1950s the two major political parties finally agreed to govern the country together. Their cooperation has led to a stable democratic government.

Growth of Cities Like other South American countries, Colombia has a rapidly growing urban population. Since the 1940s Colombian farmers, known as *campesinos,* and their families have journeyed from their villages to the cities to look for work.

Fourteen-year-old Carmen Rivera lives in the Caribbean port city of Barranquilla (BAHR•ruhn•KEE•yuh). She enjoys performing many of the regional dances of her country. Her brother, Tomás, likes to join in the Carnival procession before the Easter season. In the parade, he wears a brightly colored mask.

SECTION 1 ASSESSMENT

REVIEWING TERMS AND FACTS
1. **Define the following:** cordillera, *llanos,* cash crop, *campesino.*
2. **PLACE** Where is Colombia located?
3. **HUMAN/ENVIRONMENT INTERACTION** What is Colombia's main export?

MAP STUDY ACTIVITIES
4. Look at the physical map on page 218. What major river of Colombia flows into the Caribbean Sea?
5. Turn to the map on page 219. What Andean countries depend on fishing as part of their economies?

INCA ENGINEERING

MATH	SCIENCE	HISTORY	LITERATURE	**TECHNOLOGY**

They built thousands of miles of roads. They guaranteed one-day delivery for messages delivered up to 140 miles (225 km) away. They built a city in the sky and the first true suspension bridges. No, we're not talking about modern architects. The people who accomplished these feats—and more—lived during the 1400s and 1500s. They were the Inca. Surprisingly, these Native Americans built their structures without the aid of the wheel, wagons, or pulleys.

ROADS To control their huge empire that stretched about 2,500 miles (4,023 km) along the west coast of South America, the Inca needed roads. Eventually, they carved 14,260 miles (22,944 km) of roads throughout the Andes. To conquer the steep slopes, the Inca laid roads in a zigzag pattern. They also built stairways of long, shallow steps. On the steepest slopes, roads were just a series of stone steps thrust into the mountainside. A system of way stations provided stops for the Inca army or messengers to rest. Across deep gorges, the Inca built sturdy suspension bridges made of rope and wood.

CITY IN THE SKY Probably the most spectacular feat of ancient engineering in the

Western Hemisphere is the Inca city, Machu Picchu (MAH•choo PEE•choo), built high in

Machu Picchu, Peru

the Andes. You can see its ruins today if you visit the country of Peru. Jagged Andean cliffs hid and protected Machu Picchu, which rose 2,000 feet (about 610 m) above a narrow valley. Using only stone hammers and a polish made of wet sand, the Inca erected Machu Picchu's 143 buildings from granite blocks cut from the mountains. They transformed the steep slopes below the city into gardens by building terraces, or stepped ridges of land suitable for farming.

Making the Connection

1. How did the Inca overcome the rugged Andes in their road building?
2. What features make Machu Picchu a remarkable city?

SECTION 2 · Peru and Ecuador

PREVIEW

Words to Know
- *altiplano*
- navigable
- empire

Places to Locate
- Peru
- Lima
- Lake Titicaca
- Ecuador
- Quito
- Guayaquil
- Galápagos Islands

Read to Learn . . .
1. what landforms are found in Peru and Ecuador.
2. what mineral resources are mined in Peru and Ecuador.
3. how people live in Peru and Ecuador.

What do you suppose this lizard is thinking as it suns itself on the rocky coast of the Galápagos Islands? Is it aware that its home rises up in the Pacific Ocean about 600 miles (965 km) west of South America? Since 1832 the Galápagos Islands have belonged to Ecuador, one of the Andean countries.

Ecuador and its larger neighbor, Peru, lie in western South America along the Pacific Ocean. They are lands of enormous contrasts in landscape and climate, but they share a Native American and Spanish heritage.

Place
Peru

Peru is South America's third-largest country in area. Peru—a Native American word that means "land of abundance"—is rich in mineral resources.

The Land and Economy Dry deserts, snowtopped mountains, and hot, humid rain forests greet you in Peru. On a narrow coastal strip of plains and deserts lies Peru's many farms and cities. The cold Peru Current in the Pacific Ocean keeps coastal temperatures fairly mild.

The Andes, with their broad valleys and plateaus, sweep through the center of Peru. The southern part of Peru's Andean region contains a large plateau known as the ***altiplano.*** If you travel on the *altiplano,* you will see Lake Titicaca (TIH•tih•KAH•kuh), the highest navigable lake in the world. **Navigable** means that a body of water is wide and deep enough to allow the passage of ships.

 222

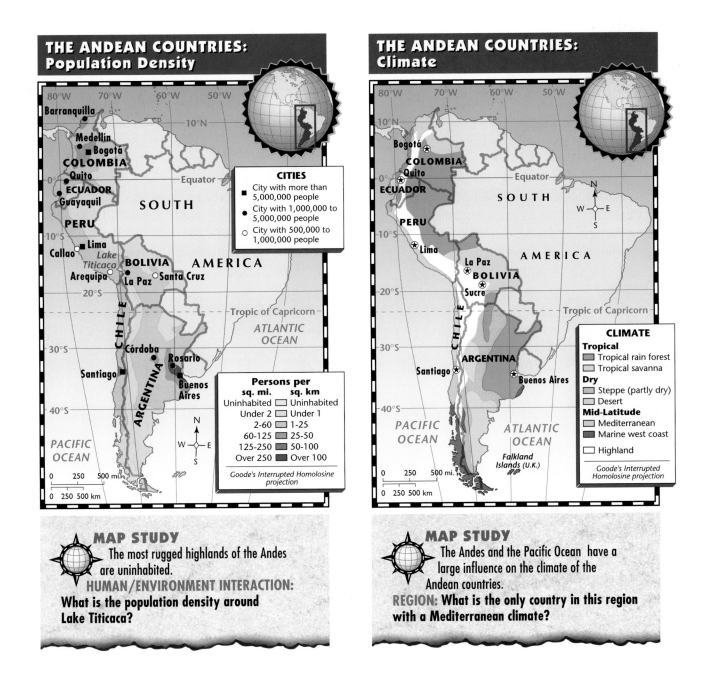

THE ANDEAN COUNTRIES: Population Density

Barranquilla
Medellín
Bogotá
COLOMBIA
Quito
ECUADOR
Guayaquil
PERU
Lima
Callao
Lake Titicaca
Arequipa
BOLIVIA
La Paz
Santa Cruz
AMERICA
SOUTH
Córdoba
Rosario
Santiago
Buenos Aires
CHILE
ARGENTINA
PACIFIC OCEAN
ATLANTIC OCEAN

CITIES
■ City with more than 5,000,000 people
● City with 1,000,000 to 5,000,000 people
○ City with 500,000 to 1,000,000 people

Persons per
sq. mi.		sq. km
Uninhabited	☐	Uninhabited
Under 2	☐	Under 1
2-60	☐	1-25
60-125	☐	25-50
125-250	☐	50-100
Over 250	■	Over 100

Goode's Interrupted Homolosine projection

0 250 500 mi.
0 250 500 km

MAP STUDY
The most rugged highlands of the Andes are uninhabited.
HUMAN/ENVIRONMENT INTERACTION:
What is the population density around Lake Titicaca?

THE ANDEAN COUNTRIES: Climate

Bogotá
COLOMBIA
Quito
ECUADOR
PERU
Lima
La Paz
BOLIVIA
Sucre
SOUTH
AMERICA
ARGENTINA
Santiago
Buenos Aires
CHILE
PACIFIC OCEAN
ATLANTIC OCEAN
Falkland Islands (U.K.)
Equator
Tropic of Capricorn

CLIMATE
Tropical
☐ Tropical rain forest
☐ Tropical savanna
Dry
☐ Steppe (partly dry)
☐ Desert
Mid-Latitude
☐ Mediterranean
☐ Marine west coast
☐ Highland

Goode's Interrupted Homolosine projection

0 250 500 mi.
0 250 500 km

MAP STUDY
The Andes and the Pacific Ocean have a large influence on the climate of the Andean countries.
REGION: What is the only country in this region with a Mediterranean climate?

East of the Andes you run into low foothills and flat plains. Thick rain forests cover almost all of the plains area, where the mighty Amazon River begins. Rainfall is plentiful, and temperatures remain high throughout the year.

The map on page 219 shows you that farming is the major economic activity of Peru. Coffee, cotton, and sugarcane are Peru's major export crops. Peru also ranks among the world's leading producers of copper, lead, silver, and zinc. These minerals come from mines in the Andean highlands.

The People During the 1400s the Inca had a powerful civilization in the area that is now Peru. Their **empire,** or group of lands under one ruler, stretched more than 2,500 miles (4,023 km). In 1530 the Spaniards defeated the Inca and made Peru a Spanish territory. Peru gained independence from Spain in 1821.

Peru's 26.1 million people live mostly in cities and towns. Lima (LEE•muh), with a population of almost 7 million, is the capital and largest city. Peru has the largest Native American population in the Western Hemisphere. People of *mestizo* and European ancestry make up the rest of Peru's population.

Place

Ecuador

Ecuador is one of the smallest countries in South America. Can you guess how Ecuador got its name? *Ecuador* is the Spanish word for "equator," which runs right through Ecuador.

The Land and Economy Swamps, deserts, and fertile plains stretch along Ecuador's Pacific coast. The Andes run through the center of the country. Nearly one-half of Ecuador's people live in the valleys and plateaus of the Andes. Quito (KEE•toh), Ecuador's capital, lies more than 9,000 feet (2,700 m) above sea level. Thick rain forests cover eastern Ecuador.

Ecuador's inland climate is hot and humid. The Peru Current in the Pacific Ocean keeps coastal temperatures moderate. In the Andes the climate gets colder as you climb higher into the mountains.

Agriculture is Ecuador's most important economic activity. Bananas, cacao, sugarcane, and other export crops grow in the coastal lowlands. Here you will find Guayaquil (gwy•uh•KEEL), Ecuador's most important port city. Farther inland, farms in the Andean highlands grow coffee, beans, corn, potatoes, and wheat. Ecuador's major mineral export is petroleum.

The People Of Ecuador's 12.2 million people, most claim Native American or *mestizo* ancestry. A small number are of African background. About 61 percent of Ecuadorians live in urban areas.

SECTION 2 ASSESSMENT

REVIEWING TERMS AND FACTS

1. Define the following: *altiplano,* navigable, empire.

2. REGION What landforms do Peru and Ecuador share?

3. HUMAN/ENVIRONMENT INTERACTION What is Ecuador's major mineral export?

MAP STUDY ACTIVITIES

4. Look at the climate map on page 223. What climate do you find in the eastern lowlands of Peru and Ecuador?

Bolivia and Chile

PREVIEW

Words to Know
- *quinoa*
- sodium nitrate

Places to Locate
- Bolivia
- La Paz
- Sucre
- Chile
- Tierra del Fuego
- Santiago

Read to Learn . . .
1. where Bolivia and Chile are located.
2. what landforms and climates are found in Bolivia and Chile.
3. how the economies of Bolivia and Chile differ.

Do you think you would want to be a farmer in the high Andean plateau of Bolivia? You would tend a flock of llamas, which are raised for wool and carrying goods. Native Americans in Bolivia live much as their ancestors did hundreds of years ago. Their way of life is common throughout much of rural Bolivia.

At first glance, Bolivia and its neighbor Chile seem very different. Bolivia lacks a seacoast, while Chile has a long, narrow coastline. The Andes, however, affect the climate and cultures of both countries.

Place
Bolivia

Bolivia lies near the center of South America, cut off from the sea. The Andes dominate Bolivia's landscape. Breathtaking scenery draws visitors from all over the world. The mountains, however, present hardships to Bolivia's people. The many Bolivians who live at high elevations must work hard to farm and extract minerals from the earth.

Land and Climate Bolivia has mountains, lowland plains, and tropical rain forests. Look at the map on page 218. You see that in western Bolivia, the Andes surround a high plateau called the *altiplano*. This region has a cool climate, kept moderate by Lake Titicaca. Few trees grow in this region, and the land is too dry to farm. Farms are more common in south central Bolivia, an area of gently sloping hills and broad valleys.

A vast lowland plain spreads over northern and eastern Bolivia. Tropical rain forests cover the northern end. Grasslands and swamps sweep across the rest of the plain. Most of this area has a hot, humid climate.

The Economy Bolivia's economy relies partly on farming. Bolivian farmers struggle to grow corn, potatoes, wheat, and a cereal grain called *quinoa* (KEEN•wah). They also raise cattle for beef and llamas for wool.

Abundant mineral resources balance Bolivia's economy. The country is rich in tin, copper, and lead. Miners extract these minerals from mines high in the Andes. The country ranks as one of the world's leading tin producers. The eastern lowlands provide gold, petroleum, and natural gas.

Landlocked Bolivia has had limits placed on its economic growth. In 1998, however, Bolivia and Peru agreed to build a highway to the Pacific coast. This link to the ocean is expected to benefit Bolivia's economy.

The People Bolivia has two capital cities—La Paz (luh•PAHZ) and Sucre (SOO•kray). They are located in the *altiplano*. La Paz—the highest capital in the world—lies at an elevation of 12,000 feet (3,660 m).

Most of Bolivia's 8 million people live in the Andean highlands. The majority claim Native American or mixed Spanish and Native American ancestry. In the cities, most people observe European or North American customs. Most people in the countryside are Native Americans and follow the traditions of their ancestors. The men carry on subsistence farming, and the women weave textiles or make pottery to earn money.

Place

Chile

Find Chile on the map on page 216. This long and narrow country runs about 2,650 miles (4,265 km) along the Pacific coast of western South America. Chile's average width, however, is only 100 miles (160 km).

The Land and Climate Chile contains many types of landforms. In the north lies the Atacama Desert, one of the world's driest places. Farther east, the Andes run along Chile's border with Argentina. Fanning out from the mountains in central Chile is the Central Valley. With its fertile soil and mild climate, the Central Valley supports most of Chile's people, industries, and farms.

The southern part of Chile is a stormy, windswept region of snow-tipped volcanoes, thick forests, and huge glaciers. In the far south, the Strait of Magellan separates mainland Chile from a group of islands known as Tierra del Fuego (fu•AY•goh)—or Land of Fire. Cold ocean waters batter the rugged coast around Cape Horn, the southernmost point of South America.

The Economy Chile has the fastest-growing economy in Latin America. It relies on the mining and export of valuable minerals. Copper is the country's most important resource and major export. Chile ranks as the world's leading

copper producer. Chile also mines and exports gold, silver, iron ore, and **sodium nitrate.** Sodium nitrate is used as a fertilizer and in explosives.

Agriculture and manufacturing are also major economic activities in Chile. Farmers produce wheat, corn, barley, rice, and oats and raise cattle, sheep, and other livestock. Chilean factory workers produce clothing, wood products, chemicals, and transportation equipment. In Chile's cities, service industries such as banking and tourism thrive.

The People Of the 14.8 million people in Chile, about 75 percent are mestizos. About 20 percent are of pure European ancestry. Nearly all the people speak Spanish, and most are Roman Catholics. About 85 percent of Chile's population live in urban areas. Santiago, the capital, has about 4.9 million people.

Antonio Pérez, a teenager from central Chile, attends rodeos to watch Chilean cowboys perform remarkable feats. He—and many other Chileans— also enjoy sports such as soccer, skiing, and volleyball. Antonio likes to vacation on the beautiful beaches along Chile's Pacific coast.

Chile has a wide variety of climates and landforms. The capital city of Santiago in central Chile *(left)* contrasts sharply with the icy southern region with massive glaciers *(right).* **PLACE: What chain of islands lies at the southern tip of Chile?**

SECTION 3 ASSESSMENT

REVIEWING TERMS AND FACTS

1. Define the following: *quinoa,* sodium nitrate.

2. PLACE What are the two capitals of Bolivia? Where are they located?

3. HUMAN/ENVIRONMENT INTERACTION What is Chile's most important mineral resource?

MAP STUDY ACTIVITIES

4. Look at the land use map on page 219. What region holds most of Chile's farms?

5. Using the same map, name the mineral resources found in Bolivia.

PREVIEW

Words to Know
- tannin
- *estancia*
- *gaucho*

Places to Locate
- Argentina
- Gran Chaco
- Patagonia
- Pampas
- Buenos Aires
- Río de la Plata

Read to Learn . . .
1. what physical regions make up Argentina.
2. what products come from Argentina.
3. where Argentina's people live.

The North American song "Home on the Range" could have been written about this vast, treeless landscape in the country of Argentina. Here cowhands drive large herds of cattle across this plains area. Beef—one of Argentina's major products—is known throughout the world for its quality and flavor.

Argentina occupies most of the southern part of South America. Uruguay, Brazil, Paraguay, and Bolivia lie on its northern borders. Argentina's eastern coastline is washed by the Atlantic Ocean. Its southern tip reaches almost to the continent of Antarctica.

Region
The Land

Argentina's land area of 1,056,640 square miles (2,736,698 sq. km) makes it South America's second-largest country, after Brazil. Within Argentina's vast spaces are a variety of landscapes: mountains, forests, plains, and deserts.

The North A roaring waterfall greets visitors to Argentina's north region. The spectacular Iguaçú (EE•gwuh•SOO) Falls, on Argentina's border with Brazil, is one of the scenic wonders of the world.

Lowland areas stretch across northern Argentina. To the west, great forests cover an area known as the Gran Chaco. As you learned in Chapter 8, the Gran

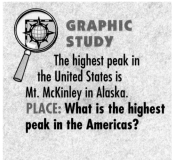

The highest peak in the United States is Mt. McKinley in Alaska.
PLACE: What is the highest peak in the Americas?

Highest mountain in:

United States	Mt. McKinley	20,320 ft. (6,194 m)
Canada	Mt. Logan	19,850 ft. (6,050 m)
Mexico	Mt. Citlaltépetl	18,701 ft. (5,700 m)
Central America	Mt. Tajumulco	13,845 ft. (4,220 m)
West Indies	Pico Duarte	10,417 ft. (3,175 m)
South America	Mt. Aconcagua	22,834 ft. (6,960 m)

Source: *World Almanac,* 1998;
Goode's World Atlas, 19th edition

Chaco also extends into neighboring Paraguay. This region, dry for most of the year, experiences sudden downpours during the summer months. The few people who live here harvest quebracho trees and practice subsistence agriculture. These hardwood trees are a source of **tannin,** a substance used in making leather.

To the east, hot, humid grasslands lie between the Paraná and Paraguay rivers. Farmers graze livestock and grow crops on this fertile soil. Along this area's northern border, the Iguaçú River empties into the Paraná River. Near where the two rivers join, about 275 waterfalls form Iguaçú Falls.

The Andes The Andes tower over the western part of Argentina. Snow-capped peaks and clear blue lakes draw tourists who come to ski and hike. Mount Aconcagua (A•kuhn•KAH•gwuh), soaring to a height of almost 23,000 feet (7,000 m), is the highest mountain in the Western Hemisphere. The chart above shows how Aconcagua compares in height with mountains in other parts of the Americas.

East of the Andes is a region of rolling hills and desert valleys. Mountain streams flow through this area. Farmers use these waters to grow sugarcane, corn, cotton, and grapes. Cities in this area are the oldest Spanish settlements in Argentina.

The Pampas In the center of Argentina are treeless plains known as the Pampas. The Pampas spread almost 500 miles (804 km) from the Atlantic coast to the Andes. Argentina's economy depends on this region's fertile soil and mild climate. Similar to the Great Plains of the United States, the Pampas are home to farmers growing grains and ranchers raising livestock. Most of Argentina's urban areas are here, where more than two-thirds of the country's people live.

Buenos Aires, Argentina's capital and largest city, lies in the area where the Pampas meet the Río de la Plata. The Río de la Plata is a funnel-shaped bay that enters the Atlantic Ocean.

The Andes tower over the Lake of Horns in southern Patagonia *(left)*. The "pink house" in Buenos Aires is the office of Argentina's president *(right)*. **HUMAN/ENVIRONMENT INTERACTION: Why do few people live in Patagonia?**

Patagonia South of the Pampas is a dry, windswept plateau called Patagonia. Most of Patagonia gets little rain and has poor soil. Sheep raising is the region's major economic activity.

Region
The Economy

Although Argentina today has one of the most developed economies in Latin America, Argentines struggled for much of the past 50 years under unstable governments. Argentina's economy depends on agriculture, manufacturing, and service industries.

Agriculture Argentina's major farm products are beef, corn, and wheat. Most of these products come from the Pampas and the northeastern part of the country. Huge *estancias,* or ranches, cover the Pampas. Owners of these estates hire *gauchos,* or cowhands, to take care of the livestock. The *gaucho* is the national symbol of Argentina. Many people in Argentina admire the *gauchos* for their independence and their horse-riding skills.

Manufacturing Argentina is one of the most industrialized countries in South America. The map on page 219 shows you that most of the country's factories are in or near Buenos Aires. Argentina's leading manufactured goods are food products, leather goods, electrical equipment, and textiles.

Petroleum is Argentina's most valuable mineral resource. The country's major oil fields are in Patagonia and the Andes. Other minerals, such as iron ore, lead, zinc, and uranium, also lie in the Andean region.

The People

The population density map on page 223 reveals that Argentina's 36.1 million people live mostly in certain areas of the country. Because of the harsh climate and landscape of the Andean region and Patagonia, settlement in these areas is sparse. More than one-third of the country's people live in or around Buenos Aires. Another one-fourth live on the Pampas.

WHAT IN THE WORLD?

Let's Play Pato

Many Argentines play a sport called *pato*. *Pato* combines polo and basketball. Players on horseback form two teams. Each team tries to keep control of a six-handled ball and toss it into the opposing team's basket.

Influences of the Past The region that is now Argentina had very few Native Americans before the arrival of the Europeans. The Spanish arrived in the 1500s from Chile and later founded a colony centered around Buenos Aires. Most of the Native Americans in the region died from disease or were killed by Europeans.

In 1816 a general named José de San Martín helped Argentina win its independence from Spain. During the late 1800s and early 1900s, some of Argentina's people grew wealthy from livestock raising and industry. Military leaders controlled Argentina's government.

In 1982 Argentina suffered defeat in a war with the United Kingdom for control of the Falkland Islands. The Falklands, known in Argentina as the Malvinas, are located in the Atlantic Ocean off the coast of Argentina. Argentina's loss led to the overthrow of its military leaders and the creation of a democracy.

Argentina Today Argentines are mostly of European ancestry. Settlers came to Argentina from Spain and Italy during the 1800s. Native Americans and mestizos form only a small part of Argentina's population. The official language of Argentina is Spanish, and the Roman Catholic faith is the major religion. About 86 percent of Argentina's people live in cities and towns. Thirteen-year-old Gabriella Marcelli is among the 12.2 million Argentines who live in Buenos Aires. She admires the city's attractive parks and elegant government buildings.

SECTION 4 ASSESSMENT

REVIEWING TERMS AND FACTS

1. **Define the following:** tannin, *estancia, gaucho.*
2. **PLACE** Where is Buenos Aires located?
3. **PLACE** What are the most important agricultural products of Argentina?
4. **REGION** How is the Andean region important to Argentina's economy?

GRAPHIC STUDY ACTIVITIES

5. Look at the chart on page 229. What is the highest mountain in Central America?
6. How much taller is Mt. Aconcagua than the highest mountain in the United States?

BUILDING GEOGRAPHY SKILLS

Building a Database

An electronic database is a collection of data—names, facts, and statistics—stored on a computer. Databases are useful for organizing large amounts of information. The information in a database can be sorted and presented in different ways.

A database organizes information in categories called *fields.* For example, a database of people you send birthday cards to might include the fields **name**, **address**, and **birthday**. Each person you enter into the database is called a *record*. After entering the records, you might create a list sorted by birthday, or use the records to create address labels.

Geographers use databases for many purposes. They often have large amounts of data that they need to analyze. For example, a geographer might want to compare and contrast certain information about the Andean countries of Latin America.

Use these steps to create a database about the Andean countries:

1. Determine what facts you want to include in your database. You might want to include facts about the land, economy, and people of each country.

2. Follow instructions in the database program that you are using to set up fields. Then enter each item of data in its assigned field.

3. Determine how you want to organize and present the facts in the database.

4. Follow the instructions in your program to sort the information.

5. Check that the information in your database is correct. If necessary, add, delete, or change information or fields.

Country	Land	Economy	People
Argentina			Mostly of European ancestry
Bolivia		Rich in tin, copper, and lead	
Chile	Contains many types of landforms		

Technology Skills Practice

For additional practice in using a database, follow the steps listed above to build a database that lists information about current political events in Latin American countries. Bring current newspapers to class and create a database using information from the newspapers. After you have completed the database, write a paragraph explaining why the database is organized the way it is and how it might be used in this class.

Chapter 9 Highlights

Important Things to Know About the Andean Countries

SECTION 1 COLOMBIA

- Colombia has coastlines on the Caribbean Sea and the Pacific Ocean.
- Coastal lowlands, central highlands, and inland rain forests make up Colombia's landscape.
- Coffee is Colombia's major farm product.

SECTION 2 PERU AND ECUADOR

- Peru's long Pacific coastline is mostly desert and plains.
- The land that is now Peru was the center of a great Native American civilization—the Inca Empire.
- Most of the people of Peru live in coastal cities and in mountain valleys.

SECTION 3 CHILE AND BOLIVIA

- Bolivia is a landlocked country near the center of South America. Chile is a long, narrow country along the Pacific coast.
- Most of Bolivia's people are Native Americans.
- The majority of Chile's population live in the fertile central valley between the mountains.
- Copper is the major mineral export of Chile; tin is Bolivia's major mineral export.

SECTION 4 ARGENTINA

- Argentina is the second-largest country in South America, after Brazil.
- A vast, treeless plain called the Pampas is home to Argentina's beef cattle industry.
- About one-third of Argentina's people live in Buenos Aires, the nation's capital.

Inca terraces, Pisac, Peru ▶

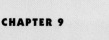

REVIEWING KEY TERMS

Match the numbered terms in Set A with their lettered definitions in Set B.

A

1. *quinoa*
2. sodium nitrate
3. tannin
4. *estancia*
5. *altiplano*
6. *campesino*
7. *gaucho*
8. *llanos*

B

A. Argentine ranch
B. substance used in making leather
C. a cereal grain
D. mineral used for fertilizer and explosives
E. farmer
F. treeless plains in Colombia
G. high, dry plateau in the Andes
H. cowhand in Argentina

Mental Mapping Activity

Draw a freehand map of the Andean countries. Label the following items:
- Buenos Aires
- Atacama Desert
- Cape Horn
- Andes
- Lake Titicaca
- Bogotá

REVIEWING THE MAIN IDEAS

Section 1

1. PLACE How do the Andes affect Colombia's climate?
2. HUMAN/ENVIRONMENT INTERACTION What part of Colombia has a lumber industry?

Section 2

3. PLACE What is the capital of Peru?
4. PLACE What body of water is the highest navigable lake in the world?

Section 3

5. REGION Describe the climate of northern Chile.
6. HUMAN/ENVIRONMENT INTERACTION What minerals are mined in Bolivia?

Section 4

7. LOCATION Where is Patagonia located?
8. PLACE What are the major economic activities in the Pampas?

CRITICAL THINKING ACTIVITIES

1. **Determining Cause and Effect** What effect does the Peru Current have on the coastal areas of Peru and Ecuador?
2. **Evaluating Information** How did the arrival of Europeans affect the Native Americans living in the Andean region of South America?

CONNECTIONS TO WORLD CULTURES

Cooperative Learning Activity Work in a group of three to learn more about daily life in one of the capital cities of the Andean countries. Choose one city and select one of the following topics to research: (a) where and how people live; (b) what people do for recreation; or (c) what holidays and festivals they celebrate. As a group prepare a script for a travel show about these countries. Share your information with the class.

GeoJournal Writing Activity

Imagine that you are working on an *estancia,* a large cattle ranch in Argentina. Write a letter to a family member or a friend describing how you begin your day, your work on the ranch, and how you end your day.

TECHNOLOGY ACTIVITY

Building a Database Create a fact sheet about the countries of Latin America. Include information under such headings as mountain ranges, bodies of water, natural resources, capitals, major cities, population, and type of government. Put this information into a database, with separate fields for each heading. Print and distribute your database to the rest of the class.

PLACE LOCATION ACTIVITY: THE ANDEAN COUNTRIES

Match the letters on the map with the places and physical features of the Andean countries. Write your answers on a separate sheet of paper.

1. Paraguay River
2. Bogotá
3. Lima
4. Strait of Magellan
5. Pampas
6. Río de la Plata
7. Bolivia
8. Ecuador
9. Lake Titicaca
10. Andes

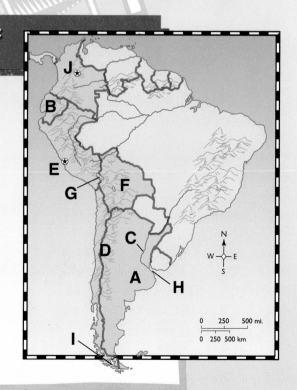

The Disappearing Rain Forest

This brilliant tree frog is one of about 4,000 species of frogs living in the forests of Central America whose habitats are threatened.

PROBLEM

The rain forests of the Amazon basin in South America are being cut or burned at a very fast rate. Why? The trees are being cut for export to countries that will pay high prices for the fine hardwoods. Farmers are burning the forest to clear the land for other profitable crops. Many people are worried that these practices—which are sometimes good for the economies of South America—will cause the rain forests to completely disappear.

SOLUTIONS

- Farmers who settle on newly cleared land are learning soil conservation.
- Governments in Latin America have set up land preserves.
- Researchers supported by a group of Brazil's largest companies are looking for less harmful ways to develop land.
- Latin America is promoting markets for goods that can be harvested without destroying trees.

RAIN FOREST FACT BANK

- Every minute of every day, between 30 and 50 acres (12 to 20 ha) of rain forest disappear.
- Every year, 27,000 species of forest plants and animals are destroyed.
- Rain forests supply one-fifth of the world's oxygen supply.
- Plants in the Amazon basin provide sources for one-fourth of the world's medicines.
- At least one-half of all plant and animal species on Earth live in the rain forests.
- Destroying the rain forests wipes out an ecosystem that does not exist anywhere else on Earth.

A family of farmers winnows dry-land rice from a plot in western Brazil. Slash-and-burn agriculture is the single largest cause of tropical forest loss around the world.

TEEN TRIBUTE

Roland Tiensuu, 13, of Sorunda, Sweden, wanted to help save the rain forests. In the early 1990s, he and fellow teenage students raised about $2 million to buy and protect 23,500 acres (9,185 ha) of forests in Costa Rica.

TAKE A CLOSER LOOK

1 What are some of the ways the rain forests affect the whole world?

2 What are the major reasons the rain forests are being cut down?

WHAT CAN YOU DO?

Plan and carry out an awareness campaign in your school or community to let people know about the destruction of the rain forests.

Help groups who are trying to protect the forests by raising money.

Plant a tree.

Write to your government representatives and encourage them to support plans that help the rain forest.

Find out what natural lands might be threatened in your community.

SAVE THE RAINFORESTS

Botanist David Neill collects plant specimens from the tops of trees felled by road builders in Ecuador. Roughly 25 percent of all prescription medicines in the U.S. are derived from rain forest plants.

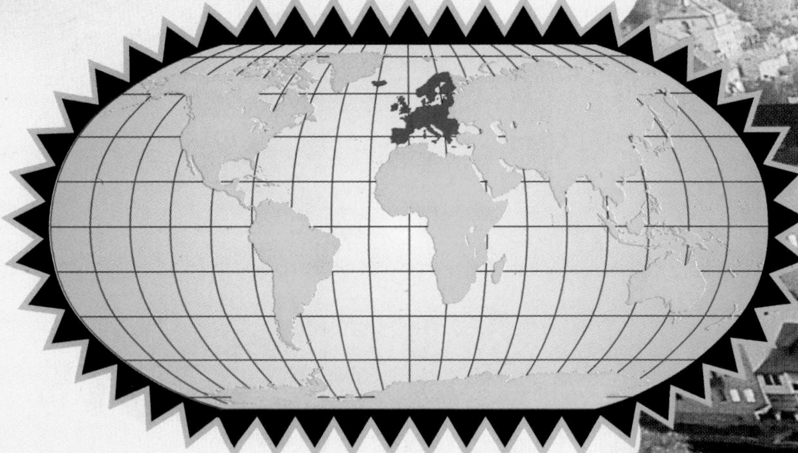

What Makes Europe a Region?

The people share . . .

- high productivity in agriculture and industry.
- densely populated cities.
- global connections in trade, communications, travel, and tourism.
- multinational cooperation since World War II.

To find out more about Europe, see the Unit Atlas on pages 240–253.

Cesky Krumlov, Czech Republic

EXPLORING THE INTERNET

To learn more about Europe, visit the Glencoe Social Studies Web site at **www.glencoe.com** for information, activities, and links to other sites.

GeoJournal Activity

Much of the world's great art comes from Europe. As you read about this region, imagine that you are an art critic. Choose at least five pieces of art or architecture shown in this unit's photographs. Then analyze each piece and comment on its style, its color, and how it reflects the environment in which it was created.

UNIT 4 ATLAS

NATIONAL GEOGRAPHIC SOCIETY

IMAGES *of the* WORLD

1. **Mowing at Stonehenge, England**
2. **Open-air barber shop, Greece**
3. **Snow-covered peaks, Lofoten Islands, Norway**
4. **Children and pony returning from a horse fair, Dublin**
5. **Grand Canal, Venice**
6. **Veteran wearing his medals of honor, Paris**
7. **Parade of bullfighters, Madrid**

All photos viewed against an aerial photo of fields and hedgerows, England.

2

3

4

5

6

Regional Focus

Europe is both a continent and a region. It is a huge peninsula that runs westward from the landmass of Asia. More than 30 countries make up Europe. Their peoples speak about 50 different languages.

Region

The Land

Europe covers about 1.9 million square miles (5.1 million sq. km). The map on page 246 shows you that several bodies of water touch Europe. The largest of these are the Arctic Ocean, the Mediterranean Sea, the Baltic Sea, and the Atlantic Ocean.

The Canals of Amsterdam in the Netherlands

Houseboats and tall, narrow apartment buildings line the banks of Amsterdam's many canals.
PLACE: What is western Europe's major river?

Seas and Coasts Europe's long, jagged coastline has many peninsulas and offshore islands. Deep bays, narrow seas, and well-protected inlets shelter fine harbors. Closeness to the sea has enabled Europeans to trade with other lands. Many Europeans also depend on the sea for food.

Mountains Europe is a continent almost completely covered by mountains. You will find low mountains in the British Isles, Scandinavia, and parts of France, Germany, and eastern Europe. Mountains elsewhere on the continent are higher and more rugged. They include the Pyrenees, which form the border between France and Spain, and the Alps. Europe's highest mountain range, the Alps sweep across southern and central parts of the continent. Joined to the Alps are the Carpathians, a chain of mountains that towers over the landscape of eastern Europe.

Plains Broad, fertile plains curve around Europe's mountains. The North European Plain stretches from the British Isles to Russia. Farms, towns, and cities dot its rolling land. You will find other plains in the peninsulas of southern and eastern Europe.

Rivers Europe's rivers provide trade links between inland areas and coastal ports. The Rhine—western Europe's major river—begins in the Alps, flows through northwestern Europe, and empties into the North Sea. Europe's other important waterway—the Danube River—flows eastward from central Europe to the Black Sea.

Region
Climate and Vegetation

Europe's small land area offers many contrasts in its climate. Europe's northern location and closeness to the sea are key to this variety of climates.

A Mild Climate Europe lies far north, but many areas of the continent enjoy mild temperatures year-round. This happens because of an ocean current known as the North Atlantic Current. It brings warm Atlantic Ocean waters and winds, which provide rainfall, to Europe's western coast.

Europe's northern location also influences the climate of the region. Northern Europe has longer, colder winters and shorter, cooler summers than southern Europe. Eastern Europe lies farther inland from the Atlantic and its mild current. Therefore, its winters are longer and colder—and its summers shorter and hotter—than those in western Europe.

Vegetation In the far north of Europe, mosses and small shrubs blanket the land. Mixed forests and grasslands are found in the milder climate regions of northwestern and eastern Europe. Air pollution from Europe's factories and cars, however, has harmed many woodlands throughout the continent. In the south the hot, dry summers of the Mediterranean area produce shrubs and short trees.

Region
The Economy

Agriculture, manufacturing, and service industries are Europe's leading economic activities. Skilled workers, a few key natural resources, and closeness to waterways have made Europe one of the economic giants of the world.

Agriculture Europe has some of the world's most productive farmland. European farmers use modern equipment to produce huge yields of grains, fruits, and vegetables on small areas of land. They also raise some of the world's finest breeds of cattle and sheep.

Industry Iron ore, coal, and other minerals are found throughout the North European Plain. Vast reserves of oil and natural gas lie beneath the North Sea and in southeastern Europe. Most of Europe's industrial centers have developed near major mineral deposits.

Many European countries are among the world's leading manufacturing centers. They produce steel, machinery, cars, textiles, electronic equipment, food products, and household goods. Service industries such as banking, insurance, and tourism are also important to Europe's economy.

Moving Products Along the Danube

Barges line the banks of the Danube River in Bratislava, Slovakia.
REGION: What factors have made Europe an economic giant?

UNIT 4 ATLAS

Region
The People

Some European countries have one main ethnic group; others are made up of two or more ethnic groups. Differences among Europe's many peoples have often led to conflicts such as civil war in the former Yugoslav republics.

Population The region of Europe has about 520 million people—a large number for a small space. It is one of the world's most densely populated areas. Most Europeans live in urban areas, while large stretches of countryside have few people. Many of the world's leading cities are in Europe. They include London, Paris, Rome, Berlin, and Warsaw. Europe's cities have historical palaces and churches as well as modern skyscrapers and shopping malls. In rural areas life still centers on villages and small towns.

History The ancient Greeks and Romans laid the foundation of European government, law, and thought. They left behind graceful works of art and architecture, such as the Parthenon in Athens. From A.D. 500 to 1500, Christianity shaped a new European civilization. The Gothic style of many cathedrals built during this time reflects Europe's Christian heritage.

Beginning in the 1400s and 1500s, Europeans advanced learning and the arts. The Renaissance—or "rebirth" of art and scholarly activities—advanced learning and the value of the individual person. After the 1700s, political changes increased freedom for the common people. An interest in science and the invention of machines during the Industrial Revolution changed the economy and, in the long run, raised standards of living. In eastern Europe once-powerful empires faced growing challenges from ethnic groups that wanted independence.

A Street in Old Vienna, Austria

People gather at sidewalk cafes in Vienna's historic Inner City to socialize and enjoy pastries and coffee.
REGION: What is the population of Europe?

Europeans also explored, settled, and conquered other lands. They brought their ways of life to every part of the globe. Competition for land and trade among Europe's nations, however, eventually led to two world wars that put an end to Europe's global power. After World War II, Europe was split into noncommunist and communist areas. The fall of communism in the late 1980s brought about a new era. Many European countries joined the European Union to bring their governments and economies closer together.

Picture Atlas of the World

Culture and Tourism

Using the CD-ROM Each European country contributes its own history, language, and culture to the continent. Some of the most fascinating countries of Europe are also the smallest ones. Monaco, Andorra, San Marino, Liechtenstein, and Vatican City claim small populations and cover relatively small areas. Find out more about the allure of these tiny European countries. Assemble a file of information that a tourist might find interesting and helpful. (See the User's Guide for information on

how to use the Collector button.) Use your findings to create a travel brochure titled "Europe: Little Places with Colossal Lure." Remember to include the following:

1. population and economic statistics
2. location of the country within Europe
3. information about the climate
4. charming photos of the people and countries
5. fascinating facts about the countries

Surfing the "Net"

Economic Unity

The goal of the European Union (EU) is to unify and strengthen the economies of member nations by removing all economic restrictions across their borders. To find out more about the reasons why the EU was created and how it benefits its member nations, search the Internet.

Getting There
Follow these steps to gather information on the EU:
1. Use a search engine. Type in the name *European Union, European Community,* or *European Common Market.*
2. Select two members of the EU on which to focus. Type in those countries' names to focus your search.

3. The search engine will provide you with a number of links to follow.

What To Do When You Are There
Click on the links to navigate through the pages of information and gather your findings. Use the information to create a chart of population and economic statistics. Note what the country produces and to whom it sells its products. Include this information in your chart. Prepare answers to the following questions: How does being a member of the EU benefit the country you selected? How do relationships among countries in the European Union compare with those among states in the United States? As a class, compare charts and discuss the questions above.

Physical Geography

EUROPE: Physical

ELEVATIONS

Feet	Meters
10,000	3,000
5,000	1,500
2,000	600
1,000	300
0	0

▲ Mountain peak

Lambert Conformal Conic projection

ARCTIC OCEAN

ICELAND

Arctic Circle

Norwegian Sea

SCANDINAVIAN PENINSULA

NORWAY

FINLAND

SWEDEN

ESTONIA

LATVIA

Shetland Is.

Orkney Is.

North Sea

Jutland Peninsula
DENMARK

Baltic Sea

LITHUANIA

UNITED KINGDOM

IRELAND

NETHERLANDS

Elbe R.

GERMANY

EUROPEAN PLAIN

Vistula R.

POLAND

Thames R.

English Channel

ATLANTIC OCEAN

BELGIUM

NORTH

CZECH REPUBLIC

SLOVAKIA

CARPATHIAN MTS.

LUXEMBOURG

Rhine R.

LIECHTEN-STEIN

Hungarian Plain

HUNGARY

ROMANIA

Seine R.

A L P S

AUSTRIA

Loire R.

FRANCE

Mt. Blanc
15,771 ft.
(4,807 m)

SWITZER-LAND

SLOVENIA

CROATIA

YUGOSLAVIA

Danube R.

Gran Paradiso
13,323 ft.
(4,061 m)

Po R.

BOSNIA-HERZEGOVINA

BALKAN MTS.

BULGARIA

Rhône R.

Adriatic Sea

125   250 mi.

0   125   250 km

IBERIAN PYRENEES

ANDORRA

SAN MARINO

MONACO

APENNINES

ITALY

FMR. YUGOSLAV
REP. OF MACE.

PORTUGAL

Douro R.

Ebro R.

ALBANIA

BALKAN

40°N

Corsica

SPAIN

PENINSULA

GREECE
PENINSULA

Aegean Sea

Sardinia

Strait of Gibraltar

Sicily

Mediterranean   Sea

MALTA

Crete

ASIA

CYPRUS

35°N

Mediterranean Sea

30°E   35°E

20°W   70°N

10°W

0°

10°E

20°E

60°N

50°N

Map Study

1. LOCATION What two countries are located on the Iberian Peninsula?

2. PLACE What sea separates the Scandinavian Peninsula from Poland?

ELEVATION PROFILE: Europe

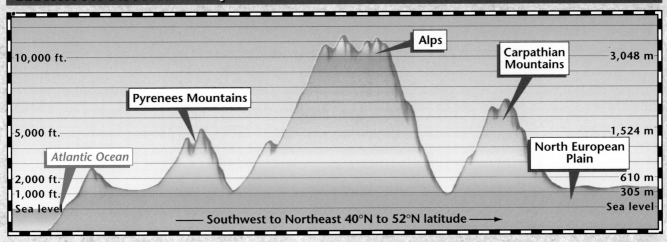

10,000 ft. — 3,048 m

Alps

Carpathian Mountains

Pyrenees Mountains

5,000 ft. — 1,524 m

North European Plain

Atlantic Ocean

2,000 ft. — 610 m
1,000 ft. — 305 m
Sea level — Sea level

← Southwest to Northeast 40°N to 52°N latitude →

GeoFacts

Highest mountain peak: Mont Blanc (France-Italy) 15,771 ft. (4,807 m) high

Longest river: Danube (central Europe) 1,776 mi. (2,858 km) long

Largest lake: Vänern (Sweden) 2,156 sq. mi. (5,584 sq. km)

Highest waterfall: Mardalsfossen [Southern] (Norway) 2,149 ft. (655 m) high

EUROPE AND THE UNITED STATES: Land Comparison

Graphic Study

1. **LOCATION** What are the highest mountains in Europe?
2. **REGION** Look at the land comparison map of Europe and the United States. Knowing that Europe's population is almost double that of the United States, what can you infer about Europe's population density?

Cultural Geography

EUROPE: Political

— National boundary
⊛ National capital

*Lambert Conformal
Conic projection*

ARCTIC OCEAN

20°W   10°W   0°   10°E   20°E
70°N

Arctic Circle

ICELAND
Reykjavík

Norwegian
Sea

SWEDEN   FINLAND

60°N

Helsinki

NORWAY

Oslo   Stockholm   ⊛Tallinn
ESTONIA

Baltic
Sea

North
Sea

LATVIA
Riga

DENMARK
Copenhagen

LITHUANIA
Vilnius

UNITED
KINGDOM

Dublin
IRELAND

NETHERLANDS

London

Amsterdam   Berlin
GERMANY

Warsaw
POLAND

50°N

ATLANTIC

Brussels
BELGIUM

Prague

Luxembourg

CZECH
REPUBLIC

SLOVAKIA
Bratislava

OCEAN

Paris
LUXEMBOURG

LIECHTEN-
STEIN

Vienna
AUSTRIA

Budapest

0   125   250 mi.

Bern

HUNGARY

ROMANIA

0   125   250 km

FRANCE

Ljubljana
SLOVENIA

Zagreb

Bucharest

SWITZER-
LAND

CROATIA

Belgrade
YUGOSLAVIA

SAN
MARINO

BOSNIA-
HERZEGOVINA

BULGARIA

MONACO

Adriatic
Sea

SERBIA

Sarajevo

MONT.

Sofia

PORTUGAL

ITALY

Skopje

40°N

ANDORRA
Madrid

Rome

FMR. YUGOSLAV
REP. OF MACE.

Tiranë
ALBANIA

Lisbon   SPAIN

GREECE

Aegean
Sea

Athens

Mediterranean   Sea

MALTA   ⊛Valletta

ASIA

Nicosia

35°N

CYPRUS

Mediterranean
Sea

30°E   35°E

N
W   E
S

Map Study

1. **LOCATION** What large island country lies about 600 miles northwest of Norway?

2. **PLACE** What is the capital of Bosnia-Herzegovina?

COMPARING POPULATION: Selected European Countries

Austria

Germany

Hungary

Ireland

Italy

Netherlands

Poland

Portugal

Romania

Slovakia

Spain

United Kingdom

= 10,000,000

Source: *Population Reference Bureau, 1998*

GeoFacts

Biggest country (land area): France 212,392 sq. mi. (550,095 sq. km)

Smallest country (land area): Vatican City 0.2 sq. mi. (.44 sq. km)

Largest city (population): Paris (1997) 9,523,000; (2015 projected) 9,500,000

Highest population density: Monaco 49,520 people per sq. mi. (18,570 people per sq. km)

Lowest population density: Iceland 7 people per sq. mi. (3 people per sq. km)

COMPARING POPULATION: Europe and the United States

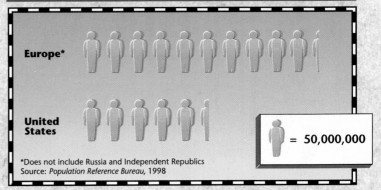

Europe*

United States

= 50,000,000

*Does not include Russia and Independent Republics
Source: *Population Reference Bureau, 1998*

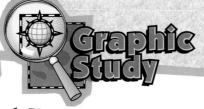

Graphic Study

1. **PLACE** About how many people live in Europe?
2. **REGION** How much difference is there between the most heavily populated and most sparsely populated countries on the graph?

Countries at a Glance

Albania

  Tiranë

LANDMASS:
10,579 sq. mi./
27,400 sq. km
MONEY:
Lek
MAJOR EXPORT:
Fuels
MAJOR IMPORT:
Machinery

CAPITAL:
Tiranë
MAJOR LANGUAGE(S):
Albanian, Greek
POPULATION:
3,300,000

Andorra

Andorra la Vella

LANDMASS:
185 sq. mi./
479 sq. km
MONEY:
French franc,
Spanish peseta
MAJOR EXPORT:
Clothing
MAJOR IMPORT:
Electronic Equipment

CAPITAL:
Andorra la Vella
MAJOR LANGUAGE(S):
Catalan
POPULATION:
54,000

Austria

Vienna

LANDMASS:
31,942 sq. mi./
82,729 sq. km
MONEY:
Schilling
MAJOR EXPORT:
Machinery
MAJOR IMPORT:
Machinery

CAPITAL:
Vienna
MAJOR LANGUAGE(S):
German
POPULATION:
8,100,000

Belgium

Brussels

LANDMASS:
11,790 sq. mi./
30,536 sq. km
MONEY:
Franc
MAJOR EXPORT:
Machinery
MAJOR IMPORT:
Machinery

CAPITAL:
Brussels
MAJOR LANGUAGE(S):
Flemish, French,
Italian, German
POPULATION:
10,200,000

Bosnia and Herzegovina

Sarajevo

LANDMASS:
16,691 sq. mi./
43,230 sq. km
MONEY:
New Yugoslav dinar
MAJOR EXPORT:
Machinery
MAJOR IMPORT:
Fuels

CAPITAL:
Sarajevo
MAJOR LANGUAGE(S):
Serbo-Croatian
POPULATION:
4,000,000

Bulgaria

 Sofia

LANDMASS:
42,683 sq. mi./
110,548 sq. km
MONEY:
Lev
MAJOR EXPORT:
Machinery
MAJOR IMPORT:
Machinery

CAPITAL:
Sofia
MAJOR LANGUAGE(S):
Bulgarian, Turkish
POPULATION:
8,300,000

Croatia

Zagreb

LANDMASS:
21,590 sq. mi./
55,918 sq. km
MONEY:
Kuna
MAJOR EXPORT:
Machinery
MAJOR IMPORT:
Machinery

CAPITAL:
Zagreb
MAJOR LANGUAGE(S):
Croatian
POPULATION:
4,200,000

Cyprus

Nicosia

LANDMASS:
3,568 sq. mi./
9,242 sq. km
MONEY:
Pound
MAJOR EXPORT:
Clothing
MAJOR IMPORT:
Transport Equipment

CAPITAL:
Nicosia
MAJOR LANGUAGE(S):
Greek, Turkish, English
POPULATION:
700,000

Czech Republic

Prague

LANDMASS:
29,838 sq. mi./
77,280 sq. km
MONEY:
Koruna
MAJOR EXPORT:
Chemicals
MAJOR IMPORT:
Chemicals

CAPITAL:
Prague
MAJOR LANGUAGE(S):
Czech
POPULATION:
10,300,000

Denmark

Copenhagen

LANDMASS:
16,320 sq. mi./
42,269 sq. km
MONEY:
Krone
MAJOR EXPORT:
Machinery
MAJOR IMPORT:
Machinery

CAPITAL:
Copenhagen
MAJOR LANGUAGE(S):
Danish
POPULATION:
5,300,000

Countries not drawn to scale.

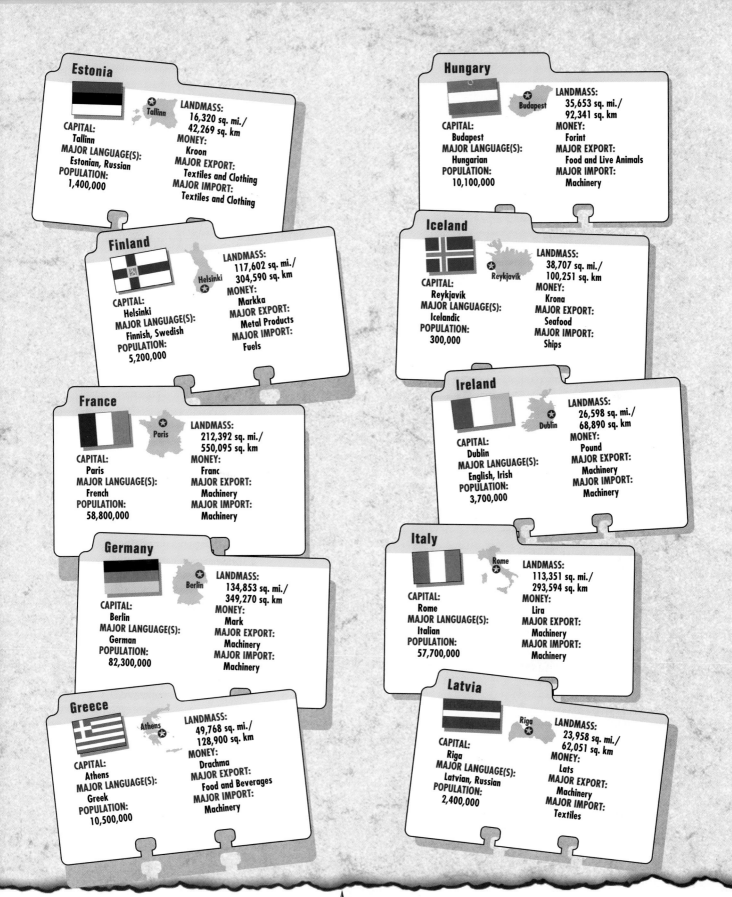

Estonia

CAPITAL:
Tallinn
MAJOR LANGUAGE(S):
Estonian, Russian
POPULATION:
1,400,000

LANDMASS:
16,320 sq. mi./
42,269 sq. km
MONEY:
Kroon
MAJOR EXPORT:
Textiles and Clothing
MAJOR IMPORT:
Textiles and Clothing

Finland

CAPITAL:
Helsinki
MAJOR LANGUAGE(S):
Finnish, Swedish
POPULATION:
5,200,000

LANDMASS:
117,602 sq. mi./
304,590 sq. km
MONEY:
Markka
MAJOR EXPORT:
Metal Products
MAJOR IMPORT:
Fuels

France

CAPITAL:
Paris
MAJOR LANGUAGE(S):
French
POPULATION:
58,800,000

LANDMASS:
212,392 sq. mi./
550,095 sq. km
MONEY:
Franc
MAJOR EXPORT:
Machinery
MAJOR IMPORT:
Machinery

Germany

CAPITAL:
Berlin
MAJOR LANGUAGE(S):
German
POPULATION:
82,300,000

LANDMASS:
134,853 sq. mi./
349,270 sq. km
MONEY:
Mark
MAJOR EXPORT:
Machinery
MAJOR IMPORT:
Machinery

Greece

CAPITAL:
Athens
MAJOR LANGUAGE(S):
Greek
POPULATION:
10,500,000

LANDMASS:
49,768 sq. mi./
128,900 sq. km
MONEY:
Drachma
MAJOR EXPORT:
Food and Beverages
MAJOR IMPORT:
Machinery

Hungary

CAPITAL:
Budapest
MAJOR LANGUAGE(S):
Hungarian
POPULATION:
10,100,000

LANDMASS:
35,653 sq. mi./
92,341 sq. km
MONEY:
Forint
MAJOR EXPORT:
Food and Live Animals
MAJOR IMPORT:
Machinery

Iceland

CAPITAL:
Reykjavik
MAJOR LANGUAGE(S):
Icelandic
POPULATION:
300,000

LANDMASS:
38,707 sq. mi./
100,251 sq. km
MONEY:
Krona
MAJOR EXPORT:
Seafood
MAJOR IMPORT:
Ships

Ireland

CAPITAL:
Dublin
MAJOR LANGUAGE(S):
English, Irish
POPULATION:
3,700,000

LANDMASS:
26,598 sq. mi./
68,890 sq. km
MONEY:
Pound
MAJOR EXPORT:
Machinery
MAJOR IMPORT:
Machinery

Italy

CAPITAL:
Rome
MAJOR LANGUAGE(S):
Italian
POPULATION:
57,700,000

LANDMASS:
113,351 sq. mi./
293,594 sq. km
MONEY:
Lira
MAJOR EXPORT:
Machinery
MAJOR IMPORT:
Machinery

Latvia

CAPITAL:
Riga
MAJOR LANGUAGE(S):
Latvian, Russian
POPULATION:
2,400,000

LANDMASS:
23,958 sq. mi./
62,051 sq. km
MONEY:
Lats
MAJOR EXPORT:
Machinery
MAJOR IMPORT:
Textiles

Countries at a Glance

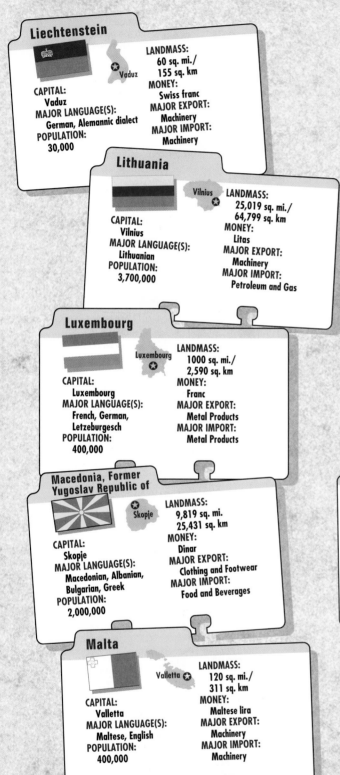

Liechtenstein

Vaduz

CAPITAL:
Vaduz
MAJOR LANGUAGE(S):
German, Alemannic dialect
POPULATION:
30,000

LANDMASS:
60 sq. mi./
155 sq. km
MONEY:
Swiss franc
MAJOR EXPORT:
Machinery
MAJOR IMPORT:
Machinery

Lithuania

Vilnius

CAPITAL:
Vilnius
MAJOR LANGUAGE(S):
Lithuanian
POPULATION:
3,700,000

LANDMASS:
25,019 sq. mi./
64,799 sq. km
MONEY:
Litas
MAJOR EXPORT:
Machinery
MAJOR IMPORT:
Petroleum and Gas

Luxembourg

Luxembourg

CAPITAL:
Luxembourg
MAJOR LANGUAGE(S):
French, German,
Letzeburgesch
POPULATION:
400,000

LANDMASS:
1000 sq. mi./
2,590 sq. km
MONEY:
Franc
MAJOR EXPORT:
Metal Products
MAJOR IMPORT:
Metal Products

Macedonia, Former Yugoslav Republic of

Skopje

CAPITAL:
Skopje
MAJOR LANGUAGE(S):
Macedonian, Albanian,
Bulgarian, Greek
POPULATION:
2,000,000

LANDMASS:
9,819 sq. mi.
25,431 sq. km
MONEY:
Dinar
MAJOR EXPORT:
Clothing and Footwear
MAJOR IMPORT:
Food and Beverages

Malta

Valletta

CAPITAL:
Valletta
MAJOR LANGUAGE(S):
Maltese, English
POPULATION:
400,000

LANDMASS:
120 sq. mi./
311 sq. km
MONEY:
Maltese lira
MAJOR EXPORT:
Machinery
MAJOR IMPORT:
Machinery

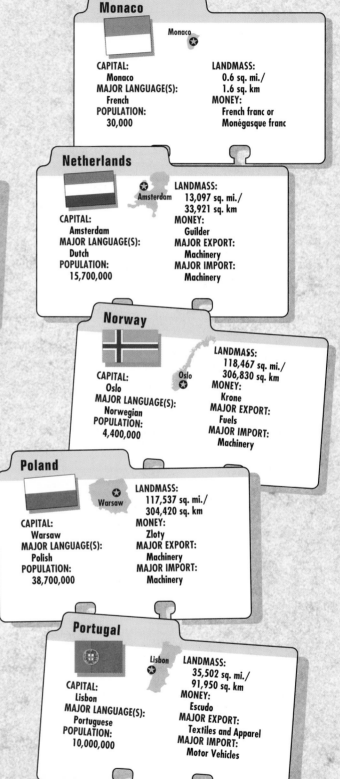

Monaco

Monaco

CAPITAL:
Monaco
MAJOR LANGUAGE(S):
French
POPULATION:
30,000

LANDMASS:
0.6 sq. mi./
1.6 sq. km
MONEY:
French franc or
Monégasque franc

Netherlands

Amsterdam

CAPITAL:
Amsterdam
MAJOR LANGUAGE(S):
Dutch
POPULATION:
15,700,000

LANDMASS:
13,097 sq. mi./
33,921 sq. km
MONEY:
Guilder
MAJOR EXPORT:
Machinery
MAJOR IMPORT:
Machinery

Norway

Oslo

CAPITAL:
Oslo
MAJOR LANGUAGE(S):
Norwegian
POPULATION:
4,400,000

LANDMASS:
118,467 sq. mi./
306,830 sq. km
MONEY:
Krone
MAJOR EXPORT:
Fuels
MAJOR IMPORT:
Machinery

Poland

Warsaw

CAPITAL:
Warsaw
MAJOR LANGUAGE(S):
Polish
POPULATION:
38,700,000

LANDMASS:
117,537 sq. mi./
304,420 sq. km
MONEY:
Zloty
MAJOR EXPORT:
Machinery
MAJOR IMPORT:
Machinery

Portugal

Lisbon

CAPITAL:
Lisbon
MAJOR LANGUAGE(S):
Portuguese
POPULATION:
10,000,000

LANDMASS:
35,502 sq. mi./
91,950 sq. km
MONEY:
Escudo
MAJOR EXPORT:
Textiles and Apparel
MAJOR IMPORT:
Motor Vehicles

Countries not drawn to scale.

Romania

Bucharest

CAPITAL:
Bucharest
MAJOR LANGUAGE(S):
Romanian, Hungarian, German
POPULATION:
22,500,000

LANDMASS:
88,934 sq. mi./
230,339 sq. km
MONEY:
Leu
MAJOR EXPORT:
Mineral Fuels
MAJOR IMPORT:
Raw Materials

San Marino

San Marino

CAPITAL:
San Marino
MAJOR LANGUAGE(S):
Italian
POPULATION:
20,000

LANDMASS:
20 sq. mi./
52 sq. km
MONEY:
Italian lira
MAJOR EXPORT:
Wine
MAJOR IMPORT:
Oil

Slovakia

Bratislava

CAPITAL:
Bratislava
MAJOR LANGUAGE(S):
Slovak
POPULATION:
5,400,000

LANDMASS:
18,564 sq. mi./
48,080 sq. km
MONEY:
New koruna
MAJOR EXPORT:
Machinery
MAJOR IMPORT:
Petroleum

Slovenia

Ljubljana

CAPITAL:
Ljubljana
MAJOR LANGUAGE(S):
Slovenian, Serbo-Croatian
POPULATION:
2,000,000

LANDMASS:
7,768 sq. mi./
20,119 sq. km
MONEY:
Tolar
MAJOR EXPORT:
Machinery
MAJOR IMPORT:
Machinery

Spain

Madrid

CAPITAL:
Madrid
MAJOR LANGUAGE(S):
Spanish, Catalan, Galician, Basque
POPULATION:
39,400,000

LANDMASS:
192,834 sq. mi./
499,440 sq. km
MONEY:
Peseta
MAJOR EXPORT:
Transport Equipment
MAJOR IMPORT:
Machinery

Sweden

Stockholm

CAPITAL:
Stockholm
MAJOR LANGUAGE(S):
Swedish
POPULATION:
8,900,000

LANDMASS:
158,927 sq. mi./
411,621 sq. km
MONEY:
Krona
MAJOR EXPORT:
Machinery
MAJOR IMPORT:
Machinery

Switzerland

Bern

CAPITAL:
Bern
MAJOR LANGUAGE(S):
German, French, Italian
POPULATION:
7,100,000

LANDMASS:
15,270 sq. mi./
39,590 sq. km
MONEY:
Franc
MAJOR EXPORT:
Chemical Products
MAJOR IMPORT:
Machinery and Electronics

United Kingdom

London

CAPITAL:
London
MAJOR LANGUAGE(S):
English, Welsh, Gaelic
POPULATION:
59,100,000

LANDMASS:
93,282 sq. mi./
241,600 sq. km
MONEY:
Pound
MAJOR EXPORT:
Machinery
MAJOR IMPORT:
Machinery

Vatican City

Vatican City

MAJOR LANGUAGE(S):
Italian, Latin
POPULATION:
1,000

LANDMASS:
0.2 sq. mi./
0.4 sq. km
MONEY:
Lira

Yugoslavia

Belgrade

CAPITAL:
Belgrade
MAJOR LANGUAGE(S):
Serbian, Macedonian, Hungarian, Albanian
POPULATION:
10,600,000

LANDMASS:
39,382 sq. mi./
102,000 sq. km
MONEY:
Dinar
MAJOR EXPORT:
Clothing
MAJOR IMPORT:
Machinery

10 The British Isles and Scandinavia

National boundary
⊛ National capital
● Other city

Lambert Conformal Conic projection

ICELAND
Reykjavik

ARCTIC OCEAN

Arctic Circle

Kiruna

Norwegian Sea

Trondheim

SWEDEN

FINLAND

Tampere

Helsinki ⊛

ATLANTIC OCEAN

NORWAY

Bergen

Oslo ⊛

Stockholm ⊛

Baltic Sea

Göteborg

Scotland

Glasgow

North Sea

DENMARK

Copenhagen ⊛

Malmö

Northern Ireland

Belfast

UNITED KINGDOM

Dublin

IRELAND

Manchester

Liverpool

England

EUROPE

Birmingham

Wales

London ⊛

English Channel

0   150   300 mi.

0   150   300 km

MAP STUDY ACTIVITY

As you read Chapter 10, you will learn about the British Isles and Scandinavia.

1. What countries in this region are islands?
2. What country lies farthest north?
3. What capital is farthest south?

The United Kingdom

Words to Know
- moor
- loch
- parliamentary democracy
- constitutional monarchy

Places to Locate
- England
- London
- Wales
- Scotland
- Northern Ireland

Read to Learn . . .
1. what landscapes and climate are found in the United Kingdom.
2. how the British earn their livings.
3. how the United Kingdom has influenced other countries of the world.

What sights would you want to see if you visited London, England? You might want to view the crown jewels or dungeons in the Tower of London. If you took a bus tour, you could see the Houses of Parliament. These stately buildings are home to Parliament, the lawmaking body of the United Kingdom.

What do you think of when you hear the words "British Isles"? Perhaps you have visions of the legendary King Arthur and his Knights of the Round Table. Maybe you hear a bagpipe echoing across the foggy highlands of Scotland. Or you might picture rock groups such as the Rolling Stones or U2.

The British Isles lie just off the Atlantic coast of western Europe. The islands form two countries: the United Kingdom and the Republic of Ireland. The United Kingdom is made up of four regions: England, Scotland, Wales, and Northern Ireland.

Region
The Land

Slightly smaller in area than the state of Oregon, the United Kingdom depends on the sea for many things. The country also lies close to the mainland of Europe. This closeness to both sea and mainland helped make the United Kingdom one of the world's major trading nations.

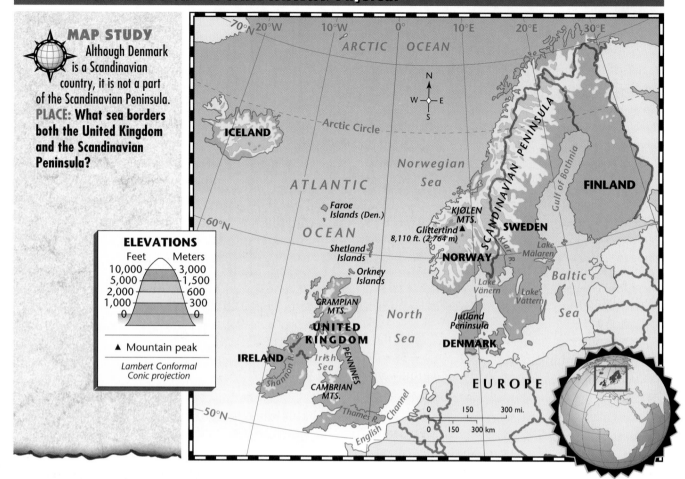

MAP STUDY
Although Denmark is a Scandinavian country, it is not a part of the Scandinavian Peninsula.
PLACE: What sea borders both the United Kingdom and the Scandinavian Peninsula?

ELEVATIONS

Feet	Meters
10,000	3,000
5,000	1,500
2,000	600
1,000	300
0	0

▲ Mountain peak

Lambert Conformal Conic projection

English Countryside

Rolling meadows surround this village in southern England.
PLACE: Why is the United Kingdom's climate mild compared to other countries at that latitude?

The Landscape

If you drew an imaginary line across the United Kingdom from the southwest to the northeast, you would find lowlands east of this line. From the air you would see rolling plains covered by a patchwork of fertile fields and meadows.

West of this imaginary line lies a region of highland areas, most of which are **moors**—treeless landscapes often swept by strong winds. In Scotland narrow bays called **lochs** cut into the highland coasts. They reach far inland and are nearly surrounded by steep mountain slopes.

The Climate

The map on page 258 shows you that the United Kingdom has a mild climate even though it lies as far north as Canada. What causes this? Winds blowing over the North Atlantic Current—an extension of the Gulf Stream—warm the United Kingdom in winter and cool it in summer. The sea winds also bring plenty of rain. If you visit the United Kingdom, take your raincoat! Foggy, damp weather is common throughout the country.

UNIT 4

Human/Environment Interaction
The Economy

More than 200 years ago, inventors and scientists in the United Kingdom sparked the Industrial Revolution. Fuel-powered machines in factories began producing a greater variety and supply of goods. The Industrial Revolution made the United Kingdom the world's leading economic power during the 1800s. Although its economic influence declined in the latter half of the 1900s, the United Kingdom is still a major industrial and trading country.

Resources and Manufacturing The United Kingdom's leading natural resource is oil, pumped from under the North Sea. Other important energy resources are coal and natural gas. All three fuels help drive industries in the United Kingdom and are exported to other countries. The United Kingdom is Europe's leading oil producer.

Heavy machinery, ships, textiles, and cars were once the United Kingdom's major industrial products. Today most of the industries making these goods have declined because of stiff competition from other countries. New high-technology industries, especially computers and electronic equipment, are replacing the older industries.

Farming In recent years farmers in the United Kingdom have adopted modern agricultural methods. Yields are high, but the United Kingdom still must import about one-third of its food supply. Why? The country does not have enough farmland to support its large population. In addition, other countries in Europe produce food at lower costs than the United Kingdom does.

The main crops grown in the United Kingdom are wheat, barley, potatoes, fruits, and vegetables. British farmers also raise cattle and sheep, and the dairy industry thrives.

Place
The People

About 59 million people live in the United Kingdom. The capital, London, ranks as the most heavily populated city in Europe. Many people in the United States and other parts of the world trace their ancestry to the United Kingdom. The British people speak English, and most are Protestants.

Let the Games Begin!

Once a year, the sound of bagpipes signals the beginning of the Highland Games in Braemar, Scotland. Brian Dunbar and Patty McClory perform traditional Highland dances for the huge crowds that attend the games. The plaid, or tartan, on their kilts represents their clan—a group of related families with the same name. Patty's favorite competition is the tossing of the caber—a 100-pound pole.

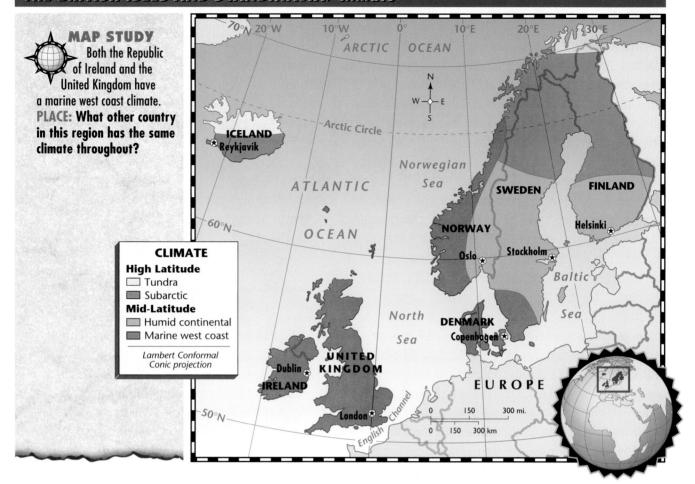

MAP STUDY
Both the Republic of Ireland and the United Kingdom have a marine west coast climate.
PLACE: What other country in this region has the same climate throughout?

CLIMATE

High Latitude
- Tundra
- Subarctic

Mid-Latitude
- Humid continental
- Marine west coast

Lambert Conformal Conic projection

From Past to Present The British are descendants of various groups. Among early arrivals were the Celts (KEHLTS), who sailed from the European mainland around 500 B.C. Their descendants live today in Scotland, Wales, and Ireland. From the A.D. 400s to 1100s, groups from northern and western Europe—the Anglo-Saxons and later, the Normans—settled England. Their descendants became known as the English.

England became a major European power in the late 1500s. In 1707 England and Wales united with Scotland to form the United Kingdom. A seafaring people, British traders, soldiers, and settlers won control of large areas of land throughout the world. By the mid-1800s, the United Kingdom governed the world's largest overseas empire. This empire was so vast that, according to a popular saying, "the sun never sets on the British Empire."

In the 1900s, two world wars weakened the United Kingdom. Nearly all of the British Empire broke into independent countries. Some of the former colonies joined the United Kingdom in a new organization called the Commonwealth of Nations. Recently the United Kingdom has joined other European countries to form the European Union. The goal of this organization is to unify and strengthen the economies of member countries.

Gifts to the World The literature of the United Kingdom ranks as one of its greatest gifts to the world. Its writers include William Shakespeare, Robert Burns, Charles Dickens, and Charlotte and Emily Brontë. Today the United Kingdom remains a center for literature, theater, and the arts.

When you vote in your first election in the United States, you will be sharing another British gift. The British form of government—**parliamentary democracy**—was copied in part by the Framers of the U.S. Constitution. In a parliamentary democracy, voters elect representatives to a lawmaking body called Parliament. The largest political party in Parliament chooses the government's leader, the prime minister.

The United Kingdom is also a **constitutional monarchy,** in which a king or queen represents the country at public events. The prime minister and other elected officials, however, actually hold the powers of government.

A Crowded Country Look at the map on page 269. Like other European countries, the United Kingdom is densely populated. About 90 percent of the British live in cities and towns. The city of London in southern England is a world center of trade, business, and banking. About 8 million people live in London and surrounding suburbs.

The ancient and the modern meet in London. Thirteen-year-old Pamela Wilson lives with her family in a small house on the outskirts of London. She enjoys going to the West End, an area of London famous for its shops, museums, theaters, and restaurants. Not far from the West End are the church of Westminster Abbey, where Britain's kings and queens are crowned, and Buckingham Palace, the London home of the British royal family. Pamela and her friends also enjoy watching soccer matches, as do many other British people.

WHAT IN THE WORLD?

The Chunnel

In 1994 England and France, separated for thousands of years by the English Channel, were once again joined. The Channel Tunnel, nicknamed the Chunnel, linked the island country to mainland Europe. By the year 2003 the Chunnel is expected to carry more than 120,000 people between England and France each day. At night, tons of freight will move through the Chunnel.

SECTION 1 ASSESSMENT

REVIEWING TERMS AND FACTS

1. Define the following: moor, loch, parliamentary democracy, constitutional monarchy.

2. REGION What landforms are found in the United Kingdom?

3. HUMAN/ENVIRONMENT INTERACTION What natural resources do the British export?

4. MOVEMENT In what ways have the British influenced other parts of the world?

MAP STUDY ACTIVITIES

5. Look at the climate map on page 258. What type of climate does the United Kingdom have?

6. Turn to the political map on page 254. What is the distance between the cities of London and Glasgow?

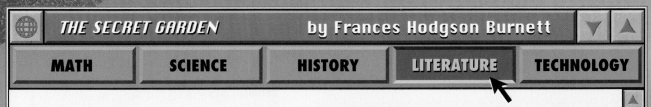

MAKING CONNECTIONS

The Secret Garden is a novel set in England in the early 1900s. The following excerpt gives us a view of the English countryside through the eyes of a young traveler.

"What is a moor?" [Mary] said suddenly to Mrs. Medlock.

"Look out of the window . . . you'll see," the woman answered. . . .

The carriage lamps shed a yellow light on a rough-looking road which seemed to be cut through bushes and low-growing things which ended in the great expanse of dark apparently spread out before and around them. A wind was rising and making a singular, wild, low, rushing sound.

"It's—it's not the sea, is it?" asked Mary, looking round at her companion.

"No, not it," answered Mrs. Medlock. "Nor it isn't fields nor mountains, it's just miles and miles and miles of wild land that nothing grows on but heather and gorse and broom, and nothing lives on but wild ponies and sheep."

"I feel as if it might be the sea, if there were water on it," said Mary. "It sounds like the sea just now."

"That's the wind blowing through the bushes," Mrs. Medlock said. "It's a wild,

Heather Blooming on the Moors

dreary enough place to my mind, though there's plenty that likes it—particularly when the heather's in bloom."

. . . Two days later . . . the rainstorm had ended and the gray mist and clouds had been swept away in the night by the wind. The wind itself had ceased and a brilliant, deep blue sky arched high over the moorland. Never, never had Mary dreamed of a sky so blue. . . . The far-reaching world of the moor itself looked softly blue instead of gloomy purple-black or awful dreary gray.

From *The Secret Garden* by Frances Hodgson Burnett. Copyright © 1938 by Verity Constance Burnett, J. B. Lippincott Company.

Making the Connection   ▼ ▲

1. How did Mrs. Medlock describe the view from the carriage?
2. To what geographic feature did Mary compare the landscape?

SECTION 2 ········ The Republic of Ireland

PREVIEW

Words to Know
- peat
- bog

Places to Locate
- Republic of Ireland
- Dublin
- Killarney

Read to Learn . . .
1. why Ireland is called the Emerald Isle.
2. how the Irish struggled to win their independence.
3. how urban and rural Irish live.

Who lives next door to you? Are they just like you, or are they different? Ireland's way of life is very different from that of its neighbor, the United Kingdom. One of the ways in which Ireland differs is in its

religion—more than 90 percent of Ireland's people are Roman Catholics. Churches such as this one dot Irish lands.

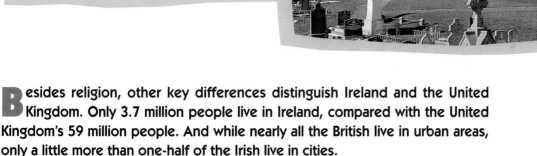

Besides religion, other key differences distinguish Ireland and the United Kingdom. Only 3.7 million people live in Ireland, compared with the United Kingdom's 59 million people. And while nearly all the British live in urban areas, only a little more than one-half of the Irish live in cities.

Place
The Emerald Isle

If you could fly over Ireland on a summer day, you would see lush green meadows and tree-covered hills. Surrounded on three sides by the Atlantic Ocean, Ireland's green color is so striking that it was named the Emerald Isle.

The Landscape The map on page 256 shows you Ireland's plains and highlands. At Ireland's center lies a wide, rolling plain dotted with low hills. Forests and farmland cover this central lowland. Much of the area is rich in **peat,** or wet ground with decaying plants that can be used for fuel. Peat is dug from **bogs,** or low swampy lands.

Two Irish workers cut slabs of peat, which will be used for heating homes.
HUMAN/ENVIRONMENT INTERACTION: On what type of land is peat usually found?

Along the Irish coast, the land rises in rocky highlands. In some places, however, the central plain spreads all the way to the sea. On the map on page 254, you will find Dublin—Ireland's capital—on an eastern stretch of the plain.

The Climate Whether plain or highland, no part of Ireland is more than 70 miles (113 km) from the sea. This nearness to the sea gives Ireland a uniform climate. Like the United Kingdom, Ireland is warmed by moist winds blowing over the North Atlantic Current. The mild weather, along with frequent rain and mist, makes Ireland's landscape green year-round.

The Economy If you look at the map on page 263, you see that Ireland has few mineral resources. The country, however, does have rich soil and pastureland.

The mild and rainy climate favors farming. In the mid-1800s, Irish farmers grew potatoes as their main food crop. When too much rain and a blight caused the potatoes to rot in the fields, famine struck, bringing hardship to the Irish. This disaster forced many Irish to emigrate to other countries, especially to the United States.

Although farming is still important to Ireland, industry now also contributes to economic development. The economy depends on the manufacturing of machinery and transportation equipment exported to the United Kingdom and the European mainland. Ships bringing mineral and energy resources to Ireland dock at the country's many ports, including Dublin and Cork.

Place
The People

Most of the Irish trace their ancestry to the Celts who settled Ireland during the 300s B.C. The English arrived in Ireland during the A.D. 1100s. Today the Celtic language, called Gaelic, and English are Ireland's two official languages. Most Irish, however, speak English as their everyday language.

Influences of the Past Stormy politics mark Ireland's history. From the 1100s to the early 1900s, the British governed Ireland. Religion and government controls mixed to cause disagreement. The Irish people resisted British rule and demands that the Roman Catholic country become Protestant. British officials seized land in Ireland and gave it to English and Scottish Protestants. At one time the British drove out Irish Catholics to make room for the new settlers.

WHAT IN THE WORLD?

What's in a Name?

Many Irish and Scottish last names begin with *Mac, Mc,* or *O',* such as MacDermott, McCormack, and O'Casey. Why? *Mac, Mc,* and *O'* mean "descendant of." *MacDermott,* then, means descendant of Dermott; *McCormack,* descendant of Cormack; and *O'Casey,* descendant of Casey.

THE BRITISH ISLES AND SCANDINAVIA: Land Use and Resources

MAP STUDY Countries bordering the North Sea harvest large amounts of seafood.
HUMAN/ENVIRONMENT INTERACTION: What resources are found in Ireland?

Agriculture
- Commercial farming
- Forestry
- Little or no activity
- Nomadic herding
- Manufacturing area

Resources
- Coal
- Copper
- Fishing
- Hydroelectric power
- Iron ore
- Lead
- Natural gas
- Petroleum
- Zinc

Lambert Conformal Conic projection

ARCTIC OCEAN

Arctic Circle

ICELAND
Sheep

ATLANTIC OCEAN

Norwegian Sea

SWEDEN

FINLAND

Helsinki

NORWAY

Turku

Oslo

Stockholm

Linköping

Baltic Sea

Göteborg

North Sea

DENMARK
Copenhagen

Malmö

Glasgow
Edinburgh
Newcastle

Belfast

Manchester

Dublin
Peat
Liverpool
Birmingham

IRELAND

Sheep

Sugar beets

EUROPE

Cardiff

London

UNITED KINGDOM

Portsmouth

English Channel

0   150   300 mi.
0   150   300 km

In 1921 the southern part of Ireland finally won its independence from the United Kingdom and became the Republic of Ireland. The northern part remained within the United Kingdom as Northern Ireland. Friction among the three regions continued—especially between Catholics and Protestants.

Disagreement over Northern Ireland exploded into violence in the 1960s. The violence continued into the 1990s. In 1998 British and Irish officials agreed to seek peace, but new violence in Northern Ireland threatened this effort.

Continuing Changes The Republic of Ireland continues to make changes to benefit its people. It is a member of the European Union (EU), which increases the number of markets for its products. In 1990 the Irish elected their first female president.

Daily Life Nearly 57 percent of Ireland's 3.7 million people live in cities and towns. Nearly one-third of its people live in the city of Dublin alone. Rural population has decreased. Life in both urban and rural areas often centers on the neighborhood or village church, where people take part in social gatherings and other activities.

Fifteen-year-old Sean Carroll lives in Dublin, Ireland's capital and largest city. He rides his bicycle to school across one of the bridges that span the River Liffey, which runs through the heart of Dublin. Sean, like most Irish boys, likes sports such as soccer and hurling, an Irish game similar to field hockey. His favorite food is Irish stew, a dish made by boiling potatoes, onions, and mutton in a covered pot.

Sean's father teaches Irish literature at the University of Dublin. Ireland is known for its writers, such as George Bernard Shaw and James Joyce. His mother works at the National Museum, which houses famous artworks from Celtic times. Sean and his family take their summer holiday in Killarney, a little town in the southwest of Ireland. The region of Killarney is popular because of its beautiful lakes, hills, and woods.

SECTION 2 ASSESSMENT

REVIEWING TERMS AND FACTS
1. **Define the following:** peat, bog.
2. **LOCATION** Where in Ireland are plains and highlands located?
3. **MOVEMENT** Why did many Irish emigrate to other countries in the mid-1800s?
4. **HUMAN/ENVIRONMENT INTERACTION** How does Irish stew reflect the rural landscape of Ireland?

MAP STUDY ACTIVITIES
5. Turn to the land use map on page 263. What is the main manufacturing area in Ireland?
6. Look at the same map on page 263. What type of farming takes place throughout Ireland?

BUILDING GEOGRAPHY SKILLS

Reading a Climate Map

Before you dress for school in the morning, you probably check the weather. You want to know how warm or cold or rainy it is today. If you're packing to go on vacation to another area, you might check a climate map. You want to know what the weather usually is like at this time of year.

A climate map reveals the usual weather patterns of a region. A climate region can be defined by its temperature and amount of precipitation. Near the Equator, for example, the climate is warm year-round. Some areas, however, have *rain forest climates* that are warm and very wet. Others have *desert climates* that are warm and very dry.

On a climate map, colors represent climate regions. The map key explains what the colors mean. To read a climate map, apply these steps:

• Identify the area shown in the map.

• Study the map key to identify the climate regions on the map.

• Locate the countries or areas in each climate region.

Geography Skills Practice

1. What two major kinds of climates are found in Norway?
2. Does the west coast of Norway have more or less snow than inland areas of the country? Why?
3. How is the climate of northern Norway different from that of southern Norway?

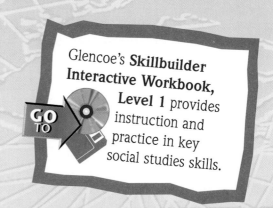

Glencoe's **Skillbuilder Interactive Workbook, Level 1** provides instruction and practice in key social studies skills.

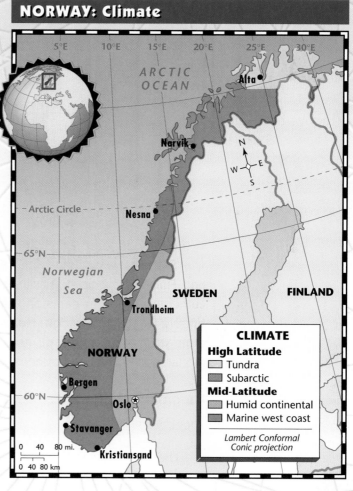

NORWAY: Climate

ARCTIC OCEAN

Alta

Narvik

Arctic Circle

Nesna

65°N

Norwegian Sea

SWEDEN

FINLAND

Trondheim

NORWAY

Bergen

60°N

Oslo ✪

Stavanger

Kristiansand

0   40   80 mi.
0   40   80 km

CLIMATE
High Latitude
☐ Tundra
☐ Subarctic
Mid-Latitude
☐ Humid continental
☐ Marine west coast

Lambert Conformal Conic projection

Scandinavia

PREVIEW

Words to Know
- fjord
- welfare state
- geyser

Places to Locate
- Norway
- Sweden
- Denmark
- Iceland
- Finland

Read to Learn . . .
1. how the Atlantic Ocean affects climate in Scandinavia.
2. where most Scandinavians live.
3. how Scandinavians work and enjoy leisure time.

Does the sun ever shine at midnight? If you ask the Sami—reindeer herders who live in northern Scandinavia—they will say, "Of course it does." Between April and August, the sun shines 24 hours a day in the northern area of Norway.

Norway is part of Scandinavia, a large region in northern Europe. Scandinavia also includes four other countries: Sweden, Denmark, Iceland, and Finland. The Scandinavian countries have democracy and the Protestant Lutheran religion in common. All share similar cultures, but Finland has a distinctive language. Sweden, Norway, and Denmark also have something in common with their neighbor, the United Kingdom. They have constitutional monarchies—kings and queens who have no real power.

Human/Environment Interaction
Norway

The Scandinavian Peninsula sweeps from the Arctic Ocean to the North Sea. Norway is a narrow kingdom that runs the length of the peninsula. The map on page 256 shows you Norway's jagged Atlantic coastline. Its many **fjords,** or steep-sided valleys, are inlets of the sea. What caused these deep valleys? Thousands of years ago glaciers scoured the land. The fjords now provide Norway with sheltered harbors.

Landscape and Climate The snowcapped Kjølen Mountains tower over most of Norway. Limited areas of land are available for farming. Rivers rushing down from the mountains provide hydroelectric power to Norway's farms, factories, and homes.

Warm winds from the North Atlantic Current give Norway's southern and western coasts a mild climate. Most of the country's 4.4 million people live in these areas. Oslo, the capital and largest city, lies at the end of a fjord on the southern coast.

People of the Sea From the icy oceans lapping its long coastline, Norway built its economy. About 1,000 years ago, the Vikings—ancestors of most modern Scandinavians—sailed west. They founded settlements in Iceland, Greenland, and even northern Canada. Today their Norwegian descendants harvest fish from the sea. The map on page 263 shows that the Norwegians also benefit from oil and natural gas found beneath the North Sea waters.

Norwegians Bring in Their Catch

Love of Sports Sonja Hamsun, an Oslo teenager, likes to travel to the mountains of Norway's interior. This part of the country has a colder climate because the mountains block the warm coastal winds. Snow covers the ground at least three months of the year. Like most Norwegians, Sonja loves outdoor sports, especially skiing. Before modern transportation linked Norway's cities, Norwegians traveled over the snow on skis. Today skiing is Norway's national sport.

A Norwegian fishing crew prepares its catch for a local fish market in Bergen, Norway.
PLACE: How were most of Norway's harbors formed?

Place

Sweden

What would you see from an airplane flying east from Norway? The many mountains of Sweden rise up to greet you. Then the shimmering peaks and valleys end abruptly at the seacoast. Sweden's long coastline of sandy beaches and rocky cliffs touches the Baltic Sea, the Gulf of Bothnia, and a narrow arm of the North Sea.

Landscape and Climate The mountains along Sweden's border with Norway block the warm winds of the North Atlantic Current. This causes northern Sweden to have cool summers and cold winters. As you descend from the snow-covered mountains, you come to forested highlands and then fertile lowlands. Many lakes, hills, and farms dot the lowland coast of southern Sweden. The North Atlantic winds reach this area, giving it a mild climate.

These people are part of a pageant on St. Lucia Day. Young people often awaken their families with a traditional song and a breakfast of buns and coffee.
PLACE: What other holiday is popular in Sweden?

Prosperous Economy Sweden has one of Europe's most prosperous economies. The Swedes have developed profitable industries based on timber, iron ore, and water power. With this economic wealth, Sweden has become a **welfare state.** In a welfare state, the government uses tax money to provide education, medical services, health insurance, and other services for its citizens.

The People The map on page 269 shows you that most of Sweden's 8.9 million people live in cities in the southern lowlands. Stockholm, on the Baltic coast, is the country's capital and largest city. Stockholm's churches, palaces, and modern buildings spread out over several islands connected by more than 40 bridges. Another city, Göteborg (YUHR•tuh•BAWR•ee), is a major port.

Like the Norwegians, the Swedes enjoy the country-side and outdoor sports. One of their favorite holidays, Midsummer's Eve, comes in late June and celebrates summer. On this holiday people dance and feast into the early morning hours—with the sky still light from the midnight sun.

Location

Denmark

Picture the Scandinavian Peninsula as the head of a huge dragon, its jaws open wide. Its morsel of food is the country of Denmark. Denmark is a small country south of Sweden between the North and Baltic seas.

Landscape and Climate The map on page 256 shows you that most of Denmark is made up of the Jutland Peninsula, which connects the country to the European mainland. Denmark also includes hundreds of nearby small islands. In addition, tiny Denmark rules the gigantic island of Greenland, 1,300 miles (2,090 km) away off the northeastern coast of Canada. Rolling, green hills and low, fertile plains run through most of Denmark. Warm North Atlantic winds give the country a mild, damp climate.

Farm Country Denmark is poor in natural resources, but it has some of the richest farmland in Scandinavia. Danish farm products include butter, cheese, bacon, and ham. Foods make up much of Denmark's exports. Food exports help pay for the fuels and metals the Danish people must import for their industries. Danish workers produce beautifully designed furniture, porcelain, and silver.

The People Like the Swedes, the 5.3 million Danes have a high standard of living and a welfare state. More than half of them live on the islands near the Jutland Peninsula. Copenhagen, Denmark's capital and largest city, is on the largest island. This city is known for its old church spires and red tile rooftops as well as its modern buildings and parks. It is also a major port. Denmark is the most densely populated Scandinavian country.

Place
Iceland

When and where was the first government legislature in the world established? In A.D. 930 in Iceland—a small island republic that lies just south of the Arctic Circle in the Atlantic Ocean. Once part of Denmark, Iceland declared its independence in 1944.

Land of Fire and Ice Most of Iceland is a rugged plateau that sits on top of a fault line, or break in the earth's crust. This makes Iceland a land of volcanoes, hot springs, and **geysers**—springs that spout hot water and steam. Icelanders use the naturally heated water to heat their buildings.

Because Iceland is so far north, it is a land of glaciers. Fast-flowing rivers, some formed by melting snow and glaciers, provide a good source of hydroelectric power. The North Atlantic Current warms most of Iceland's coast and keeps temperatures from getting too cold.

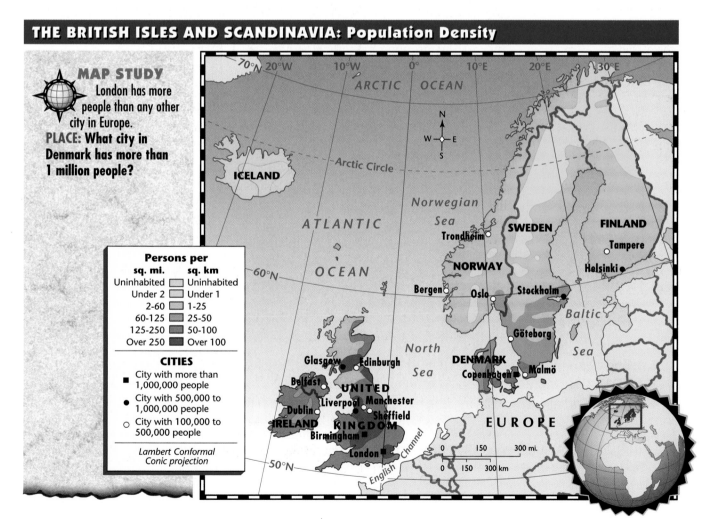

THE BRITISH ISLES AND SCANDINAVIA: Population Density

MAP STUDY
London has more people than any other city in Europe.
PLACE: What city in Denmark has more than 1 million people?

Persons per

sq. mi.	sq. km
Uninhabited	Uninhabited
Under 2	Under 1
2-60	1-25
60-125	25-50
125-250	50-100
Over 250	Over 100

CITIES
■ City with more than 1,000,000 people
● City with 500,000 to 1,000,000 people
○ City with 100,000 to 500,000 people

Lambert Conformal Conic projection

The Economy and People Iceland has few mineral resources and areas of farmland. The country's economy depends on fishing. Fish exports provide the money for Iceland to buy food and consumer goods from other countries.

Icelanders trace their heritage to Viking explorers who came from mainland Scandinavia. Today nearly all 300,000 Icelanders live near the coast. About half live in Reykjavík (RAY•kyuh•vihk), the most northern capital city in the world. Icelanders, with a 100 percent literacy rate, like reading and sports. Iceland's most famous works of literature are sagas, or long tales, written during the 1100s about Vikings.

Location

Finland

Finland is a long, narrow country tucked among Sweden, Norway, and Russia. Ruled first by Sweden and later by Russia, Finland became an independent republic in 1917. Finland lies on a flat plateau broken by small hills and valleys. Thick green forests and thousands of blue lakes cover most of the country. Finland has humid continental and subarctic climates because of its great distance from the warm North Atlantic Current.

The Economy Most of Finland's wealth comes from its huge forests. Paper and wood production are the country's major industries. Because of poor soil and a short growing season, Finland's farming areas are in the south. There, farmers raise wheat, rye, and livestock, including dairy cattle.

The People Most of Finland's people belong to an ethnic group known as Finns. The Finns' culture and language are very different from those of the other Scandinavian peoples. The rest of Finland's people are Swedes, Russians, or Sami. The Sami are a small ethnic group who live in northern Scandinavia and have their own separate culture and language. Their traditional nomadic way of life is based on herding reindeer. Most of Finland's 5.2 million people live in towns and cities on the southern coast. Helsinki, the capital and largest city, is known for its scenic harbor and modern buildings.

SECTION 3 ASSESSMENT

REVIEWING TERMS AND FACTS
1. Define the following: fjord, welfare state, geyser.
2. HUMAN/ENVIRONMENT INTERACTION How have Norwegians adapted to their environment?
3. LOCATION How does Iceland's location affect its landscape?
4. PLACE In what ways is Finland different from other Scandinavian countries?

MAP STUDY ACTIVITIES

5. Look at the population density map on page 269. What area of Norway is most sparsely populated?

Chapter 10 Highlights

Important Things to Know About the British Isles and Scandinavia

SECTION 1 THE UNITED KINGDOM

- The United Kingdom is made up of England, Scotland, Wales, and Northern Ireland.
- The landscape of the United Kingdom consists of highlands and lowlands.
- Winds blowing over the warm North Atlantic Current provide the United Kingdom with a mild climate.
- The United Kingdom is a major industrial and trading nation.
- Literature and parliamentary democracy are two of the United Kingdom's greatest gifts to the world.

SECTION 2 THE REPUBLIC OF IRELAND

- Ireland is called the Emerald Isle because of its lush vegetation that stays green year-round.
- The Irish were once dependent on the land. Now they are developing an industrial economy.
- Ireland's stormy political history has caused frequent changes in government policies.

SECTION 3 SCANDINAVIA

- The region of Scandinavia is made up of the countries of Norway, Sweden, Denmark, Iceland, and Finland.
- The warm North Atlantic Current, an extension of the Gulf Stream, keeps temperatures mild in many parts of Scandinavia.
- The economies of the Scandinavian countries depend on produce from the land and the sea.

Royal Guards in London, England

REVIEWING KEY TERMS

Match the numbered terms in Set A with their lettered definitions in Set B.

A

1. moor
2. parliamentary democracy
3. fjord
4. geyser
5. peat
6. constitutional monarchy
7. loch
8. welfare state

B

A. government in which a king or queen shares power with elected officials
B. a type of decayed plant material
C. spring that spouts hot water and steam
D. steep-sided valley filled with seawater
E. form of government in which the people rule through Parliament
F. wild, treeless land in the United Kingdom
G. system under which the government provides major services to citizens
H. narrow bay cut into the highland coasts of Scotland

REVIEWING THE MAIN IDEAS

Mental Mapping Activity

Draw a freehand map of the British Isles and Scandinavia. Label the following:
• Reykjavík
• Dublin
• Scotland
• North Sea
• Finland

Section 1

1. **HUMAN/ENVIRONMENT INTERACTION** What is the United Kingdom's major natural resource?
2. **PLACE** What kind of government does the United Kingdom have?

Section 2

3. **MOVEMENT** How does the North Atlantic Current affect Ireland's climate?
4. **MOVEMENT** How has membership in the European Union helped Ireland?

Section 3

5. **HUMAN/ENVIRONMENT INTERACTION** How has Scandinavia's northern location affected people's daily lives?
6. **LOCATION** What is the capital of Finland?

CRITICAL THINKING ACTIVITIES

1. **Evaluating Information** How has the Industrial Revolution that began in the United Kingdom affected your life?
2. **Making Comparisons** What comparisons can you make between the landscapes of Norway and Denmark? Which country's land would affect their tourist industry the most?

CONNECTIONS TO WORLD HISTORY

Cooperative Learning Activity Discover how the movement of peoples into England affected the English language. Organize into three groups. Each group will look through a dictionary to list at least 15 commonly known words that came from the language of (a) the Angles, Saxons, or Jutes (Old English); (b) Norman French (Old French); or (c) Old Norse. Write your group's complete list on a chart. Then share your words with the class through a game of charades.

GeoJournal Writing Activity

Imagine that you are working on an *estancia,* a large cattle ranch in Argentina. Write a letter to a family member or a friend describing how you begin your day, your work on the ranch, and how you end your day.

TECHNOLOGY ACTIVITY

Using the Internet Search the Internet to find a Web site presenting current information about the North Sea and how it is of major economic importance to the European economy. Use the information you find to write an article for your school newspaper or magazine.

PLACE LOCATION ACTIVITY: BRITISH ISLES AND SCANDINAVIA

Match the letters on the map with the places listed below. Write your answers on a separate sheet of paper.

1. Reykjavík
2. Baltic Sea
3. London
4. Stockholm
5. North Sea
6. Northern Ireland
7. Denmark
8. English Channel
9. Republic of Ireland
10. Kjølen Mountains

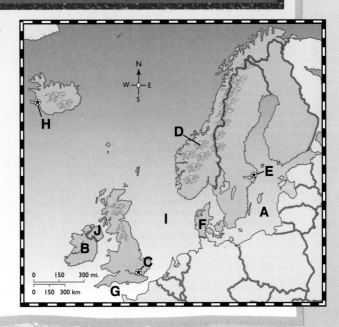

Cultural HERITAGE:
EUROPE

ARCHITECTURE ▶ ▶ ▶ ▶

The wooden stave churches in Scandinavia were built during the early 1100s. They were the first Christian churches in Scandinavia. Their steeply angled roofs are covered with scalelike shingles and are often decorated with dragons and other Viking symbols.

◀ ◀ ◀ ◀ SCULPTURE

The European Renaissance was a time of invention and creativity. Michelangelo, a brilliant Italian sculptor and painter, created the Pietá in 1500. This marble sculpture of the Virgin Mary and Christ is housed in St. Peter's Basilica in Rome.

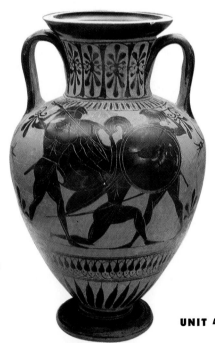

ANCIENT POTTERY ▶ ▶ ▶ ▶

The painting on this ancient Greek vase shows warriors with their shields raised in battle. Artifacts provide a glimpse of what life was like more than 2,000 years ago in Europe.

◄ ◄ ◄ ◄ ROYAL JEWELS

This diamond-studded gold crown is part of the collection of the British crown jewels. The precious ornaments and jewels of Great Britain's kings and queens are stored in an underground vault in the Tower of London.

PAINTING ► ► ► ►

The French Impressionists were a group of painters who used short brush strokes and brilliant color to show the play of light on outdoor subjects. Their goal was to give an *impression* rather than to show reality. Claude Monet painted this scene in Giverny, France, in the late 1800s.

PORCELAIN ▲ ▲ ▲ ▲

These porcelain figures were made in Germany in the 1700s. The tiny sculptures were used as table decorations. A world-famous porcelain factory still operates near Dresden, Germany.

APPRECIATING CULTURE

1. Describe what you like or do not like about Monet's Impressionist painting.

2. How is the sculpture by Michelangelo different from the miniature porcelain sculptures made in Germany?

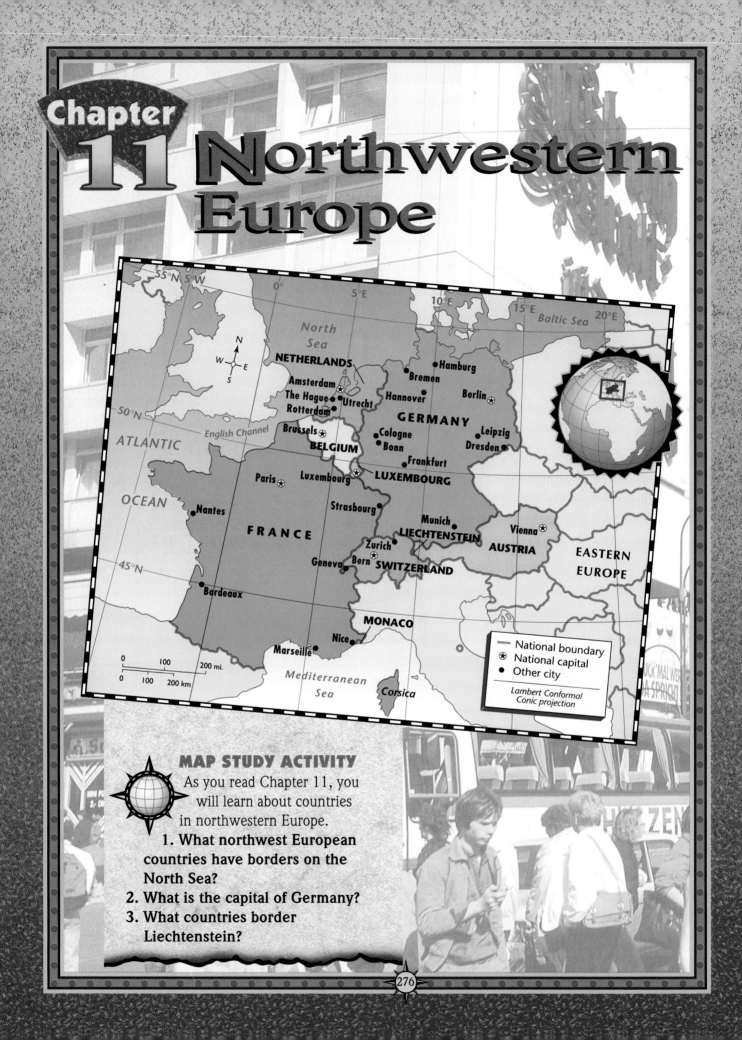

MAP STUDY ACTIVITY

As you read Chapter 11, you will learn about countries in northwestern Europe.

1. What northwest European countries have borders on the North Sea?
2. What is the capital of Germany?
3. What countries border Liechtenstein?

PREVIEW

Words to Know
- navigable
- republic

Places to Locate
- France
- Paris
- Loire River
- Seine River
- Pyrenees
- Alps

Read to Learn . . .
1. what landforms are found in France.
2. why France is able to produce huge amounts of food.
3. what French culture offers the rest of the world.

Imagine yourself on a river barge gliding between fairy-tale castles and peaceful villages. You ask the captain about this beautiful river, the Loire (LWAHR), and find out you are cruising on the longest river in France. Along its banks lie rich farmlands and prosperous manufacturing centers. France welcomes you.

France, Germany, and their smaller neighbors rank as major economic and cultural centers of Europe. In the first half of the twentieth century, warfare among these nations tore Europe apart. Since the end of World War II, however, they have put aside their differences. Joined in economic partnership in the European Union, they look forward to a peaceful and prosperous twenty-first century.

Region

The Land

Home to one of the oldest civilizations in the world, France's land area covers 212,392 square miles (550,095 sq. km). Slightly smaller in area than the state of Texas, about one-half of France's border is coastline. The Mediterranean Sea lies along its southeast coast. Along the country's west and northwest shores are the Atlantic Ocean and the English Channel. A new tunnel—the Chunnel—built under this channel connects France with Great Britain. For the first time, citizens and freight can move directly between these two nations by truck, rail, and automobile.

Windy Wonders

If you built a home in the Rhône Valley of France, you might put windows only on the southeast side of it. Why? To protect yourself from the mistral, a cold dry wind that blows through the valley about 100 days a year. So powerful is this wind—averaging 40 to 80 miles per hour (64 to 129 km per hour)—that trees grow permanently bent by its force.

Mountains Great mountain ranges also form part of France's borders. In the southwest the Pyrenees (PIHR•uh•NEEZ) separate France from Spain. In the southeast the Alps divide France from Italy and Switzerland. The soaring, icy peaks of the French Alps include Mont Blanc. At 15,771 feet (4,807 m), Mont Blanc is one of the highest mountains in Europe.

Plains and Rivers The map on page 246 shows you that most of northern France is part of the vast North European Plain. It stretches from the British Isles east to Russia. In France a variety of crops thrive in the rich soil of the plain. Livestock are raised here, too, and the area is home to many industrial cities.

An important transportation network of rivers connects major manufacturing areas within France. Most of these rivers are **navigable,** or usable by large ships. The Seine (SAYN) River flows through the capital city of Paris. The Rhône (ROHN) River lies in the southeast. The Rhine River—Europe's most important inland waterway—forms part of France's eastern border.

The Climate How many different climate regions of France do you see on the map on page 284? Note that France has three types of climate. Along the Mediterranean coast, summers are hot and winters are mild. The climate of

---

NORTHWESTERN EUROPE: Physical

MAP STUDY
Mont Blanc is one of Europe's highest mountains.
MOVEMENT: A trip from Mont Blanc to the Matterhorn would start and end in what countries?

ELEVATIONS

Feet	Meters
10,000	3,000
5,000	1,500
2,000	600
1,000	300
0	0

▲ Mountain peak

Lambert Conformal Conic projection

North Sea
Baltic Sea
NETHERLANDS
NORTH EUROPEAN PLAIN
Weser R.
Elbe R.
English Channel
BELGIUM
Meuse R.
Rhine R.
GERMANY
ATLANTIC
OCEAN
ARDENNES
LUXEMBOURG
Seine R.
SWITZERLAND
Black Forest
Danube R.
Loire R.
Lake Constance
AUSTRIA
LIECHTENSTEIN
EASTERN EUROPE
FRANCE
Lake Geneva
ALPS
Bay of Biscay
CENTRAL MASSIF
Mt. Blanc 15,771 ft. (4,807 m)
Matterhorn 14,691 ft. (4,478 m)
MONACO
PYRENEES
Corsica
Mediterranean Sea

0   100   200 mi.
0   100   200 km

western France, influenced by Atlantic winds, has cool summers and mild winters with plenty of rainfall. As you travel farther east—away from the ocean—winters are colder, summers warmer, and there is less rain.

The Seine River and Eiffel Tower are two of Paris's most famous landmarks *(left)*. On Bastille Day, soldiers parade in Paris to commemorate the French Revolution *(right)*.
PLACE: Judging from the parade, what are France's national colors?

Place
The Economy

France's well-developed economy relies on agriculture, industry, and commerce. France is the largest food producer in western Europe. French farmers grow fruits, vegetables, and grains. They also raise beef and dairy cattle. Grapes rank as one of the most important French crops. Why? The grapes are used to make famous French wines.

France is also one of the world's leading manufacturing countries, with Paris as its chief industrial center. France's natural resources include coal, iron ore, and bauxite. A variety of industrial goods—including airplanes, cars, computers, chemicals, clothing, and furniture—are produced by French workers. France is also a leading center of commerce, with an international reputation in fashion.

Place
The People

"Liberté . . . Egalité . . . Fraternité" (Liberty, Equality, Fraternity)—France's national motto—describes the spirit of the people. Although they have regional differences, the French share a strong national loyalty. Most French trace their ancestry to the Celts and Romans of early Europe. They speak French, and about 75 percent of them are Roman Catholics.

Influences of the Past Kings ruled France for nearly 800 years, establishing France as a great European power. In 1792 the French people overthrew their last king. They set up a **republic,** a government in which the people elect all important government officials. A few years later, a general named Napoleon Bonaparte seized power. He conquered much of Europe before an alliance of European countries defeated him. In the first half of the 1900s, France was a battleground during the two world wars. As strong as its motto, France has survived and holds an important role in modern world affairs.

City Life If you drew a line connecting the five largest French cities from south to north, you would form a triangle. You can see on the map on page 292 that the areas around Marseille, Bordeaux, and Paris hold most of France's people. A superb system of railroads connects these and other cities. French passenger trains are some of the fastest in the world, many averaging 132 miles per hour (212 km per hour). The population of France totals about 58.8 million people. About 74 percent of the French people live in cities and large towns.

Paris Paris itself is home to about 9.5 million people. France's literacy rate of 99 percent shows that the French prize education. Paris, a world center of art and learning, is home to many universities, museums, and other cultural sites. It is also France's leading center of industry, transport, and communications. Millions of tourists from all over the world visit Paris annually.

Thirteen-year-old Hervé Lefort lives in Paris and can see the Eiffel Tower every day. He likes to visit the tower to enjoy the wonderful view of the whole city. Among the most famous sights in Paris is the beautiful cathedral of Notre Dame, which sits on an island in the Seine River. The nearby Louvre (LOOV) is one of the greatest art museums in the world.

Hervé and his friends look forward every year to the Tour de France—Europe's premier bicycle race. Held since 1903, this three-week competition matches cyclists from all over the world.

SECTION 1 ASSESSMENT

REVIEWING TERMS AND FACTS
1. **Define the following:** navigable, republic.
2. **REGION** What is important about the region of the North European Plain?
3. **PLACE** Why is Paris an important city?
4. **HUMAN/ENVIRONMENT INTERACTION** What are France's major agricultural products?

MAP STUDY ACTIVITIES
5. Look at the physical map on page 278. What mountain ranges are located along France's borders?
6. Turn to the political map on page 276. What is France's smallest neighbor along its eastern border?

BUILDING GEOGRAPHY SKILLS

Reading a Vegetation Map

A vegetation map helps you see whole regions that have the same type of plant life. These areas often share other natural elements, too, such as climate and soil type. For example, in rain forests, broad-leaved evergreens grow consistently throughout the region. Rain forests also have the same climate and soil types. To read a vegetation map, apply these steps:

- Use the map key to match colors with vegetation regions.

- Use a dictionary to define unfamiliar vegetation terms.

- Draw conclusions about the climate and soil of the vegetation region.

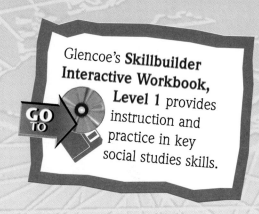

Glencoe's **Skillbuilder Interactive Workbook, Level 1** provides instruction and practice in key social studies skills.

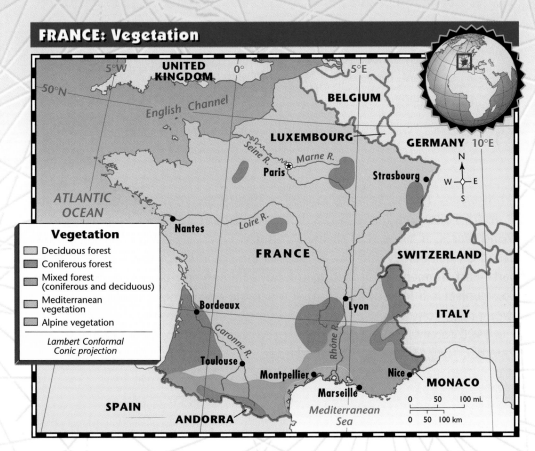

FRANCE: Vegetation

UNITED KINGDOM
5°W
0°
5°E
50°N
English Channel
BELGIUM
LUXEMBOURG
Seine R.
Marne R.
GERMANY 10°E
Paris
Strasbourg
ATLANTIC OCEAN
Nantes
Loire R.
FRANCE
SWITZERLAND
N
W   E
S
Bordeaux
Lyon
ITALY
Garonne R.
Rhône R.
Toulouse
Montpellier
Nice
MONACO
Marseille
SPAIN
Mediterranean Sea
ANDORRA

Vegetation
- Deciduous forest
- Coniferous forest
- Mixed forest (coniferous and deciduous)
- Mediterranean vegetation
- Alpine vegetation

Lambert Conformal Conic projection

0   50   100 mi.
0 50 100 km

Geography Skills Practice

1. What vegetation region covers most of France?

2. What type of vegetation is found only along France's Mediterranean coast?

3. From the map, what conclusions can you draw about the amount of rain the regions of France receive?

SECTION 2 Germany

Words to Know

- communism
- acid rain
- loess
- *autobahn*
- dialect
- Holocaust

Places to Locate

- Germany
- Berlin
- Alps
- Black Forest
- the Ruhr

Read to Learn . . .

1. what the landscape is like in Germany.
2. why the German economy is so strong.
3. how historical events affected the geography of Germany.

On October 3, 1990—in just one day—the Federal Republic of Germany grew in area by almost 50 percent. East Germany and 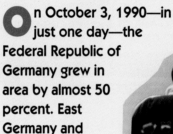 West Germany had officially reunited. Germans hurried to chip away at the Berlin Wall—the symbol of the once-divided country.

In the early 1900s, Germany ranked as one of the strongest countries in the world. Following World War I, however, it needed rebuilding. Then after World War II, it became a divided country. One part—West Germany—was based on democracy and closely tied to the countries of the West. The other part—East Germany—was tied to the Soviet Union and based on communism. **Communism** is a form of socialism in which the government, usually under one strong ruler, controls the economy and the society in general.

In 1990 communism in Europe collapsed and the two parts of Germany came together. They formed one democratic country. Modern, united Germany has a strong, free market economy.

Region

The Land

Now a much larger country, Germany covers about 134,853 square miles (349,270 sq. km). It lies in the heart of Europe, south of Scandinavia and northeast of France. Germany's northern border touches Denmark and the Baltic and North seas.

The Landscape Mountains, plateaus, and plains form the landscape of Germany. In the south the Alps rise over the German border with Switzerland and Austria. The lower slopes of these mountains—a favorite area of snow skiers—are covered with forests. Beyond the forests lie open meadows. Above the slopes are the mountain peaks, many capped with snow year-round.

Have you ever heard a cuckoo clock announce the hour? That clock probably came from Germany's Black Forest. Lying north of the Alps, this part of the Central Uplands is famous for its wood products. Is the forest really black? No, but the trees grow so close together the forest appears to be black.

In recent decades, the Black Forest has suffered severe damage from **acid rain,** or rainfall with high amounts of chemical pollution. Growing numbers of European factories and automobiles are responsible. Germans are making efforts to conserve their forests. A solution to the acid rain problem, however, has yet to be found.

From Germany's Central Uplands flows one of Europe's most important rivers—the Danube. It winds eastward for about 400 miles (650 km) across southern Germany. The Rhine, the Elbe, and the Weser rivers stretch across the North European Plain.

The Climate Westerly Atlantic Ocean winds crossing Europe help warm Germany in winter and cool it in summer. The **loess** (LEHS), or soil deposited by winds, produces good crops. Most of Germany has a marine west coast climate. Southern areas more distant from the ocean have colder winters and warmer summers.

The Bavarian Alps in Germany

The beauty of the snowcapped Bavarian Alps attracts many tourists.
LOCATION: Besides Germany, through what two countries do the Bavarian Alps extend?

Place
The Economy

The product label "Made in Germany" means superior quality and construction to consumers all over the world. Why? Germany's manufacturing economy has two major strengths: highly skilled workers and an abundance of key industrial resources.

The Ruhr Rich deposits of coal and iron ore found in the Ruhr help make this region in western Germany the industrial heart of Europe. Battles have been fought over the control of this productive area. Factories here produce machinery, cars, clothing, and chemical products. In the Ruhr, often called a "smokestack region," cities and factories stand as close together as teeth on a comb. They cover almost every square mile of the Ruhr.

Farming Germany imports about one-third of its food, although it is a leading producer of beer, wine, and cheese. German farmers raise livestock and crops

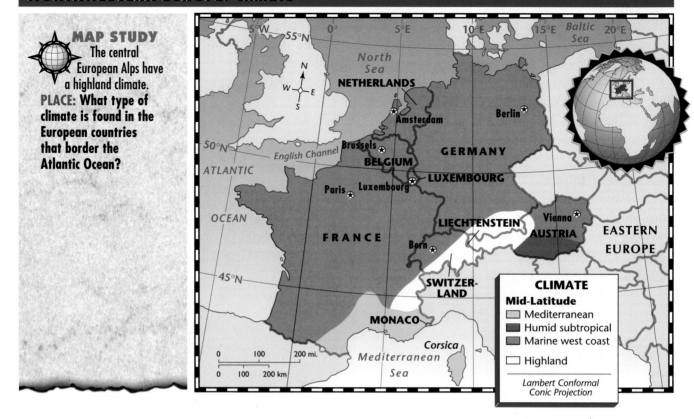

MAP STUDY
The central European Alps have a highland climate.
PLACE: What type of climate is found in the European countries that border the Atlantic Ocean?

CLIMATE
Mid-Latitude
Mediterranean
Humid subtropical
Marine west coast
Highland

Lambert Conformal Conic Projection

such as grains, vegetables, and fruits. German workers and products travel to market on superhighways called *autobahns.* These highways, along with canals and railroads, link Germany's cities.

Movement

The People

If a German said to you, *"Guten tag,"* how would you reply? You would return this greeting of "Good day" with a similar one. The German language commonly spoken within Germany has two major **dialects,** or local forms of a language. Most of Germany's 82.3 million people trace their ancestry to groups of people who settled in Europe from about the A.D. 100s to the 400s. Roman Catholics and Protestants make up most of the population and are fairly evenly represented.

Influences of the Past For hundreds of years, Germany was a collection of small territories ruled by princes. During the late 1800s, a leader named Otto von Bismarck united the territories into a single nation—Germany. The country's efforts to become a world power, in part, caused World War I.

During the 1930s dictator Adolf Hitler gained control of Germany. Under his leadership Germany increased its military power and invaded neighboring countries, setting off World War II.

One of the horrors of World War II was the **Holocaust,** the systematic murder of more than 6 million European Jews and 6 million others by Hitler and his followers. The Allies—led by the United States, Great Britain, and the Soviet Union—defeated Germany and ended the Holocaust.

Following World War II, Germany was divided into democratic West Germany and communist East Germany. West Germany built a powerful free enterprise economy. East Germany made slower progress under communism. East Germany and West Germany reunited in 1990. An expanded free enterprise system and greater European unity are goals of the new German government.

Daily Life About 85 percent of Germans live in cities and towns. Berlin, with a population of 3.5 million, is the largest city and the official capital. Hamburg, in the north, is Germany's major port. Many government offices are located in Bonn and Cologne, cities on the Rhine River. The southern cities of Munich and Stuttgart rank as the fastest-growing cities in Germany. Industrial expansion and technology have spurred this growth.

Fourteen-year-old Eva Reinhardt lives in Munich. Her favorite subjects in school are music, German literature, and mathematics. Eva is proud of what her country has contributed to the worlds of music, literature, science, and good food.

Germans excel at sports and physical activities. Eva often joins friends for a hike or a ski trip in the Bavarian Alps near Munich. Oktoberfest, a 16-day fall festival, offers an opportunity for Eva and her friends to wear traditional costumes and march in a parade. The festival includes amusement park rides, bands, folk dancers, and plenty of food—especially frankfurters.

German Rockers

By day, Sebastian Haussen and Ava Rottermich are students at a music academy in Berlin. By night, they are students of rock 'n roll! Sebastian and Ava started their own band two years ago. Someday they hope to play in one of Berlin's many cabarets. Cabarets are cafes with live entertainment. Sebastian lives in East Berlin. Now that the Berlin Wall is gone, he can travel freely in Germany.

SECTION 2 ASSESSMENT

REVIEWING TERMS AND FACTS

1. Define the following: communism, acid rain, loess, *autobahn,* dialect, Holocaust.

2. PLACE What are the three major landforms of Germany?

3. PLACE What natural resources are found in the Ruhr?

4. MOVEMENT How are goods transported within Germany?

MAP STUDY ACTIVITIES

5. Study the climate map on page 284. What is the major climate of Germany?

MAKING CONNECTIONS

| MATH | SCIENCE | **HISTORY** | LITERATURE | TECHNOLOGY |

Yelling—shouting—cheering voices rang through the streets of Berlin, Germany. It was midnight on November 9, 1989. The joyful crowd had gathered to tear down the hated Berlin Wall.

A BARRIER AND A SYMBOL The Berlin Wall—named *Schandmauer* (wall of shame) by Berliners—divided Berlin from 1961 to 1989. About 27½ miles (43 km) of concrete and barbed wire, it created East Berlin and West Berlin, dividing friends and families. The wall also became a symbol of the separation between democracy and communism. Why was the wall built?

After World War II, defeated Germany and Berlin, its capital, were divided among the victorious Allies. West Germany and West Berlin were allied to the Western democracies. They remained democratic. Communism and the Soviet Union ruled East Germany and East Berlin.

THE WALL GOES UP During the 1950s, many people escaped from East Germany by crossing into West Berlin. To stop the escapes, the East Germans built the Berlin Wall. As the world watched in shock, German soldiers laid bricks and barbed wire to seal off East Berlin. Along the wall appeared guard towers and searchlights. East German soldiers, armed with machine guns and German shepherd dogs, patrolled nearby.

Hello, Neighbor!

Crossing from East Berlin to West Berlin—without government permission—was now almost impossible. Yet hundreds of East Germans risked their lives to try.

THE WALL COMES DOWN Communism weakened in the 1980s. In 1989 East Germany announced many government changes. Included was the opening of the Berlin Wall. All that remains of the wall today are a few sections kept as reminders of a divided Germany.

Making the Connection

1. In what way was the Berlin Wall a geographical barrier?
2. In what ways was the Berlin Wall a cultural barrier?

The Benelux Countries

Words to Know
- polder
- multinational firm

Places to Locate
- Netherlands
- Amsterdam
- Belgium
- Brussels
- Luxembourg

Read to Learn . . .
1. what two ethnic groups live in Belgium.
2. how the people of the Netherlands have changed their environment.
3. why tiny Luxembourg attracts many businesses.

Tulips blossom as far as the eye can see! Tulips are not only beautiful but, in the Netherlands, they are also a major commercial crop.

The name *Benelux* comes from combining the first letters of *Bel*gium, the *Nether*lands, and *Lux*embourg. The three small Benelux countries of northwestern Europe have much in common. Much of their land area is low, flat, and densely populated. Most of their people live in cities, work in businesses or factories, and enjoy a high standard of living. Governments in the Benelux countries are parliamentary democracies with monarchs.

Location
Belgium

Belgian lace . . . Belgian diamonds . . . Belgian chocolate . . . all enjoy a worldwide reputation for excellence. Belgium is bordered by France, Germany, and the Netherlands. Because Belgium is centrally located, European wars have often been fought on its lands. Its central location has also contributed to trade and industry and a strong economy. The Belgian people import metals and fuels to produce manufactured goods for export. Rolling hills stretch through Belgium's southeast, and flat plains spread through the northwest. Traveling along Belgium's excellent road system, you are never very far from a town or city. Look at the map on page 292 to see the population density of Belgium.

The People The Walloons and the Flemings dominate the culture of Belgium. These two ethnic groups have lived in Belgium for hundreds of years. The French-speaking Walloons are the largest group of Belgians. The Flemings speak Dutch and often prefer to be called Flemish. Most of Belgium's people are Roman Catholics.

Brussels, the nation's centrally located capital city, has an international air. Many world organizations locate their headquarters there. Banks and other businesses use Brussels as a trade and commerce center.

Human/Environment Interaction
The Netherlands

The Netherlands—a country about half the size of the state of Maine—is located on the North Sea north of Belgium. At its elevation, about 40 percent of the country should be under water. That is how much of its land lies below sea level.

NORTHWESTERN EUROPE: Land Use and Resources

North Sea · NETHERLANDS · Hamburg · Bremen · Berlin · Amsterdam · Rotterdam · Cologne · GERMANY · Leipzig · Dresden · Brussels · BELGIUM · Frankfurt · Grapes · Mannheim · EASTERN EUROPE · Le Havre · Luxembourg · English Channel · Potatoes · Apples · Nantes · Paris · LUXEMBOURG · Munich · Vienna · LIECHTENSTEIN · Grapes · Grapes · AUSTRIA · ATLANTIC OCEAN · Apples · Zurich · SWITZERLAND · Grapes · FRANCE · Geneva · Tobacco · Corn · Olives · Grapes · Marseille · MONACO · Mediterranean Sea · Corsica

Agriculture
- Commercial farming
- Manufacturing area

Resources
- Bauxite
- Coal
- Copper
- Fish & other seafood
- Forest
- Iron ore
- Lead
- Natural gas
- Petroleum
- Potash
- Silver
- Uranium
- Hydroelectric power
- Zinc

0 50 100 mi.
0 50 100 km

Lambert Conformal Conic projection

MAP STUDY
The Benelux countries are members of the European Union.
PLACE: How do Belgium and Luxembourg differ from the Netherlands in terms of land use?

The Land and People The Dutch, or the people of the Netherlands, have been reclaiming their land from the sea since the A.D. 1100s. Land is valuable to the Dutch because the Netherlands is one of the most densely populated countries in the world. The Dutch method of reclaiming land is simple, although it takes hard work. First a dam is built across a coastal bay to hold back the sea. Then workers pump out the seawater to create a **polder**—an area of reclaimed land. Today, about 3,000 square miles (7,770 sq. km) of the Netherlands is made up of polders. In addition to growing tulips and other bulbs, the Dutch raise food crops and dairy cattle on polders. Amsterdam, the capital and largest city, is located on a polder.

About 90 percent of the Dutch live in towns and cities. They speak Dutch. About half of the population are Protestants and half Catholics. Because of their high standard of living and their efforts in reclaiming land from the sea, the Dutch have a reputation for hard work.

Location

Luxembourg

If you travel southeast of Belgium, you can visit Luxembourg, Luxembourg. The city of Luxembourg is the capital and largest city of one of Europe's smallest countries. The country is only about 55 miles (89 km) in length and about 35 miles (56 km) in width.

The Land and People In spite of its small size, Luxembourg is a land of contrasts. Its scenic areas of rolling hills, dense forests, and charming villages draw tourists from all over the world. Many **multinational firms,** or companies that do business in several countries, have their headquarters in Luxembourg. Ranked as one of Europe's leading steel producers, this Benelux country has a strong economy.

The people of Luxembourg have close cultural ties to both France and Germany. The country has three official languages: French, German, and Letzeburgesch—a dialect of German. The Luxembourgers, however, proudly cling to their independence, which they have enjoyed for more than 1,000 years. More than 95 percent of the people are Roman Catholics.

SECTION 3 ASSESSMENT

REVIEWING TERMS AND FACTS

1. Define the following: polder, multinational firm.

2. LOCATION How has Belgium's location affected its economic development?

3. HUMAN/ENVIRONMENT INTERACTION Why do the Dutch reclaim land from the sea?

MAP STUDY ACTIVITIES

4. Study the land use map on page 288. Which of the Benelux countries is landlocked?

5. Look again at the map on page 288. What resources are found in Belgium?

The Alpine Countries

PREVIEW

Words to Know
- neutrality
- watershed

Places to Locate
- Switzerland
- Jura Mountains
- Bern
- Austria
- Vienna
- Danube River
- Liechtenstein

Read to Learn . . .
1. what the landscape is like in the Alpine countries.
2. what languages are spoken in Switzerland and Austria.
3. where most people in Switzerland and Austria live.

A breathtaking sight, the Matterhorn soars 14,691 feet (4,478 m) into the sky! This dramatic peak has become a symbol of the Alpine countries.

The Alps form most of the landscape in Switzerland, Austria, and Liechtenstein. That is why they are called the Alpine countries. Switzerland and Austria together cover about 47,212 square miles (121,869 sq. km), an area about the size of the state of Alabama. Sandwiched between them is tiny Liechtenstein, covering only 60 square miles (156 sq. km).

Location
Switzerland

What do you think of when you imagine the country of Switzerland? Do mounds of Swiss chocolate or soaring mountaintops come to mind? The rugged Swiss Alps have always created a barrier to travel between northern and southern Europe. For centuries Switzerland guarded the few routes that cut through this barrier. Because of its location, Switzerland practiced **neutrality**—refusing to take sides in disagreements and wars between countries.

The Land and Economy The Alps make Switzerland the **watershed** of central Europe. A watershed is a region that drains into a common waterway or body of water. Several rivers, such as the Rhine and the Rhône, begin in the Swiss

290

Alps. Dams harness many of Switzerland's rivers, producing great amounts of hydroelectricity.

Another mountain range, the Jura, runs across northwest Switzerland. Between the Jura Mountains and the Alps lies a plateau called the *Mittelland,* or "Middle Land." Most of Switzerland's industries and its richest farmlands are found here. Bern, Switzerland's capital, and Zurich, its largest city, are also located here.

Switzerland's climate is strongly influenced by the Alps. Though temperatures differ with elevation, in most parts of the country winters are cold and summers are warm.

With no coastline and few natural resources, you might not expect Switzerland to be industrialized. It is, however, a thriving industrial nation. Using imported materials, Swiss workers make high-quality goods such as machine tools, electrical equipment, computers, and watches. They also produce chemicals, chocolate, cheese, and other dairy products. Zurich and Geneva are important international banking centers.

The People Multicultural describes the people of Switzerland. Switzerland has many different ethnic groups and religions. If you went to school there, you might have to learn all three of the official languages—German, French, and Italian. About half of the Swiss are Protestants and half Catholics. The Swiss have enjoyed a stable democratic government for more than 700 years.

Region
Austria

Austria's landscape looks like the setting for a fairy tale. Small when compared to some other European countries—31,942 square miles (82,729 sq. km)—Austria is slightly larger than the state of Maryland.

At one time, Austria ranked as one of the largest and most powerful countries in Europe. Like many European nations, it was once ruled by a strong monarchy that later lost power and territory. The Austrian Empire dominated politics and culture in central Europe from around 1280 until 1918.

The Land and Economy Mountains cover three-fourths of Austria. It is one of the most mountainous countries in the world. An important European river, the Danube, flows 217 miles (350 km) from east to west across the country's northern region.

Austria mines a variety of mineral resources. Iron ore, coal, copper, lead, and graphite fuel its many industries. Austrian factories produce machinery, chemical products, clothing, furniture, glass, and paper goods. The mountains cover much of the land. Austrian farmers, however, still manage to raise dairy cattle and other livestock, potatoes, sugar beets, and rye.

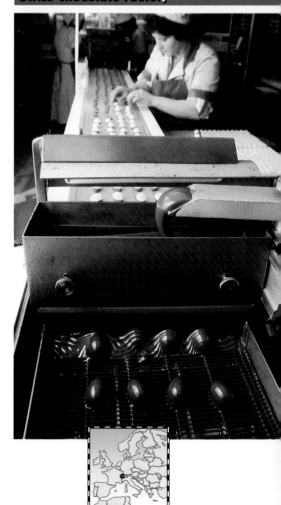

Swiss Chocolate Factory

Switzerland's factories produce some of the best chocolates in the world.
PLACE: What other products are made in Switzerland?

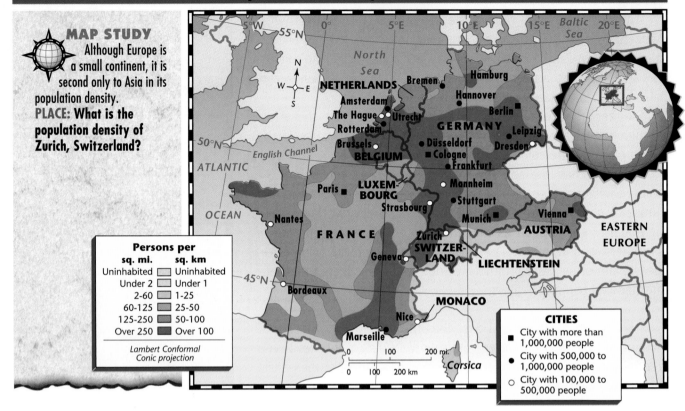

MAP STUDY
Although Europe is a small continent, it is second only to Asia in its population density.
PLACE: What is the population density of Zurich, Switzerland?

Persons per

sq. mi.	sq. km
Uninhabited	Uninhabited
Under 2	Under 1
2-60	1-25
60-125	25-50
125-250	50-100
Over 250	Over 100

Lambert Conformal Conic projection

CITIES
■ City with more than 1,000,000 people
● City with 500,000 to 1,000,000 people
○ City with 100,000 to 500,000 people

The People Austria, a republic, is home to about 8.1 million people. More than half of its population live in cities and towns. Most speak German and share cultural traditions with Switzerland. About 90 percent of Austrians are Roman Catholics. Austria is well known for its concert halls, historic palaces and churches, and its architecture.

The music of Vienna, Austria's capital, has entertained the world. Some of the world's greatest composers—Joseph Haydn, Wolfgang Amadeus Mozart, Franz Schubert—lived or performed in Vienna, the largest city in Austria.

SECTION 4 ASSESSMENT

REVIEWING TERMS AND FACTS

1. Define the following: neutrality, watershed.

2. REGION How has Switzerland's location affected its relations with neighbors?

3. PLACE What makes Switzerland a multicultural country?

4. HUMAN/ENVIRONMENT INTERACTION What minerals are found in Austria?

MAP STUDY ACTIVITIES

5. Compare the population density map above with the physical map on page 278. Where do most of the people in Switzerland live? What led them to live there?

Chapter 11 Highlights

Important Things to Know About Northwestern Europe

SECTION 1 FRANCE

- France contains mountains and fertile northern plains.
- France's well-developed economy balances agriculture and industry.
- Paris, the capital of France, is one of the world's leading cultural centers.

SECTION 2 GERMANY

- Germany is hilly and mountainous in the south and flat in the north.
- The German economy is strong because of its skilled workers and natural resources.
- Germany reunited as one country in 1990.

SECTION 3 THE BENELUX COUNTRIES

- Belgium has two major ethnic groups: the Walloons and the Flemings.
- The Dutch have reclaimed much of their land from the sea.
- Luxembourg's capital is home to the headquarters of many multi-national companies.

SECTION 4 THE ALPINE COUNTRIES

- The Alpine countries include Switzerland, Austria, and Liechtenstein.
- The Alps cover much of Switzerland and Austria.
- Switzerland has three official languages: German, French, and Italian.
- Most Austrians speak German, share cultural customs with Switzerland, and are Roman Catholic.

Berlin business district ▶

REVIEWING KEY TERMS

Match the numbered terms in Set A with their lettered definitions in Set B.

A

1. multinational firm
2. *autobahn*
3. communism
4. polder
5. watershed
6. Holocaust

B

A. German superhighway
B. an area draining into a common waterway
C. the organized killing of more than 6 million European Jews and 6 million others
D. an area of reclaimed land
E. a company that does business in several countries
F. form of government that controls society

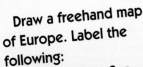

Mental Mapping Activity

Draw a freehand map of Europe. Label the following:

- Mediterranean Sea
- Alps
- Danube River
- North European Plain
- Belgium
- Rhine River
- Vienna

REVIEWING THE MAIN IDEAS

Section 1

1. PLACE What river forms France's border on the east?
2. HUMAN/ENVIRONMENT INTERACTION What makes France's land good for agriculture?

Section 2

3. LOCATION What is Germany's leading industrial region?
4. REGION What four major rivers flow through Germany?

Section 3

5. REGION Why are Belgium, the Netherlands, and Luxembourg grouped together into one region called Benelux?
6. LOCATION Where do most of the people of the Netherlands live?

Section 4

7. PLACE What is the capital of Switzerland?
8. PLACE What river runs through northern Austria?

CRITICAL THINKING ACTIVITIES

1. **Making Generalizations** Why do you think so many historic castles are found along Europe's rivers?
2. **Making Comparisons** What similarities are there among the people of Germany and Austria?

CONNECTIONS TO ECONOMICS

Cooperative Learning Activity Work in a group to write a play about one of the countries discussed in the chapter. Choose a topic that relates to the economy of the country. Each person can research and write a scene of the play. Review each scene as a group and share your play with the rest of the class. You may want to choose a cast for your play and present it to another class.

GeoJournal Writing Activity

Imagine that you used to live in communist East Germany before 1990. You have escaped to democratic West Germany. Write a letter home to a friend describing your escape and your new life.

TECHNOLOGY ACTIVITY

Using a Spreadsheet List the names of at least ten European countries in a spreadsheet, beginning with cell A2 and continuing down column A. Then use a world atlas to find the total population of each country. List the population figures in column B, beginning with cell B2. Make a bar graph that shows, at a glance, a comparison of the countries' populations. Draw conclusions about Europe's population based on your bar graph.

PLACE LOCATION ACTIVITY: NORTHWESTERN EUROPE

Match the letters on the map with the places of northwestern Europe. Write your answers on a separate sheet of paper.

1. Netherlands
2. Austria
3. Mediterranean Sea
4. English Channel
5. Danube River
6. Paris
7. Alps
8. Berlin
9. Seine River
10. Luxembourg

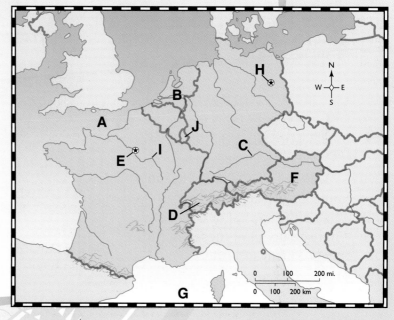

Chapter 12 Southern Europe

National boundary
⊛ National capital
● Other city

Lambert Conformal Conic projection

EUROPE

PORTUGAL
Lisbon ⊛

SPAIN
Madrid ⊛
Saragossa ●
Barcelona ●
Valencia ●
Seville ●
Malaga ●

ANDORRA

Milan ●
Venice ●
Genoa ●
Florence ●
SAN MARINO
ITALY
Rome ⊛
Naples ●
Sardinia
Balearic Islands
Tyrrhenian Sea
Palermo ●
Sicily ●

Adriatic Sea
Ionian Sea

Mediterranean Sea

GREECE
Thessaloniki ●
Patras ●
Athens ⊛
Aegean Sea
Crete

AFRICA

MALTA ⊛ Valletta

Sea

GREECE
CYPRUS
Nicosia ⊛
Mediterranean Sea
ASIA

5°W 0° 5°E 10°E 15°E 20°E 25°E 30°E
40°N 35°N

0 100 200 mi.
0 100 200 km

MAP STUDY ACTIVITY

As you read Chapter 12, you will learn about the countries in southern Europe.

1. What southern European countries are islands?
2. What is the capital of Greece?
3. What sea do all of these countries except Portugal border?

Spain and Portugal

Words to Know
- plateau
- textile
- dialect

Places to Locate
- Iberian Peninsula
- Spain
- Madrid
- Meseta
- Portugal
- Lisbon

Read to Learn . . .
1. about the landscape of Spain and Portugal.
2. how the people of Spain and Portugal earn their livings.
3. what cultural groups are found in Spain.

Past and present meet in Spain. This scene of plateaus and windmills seems untouched by the passage of time. Yet just a few hours by train over the dusty horizon is Spain's capital—Madrid—a thriving, modern urban center.

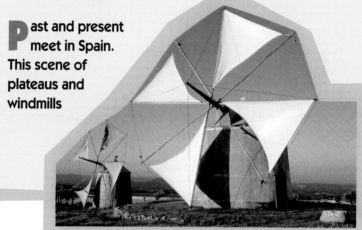

Three countries—Spain, Portugal, and Andorra—are found on the Iberian Peninsula. The map on page 298 shows you the square-shaped Iberian Peninsula. Spain takes up five-sixths of the peninsula's land area. The remaining one-sixth, on the peninsula's western edge, is Portugal. Andorra, a tiny nation of 185 square miles (479 sq. km), perches high in the Pyrenees next to France.

Region
The Land

Most of the Iberian Peninsula is high and rugged. In the northern region the Pyrenees form a wall dividing the peninsula from the rest of Europe. In the southern region another mountain chain—the Sierra Nevada—parallels the Mediterranean coast.

The Meseta Between these mountain ranges in the heart of the peninsula lies a huge central plateau called the Meseta (muh•SAY•tuh), or "tableland." A **plateau**, you will recall, is a flat landmass higher than the surrounding land.

EUROPE

ATLANTIC
OCEAN

Bay
of
Biscay

ELEVATIONS

Feet	Meters
10,000	3,000
5,000	1,500
2,000	600
1,000	300
0	0

Gran Paradiso
13,323 ft.
(4,061 m)

ALPS

Po River

SAN
MARINO

Arno R.

APENNINES

Adriatic Sea

ANDORRA

PYRENEES

PORTUGAL Douro River

Ebro R.

APENNINE PENINSULA

Tiber
River

Mt. Olympus
9,570 ft.
(2,917 m)

IBERIAN PENINSULA

MESETA

ITALY

Aliakmon
River

BALKAN

SPAIN

Tagus R.

Sardinia

Tyrrhenian
Sea

PENINSULA

**PINDUS
MTS.**

Aegean
Sea

Guadiana R.

Balearic
Islands

GREECE

Cyclades

Guadalquivir R.

Mediterranean

Peloponnesus

Ionian Islands

Ionian
Sea

Sea

Crete

Strait of
Gibraltar

▲ Mountain peak

Lambert Conformal
Conic projection

Sicily

MALTA

0   125   250 mi.

0   125   250 km

AFRICA

GREECE

ASIA

CYPRUS

Mediterranean   Sea

25°E   30°E   35°E

35°N

MAP STUDY

Italy and San Marino make up the Apennine Peninsula.
PLACE: What countries make up the Iberian Peninsula?

The Meseta covers about two-thirds of the Iberian Peninsula. Its dry climate and poor, reddish-yellow soil must be irrigated for successful farming. Farmers of the Meseta raise wheat and vegetables, and they herd sheep and goats.

Many of the Iberian Peninsula's major rivers begin in the Meseta. For example, the Tagus (TAY•guhs)—Spain's longest river—and the Guadalquivir (GWAH•duhl•kih•VIHR) River flow from the Meseta to the Atlantic Ocean. Most of the peninsula's rivers are too narrow and shallow for large ships, however. The map above shows you that several fertile basins and coastal plains reach far into the Meseta. The largest is the Guadalquivir Basin in southern Spain.

The Climate   To see how climate differs throughout the Iberian Peninsula, look at the map on page 305. Coastal areas of Spain and Portugal have a Mediterranean climate with mild winters and hot summers. Northern areas, however, enjoy a marine west coast climate. There, warm winds from the Atlantic Ocean bring much rain. The ocean's influence on the land diminishes as you travel inland. Parts of the Meseta have cold winters, hot summers, and very little rain.

The Economy

Tour buses filled with travelers from all over the world regularly crisscross Spain. In recent years tourism has brought much money to the Spanish. Visitors enjoy the sunny climate, beautiful beaches, and storybook castles. Spain also has a rapidly growing industrial and service economy. Spanish workers make machinery, trucks, cars, and clothing.

Farming and fishing form an important part of the Spanish economy. Acres and acres of olive groves make Spain the world's leading producer of olive oil. Spain also exports citrus fruits, cork, and wine.

Portugal lacks the economic development of Spain. Although Portugal benefits from **textile**, or clothing, and tourist industries, it is still mainly a subsistence farming economy. Portuguese farmers grow wine grapes, citrus fruits, olives, rice, cork, and wheat. Because of the slow growth of industry, Portugal's standard of living is one of the lowest in Europe.

Region

The People

The people of Spain and Portugal share similar histories and cultures. Many historians believe Iberians migrated to these countries from North Africa more than 5,000 years ago. They were later followed by Romans from Italy and Moors from North Africa. These earlier civilizations influenced Spanish and Portuguese culture, religion, and architecture.

Fountains and flower vendors brighten the main square of Lisbon, Portugal's capital *(left)*. Cork is an important product in both Portugal and Spain *(right)*.
MOVEMENT: **Which country—Spain or Portugal—has a more rural population?**

Urban and Rural Life in Portugal

By the late 1400s, Portugal and Spain had become independent, powerful, and Roman Catholic. A long Atlantic coastline and good harbors encouraged a spirit of adventure in both countries. Portugal and Spain sponsored explorers, such as Christopher Columbus and Vasco da Gama, who sailed abroad and set up empires in Africa, Asia, and the Americas.

The Court of Lions in the Alhambra

The Alhambra, a palace for Moorish royalty, was built when Spain was under Islamic rule.
PLACE: What is the primary religion of Spain today?

Cultural Regions After the decline of their overseas empires in the 1800s and 1900s, Spain and Portugal turned their attention to local affairs. Portugal developed a unified culture based on the Portuguese language. Spain remained in many ways a "country of different countries." Today Spain is still made up of several unique cultural regions.

The region of Castile, in the Meseta, dictated the culture of Spain for centuries. The **dialect** of the Castilian people became the major language of Spain. A dialect is a local form of a language. Today Castilian Spanish is spoken by most Spaniards.

People in other regions of Spain, however, regard themselves as separate groups. In the Mediterranean coastal region of Catalonia, the people speak Catalan, a language related to French. Galicia in northwest Spain and Andalusia in the south also have their unique cultural traditions.

The Basque people of northern Spain speak a language that is unlike any other language spoken in Europe. These people think of themselves as separate from the rest of Spain. Many Basques want complete independence from Spain in order to preserve their way of life.

A Rich Cultural Heritage The royal courts of early Spain and Portugal echoed with poetry and music. Spain and Portugal have a rich artistic heritage. In the 1500s Luiz Vaz de Camões (kuh•MOHNSH) of Portugal wrote *Os Lusídas* (OHS loo•SEE•dahs), a long poem about his country's history and heroes. In the 1600s Spain's Miguel de Cervantes (suhr•VAN•TEEZ) wrote *Don Quixote* (kee•HOH•tee), one of the world's first great novels. One of the most famous painters of the 1900s was Spain's Pablo Picasso, whose works had a major influence on modern art.

The cultures of Spain and Portugal have gained new freedoms in the past 20 years. From the 1930s until the mid-1970s, both countries were ruled by dictators. When Spanish dictator Francisco Franco died in 1975, democracy returned. Today Spain is a constitutional monarchy, and Portugal, too, has become a more open and democratic society.

Rural and City Life If you lived in Portugal today, you would most likely live in a rural village. About 52 percent of Portugal's 10 million people are villagers. Lisbon, with a population of about 2 million, is Portugal's capital and major city.

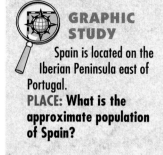

GRAPHIC STUDY

Spain is located on the Iberian Peninsula east of Portugal.
PLACE: What is the approximate population of Spain?

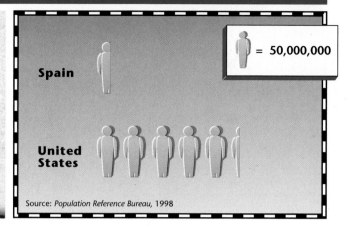

= 50,000,000

Spain

United States

Source: *Population Reference Bureau, 1998*

In contrast, almost 64 percent of Spain's nearly 40 million people live in cities and towns. A city of 4.5 million people, Madrid ranks as one of Europe's leading cultural centers. Barcelona, along the Mediterranean coast, is Spain's leading seaport and industrial center. About 3 million people live there.

Juan García, a Madrid teenager, enjoys the fast pace of life in his city. Like most Spaniards, he and his family usually do not sit down to dinner until 9 or 10 o'clock at night. The meal may not end until midnight. On special occasions, the García family enjoys *paella*, a traditional dish of shrimp, lobster, chicken, ham, and vegetables mixed with seasoned rice.

Although Juan enjoys rock music and jazz, his parents prefer folk singing and dancing. In Spain the people of each region enjoy their own special songs and dances. Musicians often accompany singers and dancers on guitars, castanets, and tambourines. Spanish dances such as the bolero and flamenco have spread throughout the world.

SECTION 1 ASSESSMENT

REVIEWING TERMS AND FACTS

1. Define the following: plateau, textile, dialect.
2. REGION What is the major landform of Spain?
3. REGION How are the economies of Spain and Portugal different?
4. LOCATION What is the capital of Spain?

 MAP STUDY ACTIVITIES

5. Study the physical map on page 298. What two major rivers flow through both Spain and Portugal?
6. Look at the political map on page 296. About how far is it from Lisbon to Seville?

BUILDING GEOGRAPHY SKILLS

Reading a Population Map

At a baseball game, most of the filled seats are in the infield. Outfield seats have fewer people, and the upper decks have few people or none at all. The infield seats, you might say, have the greatest population density in the stadium.

Population density is the average number of people living in a square mile or square kilometer. Population density maps often use color to show differences in population. Dots or squares represent cities of different population sizes. To read a population density map, follow these steps:

• Study the map key to identify the population densities on the map.

• On the map, find areas of lowest and highest population density.

• Compare the map with other information to explain the region's population pattern.

Glencoe's **Skillbuilder Interactive Workbook, Level 1** provides instruction and practice in key social studies skills.

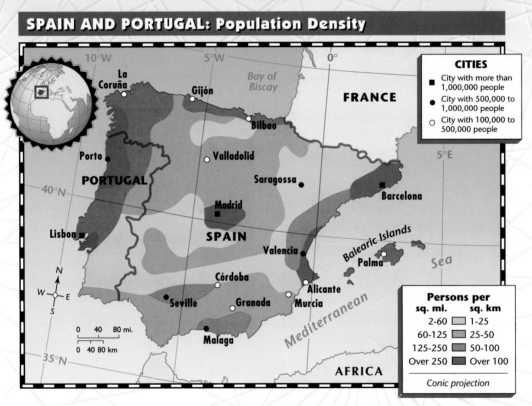

SPAIN AND PORTUGAL: Population Density

CITIES
■ City with more than 1,000,000 people
● City with 500,000 to 1,000,000 people
○ City with 100,000 to 500,000 people

Persons per
sq. mi.	sq. km
2–60	1–25
60–125	25–50
125–250	50–100
Over 250	Over 100

Conic projection

Geography Skills Practice

1. What color represents areas with 125–250 people per square mile (50–100 per sq. km)?

2. Which cities have more than 1 million people?

3. Which areas have the lowest population density? Why?

4. What color stands for over 250 people per square mile (over 100 per sq. km)?

Italy

PREVIEW

Words to Know
- sirocco
- city-state

Places to Locate
- Italy
- Apennines
- San Marino
- Rome
- Vatican City
- Po River
- Sicily
- Sardinia
- Adriatic Sea

Read to Learn . . .
1. about Italy's physical regions.
2. how the northern and southern parts of Italy differ.
3. how Italy's rich history has influenced the world.

The glistening dome of St. Peter's Basilica rises above the skyline of Rome. This historic city attracts people to Italy from all over the world.

Italy is a land of seacoasts. Look at the map on page 296. You see that Italy is a peninsula jutting out from southern Europe about 750 miles (1,207 km) into the Mediterranean Sea. The Italian Peninsula looks like a boot about to kick a triangle-shaped football. The "football" is the Italian island of Sicily. Another large Mediterranean island, Sardinia, is also part of Italy. Between Sicily and Africa lies Malta, an independent island country that has close ties to the United Kingdom as well as to Italy.

Where can you find the smallest nation in the world? Two tiny countries—San Marino and Vatican City—lie within the Italian "boot." San Marino is near the northwestern coast. Vatican City—the smallest nation—lies completely within Italy's capital, Rome. Its area totals only 0.2 square miles (0.4 sq. km).

Region
The Land

The land area of Italy—113,351 square miles (293,594 sq. km)—equals the combined areas of the states of Florida and Georgia. Mountains and highlands run down Italy's length. Plains and lowlands hug its coasts.

The Landscape Locate Italy's mountain regions on the map on page 298. The Alps tower over northern Italy, separating the country from France, Switzerland, and Austria. Another mountain range, the Apennines (A•puh•NYNZ), runs down the center of the Italian Peninsula to the toe of the boot. These mountains descend to low hills in the east and west.

The Amalfi Coast, Italy

Cliffside towns hug the jagged shoreline of Italy's south-western coast.
PLACE: What mountain range extends north to south through Italy?

The rumbling of volcanic mountains echoes through the southern part of the peninsula and the island of Sicily. Mount Etna, one of the world's most famous volcanoes, rises 11,122 feet (3,390 m) on Sicily's eastern coast.

Although mountains and hills cover most of Italy, you can also find lowland regions. In the north the largest lowland is the Po River valley, the most densely populated part of the country. The Po River is Italy's longest river, flowing 400 miles (644 km) west to east. It stretches from the French border to the Adriatic Sea. You will learn later how important the Po River valley is to Italy.

The Climate Nearly all of Italy has a mild Mediterranean climate with sunny summers and rainy winters. In spring and summer, hot dry winds called **siroccos** blow across the Italian Peninsula from North Africa. In fall and winter, cool moist air from the Atlantic Ocean replaces this hot air. Most of Italy receives enough rain to grow crops. Parts of Sicily and Sardinia, however, record little rainfall.

Human/Environment Interaction
The Economy

Italy ranks as the most prosperous country in southern Europe. The country's wealth, however, is not evenly distributed. Northern Italy is home to many large cities that thrive on industry, banking, and trade. Life in the urban north, however, contrasts sharply with farmers who carve out a living on rocky slopes in southern Italy.

The Prosperous North Fertile soil and steady, year-round rainfall have made northern Italy very productive. Most of Italy's agriculture centers on the Po River valley. Farmers there grow wheat, corn, rice, and sugar beets. In hilly areas of northern and central Italy, farmers grow grapes for wine. Italy is one of the world's major wine-producing countries.

Most manufacturing takes place in the northern Po River valley in the cities of Milan and Turin. Skilled workers there produce cars, machinery, chemicals, clothing, and leather goods. More industry is located in Genoa, an important port city. Italy has few mineral resources, but valuable natural gas deposits lie in the Po River valley. Factories in the northern region also use hydroelectric power from water sources in the nearby Alps.

304

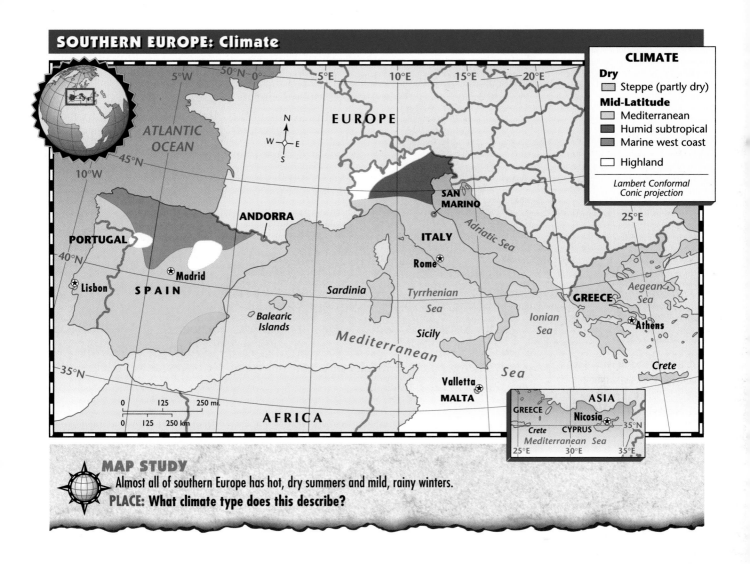

CLIMATE
Dry
☐ Steppe (partly dry)
Mid-Latitude
☐ Mediterranean
☐ Humid subtropical
☐ Marine west coast
☐ Highland

Lambert Conformal
Conic projection

MAP STUDY
Almost all of southern Europe has hot, dry summers and mild, rainy winters.
PLACE: What climate type does this describe?

The Developing South Southern Italy is poorer and less developed than northern Italy. From the map on page 309, you can tell that southern Italy lacks hydroelectric power. Southern Italy also lacks the natural gas deposits and rich soil found in northern Italy. The dry, rugged landscape of the southern region is used mainly for pastureland. The region's volcanic and clay soils are also good for growing citrus fruits, olives, and grapes. Southern Italy benefits from a steady stream of tourists.

Place
The People

Have you been on a crowded bus or in a crowded store recently? Then you know how some Italians feel living in their crowded country. About two-thirds of Italy's nearly 58 million people live on only one-fourth of the land. Why? Mountains take up so much space in the country. Many Italians from the poorer southern region have moved to northern Italian cities to find jobs. Others have emigrated to northern Europe and the United States.

Canals serve as streets for the many gondolas and motorboats in Venice, Italy.
HUMAN/ENVIRONMENT INTERACTION: What do you think it would be like to live in a city where boats are used more than cars?

Italy's Heritage For hundreds of years, Italy was the heart of Western civilization. The Roman Empire, based in Italy, influenced the government, arts, and architecture of Europe. After the fall of the Roman Empire in the A.D. 400s, Italy was divided into many small territories and **city-states**. Each city-state included an independent city and its surrounding countryside. The Renaissance developed in Italy's city-states during the 1300s and spread throughout Europe. It was a period of great achievement in the arts.

Italy was finally united as an independent country in 1861. From the 1920s to the early 1940s, dictator Benito Mussolini ruled Italy. He also became a supporter of Germany's Adolf Hitler and pulled Italy into World War II. Italy was defeated in the war.

In 1946 Italy became a democratic republic. Democracy, however, did not bring a stable government. Since the late 1940s, political power in Italy has constantly changed hands. Rivalry between the wealthy north and the poorer south has caused political tensions.

Daily Life To the rest of the world, Italy's capital—Rome—is the Eternal City. It is the keeper of historic ruins, ancient monuments, and beautiful churches and palaces. About 70 percent of Italy's population live in towns and cities. Three cities in Italy—Rome, Milan, and Naples—have populations of more than 3 million each.

The people of Italy speak Italian, which developed from Latin, the language of ancient Rome. Italian is closely related to French and Spanish. Do you know any Italian words? Many have been adopted into the English language. Pasta, made from flour and water, is a basic dish in Italy. Some pasta dishes are spaghetti, lasagna, and ravioli.

Celebrating the religious festivals of the Roman Catholic Church is a widely shared part of Italian life. More than 95 percent of Italy is Roman Catholic. Vatican City in Rome is the headquarters of the Roman Catholic Church.

SECTION 2 ASSESSMENT

REVIEWING TERMS AND FACTS

1. **Define the following:** sirocco, city-state.

2. **REGION** What are two major mountain regions of Italy?

3. **HUMAN/ENVIRONMENT INTERACTION** How is the economy of northern Italy different from the economy of southern Italy?

4. **PLACE** Why is Rome so important to Italy?

MAP STUDY ACTIVITIES

5. Using the climate map on page 305, identify the climate of Sicily.

ROMAN BUILDERS

MATH　　**SCIENCE**　　**HISTORY**　　**LITERATURE**　　**TECHNOLOGY**

Paved roads and running water—how would you live without them? You may believe that they are inventions of the modern world. They had their origins, however, centuries ago in Italy. Among other achievements, the Romans built good roads and reliable water supplies.

ROMAN ROADS Road building played a key role in Roman military conquest. It allowed Roman legions to move quickly from one trouble spot to another in the empire. Wherever Roman soldiers went, they built roads.

Engineers first analyzed the soil and marked the road edges. Then, under the watchful eyes of supervisors, teams of soldiers dug down several feet to prepare the roadbed. A flattened layer of sand was further layered with rocks, small stones, and then gravel. These layers were topped off with flat paving stones carefully cut to fit closely together.

This top pavement was raised slightly in the middle so rainwater drained off to the sides. The roads were then usable in all kinds of weather. This approach is still used today to build highways!

AQUEDUCTS The Romans also built aqueducts, or stone troughs held aloft by supports.

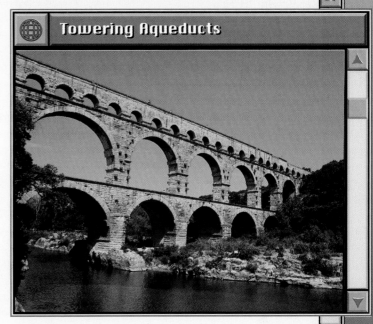

Towering Aqueducts

The aqueducts were a way to move water from its source in the mountains to communities in lower regions.

To keep the water flowing, all aqueducts were built at a constant slope from beginning to end. Each aqueduct stood about 50 feet (15.2 m) off the ground, on top of its own row of supports and foundations. The outside of the supports and foundations was made of close-fitting stones. The inside was coated with layers of concrete, which the Romans invented by mixing volcanic stone with stone rubble.

Making the Connection

1. What building materials did Romans use?
2. What is an aqueduct?

SECTION 3 · · · Greece

Words to Know
- mainland
- elevation
- service industry
- suburb
- emigrate

Places to Locate
- Greece
- Athens
- Pindus Mountains
- Plain of Thessaly
- Peloponnesus
- Crete
- Aegean Sea

Read to Learn . . .
1. how mountains and seas divide Greece.
2. how Greeks earn their livings.
3. what contributions ancient Greece made to Western civilization.

Were you named after someone—your grandmother or grandfather perhaps? The people of Athens, Greece, know how their city was named. It was called Athens in honor of Athena, the Greek goddess of wisdom. This graceful temple—the Parthenon—stands high above the city of Athens.

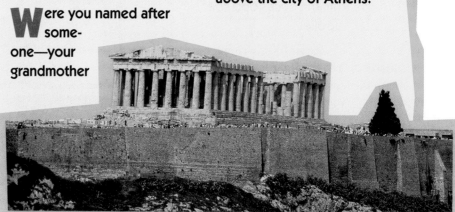

In ancient times Athens was a small town. Today the city is a modern urban center that is growing rapidly in size and population. The ancient Greeks would be amazed to see the changes that have taken place in Athens over the centuries. Yet they would find many other places in Greece that have changed very little. Outside the large cities, much of Greek life still follows traditional ways.

Region
The Land

Like Spain and Italy, much of Greece lies on a large peninsula that juts out from Europe into the Mediterranean Sea. Greece sits at the southern tip of the Balkan Peninsula.

Most of Greece is rocky and mountainous. High peaks separate valleys and plains. Long arms of the sea reach into the coast, forming many small peninsulas. Nearly 2,000 islands dot the sea around Greece. They make up about 20 percent of Greece's entire land area of 49,708 square miles (128,900 sq. km).

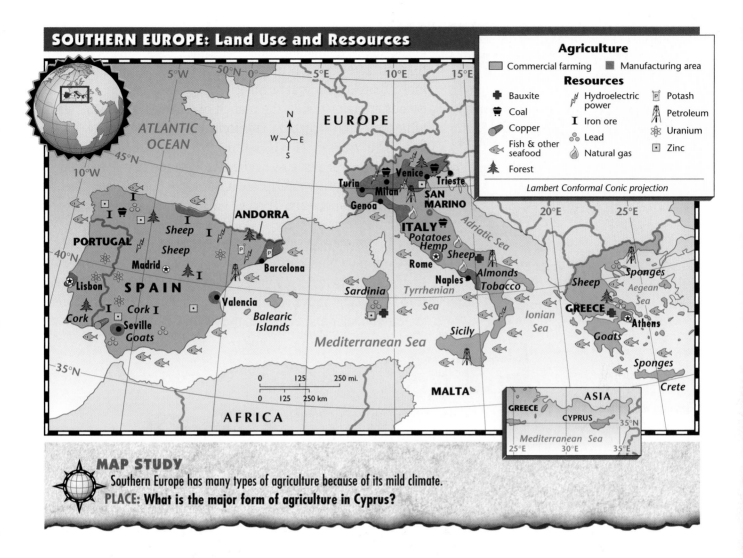

SOUTHERN EUROPE: Land Use and Resources

Agriculture

☐ Commercial farming ■ Manufacturing area

Resources

✚ Bauxite
🚃 Coal
⬭ Copper
🐟 Fish & other seafood
🌲 Forest

⚡ Hydroelectric power
I Iron ore
•• Lead
💧 Natural gas

P Potash
🏭 Petroleum
✳ Uranium
·· Zinc

Lambert Conformal Conic projection

MAP STUDY

Southern Europe has many types of agriculture because of its mild climate.

PLACE: What is the major form of agriculture in Cyprus?

Highlands As you look at the map on page 298, you can see that the Pindus Mountains run through the center of the Greek **mainland**. This part of Greece connects to the European continent. The Pindus and other smaller ranges divide Greece into many separate regions. Historically, this has kept people in one region isolated and cut off from people in other regions. Because of the poor, stony soil, most people living in the highlands graze sheep and goats.

Lowlands and Peninsulas To the east of the Pindus Mountains lie two fertile lowlands—the Plain of Thessaly and the Macedonia-Thrace Plain. These two plains form Greece's major farming areas.

At the southern end of the Pindus range is another lowland, the Plain of Attica. About one-third of the Greek population live on this plain. Athens, the Greek capital, spreads out in all directions on the Plain of Attica.

Southwest of the Plain of Attica lies the Peloponnesus (PEH•luh•puh•NEE•suhs), a large peninsula of rugged mountains and deep valleys. If you traveled through this area thousands of years ago, you might have heard the cheering crowds attending the ancient Olympic games. Among other ruins on the Peloponnesus stands Olympia. Here the first Olympic games were held more than 2,600 years ago.

Gleaming white buildings line the shore of the Greek island of Mikonos.
HUMAN/ENVIRONMENT INTERACTION: Why do you think most of the buildings on Mikonos are so close together?

Islands A large number of islands are scattered on all sides of the Greek mainland. Some islands are mere dots of land less than an acre (0.4 ha) in area. Others, like Crete, are large—more than 3,000 square miles (7,770 sq. km)—in area. The islands are named after the branch of the Mediterranean Sea in which they are located. The Ionian Islands lie in the Ionian Sea. The Aegean (ih•JEE•uhn) Islands spread across the Aegean Sea. Farther east in the Mediterranean is the independent island country of Cyprus. It has close cultural and historic ties to both Greece and Turkey.

The Climate Greece has a Mediterranean climate with hot, dry summers and mild, rainy winters. As you learned in Chapter 2, however, climate may vary, depending on **elevation,** or height above sea level. High-elevation areas in Greece are often cooler and wetter than lowlands.

Movement
The Economy

Greece has one of the least developed economies in Europe. With few mineral resources, Greece must import many manufactured goods. In recent years, however, manufacturing has grown rapidly. Greek workers produce cement, tobacco products, and clothing. Tourism and **service industries** also have grown. Service industries provide a service—such as banking, insurance, or transport—instead of making goods.

Agriculture Greece's rocky soil limits the land available for farming. Yet many Greeks manage to earn their living by tilling the soil. Small in size, most Greek farms lie in valleys where the land is fertile. There farmers cultivate citrus fruits, olives, wine grapes, and tobacco.

Teen Scene

Harvest Time

Melina Stavros and her mother have spent the day gathering ripe olives from the family's olive grove. Melina lives on a small farm with her mother, father, and three brothers. "My family is very traditional," she says, "but nothing like my grandparents' family!" Melina said her grandparents chose who her mother would marry.

310

Shipping and Tourism The sea has long been important to the Greeks. No part of Greece is more than 85 miles (137 km) from the sea. In ancient times the Greeks depended on the sea for fishing and trade. Many of their legends describe heroes taking long sea voyages. Today the sea still provides fish and trade. Greece has one of the largest merchant fleets in the world, and Greek shipping is vital to the economy.

Greek ships not only transport goods, but also carry passengers. Each year millions of tourists come to Greece to sunbathe and relax. Many visit the sunny Aegean islands and historic sites such as the Parthenon in Athens. Tourism, which has increased rapidly since the 1960s, is now one of Greece's most important industries.

Place
The People

About 59 percent of Greece's 10.5 million people live in urban areas. Athens, the capital and largest city, is home to about 700,000 people. Another 2.8 million live in its **suburbs**, or surrounding areas.

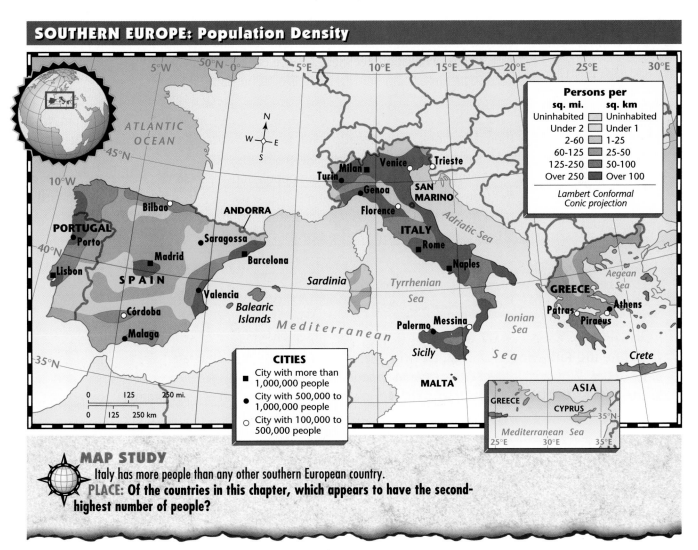

SOUTHERN EUROPE: Population Density

MAP STUDY
Italy has more people than any other southern European country.
PLACE: Of the countries in this chapter, which appears to have the second-highest number of people?

The population of Athens has mushroomed because many Greek farmers have left their villages to look for jobs in the city. Some Greeks have **emigrated**, or moved to live in other countries. Today more than 3 million people of Greek descent make their home in the United States, Australia, and western Europe.

The Ancient Greeks Many firsts mark the history of Greece. The first theories of geometry, medical science, astronomy, and physics were advanced by Greeks. The ancient Greeks developed the first European civilization.

Greek civilization reached its height in Athens during the mid-400s B.C.—the Golden Age of Greece. During this period, the city-state of Athens set up one of the world's first democratic governments. The ancient Greeks prized freedom and stressed the importance of the individual. Greek thinkers, such as Plato and Aristotle, laid the foundations of Western science and philosophy. They sought logical explanations for what happened in the world around them. Greek writers, such as Sophocles (SAH•fuh•KLEEZ), created dramas that explored human thoughts and feelings. Following the Golden Age, Greek civilization declined. Greece came under foreign rule for about 2,000 years. The Greeks did not regain their freedom until 1829.

The Modern Greeks The Greeks of today have much in common with their ancestors. They debate political issues with great enthusiasm. They love their sunny, rugged land and spend much time outdoors.

About 95 percent of the Greek population are Eastern Orthodox Christians. Religion influences much of Greek life, especially in rural areas. Easter is the most important Greek holiday. Traditional holiday foods include lamb, fish, and feta cheese—made from sheep's or goat's milk.

SECTION 3 ASSESSMENT

REVIEWING TERMS AND FACTS

1. **Define the following:** mainland, elevation, service industry, suburb, emigrate.

2. **REGION** Why is so little of Greece's land good for farming?

3. **HUMAN/ENVIRONMENT INTERACTION** How does the sea help Greece's economy?

4. **PLACE** What great contributions did ancient Greece make to Western civilization?

MAP STUDY ACTIVITIES

5. Study the land use map on page 309. What sea would you cross if you moved products by ship from Greece to Italy?

6. Using the population density map on page 311, locate Greece's main population centers.

Chapter 12 Highlights

Important Things to Know About Southern Europe

SECTION 1 SPAIN AND PORTUGAL

- Spain's dry central highland slopes down to fertile coastal lowlands. Mountains rise on its northern and southern coasts.
- Both Spain and Portugal have a mostly Mediterranean climate.
- Spain has developed an industrial economy. Portugal is still mostly agricultural.
- Portugal is more culturally uniform than Spain, which has several distinct cultural regions.

SECTION 2 ITALY

- Italy's major landform is the Apennine mountain chain that runs through the center of the Italian Peninsula.
- Fertile soil, a mild climate, and hydroelectric power help make northern Italy a productive region.
- The ancient Romans made important contributions to Western civilization in language, government, and architecture.
- Rome, Italy's capital, encompasses Vatican City, the World's smallest nation and headquarters of the Roman Catholic Church.

SECTION 3 GREECE

- Greece consists of a mountainous mainland and nearly 2,000 offshore islands.
- The Mediterranean climate found throughout Greece varies only slightly with changes in elevation.
- The Greek economy relies on tourism and shipping.
- Ancient Greece laid the foundations of Western science, art, philosophy, government, and drama.

Córdoba, Spain ▶

Chapter 12 Assessment and Activities

REVIEWING KEY TERMS

Match the numbered terms in Set A with their lettered definitions in Set B.

A

1. suburb
2. city-state
3. dialect
4. mainland
5. sirocco
6. service industry
7. emigrate

B

A. business that does not produce a product
B. an ancient city with an independent government that rules the countryside around it
C. local form of a language
D. land that is part of another large landmass
E. an area that surrounds or is next to a city center
F. move from one country to settle in another country
G. hot, dry wind from North Africa

Mental Mapping Activity

Draw a freehand map of southern Europe. Label the following:
- Meseta
- Lisbon
- Tagus River
- Adriatic Sea
- Peloponnesus
- Athens
- Ionian Islands

REVIEWING THE MAIN IDEAS

Section 1

1. **REGION** What areas of Spain and Portugal enjoy a Mediterranean climate?
2. **REGION** What are the cultural divisions of Spain?

Section 2

3. **HUMAN/ENVIRONMENT INTERACTION** How does industry in northern Italy depend on the environment?
4. **PLACE** What form of government does Italy have today?

Section 3

5. **REGION** What two plains form Greece's major farming areas?
6. **MOVEMENT** Where have Greeks emigrated to find jobs?

CRITICAL THINKING ACTIVITIES

1. **Drawing Conclusions** If you had the chance to live in southern Europe, in which country would you prefer to live? Give reasons for your choice.
2. **Making Generalizations** How did the ancient Greeks and Romans contribute to your world today?

CONNECTIONS TO GOVERNMENT

Cooperative Learning Activity Work in groups to learn more about the democratic government of ancient Athens, Greece. How was the government organized? Who had the most power and how were these people chosen? What power did citizens have? Create a chart based on your research that shows the structure of the government of this ancient city-state. Share your chart with the rest of the class. Discuss the similarities between the government of ancient Athens and the government we have today.

GeoJournal Writing Activity

Imagine you are writing an article for *National Geographic* magazine. Describe your travels through southern Europe by focusing on the region's physical features.

TECHNOLOGY ACTIVITY

Using a Computerized Card Catalog Use the electronic card catalog at your school or community library to find information about a Renaissance artist. Research the person's life and achievements. Use this information to create an oral history about that person by role-playing him or her. Have the class ask you questions about your life and your contributions. Your responses should reflect the information you collected during your research.

PLACE LOCATION ACTIVITY: SOUTHERN EUROPE

Match the letters on the map with the places and physical features of southern Europe. Write your answers on a separate sheet of paper.

1. Adriatic Sea
2. Aegean Sea
3. Athens
4. Madrid
5. Portugal
6. Sicily
7. Sardinia
8. Tagus River
9. Pindus Mountains
10. Apennines

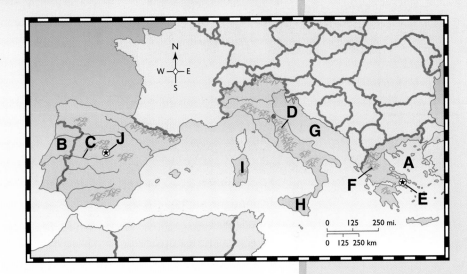

From the classroom of Rebecca A. Corley, Evans Junior High School, Lubbock, Texas.

POPULATION: A Corny Map

Background

Western Europe has some of the most densely populated areas of the world. The density and distribution of Europe's population sometimes cause problems for the governments and environments of the region. To better understand Europe's population distribution, use popcorn to create your own population map of Europe.

Shoppers Crowd London's Portobello Market

Believe it or NOT!

One of Europe's most heavily populated areas—the lowlands of Belgium, the Netherlands, and sections of Germany—holds more than 900 people per square mile (347 people per sq. km). The interior of Iceland is the most sparsely populated area in Europe with about 7 people per square mile (3 people per sq. km).

Materials

- population density map of Europe
- 1 large poster board
- newspaper
- popcorn kernels
- hot-air popcorn popper
- food coloring
- 6 small transparent cups or jars
- 1 glue stick
- 1 pencil
- 6 large mixing bowls
- 6 spoons

316

What To Do

A. Using a population density map of Europe as a reference, trace Europe's outline onto the poster board.

B. Add the lines marking the various levels of population density for the different European regions that you see on the map.

C. Most maps include 6 density levels, from uninhabited to more than 250 people per square mile. You will need 6 shades of colored popped corn, one for each level.

D. Pop your popcorn. Divide it into 6 bowls.

E. Color your popcorn. Mix food coloring drops with a small amount of water in a cup or jar. Make the color darker by adding more coloring. Then stir the colored water into a bowl of popped corn.

F. Spread the popcorn on newspaper to dry.

G. Using a glue stick, glue the colored popcorn onto your map to represent the different population densities throughout Europe.

Lab Activity Report

1. What areas of Europe are the most densely populated?

2. What areas are the most sparsely populated?

3. What inference can you make between where most people live and Europe's rivers?

4. **Drawing Conclusions** What are two physical features in Europe that contribute to low population densities in some regions?

Go A Step Further!

Activity
High population density can cause serious problems for a region. Imagine that you live in a densely populated area of Europe. Write a letter to the editor of a local newspaper in which you describe your concerns for the environment.

Chapter 13 Eastern Europe

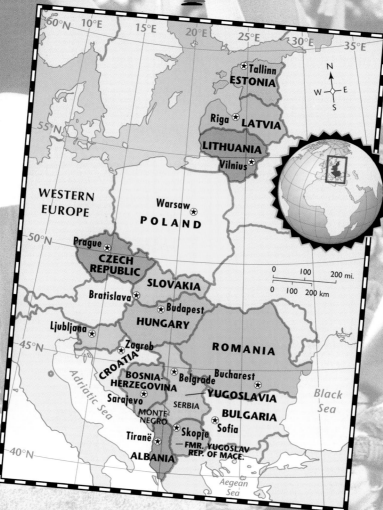

MAP STUDY ACTIVITY

As you read Chapter 13, you will learn about the countries of eastern Europe.

1. Which eastern European country has the largest area?
2. What two countries border on the Black Sea?
3. What is the capital of Poland?

The Baltic Republics

Words to Know
- communism
- oil shale

Places to Locate
- Baltic Sea
- Estonia
- Latvia
- Lithuania

Read to Learn . . .
1. how the sea affects climate in the Baltic republics.
2. what political changes occurred in the Baltic republics.
3. where most people in the Baltic republics live.

Today—with great effort— this farmer in the Baltic country of Lithuania clears land. The people of the Baltic republics are working hard to develop their economies.

Along the eastern shore of the Baltic Sea are three small republics—Estonia, Latvia, and Lithuania. In 1940 the Baltic republics came under the rule of the neighboring Soviet Union. The leaders of the Soviet Union practiced **communism**, a political and economic system in which the government controls how its citizens live and produce goods. In the early 1990s, the Soviet Union collapsed. The Baltic republics soon won back their independence. Today they are democracies with free enterprise economies.

Place
Estonia

The map on page 318 shows you that Estonia is the northernmost Baltic republic. With 16,320 square miles (42,269 sq. km), Estonia is also the smallest of the three Baltic countries.

The Land Low plains, forests, and swamps cover the land. Sandy beaches sweep along the Baltic coast. For a northern country, Estonia has a relatively mild climate. Mild Baltic Sea winds keep the weather from becoming too hot or too cold.

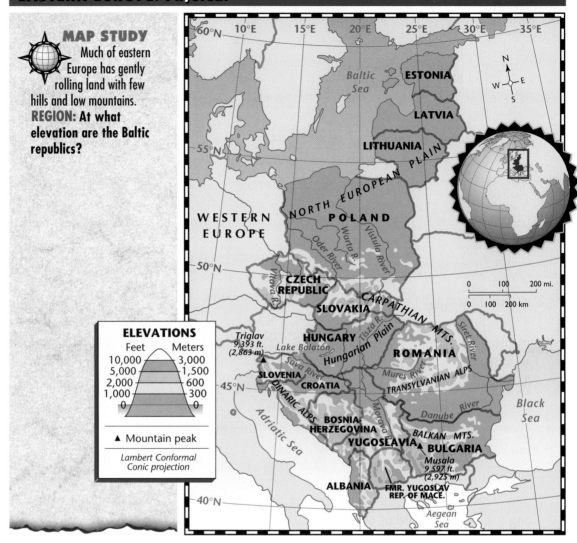

MAP STUDY
Much of eastern Europe has gently rolling land with few hills and low mountains.
REGION: At what elevation are the Baltic republics?

ELEVATIONS

Feet	Meters
10,000	3,000
5,000	1,500
2,000	600
1,000	300
0	0

▲ Mountain peak

Lambert Conformal Conic projection

Estonian farmers raise livestock and grow crops such as apples, potatoes, and sugar beets. Industry, however, provides jobs for most of Estonia's people. The country has large deposits of **oil shale**, a rock that contains oil.

The People About 64 percent of the country's people are Estonians who have close ties to the Finns. After nearly 50 years of Soviet rule, about 30 percent of Estonia's people are Russian. Most of Estonia's 1.5 million people live in cities and towns. Tallinn, the capital, is on the Baltic coast. German and Estonian builders created Tallinn's beautiful churches and castles hundreds of years ago.

Location
Latvia

South of Estonia lies the Baltic republic of Latvia. Latvia covers 23,958 square miles (62,051 sq. km)—about the area of West Virginia. Latvia's location on the Baltic Sea has made it a trading center.

320

The Land and Economy If you visited Latvia, you would find a landscape of coastal plains, low hills, and forests. The mild climate of Latvia is similar to that of Estonia.

Latvia's industrial development is greater than that of other Baltic republics. Latvia's factory workers produce electronic equipment, machinery, and household goods. Latvian farmers raise dairy cattle and grow barley, flax, oats, potatoes, and rye.

The People Only about 53 percent of Latvia's people are Latvian. Under Soviet rule, many Russians settled in Latvia. Russians make up about 33 percent of the population. The Latvians and the Russians, who have a history of conflict, often clash today. At issue between the Latvians and Russians is how one defines Latvian citizenship. Riga, the capital and largest city, is an important shipping and industrial center.

Place
Lithuania

Lithuania covers 25,019 square miles (64,799 sq. km). About 3.7 million people—at a density of 148 people per square mile (57 people per sq. km)—fill its cities and villages. In both size and population, Lithuania is the largest of the Baltic republics. It, too, struggles to develop a strong economy.

The Land and Economy Flat plains and forested hills cover most of Lithuania. Lithuania enjoys mild winter and summer temperatures. The country's economy is largely industrial with some farming. Lithuanian factory workers produce chemicals, electronic products, and machinery. Farmers here raise dairy cattle and other livestock.

The People About 80 percent of the country's people are Lithuanian, and most are city dwellers. Vilnius, the capital and largest city, is a major cultural and industrial center. Although the other Baltic republics are Protestant, Lithuania is largely Roman Catholic.

Praying in Lithuania

Worshipers gather at a Catholic church in Vilnius, Lithuania.
PLACE: Which Baltic republic is the largest in size?

SECTION 1 ASSESSMENT

REVIEWING TERMS AND FACTS

1. **Define the following:** communism, oil shale.

2. **REGION** What country ruled the Baltic countries from 1940 to 1991?

3. **PLACE** Where do most people live in the Baltic republics?

4. **PLACE** How does Lithuania compare in population to the other republics?

MAP STUDY ACTIVITIES

5. Study the physical map on page 320. What large body of water lies to the west of Latvia?

BUILDING GEOGRAPHY SKILLS

Mental Mapping

Think about how you get from place to place each day. In your mind you have a picture—or mental map—of your route. If necessary, you could probably draw sketch maps of many familiar places like the one shown below.

In the same way you can develop mental maps of places in the world. When studying a world region, picture its shape in your mind. Think about its important features and cities. Also try to imagine where in the world it is located. To develop your mental mapping skills, follow these steps:

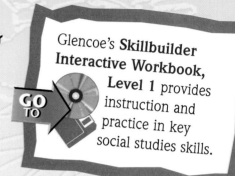

Glencoe's **Skillbuilder Interactive Workbook, Level 1** provides instruction and practice in key social studies skills.

- Picture a place in your mind. See its shape and most important features.
- Draw a sketch map of it.
- Compare your sketch to the actual place or a map. Revise your mental map and sketch.

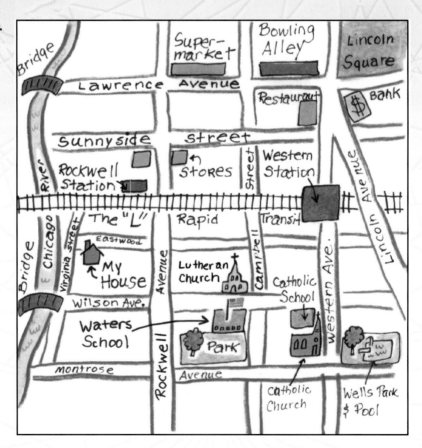

Geography Skills Practice

Study the sketch map above. On a separate sheet of paper, imagine your own neighborhood. Draw a sketch map of it from your mental map and answer these questions:

1. Which neighborhood streets or roads did you include?

2. What are the three most important features on your map?

SECTION 2 Poland

PREVIEW

Words to Know
- bog
- pope

Places to Locate
- Poland
- Carpathian Mountains
- Warsaw

Read to Learn . . .
1. how Poland's landscape differs from north to south.
2. how Poland's economy has changed in recent years.
3. what customs and beliefs the Polish people value.

Today Poland is a vigorous and visible country facing new challenges. But that was not the case in the late 1700s when Poland disappeared from world maps. This cafe in Kraków, the third-largest city in Poland, shows that the country is alive and well today!

The map on page 318 shows you that Poland lies south of the Baltic countries and east of Western Europe. In the late 1700s, Poland was swallowed up by Austria, Russia, and Germany. Poland later regained independence, but then communist rule was imposed on it after World War II.

Region
The Land

Today Poland is an independent nation. Covering 117,537 square miles (304,420 sq. km), Poland is slightly smaller than the state of New Mexico. On the north its coastline stretches for 326 miles (525 km) along the Baltic Sea. Sandy beaches and busy ports line the coast. Most of Poland lies in the North European Plain. Northern Poland also includes a hilly area with thousands of small lakes. Forests and **bogs**, or small swamps of spongy ground, run through the lakes region.

Plains and Highlands A vast plain stretches through central Poland. It forms part of the fertile North European Plain that spreads across Europe. Most of the Polish people live in this plains area.

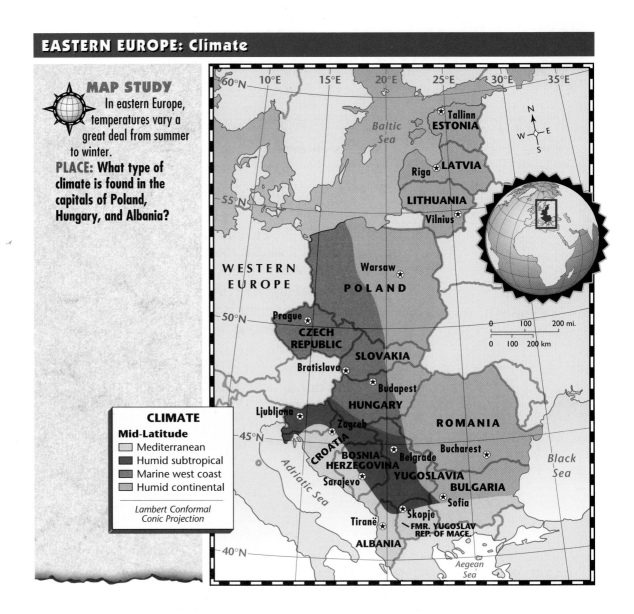

In eastern Europe, temperatures vary a great deal from summer to winter.
PLACE: What type of climate is found in the capitals of Poland, Hungary, and Albania?

CLIMATE
Mid-Latitude
- Mediterranean
- Humid subtropical
- Marine west coast
- Humid continental

Lambert Conformal Conic Projection

In southern Poland the plains gradually rise in elevation to meet the low Sudeten (soo•DAY•tuhn) Mountains and the towering Carpathian (kahr•PAY•thee•uhn) Mountains. These two mountain areas form Poland's southern border.

The source of the mighty Vistula (VIHS•chuh•luh)—the main river of Poland—lies in the Carpathians. Winding more than 680 miles (1,094 km), the Vistula River carves a huge letter *S* through the country before it empties into the Baltic Sea.

The Climate Look at the map above. You see that the climate differs from one part of Poland to another. Western Poland has a marine west coast climate. Warm winds blowing across Europe from the Atlantic Ocean bring mild weather year-round to this part of the country. Farther from the ocean, eastern Poland experiences a humid continental climate with hot summers and cold winters.

Human/Environment Interaction
The Economy

Some people joke that going to work is going "back to the salt mines." In Wieliczka (vyeh•LEECH•kah), Poland, workers say that and mean it! The world's oldest working salt mines lie beneath this city. The map on page 330 shows that the Polish economy depends on agriculture as well as industry. Polish farmers raise livestock and grow potatoes, sugar beets, wheat, and rye on the central plains. About 30 percent of the Polish population work on farms.

An Industrial Land Mining and manufacturing take place mostly in Poland's central and southern regions. Polish miners dig up coal, copper, and iron ore. Coal mining is Poland's most important industry and is concentrated near the Czech Republic. Factory workers produce machinery, cars, and trucks. Shippers at ports on the Baltic, such as Gdansk (guh•DAHNSK), send many of these products to other lands. Solidarity—the Polish labor union— struggles to protect the rights of these and other workers.

A Potato Harvest in Poland

Families work together to harvest the year's potato crop.
HUMAN/ENVIRONMENT INTERACTION: What has caused the pollution of the soil, air, and water in Poland?

A New Economy Until the early 1990s, Poland had a communist economy in which the government ran industry and set prices and wages. Since then Poland has been moving toward free enterprise.

The transition from communism to free enterprise has brought Poland new challenges. Under the communist system, many Poles had lifelong jobs in industries that were often inefficient. These jobs have disappeared as Polish companies try to compete on the world market. Even though some have lost jobs, most Poles prefer free enterprise to communism.

Environmental Challenges Poland has begun to make a serious attempt to correct its environmental problems. For years Polish factories poured wastes into the air, water, and soil. Pollution in Poland still causes severe health problems for humans and animals. The Poles are now trying to improve the environment without further loss of jobs.

Place
The People

Poland has a population of 38.7 million. Most of Poland's citizens are Poles who belong to a larger ethnic group known as Slavs. Neighboring groups such as the Russians, Ukrainians, and Czechs are also Slavs.

Many Polish couples dress in traditional clothes on their wedding day.
PLACE: To what ethnic group do most of Poland's people belong?

The Poles speak Polish, a Slavic language similar to Russian. Poles use the Latin alphabet, which English speakers also use. Some other Slavic groups, such as the Russians, use the Cyrillic (suh•RIH•lihk) alphabet. Look at the chart on page 334 to see the Slavic and other major language families of Europe.

Influences of the Past Situated on a vast plain, Poland has no natural defenses on its eastern and western borders. Over the centuries, it has been an easy target for invading armies. The ethnic term *Slav* comes from the word *slave*. Polish people have often been enslaved throughout their history. The Poles' love of country has helped them face these trials.

Daily Life About 62 percent of the Polish people live in cities and towns. Warsaw, the capital of Poland, has been an important urban center for hundreds of years. Poles still value their rural heritage and customs, however. They sometimes wear traditional costumes and often perform folk dances at weddings and other special occasions. At the same time, Polish young people enjoy rock and jazz as well as sports such as soccer, skiing, and basketball.

Religion has a strong influence on Polish life. Social life often centers around the local Roman Catholic church. Poles celebrate many religious holidays, especially Christmas and Easter. In 1978 Karol Wojtyla (voy•TEE•wah), a Polish church leader, became **pope,** or head of the Roman Catholic Church. The first Polish pope in history, Wojtyla took the name John Paul II.

SECTION 2 ASSESSMENT

REVIEWING TERMS AND FACTS
1. **Define the following:** bog, pope.
2. **REGION** What part of Poland consists of a vast plain?
3. **HUMAN/ENVIRONMENT INTERACTION** How has industry affected the environment in Poland?
4. **HUMAN/ENVIRONMENT INTERACTION** How has Poland's geography affected its history?
5. **PLACE** What traditions do most Poles have in common?

MAP STUDY ACTIVITIES

6. Look at the climate map on page 324. What two climates are found in Poland?

COPERNICUS

| MATH | SCIENCE | HISTORY | LITERATURE | TECHNOLOGY |

How many times have you stared up at the sky and wondered about the stars? A young Polish man named Nicolaus Copernicus (koh•PUHR•nih•kuhs) also wondered about the night sky. The sun, stars, and planets fascinated him throughout his life. Copernicus (1473–1543) became a scientist and astronomer and changed people's view of the solar system.

GREEK INFLUENCE The people of Copernicus's time believed, as the ancient Greeks had, that Earth was the center of the solar system. They thought that Earth remained still, while the sun, moon, planets, and stars circled it, with the planets "wandering" from place to place.

THE VIEW OF COPERNICUS Copernicus regularly watched the heavens. He carefully noted movements and measurements of stars. In 1543 he published his findings in a book called *On the Revolution of the Celestial Spheres.* Using mathematical proof, Copernicus established the sun as the center of the solar system. He showed that Earth did not stay in one place but rotated on its axis, as did all the other planets. He also proved that the

moon revolved around Earth, while Earth and the other planets revolved around the sun.

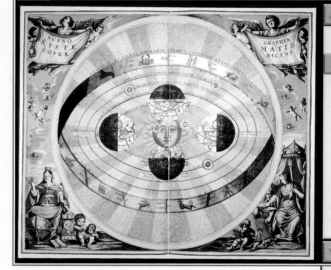

1543 Map of the Universe

MODERN ASTRONOMY Copernicus did what no other scientist had done. He proved that scientific findings and mathematical equations could predict and compare the movements of stars and planets. For the first time, he explained the solar system. For these contributions, later scientists called Copernicus the Father of Modern Astronomy.

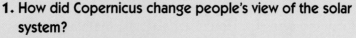

Making the Connection

1. How did Copernicus change people's view of the solar system?
2. Why is Copernicus called the Father of Modern Astronomy?

PREVIEW

Words to Know
- invest

Places to Locate
- Hungary
- Budapest
- Danube River
- Great Hungarian Plain

Read to Learn . . .
1. why the Danube River is important to Hungary.
2. how Hungary's economy changed after the fall of communism.
3. what types of foods Hungarians enjoy.

Have you ever ridden a horse or taken care of one? In Hungary raising horses has been a favorite activity in rural areas for centuries. Raised on the broad, fertile Hungarian Plain, Hungary's horses are famous throughout Europe.

Hungary lies east of Austria. Until 1918 Hungary and Austria were partners in a large central European empire. Like Austria, Hungary lost much territory after its defeat in World War I. Today it is a much smaller, landlocked country of 35,653 square miles (92,341 sq. km).

Region
The Land

Because it has no coastline, Hungary depends on the Danube River for transportation and trade. Look at the map on page 320. You see that the Danube twists and turns through several European countries, including Hungary. This river flows through so many countries that it has seven different names: Danube, Donau, Duna, Donaj, Dunarea, Dunav, and Dunai. It winds across 1,776 miles (2,858 km) before emptying into the Black Sea.

The Landscape The land of Hungary is mostly plains and highlands. The Great Hungarian Plain runs through eastern Hungary. Rivers and small hills wind across it. Dotted with farms, the plain is known for its excellent farming and grazing land.

The Danube River separates the Great Hungarian Plain from a very different region to the west. This region is called Transdanubia because it lies "across the Danube." Rolling hills, wide valleys, and forests cover Transdanubia's landscape. Lake Balaton, one of Europe's largest lakes, is in this region. Many Hungarians spend their vacations there sailing, swimming, and hiking.

In northern Hungary the Carpathian Mountains tower above the horizon. In this scenic part of Hungary you can wander through thick forests, find strange rock formations, and explore underground caves.

The Climate The map on page 324 shows you that Hungary is located far from any large body of water. As you might expect, it has a humid continental climate with cold winters and hot summers. Temperatures climb as high as 106°F (41°C) and drop as low as -29°F (-34°C). Western Hungary receives more rainfall than the eastern part of the country.

Movement
The Economy

Hungary's most important economic activities are agriculture and manufacturing. Hungarian farmers grow corn, potatoes, sugar beets, wheat, and wine grapes. They also raise dairy cattle, horses, and sheep. The country's factory workers produce chemicals, steel, textiles, and buses.

In the late 1940s, Hungary's economy came under communist control. When communism collapsed in the early 1990s, free enterprise returned. To encourage the growth of Hungarian companies, many foreign businesses now are investing in Hungary's free enterprise system. To **invest** means to put money into a company in return for a share of the profits.

Place
The People

The majority of Hungary's 10.1 million people are Hungarians, or Magyars. Unlike other eastern Europeans, Hungarians do not belong to the Slavic ethnic group. The ancestors of present-day Hungarians, the Magyars, came to eastern Europe from central Asia more than 1,000 years ago.

Hungarians are similar to other Europeans in their religious choices. About 67 percent of Hungary's people are Roman Catholic. Another 25 percent are Protestant.

Window-Shopping in Budapest

Window-shoppers look at the merchandise in Budapest's fashionable shops.
PLACE: How has Hungary's economic system changed since the collapse of communism?

MAP STUDY

Coal mining is an important industry in Poland.

PLACE: What other economic activity takes place in Poland in the areas where coal is found?

Agriculture

- Commercial farming
- Subsistence farming
- Manufacturing area

Resources

- Bauxite
- Coal
- Copper
- Fish and other seafood
- Hydroelectric power
- Iron ore
- Lead
- Natural gas
- Petroleum
- Zinc

Lambert Conformal Conic projection

ESTONIA
Riga
LATVIA
LITHUANIA
Gdansk
Potatoes
POLAND
Warsaw
Lodz
Wroclaw
Potatoes
Wheat
WESTERN EUROPE
Prague
CZECH REPUBLIC
Kraków
Tobacco
SLOVAKIA
Corn
Bratislava
Wheat
Budapest
HUNGARY
ROMANIA
SLOVENIA
Citrus fruit
Zagreb
Corn
Belgrade
Corn
CROATIA
BOSNIA-HERZEGOVINA
Wheat
Bucharest
Sarajevo
YUGOSLAVIA
Sofia
BULGARIA
Tobacco
ALBANIA
FMR. YUGOSLAV REP. OF MACE.
Adriatic Sea
Black Sea
Aegean Sea

0   100   200 mi.
0   100   200 km

City on the Danube About 63 percent of Hungarians are city dwellers. Nearly 1.9 million people live in Budapest (BOO•duh•PEHST), the capital and largest city. Fifteen-year-old Mátyás Kemény and his family live in Budapest. Mátyás always explains to visitors that Budapest is really two cities that stand on opposite banks of the Danube River. Buda, on the west bank, is a historic city of churches and palaces. Across the river, Pest is more modern and the site of many factories.

Past and Present Mátyás's father remembers when the Hungarians rose up against strict communist rule in 1956. Soviet troops crushed the uprising, and many Hungarians died. At that time, about 200,000 people left the country. Many of them moved to the United States.

Today Mátyás understands why his father does not have a job. His factory, which made parts for railroad cars, closed down because it could not compete successfully in the new free enterprise market. The family is hopeful for the future. Whatever happens, they know they do not want to return to communism.

Hungary's capital extends along both banks of the Danube River.
PLACE: Why is the Danube so important to Hungary's economy?

Daily Life Neither the grim communist past nor the present shortage of jobs can dim the Hungarians' enjoyment of life. Mátyás and his family like to celebrate. On special occasions they go to a restaurant where they can hear an orchestra play fast-paced Hungarian music.

Like most Hungarians, Mátyás enjoys good food, especially a thick stew called goulash. It is made of beef, gravy, onions, and potatoes. He also likes to flavor his food with a seasoning of red peppers called paprika. You may have seen paprika on potato salad or other foods that your family has prepared.

SECTION 3 ASSESSMENT

REVIEWING TERMS AND FACTS

1. Define the following: invest.

2. **REGION** What does Transdanubia look like?

3. **MOVEMENT** How are foreign businesses helping Hungary's economy?

4. **PLACE** What river divides Budapest into two parts?

MAP STUDY ACTIVITIES

5. Study the land use map on page 330. Where would be a good place to build an oil refining plant?

6. Look again at the map on page 330. What is Hungary's major manufacturing center?

The Czech Republic and Slovakia

PREVIEW

Words to Know
- nature preserve
- service industry

Places to Locate
- Czech Republic
- Prague
- Slovakia
- Bratislava

Read to Learn . . .
1. how people earn their livings in the Czech Republic and Slovakia.
2. what physical features dominate the Czech Republic and Slovakia.
3. why the Czech Republic and Slovakia became separate, independent countries.

Charles Bridge is a charming spot in the city of Prague, the capital of the Czech Republic. From the bridge, you can see the domes and towers of Prague. Prague is one of the best-preserved historic cities in Europe. It is also a modern city whose people look forward to a bright future.

The Czech (CHEHK) Republic and its neighbor Slovakia (sloh•VAH•kee•uh) once formed a larger country called Czechoslovakia. Like Poland and Hungary, Czechoslovakia was under communist control and closely linked to the Soviet Union. In 1989 the Czech and Slovak peoples peacefully ended communist rule. On January 1, 1993, they also peacefully split into two separate countries.

Place
The Czech Republic

The Czech Republic lies south of Poland and east of Germany. It is a land-locked country, lying deep within the continent. This interior location gives the republic warm summers and cold winters.

The Land The map on page 320 shows that mountains run through the western and northern parts of the Czech Republic. In these areas, spas famous for their healthful waters attract bathers to hot mineral springs. Tourists can also explore **nature preserves,** or lands set aside for plant and animal wildlife. The natural beauty, however, cannot hide the vast areas of bare trees ruined by industrial pollution.

Between low mountain ranges, the Czech Republic is a land of rolling hills and low fertile plains. These areas boast the nation's best farmland and major industrial centers. Rivers such as the Elbe (EHL•buh) and Vltava (VUHL•tuh•vuh) flow gracefully through the region.

The Economy Under communism the Czech economy was based on heavy industry. Today there is a free enterprise economy, and many Czechs have set up new **service industries.** They now own hotels, repair shops, and stores that sell clothing and household goods. Farming provides jobs for only about 10 percent of the population. Major crops include barley, corn, fruits, oats, and potatoes.

The People Most of the Czech Republic's 10.3 million people belong to a Slavic ethnic group called Czechs. Most Czechs live in cities and towns. Prague (PRAHG), with about 1 million people, is the capital and largest city. The Czechs have one of the highest standards of living in eastern Europe. Many of them own cars and household appliances. City dwellers, however, often have to live in crowded high-rise apartment buildings.

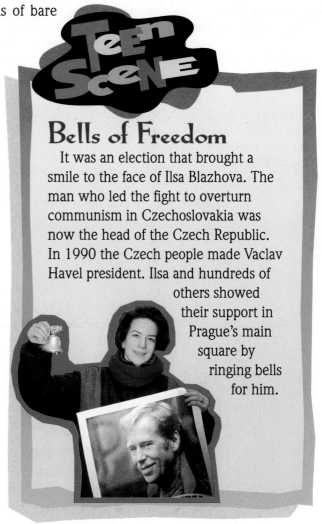

Teen Scene

Bells of Freedom

It was an election that brought a smile to the face of Ilsa Blazhova. The man who led the fight to overturn communism in Czechoslovakia was now the head of the Czech Republic. In 1990 the Czech people made Vaclav Havel president. Ilsa and hundreds of others showed their support in Prague's main square by ringing bells for him.

WHAT IN THE WORLD?

Castle Craze

Are you a victim of castle-mania? Do you find castles thrilling and fascinating? If so, you should travel to the Czech Republic and Slovakia. There you can visit a different castle every day—for seven years! Some 2,500 castles still stand in the two countries.

Place
Slovakia

Newly separated from the Czech Republic, Slovakia lies to the east. The Slovak people and the Czechs share a common Slavic heritage. The two groups, however, have different languages and cultures.

The Land and Climate A range of the Carpathian Mountains towers over most of northern Slovakia. Rugged gray peaks, thick forests, and blue lakes make this area a popular vacation spot. Farther south,

GRAPHIC STUDY

Europe has seven main language families.
MOVEMENT: From what language family did English develop?

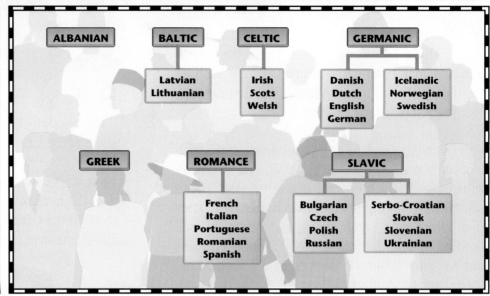

ALBANIAN	BALTIC	CELTIC	GERMANIC	
	Latvian Lithuanian	Irish Scots Welsh	Danish Dutch English German	Icelandic Norwegian Swedish

GREEK	ROMANCE	SLAVIC	
	French Italian Portuguese Romanian Spanish	Bulgarian Czech Polish Russian	Serbo-Croatian Slovak Slovenian Ukrainian

vineyards and farms spread across the fertile lowlands and plains that stretch to the Danube River. Elevation affects Slovakia's climate, but most of the country experiences cold winters and warm summers.

The Economy Slovakia has been a farming country throughout most of its history. Slovak farmers grow barley, corn, potatoes, sugar beets, and wine grapes. Under communist rule Slovakia began building factories for heavy industry. Slovakia now has a free enterprise economy with a growing number of service industries. Many inefficient factories have shut down, causing job losses.

The People An ethnic group known as Slovaks make up the majority of the country's population. People of Hungarian descent form the second-largest group. About 60 percent of Slovakia's 5.4 million people live in urban areas. Bratislava (BRA•tuh•SLAH•vuh), a port on the Danube, is Slovakia's capital and largest city. Most Slovaks are Catholics.

SECTION 4 ASSESSMENT

REVIEWING TERMS AND FACTS

1. Define the following: nature preserve, service industry.
2. LOCATION Where are the Czech Republic's major farmlands and industrial centers located?
3. PLACE What mountain range runs through northern Slovakia?

GRAPHIC STUDY ACTIVITIES

4. Study the chart of language families above. From which family does the Czech language come?

The Balkan Countries

PREVIEW

Words to Know
- consumer goods
- mosque

Places to Locate
- Slovenia
- Croatia
- Yugoslavia (Serbia and Montenegro)
- Bosnia-Herzegovina
- Former Yugoslav Republic of Macedonia (F.Y.R.O.M.)
- Romania
- Bulgaria
- Albania

Read to Learn . . .
1. why Yugoslavia broke up into separate countries.
2. how Romanians are like western Europeans.
3. what Albania needs to build its economy.

Have you ever seen such a towering wall of rock? It is one of the many mountain ranges that runs through the Balkan Peninsula in the southeast corner of Europe. The word *balkan* means "mountain" in one of the peninsula's local languages. The rugged Balkan Peninsula is home to a rich variety of cultures.

The Balkan Peninsula lies between the Adriatic Sea and the Black Sea. It also stretches into the Mediterranean Sea. The map on page 318 shows you that several countries make up this Balkan region. They are the former Yugoslav republics, plus Romania, Bulgaria, and Albania.

Place
Former Yugoslav Republics

Rugged mountains cover most of the republics that formerly made up Yugoslavia. Branches of the Alps reach into northwestern and coastal areas. Highlands at the center of the republics flatten into northern plains. The Danube River flows through the plains region.

Breakup of the Region This region once formed a communist country called Yugoslavia. After communism ended in the region, Yugoslavia broke apart because of differences among its many ethnic groups. Today the region is made up of five independent republics: Slovenia, Croatia, Yugoslavia (Serbia and Montenegro), Bosnia-Herzegovina (BAHZ•nee•uh HERT•suh•goh•VEE•nuh), and the Former Yugoslav Republic of Macedonia (F.Y.R.O.M.).

Slovenia Slovenia lies in the mountains of the northwest. It has more factories and service industries than the other republics. It also has the region's highest standard of living. Most of the 2 million Slovenians are Slavs who practice the Roman Catholic religion and use the Latin alphabet.

Croatia Beyond Croatia's island-studded Adriatic coast, mountains rise to a fertile, inland plain. An industrialized republic, Croatia also relies on agriculture. Like the Slovenians, the 4.8 million Croats (KROH•atz) are Slavs and Roman Catholics. They write in the Latin alphabet.

Serbia and Montenegro (Yugoslavia) Serbia and Montenegro, united as Yugoslavia, cover inland plains and mountains. The two republics rely on agriculture and industry. Belgrade is the capital and largest city.

Most of Yugoslavia's 10.6 million people belong to two Slavic groups—the Serbs and the Montenegrins. Eastern Orthodox in religion, they write in the Cyrillic alphabet. The largely Muslim Albanians, however, are the largest group in the province of Kosovo. In 1998 conflict erupted in Kosovo between Yugoslav forces and Albanians favoring Kosovo's independence.

Bosnia-Herzegovina Bosnia-Herzegovina lies west of Serbia. It consists of mountains, thick forests, and fertile river valleys. Many of the Bosnian people are Muslims. Others are Eastern Orthodox Serbs or Roman Catholic Croats.

Civil war ravaged this area in the early 1990s. Bosnian Serbs tried to force out Bosnian Muslims. In 1995 the Dayton Peace Treaty ended the fighting. It also divided Bosnia into two separate regions under one government.

F.Y.R.O.M. The Former Yugoslav Republic of Macedonia is the most southern of the republics. It has a developing economy largely based on agriculture. The 2 million Macedonians are a mixture of many different Balkan peoples.

Place
Romania

Romania spreads eastward from Yugoslavia to the Black Sea. It has hot summers and cold winters. The scenic Carpathian Mountains curve through northern and central Romania. Between the mountain ranges stretch a vast plateau and plains. Quaint villages and modern urban centers dot these flatlands. Transylvania, the setting for many horror novels, lies in this region.

The Economy Romania's economic activities include farming, manufacturing, and mining. The oil industry is important in the southeastern part of the country. Under communism Romania's factories produced machinery but few **consumer goods**—clothing, shoes, and other products made for people. Romania now has a free enterprise economy that turns out more consumer goods.

The People What does Romania's name tell you about its history? If you guessed that the Romans once ruled this region, you are correct. Most of Romania's 22.5 million people are descended from the Romans. The Romanian

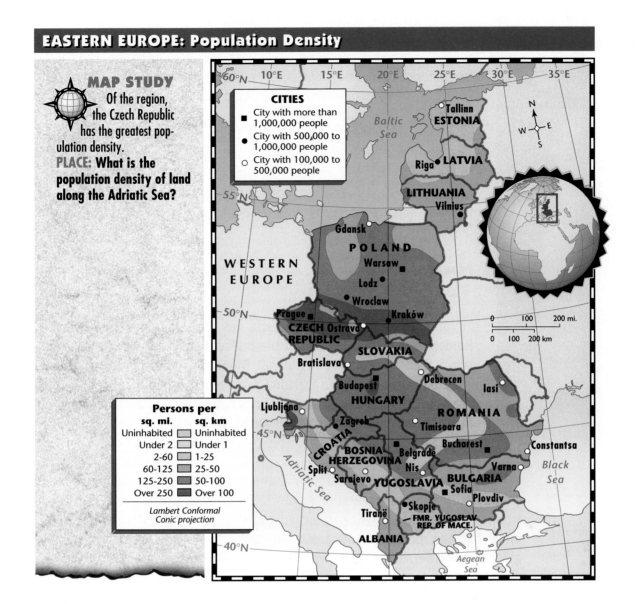

MAP STUDY
Of the region, the Czech Republic has the greatest population density.
PLACE: What is the population density of land along the Adriatic Sea?

CITIES
■ City with more than 1,000,000 people
● City with 500,000 to 1,000,000 people
○ City with 100,000 to 500,000 people

Persons per	
sq. mi.	**sq. km**
Uninhabited	Uninhabited
Under 2	Under 1
2-60	1-25
60-125	25-50
125-250	50-100
Over 250	Over 100

Lambert Conformal Conic projection

language comes from the Roman language, Latin. Romanian is closer to French, Italian, and Spanish than it is to other eastern European languages. In other ways, the Romanians are more like their Slavic neighbors. For example, many Romanians are Eastern Orthodox Christians. Half of Romania's population live in cities and towns. Bucharest, the capital and largest city, has wide streets and modern buildings.

Place

Bulgaria

Bulgaria lies south of Romania. Like Romania, Bulgaria has a coastline on the Black Sea. Two mountain ranges—the Balkan Mountains and the Rhodope (RAH•duh•pee) Mountains—span most of Bulgaria. Fertile valleys and plains separate the mountains in many areas. Bulgaria's climate varies considerably. The Black Sea coast has warmer year-round temperatures than the mountainous inland areas.

A Bulgarian woman handpicks roses to be used in making perfume.
PLACE: In what other ways do Bulgarians earn their livings?

The Economy Bulgaria is making the transition from a communist to a free enterprise economy. Bulgarian factories produce chemicals, machinery, and textiles. The country's farms grow fruits, vegetables, and grains. Roses are grown for their sweet-smelling oil, which is used in perfumes.

The People Most of Bulgaria's 8.3 million people trace their ancestry to the Slavs and other groups from central Asia. The Bulgarian language is similar to Russian and uses the Cyrillic alphabet. Most Bulgarians live in cities and towns. Sofia (SOH•fee•uh) is the country's capital and largest city.

Place
Albania

Tucked southwest of Yugoslavia lies the small country of Albania. Bordering the Adriatic coast of the Balkan Peninsula, Albania is slightly larger than the state of Maryland. Mountains cover most of Albania, contributing to Albania's isolation from neighboring countries. A small coastal plain runs along the Adriatic Sea. Most of Albania has a mild climate. Temperatures in coastal areas, however, are warmer than in mountainous inland areas.

The Economy Albania has one of Europe's least developed economies. The country is rich in mineral resources but lacks the technology to develop them. Under communist rule Albania began to develop heavy industry. Yet today, under free enterprise, most Albanians still make their living from farming. They grow corn, grapes, olives, potatoes, sugar beets, and wheat in mountain valleys.

The People About 62 percent of Albania's 3.3 million people live in the countryside. Most Albanians are Muslims. Since the fall of communism, many **mosques**, or Muslim houses of worship, have opened in Albania.

SECTION 5 ASSESSMENT

REVIEWING TERMS AND FACTS
1. Define the following: consumer goods, mosque.
2. PLACE What former Yugoslav republics have been torn by warfare?
3. LOCATION What Balkan countries have coastlines on the Black Sea?
4. HUMAN/ENVIRONMENT INTERACTION Why is Albania unable to develop its resources?

MAP STUDY ACTIVITIES
5. Look at the population density map on page 337. What city in the Yugoslav region has more than 1 million people?

Chapter 13 Highlights

Important Things to Know About Eastern Europe

SECTION 1 THE BALTIC REPUBLICS

- The Baltic republics of Estonia, Latvia, and Lithuania share a Baltic Sea coastline.
- After decades of communist rule, Estonia, Latvia, and Lithuania are now independent democracies.

SECTION 2 POLAND

- Most of Poland's farms, factories, and cities lie on a vast plain.
- Poland's economy changed from communism to free enterprise.
- Most Poles belong to the Roman Catholic Church.

SECTION 3 HUNGARY

- Landlocked Hungary depends on the Danube River for trade.
- Most Hungarians belong to an ethnic group known as Magyars.
- Hungarian businesses grow through foreign investments.

SECTION 4 THE CZECH REPUBLIC AND SLOVAKIA

- Cultural differences led Czechs and Slovaks to form two separate nations in 1993.
- Mountains, rolling hills, and plains sweep across the Czech Republic and Slovakia.
- Both republics are industrialized, with large urban populations.

SECTION 5 THE BALKAN COUNTRIES

- The Balkan Peninsula has many different ethnic groups speaking several different languages.
- The breakup of Yugoslavia has led to civil war in the region.

Celebrating independence in Tallinn, Estonia ▶

REVIEWING KEY TERMS

Match the numbered terms in Set A with their lettered definitions in Set B.

A

1. consumer goods
2. oil shale
3. bog
4. pope
5. invest
6. nature preserve
7. mosque

B

A. leader of the Roman Catholic Church
B. Muslim house of worship
C. land set aside to protect plants and animal wildlife
D. a type of rock that contains petroleum
E. small swamp of spongy ground
F. goods made for people's use, such as shoes and clothing
G. to spend money on a business, hoping to make a profit

Mental Mapping Activity

Draw a freehand map of Eastern Europe. Label the following:
- Slovakia
- Sofia
- Great Hungarian Plain
- Czech Republic
- Albania
- Poland

REVIEWING THE MAIN IDEAS

Section 1
1. PLACE How is Estonia unique among the Baltic countries?
2. LOCATION How does Latvia's location affect its economy?

Section 2
3. REGION What are the main landforms of Poland?
4. PLACE What religion do most Poles practice?

Section 3
5. MOVEMENT Where did the original Magyars come from?
6. PLACE Name a typical Hungarian food dish.

Section 4
7. REGION What two rivers flow through the Czech Republic?
8. REGION How has free enterprise changed Slovakia's economy?

Section 5
9. PLACE What is the major landform of the Balkan Peninsula?
10. PLACE What is the capital of Romania?

CRITICAL THINKING ACTIVITIES

1. **Making Comparisons** How do people earn a living in the Czech Republic? In Albania?
2. **Making Generalizations** How have mountains affected ways of life in the Balkan Peninsula?

CONNECTIONS TO WORLD HISTORY

Cooperative Learning Activity Work in a group of three to learn more about recent conflicts in the former Yugoslav republics. Research the various ethnic groups of the region. What does each group want? What land are they fighting over? Draw a map of the Balkan region with labels showing where different groups live. Propose a peace settlement that seems fair. Then share your plan with the rest of the class.

GeoJournal Writing Activity

Write a journal entry describing a trip down the Danube River from southern Germany to the Black Sea. Describe the cities and the countryside through which you would pass.

TECHNOLOGY ACTIVITY

Using a Spreadsheet Research the progress of the eastern European countries after the fall of communism. Use a spreadsheet to organize your information. Include headings such as name of country, highlights of history under communism, year of independence, how independence was attained, current economy, and present form of government. Include a map highlighting the locations of the countries and their capitals.

PLACE LOCATION ACTIVITY: EASTERN EUROPE

Match the letters on the map with the places and physical features of eastern Europe. Write your answers on a separate sheet of paper.

1. Poland
2. Budapest
3. Bosnia-Herzegovina
4. Latvia
5. Balkan Mountains
6. Danube River
7. Adriatic Sea
8. Warsaw
9. Carpathian Mountains
10. Baltic Sea

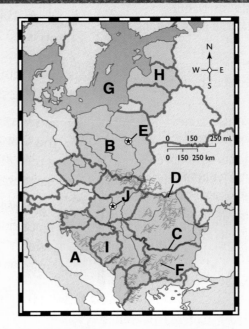

EYE ON THE ENVIRONMENT

Europe

POLLUTION

PROBLEM

Air, water, and soil pollution do not respect national boundaries. Smoke belching from one country's smokestacks becomes another country's acid rain or smog. Europe is one of the world's most polluted continents. Smoke from British factories kills forests in Scandinavia. Raw sewage and industrial toxins in the Baltic Sea spoil beaches and kill fish, birds, and other wildlife. Pollutants dumped into the Danube River end up in the drinking water of eight nations.

SOLUTIONS

- European Union members have pledged to reduce air pollution by 90 percent by the year 2005.

- Britain has established "protection zones" limiting the use of pesticides in certain areas.

- In the Czech Republic, scientists have developed a "clean coal" that burns without releasing sulfur dioxide—the cause of acid rain.

- In France, taxes on industries are being used to install machines that sniff the air. They sound a warning when pollution is detected.

- In Greece and Italy, some cities allow cars and buses to operate only on certain days.

A layer of suds pollutes the Tagus River, running beneath a stone bridge in Toledo, Spain.

POLLUTION FACT BANK

🍃 In 100 European cities, air pollution is 10 times greater than the standards set by the World Health Organization.

🍃 Air pollution and acid rain have laid waste to many German forests. In the Czech Republic and Slovakia, nearly a million acres of woodland have been damaged by acid rain.

🍃 Most of the water in Poland's rivers is undrinkable, and 50 percent of river water is so toxic it corrodes industrial machinery.

🍃 In Hungary, air pollution causes 1 in 17 deaths. An hour's stroll through Budapest's polluted streets is as bad for the lungs as smoking 20 cigarettes!

342

UNIT 4

TEEN TRIBUTE

Students in Dobra Voda in the Czech Republic hope to make their town live up to its name. *Dobra voda* means "good water" in Czech, and the town is famous for its spring water. But overuse of fertilizers and pesticides polluted the water supply. When the students tested the water, they found it unfit to drink. Now Dobra Voda imports its drinking water.

environmental activities

TAKE A CLOSER LOOK

1 What substances pollute Europe's air and water?

2 What are European countries doing to clean up their air?

CLEAN UP OUR AIR

WHAT CAN YOU DO?

🍃 Adopt a Stream! Contact this organization and support its goal to protect local waterways.

🍃 Organize a Hazardous Waste Collection Day at your school. One quart of oil can contaminate 2 million gallons of water.

🍃 Grow a plant at home—it will clean up your air by absorbing carbon dioxide.

🍃 Feed the birds! They act as natural pesticides that won't pollute our earth.

Acid rain eats away at this cherub and centuries-old statues in St. Mark's Square in Venice, Italy.

Russia and the Independent Republics

What Makes Russia and the Independent Republics a Region?

The people share . . .

• a vast land—8.5 million square miles!

• very diverse cultures within republics.

• geographic features.

• the change to a capitalist economy.

To find out more about Russia and the independent republics, see the Unit Atlas on pages 346–357.

St. Petersburg, Russia

EXPLORING THE INTERNET

To learn more about Russia and the independent republics, visit the Glencoe Social Studies Web site at **www.glencoe.com** for information, activities, and links to other sites.

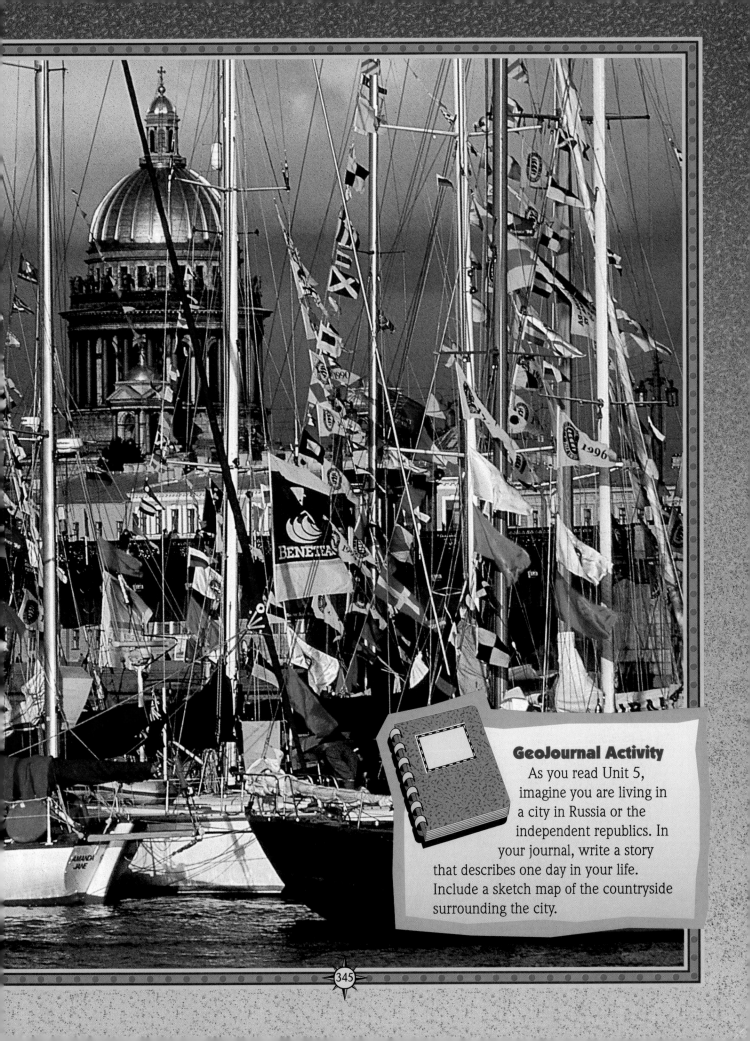

GeoJournal Activity

As you read Unit 5, imagine you are living in a city in Russia or the independent republics. In your journal, write a story that describes one day in your life. Include a sketch map of the countryside surrounding the city.

UNIT 5 ATLAS

NATIONAL
GEOGRAPHIC
SOCIETY

IMAGES
of the
WORLD

1. Child using sign language, Murmansk
2. Gold domes, St. Petersburg
3. Gas field workers, Siberia
4. Village, western Ukraine
5. Castle on the Black Sea, Ukraine
6. Open-air market, Kazan
7. Buddhist monks, Ulan-Ude

All photos viewed against the statue *Motherland Calls*, Volgograd, Russia.

Regional Focus

Russia and the other independent republics span the continents of Europe and Asia. Russia is the largest republic in area and population. The other republics in the region are Ukraine, Belarus, Moldova, Armenia, Georgia, Azerbaijan, Kazakhstan, Uzbekistan, Turkmenistan, Kyrgyzstan, and Tajikistan.

Region

The Land

The region of Russia and the independent republics covers about 8 million square miles (20.7 million sq. km). This land area is greater than the size of Canada, the United States, and Mexico combined.

Overlooking the Ural Mountains

The Ural Mountains separate European Russia from the Asian part of Russia known as Siberia. **HUMAN/ENVIRONMENT INTERACTION: Why do you think most Russians live west of the Urals?**

The North The far north of Russia is made up of low hills. Snow and ice cover the land for most of the year. The few people who live in the far north either fish, hunt, or herd reindeer. Farther south you will find deep, thick forests. In some areas farmers have cleared the woods to grow crops.

South of the forests is a vast open plain. It stretches more than 3,000 miles (4,800 km) from eastern Europe to Asia. For centuries routes across this plain brought invading armies. Today it is an important agricultural and industrial area for Russia, Ukraine, Moldova, and Belarus. The plain's fertile soil is among the best in the world for farming.

Mountains Mountains cover many areas in Russia and the independent republics. You will find two mountain ranges in areas where the continents of Europe and Asia meet. The Ural Mountains divide Russia into European and Asian parts. To the south, the Caucasus range rises along the borders of Russia, Georgia, and Azerbaijan. Find these mountains on the map on page 352.

Mountains also cross through several republics in central Asia. The Pamirs in Tajikistan have some of the region's highest peaks. The Tian Shan range in Kyrgyzstan holds some of the world's largest glaciers.

Bodies of Water Russia and the independent republics contain many bodies of water. You will find two of the world's largest lakes here. One is the Caspian Sea, which is really a salt lake. The other is Lake Baikal, the deepest lake in the world. The region also has long rivers. Some of the rivers flow eastward, like the Amur, which forms Russia's border with China. Others, like the Volga and the Dnieper, flow south through plains. Many that flow north, such as the Lena, Ob, and Yenisey, are frozen much of the year.

Region

Climate and Vegetation

The far northern location of Russia and the independent republics affects their climates. Most of the republics lie far from great bodies of water that keep land temperatures mild. Also, the region's flat lands do not provide shelter from hot summer winds and freezing winter storms.

Far northern areas have tundra and subarctic climates with cold temperatures most of the year. Most of the deep soil is always frozen. Farther south, forests stretch to a grass-covered plains area, which has a humid continental climate of short, hot summers and long, cold winters. Desert regions of central Asia have long, hot summers and short, cold winters. Plant life is very sparse there. The Black Sea coast, with its humid subtropical and Mediterranean climates, is known for subtropical plants and citrus fruits.

Vineyards in Moldova

Enough wine is produced from Moldova's vineyards to export to Russia and other independent republics.
PLACE: What crops grow well along the Black Sea?

Region

The Economy

For many years, Russia and the independent republics formed one country called the Soviet Union. It had an economy planned and run by the communist leaders. In 1991 each republic became independent and responsible for its own economy. The republics now allow people to run their own businesses and farms. This change to free enterprise has been a challenge for the people in the region.

Russia and the independent republics grow mainly wheat, rye, oats, barley, sugar beets, and cotton. The region is rich in minerals such as coal, oil, natural gas, copper, silver, manganese, and gold. Dams on rivers in Russia and Ukraine supply electric power to factories and cities. Rivers, roads, and railroads link industrial areas. Industrial growth under Soviet rule, however, led to widespread pollution of the air and water.

Place
The People

About 283 million people live in Russia and the independent republics. Russia has the region's largest population with 147 million people.

Shopping in Moscow

The GUM department store—Russia's largest shopping mall—sells everything from coats to caviar *(above)*.
REGION: How have the economies of Russia and the independent republics changed since 1991?

Population Climate and landscape affect where people live in Russia and the independent republics. Most people in the region live west of the Urals, where the climates are mild and the land is fertile. In the past most people lived in the countryside. Today most are urban dwellers. The three largest cities are Moscow, St. Petersburg, and Kiev.

Ethnic Groups Each of the republics has a major ethnic group, language, and culture. There are also many smaller groups in each republic. More than 100 different ethnic groups live throughout the entire region. Slavs (Russians, Ukrainians, and Belarusians) make up the region's largest ethnic group. They speak Slavic languages and practice Eastern Orthodox Christianity. Turkic ethnic groups (Uzbeks, Kazakhs, and Turkmenis) predominate in central Asia. They have their own languages and practice the religion of Islam. Various ethnic groups known as Caucasian inhabit Armenia and Georgia. They practice their own forms of Orthodox Christianity.

History Russia and the independent republics have a rich history. Centuries ago Slavs set up city-states along rivers in present-day Ukraine and Russia. By the 1400s powerful rulers called czars ruled Russia. Over several centuries, the Russians took over all of the region that is today Russia and the independent republics. Their territory was called the Russian Empire.

In 1917 the rule of the czars ended. A communist dictatorship emerged, and the Russian Empire became the Union of Soviet Socialist Republics. The Communists made the Soviet Union an industrial power but denied the people basic freedoms. In 1991 the Communists fell from power, and the separate parts of the Soviet Union became independent republics. Under Russian president Boris Yeltsin, Russia allied with the other former Soviet republics to form a loose union called the Commonwealth of Independent States, or CIS.

The Arts The arts of Russia and the independent republics include architecture, painting, music, and dance. Each republic has its own rich heritage. You probably have seen pictures of Russia's onion-domed churches and heard the classical music of Peter Tchaikovsky and other Russian composers. Ukraine is known for its lively folk music and colorfully decorated Easter eggs. Ancient churches with drumlike tops and bells dot the rugged countryside of Armenia and Georgia. In the central Asian republics, you will see beautiful tiles in swirling patterns decorate Islamic mosques.

Multimedia Unit 5 Activities

NATIONAL GEOGRAPHIC SOCIETY

Picture Atlas of the World

Economic Restructuring

Using the CD-ROM Today Russia and the independent republics are struggling to come to terms with the changes brought about by independence. Use the CD-ROM and other reference works to analyze the economies of these countries. Record your findings in a chart. Use the following criteria to help you assess the economic status of each country:

- number of schools
- annual per capita income
- amount of food and housing per resident

- extent of communication and transportation networks
- literacy rate
- life expectancy
- infant mortality rate
- population growth rate

Compare your chart with your classmates' charts to see how your assessments of each country differ.

Surfing the "Net"

Security, Expansion, and Geography

Throughout history countries wishing to expand their borders have targeted neighboring nations. The former Soviet Union targeted Estonia, Latvia, and Lithuania. In 1990 Estonia, Latvia, and Lithuania became the first republics to declare their independence from the Soviet Union. To find out more about the struggle for self-rule in these countries, search the Internet.

Getting There

Follow these steps to gather information on the Baltic countries:

1. Use a search engine. Type in the names of the countries.
2. Enter words like the following to focus your search: *history*, *politics*, and *independence*.

3. The search engine will provide you with a number of links to follow.

What To Do When You Are There

Complete the following activity individually or in groups. Build a presentation that explores the history and results of political revolution in Estonia, Latvia, or Lithuania. Your presentation should depict one or more of the following events: the periods of independence and control of the country, the reactions of the country's citizens to control or independence, the buildup to revolution, how and when the revolution occurred, or the results of revolution. Share your presentation with the class.

Physical Geography

RUSSIA AND THE INDEPENDENT REPUBLICS: Physical

ARCTIC OCEAN

80°N

60°N

180°

Bering
Sea

New
Siberian
Islands

Kolyma R.

Kamchatka
Peninsula

Klyuchevskaya ▲
15,584 ft.
(4,750 m)

160°E

VERKHOYANSKIY
RANGE

CHERSKOGO RANGE

Barents
Sea

EUROPE

60°N

Novaya
Zemlya

Baltic Sea

20°E

BELARUS

North

European

Plain

Dnieper R.

UKRAINE

Don R.

MOLDOVA

Volga R.

Black Sea

40°N

GEORGIA

CAUCASUS MTS.

Mt. Elbrus
18,510 ft.
(5,642 m)

40°E

ARMENIA

AZERBAIJAN

Caspian
Sea

TURKMENISTAN

60°E

Kara Kum

Kama R.

URAL MOUNTAINS

R U S S I A

West

Siberian

Plain

Ural R.

Yenisey River

Arctic Circle

Lena R.

Sea
of
Okhotsk

Sakhalin
Island

140°E

Central

Siberian

Plateau

Ob R.

Irtysh R.

KAZAKHSTAN

Aral
Sea

Lake
Balkhash

Lake
Baikal

YABLONOVY
MOUNTAINS

SAYAN
MOUNTAINS

Kyrgyz Steppe

UZBEKISTAN

KYRGYZSTAN

A S I A

N
W   E
S

100°E

120°E

TIAN SHAN
Communism Peak
▲ 24,590 ft. (7,495 m)

TAJIKISTAN

80°E

0   250   500 mi.

0   250 500 km

— National boundary
☐ Desert
▲ Mountain peak

*Lambert Equal-Area
projection*

ELEVATIONS

Feet	Meters
10,000	3,000
5,000	1,500
2,000	600
1,000	300
0	0

Map Study

1. **LOCATION** What two mountain ranges border the Central Siberian Plateau on the south?

2. **LOCATION** What large desert is found in Turkmenistan?

ELEVATION PROFILE: Russia

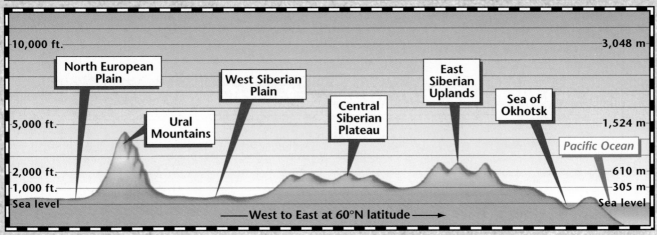

North European Plain

Ural Mountains

West Siberian Plain

Central Siberian Plateau

East Siberian Uplands

Sea of Okhotsk

Pacific Ocean

10,000 ft. — 3,048 m

5,000 ft. — 1,524 m

2,000 ft. — 610 m

1,000 ft. — 305 m

Sea level — Sea level

← West to East at 60°N latitude →

Source: *Goode's World Atlas,* 19th edition

GeoFacts

Highest point: Communism Peak (Tajikistan) 24,590 ft. (7,495 m) high

Lowest point: Caspian Sea (Russia, Azerbaijan) 92 ft. (28 m) below sea level

Longest river: Ob-Irtysh 3,362 mi. (5,409 km) long

Largest lake: Caspian Sea (Eurasia) 143,244 sq. mi. (371,002 sq. km)

Deepest lake: Lake Baikal (Russia) 5,315 ft. (1,620 m) deep

Largest desert: Kara Kum (Turkmenistan) 120,000 sq. mi. (310,800 sq. km)

RUSSIA/INDEPENDENT REPUBLICS AND THE UNITED STATES: Land Comparison

Graphic Study

1. **LOCATION** Where are the highest elevations found in Russia?
2. **REGION** About how much larger in area are Russia and the independent republics than the continental United States?

UNIT 5 ATLAS

Cultural Geography

RUSSIA AND THE INDEPENDENT REPUBLICS: Political

ARCTIC OCEAN

Barents
Sea

Bering
Sea

EUROPE

Baltic Sea

Arctic Circle

Sea
of
Okhotsk

160°E

⊛ Minsk   BELARUS

Chisinau
⊛ Kiev   ⊛ Moscow

R U S S I A

Sea
of
Japan

UKRAINE
MOLDOVA

Black Sea

GEORGIA

⊛ Astana

Lake
Baikal

40°N
⊛ Tbilisi

K A Z A K H S T A N

Yerevan
⊛ ⊛ AZERBAIJAN

Aral
Sea

Lake
Balkhash

N

Sea
of
Japan

ARMENIA   ⊛ Baku

UZBEKISTAN

W   E

A S I A

Caspian
Sea

S

Ashkhabad ⊛   Tashkent ⊛
TURKMENISTAN   Dushanbe ⊛

Bishkek ⊛
KYRGYZSTAN

60°E   TAJIKISTAN

National boundary
⊛ National capital

80°E

0   250   500 mi.

0   250   500 km

Lambert Equal-Area
projection

PACIFIC
OCEAN

100°E

120°E

140°E

Map Study

1. **PLACE** What is the largest independent republic after Russia?
2. **PLACE** What is the capital of Belarus?

Population: Russia and the Independent Republics

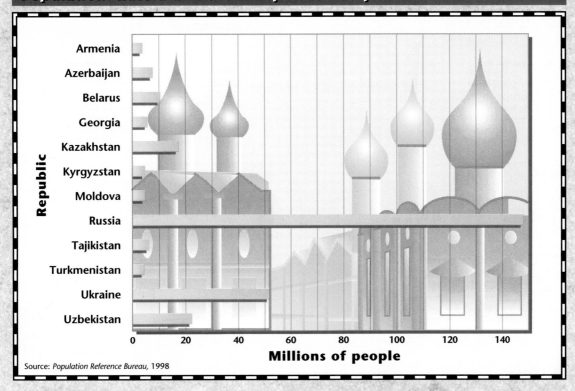

Republic: Armenia, Azerbaijan, Belarus, Georgia, Kazakhstan, Kyrgyzstan, Moldova, Russia, Tajikistan, Turkmenistan, Ukraine, Uzbekistan

Millions of people: 0 20 40 60 80 100 120 140

Source: *Population Reference Bureau, 1998*

GeoFacts

Biggest country (land area):
Russia 6,520,656 sq. mi.
(16,888,499 sq. km)

Smallest country (land area):
Armenia 10,888 sq. mi.
(28,200 sq. km)

Largest city (population):
Moscow (1998) 9,269,000
(2015 projected) 11,000,000

Highest population density:
Armenia 349 people per sq. mi.
(134 per sq. km)

Lowest population density:
Kazakhstan 15 people per sq. mi.
(6 people per sq. km)

COMPARING POPULATION: Russia/Independent Republics and the U.S.

Russia/ Independent Republics

United States

= 50,000,000

Source: *Population Reference Bureau, 1998*

Graphic Study

1. PLACE What is the combined population of Russia and the independent republics?

2. PLACE Which country in the chart above has the second-highest population?

Countries at a Glance

Armenia

Yerevan

LANDMASS:
10,888 sq. mi./
28,200 sq. km
MONEY:
Dram
MAJOR EXPORT:
Machinery
MAJOR IMPORT:
Machinery

CAPITAL:
Yerevan
MAJOR LANGUAGE(S):
Armenian
POPULATION:
3,800,000

Azerbaijan

Baku

CAPITAL:
Baku
MAJOR LANGUAGE(S):
Azeri, Turkish, Russian
POPULATION:
7,700,000

LANDMASS:
33,436 sq. mi./
86,599 sq. km
MONEY:
Manat
MAJOR EXPORT:
Food Products
MAJOR IMPORT:
Food Products

Belarus

Minsk

CAPITAL:
Minsk
MAJOR LANGUAGE(S):
Belorussian, Russian
POPULATION:
10,200,000

LANDMASS:
80,108 sq. mi./
207,480 sq. km
MONEY:
Belarus Ruble
MAJOR EXPORT:
Machinery
MAJOR IMPORT:
Machinery

Georgia

Tbilisi

CAPITAL:
Tbilisi
MAJOR LANGUAGE(S):
Georgian, Russian
POPULATION:
5,400,000

LANDMASS:
26,911 sq. mi./
69,699 sq. km
MONEY:
Lari
MAJOR EXPORT:
Food
MAJOR IMPORT:
Machinery

Kazakhstan

Astana

CAPITAL:
Astana
MAJOR LANGUAGE(S):
Kazakh, Russian, German
POPULATION:
15,600,000

LANDMASS:
1,031,170 sq. mi./
2,670,730 sq. km
MONEY:
Tenge
MAJOR EXPORT:
Raw Materials
MAJOR IMPORT:
Raw Materials

Kyrgyzstan

Bishkek

CAPITAL:
Bishkek
MAJOR LANGUAGE(S):
Kyrgyz
POPULATION:
4,700,000

LANDMASS:
74,054 sq. mi./
191,800 sq. km
MONEY:
Som
MAJOR EXPORT:
Machinery
MAJOR IMPORT:
Light-industrial Products

Moldova

Chisinau

CAPITAL:
Chisinau
MAJOR LANGUAGE(S):
Romanian, Ukrainian
POPULATION:
4,200,000

LANDMASS:
12,730 sq. mi./
32,971 sq. km
MONEY:
Leu
MAJOR EXPORT:
Food Products
MAJOR IMPORT:
Machinery

Russia

Moscow

CAPITAL:
Moscow
MAJOR LANGUAGE(S):
Russian, Ukrainian,
Belorussian, Uzbek
POPULATION:
147,000,000

LANDMASS:
6,520,656 sq. mi./
16,888,499 sq. km
MONEY:
Ruble
MAJOR EXPORT:
Fuels
MAJOR IMPORT:
Machinery

Tajikistan

Dushanbe

CAPITAL:
Dushanbe
MAJOR LANGUAGE(S):
Tadzhik, Russian
POPULATION:
6,100,000

LANDMASS:
54,286 sq. mi./
140,601 sq. km
MONEY:
Ruble
MAJOR EXPORT:
Aluminum
MAJOR IMPORT:
Chemicals

Turkmenistan

Ashkhabad

CAPITAL:
Ashkhabad
MAJOR LANGUAGE(S):
Turkmen, Russian
POPULATION:
4,700,000

LANDMASS:
181,440 sq. mi./
470,000 sq. km
MONEY:
Manat
MAJOR EXPORT:
Natural gas
MAJOR IMPORT:
Machinery

Countries not drawn to scale.

Lunch in Moscow

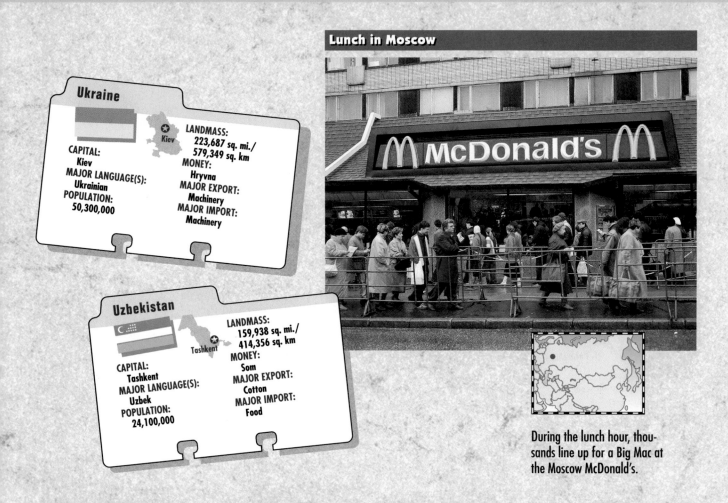

Ukraine

CAPITAL:
Kiev
MAJOR LANGUAGE(S):
Ukrainian
POPULATION:
50,300,000

LANDMASS:
223,687 sq. mi./
579,349 sq. km
MONEY:
Hryvna
MAJOR EXPORT:
Machinery
MAJOR IMPORT:
Machinery

Uzbekistan

CAPITAL:
Tashkent
MAJOR LANGUAGE(S):
Uzbek
POPULATION:
24,100,000

LANDMASS:
159,938 sq. mi./
414,356 sq. km
MONEY:
Som
MAJOR EXPORT:
Cotton
MAJOR IMPORT:
Food

During the lunch hour, thousands line up for a Big Mac at the Moscow McDonald's.

Street Scene in Lvov, Ukraine

The city of Lvov has large squares and narrow streets much like the cities of Europe.

Chapter 14 Russia

ARCTIC OCEAN

Barents Sea

Murmansk

Baltic Sea

St. Petersburg

EUROPE

Moscow ⊛

Nizhniy Novgorod

Voronezh

Kazan • Perm

Saratov

• Ufa • Yekaterinburg

Samara • Chelyabinsk

Volgograd

Astrakhan

Black Sea

Caspian Sea

Aral Sea

Omsk

Novosibirsk

R U S S I A

Arctic Circle

Krasnoyarsk

Irkutsk • *Lake Baikal*

ASIA

Yakutsk

Petropavlovsk-Kamchatski

Sea of Okhotsk

Bering Sea

Khabarovsk

Vladivostok

Sea of Japan

80°N

60°N

60°N

80°N

80°N

60°N

180°

160°E

140°E

120°E

100°E

80°E

60°E

40°N

40°E

20°E

0   250   500 mi.

0   250   500 km

N W E S

	National boundary
⊛	National capital
•	Other city

Lambert Equal-Area projection

MAP STUDY ACTIVITY

As you read Chapter 14, you will learn about the land, people, and history of Russia.

1. In what part of Russia is its capital located?

2. What sea borders Vladivostok?

3. About how many miles (km) separate Moscow and St. Petersburg?

The Land

PREVIEW

Words to Know
- tundra
- taiga
- permafrost

Places to Locate
- North European Plain
- West Siberian Plain
- Central Siberian Plateau
- East Siberian Uplands
- Ural Mountains
- Volga River
- Caspian Sea
- Lake Baikal
- Kamchatka Peninsula

Read to Learn . . .
1. where Russia is located.
2. what landforms are found in Russia.
3. what two major climates Russia has.

Brown bear, elk, and fox roam this "green ocean"—a vast forest area that stretches across the northern part of Russia. Few people visit or live here, however. Why? Although summers are warm, snow covers the ground for as long as eight months a year!

If you had to describe Russia in one word, that word would be "BIG!" Russia is the largest country in the world in area. Its almost 6.6 million square miles (17,075,400 sq. km) are spread across two continents—Europe and Asia. If you crossed Russia by train, it would take about one full week of travel and you would pass through 11 time zones!

Location
A Northern Country

The map on page 358 shows you two important facts about Russia—its far northern location and its isolation. Russia's longest coastline faces the Arctic Ocean. Along the Arctic coast, ice makes shipping difficult or impossible most of the year. Although Russia does have ports on the Baltic Sea, the Black Sea, and the Pacific Ocean, many of these are frozen for several weeks each winter.

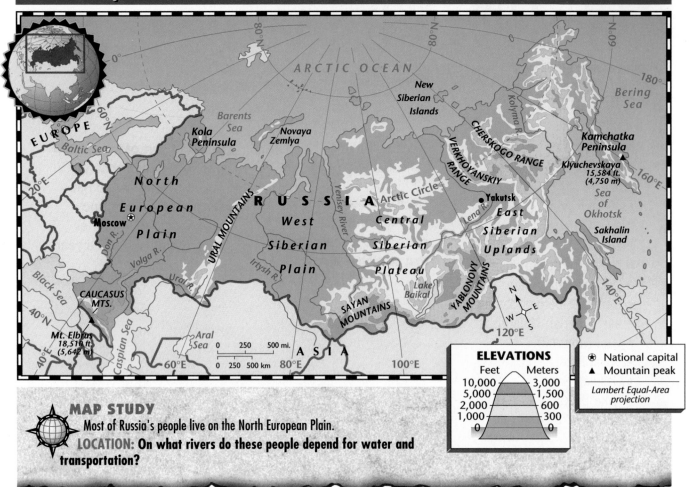

ELEVATIONS

Feet	Meters
10,000	3,000
5,000	1,500
2,000	600
1,000	300
0	0

⊛ National capital
▲ Mountain peak

Lambert Equal-Area projection

MAP STUDY

Most of Russia's people live on the North European Plain.

LOCATION: On what rivers do these people depend for water and transportation?

Ural Mountains Find the Ural Mountains on the map above. Can you see why geographers use them to mark the boundary between the continents of Europe and Asia? The Ural Mountains extend for 1,500 miles (2,414 km) from the Arctic Ocean in the north to near the Aral Sea in the south. The Urals—worn by erosion—are old mountains, rising only a few thousand feet.

Plains and Highlands West of the Urals lies a vast area known as the North European Plain. About 75 percent of the Russian people live on this plain. Russia's largest cities are located here, including its capital, Moscow, and St. Petersburg, a leading industrial and cultural center. Most of Russia's industries and farms dot this region, too.

Now look east of the Urals. This is the Asian part of Russia known as Siberia. Another plain lies on this side of the mountains—the West Siberian Plain. It is the world's largest area of flat land.

As you move farther east of the West Siberian Plain, you come to the Central Siberian Plateau. It rises in elevation from the Arctic Ocean in the north to the Sayan Mountains in the south. Another highland area—the East Siberian Uplands—forms the largest region of Siberia. This wilderness of forests, mountains, and plateaus runs from the Central Siberian Plateau to Russia's eastern coast. The endangered Siberian tiger, among other animals, makes its home here.

In Siberia's far north, large areas of **tundra,** or cold treeless plains, stretch along the Arctic shore. South of the tundra is a vast area of evergreen forests known as the **taiga.** It has so many trees that it is called a "green ocean." On its far eastern edges, a finger of land called the Kamchatka Peninsula stretches south between the Bering Sea and Sea of Okhotsk.

Caucasus Mountains In southwest Russia you can see the mighty Caucasus (KAW•kuh•suhs) Mountains. An extension of the Alps, the Caucasus are thickly covered with pines and other trees. Mount Elbrus in the Caucasus rises 18,510 feet (5,642 m). It is the tallest mountain on the continent of Europe.

Inland Water Areas Russia's landscape also includes many inland bodies of water. In the south Russia borders on the Caspian Sea—a saltwater lake. About the size of the state of California, it is the largest inland body of water in the world. The Caspian Sea lies 92 feet (28 m) *below* sea level.

RUSSIA: Climate

CLIMATE
High latitude
Subarctic
Tundra
Mid-latitude
Humid continental
Dry
Steppe (partly dry)

Lambert Equal-Area projection

MAP STUDY
The Arctic Circle passes through northern Russia.
PLACE: What two types of climates are found between the Arctic Circle and 80°N?

Now That's Cold!

In Yakutsk, a city in north-eastern Siberia, temperatures have plunged as low as −108°F (−77°C).
REGION: What kind of climate prevails in northern Siberia?

Farther east is Siberia's Lake Baikal (by•KAHL). It is the deepest lake in the world—about 5,315 feet (1,620 m) deep. It also holds more water than any other freshwater lake—about 20 percent of the world's unfrozen freshwater.

Many rivers fan across Russia's landscape. Rivers in Siberia, such as the Lena (LEE•nuh), the Ob (AHB), and the Yenisey (YEH•nuh•SAY), are among the longest in the world. Most of them flow south to north into the Arctic Ocean.

Rivers in European Russia, on the western side of the Urals, are connected to one another and to the Caspian Sea by canals. This flow of river traffic has helped unify the country. European Russia's longest river—the Volga—flows 2,193 miles (3,528 km) southward from the Moscow area to the Caspian Sea.

Region
Climate

Russia's northern location near the cold Arctic Ocean results in short, warm summers and long, cold winters for most of the country. Imagine traveling west across the forested southern parts of Siberia. These areas have a subarctic climate. Summers are cool, and snow is on the ground eight to nine months each year. Average temperatures in January plunge to -45°F (-42.8°C).

The region north of the Arctic Circle has a tundra climate with perhaps the coldest winters in the world. Across much of Siberia, a permanently frozen layer of soil called **permafrost** lies beneath the ground's surface.

Temperatures warm on the North European Plain. This area has a humid continental climate. Winters are cold, but not so brutally cold as farther north and east. Summers are rainy and warm. In the city of Moscow, snow falls only five months each year, and rainfall averages 20 inches (50 cm) in the summer.

Farmers in some areas of southern Russia are not as fortunate regarding rainfall. Here rainfall is scarce. These areas have a steppe, or partly dry, climate. The natural vegetation in this area is largely grasses.

SECTION 1 ASSESSMENT

REVIEWING TERMS AND FACTS

1. Define the following: tundra, taiga, permafrost.

2. LOCATION How does Russia's location affect its climate?

3. REGION In what area do most of Russia's people live?

MAP STUDY ACTIVITIES

4. Look at the physical map on page 360. What oceans and seas border Russia?

5. Turn to the climate map on page 361. How would you describe the climate in the far north of Siberia?

BUILDING GEOGRAPHY SKILLS

Reading a Transportation Map

Transportation maps show how people and goods move from place to place. Lines and colors represent different types of transportation. Areas with many transportation routes through them are *accessible*, or easy to reach. These areas usually have more trade, industry, and population. When reading a transportation map, apply these steps:

- Read the title and the map key to find the region and the kinds of transportation shown on the map.
- Find the areas that are most and least accessible.
- Determine what kinds of transportation are most important to this region.
- Use the map to conclude how transportation affects the region.

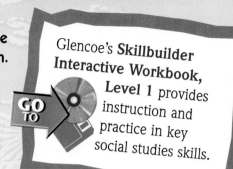

Glencoe's **Skillbuilder Interactive Workbook, Level 1** provides instruction and practice in key social studies skills.

RUSSIA: Transportation Routes

National boundary
Sea route
Railroad
Trans-Siberian Railroad
National capital
Other city
Seaport

Lambert Equal-Area projection

Geography Skills Practice

1. What region and kinds of transportation are shown on this map?

2. Which major railroad extends to the far eastern part of Russia?

3. Which parts of Russia are most accessible? Least accessible?

The Economy

PREVIEW

Words to Know
- communism
- command economy
- heavy industry
- light industry
- free enterprise system
- consumer goods

Places to Locate
- Moscow
- St. Petersburg
- Siberia

Read to Learn . . .
1. what Russia's economy was like under communism.
2. how Russia's economy has changed in recent years.
3. what economic regions make up Russia.

You may not be able to tell by looking at it, but there is something special about this electronics factory in Russia. What is it? In the past, this same factory made weapons for the military. For years the communist Russian government told workers what to produce. Today Russians are making their own decisions about what to make and sell. They are entering the growing world economy.

If you travel across Russia, you will see people farming the land and working in factories. Others take resources from the earth, forests, and rivers. Russia is rich in natural and human resources, but its economy has gone through many ups and downs in this century.

Place
Today's Challenges

From 1922 to 1991, Russia was part of an even larger empire called the Soviet Union. The Soviet Union practiced **communism,** an authoritarian political system in which the government controls the economy. Private ownership of property is not allowed.

Government Controls The Soviet Union's Communist leaders created the world's largest command economy. In a **command economy,** the national government owns all land, resources, industries, farms, and railroads. The government also makes all economic decisions.

364

UNIT 5

The colorful domes of St. Basil's Cathedral stand just outside the Kremlin in Moscow's Red Square.
PLACE: Why do you think the Kremlin was surrounded by a wall?

Communist leaders used five-year plans to run the Soviet economy. These plans set production goals, prices of goods, and workers' wages. Plans empha-sized **heavy industry,** or the making of goods such as machinery, mining equip-ment, and military weapons. They gave less attention to **light industry,** or the making of such goods as clothing, shoes, furniture, and household products.

Collapse of Communism Many workers in the Soviet Union made the change from farmers to factory workers under communism. In the 1970s, however, inefficiencies in the centrally controlled communist system gradually weakened the economy. In 1985 the Soviet leader Mikhail Gorbachev (GOR•buh•CHAWF) pushed to rebuild the economy. He reduced some of the national government's rigid controls and allowed more decisions to be made at the local level. Before his plans could bring about change, however, the Soviet Union collapsed. In 1991 Russia and the other parts of the Soviet Union became independent countries.

A New Beginning Independence turned Russia's economy upside down. The Russians are making the painfully slow change to a **free enterprise system** like that of the United States. Under free enterprise, businesses are privately owned, operating for profit with little government control. Today Russia's economy is still struggling. Goods are often scarce, and prices have steadily risen. Some Russians even favor a return to government controls.

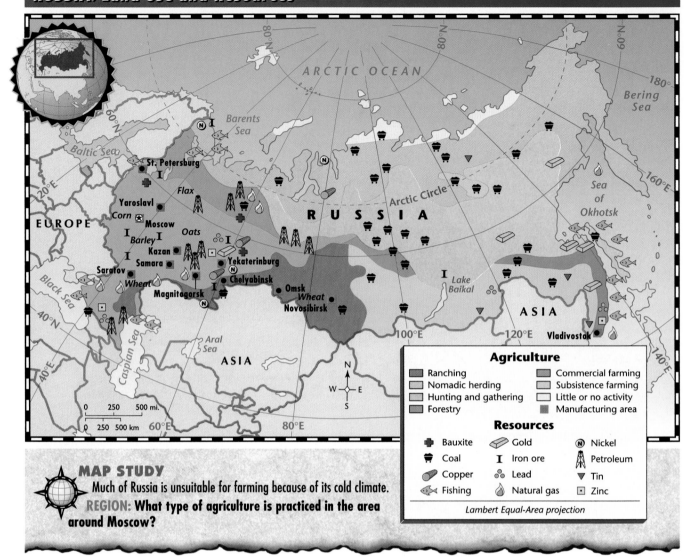

Much of Russia is unsuitable for farming because of its cold climate.
REGION: What type of agriculture is practiced in the area around Moscow?

Region
Economic Regions

Look at the map above. You see that most of Russia's industries are located west of 60°E in the European part of the country. Most farms are in this area of Russia, too. Although the country has a short growing season, its farmland is extremely productive. A dark, fertile soil spreads across part of the North European Plain into southwestern Siberia. Farmers in this area grow wheat, barley, rye, and sugar beets.

Moscow Russia's rich history shines in the city of Moscow located in the western part of Russia. About 800 years old, Moscow is the political, economic, and cultural center of Russia. This major industrial city produces textiles, electrical equipment, and automobiles. Moscow's factories recently have started to make more **consumer goods,** or household and electronics products.

Ivan Borisov, a Moscow teenager, enjoys greater freedom of choice than his parents or grandparents did. Ivan and his family—among Moscow's 9.3 million people—live in one of the city's many tall apartment houses. Ivan likes to take foreign visitors on a trip through the Metro, the Moscow subway that links every part of the city. His favorite stop is near Red Square in the center of Moscow.

St. Petersburg Northwest of Moscow lies St. Petersburg, Russia's second-largest city. Located on the Neva River near the Gulf of Finland, St. Petersburg is noted for shipbuilding. Its factories produce light machinery, textiles, and scientific and medical equipment.

St. Petersburg's 5.1 million residents take pride in their beautiful city. Built in the early 1700s, St. Petersburg spreads over about 100 islands. A network of graceful bridges and streets joins the islands.

The Volga and Urals Tucked between European Russia and Siberia lies the industrial region of the Volga River and the Ural Mountains. The Volga carries almost one-half of Russia's river traffic. It also provides water for hydroelectric power and irrigation. Both the Volga and the Urals contain large deposits of oil and natural gas. The Ural Mountains are rich in other minerals, too, such as tungsten, zinc, iron ore, and nickel.

Siberia East of the Volga and the Urals spreads Siberia, a promising economic region. Siberia has the largest supply of mineral resources in the country, including iron ore, uranium, gold, diamonds, and coal. In the Arctic region, huge deposits of oil and natural gas lie below the permafrost. About two-thirds of Siberia is covered by forests that provide lumber. In spite of this wealth in natural resources, Siberia is mostly undeveloped because of its climate and isolation. Since the late 1800s, the governments ruling Russia have built roads and railroads there to encourage development.

WHAT IN THE WORLD?

"Chizburger," Anyone?

The familiar hamburger chain, McDonald's, has introduced the people of Moscow to the "chizburger." Russians, like Americans, are enthusiastic burger fans. Daily, more than 50,000 people pack into the Moscow McDonald's to get lunch.

SECTION 2 ASSESSMENT

REVIEWING TERMS AND FACTS

1. **Define the following:** communism, command economy, heavy industry, light industry, free enterprise system, consumer goods.

2. **PLACE** Before 1991, how was the Soviet Union's economy run?

3. **REGION** What are Russia's four major economic regions?

4. **HUMAN/ENVIRONMENT INTERACTION** What region has the most promising future? Why?

MAP STUDY ACTIVITIES

5. Look at the land use map on page 366. What resources are found in Siberia? What kind of industry might locate there?

MAKING CONNECTIONS

A road that is 2,000 years old? It sounds unbelievable. Yet stretching 7,000 miles (11,263 km) across central Asia and Russia lies the ancient Silk Road. For about 14 centuries, this road was a vital trade route between Asia and Europe. Armies, merchants, missionaries, and adventurers from both eastern and western worlds traveled this road.

THE ROUTE Beginning in northern China about 100 B.C., the Silk Road wound its way from oasis to oasis across Asia's great deserts. To the south, always in plain view, were majestic snowcapped mountains. Beyond the deserts, travelers began the terrifying climb through the Himalayas. On the other side of these mountains, the road split. One branch continued westward toward Europe, winding through what is today Russia and the independent republics. Another branch turned south into India.

THE REASON Again and again, travelers on the Silk Road braved fierce desert sandstorms, howling Himalaya blizzards, and murderous thieves. Why? The only land route between

Asia and Europe, the Silk Road made it possible for costly Chinese goods to reach Europe. The most valuable of these goods was silk—the soft, shiny fabric that gave the road its name. Wealthy Europeans were willing to pay handsomely for this luxurious cloth.

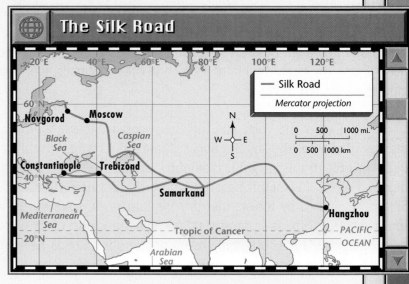

The Silk Road

THE RESULT Heavily laden camels formed caravans that transported the silk. They also carried gemstones, perfumes, tea, fine china, and gunpowder. More importantly, travelers on the Silk Road traded ideas and customs. This cultural exchange greatly enriched the peoples of central Asia, China, India, Southwest Asia, and eventually, western Europe.

Making the Connection

1. What and where was the Silk Road?
2. What products did Silk Road travelers carry from China?

The People

Words to Know

- czar
- serf
- cold war
- *glasnost*
- ethnic group

Places to Locate

- Muscovy
- Union of Soviet Socialist Republics (Soviet Union)
- Commonwealth of Independent States (CIS)

Read to Learn . . .

1. what groups influenced Russia's culture.
2. how city life differs from country life in Russia.
3. what makes up Russia's culture today.

These Russian teenagers are enjoying a weekend outing in Moscow. After eating a "chizburger" at a fast-food restaurant, they will explore the city. In recent years, they and other Russians have seen two amazing changes in their country—the fall of communism and the first stirrings of democracy. Like teenagers around the world, these young people hope for a better future.

Russia and the other countries that were once part of the Soviet Union have a dramatic history. Today these countries are independent republics, trying to build democracies and a better way of life for their people.

Movement

A Dramatic Past

To understand the challenges facing Russia today, let's look back through Russia's history. As an early Russian you would have belonged to a loose union of people called Eastern Slavs who dominated the area from the 900s to the 1200s. Then Mongols swept in from central Asia in the 1200s and ruled the Slavic territories in eastern Europe for 200 years. During this time, what is now Moscow became the center of a territory called Muscovy (MUHS•kuh•vee). As a citizen in the late 1400s, you would have rejoiced when Ivan III, a prince of Muscovy, drove out the Mongols and made the territory independent.

Out With the Old

Nadia Abramov often visits the park near her family's apartment in Moscow. Sometimes she climbs on the toppled statues of former Communist leaders. The fallen statues remind her of the many changes that Russians have experienced since the breakup of the Soviet Union in 1991. Nadia is happy with most of the changes, especially those at school. She still goes to school six days a week, but she doesn't have to wear uniforms anymore. The classes on communism have been dropped, and her new history books are filled with information that was once forbidden by the former government. For the first time, she can also take religion courses if she wants to.

Strong Rulers Muscovy slowly developed into the country we know today as Russia. The Muscovite rulers extended their power and began calling themselves **czars,** or emperors. As a Muscovite, you would have feared Czar Ivan IV, who used a secret police force to keep his people under control. He also expanded Russia's territory by conquest eastward into the Volga region and Siberia.

Over the centuries the country grew as czars conquered other lands. Many non-Russian peoples came into the Russian Empire. Czars such as Peter I and Catherine II pushed the empire's borders westward and southward. They also tried to make Russia more like Europe. A new capital—St. Petersburg—was built in the early 1700s to look like a European city. If you had been a Russian noble at this time, you would have worn European clothes and spoken French instead of Russian.

The actions of the czars, however, had little effect on ordinary citizens. Most Russians were **serfs,** or laborers who were bound to the land. They were too poor to be interested in European ways and kept their Russian culture.

Toward Revolution In 1861 Czar Alexander II freed the serfs. At about the same time, Russia began to set up industries like those of other European countries. Railroads, including the famous Trans-Siberian Railroad, spread across the country. It linked Moscow in the west with Vladivostok on Russia's Pacific coast.

Russia, however, did not progress politically. The czars clung to their power and rejected democracy. Revolution brewed. In 1917 the political leaders and workers forced Czar Nicholas II to give up the throne. At the end of the year, a group of Communists led by Vladimir Ilyich Lenin came to power. They set up a Communist government and soon moved its capital to Moscow.

The Soviet Union In 1922 the Communists formed the Union of Soviet Socialist Republics, or the Soviet Union. The new country included Russia and most of the conquered territories of the old Russian Empire. During the late 1920s, Joseph Stalin became the ruler of the Soviet Union and set out to make it a great industrial power. To reach this goal, the government took control of all industry and farming. Stalin, a cruel dictator, put down any opposition to his rule. Millions of people were either killed or sent to prison labor camps.

After World War II the Soviet Union further expanded its territory and extended communism to eastern Europe. From the late 1940s to the late 1980s, the Soviet Union and the United States waged a **cold war.** They competed for world influence without actually waging war on one another.

A New Beginning In 1985 Mikhail Gorbachev came to power in the Soviet Union. In addition to economic changes, He supported a policy of ***glasnost,*** or openness. He wanted people to speak freely about the Soviet Union's problems. Gorbachev's efforts, however, failed to stop the collapse of the Soviet Union. Many of the non-Russian nations had long resented Russian rule and wanted independence. By late 1991 the Soviet Union had broken apart, and Russia had a new leader, Boris Yeltsin. Russia and the newly independent countries formed a loose union called the Commonwealth of Independent States (CIS).

Place
City and Country Life

The teens you read about earlier are part of Russia's 147 million people. Look at the graph below to see how Russia's population compares with that of the United States. Russian Slavs make up about 80 percent of the population, and Russian is the official language. The country, however, has hundreds of smaller **ethnic groups,** or people who have a common language, culture, and history.

City Life Nearly three-fourths of Russia's people live in cities and towns. The map on page 372 shows that a large part of Russia's population lives around Moscow. Many cities are crowded, and residents face severe housing shortages. Whole families often live in one- or two-room apartments. City residents in Russia also have to deal with other problems such as food shortages, high prices, crime, and pollution from factories. The government is trying to correct such concerns and is also working to improve medical care. When people in cities relax, they spend time with family and friends, taking walks through parks, or attending concerts, movies, and the circus.

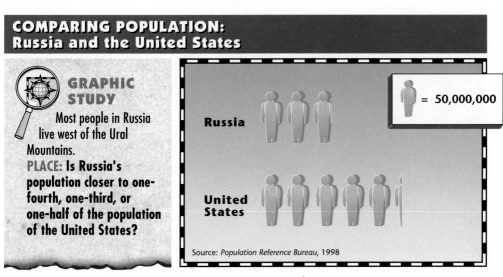

**COMPARING POPULATION:
Russia and the United States**

GRAPHIC STUDY
Most people in Russia live west of the Ural Mountains.
PLACE: Is Russia's population closer to one-fourth, one-third, or one-half of the population of the United States?

Russia

United States

= 50,000,000

Source: *Population Reference Bureau,* 1998

Country Life If you visit a village in Russia today, you will find that most people live in houses built of wood. In some rural areas, people live without heat, electricity, or plumbing. Consumer goods are scarce in the countryside, but food is more plentiful than in some cities. The quality of health care, education, and cultural activities is lower in rural areas than in urban areas. Over the years, many people have left the villages to find work in the cities.

Place
The Arts and Recreation

Russia has a rich tradition of art, music, and literature. The Russian people view their country's cultural achievements with pride. Some of their finest works of art depict religious and historical themes. Others reflect the daily lives of ordinary people.

Religion Communist rulers of the Soviet Union banned religion. They often closed houses of worship and persecuted religious people. Since the fall of communism, many Russians have returned to their religious traditions. Russians are Jews, Protestants, and Roman Catholics. Most Russians, however, follow Eastern Orthodox Christianity. Orthodox churches in Russia that have survived are excellent examples of Russian architecture. Many contain religious works of art that are hundreds of years old.

Arts and Music If you want to see Russian culture at its best, where do you go? If you enjoy dance, you might attend a ballet performed at Moscow's Bolshoi Theater. Russia has long been a world leader in ballet.

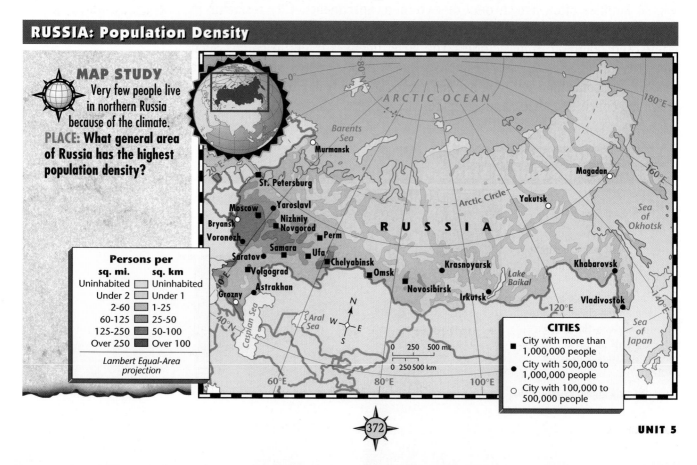

RUSSIA: Population Density

MAP STUDY Very few people live in northern Russia because of the climate.
PLACE: What general area of Russia has the highest population density?

Persons per

sq. mi.	sq. km
Uninhabited	Uninhabited
Under 2	Under 1
2-60	1-25
60-125	25-50
125-250	50-100
Over 250	Over 100

Lambert Equal-Area projection

CITIES
■ City with more than 1,000,000 people
● City with 500,000 to 1,000,000 people
○ City with 100,000 to 500,000 people

0 250 500 mi.
0 250 500 km

The wooden houses in rural Russia *(left)* have much more room than the cramped apartment buildings in Russia's cities *(right)*.
PLACE: About how many of Russia's people live in cities?

If you enjoy painting, you might stroll through the Hermitage, St. Petersburg's most famous museum. Works of noted Russian as well as western European painters hang in the Hermitage. In the early 1900s, Russia led the modern art movement. Two of the most famous Russian-born painters of this movement are Kazimir Malevich and Wassily Kandinsky.

Perhaps you prefer music. Classical music fills concert halls throughout Russia. Many of the world's greatest composers were Russian. They include Peter Tchaikovsky and Nikolay Rimsky-Korsakov. You may be familiar with Tchaikovsky's *Nutcracker Suite* and his *1812 Overture.*

The 1800s and 1900s saw a great flowering of Russian literature. Among the best Russian writers of the nineteenth century were Alexander Pushkin, Leo Tolstoy, and Fyodor Dostoyevsky. Their works vividly describe the lives of nobles, merchants, serfs, and city workers. In this century, the leading Russian writers include Maxim Gorky and Alexander Solzhenitsyn. They and other modern writers have discussed the lack of individual freedom in Russia and the hopes for a better society.

Celebrations Russians enjoy small family get-togethers as well as national celebrations. New Year's Eve is the most festive nonreligious holiday for the Russian people. On this day children decorate a fir tree and exchange presents with their families. Another important holiday is May Day. In parades and speeches, May Day celebrates Russian workers and their contributions to their country.

Food If you have dinner with a Russian family, you might begin with a big bowl of *borsch,* a soup made from beets, or *shchi,* a soup made from cabbage. Next you might be served meat turnovers called *pirozhki.* For the main course, you would have meat, poultry, or fish with boiled potatoes. If you were having a

Dedicated young dancers practice at a ballet school in St. Petersburg.
LOCATION: What famous ballet theater is located in Moscow?

special holiday dinner, you might have caviar, or fish eggs, fresh from the Caspian Sea.

Sports Russians love soccer, tennis, and ice hockey. On the lakes and in the mountains, you will find Russians skiing, hiking, camping, mountain climbing, and hunting. Reindeer races and dog-team competitions are popular sports in the cold lands of Siberia. And if you watch the Olympics, you know that Russians are major competitors in the games.

SECTION 3 ASSESSMENT

REVIEWING TERMS AND FACTS

1. Define the following: czar, serf, cold war, *glasnost,* ethnic group.

2. PLACE What was life like for most Russians under the czars?

3. PLACE How does city life differ from country life in Russia?

4. HUMAN/ENVIRONMENT INTERACTION How would you compare the practice of religion in Russia before and after 1991?

MAP STUDY ACTIVITIES

5. Turn to the population density map on page 372. Which region of Russia is sparsely populated? Why is this so?

6. Look again at the same map. About how many people live in Murmansk?

Important Things to Know About Russia

SECTION 1 THE LAND

- Russia is the largest country in the world in area.
- Russia extends across both Europe and Asia.
- Russia is a vast lowland divided or surrounded by mountains.
- Inland waterways help connect Russia's widespread regions.
- Much of Russia's climate is subarctic or humid continental.

SECTION 2 THE ECONOMY

- Russia's economy is moving from government control to a free enterprise system.
- Russia's leading industrial centers are located in Moscow, St. Petersburg, the Volga and the Urals, and Siberia.
- Oil and gas are important resources in Russia.
- Siberia, the Asian part of Russia, has vast timber resources.

SECTION 3 THE PEOPLE

- Russia was the center of a powerful empire under rulers known as czars.
- A revolution in 1917 overthrew the czars; later a communist empire called the Union of Soviet Socialist Republics, or Soviet Union, was established.
- Russia and other parts of the Soviet Union have become independent republics since the fall of communism in the early 1990s.
- Most Russians live in urban areas in the European part of Russia.
- Russia has a rich tradition of art, music, literature, and dance.

St. Petersburg, Russia ▶

REVIEWING KEY TERMS

Match the numbered terms in Set A with their lettered definitions in Set B.

A

1. taiga
2. command economy
3. czar
4. heavy industry
5. tundra
6. serf
7. permafrost
8. light industry
9. free enterprise system

B

A. permanent layer of frozen soil
B. Russian ruler
C. economy planned and directed by the communist government
D. production of household products
E. privately owned businesses with little government control
F. cold, treeless plain
G. farmers bound to the land
H. production of machinery, mining equipment, and military weapons
I. vast forest area of Siberia

REVIEWING THE MAIN IDEAS

Mental Mapping Activity

Draw a freehand map of Russia. Label the following:
• Black Sea
• Volga River
• Moscow
• Caspian Sea
• Central Siberian Plateau
• St. Petersburg

Section 1

1. **REGION** What is the world's largest region of flat land?
2. **REGION** What Russian lake is the deepest in the world?

Section 2

3. **HUMAN/ENVIRONMENT INTERACTION** What changes have come to Russia's economy since the fall of communism?
4. **HUMAN/ENVIRONMENT INTERACTION** Why is the Volga River important to Russia?

Section 3

5. **MOVEMENT** What railroad links Moscow with Russia's eastern coast?
6. **PLACE** What effect did Mikhail Gorbachev's changes have on the Soviet Union?

CRITICAL THINKING ACTIVITIES

1. **Determining Cause and Effect** How do you think climate affects the way Russians make a living?
2. **Evaluating Information** What problems did Russia face after becoming independent in the early 1990s?

CONNECTIONS TO WORLD HISTORY

Cooperative Learning Activity Work in groups to plan a trip to one of the following areas to learn more about the history and people of the area: (1) the Volga River; (2) the Ural Mountains; (3) Lake Baikal; or (4) an area of your choice. Each member of the group will do one of the following: (a) locate historic places to visit; (b) learn about the different people who live in the region; or (c) make a list of important historical events that have happened in the area. Combine your findings in a travel information packet.

GeoJournal Writing Activity

Imagine you are a friend of Ivan Borisov. He has asked you to join his family on a trip to Siberia. Write about your visit to this region of Russia. You may use the maps in your text or a reference book to help you.

TECHNOLOGY ACTIVITY

Using a Computerized Card Catalog Use a computerized card catalog to locate sources about the history of Russian czars. Create a genealogy chart of the Romanov dynasty. Include a short report explaining why the empire of the czars ended and how it affected Russian culture. Share your genealogy chart and summary with the class.

PLACE LOCATION ACTIVITY: RUSSIA

Match the letters on the map with the places and physical features of Russia. Write your answers on a separate sheet of paper.

1. Vladivostok
2. West Siberian Plain
3. Ural Mountains
4. Caspian Sea
5. Lena River
6. Kamchatka Peninsula
7. Lake Baikal
8. Caucasus Mountains
9. Moscow
10. Sea of Okhotsk

GeoLab ACTIVITY

From the classroom of
Dana Moseley,
Carl Stuart Middle School,
Conway, Arkansas

MAPMAKING: Dough It!

Background

Russia can be divided into five major geographic regions: the North European Plain, the West Siberian Plain, the Central Siberian Plateau, the East Siberian Uplands, and the lowland region east of the Caspian Sea. In this activity, you will make a salt-dough relief map of these regions.

The Ural Mountains in Russia

Verkhoyansk is a small Russian town in northeast Siberia. It has one of the world's most extreme climates. In January, the coldest month, the average temperature plunges to -58.2°F.

This is one group's salt-dough map.

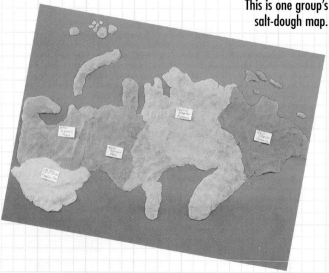

Materials

- 1 poster board
- outline map of Russia
- physical map of Russia
- salt dough (see recipe on page 379)
- glue
- food coloring

What To Do

A. Draw an outline map of Russia onto the poster board.

B. Mix the salt dough using this recipe:
- *1 cup flour*
- *1/2 cup salt*
- *1 cup water*
- *1 tablespoon vegetable oil*
- *2 teaspoons cream of tartar*

Heat all ingredients in a saucepan over medium heat, stirring about 4 minutes or until mixture forms a ball. Remove from saucepan and let dough cool on counter for 5 minutes. Knead dough about 1 minute or until it is smooth and blended. Let cool completely before using on map.

C. Separate the dough into five sections and use the food coloring to make each region a different color—brown, green, red, yellow, and blue.

D. Press the colored salt dough into the appropriate places on your outline map of Russia. Make sure you recreate the correct geographic landforms and regions.

E. Label each region.

Lab Activity Report

1. Which region covers the greatest area?

2. Which region has the highest relief overall?

3. Describe the relief of each region.

4. Drawing Conclusions
What adjustments in your life would you have to make if you lived in each region?

Go A Step Further!

Activity
Research and create a time line depicting Russia's history. Include important people, places, and events. Illustrate your time line with small, colorful pictures that focus on important events in Russia's history.

Cultural HERITAGE:

RUSSIA AND THE INDEPENDENT REPUBLICS

RELIGIOUS ART ▶ ▶ ▶ ▶

Icons hang on the walls of many homes and churches in Russia and the independent republics. Icons are paintings of Christian religious figures on wood panels. Many are hundreds of years old.

◀ ◀ ◀ ◀ MUSIC

Folk music is very popular in all the independent republics, especially Ukraine. This Ukrainian student plays the *bandura,* a multistringed instrument that is plucked like a guitar. Her brightly embroidered blouse and skirt are traditional Ukrainian clothing.

PAINTING ▶ ▶ ▶ ▶

Kazimir Malevich was one of the first painters to introduce modern art to Russia. This painting from the early 1900s does not show real objects. Instead Malevich used shapes and pure colors on plain backgrounds to represent objects.

◄ ◄ ◄ ◄ WEAVING

The people of Uzbekistan take great pride in their beautiful wool carpets and rugs. Each year, factories in the area weave more than 3 million square miles (7.9 million sq. km) of carpet. The bright colors and bold patterns of the rugs reflect the influence of Islamic ancestors.

ARCHITECTURE ▶ ▶ ▶ ▶

Churches, mosques, and synagogues reflect the many types of architecture found in Russia and the independent republics. This old church with a round tower in Msteka, Georgia, is made of stone. Others have elaborate onion-shaped domes and colorful mosaics.

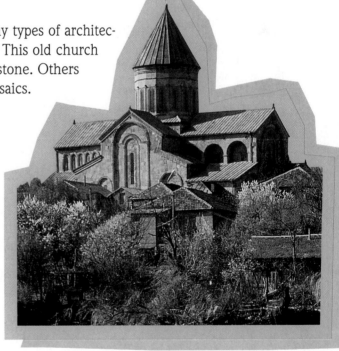

FABERGÉ EGG ▲▲▲▲

Russia's most famous jeweler, Carl Fabergé, designed this beautiful egg for a Russian czar. He crafted a series of these elaborate eggs from gold, silver, enamel, and precious stones.

APPRECIATING CULTURE

1. Russia and the independent republics have a rich and varied history. How has some of that history been recorded in artwork from this area?

2. Which type of art shown here most appeals to you? Why?

The Independent Republics

EUROPE

20°E
30°E
60°N
40°E
50°E
60°E
70°E
Baltic Sea

Minsk
BELARUS
50°N
Gomel
Lvov
Kiev
UKRAINE
Kharkov
MOLDOVA
Dnepropetrovsk
Chisinau
Odessa
Donetsk
Sevastopol
Black Sea
40°N

RUSSIA

Astana
Karaganda
KAZAKHSTAN
Lake Balkhash
Almaty
Aral Sea
Bishkek
GEORGIA
Tbilisi
UZBEKISTAN
Tashkent
KYRGYZSTAN
Yerevan
Baku
Caspian Sea
Samarkand
ARMENIA
TURKMENISTAN
Dushanbe
Mediterranean Sea
AZERBAIJAN
Ashkhabad
TAJIKISTAN
30°N

0   250   500 mi.
0   250   500 km

N
W   E
S

ASIA

⎯⎯ National boundary
✹ National capital
● Other city

Lambert Equal-Area projection

MAP STUDY ACTIVITY

As you read Chapter 15, you will learn about the independent republics that, like Russia, were once part of the former Soviet Union.

1. **How many independent republics are there?**

2. **What republics share a border with Russia?**

SECTION 1 ·
Ukraine

PREVIEW

Words to Know
- steppe

Places to Locate
- Ukraine
- Black Sea
- Dnieper River
- Dniester River
- Crimean Peninsula
- Kiev
- Odessa
- Donets Basin

Read to Learn . . .
1. what landforms are found in Ukraine.
2. why Ukraine is known as the breadbasket of Europe.
3. what challenges Ukraine faces as an independent country.

The sputtering of this tractor and the "whoosh" of harvested wheat echo across the broad plains of Ukraine. Farmers grow a variety of grains in Ukraine's rich soil and ship them abroad to foreign markets. Because of the abundance of its grain harvests, Ukraine has been called the breadbasket of Europe.

Ukraine and many of its neighbors were once part of the Soviet Union. Since the breakup of the Soviet Union in 1991, they have been independent republics. To help their economies grow, these new countries and Russia have formed the Commonwealth of Independent States (CIS).

Human/Environment Interaction
Land, Climate, Economy

Ukraine—a word meaning "frontier" in a Slavic language—was named for its location. It once marked the far western edge of the Russian empire. With 223,687 square miles (579,349 sq. km), Ukraine is about the size of Texas.

Landforms Find Ukraine on the map on page 384. Lowlands dotted with farms and forests spread across the north. Eastern highlands stretch toward the Black Sea. A coastal plain curves along the Black Sea. From the plain, a landmass juts into the water. This landmass, the Crimean Peninsula, is one of the Black Sea's most scenic areas.

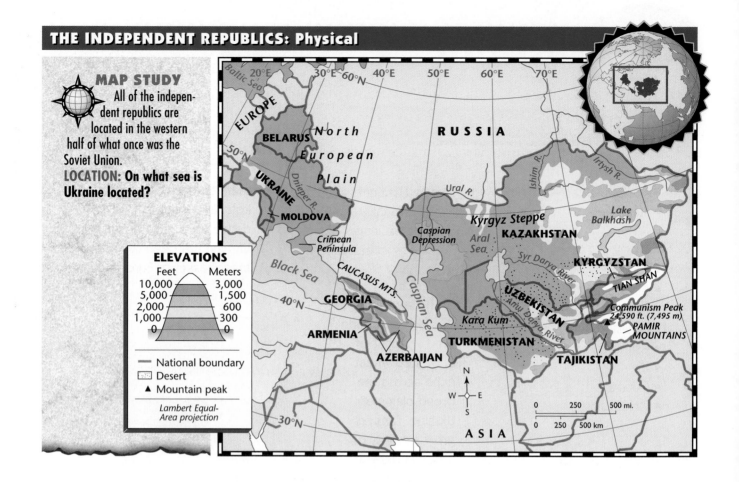

MAP STUDY
All of the independent republics are located in the western half of what once was the Soviet Union.
LOCATION: On what sea is Ukraine located?

ELEVATIONS

Feet	Meters
10,000	3,000
5,000	1,500
2,000	600
1,000	300
0	0

— National boundary
⬚ Desert
▲ Mountain peak

Lambert Equal-Area projection

Vast treeless plains called **steppes** sweep through the center of Ukraine. Rich, fertile soil and plentiful rainfall make these steppes the most productive farmland in the region. Through the steppes to the Black Sea flows Ukraine's important river—the Dnieper (NEE•puhr). The Dniester (NEES•tuhr) also flows through the area. These rivers and other waterways link Ukraine's industrial areas with the Black Sea and world markets. The city of Odessa is Ukraine's major port.

The Climate Most of Ukraine has a humid continental climate of cold winters and warm summers. Rainfall is more plentiful in the northern and central parts of the country than in the south. On the Crimean Peninsula, mountains shield the Black Sea coast from the cold northern winds.

The Economy Under Soviet rule, what Ukraine produced was controlled by the government. Today Ukraine is changing its economy to one based on free enterprise. Farming and industry are the mainstays of Ukraine's economy. Farmers harvest grains, potatoes, sugar beets, and dairy products.

Ukraine is more than a breadbasket, however. The country is rich in coal, iron ore, manganese, potash, and natural gas. The Donets Basin in eastern Ukraine contains one of the largest coal deposits in the world. It also is the country's major industrial area. A gigantic dam on the Dnieper River supplies electricity to the region's factories.

Place
People

Ukraine has a population of about 50 million. About 75 percent of the people—including Anna Lyashenko, a 13-year-old citizen of Ukraine's capital—are ethnic Ukrainians. They have their own language and culture. The rest of Ukraine's population are mostly Russian. Eastern Orthodox Christianity is the country's main religion.

From Past to Present About two-thirds of Ukraine's people live in urban areas. Kiev (KEE•EHF), the capital and largest city, has about 2.8 million people. Located on the Dnieper River, Kiev is a transportation, industrial, and cultural center. In history class, Anna has learned that, from A.D. 900 to the 1200s, Kiev was the center of a group of powerful Slavic city-states. In past centuries, outsiders such as the Poles and the Russians ruled Anna's country, and Ukrainians suffered. Independence, Anna admits, has not solved all problems. Ukraine's economy is shaky. Many people have lost their jobs as factories change ways of producing goods. Prices are high in the stores, and some goods are scarce. In addition, the growth of heavy industry during the years of Soviet rule has seriously harmed Ukraine's environment. Factory smoke and other wastes have polluted the air and water.

Culture and Celebrations The people of Ukraine take pride in a rich cultural heritage. Ukrainians are famous for their folktales and proverbs. The chart above lists several favorite proverbs of Ukraine and Russia.

During the Easter season, Ukrainian families observe the ancient tradition of decorating eggs with colorful designs. Folk music is popular, and many Ukrainian musicians perform on a stringed instrument called the *bandura*. In a popular dance—the *hopak*—male dancers make astounding acrobatic leaps!

Ukrainian and Russian Proverbs

- **Where the road is straight, don't look for a short cut.**

- **A bad peace is better than a good war.**

- **There is plenty of sound in an empty barrel.**

- **A near neighbor is better than a distant relative.**

- **When the Czar has a cold, all Russia coughs.**

Source: *The Prentice-Hall Encyclopedia of World Proverbs*

GRAPHIC STUDY
A proverb is a saying that describes a truth.
HUMAN/ENVIRONMENT INTERACTION: On this chart, which proverb might be rewritten as "don't take the easy way out"?

SECTION 1 ASSESSMENT

REVIEWING TERMS AND FACTS
1. Define the following: steppe.
2. PLACE How does the climate of Ukraine differ from north to south?
3. HUMAN/ENVIRONMENT INTERACTION How is the Donets Basin important to Ukraine's industrial growth?

MAP STUDY ACTIVITIES
4. Turn to the physical map on page 384. What major body of water lies south of Ukraine?

BUILDING GEOGRAPHY SKILLS

Reading a Historical Map

Suppose you tried to draw a map of your town as it looked 100 years ago. You would not include roads or bridges as they are today. You certainly wouldn't find any airports. Perhaps there was no town at all!

A historical map shows the cultural and political features of an area in an earlier period. Physical features of a landscape may stay much the same. Cultural features, however, change dramatically over time. To read a historical map, apply these steps:

- Read the title to identify the region and time period.
- Use the key to locate political units.
- Compare the map with a recent physical map to observe changes.
- Compare the map with a recent political map to observe changes.

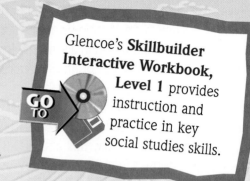
Glencoe's **Skillbuilder Interactive Workbook, Level 1** provides instruction and practice in key social studies skills.

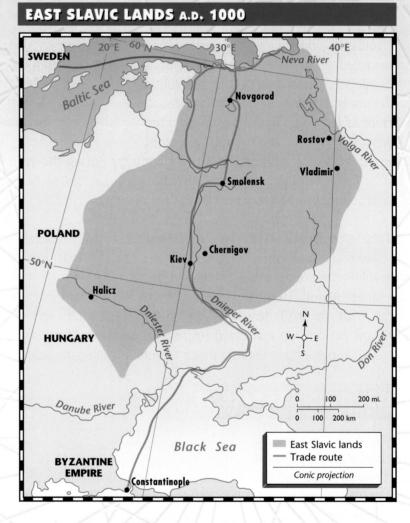

EAST SLAVIC LANDS A.D. 1000

East Slavic lands
Trade route

Conic projection

Geography Skills Practice

1. What region and time period are shown in this map?

2. Near what rivers did city-states develop?

3. What present-day independent republics were once part of the East Slavic lands?

Belarus and Moldova

PREVIEW

Words to Know
- nature preserve

Places to Locate
- Belarus
- Minsk
- Moldova
- Chisinau

Read to Learn . . .
1. where Belarus and Moldova are located.
2. what Belarus and Moldova produce.
3. what groups of people live in Belarus and Moldova.

What makes this farmer in Belarus so happy? Is it a good crop of potatoes, flax, or wheat? Or is he feeling successful because he tends his own land? While most rural families in Belarus work on government-owned farms, many have been encouraged to run their own farms and businesses.

Belarus (BEE•luh•ROOS) lies north of Ukraine, and the country of Moldova (mahl•DOH•vuh) lies south. Like Ukraine, Belarus and Moldova—both land-locked countries—were once part of the Soviet Union. Today they are independent but have joined other former Soviet republics in the CIS.

Place
Belarus

Belarus covers 80,108 square miles (207,480 sq. km)—roughly the area of the state of Kansas. The people of Belarus, or Belarusians, are Slavs who are closely related to the Ukrainians and Russians. They also have close cultural ties to Poland, another neighboring Slavic country.

The Land and Climate Dense green undergrowth, towering trees, and wide marshes mark an area Belarus and Poland also share—Belovezha (BEHL•oh•VYEH•zah) Forest. This unique stretch of land is a famous **nature preserve,** or land set aside by the government to protect plants and wildlife. Rare European bison, similar to the American buffalo, graze in the woods and meadows of the forest. Low flatlands cover the rest of Belarus in the north. Marshes, swamps, and forests are found in the south.

The map below shows that Belarus has a humid continental climate with cold winters and warm summers. The temperature averages about 22°F (-6°C) in January, the coldest month. In July—the hottest month—the temperature averages about 65°F (18°C).

The Economy Most Belarusians earn their livings as factory workers and farmers. They also work in service industries. Factory workers in Belarus make trucks, tractors, engineering equipment, and consumer goods such as television sets and refrigerators. The country's farmers produce barley, flax, potatoes, rye, sugar beets, dairy products, and livestock. Forests supply wood products that include furniture, matches, and paper goods.

The People Belarus's 10.2 million people are mostly Slavs who are proud of their separate language and culture. Smaller ethnic groups in the country include Russians, Ukrainians, and Poles. Two-thirds of the population live in cities. Minsk, the capital and largest city, has a population of 1.8 million. Belarus became independent in 1991. Minsk was chosen as the headquarters of the CIS.

Nina Muzychenko lives in Neglyubka—a village known for its textiles woven in elaborate patterns. Like most Belarusians, Nina follows the Eastern Orthodox Christian religion. Her family lives in a wooden home decorated with roof carvings.

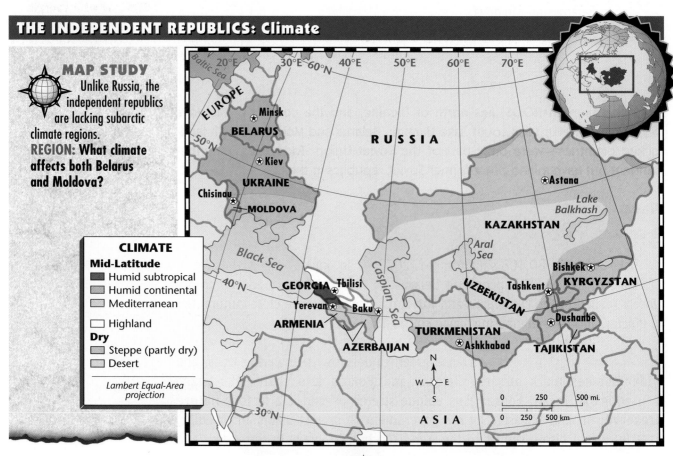

THE INDEPENDENT REPUBLICS: Climate

MAP STUDY
Unlike Russia, the independent republics are lacking subarctic climate regions.
REGION: What climate affects both Belarus and Moldova?

CLIMATE
Mid-Latitude
- Humid subtropical
- Humid continental
- Mediterranean
- Highland

Dry
- Steppe (partly dry)
- Desert

Lambert Equal-Area projection

Place
Moldova

A small country of 12,730 square miles (32,971 sq. km), landlocked Moldova broke away from the Soviet Union and became independent in 1991. Most of northern and central Moldova is hilly. Many rivers and streams flow through the rolling countryside. The river valleys of Moldova boast dark, fertile soil. Grassy plains sweep across much of the country's southern area. The map on page 388 shows you that northern Moldova has a humid continental climate of long, cold winters and short, hot summers. The southern part of the country has a steppe climate of cold, snowy winters and hot, dry summers.

Human/Environment Interaction Moldova relies on farming as its major economic activity. Farmers pluck grapes to make the country's popular wines. On the plains of the south, farmers harvest corn and wheat. In addition to farming, Moldova's light industries produce a variety of consumer and household goods. Like the other countries of the former Soviet Union, however, Moldova is struggling to move from a government-run economy to a privately owned economy with an expanded service sector.

The population of Moldova totals about 4.2 million people. It is one of the most densely populated countries in this region. Most of Moldova's people trace their language and culture to neighboring Romania. The rest of Moldova's population are Ukrainians, Russians, and Turks. Such diverse ethnic and national backgrounds often lead to conflict.

Moldova's capital and largest city is Chisinau (KEE•shih•NOW). More than half of Moldova's people, however, live in the countryside. Much of Moldova's culture is based on rural ways of life. Villagers celebrate special occasions with folk music and good food. Moldovan feasts include lamb, cornmeal pudding, and cheese made from goat's milk.

Bustling Minsk

Minsk, the capital of Belarus, is an important center of science, industry, and culture.
PLACE: In what year did Moldova and Belarus gain their independence from the Soviet Union?

SECTION 2 ASSESSMENT

REVIEWING TERMS AND FACTS
1. **Define the following:** nature preserve.
2. **PLACE** Why is the Belovezha Forest unique?
3. **PLACE** How do the economies of Belarus and Moldova differ?

MAP STUDY ACTIVITIES

4. Turn to the map on page 388. What republics share a border with Moldova?
5. Look at the map on page 388. In what climate region is the city of Minsk?

MAKING CONNECTIONS

THE FIRST SONG — BY SEMYON E. ROSENFELD

MATH | **SCIENCE** | **HISTORY** | **LITERATURE** | **TECHNOLOGY**

In this excerpt from *The First Song,* a Russian boy tells about moving to a new home. He and his family are leaving Mariopol on the Azov Sea to live in the port city of Odessa on the Black Sea.

I had spent most of my life here in Mariopol. In the summer, I would almost live on the beach, baking myself on the hot light-yellow sand, and when I was overcome with the heat, I would throw myself into the green transparent water of the Azov Sea, swim and dive. Here, also, [about two-thirds of a mile] from the city, on the Kalka—a river associated with many legends about the fighting between the Russians and the invading Mongols, centuries ago—we boys would often make our way across in an old flat-bottomed rowboat. Tall reeds grew thickly on the opposite shore—this was almost the beginning of the Don Cossack Region. On this spot we had made a momentous decision: to explore this land of the Cossacks for buried Tartar [Mongol] treasure, hoping to find a cauldron of gold, a trunkful of daggers, and a chestful of ornamented saddles and harness. Yes, I was leaving behind me a great deal, and my spirit was heavy.

. . . We soon came to the end of the Azov Sea and entered the Kerchinsky Strait, and after passing one more floating beacon, we were on the Black Sea. But I experienced a

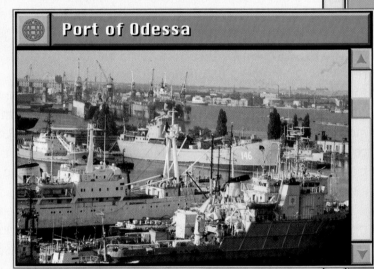

Port of Odessa

real disappointment now—this sea was not at all black. And no one was able to explain why it was called black. I thought even the Azov was darker. However, when I found out that the Black Sea was bigger and deeper, and that its water was bitter and salty and never froze over, while the water of the Azov Sea was sweet and fresh and froze over for several months in the year, I unwillingly felt respect for its vastness and depth.

From *The First Song* by Semyon E. Rosenfeld, in *A Harvest of Russian Children's Literature,* University of California Press, 1967. Reprinted by permission of the Estate of Miriam Morton.

Making the Connection

1. What activities with friends was this boy leaving behind?
2. How is the Black Sea different from the Azov Sea?

SECTION 3 ··· The Caucasus Republics

PREVIEW

Words to Know
- fault
- food processing
- ethnic group

Places to Locate
- Caucasus Mountains
- Armenia
- Yerevan
- Georgia
- Tbilisi
- Azerbaijan

Read to Learn . . .
1. what major landform is found in the Caucasus republics.
2. what farm products are grown in this region.
3. how people live in the republics of the Caucasus.

Imagine living alongside a mountain range that seems to touch the sky! You can enjoy such a view in the country of Armenia. There snow-topped peaks form part of an enormous chain of mountains called the Caucasus, which stretches about 750 miles (1,210 km) between the Black and Caspian seas.

Armenia shares the Caucasus Mountains with two neighboring countries: Georgia and Azerbaijan (A•zuhr•BY•JAHN). Together these three lands are known as the Caucasus republics. All of them lie south of Russia. Once part of the Soviet Union, they are now independent countries.

Place
Armenia

Tiny landlocked Armenia takes pride in a history that reaches back to the ancient world. The map on page 382 shows that today Armenia is without a coastline. Centuries ago—as early as 800 B.C.—Armenia ruled a large empire that reached from the Caspian Sea to the Mediterranean Sea.

Rugged Land Most of Armenia is a rugged plateau of ridges, narrow valleys, and lakes. Formed millions of years ago by volcanic eruptions, this plateau today sits across many **faults,** or openings in the earth's crust. What occurs near faults? If you guessed earthquakes, you are correct. Earthquakes occur quite frequently in Armenia. In 1988 a severe quake shook the country, killing thousands of people and destroying many villages and towns.

The Economy How do people in Armenia earn a living? In years past, many Armenians left their own country to work in the oil fields of neighboring Azerbaijan. Today they work in their own factories, making chemicals, electronic goods, machinery, and textiles. Armenia's miners take copper, gold, lead, and zinc out of the earth. Farmers—although few in number—grow wheat, barley, grapes, and other fruits in the hot, dry climate of the valleys.

Overlooking Tbilisi, Georgia

The ruins of an ancient castle and fort stand guard over the Georgian capital of Tbilisi. **PLACE: Why is Georgia popular with tourists?**

The People Armenia's 3.8 million people are mostly ethnic Armenians who share a unique language, culture, and the Christian religion. In recent centuries Turkey and Russia ruled Armenia. The Armenians suffered under foreign rule, and many settled in other countries.

During the years of foreign rule, the Armenians protected their culture. They now value their newly won independence. Today about 70 percent of Armenia's people live in urban areas. Yerevan (YEHR•uh•VAHN), the capital, has broad streets, attractive fountains, and colorful buildings made of volcanic stone. Founded in 782 B.C., Yerevan is one of the ancient cities of the world.

Victor Basmajian is a teenager who lives in Yerevan. Like most Armenians, Victor and his family share close ties and like to entertain friends at home. Victor enjoys foods such as shish kebab, or meat and vegetables cooked on an iron skewer, and cabbage leaves stuffed with rice and meat. He also likes to play chess and a board game called backgammon.

Place

Georgia

A Georgian poet once said, "Three divine gifts have been bestowed on us by our ancestors: language, homeland, faith." The homeland of Georgia, as you can see on the map on page 384, lies north of Armenia. Georgia's west coast touches the Black Sea.

Mountains and Highlands Sandwiched between two mountain ranges, Georgia is a rugged land said to have been named for St. George, a Christian saint. Fertile mountain valleys and lowlands along the Black Sea coast are home to most of Georgia's people.

392

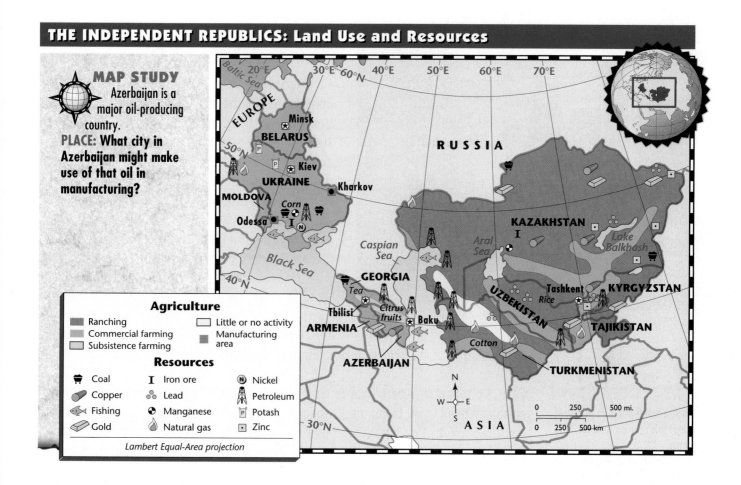

MAP STUDY
Azerbaijan is a major oil-producing country.
PLACE: What city in Azerbaijan might make use of that oil in manufacturing?

Agriculture
■ Ranching
■ Commercial farming
■ Subsistence farming
□ Little or no activity
■ Manufacturing area

Resources
⛏ Coal   I Iron ore   Ⓝ Nickel
🛢 Copper   ⚛ Lead   🗼 Petroleum
🐟 Fishing   ◉ Manganese   P Potash
🪙 Gold   💧 Natural gas   ⊡ Zinc

Lambert Equal-Area projection

Farms and Industries The map above shows you that farming ranks as Georgia's major economic activity. What would you grow if you were a Georgian farmer? You would probably harvest citrus fruits, tea, grapes, or wheat.

Georgia's major industries are wine making and **food processing,** or the preparing of foods as products for sale. Another leading industry is tourism. Resorts along the mild, scenic Black Sea coast draw thousands of visitors each year.

The People About 70 percent of Georgia's 5.4 million people are ethnic Georgians who are proud of their language, unique alphabet, and Christian heritage. The rest of Georgia's people belong to a number of other **ethnic groups**, or people having a common language, culture, and history. Since independence in 1991, conflict has broken out between Georgians and other ethnic groups. These groups want to separate from Georgia and create their own countries. Georgia first rejected membership in the Commonwealth of Independent States, but later joined in 1993.

More than half of Georgia's people live in cities and towns. Tbilisi (tuh•BEE•luh•see), the capital and largest city, prizes its warm mineral springs and scenic location. Georgians are known for their business skills, love of festivals, and friendliness to foreign visitors. They also maintain close family ties. Popular Georgian foods at family gatherings include shish kebab, pressed fried chicken, cheese, olives, and fruit.

An Azerbaijani family prepares vegetables for an outdoor lunch.
PLACE: What is Azerbaijan's most important natural resource?

Place
Azerbaijan

Also bordering Armenia is Azerbaijan, a country about the size of the state of Maine. It lies on the Caspian Sea. A narrow piece of Azerbaijan is separated from the rest of the country by Armenia. Since winning independence in 1991, Azerbaijan and Armenia have battled for control of territory and people.

The Land and Climate Like Armenia and Georgia, Azerbaijan is a mountainous country. The Caucasus ranges, having a highland climate, spread through the northern and western parts of Azerbaijan. A Mediterranean climate runs along the Caspian Sea. A dry coastal plain stretches to the north. Most of Azerbaijan is partly dry, with hot summers and mild winters.

The Economy Azerbaijan is largely agricultural. The country's farmers grow cotton, wheat, rice, fruits, and tobacco. Yet, along the shore of the Caspian Sea, countless oil-drilling rigs rise above the water. As you can see on the map on page 393, petroleum, lumber, and gold are Azerbaijan's main natural resources.

The People The 7.7 million people of Azerbaijan are mostly Azeris, a people distantly related to the people of Turkey in Southwest Asia. Armenians and Russians make up a small part of the population. The major faith of Azerbaijan is the religion of Islam.

About 55 percent of Azerbaijan's people live in cities. Baku (bah•KOO), the capital and largest city, has a population of 1.8 million. Founded more than 1,000 years ago, Baku got its name from Persian words meaning "windy town." Baku lies in one of the world's major oil-producing areas.

SECTION 3 ASSESSMENT

REVIEWING TERMS AND FACTS
1. Define the following: fault, food processing, ethnic group.
2. REGION What two bodies of water lie on either side of Armenia, Georgia, and Azerbaijan?
3. PLACE What kind of landscape is found in Armenia?
4. PLACE What is Azerbaijan's major religion?

MAP STUDY ACTIVITIES
5. Look at the land use map on page 393. What is Georgia's major economic activity?

The Central Asian Republics

PREVIEW

Words to Know
- nomad
- oasis

Places to Locate
- Kazakhstan
- Astana
- Kyrgyzstan
- Bishkek
- Uzbekistan
- Tashkent
- Tajikistan
- Dushanbe
- Turkmenistan
- Ashkhabad
- Aral Sea
- Kara Kum

Read to Learn . . .
1. what landforms and climates are found in the Central Asian republics.
2. how people earn a living in these republics.
3. how cultures here blend traditions with modern ways.

Have you ever bargained for goods at a market? This seller displays tomatoes and dill at a market in Almaty, a city in Kazakhstan. Taking its first steps toward capitalism, the country is building a free market economy. Many people in Kazakhstan's cities and towns now own small businesses.

The Central Asian republics include Kazakhstan (KA•zak•STAN), Kyrgyzstan (KIHR•gih•STAN), Tajikistan (tah•JIH•kih•STAN), Turkmenistan (turk•MEH•nuh•STAN), and Uzbekistan (uz•BEH•kih•STAN). Reaching from east of the Caspian Sea to the borders of China, these republics cover an immense area. Once part of the Soviet Union, these countries became independent in 1991.

Place
Kazakhstan

Kazakhstan is a large country covering 1,031,170 square miles (2,670,730 sq. km). About one-third the size of the United States, this country lies south of Russia in central Asia.

A Variety of Landscapes Kazakhstan has lowlands, steppes, deserts, and mountains. The map on page 384 shows you that the Tian Shan (tee•AHN SHAHN) mountain range extends through the eastern part of Kazakhstan. Major rivers, such as the Irtysh (ihr•TISH), provide water to farms in dry areas.

Kazakhstan as a whole receives very little rainfall. Its people endure very cold winters and long, hot summers. Wheat farming and sheep raising are the country's major economic activities. Kazakhstan is also rich in minerals such as coal and oil. Some 9 billion barrels of oil may lie beneath its soil. The map on page 393 shows natural resources of Kazakhstan. The government recently opened its border with China, hoping to increase trade with Asia.

The People Many Kazakhs and other people of the region trace their ancestry to the Mongols. Kazakhs make up about 40 percent of Kazakhstan's population. Most follow Islam and live in rural areas. Russians, who form the second-largest group, live in the major cities. In 1998 the Kazakh government moved the country's capital from Almaty in the south to Astana in the north.

Kazakh culture is rich in folk songs and legends. At one time Kazakhs were **nomads,** or people who moved with their flocks in search of pasture. Today Kazakhs live in rural houses or city apartments.

Place

Kyrgyzstan

The country of Kyrgyzstan—one of the most southern republics—lies southeast of Kazakhstan. Kyrgyzstan's 74,054 square miles (191,800 sq. km) make it about the size of the state of Minnesota.

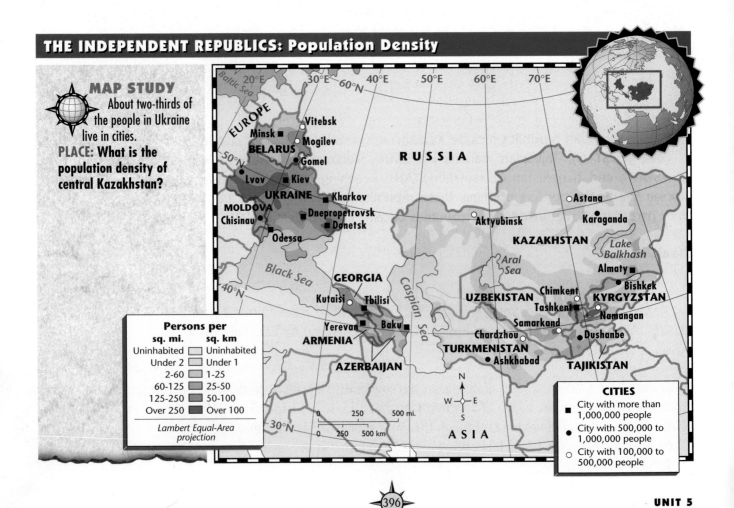

THE INDEPENDENT REPUBLICS: Population Density

MAP STUDY About two-thirds of the people in Ukraine live in cities.
PLACE: What is the population density of central Kazakhstan?

Persons per

sq. mi.	sq. km
Uninhabited	Uninhabited
Under 2	Under 1
2-60	1-25
60-125	25-50
125-250	50-100
Over 250	Over 100

Lambert Equal-Area projection

CITIES
■ City with more than 1,000,000 people
● City with 500,000 to 1,000,000 people
○ City with 100,000 to 500,000 people

A Land of Mountains The Tian Shan mountain range covers a large area of Kyrgyzstan. Nestled between the mountains lie high grassy plains and low river valleys. As in other highland areas, the climate of Kyrgyzstan differs with elevation. Lower valleys and plains have warm, dry summers and chilly winters. Highland areas have cool summers and bitterly cold winters. Most of Kyrgyzstan's people raise livestock or grow cotton, vegetables, and fruits.

The People More than half of Kyrgyzstan's 4.7 million people belong to the Kyrgyz (kihr•GEEZ) ethnic group. Russians make up the rest of the country's population. The Kyrgyz tend to live in rural areas, and the Russians live in the cities. Bishkek, with a population of about 616,000, is the capital and largest city.

Place
Tajikistan

Southwest of Kyrgyzstan lies the mountainous country of Tajikistan. Covering an area of 54,286 square miles (140,601 sq. km), Tajikistan is about the size of the state of Iowa.

Mountains and Valleys "Roof of the World" is what some people call Tajikistan's mountains. In one spot there, four major Asian mountain ranges come together. The map on page 384 shows that the Pamir Mountains rise in southeastern Tajikistan. Here Communism Peak reaches 24,590 feet (7,495 m).

Tajikistan is a farming country that depends on irrigation. In the river valleys, Tajik farmers grow cotton, rice, and fruits. In highland areas, coal, lead, and zinc are mined.

The People Most Tajiks are Muslims who speak a language similar to that spoken in Iran. About two-thirds of Tajikistan's people live in villages along the rivers. The largest city is Dushanbe (doo•SHAM•buh), the capital.

Place
Turkmenistan

Turkmenistan has some of the most isolated areas in the Central Asian republics. Covering 181,440 square miles (470,000 sq. km), it is slightly larger than the state of California.

Desert Country More than 85 percent of Turkmenistan is a vast desert known as the Kara Kum (KAHR•uh•KOOM), which means "black sand." For their water supply, the people of Turkmenistan rely on a few rivers and **oases**, or fertile desert areas watered by underground springs.

Teen Scene

Making His Mark

Kasim Ben Amir has found a way to be creative and earn money at the same time. Fifteen-year-old Kasim lives near Khodzent, Tajikistan. Much of the cotton and silk produced on Tajikistan farms is made into textiles here. Kasim buys plain shirts for a few rubles and then stamps them with designs that he creates. He sells his creations at a nearby market. Kasim gives some of the money he earns to his family. He's saving the rest to pay for a trip to Iran to visit his cousins.

The Shrinking Aral Sea

Stranded ships lie aground on the shoreline of the shrinking Aral Sea.
HUMAN/ENVIRONMENT INTERACTION: What two countries share the Aral Sea?

Cotton, Turkmenistan's major crop, covers acres of irrigated land. Other farm products include grains, grapes, and potatoes. Turkmenistan's natural resources include natural gas, petroleum, copper, gold, and lead.

The People Most of Turkmenistan's 4.7 million people belong to the Turkmen ethnic group. Islam is Turkmenistan's major religion. The population is fairly evenly divided between rural and urban dwellers. Ashkhabad (ASH•kuh•BAD), with about 518,000 people, is the capital and largest city.

Place

Uzbekistan

Like Turkmenistan, Uzbekistan is slightly larger than California—about 159,938 square miles (414,356 sq. km). But Uzbekistan has about 8 times as many people as Turkmenistan.

Cotton Country Deserts and high plains cover much of Uzbekistan. In the west are dry lowlands around the Aral Sea, a large inland salt lake. In the east are fertile valleys that receive plenty of water from the Tian Shan mountain range. Uzbekistan's economy relies on cotton, grains, fruits, and vegetables that are grown in irrigated areas.

Cotton growing in Uzbekistan has taken its toll on the environment. Under the Soviets, farmers were ordered to irrigate crops with water drained from the rivers flowing into the Aral Sea. Since the 1960s the Aral Sea has shrunk drastically in size, and its salt content has increased. Fish and other wildlife have disappeared. Salt particles carried by wind from the lake pollute the air and soil.

The People Most of Uzbekistan's 24.1 million people are Uzbeks. More than 2.2 million people live in the capital city of Tashkent—the largest city in all of central Asia. Most Uzbeks practice Islam and live in rural areas.

SECTION 4 ASSESSMENT

REVIEWING TERMS AND FACTS

1. Define the following: nomad, oasis.
2. REGION What type of landscape covers most of Turkmenistan?
3. PLACE What is the capital of Uzbekistan?
4. HUMAN/ENVIRONMENT INTERACTION What caused the Aral Sea to shrink?

MAP STUDY ACTIVITIES

5. Look at the population density map on page 396. What Central Asian republic is the most densely populated?

Chapter 15 Highlights

Important Things to Know About the Independent Republics

SECTION 1 UKRAINE

- Ukraine's landscape consists of lowlands, highlands, a coastal plain, and a central plains area.
- Ukraine is called the breadbasket of Europe because of its fertile farmland.
- Ukraine's capital, Kiev, was the center of an early Slavic civilization.

SECTION 2 BELARUS AND MOLDOVA

- The major economic activities of Belarus are farming, manufacturing, and service industries.
- Moldovans have strong ties to Romanians in language and culture.

SECTION 3 THE CAUCASUS REPUBLICS

- The Caucasus Mountains cover most of Armenia, Georgia, and Azerbaijan.
- The people of Armenia and Georgia have separate cultures, languages, and heritages.
- Located on the Caspian Sea, Azerbaijan is a major oil-producing country.

SECTION 4 THE CENTRAL ASIAN REPUBLICS

- Mountains, deserts, and plains form the landscapes of the Central Asian republics.
- Most of the peoples of the Central Asian republics practice the Islamic religion.
- Farmers in these republics depend on irrigation to grow cotton, grains, fruits, and vegetables.

Ukrainian folk dancers ▶

REVIEWING KEY TERMS

Match the numbered terms in Set A with their lettered definitions in Set B.

A

1. nomad
2. steppe
3. landlocked
4. oasis
5. food processing
6. fault

B

A. having no coastline
B. break in the earth's crust
C. fertile place in a desert area supplied by underground water
D. vast treeless plain
E. person who travels in search of pasture for grazing animals
F. the preparing of food for sale as products

REVIEWING THE MAIN IDEAS

Mental Mapping Activity

Draw a freehand map of the independent republics. Label the following:
- Kiev
- Black Sea
- Tian Shan Mountains
- Dnieper River
- Aral Sea
- Georgia

Section 1
1. LOCATION Describe the location of Ukraine.
2. REGION Why is Ukraine called a breadbasket?

Section 2
3. REGION What landforms are found in Belarus?
4. HUMAN/ENVIRONMENT INTERACTION What agricultural products come from Moldova?

Section 3
5. MOVEMENT Why did many Armenians migrate to other countries?
6. REGION What are the two major religions of the Caucasus republics?

Section 4
7. REGION What countries make up the Central Asian republics?
8. HUMAN/ENVIRONMENT INTERACTION What is the role of rivers in the Central Asian republics?

CRITICAL THINKING ACTIVITIES

1. **Drawing Conclusions** Why do you think Moldova is a densely populated country?
2. **Evaluating Information** Did years of Soviet rule help or harm the development of the independent republics? Explain.

CONNECTIONS TO ECONOMICS

Cooperative Learning Activity Organize into 11 groups, and then choose one of the republics discussed in this chapter. Each group will research the economy of its republic. Consider the following: What industries are strongest? What economic problems does the republic face? Groups should combine their information to present to another class.

TECHNOLOGY ACTIVITY

Developing Multimedia Presentations
Conduct research to find information about the Chernobyl power station. Then imagine that you are a news reporter covering the nuclear disaster. Create a short multimedia presentation about the disaster and its effects on Ukraine and surrounding areas. Incorporate images, mock interviews, and maps in your presentation.

GeoJournal Writing Activity
Imagine you are vacationing in Ukraine and the other independent republics. Write a letter home describing what you are seeing and where you are staying. Use travel books and your text to describe in detail each of your stops. In your letter, describe the scenery, climate, and the activities of the people you see.

PLACE LOCATION ACTIVITY: THE INDEPENDENT REPUBLICS

Match the letters on the map with the places of the independent republics. Write your answers on a separate sheet of paper.

1. Black Sea
2. Caucasus Mountains
3. Ukraine
4. Crimean Peninsula
5. Aral Sea
6. Dnieper River
7. Kazakhstan
8. Kara Kum
9. Baku
10. Minsk

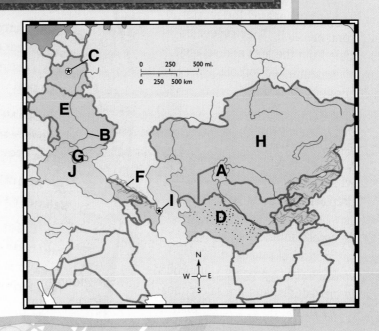

CHERNOBYL
NUCLEAR DISASTER

PROBLEM

On April 26, 1986, Reactor Number 4 of the Chernobyl Nuclear Power Plant exploded. Radioactive material flew more than 3 miles into the air. Blown by the wind, radioactive clouds contaminated thousands of square miles. Nearly 5 million people in Ukraine, Belarus, and Russia were affected.

Authorities quickly encased the reactor in a steel-and-concrete shell. They evacuated 116,000 people from the area and scraped away tons of contaminated topsoil. The cleanup continues to this day.

SOLUTIONS

● Ukraine authorities have agreed to shut down the 3 nuclear reactors remaining in Chernobyl and to work with the United States to replace electricity generated by the plants.

● Russia has agreed to encourage the use of natural gas instead of nuclear energy.

● Better worker-training programs, improved operating procedures, and fire safety measures have been started at nuclear plants.

Reindeer move up a ridge in Lapland. Radiation from the Chernobyl explosion contaminated reindeer herds, making the meat and milk unsafe for human consumption.

CHERNOBYL FACT BANK

🍃 The Chernobyl accident was the worst nuclear-power-plant disaster of all time. An estimated 5,000 people died and 30,000 were disabled.

🍃 More than 30,000 square miles of farmland were contaminated. Nothing can be grown there again for 100 years.

🍃 About 23 percent of the land in Belarus was contaminated.

🍃 Cases of thyroid cancer in children and birth defects have increased in the area near the reactor and in the path of the radioactive cloud.

🍃 Penguins in Antarctica, reindeer in Lapland, and countless other animals have suffered the effects of the radioactive clouds that encircled the planet.

USE ALTERNATIVE ENERGY

At 1:23 A.M. on April 26, 1986, a test on emergency systems in the Chernobyl Nuclear Power Plant went wrong—and led to the explosion of Reactor Number 4. As pictured in this illustration, the explosion ripped apart the reactor and sent radioactive material high into the atmosphere on a plume of intense heat.

go by bike!

environmental activities

TEEN TRIBUTE

In April 1986, a group from Ramapo High School in Spring Valley, New York, was visiting Leningrad (now St. Petersburg) when the Chernobyl reactor blew up. The visitors returned home—their trip cut short by the disaster. Since then, Ramapo students and their teacher, working with community leaders and drug companies, have provided more than $11 million worth of medicine, vitamins, and toys to hospitals in Belarus and Russia.

WHAT CAN YOU DO?

TAKE A CLOSER LOOK

1 What disaster occurred in Ukraine at the Chernobyl Nuclear Power Plant?

2 What were the lasting effects of the Chernobyl disaster?

🍃 SAVE ENERGY–ride a bike!

🍃 Support a group working for safe disposal of toxic and nuclear wastes.

🍃 Don't add to toxic waste–use fewer batteries and buy only rechargeable ones.

🍃 SAVE ENERGY– encourage your parents to lower the thermostat 6 degrees in winter. If all Americans did, we'd save more than 500,000 barrels of oil a day!

403

БЪЛГАРИЯ ст. 1

This Ukrainian couple had to resettle after the accident, with only their colorful wall hangings to remind them of their lost home.

Unit 6

Southwest Asia and North Africa

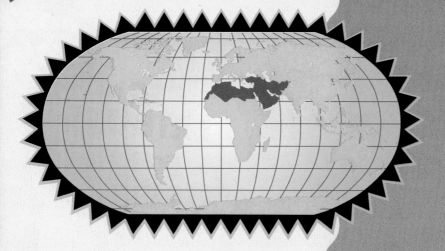

What Makes Southwest Asia and North Africa a Region?

The people share . . .

• a constant quest for water.

• Mediterranean, steppe, and desert climates.

• an abundance of oil and natural gas reserves.

• the historical beginnings of the Islamic, Jewish, and Christian religions.

To find out more about Southwest Asia and North Africa, see the Unit Atlas on pages 406–417.

The Great Sphinx near Giza, Egypt

EXPLORING THE INTERNET

To learn more about Southwest Asia and North Africa, visit the Glencoe Social Studies Web site at **www.glencoe.com** for information, activities, and links to other sites.

GeoJournal Activity

As you read about this region, imagine that you are a journalist writing a book about Southwest Asia and North Africa. Write descriptions of at least five cities or areas. Then use the descriptions to write an article for your local newspaper.

UNIT 6 ATLAS

NATIONAL GEOGRAPHIC SOCIETY

IMAGES of the WORLD

1. **Portrait of a child, Turkey**
2. **Boy in a poppy field, Tunisia**
3. **Dome of the Rock, Jerusalem**
4. **Hobbled camels, Egypt**
5. **Woman and children, Saudi Arabia**
6. **Pyramid, Egypt**
7. **Fishermen with their catch, Yemen**

All photos viewed against a desert landscape, Tunisia.

UNIT 6 ATLAS

Regional Focus

Southwest Asia and North Africa lie where the continents of Asia, Africa, and Europe meet. You will find the world's largest desert and part of the world's longest river in this region. Some of the world's earliest civilizations and three great world religions started here. Today this region is one of the world's major suppliers of oil.

Region
The Land

The region of Southwest Asia and North Africa covers about 5.5 million square miles (14.2 million sq. km)—about 10 percent of the earth's total land area. Look at the physical map on page 412. The region stretches from sandy beaches along the Atlantic Ocean on the west to the towering Hindu Kush on the east. North to south, the region reaches from the deep blue waters of the Mediterranean Sea to the Sahara.

Euphrates River in Iraq

A thick stand of palm trees lines the banks of the Euphrates River.
REGION: Why do you think most people in Iraq live near the Tigris and Euphrates rivers?

Mountains Towering mountains rise over much of Southwest Asia and North Africa. The Taurus Mountains run east and west through the southern edge of the country of Turkey. Another range—the Zagros—stretches between Iran and Iraq. Other mountain chains are the Elburz Mountains of Iran and the Asir Mountains of Saudi Arabia. Farther west, in North Africa, stretch the Atlas Mountains—Africa's longest mountain range.

Deserts Mountains and prevailing winds have created vast deserts in many parts of Southwest Asia and North Africa. The Rubal Khali, or Empty Quarter, of the Arabian Peninsula is made up of shifting sand dunes. The Sahara—the largest desert in the world—covers much of North Africa.

Water Several bodies of water border the region of Southwest Asia and North Africa. These include the Mediterranean Sea, the Black and Caspian seas, the Persian Gulf, and the Arabian Sea. Freshwater, however, is scarce. Land near rivers and underground springs is green and fertile. With the help of irrigation, crops grow in these areas.

The Nile River, more than 4,000 miles (6,400 km) long, flows north through Egypt to the Mediterranean Sea. It is the world's longest river. Two other major rivers—the Tigris and the Euphrates—flow through southeastern Turkey, Syria, and Iraq into the Persian Gulf.

Natural Resources Oil and natural gas are the most important natural resources in Southwest Asia and North Africa. You will find these minerals in the countries of North Africa and those bordering the Persian Gulf. Southwest Asia and North Africa are also rich in iron ore, copper, lead, manganese, zinc, and phosphate. Phosphate is used for making fertilizers.

Region
Climate and Vegetation

The many mountains of Southwest Asia and North Africa block the flow of rain clouds to the inland side of the region. Only coastal areas enjoy regular and adequate rainfall. Rainfall averages about 10 inches (25 cm) or less each year. This amount is inadequate to grow most food crops, and irrigation is necessary.

Desert and steppe climates cover much of Southwest Asia and North Africa. During the day, temperatures in the desert may soar above 100°F (38°C). In steppe areas, enough rain falls for grasses to grow. This allows farmers to raise livestock. A Mediterranean climate of hot, dry summers and cool, rainy winters is found on coastal plains.

Human/Environment Interaction
Economy

Economies differ from country to country in Southwest Asia and North Africa. The people living in countries that have oil, manufacturing, and trade generally enjoy a high standard of living. People living in countries where the economies are based on farming have a low standard of living.

Agriculture Most people in Southwest Asia and North Africa are farmers or herders. Cereals, citrus fruits, grapes, and dates grow in river valleys and coastal areas. Another source of food for the people of the region is livestock—mostly cattle and sheep. Farmers in river valleys or in irrigated areas also grow cotton, one of the region's major exports.

Industrial Growth Oil and natural gas are important sources of wealth for many Southwest Asian and North African countries. These resources are major exports and have allowed industries to develop. Factories in the region produce textiles, fertilizers, medicines, plastics, and paints. Service industries such as banking, insurance, and tourism have also grown in recent years.

Transportation and Communication In the past, mountains and deserts made transportation costly in the region. Today roads and railroads link cities, oil fields, and seaports. Inland waterways such as the Nile River and the Suez Canal also move people and goods.

Damascus, Syria

People gather in marketplaces to shop for items made by local workers.
PLACE: What mineral is an important source of wealth for many countries in this region?

UNIT 6 ATLAS

Place
The People

For centuries Southwest Asia and North Africa have served as the meeting place for the peoples of Asia, Africa, and Europe. The region today has people from many different ethnic backgrounds.

History and Culture Southwest Asia and North Africa have a rich culture and history. Centuries ago the region's farmers were the first to raise many of the grains, vegetables, and animals still used as basic foods in much of the world. Some of the world's oldest civilizations developed in the Nile and Tigris-Euphrates river valleys. In Egypt you can still see the huge pyramids and temples left by the ancient Egyptians.

Ancient Thebes

Giant stone figures guard the 3,000-year-old Temple of Luxor.
LOCATION: In what country can you see ancient pyramids?

The religions of Judaism, Christianity, and Islam began in Southwest Asia. All of them share the belief in one God. From the 600s to the 1500s, Islamic empires flourished in Southwest Asia and North Africa and passed on much of their knowledge to other parts of the world. By the 1800s, Europeans ruled most of the region. After World War II, the peoples of Southwest Asia and North Africa set up independent states.

Ethnic Groups Most of the people of Southwest Asia and North Africa are Arabs. About 90 percent are Muslims who practice the religion of Islam. The Arab culture, especially the Arabic language, has had a lasting influence on the region. Two other groups—the Iranians and the Turks—have contributed to the region's cultures. Neither group is Arab, but many in each group are Muslims. Living among the Turks, Iranians, and Arabs are two smaller ethnic groups—Armenians and Kurds. Most people in the country of Israel are Jewish. Israel was founded as a Jewish state in 1948. For many years conflicts have taken place between Jews and Arabs. Today, however, peace efforts are underway between Israeli and Arab leaders.

Population About 400 million people live in Southwest Asia and North Africa. Because of deserts, most people live in rural areas along seacoasts or rivers, or near highlands. The environment in these areas makes it possible to grow crops and raise animals. For centuries the people of Southwest Asia and North Africa lived in rural areas. Today many of them are moving to large cities such as Cairo or Tehran. Increasing population has raised concerns about food and housing shortages as well as air and water pollution.

Multimedia Unit 6 Activities

 **NATIONAL GEOGRAPHIC SOCIETY**

Picture Atlas of the World

Cultural Differences

Using the CD-ROM People of many different ethnic backgrounds and religions make up the human environment of Southwest Asia and North Africa. Put together a cultural reference file on the region. (See the User's Guide on how to use the Collector button.) Collect photographs on the CD-ROM for the following countries: Afghanistan—nomads and Kirghiz tribespeople; Iraq—Bedouins and a Kurdish family; Israel—the Dome of the Rock and a Palestinian man; Saudi Arabia—Saudi women and Muslim pilgrims. Read the photo captions and answer the following:

1. Describe the life of nomads in Afghanistan.
2. Why do Kurds live at the mercy of often hostile neighbors?
3. To which three religions is the Dome of the Rock sacred?
4. How do women in Saudi Arabia show their devotion to Islam?

Surfing the "Net"

Libya's Need for Water

Most of Libya lies in the Sahara, which makes it one of the driest countries in the world. Thus, there are very few areas in the country that are suitable for farming. However, in 1996 the Libyan government opened a series of long irrigation pipelines to carry water from underground, water-bearing layers of rock. These aquifers in the Sahara will bring water to areas along the coast. To find out more about this recent development, search the Internet.

Getting There

Follow these steps to gather information on Libya's new irrigation project.
1. Use a search engine. Type in the word *Libya*.
2. After typing *Libya*, enter words like the following to focus your search: *environment, agriculture, water, irrigation,* and *pipeline.*
3. The search engine will provide you with a number of links to follow.

What To Do When You Are There

Click on the links to navigate through the pages of information. Find information about the location of aquifers in the Sahara and the routes taken by the pipelines north to the coastal areas. Then use a large wall map of North Africa to trace the routes of the pipelines. Place pins on the map to represent the location of cities and towns to which the pipelines will deliver water. Write a paragraph explaining how this new development will help change the Libyan landscape.

Physical Geography

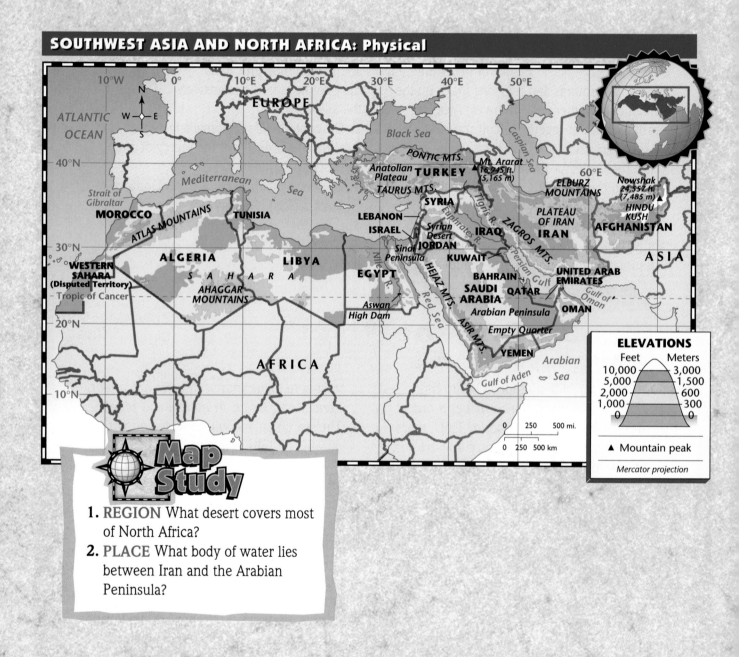

SOUTHWEST ASIA AND NORTH AFRICA: Physical

Map Study

1. **REGION** What desert covers most of North Africa?
2. **PLACE** What body of water lies between Iran and the Arabian Peninsula?

ELEVATION PROFILE: Southwest Asia and North Africa

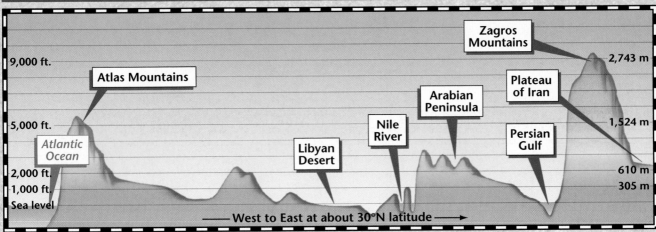

Zagros Mountains

Atlas Mountains

Plateau of Iran

Arabian Peninsula

Persian Gulf

Nile River

Libyan Desert

Atlantic Ocean

9,000 ft.
5,000 ft.
2,000 ft.
1,000 ft.
Sea level

2,743 m
1,524 m
610 m
305 m

← West to East at about 30°N latitude →

Source: *Goode's World Atlas,* 19th edition

GeoFacts

Highest point: Nowshak (Afghanistan) 24,557 ft. (7,485 m) high

Lowest point: Dead Sea (Israel-Jordan) 1,312 ft. (400 m) below sea level

Longest river: Nile River (Africa) 4,160 mi. (6,693 km) long

Largest lake: Caspian Sea (Europe-Asia) 143,244 sq. mi. (371,002 sq. km)

Largest desert: Sahara (Northern Africa) 3,500,000 sq. mi. (9,065,000 sq. km)

SOUTHWEST ASIA/NORTH AFRICA and the UNITED STATES: Land Comparison

Graphic Study

1. HUMAN/ENVIRONMENT INTERACTION How might elevation affect the population in the Zagros and Atlas mountains?

2. PLACE How would you describe the physical size of the United States compared to the combined sizes of Southwest Asia and North Africa?

Cultural Geography

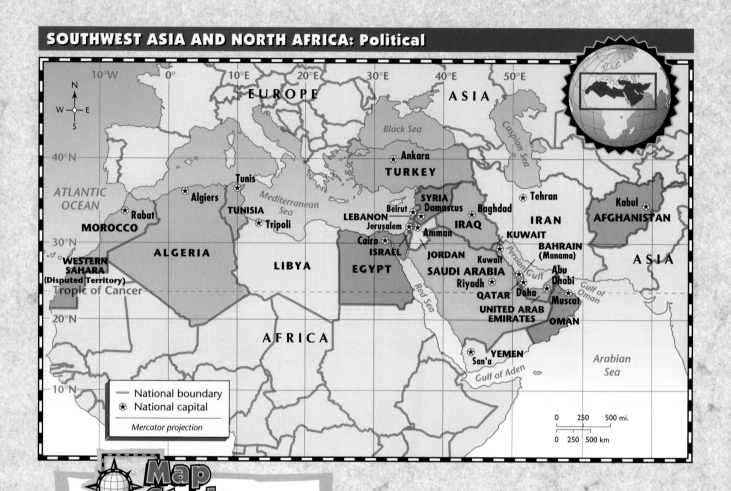

SOUTHWEST ASIA AND NORTH AFRICA: Political

1. **LOCATION** What country lies north of Syria?
2. **PLACE** What is the capital of Algeria?

SOUTHWEST ASIA AND NORTH AFRICA: Largest Urban Areas

Tehran, Iran

Cairo, Egypt

Istanbul, Turkey

Baghdad, Iraq

Alexandria, Egypt

Casablanca, Morocco

Ankara, Turkey

 = 1,000,000

Source: *The World Almanac*, 1998

COMPARING POPULATION:
Southwest Asia/North Africa and the United States

Southwest Asia/
North Africa

United States

 = 50,000,000

Source: *Population Reference Bureau*, 1998

Biggest country (land area): Algeria
919,590 sq. mi.
(2,381,738 sq. km)

Smallest country (land area): Bahrain
266 sq. mi. (689 sq. km)

Largest city (population): Tehran
(1997) 11,681,000
(2015 projected) 14,000,000

Highest population density: Bahrain
2,387 people per sq. mi. (922
people per sq. km)

Lowest population density: Libya
8 people per sq. mi. (3 people
per sq. km)

Graphic Study

1. PLACE About how many people live in
Southwest Asia and North Africa?
2. PLACE What is the largest city in Southwest
Asia and North Africa?

UNIT 6 ATLAS

Countries at a Glance

Afghanistan

Kabul

LANDMASS:
251,772 sq. mi./
652,089 sq. km
MONEY:
Afghani
MAJOR EXPORT:
Dried Fruits and Nuts
MAJOR IMPORT:
Machinery

CAPITAL:
Kabul
MAJOR LANGUAGE(S):
Pushtu, Dari Persian,
Uzbek
POPULATION:
24,800,000

Algeria

Algiers

LANDMASS:
919,590 sq. mi./
2,381,738 sq. km
MONEY:
Dinar
MAJOR EXPORT:
Petroleum
MAJOR IMPORT:
Machinery

CAPITAL:
Algiers
MAJOR LANGUAGE(S):
Arabic, Berber
POPULATION:
30,200,000

Bahrain

Manama

LANDMASS:
266 sq. mi./
689 sq. km
MONEY:
Dinar
MAJOR EXPORT:
Petroleum Products
MAJOR IMPORT:
Petroleum Products

CAPITAL:
Manama
MAJOR LANGUAGE(S):
Arabic, Farsi, Urdu
POPULATION:
600,000

Egypt

Cairo

LANDMASS:
384,344 sq. mi./
995,450 sq. km
MONEY:
Pound
MAJOR EXPORT:
Petroleum
MAJOR IMPORT:
Machinery

CAPITAL:
Cairo
MAJOR LANGUAGE(S):
Arabic, English
POPULATION:
65,500,000

Iran

Tehran

LANDMASS:
631,660 sq. mi./
1,635,999 sq. km
MONEY:
Rial
MAJOR EXPORT:
Petroleum
MAJOR IMPORT:
Motor Vehicles

CAPITAL:
Tehran
MAJOR LANGUAGE(S):
Farsi, Turkish, Kurdish,
Arabic
POPULATION:
64,100,000

Iraq

Baghdad

LANDMASS:
168,870 sq. mi./
437,373 sq. km
MONEY:
Dinar
MAJOR EXPORT:
Fuels
MAJOR IMPORT:
Machinery

CAPITAL:
Baghdad
MAJOR LANGUAGE(S):
Arabic, Kurdish
POPULATION:
21,800,000

Israel

Jerusalem

LANDMASS:
7,961 sq. mi./
20,619 sq. km
MONEY:
Shekel
MAJOR EXPORT:
Machinery, Polished
Diamonds
MAJOR IMPORT:
Rough Diamonds

CAPITAL:
Jerusalem
MAJOR LANGUAGE(S):
Hebrew, Arabic
POPULATION:
6,000,000

Jordan

Amman

LANDMASS:
34,336 sq. mi./
88,930 sq. km
MONEY:
Dinar
MAJOR EXPORT:
Phosphate Fertilizers
MAJOR IMPORT:
Food and Live Animals

CAPITAL:
Amman
MAJOR LANGUAGE(S):
Arabic
POPULATION:
4,600,000

Kuwait

Kuwait

LANDMASS:
6,880 sq. mi./
17,819 sq. km
MONEY:
Dinar
MAJOR EXPORT:
Petroleum
MAJOR IMPORT:
Machinery

CAPITAL:
Kuwait
MAJOR LANGUAGE(S):
Arabic
POPULATION:
1,900,000

Lebanon

Beirut

LANDMASS:
3,950 sq. mi./
10,231 sq. km
MONEY:
Pound
MAJOR EXPORT:
Jewelry
MAJOR IMPORT:
Machinery

CAPITAL:
Beirut
MAJOR LANGUAGE(S):
Arabic, French
POPULATION:
4,100,000

Countries not drawn to scale.

Libya

CAPITAL:
Tripoli
MAJOR LANGUAGE(S):
Arabic
POPULATION:
5,700,000

LANDMASS:
679,360 sq. mi./
1,759,542 sq. km
MONEY:
Dinar
MAJOR EXPORT:
Petroleum
MAJOR IMPORT:
Food

Morocco*

CAPITAL:
Rabat
MAJOR LANGUAGE(S):
Arabic, Berber
POPULATION:
28,600,000

LANDMASS:
172,320 sq. mi./
446,309 sq. km
MONEY:
Dirham
MAJOR EXPORT:
Food
MAJOR IMPORT:
Crude Oil

Oman

CAPITAL:
Muscat
MAJOR LANGUAGE(S):
Arabic
POPULATION:
2,500,000

LANDMASS:
82,030 sq. mi./
212,458 sq. km
MONEY:
Rial
MAJOR EXPORT:
Petroleum
MAJOR IMPORT:
Machinery

Qatar

CAPITAL:
Doha
MAJOR LANGUAGE(S):
Arabic, English
POPULATION:
500,000

LANDMASS:
4,250 sq. mi./
11,008 sq. km
MONEY:
Riyal
MAJOR EXPORT:
Petroleum
MAJOR IMPORT:
Machinery

Saudi Arabia

CAPITAL:
Riyadh
MAJOR LANGUAGE(S):
Arabic
POPULATION:
20,200,000

LANDMASS:
830,000 sq. mi./
2,149,700 sq. km
MONEY:
Riyal
MAJOR EXPORT:
Petroleum
MAJOR IMPORT:
Machinery

Syria

CAPITAL:
Damascus
MAJOR LANGUAGE(S):
Arabic, Kurdish,
Armenian, Turkish
POPULATION:
15,600,000

LANDMASS:
70,958 sq. mi./
183,781 sq. km
MONEY:
Pound
MAJOR EXPORT:
Petroleum
MAJOR IMPORT:
Food Products

Tunisia

CAPITAL:
Tunis
MAJOR LANGUAGE(S):
Arabic, French
POPULATION:
9,500,000

LANDMASS:
59,980 sq. mi./
155,348 sq. km
MONEY:
Dinar
MAJOR EXPORT:
Clothing
MAJOR IMPORT:
Textiles

Turkey

CAPITAL:
Ankara
MAJOR LANGUAGE(S):
Turkish, Kurdish,
Arabic, Greek
POPULATION:
64,800,000

LANDMASS:
297,150 sq. mi./
769,619 sq. km
MONEY:
Lira
MAJOR EXPORT:
Textiles
MAJOR IMPORT:
Machinery

United Arab Emirates

CAPITAL:
Abu Dhabi
MAJOR LANGUAGE(S):
Arabic
POPULATION:
2,700,000

LANDMASS:
32,280 sq. mi./
83,605 sq. km
MONEY:
Dirham
MAJOR EXPORT:
Petroleum
MAJOR IMPORT:
Machinery

Yemen

CAPITAL:
San'a
MAJOR LANGUAGE(S):
Arabic
POPULATION:
15,800,000

LANDMASS:
203,850 sq. mi./
527,972 sq. km
MONEY:
Rial (also the Dinar)
MAJOR EXPORT:
Coffee
MAJOR IMPORT:
Food and Live Animals

* Morocco claims the Western Sahara area but
other countries do not accept this claim.

Chapter 16 Southwest Asia

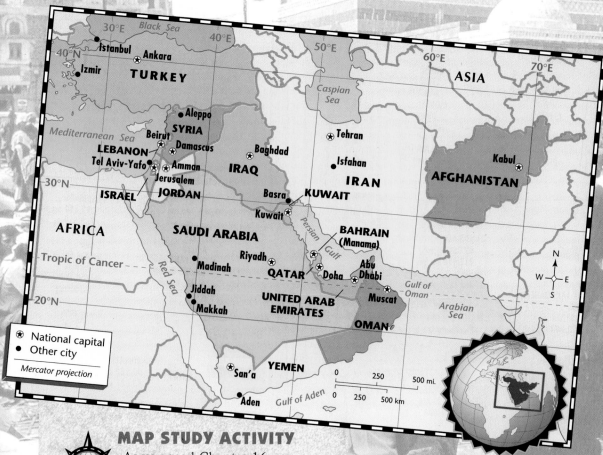

National capital
Other city
Mercator projection

Black Sea
30°E
40°E
50°E
60°E
70°E
40°N
Istanbul ⊛Ankara
Izmir
TURKEY
ASIA
Caspian Sea
Mediterranean Sea
Aleppo
SYRIA
Beirut
Damascus
⊛Tehran
Isfahan
Kabul ⊛
AFGHANISTAN
LEBANON
Tel Aviv-Yafo
Amman
Baghdad
IRAQ
IRAN
Jerusalem
JORDAN
Basra
KUWAIT
30°N
ISRAEL
Kuwait
AFRICA
SAUDI ARABIA
Persian Gulf
BAHRAIN
(Manama)
Tropic of Cancer
Riyadh
Madinah
QATAR
Doha
Abu Dhabi
Gulf of Oman
Red Sea
Jiddah
UNITED ARAB
EMIRATES
Muscat
20°N
Makkah
OMAN
Arabian Sea
YEMEN
0   250   500 mi.
San'a
0   250   500 km
Aden
Gulf of Aden

MAP STUDY ACTIVITY

As you read Chapter 16, you will learn about Southwest Asia. The countries in this region lie in the southwest corner of Asia.

1. How many countries make up Southwest Asia?
2. What are the two largest countries?
3. Which country connects Europe to Asia?

Turkey

PREVIEW

Words to Know
- mosque
- migrate

Places to Locate
- Turkish Straits
- Anatolian Plateau
- Pontic Mountains
- Taurus Mountains
- Istanbul
- Ankara

Read to Learn . . .
1. how Turkey's location has affected its history and development.
2. how Turkey blends its ancient heritage with modern ways.

Flocks of grazing sheep and goats are common sights in Turkey and other Southwest Asian countries. This shepherd watches a flock of sheep in

Turkey, one of the largest countries in Southwest Asia.

A little larger than the state of Texas, Turkey has a unique location—it lies on two continents: Europe and Asia. Because of its location, Turkey has links to both Asian and European cultures.

Location

The Land

The European and Asian parts of Turkey are separated by three important waterways—the Bosporus (BAHS•puh•ruhs), the Sea of Marmara (MAHR•muh•ruh), and the Dardanelles (DAHR•duhn•EHLZ). Together, these waterways are known as the Turkish Straits. The Turkish Straits join the Black Sea and the Mediterranean Sea. Find these waterways on the map on page A20.

The Land and Climate Highlands and plateaus cover much of Turkey. The heart of Turkey is the Anatolian (A•nuh•TOH•lee•uhn) Plateau. Mountain ranges surround this broad sweep of dry highlands. The Pontic Mountains border the plateau on the north. The Taurus Mountains tower over it on the south.

Grassy plains cover the northern part of Turkey along the Black Sea. On the western coast, you find broad, fertile river valleys running inland from the Aegean Sea. In the south, coastal plains extend along the Mediterranean Sea.

MAP STUDY
Mountains cover much of Southwest Asia.
LOCATION: What plateau covers central Turkey?

Black Sea
30°E   40°E   50°E   60°E   70°E

PONTIC MTS.
40°N
Anatolian Plateau   **TURKEY**
TAURUS MTS.
Caspian Sea

Mt. Ararat
16,945 ft.
(5,165 m)   ELBURZ MTS.
Mt. Nowshak
24,557 ft.
(7,485 m)   HINDU KUSH

SYRIA
Damavand
18,386 ft.
(5,604 m)   Plateau of Iran   **AFGHANISTAN**

LEBANON
Mediterranean Sea
ISRAEL   SYRIAN DESERT
IRAQ   ZAGROS MTS.   **IRAN**

JORDAN   Euphrates R.   Tigris R.

30°N   **ASIA**

KUWAIT

AFRICA   **SAUDI ARABIA**   Persian Gulf

HEJAZ MTS.   **BAHRAIN**   Gulf of Oman
Tropic of Cancer   **QATAR**
Red Sea   **UNITED ARAB EMIRATES**
Arabian Peninsula
20°N   Empty Quarter   **OMAN**
ASIR MTS.   Arabian Sea

YEMEN   0   250   500 mi.
0   250   500 km
Gulf of Aden
10°N

ELEVATIONS

Feet		Meters
10,000		3,000
5,000		1,500
2,000		600
1,000		300
0		0

▲ Mountain peak

Mercator projection

The map on page 424 shows you that Turkey's climate varies throughout the country. If you lived on the Anatolian Plateau, you would experience the hot, dry summers and cold, snowy winters of the steppe climate. The coastal areas enjoy a Mediterranean climate—hot, dry summers and mild, rainy winters.

Human/Environment Interaction
The Economy

Agriculture plays an important role in Turkey's economy. Wheat and livestock are the country's leading farm products. The best farmlands are found in the coastal areas. In recent years, Turkish industry has grown tremendously, especially in the making of textiles. The map on page 429 shows that Turkey is rich in mineral deposits, including iron ore, coal, and copper.

Region
The People

Most of Turkey's 65 million people live in the northern part of the Anatolian Plateau, on coastal plains, or in valleys. Turkish is the major language, but Arabic, Greek, and Kurdish are also spoken. About 98 percent of the Turkish people are Muslims. They follow Islam, a religion that teaches belief in one God. Islam is the major faith of most of Southwest Asia. Turkey also has small groups of Christians and Jews.

UNIT 6

City and Country About 64 percent of Turkey's people live in cities or towns. Ahmed ul-Hamid, a 14-year-old Turkish student, lives among the 8 million people of Istanbul (IHS•tuhn•BOOL), Turkey's largest city. Istanbul is the only city in the world located on two continents. Ahmed has the opportunity to visit Istanbul's beautiful palaces, museums, and **mosques**, or Muslim places of worship. Only once has Ahmed visited Turkey's capital and second-largest city, Ankara (ANG•kuh•ruh). It has about 2.8 million people and is located on the Anatolian Plateau.

Roughly 35 percent of Turkey's people live on farms or in small villages. Since the mid-1900s, hundreds of thousands of people have left their villages to seek work in the cities. The cities, however, can't supply enough jobs for everyone. This results in many searching for work in other parts of Southwest Asia and in western Europe and Australia.

Influences of the Past Most of Turkey's people are descendants of an Asian people called Turks who **migrated**, or moved, to the Anatolian Plateau during the A.D. 900s. One group of Turks—the Ottomans—established a Muslim empire centered in Turkey.

During World War I, the Ottomans were defeated. Kemal Atatürk, a military hero, became the first president of the new country—the modern republic of Turkey. Atatürk introduced many political and social changes to the country.

Turkey soon began to consider itself European as well as Asian. Modern Turkish people, however, remain proud of their Muslim faith. In the 1990s the Muslim party played an important role in Turkey's government.

Istanbul, Turkey

Domed roofs and graceful towers, called minarets, rise above the many mosques of Istanbul.
PLACE: What is the major religion of Turkey?

SECTION 1 ASSESSMENT

REVIEWING TERMS AND FACTS
1. **Define the following:** mosque, migrate.
2. **LOCATION** What makes Turkey's location unique?
3. **PLACE** What areas of Turkey are best for farming? coogie
4. **MOVEMENT** Where do Turkish workers move to find work?

MAP STUDY ACTIVITIES
5. Look at the physical map on page 420. What feature dominates central Turkey?
6. Turn to the political map on page 418. About how far is Ankara from Istanbul?

BUILDING GEOGRAPHY SKILLS

Reading a Time Zone Map

Earth's surface is divided into 24 time zones. The 0° line of longitude—the Prime Meridian—is the starting point for figuring out time around the world.

To read a time zone map, follow these steps:

- Locate a place where you know what time it is and select another place where you wish to know the time.

- Notice the time zones you cross between these places.

- If the second place lies east of the first, add an hour for each time zone. If it lies west, subtract an hour for each zone. If you must cross the International Date Line—the 180° meridian—add or subtract one day.

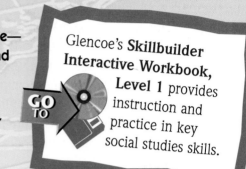

Glencoe's **Skillbuilder Interactive Workbook, Level 1** provides instruction and practice in key social studies skills.

World Time Zones

Geography Skills Practice

1. If it is 7 A.M. in New York City, New York, what time is it in Istanbul, Turkey?

2. If it is 6 A.M. on Wednesday in Mumbai, India, what day and time is it in Rio de Janeiro, Brazil?

Israel

PREVIEW

Words to Know
- potash
- phosphate
- monotheism
- Holocaust

Places to Locate
- Mediterranean Sea
- Sea of Galilee
- Dead Sea
- Negev
- Tel Aviv

Read to Learn . . .
1. how Israelis have developed their resources.
2. what groups of people live in Israel.
3. how the past affects Israel today.

I srael is the only Jewish nation in the world. Established as a nation in 1948, Israel celebrates its Independence Day just as Americans do—with parades and flag-waving.

I srael lies at the eastern end of the Mediterranean Sea. It is 256 miles (412 km) long from north to south and only 68 miles (109 km) wide from east to west. An hour's drive takes you from the eastern boundary of the country almost to the Mediterranean Sea.

Place
The Land

Israel's landscape includes plains, deserts, and highlands. In Israel's far north lie the mountains of Galilee. East of these mountains is a plateau called the Golan Heights. The Judean (ju•DEE•uhn) Hills lie south.

The Dead Sea lies between Israel and Jordan. At 1,312 feet (400 m) *below* sea level, the rim of the Dead Sea is the lowest place on the earth! It is also the saltiest body of water in the world—about 9 times as salty as the ocean. A desert landform, the Negev (NEH•gehv) in southern Israel, covers almost one-half of the country. It is a triangular area of dry hills, valleys, and plains. Not all of southern Israel is desert, however. A narrow fertile plain—only 20 miles

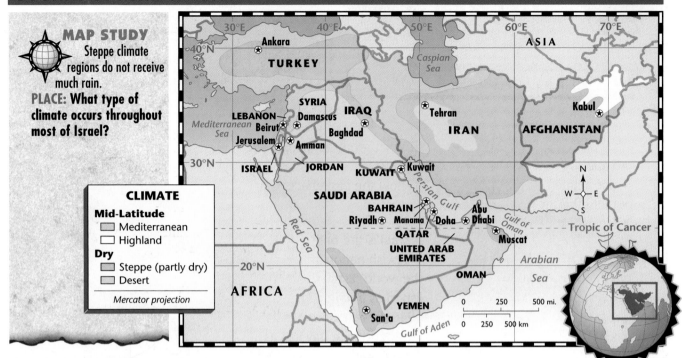

MAP STUDY
Steppe climate regions do not receive much rain.
PLACE: What type of climate occurs throughout most of Israel?

CLIMATE

Mid-Latitude
- Mediterranean
- Highland

Dry
- Steppe (partly dry)
- Desert

Mercator projection

(33 km) wide at its widest point—lies along Israel's Mediterranean coast. To the east, the Jordan River cuts through the floor of a long, narrow valley. The longest of Israel's few rivers, the Jordan flows south into the Dead Sea.

The Climate As you can see on the map above, Israel's climate varies from north to south. Northern Israel has a Mediterranean climate with hot, dry summers and mild winters. About 40 inches (100 cm) of rain fall in the north each year. Southern Israel has a desert climate, with summer temperatures soaring to 120°F (49°C) or higher. Less than 1 inch of rain (2.5 cm) falls in the south.

Place
The Economy

Israel has limited farmland and natural resources. Israelis, however, make good use of their few resources and have built a strong economy. A developing system of transportation and communication supports this economy.

Mining and Manufacturing Israel is the most industrialized country in Southwest Asia. Its factories produce food products, clothing, chemicals, building materials, electronic appliances, and machinery. Diamond cutting is also a major Israeli industry. Tel Aviv is Israel's largest manufacturing center.

Mining is important to Israel's economy. The Dead Sea area is rich in deposits of **potash**, a type of mineral salt. Potash and other minerals support a growing chemical industry in Israel. The Negev area is also a source of copper and **phosphate**, a mineral used in making fertilizer.

Agriculture The Israelis have drained swamps, built irrigation systems, and used fertilizers to improve their soil. Israel's major farming area, which produces citrus fruits such as oranges, grapefruits, and lemons, is the coastal plain. Citrus fruits are Israel's chief farm export. Farming also occurs in parts of the Negev but only with the help of irrigation.

Place

The People

A single Israeli law—The Law of Return—increased the country's population more than any other factor. Passed in 1950, the law states that settlement in Israel is open to all Jews from anywhere in the world. Today Israel has about 6 million people. About 80 percent of Israel's population are Jews, but Israel's people also include Muslims and Christian Arabs. Most of Israel's Arabs are part of a group called the Palestinians. Both Israeli Jews and Arabs trace their roots to groups that lived in the area centuries ago.

Because of their country's industrial growth, about 90 percent of Israelis live in urban areas. The largest cities are Tel Aviv and the neighboring cities of Jerusalem, Yafo (yah•FOH), and Haifa (HY•fuh).

Jerusalem is a holy city for Jews, Christians, and Muslims, whose religions began in Southwest Asia centuries ago. All three groups practice **monotheism**, or a belief in one God.

Teen Scene

A Special Birthday

If you see a look of pride on Shamir Mazar's face, it comes from his many days of preparation for his bar mitzvah. A bar mitzvah is a special ceremony for Jewish boys when they turn 13. A similar ceremony called a bat mitzvah is held for girls at the same age. At this time young people are accepted as members of the Jewish community. Dressed in a traditional prayer shawl and cap, Shamir carried the scrolls of the Torah, the Jewish holy book. As part of the ceremony, he read a passage from the scrolls.

Jerusalem, Israel

The Muslim Dome of the Rock and the Jewish Western Wall are holy sites in Jerusalem's old city.
PLACE: Jerusalem is considered a holy city for followers of what three religions?

425

Early History Many Israeli Jews today come from such places as Europe, North Africa, and the Americas. They trace their origins, however, to the Hebrews who set up a kingdom with its capital at Jerusalem in about 1000 B.C. By 922 B.C. the Hebrew kingdom had split into two states—Israel and Judah. The people of Judah were known as Jews. Israel and Judah were eventually destroyed by invaders.

Over time, the area that is now Israel was ruled by a series of peoples—Greeks, Romans, Byzantines, Arabs, and Ottoman Turks. Under the Romans, Israel and Judah became the country of Palestine. In the late 1800s, the people in Palestine included Arabs and a small community of Jews. At that time many Jews from Europe moved to Palestine. These Jewish settlers, known as Zionists, wanted to set up a homeland for Jews.

Benjamin Netanyahu (right) became Israel's leader in 1996. He firmly defended Israeli interests but was willing to meet with Palestinian leader Yasir Arafat (left) and sign a historic agreement in 1998. **REGION: What happened to Israel and the Arab countries between 1948 and 1974?**

Modern Israel After World War I, Palestine came under British control. During World War II, millions of Jews in Europe were killed by German Nazis. This mass slaughter known as the **Holocaust** brought worldwide attention to the Zionist cause. In 1947 the United Nations voted to divide Palestine into a Jewish and an Arab state. In 1948 the Jews of Palestine proclaimed an independent republic called the State of Israel. The Arabs of Palestine, however, rejected this action.

A series of Arab-Israeli wars in which Israel gained Arab land were fought between 1948 and 1974. A 1978 treaty between Israel and Egypt was a first step toward peace. Agreements made in 1993 and 1994 between Israel and Palestinian Arab leaders and between Israel and Jordan also gave hope that peace may come to the region.

SECTION 2 ASSESSMENT

REVIEWING TERMS AND FACTS

1. **Define the following:** potash, phosphate, monotheism, Holocaust.

2. **LOCATION** Where is most of Israel's industry located?

3. **PLACE** What is Israel's major crop?

4. **MOVEMENT** Why do people around the world visit Jerusalem?

MAP STUDY ACTIVITIES

5. Turn to the map on page 424. What body of water borders Israel on the west?

6. Look at the same map on page 424. What climates are found in Israel?

IT'S TIME FOR DINNER

| MATH | SCIENCE | HISTORY | LITERATURE | TECHNOLOGY |

Your friend Bashir from Southwest Asia has invited you to dinner. Some of the foods you're served look familiar. The table holds dates and other fruit, a bowl of rice, and the pita bread your own mother buys to make sandwiches. Whether you are eating Israeli matzo bread or Lebanese pita bread, you are enjoying the foods of Southwest Asia.

A family meal in Jordan

ANCIENT ORIGINS The foods of this region date to ancient times. Olives—a key ingredient in the olive oil used in the region's foods—have always grown well in a Mediterranean climate. About 5,000 years ago the Phoenicians taught the Greeks to burn olive oil in their clay lamps. The Greeks quickly learned to cook with the oil. Date palms were also cultivated in the region as early as 400 B.C. Wheat was grown in the Fertile Crescent by the earliest known civilizations. Oranges and spices came to the region from India in the 1200s.

SHARED REGIONAL FOODS The foods common to all of Southwest Asia include dishes made from wheat, fruits and vegetables, and flavorful spices. Dishes prepared with eggplant, chickpeas and other legumes, celery, tomatoes, onions, chicken, and lamb appear in most countries. A generous dash of saffron, cinnamon, cardamom, cumin, or cilantro adds flavor. Fruits such as dates, raisins, lemons, figs, pomegranates, and apricots are also popular, especially as desserts.

Lack of refrigeration in some places led people to develop a taste for yogurt—fermented camel's milk—rather than fresh milk products. In your friend's home you might be offered fruit juices or tea to end your meal.

Making the Connection

1. What are some of the oldest foods grown in the Southwest Asian region?
2. Of the foods described here, which have you tasted? How did you like them?

Syria, Lebanon, and Jordan

Words to Know
- Bedouin
- civil war
- constitutional monarchy

Places to Locate
- Syria
- Lebanon
- Syrian Desert
- Damascus
- Beirut
- Jordan
- Jordan River
- Amman

Read to Learn . . .
1. what landforms make up Syria.
2. why Lebanon was torn by a fierce civil war.
3. how Jordan is governed.

From the sight of modern buildings in busy Amman, Jordan, you might not guess that the city is more than 3,500 years old. Today Amman is the capital and major economic center of Jordan, a small country that lies east of Israel in Southwest Asia.

Syria, Lebanon, and Jordan share a stretch of fertile land in the middle of Southwest Asia's mountains and deserts. This area is called the Fertile Crescent, because it is shaped like a half-moon. The Fertile Crescent runs from the Mediterranean coast in the west to the Persian Gulf in the east.

Place

Syria

Syria lies at the eastern end of the Mediterranean Sea and has been a center of commerce and trade for centuries. Throughout its early history, Syria was a part of many empires, but in 1946 it became an independent republic.

The Land and Economy Syria's land includes fertile coastal plains and valleys along the Mediterranean Sea. Inland mountains running north and south keep moist sea winds from reaching the eastern part of Syria. The map on page 420 shows that the vast Syrian Desert covers this region.

Agriculture is Syria's major economic activity. Syrian farmers raise mostly cotton and wheat in the rich soil of mountain valleys and coastal plains. In

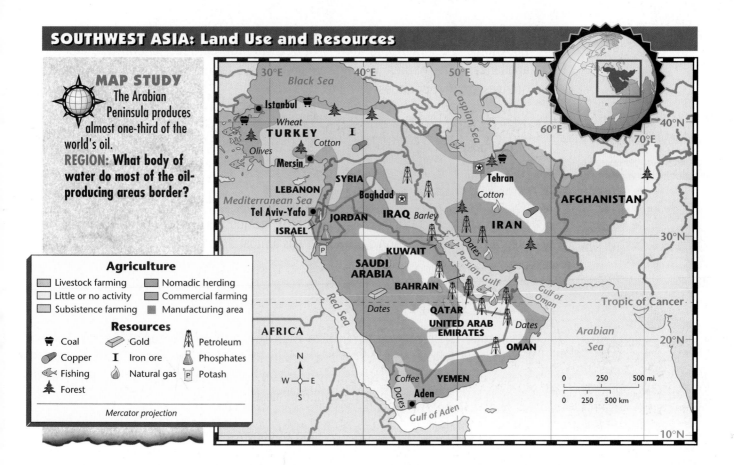

MAP STUDY
The Arabian Peninsula produces almost one-third of the world's oil.
REGION: What body of water do most of the oil-producing areas border?

Agriculture
- Livestock farming
- Little or no activity
- Subsistence farming
- Nomadic herding
- Commercial farming
- Manufacturing area

Resources
- Coal
- Copper
- Fishing
- Forest
- Gold
- Iron ore
- Natural gas
- Petroleum
- Phosphates
- Potash

Mercator projection

many dry areas, farmers rely on irrigation to grow crops. Syria's primary industry is textile making. Syrian cloth and fabrics have been highly valued since ancient times. The country also produces cement, chemicals, and glass.

The People About half of Syria's 15.6 million people live in rural areas, and a few are **Bedouins**, or people who move through the deserts. The word *bedouin* is Arabic for "desert dweller." The other half of the population live in cities. The city of Damascus, Syria's capital, is home to more than 2 million people. It was founded more than 5,000 years ago as a trading center.

Place
Lebanon

Lebanon, one of the smallest countries of Southwest Asia, lies west of Syria on the eastern coast of the Mediterranean Sea. Its land area is slightly smaller than the land area of the state of Connecticut.

The Land and Economy Lebanon consists of plains and mountains. Sandy beaches and a narrow plain run along the Mediterranean coast. Rugged mountains rise east of the plain and along Lebanon's eastern border. A fertile valley lies between the two mountain ranges. People living on the coast enjoy a mild, humid climate. Inland areas receive less rainfall and have lower winter and higher summer temperatures.

Service industries such as banking are Lebanon's major source of income. Lebanon's chief manufactured products include cement, chemicals, electric appliances, and textiles. If you were a Lebanese farmer, you might grow apples, oranges, grapes, potatoes, or olives.

The People Almost 70 percent of Lebanon's 4.1 million people are Muslims, while most of the rest are Christians. About 86 percent of Lebanon's people live in urban areas. Beirut (bay•ROOT), located on the coast, is the nation's capital and largest city.

From 1975 to 1991, groups of Muslims and Christians in Lebanon fought a **civil war,** or a conflict among different groups within a country. Many lives were lost, and the nation's economy was almost destroyed.

The stately cedar tree is Lebanon's national symbol and appears on its flag, coins, and stamps. LOCATION: How would you describe the landscape where these trees grow?

Place
Jordan

Jordan is slightly larger in size than the state of Kentucky. Before 1967 Jordan also held land west of the Jordan River. This area—the West Bank—came under Israeli control during the 1967 Arab-Israeli war. Today the West Bank is home to more than 1.8 million Palestinian Arabs.

The Land and Economy A land of contrasts, Jordan reaches from the fertile Jordan River valley in the west to dry, rugged country in the east. The northern and western parts of Jordan have mild, rainy winters and dry, hot summers. Southern and eastern parts of the country have a hot, dry climate year-round.

Most workers in Jordan are employed in agriculture and manufacturing. The leading manufactured goods are cement, chemicals, and petroleum products. Jordan's farmers in the Jordan River valley grow citrus fruits, tomatoes, and melons.

The People Most of Jordan's 4.6 million people are Arab Muslims and Christians. They include about 1 million Palestinian Arabs who fled Israel or the West Bank during the Arab-Israeli wars. Amman, with 1.2 million people, is the leading city and capital. Jordan is a **constitutional monarchy,** a form of government in which the monarch shares power with elected officials.

SECTION 3 ASSESSMENT

REVIEWING TERMS AND FACTS
1. Define the following: Bedouin, civil war, constitutional monarchy.
2. PLACE What city in Syria is considered one of the oldest cities in the world?
3. PLACE How has civil war affected Lebanon?

MAP STUDY ACTIVITIES
4. Look at the map on page 429. What is the primary economic activity of Syria?
5. Look at the same map on page 429. What countries border Lebanon?

UNIT 6

The Arabian Peninsula

PREVIEW

Words to Know
- oasis
- *hajj*

Places to Locate
- Saudi Arabia
- Kuwait
- Bahrain
- United Arab Emirates
- Qatar
- Yemen
- Oman
- Makkah
- Riyadh

Read to Learn . . .
1. why Saudi Arabia is important to the world's Muslims.
2. how oil affects the lives of the people of the Arabian Peninsula.

The hot desert sun rises and sets over the Arabian Peninsula. Under these windswept dunes, oil constantly flows through the country's network of pipelines.

The Arabian Peninsula is like a giant platform that tilts toward the east and the north. Its highest elevations are in the south. The mostly desert land in the north slopes toward flat, sandy beaches along the Persian Gulf. The country of Saudi Arabia takes up about 80 percent of the Peninsula.

Place

Saudi Arabia

Saudi Arabia, the largest country in Southwest Asia, is about the size of the eastern half of the United States. Saudi Arabia holds a major share of the world's oil. The graph on page 432 compares the amount of oil that Southwest Asia pumps to that of other oil-producing nations.

The Land Vast deserts cover Saudi Arabia. The largest desert is the Empty Quarter in the southern part of the country. The Empty Quarter has mountains of sand that reach heights of more than 1,000 feet (305 m). In the west, highlands stretch along the Red Sea. Valleys among these highlands provide fertile farmland. Eastern Saudi Arabia is a coastal plain along the Persian Gulf.

GRAPHIC STUDY

Southwest Asia has more oil than all other regions of the world combined.
REGION: What is the total percentage of oil found in all other regions of the world?

Percentages of World Oil Reserves Found in Each Region

Southwest Asia — 66.9%

Latin America — 12.1%

Europe* — 7.4%

Africa — 6.1%

Asia and Oceania — 4.4%

U.S. and Canada — 3.1%

Source: *The World Almanac*, 1998

*including Russia and the independent republics

The Economy The economy of Saudi Arabia is based on oil. Money from oil exports has built schools, hospitals, roads, and airports. Agriculture is less important than oil because of the limited amount of water and farmland. To meet the need for freshwater, the country's engineers have built industrial plants to remove salt from the seawater of the Persian Gulf and the Red Sea.

The People The people of Saudi Arabia once were divided into numerous family groups. In 1932 a monarchy led by the Saud family unified the country. The Saud family still rules Saudi Arabia today.

Most of the 20.2 million Saudi Arabians live in towns and villages. These are found along the oil-rich Persian Gulf coast or around **oases**—lush, green places in dry areas that have enough water for crops to grow. Some of the oases have grown into large cities such as Riyadh (ree•YAHD).

The Islamic religion affects almost all parts of life in Saudi Arabia—from government to the everyday lives of the people. The holy city of Makkah (MAH•kah) was very important in the life of Muhammad (moh•HA•muhd), the man who brought Islam to the Arabs in the A.D. 600s. Today hundreds of thousands of Muslims from around the world visit Makkah each year. Each Muslim tries to make a *hajj,* or religious journey to Makkah, at least once.

Saudi Women at Work

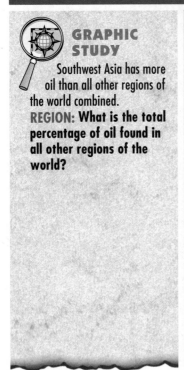

A female doctor in Saudi Arabia examines her patient.
PLACE: Income from what resource has helped build Saudi Arabia's hospitals?

Place
Other Arabian Peninsula Nations

Saudi Arabia shares the Arabian Peninsula with six other Arab Muslim countries. They are Kuwait (ku•WAYT), Bahrain (bah•RAYN), Qatar (KAH•tuhr), the United Arab Emirates, Oman (oh•MAHN), and Yemen (YEH•muhn).

Persian Gulf States Kuwait, Bahrain, Qatar, and the United Arab Emirates are located along the Persian Gulf. Beneath their flat deserts and offshore areas lie vast deposits of oil. The Persian Gulf states have used oil profits to build prosperous economies. Their people—for the most part—enjoy a high standard of living with excellent health care, education, and housing. Many workers from other countries have settled in the Persian Gulf countries to work in the modern cities and oil fields.

The Kaaba, the most sacred Islamic shrine, is in the courtyard of Makkah's Grand Mosque.
PLACE: Who brought the Islamic religion to the Arabs?

Oman and Yemen At the southern end of the Arabian Peninsula are the countries of Oman and Yemen. Oman is largely desert, but its bare land yields oil—the basis of the country's economy. Most of the people of Oman live in rural villages. The oil industry, however, has drawn many rural dwellers as well as foreigners to Oman's growing urban areas.

The country of Yemen is made up of a narrow coastal plain and inland mountains. It has almost no natural mineral resources, and most of the people are farmers or herders of sheep and cattle. They live in the high fertile interior where the capital, San'a (sa•NAH), is located. Farther south lies Aden (AH•duhn), a major port for ships traveling between the Arabian and Red seas.

SECTION 4 ASSESSMENT

REVIEWING TERMS AND FACTS
1. Define the following: oasis, *hajj.*
2. PLACE What country takes up most of the Arabian Peninsula?
3. LOCATION Where is the Empty Quarter located?
4. MOVEMENT Why is the port of Aden important to Yemen?

GRAPHIC STUDY ACTIVITIES
5. Turn to the graph on page 432. What region comes closest to matching Southwest Asia's oil production?
6. Look at the graph on page 432. What two regions together produce about as much oil as Europe does?

Iraq, Iran, and Afghanistan

PREVIEW

Words to Know
- alluvial plain
- shah
- ethnic group

Places to Locate
- Iraq
- Baghdad
- Iran
- Caspian Sea
- Tehran
- Afghanistan
- Hindu Kush Mountains
- Kabul

Read to Learn . . .
1. why the Tigris and Euphrates rivers are important to the people of Iraq.
2. how a religious government came to power in Iran.

Poems, stories, and songs describe the ancient beauty of Baghdad—the heart of Iraq. Today Baghdad boasts modern buildings as well as old neighborhoods and busy marketplaces.

Baghdad, with a population of about 4.3 million, is one of the largest cities in Southwest Asia. It is the capital of Iraq, a country in the northern part of Southwest Asia. The other countries in this region are Iran and Afghanistan.

Location

Iraq

Iraq lies at the head of the Persian Gulf and north of the Arabian Peninsula. Some of the world's oldest civilizations developed there between the Tigris and Euphrates rivers in an area called Mesopotamia. As a nation, however, modern Iraq has been in existence only since the early 1900s. Today it is one of the important oil-producing countries of Southwest Asia.

The Land The Tigris and Euphrates rivers are the major geographic features of Iraq. They flow through Iraq's northern highlands and central plain before joining to enter the Persian Gulf. West of the Tigris-Euphrates area, the landscape is mostly desert. What types of climates can you find in Iraq on the map on page 424?

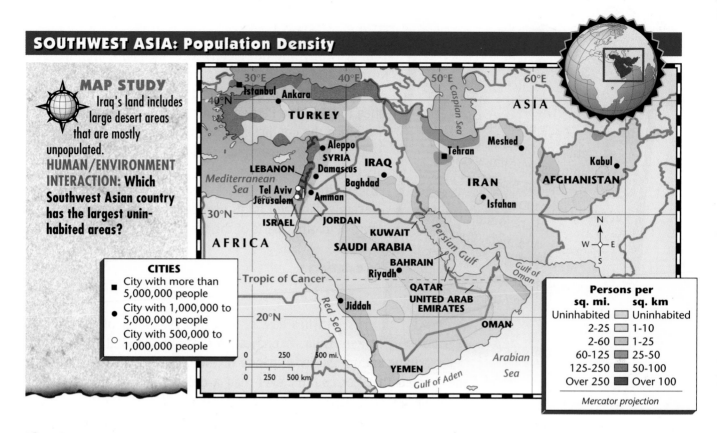

MAP STUDY
Iraq's land includes large desert areas that are mostly unpopulated.
HUMAN/ENVIRONMENT INTERACTION: Which Southwest Asian country has the largest uninhabited areas?

CITIES
- ■ City with more than 5,000,000 people
- ● City with 1,000,000 to 5,000,000 people
- ○ City with 500,000 to 1,000,000 people

Persons per	
sq. mi.	**sq. km**
Uninhabited	Uninhabited
2-25	1-10
2-60	1-25
60-125	25-50
125-250	50-100
Over 250	Over 100

Mercator projection

The Economy Iraq's economy is based on oil and agriculture as you can see on the map on page 429. Oil is the country's major export. Manufacturing also has developed in recent years, with Iraqi factories making food products, cloth, soap, and leather goods.

Most farming in Iraq takes place near the Tigris and Euphrates rivers. Find this area on the map on page 420. The plain between the rivers is called an **alluvial plain** because it is built up by rich fertile soil left by river floods. Iraq's farmers grow crops such as barley, dates, grapes, melons, citrus fruits, and wheat.

The People The land lined with date palms along the Tigris and Euphrates rivers is home to most of Iraq's 21.8 million people. About 70 percent live in urban areas such as Baghdad. Muslim Arabs make up the largest group in Iraq's population. A Muslim people, Kurds, are the second-largest group. The Kurds have their own culture and live in the mountains of northeast Iraq. Many Kurds in Iraq and in neighboring Turkey want to form their own independent country.

Past and Present Throughout the centuries, many empires have risen and fallen on the Tigris-Euphrates plain. From the A.D. 700s to 1200s, Baghdad was the center of a large Muslim empire that made many advances in the arts and sciences. Modern Iraq, with its wealth in oil, seeks to become a leading power in Southwest Asia. Over the past few years, Iraq has fought wars with some of its neighbors to gain more territory.

In 1990 Iraqi armies invaded Kuwait, Iraq's southern neighbor. In early 1991 a United Nations force led by the United States launched air and missile attacks on Iraq and its forces in Kuwait. The war against Iraq became known as the Persian Gulf War. Iraqi forces were forced to withdraw from Kuwait, but not before they set afire hundreds of Kuwait's oil wells. Oil spilling from the wells created huge, explosive lakes of petroleum and fires and smoke that threatened much of the Persian Gulf's plant and animal life.

Since the conflict, Iraq's people have faced many hardships. Iraq's government continues to threaten to invade its neighbors. In response many foreign nations, including the United States, refuse to trade with Iraq. This has severely damaged Iraq's economy, which depends on selling oil to other parts of the world.

Location
Iran

Iran lies east of Iraq and northeast of the Arabian Peninsula. Zapha, a 14-year-old schoolgirl in modern Iran, knows she lives in one of the world's oldest countries. In school she learned that her country even had a different name at one time—Persia. What has not changed is the land of Iran itself.

The Land Iran is about the size of the state of Alaska. A central, dry plateau stretches across most of Iran. Two vast mountain ranges—the Elburz and the Zagros—surround most of this plateau. Coastal plains lie outside the mountains along the shore of the Caspian Sea in the north and along the Persian Gulf coast in the south.

Iran has a variety of climates. Mountainous areas experience severely cold winters and mild summers. Most of the central plateau has a dry climate, with cold winter temperatures and mild to hot summers.

A Science Lesson in an Iranian School

Iranian students and their teacher work in an all-male science class.
PLACE: How is this science classroom similar to science classrooms in the United States?

The Economy Iran is one of the world's leading oil-producing countries. A major part of its income results from the sale of oil to foreign countries. In recent years, however, the falling price of oil on world markets has brought less money into Iran. Other Iranian industries produce a variety of goods, including leather goods, machine tools, and tobacco products.

As in much of Southwest Asia, water is scarce in Iran. Less than 12 percent of Iran's land can be used for farming. In dry areas, you see Iranian farmers bringing water to their fields from wells or through underground tunnels. Iran's chief agricultural products are wheat, barley, corn, cotton, dates, nuts, tea, and tobacco.

Tehran, Iran

Iran's capital and largest city, Tehran, lies at the base of the majestic Elburz Mountains.
MOVEMENT: Why are many Iranians moving from rural areas to the cities?

The People Today about 64.1 million people live in Iran, most of whom speak Persian, or Farsi, a language that is different from Arabic. About 98 percent of Iranians are Muslims who belong to a group within Islam known as Shiites (SHEE•YTES). Most of the world's Muslims belong to another group known as Sunnis (SU•NEES).

About 61 percent of Iran's people live in urban areas. In recent years towns and cities have grown tremendously because many people from rural areas have moved to urban areas in search of jobs. Find Tehran, Iran's capital and largest city, on the map on page 435. It has a population of about 11.7 million.

Iran was part of the powerful Persian Empire about 3,500 years ago. The descendants of ancient Persians make up the majority of Iran's population today. For hundreds of years Iran was ruled by kings known as **shahs**. Mohammed Reza Pahlavi (PA•luh•vee) became shah during World War II. In the 1960s and 1970s, he used money from oil sales to promote industry and to build Iran's military forces. The shah, however, was accused of ruling harshly, and many Iranians turned against him. In 1979 a revolution forced the shah out of Iran. In place of the shah's government, the Iranians set up a government run by religious leaders. Iran's Muslim government has placed strict laws on the country, and many non-Muslim customs are now forbidden.

Place
Afghanistan

Afghanistan is the easternmost country in Southwest Asia. It is a land of rugged landscapes and has no seacoast. Because of these geographic features, Afghanistan has had difficulty developing its resources.

The Land The Hindu Kush range covers most of Afghanistan. It has peaks as high as 25,000 feet (7,620 m). The Khyber (KY•buhr) Pass, which cuts through the Hindu Kush, for centuries has been used as a major trade route linking Southwest Asia with other parts of Asia. Rolling grasslands and low deserts make up the nonmountainous areas of Afghanistan. Highland and desert climates prevail in most of the country.

The Economy Afghanistan's major economic activity is farming. You see such crops as wheat, barley, corn, cotton, nuts, fruits, and vegetables growing there. Afghanistan is rich in minerals, but it has little industry. A few mills produce textiles, and skilled craftspeople produce beautiful rugs and jewelry in their homes.

The People The 24.8 million people of Afghanistan are divided into about 20 different **ethnic groups,** with each group having its own ancestry, language, and culture. The two largest ethnic groups are the Pushtuns (PUHSH•TOONZ) and the Tajiks (tah•JIHKS). Afghanistan's people practice Islam and live in the fertile valleys of the Hindu Kush range. The capital city, Kabul, is located in one of these valleys.

Afghanistan has had a long and troubled history. From ancient times to the present, a series of foreign powers have won or tried to win control of the country. The country also has been torn apart by continuing civil wars among its many ethnic groups.

SECTION 5 ASSESSMENT

REVIEWING TERMS AND FACTS

1. Define the following: alluvial plain, shah, ethnic group.

2. HUMAN/ENVIRONMENT INTERACTION How did the Persian Gulf War affect the environment of the Persian Gulf area?

3. PLACE What type of government rules Iran today?

4. PLACE What is the major landform of Afghanistan?

MAP STUDY ACTIVITIES

5. Using the map on page 435, name the large body of water that lies southeast of Iraq and south of Iran.

6. Look at the map on page 435. Around what body of water do most Iranians live?

Chapter 16 Highlights

Important Things to Know About Southwest Asia

SECTION 1 TURKEY

- Turkey is located in both Europe and Asia.
- Highlands and plateaus make up most of Turkey's land area.
- Most of Turkey's people live in cities and towns.

SECTION 2 ISRAEL

- Israel is the only Jewish nation in the world.
- Despite its small size, Israel has a well-developed economy.
- Israel and its Arab neighbors are moving toward peace after a long period of tension and warfare.

SECTION 3 SYRIA, LEBANON, AND JORDAN

- Syria has a coastal plain, mountains, and deserts.
- Lebanon is one of the smallest countries of Southwest Asia.
- Jordan's farmers grow crops in the rich Jordan River valley.

SECTION 4 THE ARABIAN PENINSULA

- Oil-rich Saudi Arabia is the largest country of Southwest Asia.
- Most of Saudi Arabia's people live in coastal areas.
- Persian Gulf countries have strong economies based on oil.

SECTION 5 IRAQ, IRAN, AND AFGHANISTAN

- Iraq has recently fought wars to increase its power in Southwest Asia.
- Iran is governed by Muslim religious leaders.
- Afghanistan's land area is largely mountainous.

A market in San'a, Yemen ▶

REVIEWING KEY TERMS

Match the numbered terms in Set A with their lettered definitions in Set B.

A

1. potash
2. Bedouin
3. *hajj*
4. shah
5. monotheism
6. oasis
7. alluvial plain

B

A. area of fertile soil left by river floods
B. mineral used in making fertilizer
C. Muslim pilgrimage to Makkah
D. area in dry places that has enough water to grow crops
E. Iranian king
F. nomadic desert people
G. belief in one God

Mental Mapping Activity

Draw a freehand map of the Arabian Peninsula. Label the following:
• Persian Gulf
• Kuwait
• Saudi Arabia
• Yemen
• Red Sea

REVIEWING THE MAIN IDEAS

Section 1

1. LOCATION In what part of Turkey is the Anatolian Plateau?
2. PLACE What is a major industrial product of Turkey?

Section 2

3. LOCATION What river flows into the Dead Sea?
4. HUMAN/ENVIRONMENT INTERACTION How have Israelis made their land suitable for farming?

Section 3

5. PLACE What kind of climate does Lebanon have?
6. LOCATION Where is Jordan's major farming region?

Section 4

7. PLACE What is the major industry of Saudi Arabia?
8. MOVEMENT What has drawn foreigners to Oman's cities?

Section 5

9. PLACE What major rivers flow through Iraq?
10. LOCATION What Southwest Asian country is located east of Iran?

CRITICAL THINKING ACTIVITIES

1. **Making Inferences** From what you know about a desert climate, what could you expect to be true about farming in the Negev region of Israel?
2. **Determining Cause and Effect** How has the discovery of oil changed the lives of people living on the Arabian Peninsula?

CONNECTIONS TO ECONOMICS

Cooperative Learning Activity Working in a small group, select a country of Southwest Asia and research how the people have adapted their environment to help develop their country's economy. Each person in your group can research a different part of the economy and then share findings with the group. Prepare charts, storyboards, and posters that will illustrate your presentation to the rest of the class.

GeoJournal Writing Activity

In this chapter on Southwest Asia, you were introduced to several important cities. Review the chapter to see how these cities are similar and how they are different. Write a short story or letter describing a trip to two of these cities.

TECHNOLOGY ACTIVITY

Using the Internet Search the Internet for an on-line newspaper with current articles about Southwest Asia. Find a recent article about a Southwest Asian country that relates to its geography. Write a brief summary of the article, noting the source of the information. Then write a statement about how the issue described in the article might affect life in your community.

PLACE LOCATION ACTIVITY: SOUTHWEST ASIA

Match the letters on the map with the places in Southwest Asia. Write your answers on a separate sheet of paper.

1. Zagros Mountains
2. Arabian Sea
3. Istanbul
4. Persian Gulf
5. Afghanistan
6. Tel Aviv
7. Euphrates River
8. Makkah
9. Tehran
10. Saudi Arabia

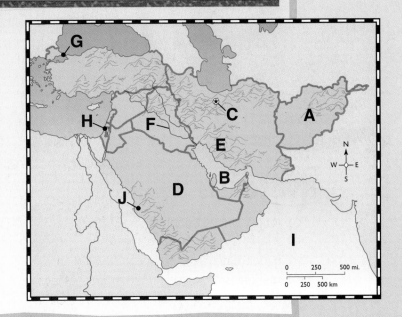

GeoLab ACTIVITY

From the classroom of Eleanor Bloom, Bloom-Carroll Schools, Lithopolis, Ohio

SUNDIALS: Shadow Time

Background

The Chinese used water clocks to measure time more than 3,000 years ago. Egyptians also made water clocks, in which water spilled from one vessel to another. Early civilizations in Southwest Asia and North Africa measured time, too, but instead of water they used the sun as their timekeeper. Sundials measure the shadows an object casts as the sun moves from east to west during the day. Find out more about how these early civilizations measured time by making your own sundial.

Modern Sundial

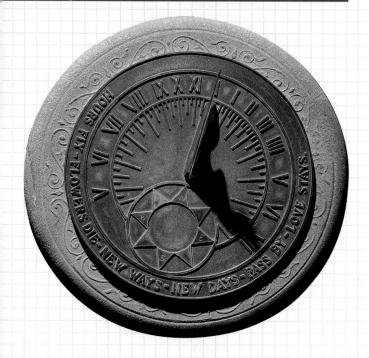

In the 1300s Henry de Vick invented a clock that contained many of the parts that modern clocks have. Atomic clocks are the world's most accurate timepieces today, gaining or losing only a few seconds every 100,000 years!

A white or light-colored paper plate makes the best base for your sundial.

Materials

- 1 empty thread spool
- 1 ruler
- 1 paper plate
- 1 marker
- glue
- 1 new, sharpened pencil

What To Do

A. Glue the spool to the center of the paper plate.
B. Stand the pencil, with the point up, in the top hole of the spool.
C. Draw an arrow on the paper plate.
D. Place your sundial in a sunny window where it gets sunlight all day long. Make sure the arrow is pointing south.
E. Every hour, trace a line along the shadow's line. Use pictograms or Roman numerals to mark the hours.

Lab Activity Report

1. Are the shadow lines you made an equal distance apart?

2. As you made your shadow lines, did the same thing happen every hour?

3. What are the advantages and disadvantages of using a sundial to measure time?

4. **Drawing Conclusions**
   Does the space between the shadow lines have anything to do with telling time? Explain.

Go A Step Further!

Activity
Invent a new way to tell time. Think of using everyday items and materials that you usually do not associate with telling time. Write a news report explaining your new invention.

Cultural HERITAGE:
SOUTHWEST ASIA AND NORTH AFRICA

CALLIGRAPHY ▶ ▶ ▶ ▶

A page from the Quran, the Islamic holy book, shows the flowing curves of Arabic script called calligraphy. Artists and architects use calligraphy to decorate the walls of buildings, metalware, and pottery.

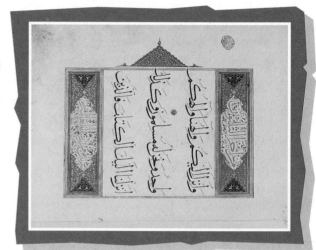

◀ ◀ ◀ ◀ RELIGIOUS ART

Religion is a common subject in this region's artwork. This stained glass window in Israel shows the 12 original Hebrew tribes. The window was created by world-famous painter Marc Chagall, a Russian Jew born in 1887.

ARTIFACTS ▶ ▶ ▶ ▶

The burial tomb of Tutankhamen, discovered in Egypt in 1922, was filled with jewelry, statues, clothing, and even toys. This funeral mask is pure gold and inlaid with semiprecious stones.

A blue-domed mosque in Isfahan, Iran, shows typical calligraphy and geometric patterns. Most mosques have a *mihrab*—an arch or other structure that marks the wall closest to the holy city of Makkah.

MUSIC ▲ ▲ ▲ ▲

An Egyptian musician draws a bow across this instrument called a *kamanja.* The music of North Africa and Southwest Asia is most often played on stringed instruments, flutes, drums, and tambourines.

JEWELRY ▲ ▲ ▲ ▲

This elaborate head covering is called a *gargush.* It is made by knitting tiny silver wires together with multicolored beads. Some Jewish women of Yemen still wear this traditional headdress.

APPRECIATING CULTURE

1. What does the mask of Tutankhamen tell us about the king and the time in which he lived?

2. In what ways is calligraphy used as art in our country?

Chapter 17 North Africa

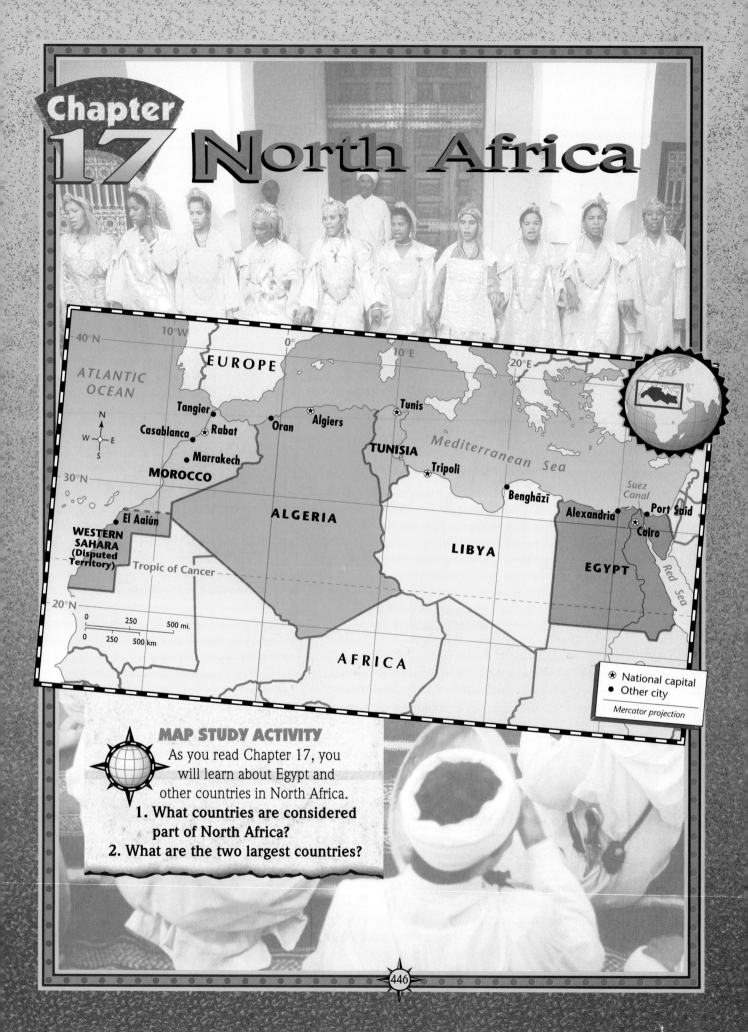

MAP STUDY ACTIVITY

As you read Chapter 17, you will learn about Egypt and other countries in North Africa.

1. What countries are considered part of North Africa?
2. What are the two largest countries?

Egypt

PREVIEW

Words to Know
- silt
- delta
- hydroelectric power
- consumer goods
- *fellahin*
- bazaar
- hieroglyph

Places to Locate
- Nile River
- Sinai Peninsula
- Suez Canal
- Libyan Desert
- Arabian Desert
- Cairo
- Alexandria

Read to Learn . . .
1. why the Nile River is important to Egypt's people.
2. how the Aswan High Dam has affected Egypt's environment.

You live along the lush, green banks of the longest river in the world. Where are you living? You live along the Nile River in the North African country of Egypt.

For centuries the people of Egypt have called their country the "gift of the Nile." The river gives them water, fertile soil, and transportation. Egypt receives little rain, and almost all of it is desert. Deserts also cover most of the lands of Egypt's neighbors in North Africa—Libya, Tunisia, Algeria, and Morocco. Egypt has the most people of any country in the region. Look at the graph on page 451 to see how Egypt's population compares with the population of the United States.

Place
The Land

Egypt has three major land areas. They are the Nile River valley, the Sinai (SY•NY) Peninsula, and desert areas. Most of Egypt has a desert climate, with hot summers and mild winters.

Nile River Valley The lifeline of Egypt is the Nile River, which supplies 85 percent of the country's water. The Nile River begins in east Africa. It flows north through the countries of Sudan and Egypt to the Mediterranean Sea. The

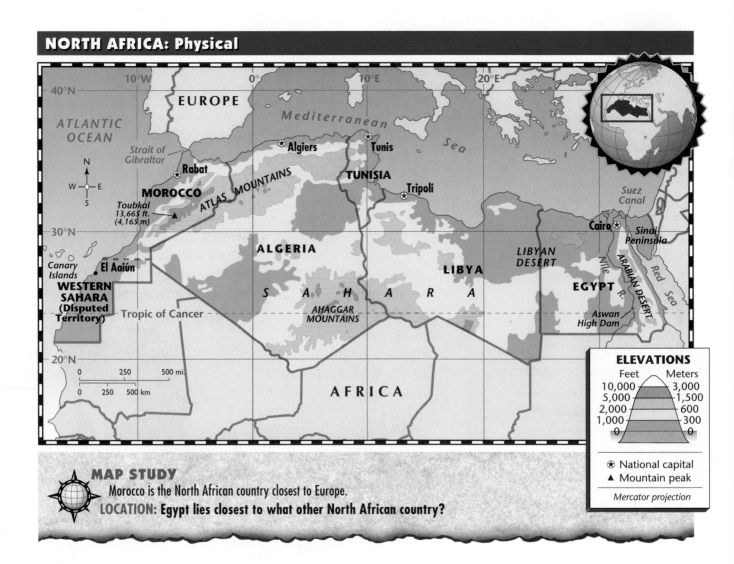

MAP STUDY
Morocco is the North African country closest to Europe.

LOCATION: Egypt lies closest to what other North African country?

Nile River's length of 4,180 miles (6,690 km) far surpasses the second-longest river in the world—the Amazon—which is 3,912 miles long (6,296 km).

Imagine you are looking down on Egypt from an airplane flying from south to north. Below you the Nile River and its banks appear as a narrow, green ribbon cutting through the vast desert.

Flood Control The Nile River valley has rich soils formed by **silt,** or particles of earth deposited by the river. In ancient times, floodwaters left heavy deposits of silt every summer. Today dams control the river's flow, and flooding no longer occurs regularly.

As you continue your flight, you can see the Nile River spread out into a broad wedge of farmland about 150 miles (240 km) from the Mediterranean Sea. You are viewing a **delta,** or triangle-shaped area of land at a river's mouth.

Sinai Peninsula The Sinai Peninsula lies southeast of the Nile Delta. This section of Egypt is not in Africa but is actually part of Southwest Asia. If you look at the map above, you can see why this area is a major crossroads between these two continents.

A human-made waterway called the Suez Canal separates the Sinai Peninsula from the rest of Egypt. Egyptians and Europeans built this important canal in the mid-1860s to allow ships to pass between the Red and the Mediterranean seas.

Deserts From the airplane you can see that most of Egypt is desert. West of the Nile River lies the Libyan (LIH•bee•uhn) Desert. It covers two-thirds of Egypt's land area. Oases are found throughout the Libyan Desert. East of the Nile River and west of the Red Sea is the Arabian Desert.

The Libyan and Arabian deserts are parts of the Sahara—one of the largest deserts in the world. *Sahara* is from the Arabic word meaning "desert." The Sahara—about the size of the United States—runs from Egypt westward through North Africa to the Atlantic Ocean.

Human/Environment Interaction
The Economy

Egypt has a developing economy that has grown considerably in recent years. Since the 1950s government and business leaders have started many new industries. Agriculture, however, remains Egypt's main economic activity.

Agriculture Only about 4 percent of Egypt's land is used for farming. The map on page 450 shows that Egypt's best farmland lies in the fertile Nile River valley. Most Egyptian farmers tend small plots and grow cotton, dates, vegetables, sugarcane, and wheat. They use irrigation, modern tools, and farming methods developed by their ancestors. Cotton is Egypt's leading agricultural export.

Aswan High Dam Until the 1960s Egyptian farmers could plant only once a year—after the Nile River finished flooding in the summer. In 1968 engineers completed the construction of the Aswan High Dam. It holds back the Nile's floodwater. Now Egyptian farmers plant crops two or three times a year.

The Aswan High Dam brings problems as well as blessings. Why? It now stops much of the rich silt flow that fertilized Egypt's fields each year. Egypt's farmers must rely on expensive artificial fertilizers. The dam also restricts the flow of water into the river valley and the Mediterranean Sea. Salty seawater now moves farther inland into the delta.

Industry The Aswan High Dam provides **hydroelectric power,** or electric energy from moving water, to Egypt's growing industries. The largest industrial centers are the capital city of Cairo and the seaport of Alexandria. Egyptian factories there produce mainly food products, textiles, and **consumer goods,** or household goods, clothing, and shoes that people buy.

Harvesting Cotton in Egypt

An Egyptian woman gathers cotton, Egypt's most valuable crop.
LOCATION: Where do you think most of Egypt's cotton fields are located?

Egyptian Music

Can you imagine keeping an orchestra in your home? In ancient Egypt, some wealthy households had their own household orchestras. Ancient Egyptians also put musical instruments in tombs. They believed that this would guarantee them entertainment in their lives after death.

Oil ranks as the country's most important natural resource and a major export. The map below shows that many of Egypt's oil wells are in and around the Red Sea. Another growing industry is tourism.

Place
The People

Most Egyptians live within 20 miles (33 km) of the Nile River. This means that 99 percent of Egypt's population live on only 3.5 percent of the land. Look at the population density map on page 457. You can see that the Nile River valley is one of the most densely populated areas in the Southwest Asia/North Africa region. It is also one of the most densely populated areas in the world.

Rural Life About 43 percent of Egypt's people live in rural areas. Most rural Egyptians are farmers called **fellahin** (FEHL•uh•HEEN). They live in villages and work on small plots of land rented from landowners. Many *fellahin* raise only

NORTH AFRICA: Land Use and Resources

Agriculture

☐ Livestock farming	☐ Nomadic herding
☐ Little or no activity	☐ Commercial farming
☐ Manufacturing areas	

Resources

Coal	Lead	Phosphates
Fishing	Natural gas	Salt
I Iron ore	Petroleum	Zinc

Mercator projection

MAP STUDY

Phosphates and iron ore are mined in the mountains of Morocco and Algeria.

PLACE: What are the major mineral resources in Egypt?

UNIT 6

GRAPHIC STUDY

Most people in Egypt live along the Nile River and its delta.
PLACE: What is the approximate population of Egypt?

Egypt

United States

= 50,000,000

Source: *Population Reference Bureau, 1998*

enough crops to feed their families. Any food left over is sold in towns at a **bazaar,** or marketplace.

City Life Egypt's large cities offer a different kind of life. Skilled and unskilled workers make up the largest groups of Egyptians in cities. Cairo is a huge and rapidly growing city, with more than 11 million people. It is the largest city in Egypt as well as in Africa. Cairo's mosques, schools, and universities make it a leading center of the Muslim world.

Why is Cairo's population increasing? First, Egypt is a country with a high birthrate. Second, many *fellahin* from rural areas have moved into Cairo to find work. As Cairo's population grows, the city cannot provide enough living space for all of its people. If you were a student in one of Cairo's schools, you might attend crowded classes on a split shift.

Influences of the Past What do you think of when you hear the word "Egypt"? Pyramids? Pharaohs and kings? Egypt has a long and fascinating history. One of the ancient world's most advanced civilizations developed along the Nile River. Ancient Egyptians used **hieroglyphs,** or picture symbols, for writing. One of the first civilizations to make paper from the reedlike papyrus plant, the Egyptians also created a calendar to keep track of the growing season. They became skilled in building such lasting monuments as the Great Sphinx and the Pyramids.

In later centuries, Egypt was ruled by outsiders—Romans, Persians, Arabs, Turks, and the British. Under Arab rule, Islam—its followers are called Muslims—and the Arabic language came to Egypt. Both Islam and Arabic have been an important part of Egyptian life ever since. Today almost 90 percent of Egyptians are Muslims and about 10 percent are Christians.

Government Egypt became a republic in 1953. Gamal Abdel Nasser was president of Egypt from 1954 to 1970. Under Nasser, Egypt became the most powerful country in the Arab world. Nasser, supported by other Arab countries,

Wide avenues wind through modern Cairo.
PLACE: Why is Cairo so densely populated?

opposed Israel. He also angered many Europeans by taking control of the British-run Suez Canal, which was vital to European trade.

To regain control of the canal, Britain, France, and Israel invaded Egypt in 1956. The United Nations persuaded the Europeans and the Israelis to withdraw their troops. Tensions between the countries remained, however. In the 1970s Egypt and Israel decided to settle their differences. In recent years, Egypt has also strengthened its ties with the United States.

SECTION 1 ASSESSMENT

REVIEWING TERMS AND FACTS

1. **Define the following:** silt, delta, hydroelectric power, consumer goods, *fellahin,* bazaar, hieroglyph.
2. **PLACE** Why is Egypt called "the gift of the Nile"?
3. **HUMAN/ENVIRONMENT INTERACTION** How has the Aswan High Dam affected the lives of Egyptians?
4. **LOCATION** Where is Cairo located?

MAP STUDY ACTIVITIES

5. Turn to the physical map of North Africa on page 448. Where is the Aswan High Dam located?
6. Look at the political map of North Africa on page 446. (a) What direction is Alexandria from Cairo? (b) About how many miles (km) separate the two cities?

BUILDING GEOGRAPHY SKILLS

Using Technology

Using an Electronic Spreadsheet

A spreadsheet is an electronic worksheet that can manage numbers quickly and easily. Spreadsheets are powerful tools because you can change or update information and the spreadsheet automatically performs the calculations.

All spreadsheets follow a basic design of rows and columns. Each column is assigned a letter and each row is assigned a number. Each point where a column and a row intersect is called a *cell*. The cell's position on the spreadsheet is labeled according to its corresponding column and row—*A1* in column A, row 1; *B2* in column B, row 2, and so on. The formula (C1+C2) tells the spreadsheet to add the numbers in cells C1 and C2.

Spreadsheets are important tools for geographers. For example, a geographer might want to compare how much oil North African countries have produced over time.

Use these steps to create a spreadsheet that will provide this information.

1. In cells B1 and C1, type the years 1980 and 1994. In cell D1, type the word *total*.
2. In cells A2–A6, type the names of North Africa's countries. In cell A7, type the word *total*.
3. In row 2, enter the number of metric tons of oil produced in 1980 and 1994 by the country named in cell A2.
4. Repeat step 3 in rows 3–6. You can find the information you need for each country in a world almanac or encyclopedia.

	A	B	C	D
1		1980	1994	total
2	Algeria			
3	Egypt			
4	Libya			
5	Morocco			
6	Tunisia			
7	total			

5. Create a formula that tells which cells (B2+C2) to add together so the computer can calculate the number of tons of oil for that country.
6. Copy the formula down in the cells for the other four countries.
7. Use the process in steps 5 and 6 to create and copy a formula to calculate the total amount of oil produced each year.

Technology Skills Practice

For additional practice in using an electronic spreadsheet, follow the steps listed above to create a spreadsheet that tracks the number of students in your school over time. Find out the number of students in each grade in your school in the years 1995 and 2000 and enter these numbers in your spreadsheet. After you have completed the spreadsheet, write a conclusion about how the numbers changed between 1995 and 2000.

Other North African Countries

PREVIEW

Words to Know
- dictatorship
- erg
- *casbah*

Places to Locate
- Sahara
- Libya
- Tunisia
- Atlas Mountains
- Maghreb
- Algeria
- Morocco
- Casablanca

Read to Learn . . .
1. what natural resources are found in North Africa.
2. why most people in North Africa live in coastal areas.
3. how urban and rural life differ in North Africa.

The busy sounds of this modern oil-refining plant echo across the Sahara in the country of Libya. You probably think the Sahara is nothing but miles and miles of shifting sand. The Sahara covers major oil reserves.

Like Egypt, the other countries of North Africa—Libya, Tunisia, Algeria, and Morocco—have established economies based on oil and other natural resources found in the Sahara. Look at the map on page 448. You see that the vast stretches of the Sahara lie to the south of North Africa, and the deep blue waters of the Mediterranean Sea lie to the north.

Place
Libya

Libya is a huge country about one-fifth the size of the United States. Until the mid-1900s Libya was a poor country with few natural resources. The discovery of oil in 1959 brought Libya great wealth.

The Land More than 90 percent desert, Libya is one of the world's driest countries. Its need for water increases as its population grows. In 1996 the Libyan government opened a long irrigation pipeline to bring water to coastal areas from aquifers in the south.

The Economy Although most of Libya's land is too dry for farming, Libya is the richest country in North Africa. It earns an enormous income from its oil exports. A Libyan oil worker earns more in one month than could be earned in one year of farming. The government uses profits from oil to improve farming and to build new schools, houses, and hospitals.

The People The 5.7 million people of Libya are of mixed Arab and Berber ancestry. The Berbers were the first people known to live in North Africa. During the A.D. 600s, the Arabs brought Islam and the Arabic language to North Africa, including Libya.

If you lived in Libya today, you would probably live in an urban area along the Mediterranean coast, as about 86 percent of Libyans do. Most live in two modern cities—Tripoli, which is the capital, and Benghazi.

For many centuries Libya was a part of either Muslim or European empires. It finally became an independent country in 1951 under a monarchy. In 1969 a military officer named Muammar al-Qaddhafi (kuh•DAH•fee) came to power. He set up a **dictatorship,** or a government under the control of one all-powerful leader.

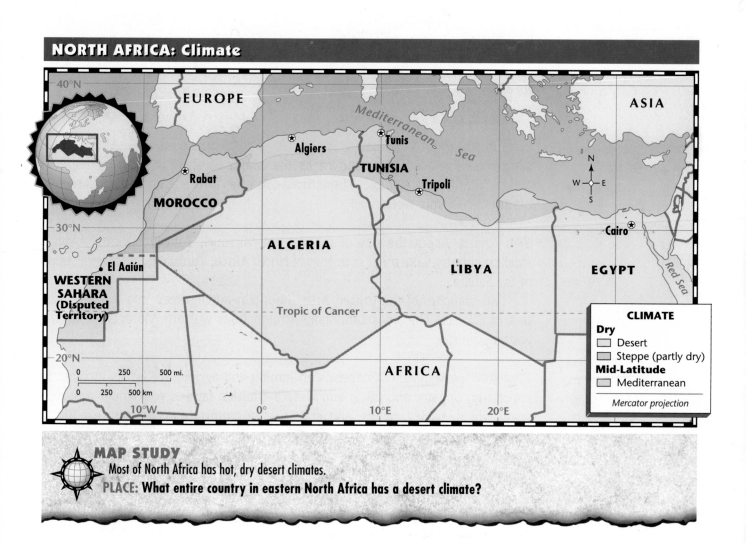

NORTH AFRICA: Climate

CLIMATE
Dry
☐ Desert
☐ Steppe (partly dry)
Mid-Latitude
☐ Mediterranean

Mercator projection

MAP STUDY
Most of North Africa has hot, dry desert climates.
PLACE: What entire country in eastern North Africa has a desert climate?

Shepherds tend their flocks on the rocky slopes of North Africa's Atlas Mountains. **LOCATION: Through what three countries do the Atlas Mountains extend?**

Region
Tunisia

Tunisia and two other western North African countries—Algeria and Morocco—form a region known as the Maghreb (MAH•gruhb). *Maghreb* means "the West" in Arabic. These three countries form the westernmost region of the Arab-Islamic world.

The Land About the size of the state of Missouri, Tunisia is North Africa's smallest country. Like other countries of North Africa, Tunisia includes large areas of the Sahara.

Two branches of the rugged Atlas mountain range sweep into the northern part of Tunisia. The Atlas Mountains stretch from western Morocco to northeastern Tunisia.

The Economy Tunisia depends on mining and agriculture for its income. It exports phosphates and oil. If you were a Tunisian farmer, you would probably grow wheat, barley, olives, and grapes—the country's major farm products. Most of Tunisia's farms are located along the fertile east coast. As the map on page 455 shows, this coastal area has a mild Mediterranean climate that favors the growing of crops.

The People Almost all of Tunisia's 9.5 million people are of Arab and Berber ancestry and speak Arabic. Look at the population density map on page 457.

Notice that most Tunisians live in the northern and eastern areas of the country. About 60 percent of Tunisia's people live in urban areas. Tunis, with a population of about 1 million, is the capital and largest city.

In ancient times the city of Carthage arose in the area that is now Tunisia. Carthage fought unsuccessfully with Rome for control of the Mediterranean world. In modern times France ruled Tunisia as a colony. Although Tunisia became an independent republic in 1956, French influences can still be seen.

Muslims gather at a mosque for daily prayers.
REGION: What is the major religion of Tunisia?

Place
Algeria

Crossing Algeria is no quick trip. Algeria is the largest country in the region of Southwest Asia and North Africa. It is more than three times bigger than the state of Texas.

The Land The landscape of Algeria varies from plains to mountains to deserts. Along the Mediterranean coast is a narrow strip of land known as the Tell. The word *tell* is an Arabic term meaning "hill." The hills and plains of the

NORTH AFRICA: Population Density

CITIES
- City with more than 5,000,000 people
- City with 1,000,000 to 5,000,000 people
- City with 500,000 to 1,000,000 people

Persons per	
sq. mi.	**sq. km**
Uninhabited	Uninhabited
Under 2	Under 1
2-60	1-25
60-125	25-50
125-250	50-100
Over 250	Over 100

Mercator projection

EUROPE

Mediterranean Sea

Tunis
Oran Algiers
Rabat TUNISIA
Casablanca Tripoli
MOROCCO

Alexandria
Giza Cairo

30°N

ALGERIA

WESTERN
SAHARA
(Disputed
Territory)
Tropic of Cancer

LIBYA

EGYPT

Red Sea

20°N

AFRICA

N
W—E
S

0   250   500 mi.
0   250   500 km

MAP STUDY
Most people in North Africa live near the Mediterranean coast and in river valley areas where water is available.
HUMAN/ENVIRONMENT INTERACTION: What North African country has the highest population density?

Tell contain Algeria's best farmland. A Mediterranean climate with hot, dry summers and mild, rainy winters helps crops grow.

South of the Tell are the Atlas Mountains. In Algeria the Atlas Mountains form two major chains with an average height of about 7,000 feet (2,134 m). Another mountain range known as the Ahaggar (uh•HAH•guhr) lies in southern Algeria. Between the Atlas and Ahaggar mountains are parts of the Sahara known as **ergs,** huge areas of shifting sand dunes. You find a hot, dry climate in most parts of Algeria south of the Tell.

The Economy Agriculture, manufacturing, and mining are the basis of Algeria's developing economy. Algeria has large deposits of natural gas and petroleum. These two natural resources make up 30 percent of the country's income.

The People About 30 million people live in Algeria. Like Libya and Tunisia, Algeria has a population of mixed Arab and Berber ancestry. Most Algerians practice Islam and speak Arabic. People in the countryside raise livestock or farm small pieces of land. Algiers, the capital and largest city, has about 1.5 million people.

What is it like to be a teenager living in Algiers? Fatimah Malek is an Algerian teen who enjoys visiting the old areas of Algiers called **casbahs.** In these sections, shops, mosques, bazaars, and homes line narrow streets. Fatimah lives in a new section of Algiers that has broad streets and modern buildings.

In school Fatimah has learned about her country's independent spirit. Algeria came under French rule in the 1800s. The Algerians never accepted French control of their country, however. In 1962 Algeria became an independent republic. Since the early 1990s, Fatimah and other Algerians have faced hardships as a result of a conflict between the Algerian government and opposing political groups.

Teen Scene

It's Festival Time!

Amineh and Jamila are wearing their nicest *caftans* and head scarves. Tonight, the Berber folklore festival will begin. Dancers, singers, and acrobats from all over Morocco come to entertain the large crowds. The 15-day folklore festival takes place every year in May. Amineh and Jamila like to share in all the activities, but their favorite part is meeting new people from other Berber villages.

Location
Morocco

The country of Morocco lies in the northwestern corner of Africa. Its long seacoast is bordered by the Mediterranean Sea on the north and the Atlantic Ocean on the west. The northern tip of Morocco meets the Strait of Gibraltar.

The Land Morocco's landforms are similar to those of neighboring Algeria. A fertile coastal plain borders the Mediterranean Sea and the Atlantic Ocean. This area has a Mediterranean climate of hot, dry summers and mild, rainy winters. Farther inland rise the Atlas Mountains. You can see from the map on page 448 that the Sahara lies east and south of the Atlas Mountains.

The Economy The major economic activities of Morocco are agriculture, mining, fishing, and tourism. Morocco's farms produce citrus fruits, potatoes, olives, sugar beets, and wheat. The country's mines yield many resources, such as phosphates, iron ore, lead, and natural gas. Morocco leads the world in the export of phosphate rock and is a leading producer of phosphates. Tourism ranks as one of the country's major service industries.

The People Morocco has a population of about 29 million. Most Moroccans are Muslims of mixed Arab and Berber ancestry. They speak the Arabic language or one of several Berber dialects.

About half of Morocco's population live in rural areas. Some live in villages and farm the land; others are herders in desert areas. The rest live in cities. Rabat, with a population of more than 1 million, is Morocco's capital.

The country's largest city and port is Casablanca (KA•suh•BLANG•kuh), home to more than 3 million people. Both Rabat and Casablanca are located on the Atlantic coast. Moroccan cities face the challenge of rapidly growing populations.

History From the A.D. 1000s to the early 1900s, Morocco was a Muslim kingdom that ruled much of northwestern Africa. France and Spain, however, gained control of Morocco's affairs at the beginning of this century. In 1956 Morocco became an independent kingdom once again. At this time, the Moroccan government laid claim to Western Sahara, a large stretch of African territory south of Morocco. Many countries, however, do not recognize this claim.

SECTION 2 ASSESSMENT

REVIEWING TERMS AND FACTS
1. **Define the following:** dictatorship, erg, *casbah.*
2. **PLACE** Where do most of Libya's people live?
3. **LOCATION** What body of water lies north and east of Tunisia?
4. **MOVEMENT** What is Morocco's major export?

MAP STUDY ACTIVITIES
5. Look at the population map of North Africa on page 457. What is the population density around Algiers, Algeria?
6. Using the climate map on page 455, name the North African countries that have both steppe and Mediterranean climates.

MAKING CONNECTIONS

MATH	SCIENCE	HISTORY	LITERATURE	TECHNOLOGY

By A.D. 750 Arabs united by the religion of Islam had created a huge empire. Their empire's territory stretched across Southwest Asia and North Africa. The Arabs came into contact with Persian, Indian, Christian, and Jewish scholars living within their empire.

From the A.D. 700s to the 1200s, the city of Baghdad was the Muslim empire's center of learning. The Arabs not only learned Persian astronomy, Indian mathematics, and Greek science, but also added new knowledge to these subjects. Indeed, Muslim mathematicians gave the Western world two kinds of mathematics: algebra and trigonometry.

ALGEBRA About A.D. 825 al-Khwarizmi (ahl•KWAHR•eez•mee), a teacher of mathematics in Baghdad, wrote a book titled *Kitab al-jabr wa al-muqabalah.* His work was a presentation of what was known about algebra at the time. The Arab mathematician improved and expanded algebra, the mathematics that uses symbols to solve complex mathematical problems. Al-Khwarizmi—and readers of his book—invented so many things that scholars consider algebra an Arab creation. Even the name *algebra* comes from part of the title of al-Khwarizmi's book—*al-jabr wa.*

TRIGONOMETRY During the A.D. 900s Muslim mathematicians developed trigonometry, a type of mathematics that deals with the measurements of triangles. They studied and described the spheres in which the sun, moon, and planets were thought to move. Their studies led them to invent the basic ratios, or comparisons, that allow engineers and scientists to measure triangles and distances.

Astronomers and Scientists in Istanbul

Making the Connection

1. From whom did the Arabs learn basic science and mathematics?

2. What contributions did Muslim mathematicians make to Western culture?

Chapter 17 Highlights

Important Things To Know About North Africa

SECTION 1 EGYPT

- The Libyan Desert covers two-thirds of Egypt's land area.
- The Aswan High Dam holds back the Nile's floodwater. The dam provides hydroelectric power to Egypt's growing industries.
- Most of Egypt's people live near the Nile River and depend on it for water and transportation.
- About 43 percent of Egypt's people live in rural areas and farm small plots of land.
- Cairo, Egypt's capital, is the largest city in Africa and a leading center of the Muslim world.
- One of the ancient world's most advanced civilizations developed in the Nile River valley.
- Today Egypt is a republic. Almost 90 percent of Egyptians are Muslims and about 10 percent are Christians.

SECTION 2 OTHER NORTH AFRICAN COUNTRIES

- North Africa also includes the countries of Libya, Tunisia, Algeria, and Morocco. The people of all four countries are of mixed Arab and Berber ancestry.
- The landscape of North Africa is mostly desert or mountains.
- The region's people live on fertile coastal plains or near oases.
- Oil, natural gas, and phosphate rock are among the most important natural resources of North Africa.
- All countries of the region border on the Mediterranean Sea or Atlantic Ocean. This affects climate.

A *casbah* near Duarzazate, Morocco ▶

REVIEWING KEY TERMS

Match the numbered terms in Set A with their lettered definitions in Set B.

A

1. delta
2. *fellahin*
3. hydroelectric power
4. hieroglyphs
5. bazaar
6. consumer goods
7. dictatorship
8. erg
9. *casbah*

B

A. picture symbols
B. government ruled by one all-powerful leader
C. Egyptian farmers
D. sand dune
E. broad swampy triangle where a river runs into an ocean or sea
F. old section of a North African city
G. household products, clothes, and shoes that people buy
H. energy produced from moving water
I. marketplace

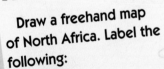

Mental Mapping Activity

Draw a freehand map of North Africa. Label the following:
- Egypt
- Nile River
- Cairo
- Libya
- Sahara
- Atlas Mountains

REVIEWING THE MAIN IDEAS

Section 1

1. LOCATION Where is Egypt located?
2. HUMAN/ENVIRONMENT INTERACTION Why did the Egyptians build the Aswan High Dam?
3. PLACE In what direction does the Nile River flow?

Section 2

4. LOCATION Which North African countries border Libya?
5. PLACE What landforms are found in the Sahara?
6. REGION What region includes Tunisia, Morocco, and Algeria?
7. HUMAN/ENVIRONMENT INTERACTION How does Morocco's location help it have a prosperous fishing industry?

CRITICAL THINKING ACTIVITIES

1. **Determining Cause and Effect** Why do you think an early Egyptian civilization developed in a river valley?
2. **Analyzing Information** How have the people of North Africa adapted to their desert environment?

CONNECTIONS TO WORLD HISTORY

Cooperative Learning Activity Work in small groups to discover what life was like in ancient Egypt. Different groups can research the pharaohs, the building of the pyramids, and the daily lives of people. Collect photographs, posters, or real objects to represent what you discover. Put all information together to make a visual presentation of "Life in Ancient Egypt."

GeoJournal Writing Activity

Imagine that you are taking a boat trip along the Nile River. Write a daily diary for seven days, describing what you see and telling about your experiences.

TECHNOLOGY ACTIVITY

Using the Internet Search the Internet for information on the hydroelectric power provided by Egypt's Aswan High Dam. Based on the information, write an article for the school newspaper or magazine about your topic. Be sure to provide a diagram, including labels, of the dam and power plant to explain how power is produced.

PLACE LOCATION ACTIVITY: NORTH AFRICA

Match the letters on the map with the places and physical features of North Africa. Write your answers on a separate sheet of paper.

1. Tripoli
2. Sinai Peninsula
3. Tunis
4. Atlas Mountains
5. Libya
6. Strait of Gibraltar
7. Mediterranean Sea
8. Morocco
9. Cairo

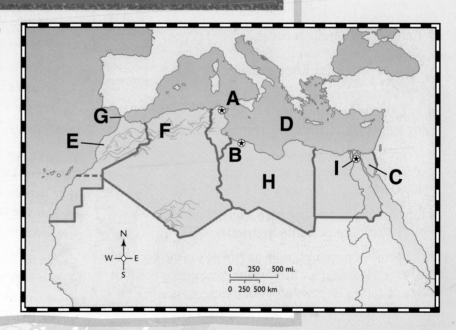

Water: A Precious Resource

PROBLEM

Most of the usable water in **Southwest Asia and North Africa** comes from aquifers and from three river basins: the Jordan, the Tigris-Euphrates, and the Nile. Despite these large river systems, water is scarce. Each of the rivers and its tributaries flows through several countries—and each country's needs are growing. Drought, industrialization, irrigation needs, and mushrooming populations—all strain the limited water supply.

SOLUTIONS

- To resolve water-rights issues, Israel is negotiating with Jordan and the Palestinians over rights to the Jordan River and West Bank aquifers.

- To promote conservation, Tunisia is introducing higher prices for water used for irrigation.

- A recycling project in Tel Aviv, Israel, generates enough recycled wastewater to cultivate 20,000 acres (8,100 hectares) of farmland.

- Advances in technology have lowered the cost of converting seawater to freshwater.

- Cooperative measures, such as Turkey's plans for a Peace Pipeline, may carry water from water-rich countries to water-poor countries.

Drip irrigation, which minimizes water loss, delivers a precise drink of water to a seedling in the desert. Drip irrigation may use 25 percent less water than sprinkler systems.

WATER FACT BANK

- By the year 2015, most of the nations of North Africa will not have enough water to meet basic human needs.

- Crops growing in the desert require huge amounts of water. It takes 3 cubic meters of water (780 gallons) to produce about 2 pounds (1 kilogram) of wheat.

- Turkey's Atatürk Dam, on the Euphrates River, holds back water that once reached other nations downstream. Soon, Syria's share of the Euphrates will shrink by 40 percent, and Iraq's share by 80 percent.

- Saudi Arabia leads the world in producing freshwater from salt water. Its 22 desalination plants produce 30 percent of all the desalinated water in the world.

environmental activities

TEEN TRIBUTE

Hamidreza Modaberi, age 12, lives in the desert country of Iran. In a letter to an environmental publication, she wrote: "I dreamed the whole world had changed. All these dry lands had been turned into a beautiful nature full of trees, full of rivers ... war and bloodshed had ended.... I wish the world would be like my dream."

TAKE A CLOSER LOOK

1 What sources provide Southwest Asia and North Africa with water?

2 What human activities have caused water shortage problems in this region?

WHAT CAN YOU DO?

🌿 Put "Water Watch" posters around your school giving tips about conserving water.

🌿 Improve your own water-saving habits and avoid using products that are polluters.

🌿 Write a letter to the editor of your local newspaper to protest industrial pollution in your community.

🌿 Educate yourself — find out where your water comes from and what is being done to protect that source.

🌿 Fix that dripping faucet!

good planets are hard to find

Unit 7

Africa South of the Sahara

What Makes Africa South of the Sahara a Region?

Most people share...

- a location almost entirely in the tropics.
- the world's fastest-growing and youngest population.
- challenges to the environment, especially natural resources and wildlife.
- a struggle to improve the quality of life.

To find out more about Africa south of the Sahara, see the Unit Atlas on pages 468–483.

Cape Town, Republic of South Africa

EXPLORING THE INTERNET

To learn more about Africa south of the Sahara, visit the Glencoe Social Studies Web site at **www.glencoe.com** for information, activities, and links to other sites.

GeoJournal Activity
As you read this unit, think about why many geographers call Africa an unopened treasure chest. In your journal, choose the three most important treasures you think Africa has to share with the rest of the world.

UNIT 7 ATLAS

NATIONAL GEOGRAPHIC SOCIETY

IMAGES *of the* WORLD

1. African elephant, Kenya
2. Woman in Tombouctou, Mali
3. Victoria Falls, Zimbabwe
4. Shoppers at a market, Madagascar
5. Mosque at Omdurman, Sudan
6. Storefronts, Kenya
7. Oil company supervisor, off Gabon

All photos viewed against a village in the Tibesti Mountains, Chad.

Regional Focus

Africa south of the Sahara covers the central and southern parts of the continent. With the Equator crossing its middle, Africa south of the Sahara is the only region in the world that lies almost entirely in the tropics.

Region
The Land

The region of Africa south of the Sahara is about 9.5 million square miles (24.6 sq. km) in size. Its center is mostly made up of large, rolling plateaus. Narrow coastal plains run along the Atlantic and Indian oceans.

Plateaus The region's plateaus rise from west to east like a stairway. Because of these plateaus, Africa has a higher average elevation than any other continent in the world. The map on page 474 shows you that the average elevation in Africa is more than 2,000 feet (610 m) above sea level. Separating the plateaus are steep cliffs. Rivers often spill over the cliffs in thundering waterfalls. The cliffs and waterfalls make land and river travel difficult in many parts of Africa.

Mountains and Valleys Even with its high elevation, Africa south of the Sahara has few tall mountains. Wind and water wore away many of them over millions of years. At 19,340 feet (5,895 m) high, Kilimanjaro is the tallest peak in Africa.

Crossing the plateaus is the Great Rift Valley. This long, narrow break in the earth's surface runs more than 3,000 miles (4,800 km) through eastern Africa. Movements of the earth's crust formed the Great Rift Valley millions of years ago. A chain of deep lakes lies in the Great Rift Valley near the Equator.

Rivers Four great rivers slice through Africa: the Nile, the Congo, the Niger, and the Zambezi. The largest river system in Africa south of the Sahara is the Congo. It twists and turns for almost 2,800 miles (4,505 km) through the region. A number of small rivers flow into the Congo and the other three large rivers. Africa's major river systems are important means of transportation that link inland areas with oceans.

A Herder in Burkina Faso

Most people in the country of Burkina Faso make their living by herding cattle.
PLACE: What is the average elevation of land in Africa?

470

Region
Climate and Vegetation

Africa south of the Sahara has many types of climate and vegetation. Differences in elevation create much of this variety. If you travel in Africa toward the Equator from either the north or the south, you will find four main climate regions: desert, steppe, tropical savanna, and tropical rain forest. Only the highlands in eastern Africa and the southern tip of Africa have moderate climates.

More of Africa is covered by deserts than any other continent. This occurs because many high plateaus block rain clouds from reaching inland areas. The largest African deserts are the Sahara in the north and the Namib and Kalahari in the south. Around Africa's deserts stretch steppe areas, or partly dry grasslands. Because of drought and human activity, the Sahel steppe area in western Africa is gradually becoming a desert.

South of the Sahara, much of Africa is made up of savannas—tropical grasslands with scattered trees. A wide variety of animal life live in the savannas. Steamy rain forests, however, cover land near the Equator. They have many hardwood trees that provide timber.

Volcanoes National Park, Rwanda

Volcanic peaks tower over the vast plateaus of northwestern Rwanda.
HUMAN/ENVIRONMENT INTERACTION: What do Africa's rain forests provide?

Human/Environment Interaction
The Economy

The economy of Africa south of the Sahara depends mainly on farming. Another important economic activity is the mining and export of mineral resources.

Agriculture and Mining Farming and herding are Africa's leading economic activities. Some Africans grow only enough food to feed their families. Others work on plantations that grow crops for export. Africa is a major producer of cacao, sweet potatoes, coffee, bananas, cotton, peanuts, tea, and sugar.

Africans also work in mining. In central and southern Africa, miners work some of the world's largest deposits of gold, diamonds, and copper. Oil is found in western and central Africa.

Manufacturing In spite of many resources, manufacturing plays only a small role in Africa's economy. In the past, European nations ruled most of the continent. Today many countries in Africa south of the Sahara are building industries and improving transportation. Few paved roads exist outside of urban areas. Railroads and airplanes are best for traveling long distances in the region.

Region
The People

Johannesburg, South Africa

Johannesburg is South Africa's largest urban area and most important industrial and commercial center.
PLACE: About how many people live in Africa south of the Sahara?

South of the Sahara, Africa is divided into 48 nations. Each nation has many ethnic groups, languages, religions, and customs.

Population More than 625 million people live in this region, which is growing rapidly in population. Because so much of Africa is desert or partly dry grasslands, most people live along coasts. Some African governments have encouraged their people to move from the crowded coasts to inland areas. Three countries—Côte d'Ivoire, Nigeria, and Tanzania—have moved or will move their capitals from coastal cities to newly built inland cities.

About 73 percent of Africans south of the Sahara live in rural villages. Economic hardships and the desire for a better life, however, are drawing many people to the cities. African cities are among the fastest-growing urban areas. The three largest cities are Lagos in Nigeria, Kinshasa in the Democratic Republic of the Congo, and Abidjan in Côte d'Ivoire.

Culture People in this region are devoted to family life. Households often are made up of extended families—grandparents and other relatives as well as parents and children. Africans also have close ties to their ethnic groups. In many countries, loyalty to an ethnic group is more important than loyalty to the national government.

History Early Africa saw the rise of powerful empires and city-states that became important cultural centers. From the 1500s to the 1800s, Europeans explored the continent, enslaved many Africans, and sent them to other parts of the world. By 1914, Europeans had divided nearly all of Africa among themselves. They set up borders without regard to the different peoples living in the area.

In many parts of Africa, however, Africans resisted European rule. By the 1960s, most of Africa was made up of independent nations. Descendants of European settlers, however, continued to rule the country of South Africa. They withheld many rights from black Africans and other non-Europeans. In the early 1990s, South Africa extended its democratic form of government to all of its citizens.

The Arts Africans have developed many art forms. African artists are known for their decorative cloth weavings, wooden sculptures, metal jewelry, and pottery. Through the centuries African storytellers have passed down myths, history, and poems from early ancestors. African authors today blend these traditions with modern writing styles. African music has influenced popular music throughout the world.

472

Picture Atlas of the World

Population Geography

Using the CD-ROM Calculate the population density of countries in Africa south of the Sahara. Population density refers to the number of people per square mile (or kilometer). To do this you must first go to the *Country* screen. Select 10 African nations. Then calculate the population density for each country by dividing the country's total population by its area. Record your findings on a chart and discuss the following questions:

1. Is the most populous country also the most densely populated country? If not, determine the most densely populated country.

2. How do the following characteristics affect population density?

    climate   literacy rate
    access to water   fertility rate
    elevation   population growth rate

3. What might be the advantages and disadvantages of living in a densely populated city?

Surfing the "Net"

Snapshot of Africa

Imagine that you are a journalist and your editor has assigned you to prepare a feature story on a country of Africa. Your article will focus on the politics, climatic changes, tourist sites, and/or recent events of an African nation you choose. The theme of your feature story should be the changing face of Africa. The article should describe how traditions, popular culture, and the modern world meet in the nations of Africa today.

Getting There

Follow these steps to gather information on African nations:

1. Use a search engine. Type in *Africa*.

2. Enter names of individual African nations and words like the following to focus your search: *politics, news, culture, history*.

3. The search engine will provide you with a number of links to follow.

What To Do When You Are There

Begin your assignment by selecting one African nation. This will be the topic of your feature article. Devise a plan by determining an aspect of this nation on which to focus your article. Create an outline for your article. Navigate through the Web pages to collect information to build your feature story. Prepare your outline along with headline and photo ideas to present to your editor (your teacher) for approval. Once your outline is approved, work to complete your feature story. Share the final article with the class.

Physical Geography

AFRICA SOUTH OF THE SAHARA: Physical

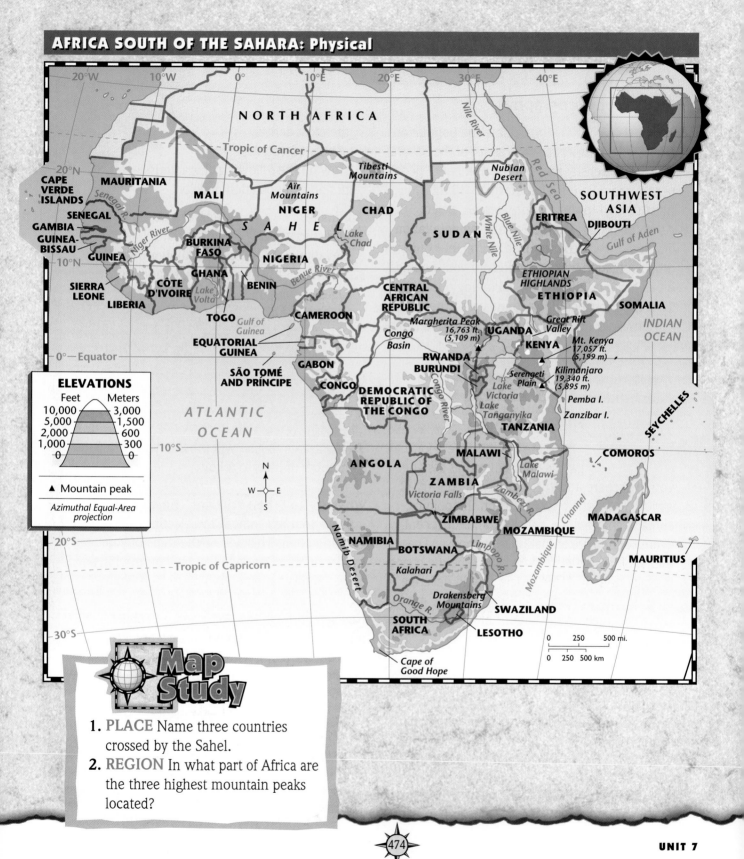

ELEVATIONS

Feet	Meters
10,000	3,000
5,000	1,500
2,000	600
1,000	300
0	0

▲ Mountain peak

Azimuthal Equal-Area projection

Map Study

1. **PLACE** Name three countries crossed by the Sahel.
2. **REGION** In what part of Africa are the three highest mountain peaks located?

ELEVATION PROFILE: Africa

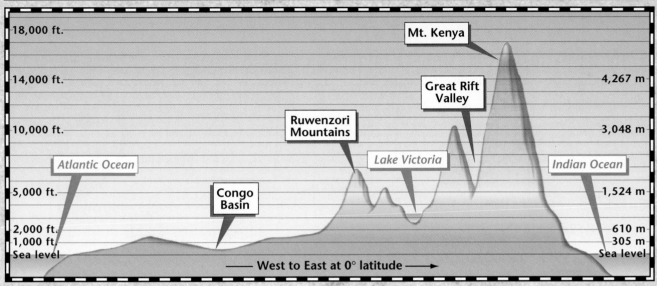

- 18,000 ft.
- 14,000 ft. — 4,267 m
- Mt. Kenya
- Great Rift Valley
- 10,000 ft. — 3,048 m
- Ruwenzori Mountains
- Atlantic Ocean
- Lake Victoria
- Indian Ocean
- Congo Basin
- 5,000 ft. — 1,524 m
- 2,000 ft. — 610 m
- 1,000 ft. — 305 m
- Sea level — Sea level

⟶ West to East at 0° latitude ⟶

Source: *Goode's World Atlas,* 19th edition

GeoFacts

Highest point: Kilimanjaro (Tanzania) 19,340 ft. (5,895 m) high

Lowest point: Lake Assal (Djibouti) 512 ft. (156 m) below sea level

Longest river: Nile River 4,160 mi. (6,693 km) long

Largest lake: Lake Victoria (Uganda/Tanzania) 26,828 sq. mi. (69,485 sq. km)

Largest desert: Sahara 3,500,000 sq. mi. (9,065,000 sq. km)

AFRICA SOUTH OF THE SAHARA AND THE UNITED STATES: Land Comparison

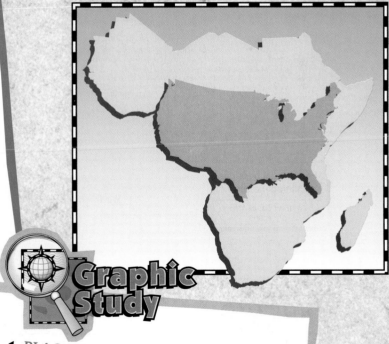

Graphic Study

1. PLACE What is the elevation of Lake Victoria?

2. REGION About how many times larger than the continental United States is Africa south of the Sahara?

Cultural Geography

AFRICA SOUTH OF THE SAHARA: Political

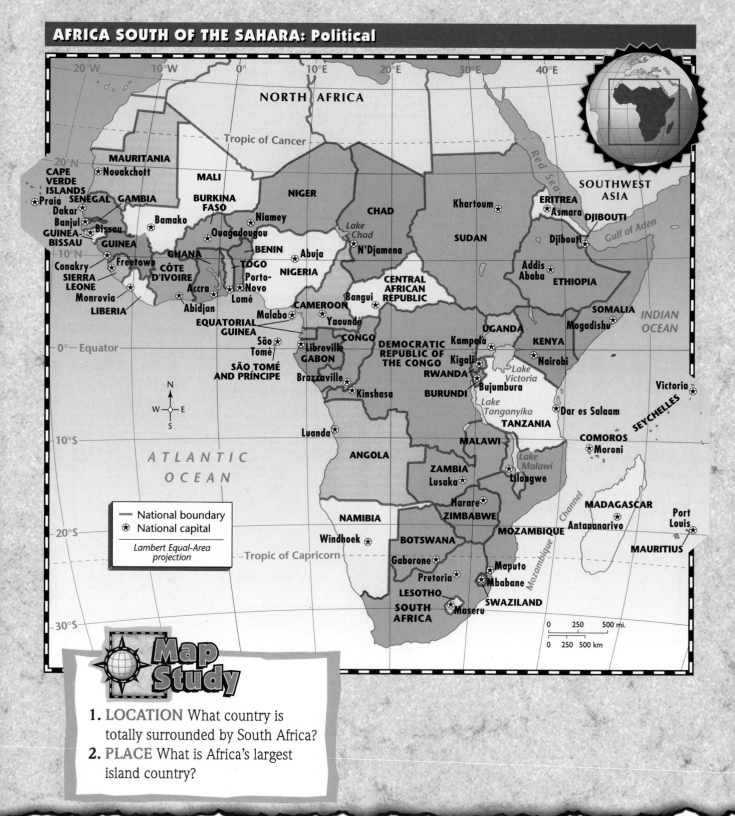

NORTH AFRICA

Tropic of Cancer

MAURITANIA
⊛Nouakchott

CAPE VERDE ISLANDS
⊛Praia

SENEGAL
Dakar⊛
Banjul⊛
GAMBIA

MALI
BURKINA FASO

NIGER

CHAD

Khartoum⊛

Red Sea

SOUTHWEST ASIA

ERITREA
⊛Asmara
DJIBOUTI
⊛Djibouti

Gulf of Aden

GUINEA-BISSAU
⊛Bissau
GUINEA

Bamako⊛
Niamey⊛
Ouagadougou⊛

BENIN

⊛Abuja

N'Djamena⊛

SUDAN

Addis Ababa⊛
ETHIOPIA

Conakry⊛
Freetown⊛
SIERRA LEONE
Monrovia⊛
LIBERIA

GHANA
CÔTE D'IVOIRE
Accra⊛
Abidjan⊛

TOGO
Porto-Novo⊛
Lomé⊛

NIGERIA

CENTRAL AFRICAN REPUBLIC

Bangui⊛

SOMALIA
Mogadishu⊛

INDIAN OCEAN

EQUATORIAL GUINEA
Malabo⊛
São Tomé⊛
SÃO TOMÉ AND PRÍNCIPE

CAMEROON
⊛Yaoundé

Libreville⊛
GABON
CONGO

Brazzaville⊛

Kinshasa⊛

DEMOCRATIC REPUBLIC OF THE CONGO

UGANDA
Kampala⊛
Kigali⊛
RWANDA
Bujumbura⊛
BURUNDI

KENYA
⊛Nairobi

Lake Victoria

Lake Tanganyika

TANZANIA
Dar es Salaam⊛

Victoria⊛
SEYCHELLES

0°—Equator

N
W E
S

Luanda⊛

ATLANTIC OCEAN

ANGOLA

MALAWI

ZAMBIA
Lusaka⊛

Lake Malawi
Lilongwe⊛

COMOROS
⊛Moroni

10°S

National boundary
⊛ National capital

Lambert Equal-Area projection

NAMIBIA

Windhoek⊛

Harare⊛
ZIMBABWE

MOZAMBIQUE

Mozambique Channel

MADAGASCAR
Antananarivo⊛

Port Louis⊛
MAURITIUS

BOTSWANA
Gaborone⊛

Pretoria⊛
LESOTHO
SOUTH AFRICA
Maseru⊛

Maputo⊛
Mbabane⊛
SWAZILAND

Tropic of Capricorn

0   250   500 mi.
0   250   500 km

30°S

Map Study

1. **LOCATION** What country is totally surrounded by South Africa?
2. **PLACE** What is Africa's largest island country?

AFRICA SOUTH OF THE SAHARA: Selected Rural and Urban Populations

	Rural	Urban
WEST AFRICA		
Niger	85%	15%
Cape Verde	56%	44%
CENTRAL AFRICA		
Angola	58%	42%
Central African Republic	61%	39%
EAST AFRICA		
Rwanda	95%	5%
Djibouti	19%	81%
SOUTHERN AFRICA		
Lesotho	84%	16%
South Africa	43%	57%

Source: *Population Reference Bureau, 1998*

GeoFacts

Biggest country (land area): Sudan 917,375 sq. mi. (2,376,000 sq. km)

Smallest country (land area): Seychelles 174 sq. mi. (451 sq. km)

Largest city (population): Lagos (1997) 10,287,000; (2015 projected) 24,600,000

Highest population density: Mauritius 1,483 people per sq. mi. (573 people per sq. km)

Lowest population density: Namibia 5 people per sq. mi. (2 people per sq. km)

COMPARING POPULATION: Africa South of the Sahara and the United States

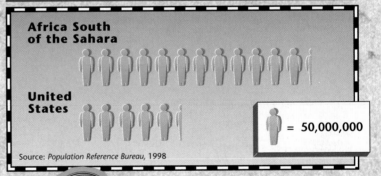

Africa South of the Sahara

United States

= 50,000,000

Source: *Population Reference Bureau, 1998*

Graphic Study

1. **PLACE** About how many people live in Africa south of the Sahara?
2. **REGION** What countries in Africa south of the Sahara have a greater urban population than a rural population?

Countries at a Glance

Angola

Luanda

LANDMASS:
481,350 sq. mi./
1,246,697 sq. km
MONEY:
New Kwanza
MAJOR EXPORT:
Mineral Fuels
MAJOR IMPORT:
Transport Equipment

CAPITAL:
Luanda
MAJOR LANGUAGE(S):
Portuguese, Bantu languages
POPULATION:
12,000,000

Benin

Porto-Novo

LANDMASS:
42,710 sq. mi./
110,619 sq. km
MONEY:
CFA Franc
MAJOR EXPORT:
Cotton
MAJOR IMPORT:
Yarn and Fabric

CAPITAL:
Porto-Novo
MAJOR LANGUAGE(S):
French, Fon, Yoruba, Somba
POPULATION:
6,000,000

Botswana

Gaborone

LANDMASS:
218,810 sq. mi./
566,718 sq. km
MONEY:
Pula
MAJOR EXPORT:
Diamonds
MAJOR IMPORT:
Transport Equipment

CAPITAL:
Gaborone
MAJOR LANGUAGE(S):
English, Setswana
POPULATION:
1,400,000

Burkina Faso

Ouagadougou

LANDMASS:
105,637 sq. mi./
273,763 sq. km
MONEY:
CFA Franc*
MAJOR EXPORT:
Cotton
MAJOR IMPORT:
Machinery

CAPITAL:
Ouagadougou
MAJOR LANGUAGE(S):
French, Sudanic languages
POPULATION:
11,300,000

Burundi

Bujumbura

LANDMASS:
9,900 sq. mi./
25,641 sq. km
MONEY:
Burundi Franc
MAJOR EXPORT:
Coffee
MAJOR IMPORT:
Machinery

CAPITAL:
Bujumbura
MAJOR LANGUAGE(S):
Rundi, French
POPULATION:
5,500,000

Cameroon

Yaoundé

LANDMASS:
179,690 sq. mi./
465,397 sq. km
MONEY:
CFA Franc*
MAJOR EXPORT:
Petroleum
MAJOR IMPORT:
Machinery

CAPITAL:
Yaoundé
MAJOR LANGUAGE(S):
English, French
POPULATION:
14,300,000

Cape Verde Islands

Praia

LANDMASS:
1,560 sq. mi./
4,040 sq. km
MONEY:
Escudo
MAJOR EXPORT:
Bananas
MAJOR IMPORT:
Food

CAPITAL:
Praia
MAJOR LANGUAGE(S):
Portuguese
POPULATION:
400,000

Central African Republic

Bangui

LANDMASS:
240,530 sq. mi./
622,973 sq. km
MONEY:
CFA Franc*
MAJOR EXPORT:
Diamonds
MAJOR IMPORT:
Food Products

CAPITAL:
Bangui
MAJOR LANGUAGE(S):
French, ethnic languages
POPULATION:
3,400,000

Chad

N'Djamena

LANDMASS:
486,180 sq. mi./
1,259,206 sq. km
MONEY:
CFA Franc*
MAJOR EXPORT:
Cotton
MAJOR IMPORT:
Petroleum Products

CAPITAL:
N'Djamena
MAJOR LANGUAGE(S):
French, Arabic
POPULATION:
7,400,000

Comoros

Moroni

LANDMASS:
860 sq. mi./
2,227 sq. km
MONEY:
CFA Franc*
MAJOR EXPORT:
Vanilla
MAJOR IMPORT:
Rice

CAPITAL:
Moroni
MAJOR LANGUAGE(S):
Arabic, French
POPULATION:
500,000

Countries not drawn to scale.

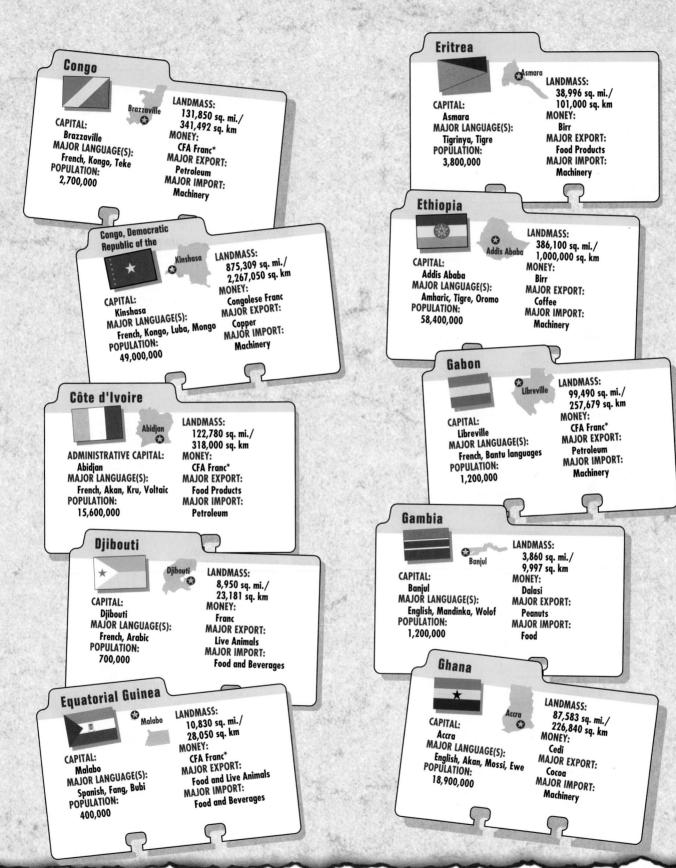

Congo

CAPITAL:
Brazzaville
MAJOR LANGUAGE(S):
French, Kongo, Teke
POPULATION:
2,700,000

LANDMASS:
131,850 sq. mi./
341,492 sq. km
MONEY:
CFA Franc*
MAJOR EXPORT:
Petroleum
MAJOR IMPORT:
Machinery

Congo, Democratic Republic of the

CAPITAL:
Kinshasa
MAJOR LANGUAGE(S):
French, Kongo, Luba, Mongo
POPULATION:
49,000,000

LANDMASS:
875,309 sq. mi./
2,267,050 sq. km
MONEY:
Congolese Franc
MAJOR EXPORT:
Copper
MAJOR IMPORT:
Machinery

Côte d'Ivoire

ADMINISTRATIVE CAPITAL:
Abidjan
MAJOR LANGUAGE(S):
French, Akan, Kru, Voltaic
POPULATION:
15,600,000

LANDMASS:
122,780 sq. mi./
318,000 sq. km
MONEY:
CFA Franc*
MAJOR EXPORT:
Food Products
MAJOR IMPORT:
Petroleum

Djibouti

CAPITAL:
Djibouti
MAJOR LANGUAGE(S):
French, Arabic
POPULATION:
700,000

LANDMASS:
8,950 sq. mi./
23,181 sq. km
MONEY:
Franc
MAJOR EXPORT:
Live Animals
MAJOR IMPORT:
Food and Beverages

Equatorial Guinea

CAPITAL:
Malabo
MAJOR LANGUAGE(S):
Spanish, Fang, Bubi
POPULATION:
400,000

LANDMASS:
10,830 sq. mi./
28,050 sq. km
MONEY:
CFA Franc*
MAJOR EXPORT:
Food and Live Animals
MAJOR IMPORT:
Food and Beverages

Eritrea

CAPITAL:
Asmara
MAJOR LANGUAGE(S):
Tigrinya, Tigre
POPULATION:
3,800,000

LANDMASS:
38,996 sq. mi./
101,000 sq. km
MONEY:
Birr
MAJOR EXPORT:
Food Products
MAJOR IMPORT:
Machinery

Ethiopia

CAPITAL:
Addis Ababa
MAJOR LANGUAGE(S):
Amharic, Tigre, Oromo
POPULATION:
58,400,000

LANDMASS:
386,100 sq. mi./
1,000,000 sq. km
MONEY:
Birr
MAJOR EXPORT:
Coffee
MAJOR IMPORT:
Machinery

Gabon

CAPITAL:
Libreville
MAJOR LANGUAGE(S):
French, Bantu languages
POPULATION:
1,200,000

LANDMASS:
99,490 sq. mi./
257,679 sq. km
MONEY:
CFA Franc*
MAJOR EXPORT:
Petroleum
MAJOR IMPORT:
Machinery

Gambia

CAPITAL:
Banjul
MAJOR LANGUAGE(S):
English, Mandinka, Wolof
POPULATION:
1,200,000

LANDMASS:
3,860 sq. mi./
9,997 sq. km
MONEY:
Dalasi
MAJOR EXPORT:
Peanuts
MAJOR IMPORT:
Food

Ghana

CAPITAL:
Accra
MAJOR LANGUAGE(S):
English, Akan, Mossi, Ewe
POPULATION:
18,900,000

LANDMASS:
87,583 sq. mi./
226,840 sq. km
MONEY:
Cedi
MAJOR EXPORT:
Cocoa
MAJOR IMPORT:
Machinery

* Communauté Financière Africaine
(African Financial Community)

Countries at a Glance

Guinea

Conakry

LANDMASS:
94,873 sq. mi./
245,721 sq. km
MONEY:
Franc
MAJOR EXPORT:
Bauxite
MAJOR IMPORT:
Petroleum

CAPITAL:
Conakry
MAJOR LANGUAGE(S):
French, Peul, Mande
POPULATION:
7,500,000

Guinea-Bissau

Bissau

LANDMASS:
10,860 sq. mi./
28,127 sq. km
MONEY:
Peso
MAJOR EXPORT:
Cashews
MAJOR IMPORT:
Transport Equipment

CAPITAL:
Bissau
MAJOR LANGUAGE(S):
Portuguese, Crioulo,
ethnic languages
POPULATION:
1,100,000

Kenya

Nairobi

LANDMASS:
219,745 sq. mi./
569,139 sq. km
MONEY:
Shilling
MAJOR EXPORT:
Tea
MAJOR IMPORT:
Machinery

CAPITAL:
Nairobi
MAJOR LANGUAGE(S):
Swahili, English, Kikuyu,
Luhya
POPULATION:
28,300,000

Lesotho

Maseru

LANDMASS:
11,720 sq. mi./
30,355 sq. km
MONEY:
Loti
MAJOR EXPORT:
Machinery
MAJOR IMPORT:
Clothing

CAPITAL:
Maseru
MAJOR LANGUAGE(S):
English, Sotho
POPULATION:
2,100,000

Liberia

Monrovia

LANDMASS:
37,190 sq. mi./
96,322 sq. km
MONEY:
Dollar
MAJOR EXPORT:
Iron Ore
MAJOR IMPORT:
Machinery

CAPITAL:
Monrovia
MAJOR LANGUAGE(S):
English, ethnic languages
POPULATION:
2,800,000

Madagascar

Antananarivo

LANDMASS:
224,530 sq. mi./
581,533 sq. km
MONEY:
Franc
MAJOR EXPORT:
Coffee
MAJOR IMPORT:
Machinery

CAPITAL:
Antananarivo
MAJOR LANGUAGE(S):
Malagasy, French
POPULATION:
14,000,000

Malawi

Lilongwe

LANDMASS:
36,320 sq. mi./
94,069 sq. km
MONEY:
Kwacha
MAJOR EXPORT:
Tobacco
MAJOR IMPORT:
Machinery

CAPITAL:
Lilongwe
MAJOR LANGUAGE(S):
English, Chewa,
Lomwe, Yao
POPULATION:
9,800,000

Mali

Bamako

LANDMASS:
471,120 sq. mi./
1,220,201 sq. km
MONEY:
CFA Franc*
MAJOR EXPORT:
Cotton Products
MAJOR IMPORT:
Machinery

CAPITAL:
Bamako
MAJOR LANGUAGE(S):
French, Bambara, Senufo
POPULATION:
10,100,000

Mauritania

Nouakchott

LANDMASS:
395,840 sq. mi./
1,025,226 sq. km
MONEY:
Ouguiya
MAJOR EXPORT:
Fish
MAJOR IMPORT:
Machinery

CAPITAL:
Nouakchott
MAJOR LANGUAGE(S):
Arabic, French,
Hassanya Arabic
POPULATION:
2,500,000

Mauritius

Port Louis

LANDMASS:
784 sq. mi./
2,030 sq. km
MONEY:
Rupee
MAJOR EXPORT:
Clothing and Textiles
MAJOR IMPORT:
Machinery

CAPITAL:
Port Louis
MAJOR LANGUAGE(S):
English, French Creole,
Bhojpuri
POPULATION:
1,200,000

Countries not drawn to scale.

UNIT 7

Mozambique

CAPITAL:
Maputo
MAJOR LANGUAGE(S):
Portuguese, Makua, Malawl
POPULATION:
18,600,000
LANDMASS:
302,740 sq. mi./
784,097 sq. km
MONEY:
Metical
MAJOR EXPORT:
Shrimp
MAJOR IMPORT:
Food

Namibia

CAPITAL:
Windhoek
MAJOR LANGUAGE(S):
Afrikaans, English
POPULATION:
1,600,000
LANDMASS:
317,870 sq. mi./
823,283 sq. km
MONEY:
South African Rand
MAJOR EXPORT:
Minerals
MAJOR IMPORT:
Petroleum Products

Niger

CAPITAL:
Niamey
MAJOR LANGUAGE(S):
French, Hausa, Fulani
POPULATION:
10,100,000
LANDMASS:
489,070 sq. mi./
1,266,691 sq. km
MONEY:
CFA Franc*
MAJOR EXPORT:
Uranium
MAJOR IMPORT:
Machinery

Nigeria

CAPITAL:
Abuja
MAJOR LANGUAGE(S):
English, Hausa,
Yoruba, Ibo
POPULATION:
121,800,000
LANDMASS:
351,650 sq. mi./
910,774 sq. km
MONEY:
Naira
MAJOR EXPORT:
Petroleum
MAJOR IMPORT:
Machinery

Rwanda

CAPITAL:
Kigali
MAJOR LANGUAGE(S):
French, Rwanda
POPULATION:
8,000,000
LANDMASS:
9,525 sq. mi./
24,670 sq. km
MONEY:
Franc
MAJOR EXPORT:
Coffee and Tea
MAJOR IMPORT:
Machinery

São Tomé and Príncipe

CAPITAL:
São Tomé
MAJOR LANGUAGE(S):
Portuguese
POPULATION:
200,000
LANDMASS:
293 sq. mi./
759 sq. km
MONEY:
Dobra
MAJOR EXPORT:
Cocoa
MAJOR IMPORT:
Food

Senegal

CAPITAL:
Dakar
MAJOR LANGUAGE(S):
French, Wolof, Serer, Peul
POPULATION:
9,000,000
LANDMASS:
74,340 sq. mi./
192,541 sq. km
MONEY:
CFA Franc*
MAJOR EXPORT:
Peanut Oil
MAJOR IMPORT:
Machinery

Seychelles

CAPITAL:
Victoria
MAJOR LANGUAGE(S):
English, French
POPULATION:
100,000
LANDMASS:
174 sq. mi./
451 sq. km
MONEY:
Rupee
MAJOR EXPORT:
Food Products
MAJOR IMPORT:
Machinery

Sierra Leone

CAPITAL:
Freetown
MAJOR LANGUAGE(S):
English, ethnic languages
POPULATION:
4,600,000
LANDMASS:
27,650 sq. mi./
71,614 sq. km
MONEY:
Leone
MAJOR EXPORT:
Food Products, Diamonds
MAJOR IMPORT:
Food and Live Animals

Somalia

CAPITAL:
Mogadishu
MAJOR LANGUAGE(S):
Somali, Arabic
POPULATION:
10,700,000
LANDMASS:
242,220 sq. mi./
627,350 sq. km
MONEY:
Shilling
MAJOR EXPORT:
Live Animals
MAJOR IMPORT:
Petroleum

*** Communauté Financière Africaine
(African Financial Community)**

Countries at a Glance

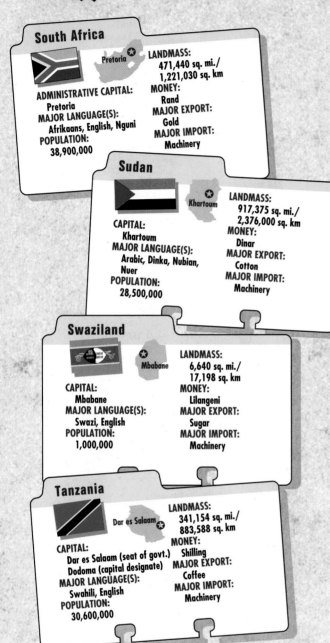

South Africa

ADMINISTRATIVE CAPITAL:
Pretoria
MAJOR LANGUAGE(S):
Afrikaans, English, Nguni
POPULATION:
38,900,000

Pretoria ★

LANDMASS:
471,440 sq. mi./
1,221,030 sq. km
MONEY:
Rand
MAJOR EXPORT:
Gold
MAJOR IMPORT:
Machinery

Sudan

CAPITAL:
Khartoum
MAJOR LANGUAGE(S):
Arabic, Dinka, Nubian,
Nuer
POPULATION:
28,500,000

★ Khartoum

LANDMASS:
917,375 sq. mi./
2,376,000 sq. km
MONEY:
Dinar
MAJOR EXPORT:
Cotton
MAJOR IMPORT:
Machinery

Swaziland

CAPITAL:
Mbabane
MAJOR LANGUAGE(S):
Swazi, English
POPULATION:
1,000,000

★ Mbabane

LANDMASS:
6,640 sq. mi./
17,198 sq. km
MONEY:
Lilangeni
MAJOR EXPORT:
Sugar
MAJOR IMPORT:
Machinery

Tanzania

CAPITAL:
Dar es Salaam (seat of govt.)
Dodoma (capital designate)
MAJOR LANGUAGE(S):
Swahili, English
POPULATION:
30,600,000

Dar es Salaam ★

LANDMASS:
341,154 sq. mi./
883,588 sq. km
MONEY:
Shilling
MAJOR EXPORT:
Coffee
MAJOR IMPORT:
Machinery

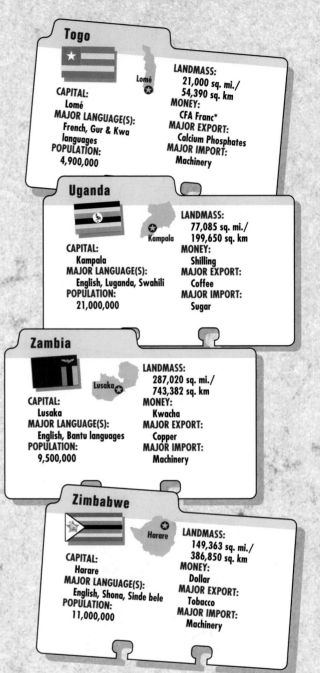

Togo

CAPITAL:
Lomé
MAJOR LANGUAGE(S):
French, Gur & Kwa
languages
POPULATION:
4,900,000

Lomé
★

LANDMASS:
21,000 sq. mi./
54,390 sq. km
MONEY:
CFA Franc*
MAJOR EXPORT:
Calcium Phosphates
MAJOR IMPORT:
Machinery

Uganda

CAPITAL:
Kampala
MAJOR LANGUAGE(S):
English, Luganda, Swahili
POPULATION:
21,000,000

★ Kampala

LANDMASS:
77,085 sq. mi./
199,650 sq. km
MONEY:
Shilling
MAJOR EXPORT:
Coffee
MAJOR IMPORT:
Sugar

Zambia

CAPITAL:
Lusaka
MAJOR LANGUAGE(S):
English, Bantu languages
POPULATION:
9,500,000

Lusaka ★

LANDMASS:
287,020 sq. mi./
743,382 sq. km
MONEY:
Kwacha
MAJOR EXPORT:
Copper
MAJOR IMPORT:
Machinery

Zimbabwe

CAPITAL:
Harare
MAJOR LANGUAGE(S):
English, Shona, Sinde bele
POPULATION:
11,000,000

Harare ★

LANDMASS:
149,363 sq. mi./
386,850 sq. km
MONEY:
Dollar
MAJOR EXPORT:
Tobacco
MAJOR IMPORT:
Machinery

Countries not drawn to scale.

* Communauté Financière Africaine
(African Financial Community)

UNIT 7

The Sahel

A dust storm kicks up a huge wall of sand on the Sahel in northern Burkina Faso.

Bringing in the Nets in Accra, Ghana

A fisher pulls in his fishing nets from a harbor in Accra, Ghana's capital.

Drilling for Gold in South Africa

South African gold miners work with large drills and dynamite to extract gold ore.

Chapter 18 West Africa

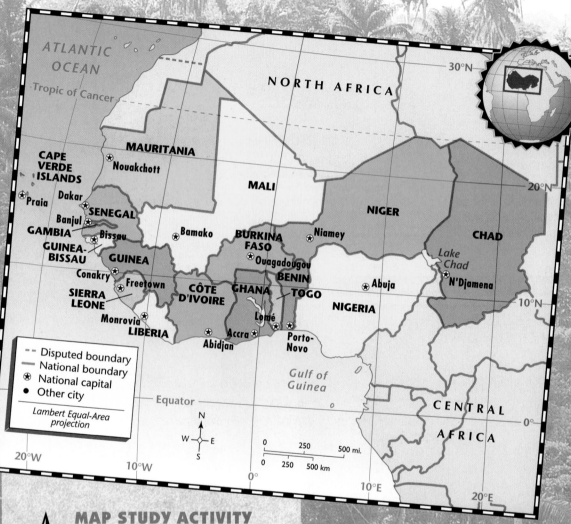

ATLANTIC OCEAN

Tropic of Cancer

NORTH AFRICA

30°N

MAURITANIA

Nouakchott

MALI

20°N

NIGER

CHAD

CAPE VERDE ISLANDS

Praia

Dakar

SENEGAL

Banjul

GAMBIA

Bissau

GUINEA-BISSAU

GUINEA

Bamako

BURKINA FASO

Niamey

Ouagadougou

BENIN

Abuja

Lake Chad

N'Djamena

Conakry

Freetown

CÔTE D'IVOIRE

GHANA

TOGO

NIGERIA

10°N

SIERRA LEONE

Monrovia

LIBERIA

Lomé

Accra

Abidjan

Porto-Novo

Gulf of Guinea

CENTRAL AFRICA

0°

Equator

N
W E
S

0   250   500 mi.
0 250 500 km

20°W   10°W   0°   10°E   20°E

Legend:
- - - Disputed boundary
—— National boundary
⊛ National capital
• Other city

Lambert Equal-Area projection

MAP STUDY ACTIVITY

In this chapter you will read about Nigeria and the other countries of West Africa.

1. **How many countries make up West Africa?**
2. **What West African country is made up of islands?**
3. **What body of water do the coastal countries border?**

Nigeria

Words to Know
- mangrove
- savanna
- harmattan
- cacao
- ethnic group
- compound

Places to Locate
- Nigeria
- Gulf of Guinea
- Niger River
- Lagos
- Abuja

Read to Learn . . .
1. what landforms are found in Nigeria.
2. what mineral supports Nigeria's economy.
3. why Nigeria has many ethnic groups.

The mighty Niger River flows about 2,600 miles (4,180 km) through the western part of Africa. It is the third-longest river in Africa and drains an area larger than the state of Alaska! The Niger River is a major "highway" for those who live along it.

The West African country of Nigeria takes its name from the Niger River, which flows through western and central Nigeria. One of the largest nations in Africa, Nigeria covers 351,650 square miles (910,774 sq. km). It is more than twice the size of the state of California.

Place
The Land

Imagine taking a trip through Nigeria from south to north. Nigeria has a long coastline on the Gulf of Guinea. The map on page 484 shows you that this gulf is part of the Atlantic Ocean. Along Nigeria's coast, the land is laced with many rivers and creeks and covered with mangrove swamps. A **mangrove** is a tropical tree with roots extending above and beneath the water.

As you travel north, the land becomes a vast tropical rain forest. Small villages and farms appear in only a few clearings. The forests gradually thin into woodlands and savannas in central Nigeria. **Savannas** are tropical grasslands with scattered trees. Highlands and plateaus also blanket this area. Farther north, a stretch of partly dry grasslands lies on the edge of the vast Sahara.

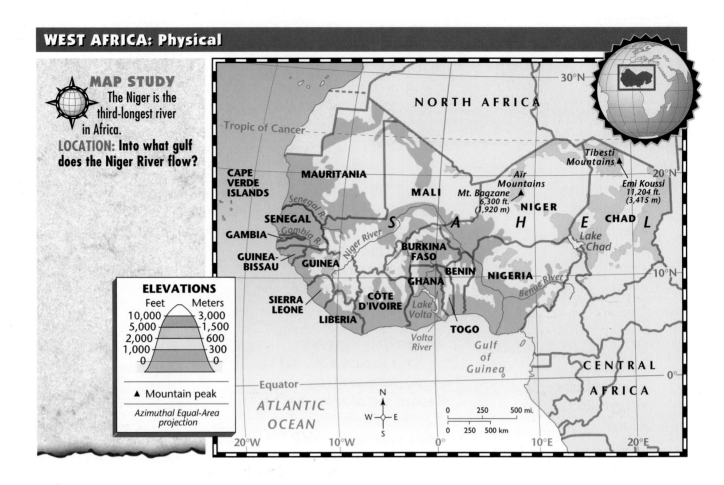

MAP STUDY The Niger is the third-longest river in Africa.

LOCATION: Into what gulf does the Niger River flow?

ELEVATIONS

Feet	Meters
10,000	3,000
5,000	1,500
2,000	600
1,000	300
0	0

▲ Mountain peak

Azimuthal Equal-Area projection

The Climate The map on page 491 shows you that Nigeria has several climate regions. Most of the country, however, is tropical with high average temperatures. Tropical areas in the south receive plenty of rainfall year-round. Steppe regions to the north have a distinct dry season. In the winter months, a dry, dusty wind called the **harmattan** blows south from the Sahara.

Human/Environment Interaction

The Economy

Nigeria has a developing economy based on mining and farming. It is one of the world's major oil-producing countries and one of the few industrialized African countries. In recent years oil has become the leading source of income for the government. The Nigerian government has used money from oil exports to build factories that turn out textiles, food products, and chemicals. Coal and tin deposits also help boost the country's industrial growth.

Boom and Bust In the 1980s Nigeria's oil-rich economy faced hard times. World oil prices fell, and Nigeria's income declined. Nigerians also realized that, in focusing on industry, they had neglected agriculture. Many farmers had left their farms in search of jobs in the cities. Food production fell so much that Nigeria had to import food. Today Nigeria is using income from oil to improve farming and to raise more food products.

Agriculture If you like chocolate, you may have Nigeria to thank. Nigeria's major crops include peanuts, cotton, rubber, palm products, and **cacao.** Beans from the cacao, or cocoa, pod are used to make chocolate. Despite their country's mineral riches, most Nigerians earn their livings as farmers. More than one-half of Nigeria's land is good for farming. Some Nigerian farmers grow subsistence crops, or crops only for their own use. Others produce cash crops, or crops to sell for export.

Place
The People

Like many African countries, Nigeria has a high birthrate and a rapidly growing population. About 98 million people live in Nigeria—more people than in any other country in Africa.

Ethnic Groups Nigeria's people belong to more than 250 ethnic groups. An **ethnic group** is a group of people that has a common language, culture, and history. The four largest ethnic groups are the Hausa (HOW•suh), Fulani (FOO•LAH•nee), Yoruba (YAWR•uh•buh), and Ibo (EE•boh).

Nigerians speak many different African languages. They use English, however, in business and government affairs. About half of Nigeria's people are Muslims and another 40 percent are Christians. The remaining 10 percent practice traditional African religions.

Lagos, Nigeria

Lagos is Nigeria's chief port and commercial center.
PLACE: What language do most Nigerians use when conducting business?

Leading Cocoa-Producing Countries

GRAPHIC STUDY
Half of the world's top cocoa-producing countries are in Africa south of the Sahara.
PLACE: What African country leads the world in cocoa production?

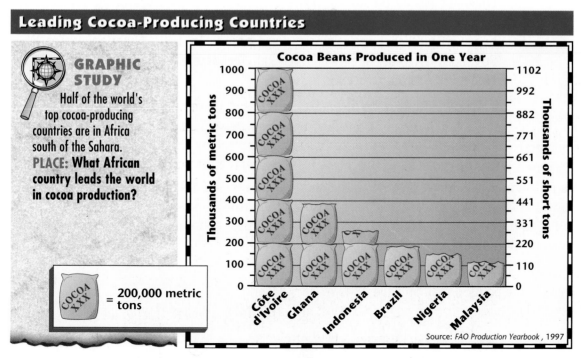

= 200,000 metric tons

Cocoa Beans Produced in One Year

Thousands of metric tons: 0, 100, 200, 300, 400, 500, 600, 700, 800, 900, 1000

Thousands of short tons: 0, 110, 220, 331, 441, 551, 661, 771, 882, 992, 1102

Côte d'Ivoire, Ghana, Indonesia, Brazil, Nigeria, Malaysia

Source: *FAO Production Yearbook*, 1997

Influences of the Past The history of Nigeria reaches farther back than that of almost any other in Africa. About 500 B.C., the Nok people hammered iron into tools and made baked clay sculptures. During the following centuries, powerful city-states and kingdoms arose throughout the region and became centers of trade and the arts. Territories in the north came under the influence of Islamic cultures from North Africa. Peoples in the south developed cultures based on traditional African religions.

Europeans looking for gold and slave labor arrived in Africa in the 1400s. They set up trading forts in Nigeria as early as 1498. By the early 1900s, the British controlled the area that is now Nigeria. The many ethnic groups there resisted British rule. On October 1, 1960, the colony of Nigeria finally became an independent nation. The ethnic differences that intensified during the time of European rule later erupted in civil war. Today Nigeria is moving from harsh military rule to democracy.

Ways of Life About 84 percent of Nigeria's population live in rural villages. Fourteen-year-old Elizabeth Tofa lives in a small village in western Nigeria. Like most rural Nigerians, Elizabeth and her family live in a **compound,** or a group of houses surrounded by walls. Every four days there is a market in her village. Run by women, markets are important in Nigeria and the rest of Africa. They not only provide a variety of goods but also a chance for friends to meet. Elizabeth, her mother, and her sisters sell meat, palm oil, cloth, yams, and nuts.

Nigeria has several large cities bustling with activity. Lagos, located on the coast, is the largest city and commercial center. Other cities—such as Ibadan (ih•BAH•duhn) and Abuja (ah•BOO•jah), the national capital—lie inland.

Nigerians take pride in their art. They make elaborate wooden masks, metal sculptures, and colorful cloth. Playing drums, horns, and other instruments, musicians create rhythms with two or more patterns going at once. In the past, Nigerians passed on stories, proverbs, and riddles by word of mouth from one generation to the next. During the mid-1900s, however, many Nigerian authors began to publish novels, stories, and poetry. In 1986 Nigerian writer Wole Soyinka (WAH•lay shaw•YIHNG•kah) became the first African to win the Nobel Prize for literature.

SECTION 1 ASSESSMENT

REVIEWING TERMS AND FACTS

1. Define the following: mangrove, savanna, harmattan, cacao, ethnic group, compound.

2. PLACE How does Nigeria's landscape change from south to north?

3. HUMAN/ENVIRONMENT INTERACTION What is Nigeria's most important mineral resource?

4. PLACE Why are village markets important in Nigeria?

MAP STUDY ACTIVITIES

5. Turn to the political map on page 484. What is Nigeria's capital?

6. Look at the physical map on page 486. Most of northern Nigeria lies at what elevation?

BUILDING GEOGRAPHY SKILLS

Interpreting a Diagram

Have you ever tried to put something together without directions or even a picture to follow? A diagram can be very helpful in cases like this. A *diagram* is a drawing that shows how something works or how its parts fit together.

In most diagrams, labels point out parts or steps in a process. Labels such as words, letters, colors, or numbers identify different parts. Arrows sometimes show the order in which steps follow one another or the direction of an object's movement.

To interpret a diagram, apply the following steps:

- Read the title or caption to find out what the diagram shows.
- Read all labels to determine their meanings.
- Look for arrows that show movement or the order of the steps to be completed.

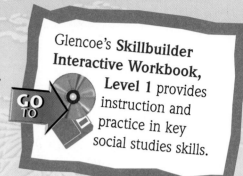

Glencoe's **Skillbuilder Interactive Workbook, Level 1** provides instruction and practice in key social studies skills.

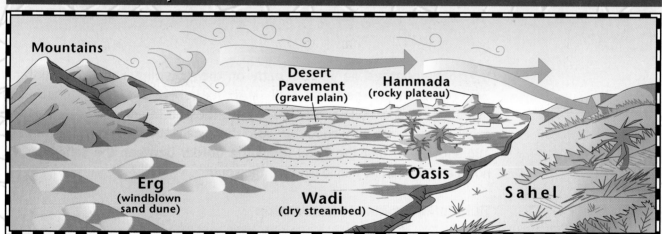

A Saharan Landscape

Mountains

Desert Pavement (gravel plain)

Hammada (rocky plateau)

Erg (windblown sand dune)

Wadi (dry streambed)

Oasis

Sahel

Geography Skills Practice

1. What is the title of this diagram?
2. What labels are used?
3. What do the arrows in the diagram show?

4. According to the diagram, how is the Sahel landscape different from the desert landscape?

SECTION 2

The Sahel Countries

PREVIEW

Words to Know
- drought
- desertification

Places to Locate
- the Sahel
- Niger River
- Mauritania
- Mali
- Niger
- Burkina Faso
- Chad

Read to Learn . . .
1. where the Sahel is located.
2. why Sahel grasslands are turning into desert areas.
3. how people in the Sahel countries live.

Slowly but surely, the desert is creeping into grassy inland areas of West Africa north of Nigeria. Over the past 100 years, a stretch of the Sahara more than 93 miles (150 km) wide has swallowed parts of countries in West Africa. This growing desert is like an invading army slowly taking over the countries.

What countries lie in the path of the creeping Sahara? Mauritania (MAWR•uh•TAY•nee•uh), Mali (MAH•lee), Niger (NY•juhr), Chad, and Burkina Faso (bur•KEE•nuh FAH•soh) are located in an area known as the Sahel. *Sahel* comes from an Arabic word that means "border." The map on page 486 shows you that the Sahel forms the border between the Sahara to the north and the fertile, humid lands to the south.

Human/Environment Interaction
The Growing Desert

The Sahel receives little rainfall, so only short grasses and small trees are able to grow here. As you might guess, this type of vegetation is good for grazing animals. Most people in the Sahel are livestock herders. Their flocks, unfortunately, have overgrazed the land, stripping areas so bare that plant life does not grow back. Without plants, the bare soil is blown away by winds.

To make matters worse, the Sahel has entered a period of **drought.** A drought is an extreme shortage of water. Droughts are common in the Sahel,

490

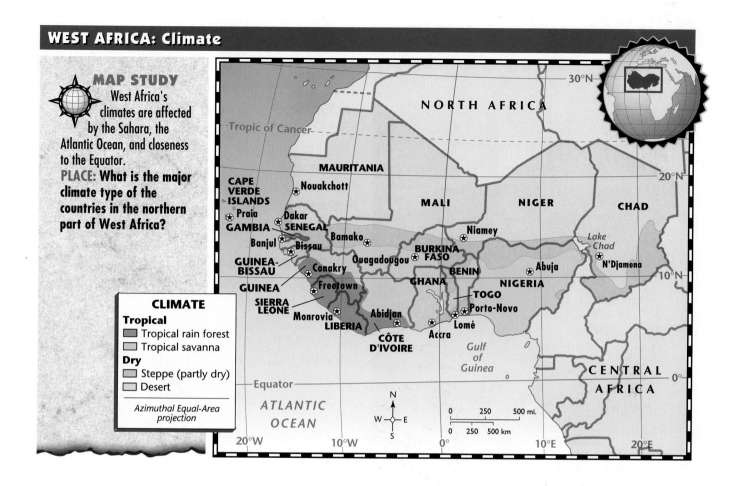

MAP STUDY
West Africa's climates are affected by the Sahara, the Atlantic Ocean, and closeness to the Equator.

PLACE: What is the major climate type of the countries in the northern part of West Africa?

CLIMATE
Tropical
■ Tropical rain forest
□ Tropical savanna
Dry
▨ Steppe (partly dry)
□ Desert

Azimuthal Equal-Area projection

where dry and wet periods regularly follow each other. The latest period of drought began in the 1980s. Since then rivers have dried up, crops have failed, and millions of cattle and other animals have died. Thousands of people have died of hunger. Millions more have fled to fertile southern areas in search of food.

Over the years, both overgrazing and drought have ruined once-productive areas of the Sahel. Many grassland areas have become desert—a change called **desertification.** Scientists believe that the increasing human use of land in the Sahel will only increase the spread of desert. You can read more about desertification south of the Sahara on page 568.

Place
People of the Sahel

The countries of the Sahel are large in size but have small populations. Look at the population density map on page 496. It shows you that most people live in the southern areas of these countries. Can you guess why? Rivers flow here, and the land can be farmed or grazed. Even these areas do not have enough water and fertile land to support large numbers of people, however.

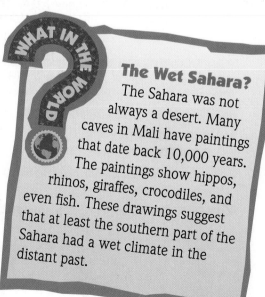

The Wet Sahara?
The Sahara was not always a desert. Many caves in Mali have paintings that date back 10,000 years. The paintings show hippos, rhinos, giraffes, crocodiles, and even fish. These drawings suggest that at least the southern part of the Sahara had a wet climate in the distant past.

CHAPTER 18

Djenné, Mali

Villagers set up shop at a street market outside the Great Mosque in Djenné.
MOVEMENT: How is the Arab influence in Mali shown in this photo?

Influences of the Past You may be surprised to learn that the Sahel was an early trading and cultural area. At a time when the climate was favorable, three great African empires—Ghana (GAH•nuh), Mali, and Songhai (SONG•hy)—arose in the Sahel. From the A.D. 300s to the 1500s, these empires controlled the trade in gold, salt, and other resources between West Africa and the Arab lands of the Mediterranean. The city of Tombouctou (tohn•book•too) on the Niger River was a leading center of learning, drawing students from Europe, Asia, and Africa.

During the 1800s, the countries of the Sahel came under French rule. They finally won their independence in 1960.

A Rural Population Most people in the Sahel live in small towns and villages. They raise only enough food for their own use. Porridges made with sorghum, millet, and other grains are popular dishes.

For years, nomadic groups such as the Tuareg (TWAH•REHG) and Fulani crossed northern desert areas with herds of camels, cattle, goats, and sheep. The recent droughts, however, have forced many of them to give up their traditional way of life and move to the towns and villages.

The people of the Sahel practice a mix of African, Arab, and European traditions. Most follow the religion of Islam. They speak Arabic as well as a variety of African languages. French is also spoken in such cities as Nouakchott (nu•AHK•SHAHT) in Mauritania, Bamako (BAH•muh•KOH) in Mali, and Niamey (nee• AH•MAY) in Niger.

SECTION 2 ASSESSMENT

REVIEWING TERMS AND FACTS

1. Define the following: drought, desertification.

2. LOCATION Where is the Sahel located?

3. HUMAN/ENVIRONMENT INTERACTION How have humans affected the spread of desert in the Sahel?

4. PLACE How do most people make their livings in the Sahel countries?

MAP STUDY ACTIVITIES

5. Look at the climate map on page 491. What climates do you find in the Sahel countries?

MAKING CONNECTIONS

Wole Soyinka of Nigeria is perhaps Africa's best-known playwright and writer. In this excerpt from *The Interpreters,* friends talk as they journey in their canoe. They discuss how the landscape makes them feel.

Two paddles clove [moved through] the still water of the creek, and the canoe trailed behind it a silent groove, between gnarled tears of mangrove; it was dead air, and they came to a spot where an old rusted cannon showed above the water. . . . The paddlers slowed down and held the boat against the cannon. Egbo put his hand in the water and dropped his eyes down the brackish stillness, down the dark depths to its bed of mud.

. . . The canoe began to move off.

. . . From the receding cannon a quizzical crab emerged, seemed to stretch its claws in the sun and slipped over the edge, making a soft hole in the water. The mangrove arches spread seemingly endless and Kola broke the silence saying, 'Mangrove depresses me.'

'Me too,' said Egbo. . . . I remember when I was in Oshogbo I loved [the] grove and would lie there for hours listening at the edge of the water. It has a quality of this part of the creeks, peaceful and comforting. . . .' He

trailed his hand in the water as he went, pulling up lettuce and plaiting [braiding] the long whitened root-strands.

Mangrove Trees

'That was only a phase of course, but I truly yearned for the dark. I loved life to be still, mysterious. I took my books down there to read, during the holidays. But later, I began to go further, down towards the old suspension bridge where the water ran freely, over rocks and white sand. And there was sunshine. . . . It was so different from the grove where depth swamped me. . . .

From Wole Soyinka's *The Interpreters.* Copyright © 1965 by Wole Soyinka, New York, N.Y.: Africana Publishing Corporation.

Making the Connection

1. What type of physical environment surrounds Kola and Egbo?
2. How do they feel when they are in this area?

SECTION 3 Coastal Countries

PREVIEW

Words to Know
- cassava
- bauxite

Places to Locate
- Senegal
- Dakar
- Cape Verde Islands
- Liberia
- Côte d'Ivoire
- Abidjan
- Ghana

Read to Learn . . .
1. what climates are found in coastal West Africa.
2. how people earn their livings in this region.
3. how coastal West African countries won their independence.

Do you like peanuts? The next time you bite into a peanut butter sandwich, think about West Africa. Workers in the West African country of Gambia dry peanuts to send to other countries. Gambia began exporting peanuts in the 1800s. Today it depends on this one crop for about 95 percent of its income.

Besides Nigeria and the Sahel countries, West Africa includes 11 coastal countries. One—Cape Verde—is a group of 15 islands in the Atlantic Ocean. The other countries spread along West Africa's Atlantic bend, north of the Equator, from Senegal in the west along the coast to Benin. Find these countries on the map on page 484.

Human/Environment Interaction
The Land

Coastal lowlands sweep along the Atlantic shore of West Africa's coastal countries. Sandy beaches, rain forests, and thick mangrove swamps cover most of this area. Inland from the coast lie highland areas with grasses and trees. Several major rivers—the Senegal, Gambia, Volta, and Niger—flow from these highlands to the coast. You won't see many ships on these rivers. Why? Rapids and shallows prevent ships from traveling far into inland areas.

The Climate Because they border the ocean, the coastal countries of West Africa receive plenty of rainfall. Warm currents in the Gulf of Guinea create a moist, tropical climate in the coastal lowlands year-round. Some coastal areas receive more than 170 inches (432 cm) of rain a year! Farther inland, rainfall is not as plentiful. These highland areas have a tropical savanna climate with dry and wet seasons.

The Economy Because of the wet climate, most people in coastal West Africa are farmers. On small plots of land, farmers grow yams, corn, rice, cassava, and other foods for their families. **Cassava** is a plant root that looks like a potato. It is used to make flour for bread.

West Africa also has large plantations that grow coffee, rubber, cacao, palm oil, and kola nuts for export. Can you guess what the kola nut is used for? This sweet-tasting nut is used in making soft drinks. Until industries become more developed in the region, these cash crops will continue to be the major source of income.

Region
The People

West Africa's coastal countries have a long and rich history. In early times, powerful kingdoms such as the Ashanti and Benin ruled the region. These kingdoms were centers of trade, learning, and the arts.

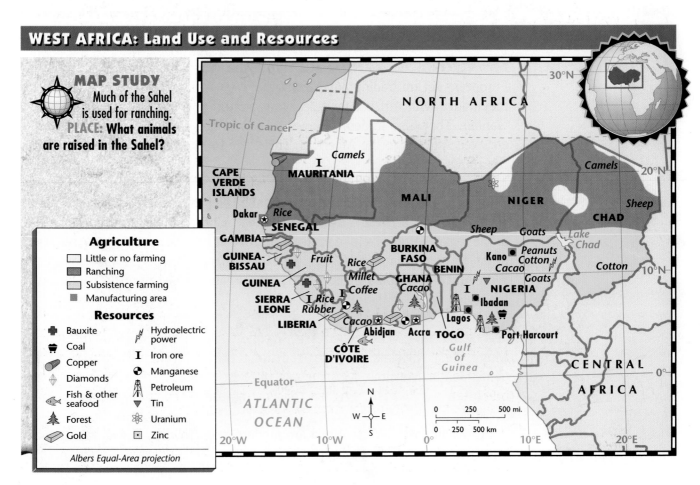

WEST AFRICA: Land Use and Resources

MAP STUDY
Much of the Sahel is used for ranching.
PLACE: What animals are raised in the Sahel?

Agriculture
☐ Little or no farming
■ Ranching
☐ Subsistence farming
■ Manufacturing area

Resources
✚ Bauxite
🚃 Coal
📦 Copper
♦ Diamonds
🐟 Fish & other seafood
🌲 Forest
▭ Gold

⚡ Hydroelectric power
I Iron ore
☻ Manganese
⚒ Petroleum
▽ Tin
☢ Uranium
⊡ Zinc

Albers Equal-Area projection

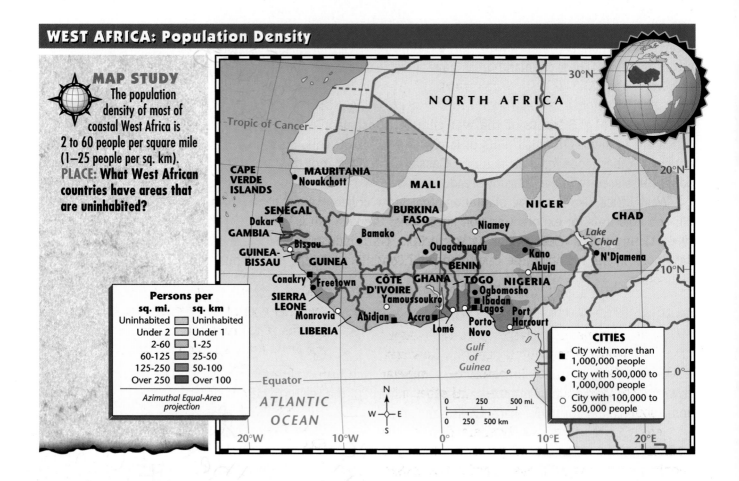

MAP STUDY The population density of most of coastal West Africa is 2 to 60 people per square mile (1–25 people per sq. km). **PLACE: What West African countries have areas that are uninhabited?**

Persons per

sq. mi.	sq. km
Uninhabited	Uninhabited
Under 2	Under 1
2-60	1-25
60-125	25-50
125-250	50-100
Over 250	Over 100

Azimuthal Equal-Area projection

CITIES
- ■ City with more than 1,000,000 people
- ● City with 500,000 to 1,000,000 people
- ○ City with 100,000 to 500,000 people

The Arrival of Europeans From the late 1400s to the early 1800s, Europeans set up trading posts along the West African coast. From these posts they traded with Africans for gold, ivory, and other goods that people in Europe wanted. They also traded enslaved Africans. The traders shipped millions of enslaved Africans to the Americas. This trade in human beings finally ended in the 1800s.

Toward Freedom The British, French, and Portuguese eventually set up colonies in coastal West Africa. They brought Christianity and European culture to the region. After several decades of European rule, the peoples of West Africa began to demand independence. In 1957 Ghana became the first country in West Africa to become independent. By the late 1970s, all of the countries in the region had won their independence.

Ways of Life The cultures of West Africa show European influences. Many West Africans practice traditional African religions, but some are Christians or Muslims. Most people in West Africa speak African languages but often use French, English, or Portuguese in business or government activities.

Cities in coastal West Africa are modern and growing. If you visited a city like Ghana's capital—Accra (uh•KRAH)—or Côte d'Ivoire's capital—Abidjan (A•bih•JAHN)—you would find downtown areas with modern government and office buildings. You would also see people dressed in business clothes or in traditional African clothing and wide streets with busy traffic. In rural areas, people live in small villages. Many village homes have mud walls and straw or metal roofs. People here follow many of their ancestors' traditions.

Region
Country Profiles

The 11 countries of coastal West Africa have many things in common. Each country, however, has certain features that make it stand out from the others.

Senegal, Gambia, and Guinea These countries all depend on farming as their major source of income. The map on page 495, however, shows that Guinea is rich in **bauxite**—a mineral from which we make aluminum—and other mineral resources.

Senegal has more city dwellers than either Gambia or Guinea. About 42 percent of Senegal's people live in urban areas. Dakar (DA•KAHR), Senegal's capital and largest city, also is a major port on the Atlantic Ocean.

Village Life

Iyabo Ayinde lives in a small village outside Abidjan. The people in her village are related, and each family has its own group of dwellings. Her family's homes have cone-shaped thatched roofs. Villagers share a small plot of land where they raise corn and yams. Today after school, Iyabo will meet her cousins in the central square of her village to plan a trip to Abidjan. Iyabo's cousins want to go to the art museum. Iyabo has another idea. "Let's go to the ice skating rink!" she says.

Guinea-Bissau and Cape Verde Guinea-Bissau lies between Senegal and Guinea. The islands of Cape Verde are located about 375 miles (600 km) offshore in the Atlantic Ocean. Guinea-Bissau and Cape Verde were Portuguese colonies until they won independence in 1975. Portuguese influences are widespread in the languages and cultures of both countries. Most of the people here make their livings by farming or fishing.

Liberia and Sierra Leone At the "bend" of Africa's west coast are Liberia and Sierra Leone. Liberia is the only West African country that was never a colony. African Americans freed from slavery founded Liberia in 1822. Monrovia, Liberia's capital and largest city, was named for James Monroe—American President when Liberia was founded. From 1989 to 1996, fighting among Liberian political groups cost many lives and left thousands homeless.

Like Liberia, Sierra Leone was established as a home for people freed from enslavement. The British ruled Sierra Leone from 1787 until 1961, when Sierra Leone became independent. Most of the country's land is used for agriculture, but minerals also provide wealth.

Côte d'Ivoire and Ghana Côte d'Ivoire was a French colony until winning its independence in 1960. Many ethnic groups live in Côte d'Ivoire's rural

Abidjan, Côte d'Ivoire

Hotel Sofitel

The modern port of Abidjan handles many of the imports and exports of West Africa. **PLACE: What European country controlled Côte d'Ivoire until 1960?**

villages. They cultivate coffee, cacao, and forest products for export. In recent years some farmers have moved from their villages to find factory jobs in Abidjan and other cities.

Ghana—like Côte d'Ivoire—produces cacao and forest products for export. Ghana's people belong to about 100 different ethnic groups. The Ashanti and the Fante are the largest groups. About 35 percent of Ghanaians live in cities. Accra, located on the coast, is the capital and largest city.

Togo and Benin The countries of Togo and Benin lie between Ghana and Nigeria. The map on page 484 shows you that both countries are long and narrow. Togo and Benin—once French colonies—became independent in 1960. Many ethnic groups whose dress, language, and other ways of life differ a great deal make up the cultures of Togo and Benin.

SECTION 3 ASSESSMENT

REVIEWING TERMS AND FACTS

1. **Define the following:** cassava, bauxite.

2. **LOCATION** Where is Cape Verde located?

3. **HUMAN/ENVIRONMENT INTERACTION** What food crops are grown in coastal West Africa?

4. **REGION** How did the arrival of Europeans influence life in coastal West Africa?

MAP STUDY ACTIVITIES

5. Study the population density map on page 496. Where do most people in coastal West Africa live?

Chapter 18 Highlights

Important Things to Know About West Africa

SECTION 1 NIGERIA

- Nigeria's major landforms are coastal lowlands, savanna highlands, and partly dry grasslands.
- Most Nigerians farm the land, but oil is Nigeria's major export.
- Many different ethnic groups live in Nigeria.
- Nigeria is the most highly populated African country.

SECTION 2 THE SAHEL COUNTRIES

- The Sahel countries are Mauritania, Mali, Niger, Chad, and Burkina Faso.
- The Sahel lies between the Sahara in the north and fertile, humid lands in the south.
- Most people in the Sahel countries are herders or farmers.
- Desertification has destroyed much of the Sahel's most productive land.

SECTION 3 COASTAL COUNTRIES

- The 11 countries that make up coastal West Africa are Senegal, Gambia, Guinea, Guinea-Bissau, Cape Verde, Liberia, Sierra Leone, Ghana, Côte d'Ivoire, Togo, and Benin.
- Because of the wet climate, farming is the major economic activity in coastal West Africa.
- Most people in the coastal countries of West Africa live in the coastal lowlands.
- People freed from slavery helped establish the countries of Liberia and Sierra Leone.

Rain forest in Liberia ▶

REVIEWING KEY TERMS

Match the numbered terms in Set A with their lettered definitions in Set B.

A
1. cassava
2. savanna
3. mangrove
4. cacao
5. desertification
6. harmattan
7. bauxite

B
A. change of grasslands to desert
B. dusty wind blowing south from the Sahara
C. plant root used in making flour
D. tropical grassland with scattered trees
E. mineral from which aluminum is made
F. beans used in making chocolate
G. tropical tree with roots above and below water

Mental Mapping Activity

Draw a freehand map of West Africa. Label the following:
- Gulf of Guinea
- Atlantic Ocean
- Lagos
- Senegal River
- Gambia
- Sahara

REVIEWING THE MAIN IDEAS

Section 1
1. HUMAN/ENVIRONMENT INTERACTION What challenges has Nigeria faced because of its dependence on oil?
2. MOVEMENT Why are Nigeria's cities growing?

Section 2
3. HUMAN/ENVIRONMENT INTERACTION Why is farming difficult in the Sahel?
4. PLACE Why was the city of Tombouctou important?
5. REGION What is the major religion in the Sahel countries?

Section 3
6. REGION Why does the coastal strip in coastal West Africa receive plenty of rainfall?
7. PLACE What West African country was the first to win independence?

CRITICAL THINKING ACTIVITIES

1. **Drawing Conclusions** Why do you think people in many areas of the Sahel have followed a nomadic way of life?
2. **Determining Cause and Effect** How has location on the coast affected the history and culture of coastal West Africa?

CONNECTIONS TO WORLD CULTURES

Cooperative Learning Activity Work in a group of three to learn more about the arts of West Africa. As a group, choose one country in West Africa. Each member in the group will then research that country's sculpture, music, or dance. After your research is complete, share your information with your group. Present your country's art to the class as a whole.

GeoJournal Writing Activity

Imagine that you are on a journey down the Niger River. Write several entries in a diary describing how the landscape changes as you travel down the river from inland areas to the coast.

TECHNOLOGY ACTIVITY

Using a Computerized Card Catalog

Choose a West African country to research, then use a computerized card catalog to find information about its early history to the present. Create a bulletin board display about that country, including an illustrated time line of significant events in the country's history. Be sure to include current information about natural resources and environmental concerns in your display.

PLACE LOCATION ACTIVITY: WEST AFRICA

Match the letters on the map with the places and physical features of West Africa. Write your answers on a separate sheet of paper.

1. Côte d'Ivoire
2. Monrovia
3. Tibesti Mountains
4. Senegal
5. Mali
6. Gulf of Guinea
7. Lake Chad
8. Niger River
9. Cape Verde Islands
10. Lagos

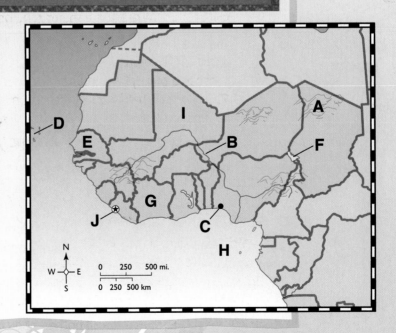

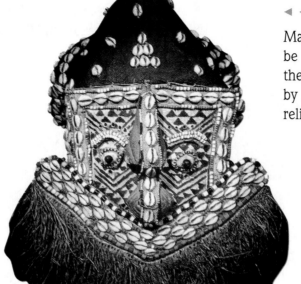

◄ ◄ ◄ ◄ MASKS

Many of Africa's ethnic groups can be identified by the types of masks they create. Masks are usually worn by African dancers in social and religious ceremonies.

SCULPTURE ► ► ► ►

Europeans often paid African artists to create intricate carvings for their personal collections. Skilled ivory carvers created this salt container from a single elephant tusk in the 1600s.

◄ ◄ ◄ ◄ DANCE

Many African ethnic groups use dance to mark important events in their lives. The rhythms in African music are reproduced by dancers' bodies. Their heads may be moving to one rhythm, their shoulders to another, and their arms and feet to yet another.

◄ ◄ ◄ ◄ ARCHITECTURE

Much of the architecture in Africa reflects the influence of the Arab world. This modern mosque with its towering minarets is located in Nigeria.

MUSIC ▶ ▶ ▶ ▶

African music can have as many as 5 to 12 different rhythms. The complicated rhythms are often produced on drums, horns, and trumpets made from natural materials. Jazz and other music popular in the United States have been greatly influenced by African music.

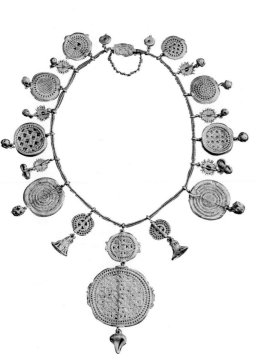

JEWELRY ▲ ▲ ▲ ▲

The gold trade in the Ashanti empire in the 1600s brought great riches to the Ashanti kings. Many kings even had their own goldsmiths. This breast ornament was made by twisting delicate gold wire into lacelike patterns.

APPRECIATING CULTURE

1. How would you describe this African mask to someone who had not seen it?
2. What are some characteristics of African music?
3. Why do you think gold was a symbol of power and status in the Ashanti empire?

Chapter 19 Central Africa

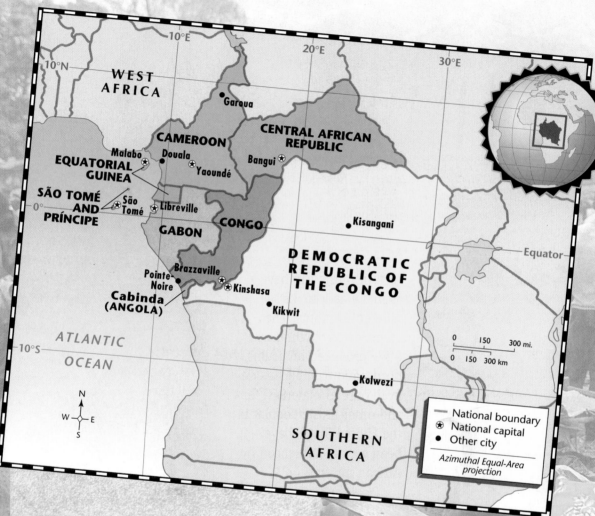

WEST AFRICA

10°N

10°E

20°E

30°E

Garoua

CAMEROON

CENTRAL AFRICAN REPUBLIC

Malabo

Douala

Yaoundé

Bangui

EQUATORIAL GUINEA

SÃO TOMÉ AND PRÍNCIPE

0°

São Tomé

Libreville

CONGO

Kisangani

GABON

DEMOCRATIC REPUBLIC OF THE CONGO

Equator

Brazzaville

Pointe-Noire

Kinshasa

Cabinda (ANGOLA)

Kikwit

ATLANTIC

10°S

OCEAN

Kolwezi

N
W E
S

SOUTHERN AFRICA

0   150   300 mi.
0   150   300 km

—— National boundary
⊛ National capital
● Other city

Azimuthal Equal-Area projection

MAP STUDY ACTIVITY

As you read Chapter 19, you will learn about the countries of Central Africa.

1. How many countries are in Central Africa?

2. What is the largest country in Central Africa?

3. What is the capital of Gabon?

Democratic Republic of the Congo

PREVIEW

Words to Know
- canopy
- hydroelectricity

Places to Locate
- Congo River
- Kinshasa
- Lake Tanganyika

Read to Learn . . .
1. why the Congo River is a highway for the Democratic Republic of the Congo's people.
2. about the potential wealth of the Democratic Republic of the Congo.
3. how large movements of people have shaped the history of the Democratic Republic of the Congo.

This lush countryside lies in the Democratic Republic of the Congo, one of the countries in the region of Central Africa. Water is abundant in the rainy and green stretches of this region.

Central Africa's many rivers are a source of life for the people of the region. Africa's second-longest river—the Congo—flows through the Democratic Republic of the Congo. This country is located on the Equator in the very heart of Africa. A large country, this republic is bordered by nine other African countries.

Place
The Land

The Democratic Republic of the Congo covers an area of 875,309 square miles (2,267,050 sq. km)—about one-fourth the size of the United States. In Africa, only Sudan and Algeria are larger. The Democratic Republic of the Congo encompasses many landscapes, including rain forests, savannas, and highlands. Its only coastline is a 25-mile (41-km) strip on the Atlantic Ocean.

Rain Forests Vast tropical rain forests cover about one-third of the Democratic Republic of the Congo. It is one of the largest rain forest areas in the world. The treetops of the rain forests form a **canopy,** or umbrella-like forest covering. This canopy is so thick that sunlight seldom reaches parts of the forest

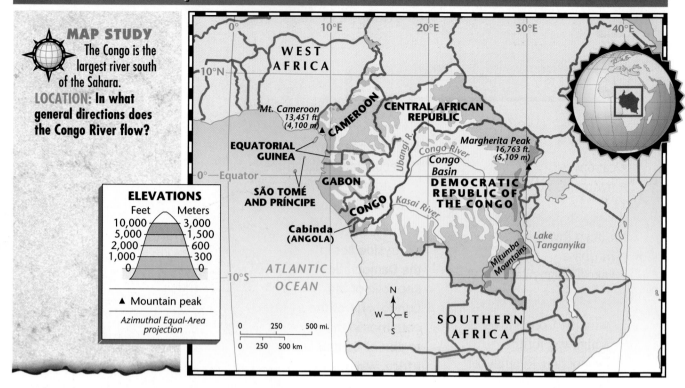

MAP STUDY
The Congo is the largest river south of the Sahara.
LOCATION: In what general directions does the Congo River flow?

ELEVATIONS

Feet	Meters
10,000	3,000
5,000	1,500
2,000	600
1,000	300
0	0

▲ Mountain peak

Azimuthal Equal-Area projection

floor, which is almost clear of plants. Many monkeys, snakes, birds, and other small animals spend their entire lives in the canopy.

More than 750 different kinds of trees grow in the rain forests. Mahogany, ebony, and other woods prized by furniture makers are harvested in abundance. Like South America's Amazon rain forest, Central Africa's rain forests are being destroyed at a fast rate.

Highlands and Savannas Spectacular mountains rise along the eastern border of the Democratic Republic of the Congo. Savannas cross the northern and southern parts of the country. These vast open grasslands are home to some of the country's most valuable natural resources—its wildlife. Antelopes, leopards, lions, rhinoceroses, zebras, and giraffes roam freely on government preserves. An animal similar to a giraffe—the okapi—is found only in the Democratic Republic of the Congo and is the country's national symbol.

Rivers and Lakes The mighty Congo River—above 2,800 miles (4,505 km) long—courses its way through the rain forests of Central Africa to the Atlantic Ocean. Many smaller rivers branch out from the Congo River.

One of the world's longest rivers, the Congo River is the country's highway for trade and travel. Dugout canoes, steamers carrying passengers to various port cities, and cargo ships travel the river to the Atlantic Ocean.

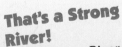

That's a Strong River!
The Congo River doesn't stop when it meets the Atlantic Ocean. The river's current carries water about 100 miles (161 km) into the ocean. The force of the current has gouged a 4,000-foot (1,200-m) canyon into the ocean floor!

In the eastern mountainous region of the Democratic Republic of the Congo, you will find Lake Albert, Lake Edward, Lake Kivu, and Lake Tanganyika (TAN•guh•NYEE•kuh). Lake Tanganyika is the world's longest freshwater lake and the second-deepest, after Lake Baikal in Russia.

The Climate Because of its location on the Equator, the Democratic Republic of the Congo generally has a tropical climate. Rain forest areas are hot and humid all year. Average daytime temperatures here reach about 90°F (32°C). Heavy rainstorms bring 80 inches (203 cm) or more of rain each year. Savanna and highland areas are cooler and drier.

Human/Environment Interaction
The Economy

The Democratic Republic of the Congo has the potential to become wealthy. The map on page 512 shows you its mineral resources. Mining is the major economic activity. The Congolese are among world leaders in copper production. They also lead the rest of Africa in the mining of industrial diamonds, which are used in precision tools.

Energy Sources The Democratic Republic of the Congo has oil and natural gas deposits, but rivers provide the main source of energy. Experts believe that the country's rivers have the ability to produce about 13 percent of the world's **hydroelectricity,** electricity created by moving water. The Congo River carries

A ferry shuttles passengers and vehicles across the Congo River.
HUMAN/ENVIRONMENT INTERACTION: What is hydroelectricity?

Leading Diamond-Producing Countries

GRAPHIC STUDY

Three of the world's top diamond-producing countries are in Africa south of the Sahara.
PLACE: What two countries in Africa produce the most diamonds?

 = 5,000 carats (2,204 pounds)

Diamonds Produced in One Year

Thousands of carats: 50,000 — 40,000 — 30,000 — 20,000 — 10,000 — 0

Kilograms: 10,000 — 8,000 — 6,000 — 4,000 — 2,000 — 0

Country: Australia, Russia, Botswana, Dem. Rep. of the Congo, South Africa

Source: *Encyclopedia Britannica Book of the Year, 1997*

more water than any other river in the world except the Amazon. About 10 million gallons (38 million l) of water rush down the Congo River every second!

Farming Despite rich resources, most people in the Democratic Republic of the Congo make a living by subsistence farming. Farmers grow crops such as corn, rice, and cassava for their families. Coffee, cotton, and palm oil are grown to sell. Along with farming, the Congolese also hunt wild animals in the forests and fish in the rivers, sometimes with traps woven from vines.

Movement

The People

The Democratic Republic of the Congo has a population of about 49 million. Like other African countries, it is made up of different ethnic groups with their own separate languages. At times, tensions between ethnic groups have led to fighting. In recent years, the country has been torn apart by warfare that has cost many lives.

Influences of the Past Groups of people began moving into the Congo region from other parts of Africa about 2,000 years ago. Before the A.D. 1400s, several powerful kingdoms developed in the savanna area south of the rain forests. The largest kingdom was the Kongo.

In the late 1400s, Portuguese and other European traders arrived in Central Africa. During the next 300 years, they enslaved thousands of Central Africans, shipping them to the Americas.

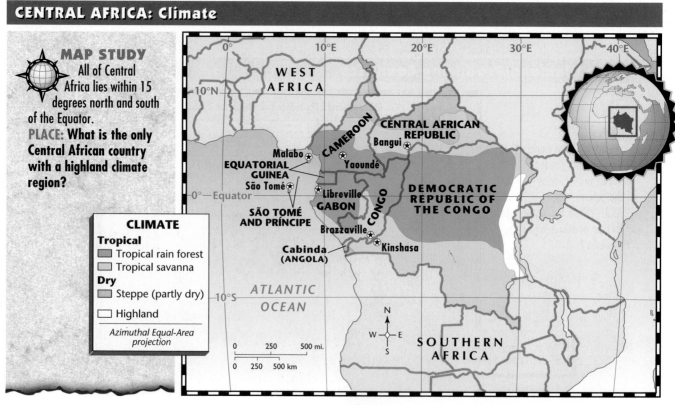

CENTRAL AFRICA: Climate

MAP STUDY All of Central Africa lies within 15 degrees north and south of the Equator.
PLACE: What is the only Central African country with a highland climate region?

CLIMATE
Tropical
- Tropical rain forest
- Tropical savanna

Dry
- Steppe (partly dry)
- Highland

Azimuthal Equal-Area projection

0   250   500 mi.

0   250   500 km

In 1878 Belgium's King Leopold II took over a large area of Central Africa. He considered the entire region his own plantation and treated the people harshly. When these outrages were publicized in 1908, the Belgian government took control of the colony and named it the Belgian Congo. In 1960 the Belgian Congo became an independent nation known as Congo. Nine years later, the country changed its name to Zaire. In 1997, Congolese troops overthrew longtime leader Mobutu Sese Seko. The new military government then renamed the country the Democratic Republic of the Congo.

Ways of Life The culture of the Democratic Republic of the Congo is largely African with European influences. More than 75 percent of the Congolese are Christians. Most of these are Roman Catholics. Other people in the country practice traditional African religions.

About 70 percent of the Congolese live in rural villages. Since the 1960s, however, many of them have migrated to cities in search of jobs. This movement of people has turned Kinshasa, the capital, into a city of about 4.2 million people. Kinshasa is a major port on the Congo River.

Fifteen-year-old Kilundu Basese lives in Kinshasa. During his free time, Kilundu strolls along Kinshasa's streets lined with trees, elegant shops, and outdoor cafés. Many areas of the city are crowded with new residents looking for work.

Market Day

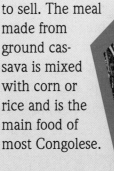

Masika Mbwana lives in a village on the banks of the Congo River. This morning, with other girls from her village, she will walk to the market in a nearby town. Masika's family grows its own food. Any extra food is gathered to sell at the market. Today Masika has a basket of cassava roots to sell. The meal made from ground cassava is mixed with corn or rice and is the main food of most Congolese.

SECTION 1 ASSESSMENT

REVIEWING TERMS AND FACTS

1. Define the following: canopy, hydroelectricity.

2. PLACE Why is the Congo River important to the Congolese people?

3. HUMAN/ENVIRONMENT INTERACTION The Democratic Republic of the Congo is among world leaders in the production of what mineral resource?

4. MOVEMENT How was the Congo region settled?

MAP STUDY ACTIVITIES

5. Turn to the physical map on page 506. What large body of water forms part of the southeast boundary of the Democratic Republic of the Congo?

6. Look at the climate map on page 508. What climates are found in Central Africa?

BUILDING GEOGRAPHY SKILLS

Reading a Bar Graph

You want to compare your running time in the 100-yard dash to that of the other sprinters on your team. Putting your running times on a bar graph would allow you to visually compare all the runners' times at once.

A *bar graph* presents numerical information in a visual way. Bars of various lengths stand for different quantities. Bars may be drawn vertically—up and down—or horizontally—left to right. Labels along the axes, or the left side and bottom of the graph, explain what the bars represent. Some bar graphs show changes over time. Others compare quantities during the same time period, but in different locations. To read a bar graph, apply these steps:

- Read the title to find out what the graph is about.
- Study the information on both axes to figure out what the bars represent.
- Compare the lengths of the bars to draw conclusions about the graph's topic.

Glencoe's **Skillbuilder Interactive Workbook, Level 1** provides instruction and practice in key social studies skills.

Literacy Rate in Selected African Countries

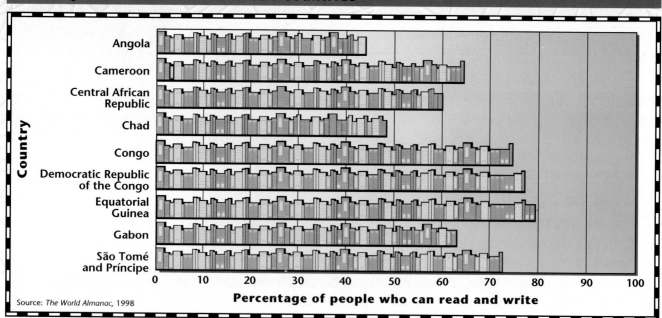

Source: *The World Almanac, 1998*

Percentage of people who can read and write

Country: Angola, Cameroon, Central African Republic, Chad, Congo, Democratic Republic of the Congo, Equatorial Guinea, Gabon, São Tomé and Príncipe

Geography Skills Practice

1. Which countries are compared on this graph?
2. What quantities appear on the horizontal axis? The vertical axis?
3. Which country has the highest literacy rate? Which has the lowest?

Other Countries of Central Africa

PREVIEW

Words to Know
- basin
- tsetse fly

Places to Locate
- Central African Republic
- Cameroon
- Gabon
- Congo
- Equatorial Guinea
- São Tomé and Príncipe

Read to Learn . . .
1. where the Central African Republic and Cameroon are located.
2. what resources are found in Gabon and Congo.
3. how people in Equatorial Guinea and São Tomé and Príncipe make their livings.

The sound of buzz saws rings through Gabon's rain forests. These rain forests hold vast resources, including the ebony trees shown here.

Ebony and mahogany hardwoods provide Central Africa with one of its most valuable sources of income.

In addition to the Democratic Republic of the Congo, Central Africa includes Gabon (ga•BOHN), the Central African Republic, Cameroon, Congo, Equatorial Guinea, and São Tomé (SOWN tuh•MAY) and Príncipe (PRIHN•suh•puh). All of these countries are only beginning to develop their natural resources.

Location

The Central African Republic and Cameroon

The Central African Republic and Cameroon became independent from France in 1960. Both countries are working to develop their economies. Cameroon, with a coastline on the Gulf of Guinea, enjoys greater natural advantages than the landlocked Central African Republic.

The Land The Central African Republic lies deep in the middle of Africa, just north of the Equator. Most of the country lies on a vast plateau bordered on both sides by basins. A **basin,** you will remember, is a broad flat valley. Savannas cover most of the plateau, but tropical rain forests are found in the

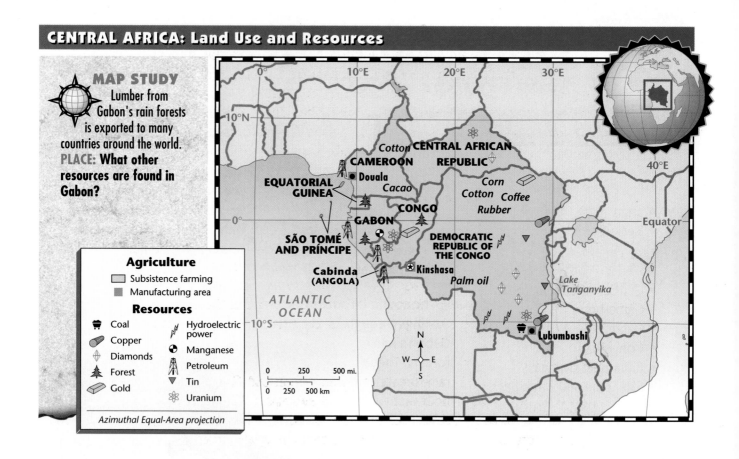

MAP STUDY Lumber from Gabon's rain forests is exported to many countries around the world. **PLACE: What other resources are found in Gabon?**

Agriculture
- ☐ Subsistence farming
- ■ Manufacturing area

Resources
- 🛒 Coal
- 🪵 Copper
- ◈ Diamonds
- 🌲 Forest
- ▭ Gold
- Ⓝ Hydroelectric power
- ◍ Manganese
- ⚒ Petroleum
- ▽ Tin
- ⚛ Uranium

Azimuthal Equal-Area projection

southwestern part of the country. The map on page 508 shows you that the Central African Republic has a tropical savanna climate and a steppe climate.

Cameroon lies on the west coast of Central Africa. In the south hot, humid lowlands stretch along the Gulf of Guinea. The central part of Cameroon is a forested plateau with cooler temperatures and less rainfall. To the north lies a partly dry grasslands area. Mountains and hills run along Cameroon's western border.

The Economy Most people in the Central African Republic and Cameroon depend on farming for a living. In the Central African Republic, most farmers grow only enough food to feed their families. A few large plantations raise coffee, cotton, and rubber for export. Farmers in Cameroon raise cassava, corn, millet, and yams. The chief cash crops are bananas, cacao, coffee, cotton, and peanuts.

The Central African Republic and Cameroon are only beginning to industrialize. Cameroon, however, has had greater success in this effort. It has coastal ports and many natural resources such as petroleum, bauxite, and forest products. With no seaports and limited resources, the Central African Republic can claim only diamond mining as an important industry.

Some people in these two countries herd livestock. They raise their animals in regions that are safe from tsetse flies. The bite of the **tsetse fly** causes a deadly disease called sleeping sickness in cattle. Tsetse flies can also spread this disease to humans.

The People Most of the people in the Central African Republic and Cameroon live in rural areas. In recent years, however, many people from the countryside have moved to cities such as Yaoundé (yown•DAY) in Cameroon and Bangui (bahn•gee) in the Central African Republic in search of work.

Many languages are spoken in these cities. The people of the Central African Republic and Cameroon belong to many ethnic groups. In both countries, most people follow traditional African religions. Smaller numbers are Christians or Muslims.

Place
Gabon and Congo

Gabon and Congo—once French colonies—became independent countries in 1960. Both have many natural resources and excellent waterways. The map on page 508 shows you that Gabon and Congo lie on the Equator along the west coast of Central Africa. Because of this location, both countries have hot, humid climates.

The Land Although they share the same type of climate, Gabon and Congo have different landforms. In Gabon, palm-lined beaches and swamps run along the Gulf of Guinea. As you move inland, you see that the land rises to become rolling hills and low mountains. Thick rain forests crossed by rivers cover most of Gabon.

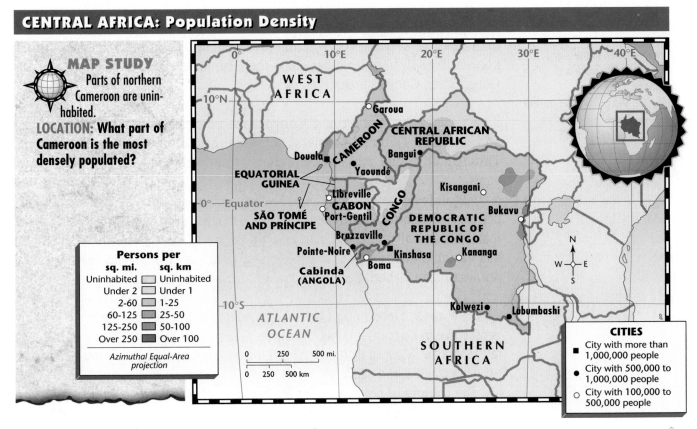

CENTRAL AFRICA: Population Density

MAP STUDY
Parts of northern Cameroon are uninhabited.
LOCATION: What part of Cameroon is the most densely populated?

Persons per	
sq. mi.	**sq. km**
Uninhabited	Uninhabited
Under 2	Under 1
2-60	1-25
60-125	25-50
125-250	50-100
Over 250	Over 100

Azimuthal Equal-Area projection

0   250   500 mi.

0   250   500 km

CITIES
■ City with more than 1,000,000 people
● City with 500,000 to 1,000,000 people
○ City with 100,000 to 500,000 people

Boats are used for fishing and transportation in the island country of São Tomé and Príncipe.
PLACE: Why does the nation of São Tomé and Príncipe have such fertile soil?

In Congo, a low treeless plain stretches along the Atlantic coast. Just inland from this coastal plain, you will find low mountain ranges and plateaus. To the north a large swampy area lies along the Ubangi River. Farther south flows the Congo River, Congo's major waterway.

The Economy Gabon and Congo have economies based on agriculture and mining. Gabon's farmers cultivate cacao and coffee for export. Throughout Congo subsistence farmers grow food crops such as cassava and yams.

Gabon is rich in natural resources. Its rain forests provide high-quality lumber that is exported throughout the world. Gabon also has oil reserves on the mainland and off the coast. The map on page 512 shows you that Gabon also has valuable deposits of manganese and uranium. Congo has fewer natural resources than Gabon. Petroleum and lumber are its main exports.

The People Gabon and Congo have relatively few people. Only 1.2 million people live in Gabon—mainly in villages along rivers or on the coast at the capital, Libreville. Most of Congo's 2.7 million people live along the Atlantic coast or near Brazzaville, the capital.

The cultures of both countries are made up of many ethnic groups. In Gabon the Fang group controls the national government. Two groups—the Kongo and

514

the Bateke (bah•TEH•keh)—make up most of Congo's population. Most Kongo farm for a living. The Bateke are primarily hunters and fishers.

Place
The Island Countries

The map on page 504 shows you that Equatorial Guinea and São Tomé and Príncipe are island nations. Equatorial Guinea includes mainland territory on the west coast of Africa, plus five offshore islands. São Tomé and Príncipe lies in the Gulf of Guinea, about 180 miles (290 km) west of the Central African coast. This country consists of two main islands and several tiny islands.

Equatorial Guinea Equatorial Guinea was a Spanish colony until 1968. Today the country is home to about 400,000 people from different ethnic groups. Most of Equatorial Guinea's people live on the mainland. They live in rural areas where they grow bananas and coffee. They also harvest timber from the country's thick rain forests. Equatorial Guinea's largest island, Bioko (bee•OH•koh), lies in the Gulf of Guinea. Farmers grow cacao in Bioko's rich volcanic soil. Malabo, the nation's capital and largest city, is on Bioko.

São Tomé and Príncipe At one time a Portuguese colony, the island country of São Tomé and Príncipe became independent in 1975. The Portuguese first settled the islands in 1470. At that time, the islands had no humans living on them. Today São Tomé and Príncipe has about 200,000 people. Most are of mixed African and Portuguese ancestry. Nearly all of the population live on São Tomé, the largest island.

São Tomé and Príncipe lies on a chain of inactive volcanoes, which no longer erupt. Over the years, volcanic ash formed deep layers of fertile soil. Today plantation workers make use of this rich soil to grow bananas, cacao, coconuts, and coffee for export.

SECTION 2 ASSESSMENT

REVIEWING TERMS AND FACTS
1. **Define the following:** basin, tsetse fly.
2. **LOCATION** What Central African country is landlocked?
3. **HUMAN/ENVIRONMENT INTERACTION** What is Gabon's most important natural resource?
4. **HUMAN/ENVIRONMENT INTERACTION** How do most people in Equatorial Guinea and São Tomé and Príncipe make their livings?

MAP STUDY ACTIVITIES
5. Turn to the land use map on page 512. What natural resources are found in Congo?
6. Look at the population density map on page 513. What is the population density of most of Central Africa?

MAKING CONNECTIONS

Around 500 B.C., a people known as the Bantu lived in West Africa. Bantu, meaning "the people," is the name of a major African language group. The Bantu lived simply—farming, hunting, and fishing—probably in the Benue Valley of Nigeria. Within 500 years the Bantu had learned to mine and to work metals.

POPULATION GROWTH The development of metalworking allowed Bantu farmers to make iron tips for their tools and Bantu hunters to fashion weapons. These advancements made farming easier and hunting more successful. As food supplies increased, more people could be supported. The population in Bantu settlements thrived and grew.

By about 2,000 years ago, the Bantu population had increased so much that most settlements could no longer feed all their members. As a result, small groups split off to search for new land.

CENTURIES OF MOVEMENT Anthropologists have come to believe that Bantu-speaking groups gradually spread throughout central, eastern, and southern Africa. As they pushed into new areas, the Bantu shared their farming and metalworking knowledge—as well as their language—with other African groups.

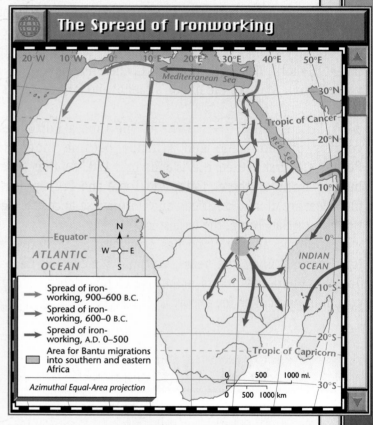

The Spread of Ironworking

→ Spread of iron-working, 900–600 B.C.

→ Spread of iron-working, 600–0 B.C.

→ Spread of iron-working, A.D. 0–500

▨ Area for Bantu migrations into southern and eastern Africa

Azimuthal Equal-Area projection

By A.D. 1500, nearly 15 centuries of migration had produced hundreds of ethnic groups, each culturally different from the others. Today between 60 and 80 million African people speak Bantu languages.

Making the Connection

1. Why did Bantu-speaking settlements increase in population?
2. To where did Bantu groups migrate?

Chapter 19 Highlights

Important Things to Know About Central Africa

SECTION 1 DEMOCRATIC REPUBLIC OF THE CONGO

- The land area of the Democratic Republic of the Congo is about one-fourth the size of the United States.
- The Congo River is the major "highway" for the Congo region.
- The Democratic Republic of the Congo has one of the largest tropical rain forest areas in the world.
- Vast savannas in the Congo region are home to valuable natural resources—antelopes, leopards, lions, zebras, giraffes, and other wildlife.
- Because of its location on the Equator, the Democratic Republic of the Congo generally has a tropical climate.
- Many important mineral resources, including copper and industrial diamonds, are found in the Democratic Republic of the Congo.
- The 49 million people of the Democratic Republic of the Congo belong to many different ethnic groups.

SECTION 2 OTHER COUNTRIES OF CENTRAL AFRICA

- The Central African Republic's landlocked location makes economic development difficult.
- Cameroon's coastal location has boosted its economic growth.
- Gabon and Congo export lumber and petroleum.
- Most of the people in Equatorial Guinea live on the mainland where they farm and harvest wood from the rain forests.
- São Tomé and Príncipe is an island nation with rich volcanic soil good for farming.

A market in the Democratic Republic of the Congo ▶

REVIEWING KEY TERMS

Match the numbered terms in Set A with their lettered definitions in Set B.

A
1. hydroelectricity
2. tsetse fly
3. basin
4. canopy

B
A. forest covering
B. insect that spreads tropical diseases to humans and animals
C. power developed from the force of moving water
D. broad, flat valley

Mental Mapping Activity

Draw a freehand map of Central Africa. Label the following:
- Democratic Republic of the Congo
- Congo River
- Gulf of Guinea
- Gabon
- Central African Republic
- Cameroon
- Congo

REVIEWING THE MAIN IDEAS

Section 1

1. **HUMAN/ENVIRONMENT INTERACTION** What challenges to the environment does the Democratic Republic of the Congo face?
2. **PLACE** How do the many ethnic groups in the Democratic Republic of the Congo affect its national unity?
3. **REGION** Why are the savannas of the Democratic Republic of the Congo important?

Section 2

4. **LOCATION** Describe how the locations of the Central African Republic and Cameroon have influenced their economic development.
5. **PLACE** What mineral resources are found in Gabon?
6. **PLACE** What Central African nation is made up entirely of islands?

CRITICAL THINKING ACTIVITIES

1. **Determining Cause and Effect** How do you think the presence of thick rain forests has affected the development of Central Africa?
2. **Drawing Conclusions** As the nations of Central Africa make use of their resources, what challenges do they face in developing their economies?

CONNECTIONS TO WORLD HISTORY

Cooperative Learning Activity Work in groups of three to learn about the colonial history of one of the Central African countries. Then create a documentary showing your findings. Each group should choose one country and do research for a documentary. Your findings should be presented in storyboard form—showing still pictures and dialogue that go together to form the documentary. Research as a group, then assign the following individual roles: (a) author, (b) editor, and (c) critic.

GeoJournal Writing Activity

Imagine that you work in a large corporation. You have been sent by your company to open a branch office in Kinshasa. Write a letter to your employer back in the United States describing how life in the Democratic Republic of the Congo is different from and similar to life in the United States.

TECHNOLOGY ACTIVITY

Using a Database Create a database about countries in Africa south of the Sahara. Your database should have a separate record for each country. Each record should have a separate field for the following: capital, languages, religions, population, land area, principal export, and principal import. After you have completed the database, write a short summary pointing out the similarities and differences among the countries.

PLACE LOCATION ACTIVITY: CENTRAL AFRICA

Match the letters on the map with the places and physical features of Central Africa. Write your answers on a separate sheet of paper.

1. Congo
2. Lake Tanganyika
3. Libreville
4. Cameroon
5. São Tomé and Príncipe
6. Kinshasa
7. Congo River

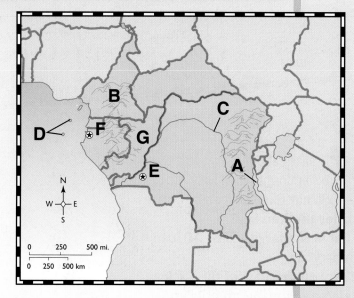

National boundary
National capital
Other city

Azimuthal Equal-Area projection

20°E
Tropic of Cancer
30°E
40°E
50°E

SOUTHWEST ASIA

20°N

Red Sea

Port Sudan

Omdurman
Khartoum

ERITREA
Asmara

DJIBOUTI

SUDAN

Djibouti
Gulf of Aden

10°N

ETHIOPIA

CENTRAL

Addis Ababa

AFRICA

SOMALIA

0°
Equator

UGANDA
Kampala

KENYA

Mogadishu

INDIAN OCEAN

Kisumu

RWANDA
Kigali
Nairobi

Mwanza

BURUNDI
Bujumbura
Lake Victoria

0   250   500 mi.

Mombasa

0   250   500 km

Lake Tanganyika

Dodoma

Victoria

TANZANIA

Dar es Salaam

SEYCHELLES

10°S

MAP STUDY ACTIVITY

As you read Chapter 20, you will learn about Kenya and the other countries in East Africa.

1. What East African countries border the Red Sea?

2. What is the largest East African country?

3. What country east of the African mainland is made up of islands?

SECTION 1

Kenya

PREVIEW

Words to Know
- coral
- reef
- fault
- escarpment
- poacher

Places to Locate
- Kenya
- Great Rift Valley
- Indian Ocean
- Nairobi
- Mombasa

Read to Learn . . .
1. what landforms are found in Kenya.
2. why most Kenyans live in highland areas.
3. what languages are spoken in Kenya.

The Great Rift Valley cuts a deep gash through East Africa. In places, the valley's sides are about 1 mile (1.6 km) high, and the valley floor is more than 50 miles (80 km) wide. The Great Rift Valley was formed when two of the plates that make up the earth's crust moved apart millions of years ago.

The Great Rift Valley is only one of East Africa's many geographic features. The region of East Africa includes 11 countries. Find these countries on the map on page 520. East Africa begins in the north with the country of Sudan. The region then runs along the Red Sea and the Indian Ocean to Tanzania (TAN•zuh•NEE•uh) in the south. About 1,000 miles (1,600 km) east of the African mainland lies the island country of Seychelles (say•SHEHLZ). On the mainland, Kenya lies close to the center of East Africa.

Region

The Land

Kenya's land area of 219,745 square miles (569,139 sq. km) makes it slightly smaller than the state of Texas. Like Texas's landforms, Kenya's landforms include coastal areas, plains, and highlands.

The Coast The blue Indian Ocean borders Kenya on the east. Kenya's long coastline has stretches of white beaches lined with palm trees. Not far offshore is a coral reef. **Coral** is a hard, rocklike material made of the skeletons of small

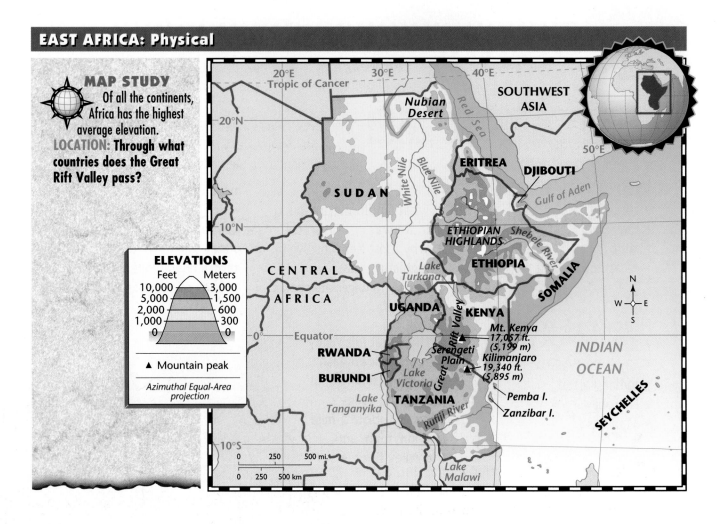

MAP STUDY Of all the continents, Africa has the highest average elevation.
LOCATION: Through what countries does the Great Rift Valley pass?

sea animals. A **reef** is a narrow ridge of coral, rock, or sand at or near the water's surface. Beaches near Kenya's coral reef area draw tourists from around the world.

Plains If you traveled about 20 miles (32 km) inland, you would find a vast plains area covering about three-fourths of Kenya. Bushes, shrubs, and low thorn trees thrive on the plains. Except for groups of nomads who herd their cattle through the region, few people make their homes here. Many animals, however, roam these vast plains. These include antelopes, water buffaloes, elephants, giraffes, lions, zebras, and other animals.

Highlands Southwestern Kenya is a highlands area made up of mountains, valleys, and plateaus. Forests and grasslands sweep through the highlands. This region, which boasts Kenya's most fertile soil, is home to about 75 percent of Kenya's people.

Look at the map above. You see that the Great Rift Valley **fault,** or crack in the earth, divides Kenya's highlands into eastern and western areas. Mountain ranges and **escarpments,** or steep cliffs, tower over both sides of the valley. Mount Kenya—Kenya's highest peak—rises 17,057 feet (5,199 m) in the eastern highlands. Many lakes are found in the western highlands.

The Climate The map on page 528 shows you that the Equator passes through the middle of Kenya. Almost all of Kenya has a savanna or steppe climate with hot or warm temperatures. The coast is hot and humid year-round. This area gets enough rain to support a few small rain forests. The plains region is the driest part of Kenya and has a desert climate.

In the highlands, the climate is mild with plenty of rainfall. Temperatures here average about 67°F (19°C), and rainfall may reach about 50 inches (130 cm) each year. The mild climate and fertile soil make the highlands Kenya's most important farming area.

Human/Environment Interaction
The Economy

Kenya has a developing economy based on free enterprise. Its chief economic activity is farming. About 50 percent of Kenya's farm products are subsistence food crops, the most important of which is corn. Kenyans grind corn into a porridge and mix it with vegetables to make a stew. Other subsistence food crops include bananas, beans, and cassava.

Cash Crops The other half of Kenya's farm products are cash crops cultivated for export. The country's leading cash crops—coffee and tea—are Kenya's main source of income. You can find both of these products growing on the slopes of the highlands.

Tsavo National Park, Kenya

Elephants and other animals roam the vast plains in Kenya's parks and game reserves.
REGION: What is the climate of Kenya's plains region?

Industries Although Kenya has no major mineral deposits to develop, the government has encouraged the growth of manufacturing in recent years. Kenya's chief factory-made products are cement, chemicals, light machinery, and household appliances.

Tourism adds to Kenya's treasury. Thousands of tourists visit Kenya each year to take trips called safaris. On safaris, visitors tour through national parks to see the country's spectacular wildlife in natural surroundings. The government has set up these parks to protect endangered animals from **poachers,** or people who hunt and kill animals illegally.

Place
The People

Harambee, which means "pulling together," is an important word in Kenya. Kenyans come from many different ethnic groups and speak many languages. Since their independence, Kenyans have tried to strengthen their country by working together. *Harambee,* however, is being tested as ethnic disputes become more frequent.

Impact of the Past Scientists have found some of the earliest-known remains of humans in the Great Rift Valley. About 3,000 years ago, people from other parts of Africa began to move into Kenya. These people farmed and herded animals in the highlands. They became the ancestors of today's Kenyans.

In the A.D. 700s, Arabs from Southwest Asia set up trading settlements along Kenya's coast. Some of these Arabs married Africans. From these marriages came a new people, culture, and language known as Swahili.

In the late 1800s, Kenya came under British control. Many British people moved to Kenya's highlands because of the mild climate and fertile soil. They took land away from Africans and established their own farms on that land. By the 1940s Kenya's African population had organized to reclaim their land. They opposed the British until Kenya finally won its independence in 1963.

After independence, Kenya became the political and economic leader of East Africa. Since the late 1970s, Kenya has struggled to overcome ethnic conflicts and to build a strong and lasting democracy.

Kenyans Today Kenya's population of 28.3 million people is growing rapidly. Kenya has a high population growth rate—over 2 percent a year. Experts predict that the Kenyan population will reach about 33 million by the year 2015. One of the biggest challenges Kenya faces is providing enough food and jobs for its people.

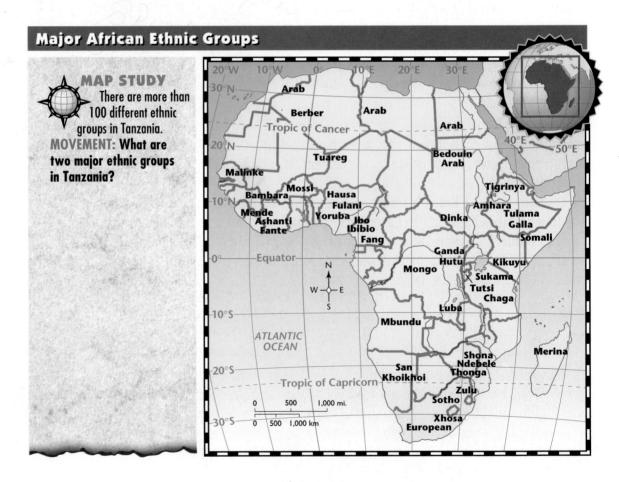

Major African Ethnic Groups

MAP STUDY There are more than 100 different ethnic groups in Tanzania. **MOVEMENT: What are two major ethnic groups in Tanzania?**

About 40 different ethnic groups are found in Kenya. The map on page 524 shows several of the many ethnic groups of Africa. The largest of Kenya's ethnic groups is the Kikuyu (kee•KOO• yoo), who live mostly in the highlands. What other major groups are found in East Africa?

Although Kenya's many ethnic groups speak many different African languages, Swahili—a blend of African languages and Arabic—is the most widely spoken. About 65 percent of Kenya's population are Christians, while another 25 percent of the people practice traditional African religions.

Most Kenyans—about 73 percent— live in rural villages. Only 27 percent of the population live in cities such as Nairobi (ny•ROH•bee)—Kenya's modern capital— and the Indian Ocean port of Mombasa (mahm•BAH•suh). Find these cities on the map on page 520.

In recent years, many Kenyans have moved from the countryside to the cities. Lucy Ombasa, a Nairobi teenager, and her family recently moved to the city from their village in western Kenya. Their village specialized in stone carvings of animals, fish, and birds. Lucy's parents and sister now sell carvings to tourist shops in the Kenyan capital.

Nairobi, Kenya

Kenya's capital and largest city has many parks and modern office buildings.
MOVEMENT: Why do you think many rural Kenyans have moved to cities?

SECTION 1 ASSESSMENT

REVIEWING TERMS AND FACTS
1. Define the following: coral, reef, fault, escarpment, poacher.
2. LOCATION Where is Kenya's most important farming area?
3. HUMAN/ENVIRONMENT INTERACTION What draws tourists to Kenya?
4. PLACE What is the most widely spoken language in Kenya?

MAP STUDY ACTIVITIES
5. Turn to the physical map on page 522. What large lake lies entirely within Kenya?
6. Look at the political map on page 520. About how far is Kenya's capital from the Ugandan border?

BUILDING GEOGRAPHY SKILLS

Reading a Line Graph

A **line graph** is a good way to show how things change. On a line graph, time intervals appear on the bottom of the graph—the horizontal axis. The information being compared usually appears on the left side of the graph—the vertical axis. Dots or other symbols mark specific quantities. These dots are connected with a line to show relationships and trends. To read a line graph, apply these steps:

• Read the title to find out the subject of the graph.

• Familiarize yourself with the information on both axes.

• Examine where the dots are placed on the graph.

• Determine what the lines or curves mean.

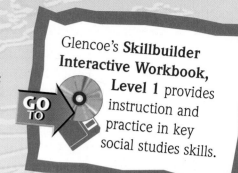

Glencoe's **Skillbuilder Interactive Workbook, Level 1** provides instruction and practice in key social studies skills.

GO TO

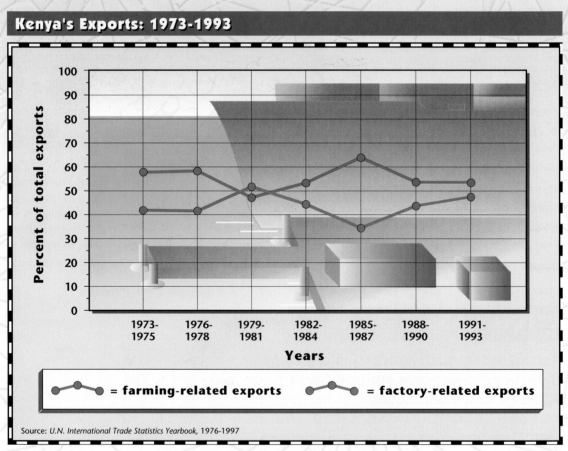

Kenya's Exports: 1973-1993

Percent of total exports

Years

●—●—● = farming-related exports   ●—●—● = factory-related exports

Source: *U.N. International Trade Statistics Yearbook, 1976-1997*

Geography Skills Practice

1. What is measured on this graph? Over what time period?

2. In what years did Kenya have the fewest farming-related exports?

3. In what years did Kenya have the most factory-related exports?

Tanzania

PREVIEW

Words to Know
- sisal
- cloves

Places to Locate
- Tanzania
- Lake Victoria
- Lake Tanganyika
- Dar es Salaam
- Zanzibar
- Dodoma

Read to Learn . . .
1. how many ethnic groups make up Tanzania's population.
2. how most people in Tanzania earn a living.

The gleaming, snow-capped peak of Kilimanjaro towers 19,340 feet (5,895 m) over grassy plains in northern Tanzania. For centuries, Africans have gazed at Africa's tallest mountain with awe and wonder. An old African story says that God built Kilimanjaro as a throne from which to view the world.

Kilimanjaro is probably the best-known sight in Tanzania. One of the largest countries in East Africa, Tanzania has a land area of 341,154 square miles (883,588 sq. km). Look at the map on page 520. You see that Tanzania lies south of Kenya and below the Equator.

Place
The Land

Most of Tanzania lies on the mainland of East Africa. The country also includes several small coral islands just off the coast, in the Indian Ocean. The largest of these islands is called Zanzibar.

Coast, Plateaus, and Lakes Like the coast of Kenya, Tanzania's coastline boasts white beaches and palm trees. As you travel inland, the country's elevation rises gradually from humid lowlands to partially dry plateaus. Huge grasslands with patches of trees and shrubs cover the plateaus. To the north, near the Kenyan border, lies a mountainous area that includes Kilimanjaro.

Much of western Tanzania is part of the Great Rift Valley. A number of lakes lie in this area, including Lake Victoria and Lake Tanganyika. Lake Victoria is the largest lake in Africa. Lake Tanganyika's floor is the deepest point on the African continent. Unusual fish found nowhere else in the world live in Lake Tanganyika's deep, dark waters.

Wildlife Like Kenya, Tanzania has many kinds of animals. The Tanzanian government has set aside thousands of square miles to protect its wildlife. Serengeti National Park covers about 5,600 square miles (14,500 sq. km). It is home to many lions and huge herds of antelopes and zebras. During the dry season, thousands of animals roam the plains in search of water.

Place

The Economy and People

Tanzania has a developing economy based on agriculture. Service industries are growing, but manufacturing plants are still small. Tanzania is rich in mineral resources such as gold and diamonds. These riches, however, have not yet been developed.

Most Tanzanians raise livestock or farm small plots of land. Farmers grow only enough bananas, cassava, corn, millet, and rice to feed their families. Large, government-run farms grow many of the crops that Tanzania exports. These cash crops include coffee, cotton, tea, and tobacco.

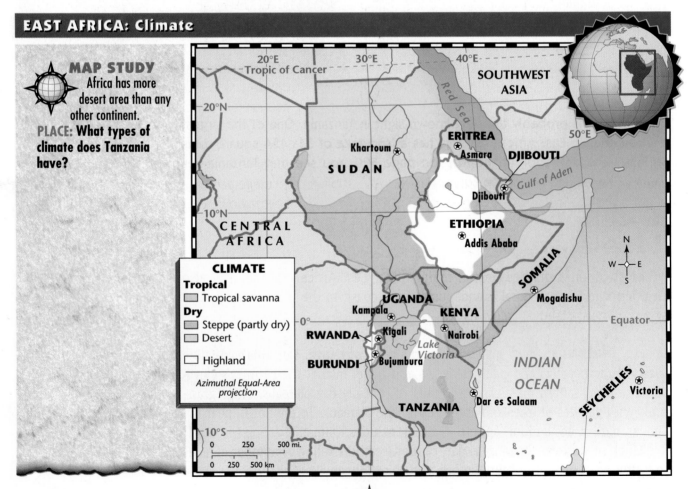

EAST AFRICA: Climate

MAP STUDY Africa has more desert area than any other continent.
PLACE: What types of climate does Tanzania have?

CLIMATE
Tropical
◻ Tropical savanna
Dry
◻ Steppe (partly dry)
◻ Desert
◻ Highland

Azimuthal Equal-Area projection

Tanzania is the world's largest producer of **sisal,** a plant fiber used in making rope and twine. Farmers on the island of Zanzibar raise the world's largest supply of **cloves,** a spice made from the buds of clove trees.

National Unity About 31 million people live in Tanzania. Most of the population belong to one of about 120 African ethnic groups. A small number of the country's people are Asians or Europeans. Although each ethnic group has its own language, many people speak Swahili, the national language. The population is about evenly divided among Christian, Muslim, and traditional African religions. Because there are so many ethnic groups and religions in Tanzania, no single group controls the country. This ethnic and religious balance has helped Tanzania achieve national unity.

The Past and Present Scientists have found the remains of some of the earliest human settlements in Tanzania. By about A.D. 500, Bantu-speaking peoples had settled in this area. Nearly 600 years later, Arabs from Southwest Asia set up major trading centers on Zanzibar and other islands. The region that is now Tanzania came under European control beginning in the early 1500s. After periods of Portuguese and German rule, the area came under British control following World War I. In the early 1960s, the Tanzanians finally won their independence as the United Republic of Tanzania.

Today about 80 percent of Tanzania's people live in rural villages. Dar es Salaam, on the Indian Ocean, is Tanzania's capital, largest city, and major port. The central area of Tanzania has few people. In an effort to encourage people to move there, the Tanzanian government is planning to build a new national capital—Dodoma—in the area.

Teen Scene

Olympic Hopeful

Sadiki Abasi hopes to be in the Olympics someday. Soccer is his favorite hobby, but his real talent is running. Sadiki is competing with long-distance runners from all over Tanzania for a spot on the national team. He thinks he's got a chance. "I started running with my uncle when I was seven," Sadiki says. "I try to run at least 100 miles a week to stay competitive." Sadiki's uncle, a former national champion, is Sadiki's coach.

SECTION 2 ASSESSMENT

REVIEWING TERMS AND FACTS

1. **Define the following:** sisal, cloves.

2. **LOCATION** In what area of Tanzania is Kilimanjaro located?

3. **PLACE** What agricultural products does Tanzania export?

4. **MOVEMENT** Why will Tanzania move its capital from Dar es Salaam to Dodoma?

MAP STUDY ACTIVITIES

5. Turn to the climate map on page 528. What type of climate is found in southern Tanzania?

MAKING CONNECTIONS

A REMARKABLE FIND

MATH | **SCIENCE** | HISTORY | LITERATURE | TECHNOLOGY

On July 17, 1959, archaeologist Mary Leakey discovered fossils of teeth buried in the ground. An expert on prehistoric bones, she had never seen teeth quite like these before. Hopeful and excited, Mary Leakey ran to tell her husband, Louis, also an archaeologist.

EARLY DIGS Louis and Mary Leakey worked at the Olduvai (OHL•duh•vy) Gorge in northern Tanzania. The Gorge—a 25-mile-long (41km) canyon—had proven to be a rich treasure of human and animal fossils. The Leakeys had been digging at various places along the Gorge since the early 1930s. They had already unearthed thousands of remains.

In 1948 Mary Leakey had found a fossilized skull and other bones. These remains were between 25 and 40 million years old. Putting the bones together, the Leakeys produced a creature that had human-like jaws and walked upright on hind legs. It was the first significant evidence that the human race may have started in Africa rather than in Asia, as most archaeologists believed. As the Leakeys ran back to excavate the teeth on that July day in 1959, they thought the teeth might provide further evidence of humankind's beginnings in Africa.

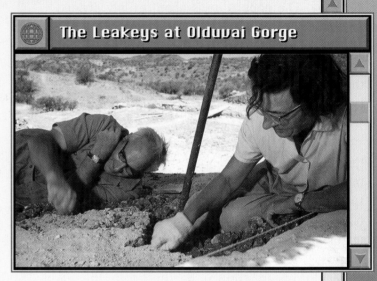

The Leakeys at Olduvai Gorge

TWICE AS OLD With dental picks and brushes, the Leakeys slowly uncovered the teeth, which turned out to be part of an almost-complete human skull. When dated, the skull proved to be about 1.75 million years old. It was twice as old as scientists had judged the human race to be.

The skull and other Leakey finds forced archaeologists to change their search for human origins from Asia to Africa. In addition, scientists were forced to hypothesize that the development of human beings took much longer than previously thought.

Making the Connection

1. Where is Olduvai Gorge?
2. How did the Leakeys' discoveries change scientists' thinking about human origins?

SECTION 3 ·
Inland East Africa

PREVIEW

Words to Know
- autonomy
- watershed
- civil war
- refugee

Places to Locate
- Uganda
- Rwanda
- Burundi
- Lake Victoria
- Nile River
- Kampala

Read to Learn . . .
1. why inland East Africa has a mild climate.
2. what farm products are grown in inland East Africa.
3. how conflict between ethnic groups has divided the people of inland East Africa.

The sparkling waters of Lake Victoria

cover 26,828 square miles (69,485 sq. km). About half of this magnificent lake lies inside the landlocked East African country of Uganda. Large lakes are also found in two other inland countries of East Africa—Rwanda and Burundi.

Great lakes, snow-covered mountains, deep tropical valleys, and thundering rivers are all found in Uganda, Rwanda, and Burundi. Each of these three countries of inland East Africa is also home to many different peoples.

Place
Uganda

Uganda lies in the highlands region of East Africa. Uganda's 77,085 square miles (199,650 sq. km) make it slightly smaller than Oregon. A plateau covers most of northern Uganda. In the center of the country is a large area of marshes and lakes. Here, waters flowing out of Lake Victoria gather to form a source of the Nile River. In southern Uganda, thick forests spread across the landscape. To the east and west, mountains form Uganda's borders with Kenya and the Democratic Republic of the Congo.

The Equator crosses southern Uganda. The country's generally high elevation, however, keeps temperatures mild. In most areas, at least 40 inches (100 cm) of rain falls each year.

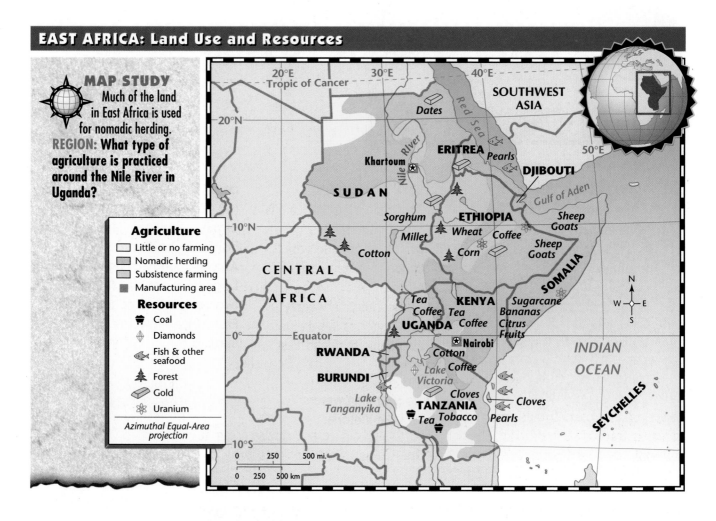

MAP STUDY
Much of the land in East Africa is used for nomadic herding.
REGION: What type of agriculture is practiced around the Nile River in Uganda?

Agriculture
- ☐ Little or no farming
- ☐ Nomadic herding
- ☐ Subsistence farming
- ■ Manufacturing area

Resources
- ⛏ Coal
- ◇ Diamonds
- 🐟 Fish & other seafood
- 🌲 Forest
- ▭ Gold
- ✿ Uranium

Azimuthal Equal-Area projection

The Economy

Agriculture is the most important economic activity in Uganda. Fertile soil and a mild climate help crops to grow well. The most productive areas lie along the western and northern sides of Lake Victoria and in the western highlands.

Uganda's farmers grow bananas, beans, cassava, corn, and sweet potatoes to feed their families. They sell any extra food at local markets. Coffee is Uganda's leading export crop. Cotton, sugarcane, and tea are valuable exports, too. These cash crops support Uganda's economy.

The People

Uganda has about 21 million people. You will find almost two-thirds of Uganda's people living in the fertile south. There you will find Kampala, the capital, on the shores of Lake Victoria.

The people of Uganda belong to more than 20 different ethnic groups. The Ganda are the largest and most prosperous group. Until 1967 the Ganda enjoyed local **autonomy,** or self-government. Today they are part of a united Uganda.

For most of this century, the British ruled Uganda. After Uganda won its independence in 1962, fighting broke out between ethnic groups in the north and the south. Ethnic conflict, plus a period of cruel dictatorship, hurt Uganda throughout much of the 1980s. Today Ugandans are making great strides in rebuilding their country after this time of suffering and warfare.

Place
Rwanda and Burundi

Rwanda and Burundi are located deep in inland East Africa. Although these two countries lie just south of the Equator, high altitudes give them a mild climate. They also lie on the ridge that separates the Nile and Congo **watersheds.** A watershed is an area drained by a river. To the west of the ridge, waters eventually run into the Congo River watershed. To the east, they run into the Nile watershed.

The Economy Most people in Rwanda and Burundi are farmers. They grow tea, coffee, beans, bananas, cotton, and grains on gently rolling plateaus. In both countries, some people raise livestock. Along lakes, such as Lake Tanganyika and Lake Kivu, fishing is an important economic activity. Some tin mining takes place in mountainous areas, but neither country has valuable minerals.

Because Burundi and Rwanda are landlocked, they have trouble getting their exports to foreign buyers. There are few paved roads and no railroads. Most goods must be transported by road to Lake Tanganyika, where boats take them to Tanzania and the Democratic Republic of the Congo. Another route is by dirt road to Tanzania and then by rail to Dar es Salaam.

Ethnic Conflict Burundi and Rwanda are two of the smallest and most crowded nations in Africa. Rwanda, for example, has an average of 835 persons per square mile (322 persons per sq. km). This density is about 13 times as great

WHAT IN THE WORLD?

Glaciers in Africa?

"Impossible!" thought British geographers in the mid-1800s. They were proved wrong in 1889, when British explorer Henry Morton Stanley saw glaciers in the Ruwenzori (ROO•uhn•ZOHR•ee) Mountains. The Ruwenzori range lies between Uganda and the Democratic Republic of the Congo. Even harder to believe is the fact that the Nile—the river that flows through rain forests and deserts—begins as a trickle in these frozen cliffs.

Volcanoes National Park, Rwanda

A small village sits in the shadow of one of several inactive volcanoes in Rwanda.
MOVEMENT: How does the lack of a coastline affect the economy of Rwanda?

Tea pickers gather leaves in the highland plantations of Burundi.
PLACE: What is the occupation of most people in Burundi?

as Africa's population density as a whole! Less than 10 percent of these people live in towns. Most live in villages surrounded by their fields.

The majority of the population in Rwanda and Burundi belong to two ethnic groups—the Hutu and the Tutsi. The Hutu are the largest group in both countries. The Tutsi, however, have controlled the countries' governments and economies. Since Rwanda's and Burundi's independence from Belgium in 1962, the Hutu have tried to gain part of this control.

In 1994 **civil war,** or fighting within a country, broke out in Rwanda between the Hutu and the Tutsi. About 500,000 people were killed. More than 2 million refugees fled Rwanda and settled in camps on the borders of neighboring countries. A **refugee** is a person who must flee his or her home and seek safety elsewhere. In 1996, many refugees returned to Rwanda. By this time, however, the ethnic conflict had spread to Burundi.

SECTION 3 ASSESSMENT

REVIEWING TERMS AND FACTS

1. Define the following: autonomy, watershed, civil war, refugee.

2. PLACE How does elevation affect climate in inland East Africa?

3. HUMAN/ENVIRONMENT INTERACTION What is Uganda's most important economic activity?

4. PLACE What two major ethnic groups live in Rwanda and Burundi?

MAP STUDY ACTIVITIES

5. Turn to the land use map of East Africa on page 532. What type of farming takes place in Rwanda and Burundi?

The Horn of Africa

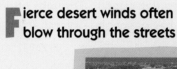

PREVIEW

Words to Know
- drought
- clan

Places to Locate
- Sudan
- Ethiopia
- Eritrea
- Somalia
- Djibouti
- Gulf of Aden
- Indian Ocean

Read to Learn . . .
1. what ethnic groups live in Sudan.
2. why Ethiopia has been independent for thousands of years.
3. how civil war has affected countries in the Horn of Africa.

Fierce desert winds often blow through the streets of Khartoum (kahr•TOOM), the capital of Sudan. Yet water flows nearby. Two branches of the Nile River meet at Khartoum. As in Egypt, the Nile River is the giver of life to people in Sudan and other countries of northeastern Africa.

A large area of northeastern Africa is called the Horn of Africa. The map on page 520 shows you how this region got its name. The land is shaped like a horn as it juts out into the Indian Ocean. The countries that lie in the Horn of Africa region are Sudan, Ethiopia, Eritrea (EHR•uh•TREE•uh), Somalia, and Djibouti (juh•BOO•tee).

Place
Sudan

Sudan is the largest country in Africa. Its 917,375 square miles (2,376,000 sq. km) cover an area about one-third the size of the contiguous United States.

Northern Sudan is mostly a desert of bare rocks or sand dunes. Grassy plains cross the central part of Sudan. Here, two great branches of the Nile River—the Blue Nile and the White Nile—meet. This area is the most fertile part of the country. To the south, Sudan has humid tropical rain forests and swamps.

Sudan is one of the world's leading producers of cotton. Most of Sudan's 28.5 million people live along the Nile River or one of its branches. They use water from the Nile to irrigate huge fields of cotton. Some work on large farms owned by the government.

The People If you visited Sudan, you would find that the people differ greatly from north to south. Most people who live in the northern two-thirds of the country are Muslim Arabs. People who live in the south come from several different African ethnic groups. They are Christians or practice traditional African religions.

Since its independence in 1956, Sudan has been torn by civil war between the north and the south. The fighting has disrupted food production and caused widespread hunger. In the early 1990s, a **drought**—an extended dry period—occurred in Sudan and caused even more suffering.

Place

Ethiopia

To the east of Sudan lies Ethiopia. If you look at the map on page 522, you will see that Ethiopia is a rugged, landlocked, mountainous country with a high plateau. The Great Rift Valley cuts through this plateau, forming deep river gorges and sparkling waterfalls. Mild temperatures and fertile soil make areas of the plateau excellent for farming. Ethiopia's farmers, however, cannot always depend on regular rainfall. Drought often occurs, destroying crops and bringing famine. This has happened several times since the 1970s.

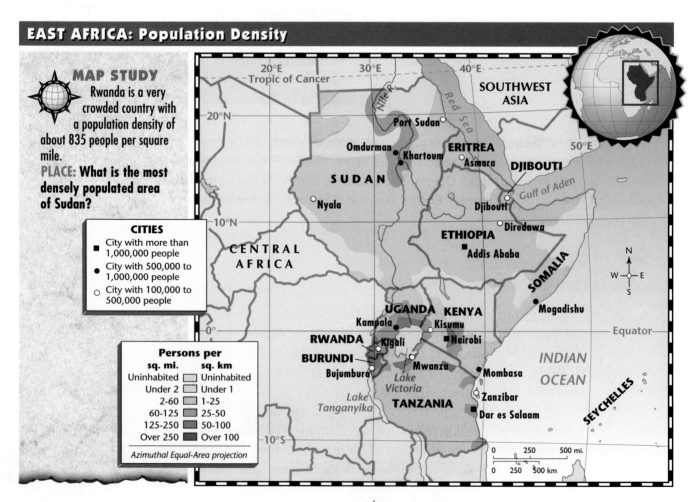

EAST AFRICA: Population Density

MAP STUDY Rwanda is a very crowded country with a population density of about 835 people per square mile.
PLACE: What is the most densely populated area of Sudan?

CITIES
■ City with more than 1,000,000 people
● City with 500,000 to 1,000,000 people
○ City with 100,000 to 500,000 people

Persons per	
sq. mi.	sq. km
Uninhabited	Uninhabited
Under 2	Under 1
2-60	1-25
60-125	25-50
125-250	50-100
Over 250	Over 100

Azimuthal Equal-Area projection

Most Ethiopians live on the high plateau and grow wheat, corn, and other grains on subsistence farms. Some farmers grow coffee to sell. Others also raise livestock.

The People Ethiopia—one of the world's oldest countries—sent government representatives to meet Egyptian pharaohs centuries ago. Ethiopia's mountains kept it isolated and independent for thousands of years. During most of its history, emperors and empresses ruled. In 1974, military leaders overthrew the last emperor. Today Ethiopia is struggling to build a democracy.

About 58.4 million people live in Ethiopia. Only 16 percent of them live in urban areas. Addis Ababa (A•duhs A•buh•buh) is Ethiopia's capital and largest city. About 70 different languages are spoken in Ethiopia. Amharic, similar to Hebrew and Arabic, is the official language. Most Ethiopians are either Christians or Muslims. A smaller number practice traditional African religions.

The walls and ceiling of this church tell the story of Ethiopian Christianity *(left)*. Many residents of Ethiopia's capital live and work in modern buildings *(right)*.
PLACE: What are the two main religions of Ethiopia?

Place
Eritrea

Ethiopia is one of the oldest countries in Africa, and Eritrea is the newest! In 1993, after 30 years of war, Eritrea won its independence from Ethiopia.

Look at the map on page 522. It shows you that Eritrea lies on the Red Sea. Along this Red Sea coast runs a wide plain—one of the hottest and driest areas in Africa. As you travel inland, you will discover that mountains cover the rest of Eritrea.

Most of Eritrea's 3.8 million people farm the land, although farming is uncertain in this dry land. Eritrea's long war with Ethiopia has also made farming difficult. Soldiers cut down the forests, which led to soil erosion. Now that an unstable peace prevails, Eritreans are rebuilding their country.

Asmara is one of the few large cities in the tiny country of Eritrea.
PLACE: What country was Eritrea a part of until 1993?

Place
Somalia

Like Eritrea, Somalia is a hot, dry country. Grassy plains cover most of Somalia. About half of the country's 10.7 million people are nomads. Southern Somalia, however, has rivers that provide water for irrigation. Farmers there grow sugarcane and citrus fruits.

The People Nearly all Somalis are Muslims and speak either Somali or Arabic. Yet they are deeply divided. They belong to **clans,** groups of people related to one another. In the late 1980s, clan disputes led to civil war. When a drought struck a few years later, hundreds of thousands of Somalis starved. Other countries sent food, but the fierce fighting kept much of the food from reaching the starving people.

In 1992 the United States and other countries sent troops to protect food supplies. The troops withdrew in 1995 after the worst of the famine ended. Armed Somali groups, however, still control parts of the country.

Place
Djibouti

Djibouti is a tiny land wedged between Eritrea and Somalia. Djibouti's 700,000 people are mostly Muslims who, in the past, lived a nomadic life. Now many of them have settled permanently in the capital city, also called Djibouti.

Almost all of the country's income derives from shipping. Farming is difficult on its dry lands. The city of Djibouti on the Gulf of Aden is an important port. A railroad line connecting it to Ethiopia makes Djibouti an important outlet for Ethiopia's products.

SECTION 4 ASSESSMENT

REVIEWING TERMS AND FACTS
1. **Define the following:** drought, clan.
2. **PLACE** Where do most of Sudan's people live?
3. **PLACE** What is the newest country in Africa?
4. **HUMAN/ENVIRONMENT INTERACTION** What challenges do farmers face in the Horn of Africa?

MAP STUDY ACTIVITIES
5. According to the population density map on page 536, what country in the Horn of Africa is the most densely populated?

Chapter 20 Highlights

Important Things to Know About East Africa

SECTION 1 KENYA

- Kenya has coastal lowlands as well as inland plains and highlands.
- The Great Rift Valley is a huge gash in the earth that runs through most of East Africa.
- Two important economic activities in Kenya are farming and tourism.
- Coffee and tea exports are Kenya's main source of income.
- Kenya's population of about 28 million is growing rapidly.

SECTION 2 TANZANIA

- Tanzania is made up of a mainland area and a group of offshore islands.
- Kilimanjaro, Africa's tallest mountain, lies in the northern area of Tanzania.
- Serengeti National Park in Tanzania is home to many animal species, including lions, antelopes, and elephants.
- Tanzania's population consists of about 120 ethnic groups.

SECTION 3 INLAND EAST AFRICA

- Uganda has a fertile plateau that is good for farming.
- Uganda, Rwanda, and Burundi lie close to the Equator, but their generally high elevation keeps their temperatures mild.
- The two main ethnic groups in Rwanda and Burundi are the Hutu and the Tutsi.

SECTION 4 THE HORN OF AFRICA

- Most of Sudan's people live along the Nile River and farm for a living.
- Ethiopia has been an independent country for thousands of years.
- Eritrea, Somalia, and Djibouti are hot, dry countries where farming is difficult.

Goma cane dancers in Kenya ▶

REVIEWING KEY TERMS

Match the numbered terms in Set A with their lettered definitions in Set B.

A

1. clan
2. poacher
3. autonomy
4. refugee
5. civil war
6. cloves
7. reef
8. sisal

B

A. person who kills animals illegally
B. plant fiber used to make rope
C. person who flees home for safety elsewhere
D. sea ridge rising above or below the water's surface
E. spice made from dried flower buds
F. self-government
G. conflict within a country
H. large group of related people

Mental Mapping Activity

Draw a freehand map of East Africa. Label the following:
- Kenya
- Lake Tanganyika
- Rwanda
- Nairobi
- Eritrea
- Indian Ocean
- Somalia

REVIEWING THE MAIN IDEAS

Section 1

1. MOVEMENT What group of people set up trading settlements along East Africa's coast?
2. HUMAN/ENVIRONMENT INTERACTION What major challenge does Kenya face today?

Section 2

3. PLACE What is the largest lake in Africa?
4. PLACE Where are cloves grown in Tanzania?

Section 3

5. LOCATION In what part of Uganda do most Ugandans live?
6. MOVEMENT Why did more than 2 million refugees flee Rwanda during the early 1990s?

Section 4

7. MOVEMENT Why were troops sent to Somalia in the early 1990s?
8. LOCATION How does Djibouti's location affect its economy?

CRITICAL THINKING ACTIVITIES

1. **Identifying Central Issues** How do ethnic differences create problems for the countries of the East African region?
2. **Predicting Consequences** What effect do you think moving from the countryside to cities has on the way people live in East Africa?

540

CONNECTIONS TO WORLD CULTURES

Cooperative Learning Activity Work in a group of four to learn about an ethnic group in East Africa. Choose one ethnic group, then have each student research one of the following topics: (a) family life; (b) education; (c) way of earning a living; or (d) the arts. After your research is complete, share your information with your group. Brainstorm ways to present your ethnic group to the entire class.

GeoJournal Writing Activity

Imagine you have been sent to Somalia, Sudan, or Rwanda as part of a relief agency such as the American Red Cross, CARE, or Doctors Without Borders. Write a short article for your hometown newspaper describing your experience.

TECHNOLOGY ACTIVITY

Using the Internet Search the Internet to find information about organizations dedicated to the preservation of animals in Africa south of the Sahara. Choose one organization and then research and write a report about the organization and what it is currently doing to protect endangered animals. Include photographs, clip art, and a map in your report.

PLACE LOCATION ACTIVITY: EAST AFRICA

Match the letters on the map with the places and physical features of East Africa. Write your answers on a separate sheet of paper.

1. Zanzibar
2. Tanzania
3. Lake Victoria
4. Nile River
5. Mogadishu
6. Red Sea
7. Khartoum
8. Uganda
9. Ethiopia

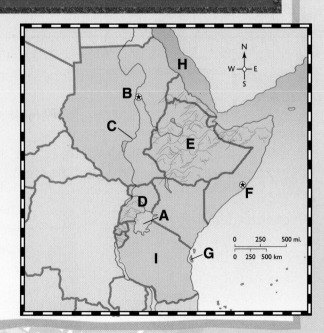

GeoLab ACTIVITY

**From the classroom of
Carole Mayrose
Clay City, Indiana**

LANDFORMS: Contour Mapping

Background

Africa south of the Sahara rises from west to east in a series of plateaus. How can you show plateaus or a rise in elevation on a map? In this activity you will create a contour map by drawing lines that follow the shape of a landform at different elevations.

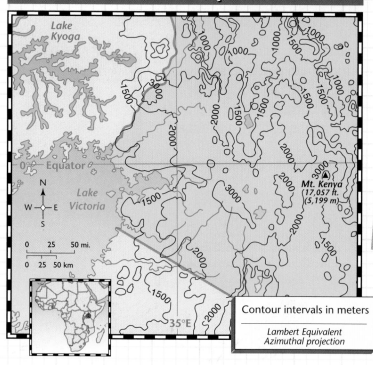

WEST KENYA: Contour Map

Lake Kyoga

1000

1500

3000

2000

0° — Equator

Lake Victoria

N
W — E
S

0   25   50 mi.
0   25   50 km

35°E

1500

2000

3000

Mt. Kenya
(17,057 ft.
(5,199 m)

2000

1500

Contour intervals in meters

*Lambert Equivalent
Azimuthal projection*

Believe it or NOT!

Most of the African continent is a plateau, with the exception of the Great Rift Valley, which slices through eastern Africa. This valley has sides up to 1 mile (1.6 km) high and a floor up to 50 miles (80 km) wide. The Great Rift Valley is like a huge scar on the earth—and can be seen from space.

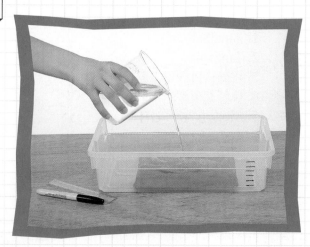

Materials

- a metric ruler
- 1 clear plastic box and lid
- modeling clay
- a beaker of water
- 1 transparency (clear plastic)
- a transparency marker
- tape

What To Do

A. Using the ruler and the transparency marker, measure and mark 1-cm lines up the side of the box. (The bottom of the box will be zero elevation, or sea level.)

B. Tape the transparency to the top of the box lid.

C. With the modeling clay, mold a landform with mountains, plateaus, and valleys.

D. Place the model in the box.

E. Using the beaker, pour water into the box to a height of 1 cm. Place the lid on the box.

F. Looking down through the transparency on top of the box, use the marker to trace the top of the water line as it surrounds the landform.

G. Using the scale 1 cm=5 feet, mark the elevation on the line.

H. Repeat steps D through F, adding water to the next 1-cm level and tracing until you have mapped the landform by means of contour lines.

I. Transfer the tracing of the contours of the landform onto paper.

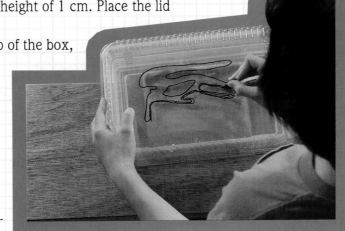

Lab Activity Report

1. What is the difference in elevation between each contour line on your map?

2. What is the average elevation of your landform?

3. Making Comparisons How would the number of contour lines on an area of steep mountains compare with the number on an area of flat plains?

Go A Step Further

Activity

Describe the kind of landform you think occurs where the contour lines are close together. What kind of landform occurs where the lines are far apart?

Chapter 21 South Africa and Its Neighbors

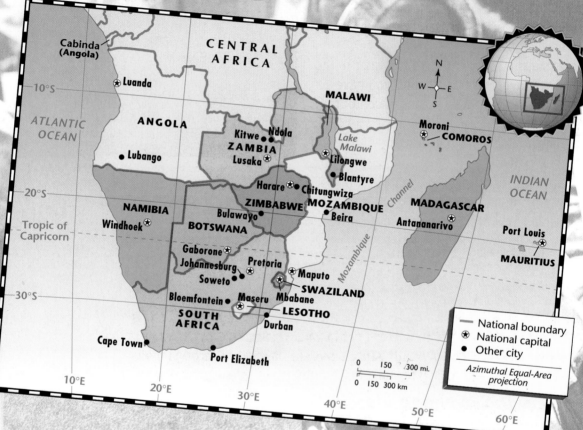

Cabinda (Angola)

CENTRAL AFRICA

10°S

Luanda

ATLANTIC OCEAN

ANGOLA

MALAWI

Moroni

COMOROS

Kitwe Ndola

ZAMBIA

Lubango

Lusaka

Lilongwe

Lake Malawi

Blantyre

Harare Chitungwiza

20°S

NAMIBIA

Bulawayo

ZIMBABWE MOZAMBIQUE

Mozambique Channel

MADAGASCAR

INDIAN OCEAN

Tropic of Capricorn

Windhoek

BOTSWANA

Beira

Antananarivo

Port Louis

Gaborone

MAURITIUS

Johannesburg

Pretoria

Maputo

Soweto

SWAZILAND

30°S

Bloemfontein

Maseru Mbabane

LESOTHO

SOUTH AFRICA

Durban

Cape Town

Port Elizabeth

10°E 20°E 30°E 40°E 50°E 60°E

National boundary
National capital
Other city

0 150 300 mi.
0 150 300 km

Azimuthal Equal-Area projection

MAP STUDY ACTIVITY

As you read Chapter 21, you will learn about South Africa and its neighbors.

1. What three island countries lie off Africa's southeastern coast?
2. What country forms the southern tip of the African continent?
3. Through what countries does the Tropic of Capricorn pass?

Republic of South Africa

Words to Know
- enclave
- high veld
- escarpment
- apartheid
- township

Places to Locate
- Republic of South Africa
- Cape Town
- Drakensberg Mountains
- Johannesburg
- Durban
- Pretoria

Read to Learn . . .
1. what landscapes are found in South Africa.
2. what mineral resources South Africa has.
3. what changes have occurred recently in South Africa's government.

This huge natural wonder is called Table Mountain. It looms over the city of Cape Town—a major port in the Republic of South Africa.

The Republic of South Africa spreads across the southern end of Africa. It is a land of breathtaking scenery and great mineral wealth. It is also a land of great change. In recent years, South Africa's people have gone through many changes in their lives and their government.

Region
The Land

South Africa sprawls across 471,440 square miles (1,221,030 sq. km)—an area nearly three times the size of California. South Africa has many different landscapes: winding coastlines, tall mountains, deep valleys, and a high plateau. The map on page 544 shows you that South Africa's large land area also swallows up two small independent African nations—Lesotho (luh•SOH•toh) and Swaziland. These are **enclaves,** or small countries surrounded or nearly surrounded by a larger country.

Coasts The west coast of South Africa borders the Atlantic Ocean, and the south and east coasts border the Indian Ocean. Northwest along the Atlantic coast stretches the vast Namib Desert. Farther south lies the Cape of Good Hope, the southernmost point of Africa. A group of long, narrow mountain ranges runs through this southwestern area. Between the ranges lies a dry, flat land called the Great Karroo. Farther east, along the Indian Ocean, are high cliffs, sandy beaches, rolling hills, and a lowland plain.

A Plateau and Mountains A large plateau spreads through the center of South Africa. It ranges from about 2,000 to 8,000 feet (610 to 2,440 m) above sea level. Part of the South African plateau is made up of flat, grass-covered plains called the **high veld.** Isolated rocky hills rise as high as 100 feet (30 m) above the surrounding land.

A group of mountains and cliffs circle the plateau and divide it from the coastal areas. They are known as the Great Escarpment. An **escarpment,** you will recall, is a steep cliff or slope that divides high and low ground. The Great Escarpment rises to its highest elevations in the eastern Drakensberg Mountains. Many peaks in this range are more than 10,000 feet (3,000 m) high.

The Climate South Africa lies south of the Equator. Its seasons are opposite those in the Northern Hemisphere. When it is winter north of the Equator, it is summer south of the Equator. In winter the South African plateau is cool and sunny—with some rainfall—during the day. Temperatures sometimes drop below freezing at night. Summers are mild because of the high elevation.

Cape Town—a major port city—has a Mediterranean climate of cool, rainy winters and hot, often dry summers. Along the eastern coast warm winds from the Indian Ocean bring a humid subtropical climate.

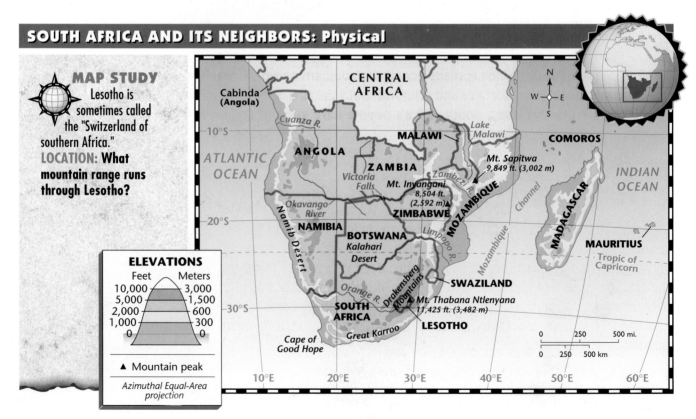

SOUTH AFRICA AND ITS NEIGHBORS: Physical

MAP STUDY
Lesotho is sometimes called the "Switzerland of southern Africa."
LOCATION: What mountain range runs through Lesotho?

ELEVATIONS

Feet	Meters
10,000	3,000
5,000	1,500
2,000	600
1,000	300
0	0

▲ Mountain peak

Azimuthal Equal-Area projection

Cabinda (Angola)
Cuanza R.
10°S
ATLANTIC OCEAN
ANGOLA
CENTRAL AFRICA
MALAWI
Lake Malawi
COMOROS
ZAMBIA
Victoria Falls
Zambezi R.
Mt. Sapitwa 9,849 ft. (3,002 m)
Mt. Inyangani 8,504 ft. (2,592 m)▲
MOZAMBIQUE
INDIAN OCEAN
Okavango River
ZIMBABWE
Mozambique Channel
MADAGASCAR
20°S
NAMIBIA
Namib Desert
BOTSWANA
Kalahari Desert
Limpopo R.
MAURITIUS
Tropic of Capricorn
Orange R.
Drakensberg Mountains
SWAZILAND
30°S
SOUTH AFRICA
Mt. Thabana Ntlenyana 11,425 ft. (3,482 m)
LESOTHO
Cape of Good Hope
Great Karroo
0   250   500 mi.
0   250   500 km
10°E   20°E   30°E   40°E   50°E   60°E

N W E S

Many people raise livestock in South Africa's high veld *(left)* and in the small enclave of Swaziland *(right).*
PLACE: What is an enclave?

Human/Environment Interaction
The Economy

South Africa has the most developed economy in Africa. Its workers mine one-half of Africa's minerals and produce two-fifths of Africa's manufactured goods. South Africa is also the continent's largest exporter of farm products.

Not all South African workers take part in this prosperous economy. Workers in modern industries and businesses live in urban areas. Yet many people in rural South Africa are poor and depend on subsistence farming.

Land of Gold and Diamonds In terms of mineral wealth, South Africa is one of the richest countries in the world. South Africa produces about one-third of all the gold mined in the world each year. The Witwatersrand (WIHT•WAW•tuhrz•RAND)—an area around the city of Johannesburg—holds the world's largest and richest goldfield. South Africa also has the world's largest deposits of diamonds.

Manufacturing and Farming South Africa's industrial workers produce many manufactured goods. Many of the products South Africans use are made in their own factories. South Africa also exports metal products, chemicals, clothing, and processed foods.

Most of South Africa is either too dry or too hilly to farm. On the land that is available, South Africa's farmers grow enough food for the country's needs and for export. Among the crops they cultivate are corn, fruits, potatoes, and wheat. Herding sheep and livestock is a major economic activity on the plateau.

WHAT IN THE WORLD?

The Biggest Diamond

The largest diamond ever found was dug from Premier Mine near the city of Pretoria in South Africa. It was about the size of a person's fist and weighed 3,106 carats—about 1.4 pounds (.64 kg). When it was cut, it made 9 large jewels and more than 100 smaller ones. The largest jewel became known as the Star of Africa.

Place
Influences of the Past

About 39 million people live in South Africa. African ethnic groups make up 74 percent of South Africa's population. Africans of European origin represent 14 percent, and Asians 3 percent. People of mixed European, African, and Asian backgrounds make up the remaining 9 percent.

Africans Most Africans in South Africa trace their ancestry to Bantu-speaking peoples that settled throughout Africa between A.D. 100 and 1000. The largest ethnic groups in South Africa today are the Sotho, Zulu, and Xhosa (KOH•suh).

Europeans In the 1600s the Dutch became the first Europeans to settle in South Africa. Over time German and French settlers joined them. Together these European groups came to speak their own language—Afrikaans (A•frih•KAHNS). The Afrikaners pushed Africans off the best land and set up farms.

The British settled in South Africa in the early 1800s. By the end of the century, diamonds and gold drew many British people to South Africa. War broke out between the British and the Afrikaners in 1899. After three years of fighting, the British won the conflict.

African Independence Dates

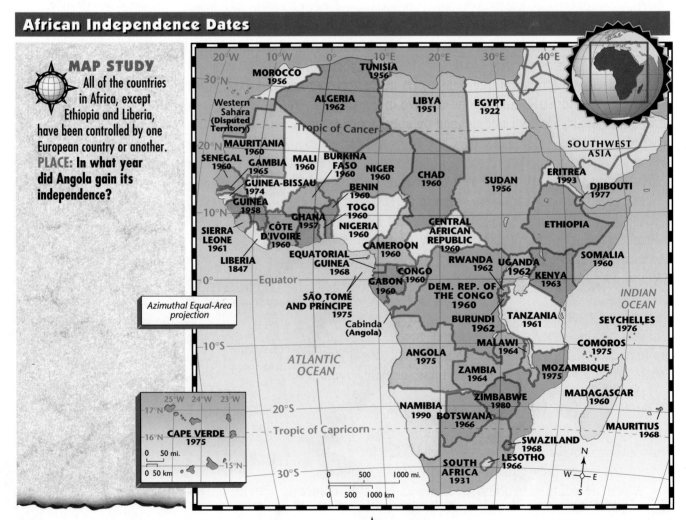

MAP STUDY All of the countries in Africa, except Ethiopia and Liberia, have been controlled by one European country or another. **PLACE: In what year did Angola gain its independence?**

Azimuthal Equal-Area projection

A New Nation In 1910 Afrikaner and British territories united into one country—the Union of South Africa—within the British Empire. Black South Africans founded the African National Congress in 1912 in hopes of gaining power. In 1931 the United Kingdom gave South Africa full independence.

In 1948 the Afrikaners set up **apartheid**—"apartness"—or practices that separated South Africans of different ethnic groups. For example, laws forced black South Africans to live in separate areas and attend different schools from European South Africans. People of non-European background could not vote. They had virtually no political rights enjoyed by those of European ancestry.

For more than 40 years, people inside and outside South Africa protested against the practice of apartheid. In 1991, the South African government finally agreed to end apartheid. In April 1994, South Africa had its first-ever election in which people of all ethnic groups could vote. South Africans elected their first black African president, Nelson Mandela.

In 1994, for the first time, South Africa's elections were open to all races *(left).* Nelson Mandela became the country's first black president *(right).* **PLACE: What group of people were favored under apartheid?**

COMPARING POPULATION: South Africa and the United States

GRAPHIC STUDY
Cape Town is South Africa's largest city. **PLACE: What is South Africa's total population?**

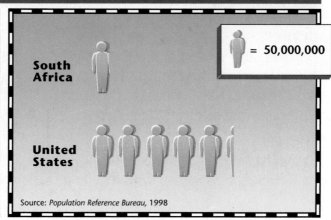

South Africa

United States

= 50,000,000

Source: *Population Reference Bureau, 1998*

The People

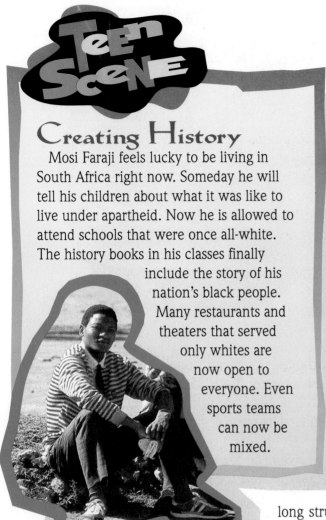

TEEN SCENE

Creating History

Mosi Faraji feels lucky to be living in South Africa right now. Someday he will tell his children about what it was like to live under apartheid. Now he is allowed to attend schools that were once all-white. The history books in his classes finally include the story of his nation's black people. Many restaurants and theaters that served only whites are now open to everyone. Even sports teams can now be mixed.

About 57 percent of South Africans live in urban areas. The map on page 562 shows you that South Africa has five large cities with more than 500,000 people. These cities are Johannesburg, Cape Town, Durban, Soweto, and Pretoria. Cape Town and Durban are port cities. Johannesburg is an inland industrial and commercial center. North of Johannesburg is Pretoria, which is one of the national capitals along with Cape Town and Bloemfontein (BLOOM•fuhn•TAYN).

One of the challenges facing South Africa is to develop a better standard of living for all of its people. Most European South Africans live in modern homes or apartments, own cars, and enjoy foods similar to those common in the United States and Europe. Most African, Asian, and mixed-group South Africans, however, have a much harder life. Many live in crowded **townships,** or neighborhoods outside cities. Others live in rural areas where the land is difficult to farm.

The arts in South Africa reflect the country's long struggle for justice and equality. In recent years, African musicians have combined traditional African dance rhythms and modern rock music. This style of music has become popular in many countries, especially the United States. You may be familiar with South African groups such as Mahlathini and the Mahotella Queens and Ladysmith Black Mambazo.

SECTION 1 ASSESSMENT

REVIEWING TERMS AND FACTS

1. Define the following: enclave, high veld, escarpment, apartheid, township.

2. PLACE What landform covers most of inland South Africa?

3. HUMAN/ENVIRONMENT INTERACTION What are South Africa's two main mineral resources?

4. PLACE How did South Africa's government change in the early 1990s?

MAP STUDY ACTIVITIES

5. Look at the political map on page 544. About how far is Cape Town from Pretoria?

6. Turn to the physical map on page 546. At what elevation does Lesotho lie?

BUILDING GEOGRAPHY SKILLS

Using Technology

Developing Multimedia Presentations

You have been assigned a research report to present to your class. You want to make your presentation informative, but also interesting and fun. How can you do this? One way is to combine several types of media into a multimedia presentation.

Suppose you plan to give a multimedia presentation on life in South Africa. You might show photographs of the country, play a recording of local music, or present a video showing the people at work and at play. You could also develop your multimedia presentation on a computer. Computer multimedia programs can allow you to combine text, video, audio, graphics, and other media to create dynamic presentations.

In order to create multimedia presentations on a computer, you need to have certain tools. These may include computer graphics tools and drawing programs, animation programs that make certain images move, and presentation programs that tie everything together. Your computer manual will tell you which programs your computer supports.

Ask questions like the following to develop a multimedia presentation:

- Which forms of media do I want to include: Video? Sound? Animation? Photographs? Graphics? Other media?
- Which of these media forms does my computer support?
- What kind of software programs do I need?
- Is there a "do-it-all" program that I can use to develop the kind of presentation I want?

Women sell food in a village market in the Democratic Republic of the Congo. Most farming in this country centers around small family plots. People produce food for their own needs, as well as food to sell and trade for other products.

Technology Skills Practice

For additional practice in developing multimedia presentations, use the questions above to help you develop a presentation on life in one of the African countries you have studied in this unit. Write a plan describing a multimedia presentation you would like to develop. Indicate what tools you will need and what steps you must take to make the presentation a reality.

Atlantic Countries

Words to Know
- exclave

Places to Locate
- Namibia
- Angola
- Windhoek
- Luanda
- Namib Desert
- Kalahari Desert

Read to Learn . . .
1. what landforms Angola and Namibia share.
2. why Portuguese is the official language of Angola.
3. why Namibia has difficulty feeding its people.

You would not want to get a flat tire on the vast Namib Desert! This empty stretch of sand and rock is one of the loneliest places on the earth. The

Namib Desert stretches north to south about 1,200 miles (1,931 km) through southwestern Africa.

The Namib Desert extends most of the length of the country of Namibia. In addition to desert, Namibia and its northern neighbor—Angola—have long coastlines on the Atlantic Ocean. For this reason, they are known as southern Africa's Atlantic countries.

Place
Angola

Angola is a huge, nearly square mass of land. It covers 481,350 square miles (1,246,697 sq. km)—an area larger than Texas and California put together. Angola also includes a tiny exclave called Cabinda. An **exclave** is a tiny area of a country that is separated from the main part. The map on page 544 shows you that Cabinda is separated from Angola.

The Land Most of Angola is part of the same huge inland plateau that sweeps through the Republic of South Africa. Many rivers cross this area in Angola. Some flow into the Congo River in the north; others flow into the Atlantic Ocean or the Indian Ocean.

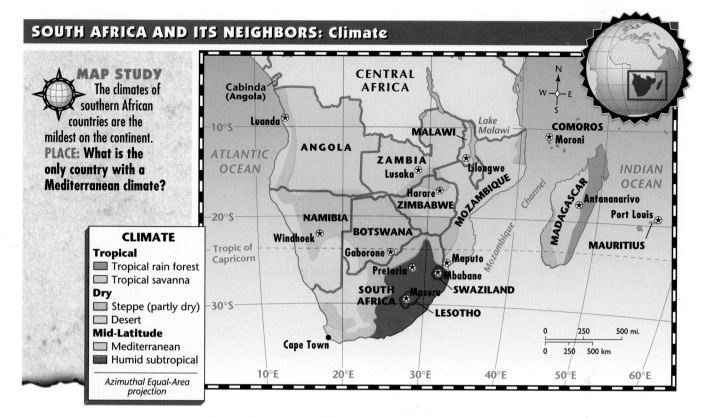

MAP STUDY
The climates of southern African countries are the mildest on the continent.
PLACE: What is the only country with a Mediterranean climate?

CENTRAL AFRICA

Cabinda (Angola)

Luanda

10°S

ATLANTIC OCEAN

ANGOLA

MALAWI

Lake Malawi

ZAMBIA

Lusaka

Lilongwe

MOZAMBIQUE

COMOROS
Moroni

INDIAN OCEAN

Harare
ZIMBABWE

Mozambique Channel

MADAGASCAR

Antananarivo

Port Louis

20°S

NAMIBIA

Windhoek

BOTSWANA

Gaborone

MAURITIUS

Tropic of Capricorn

Maputo

Pretoria

Mbabane

Mozambique

SOUTH AFRICA

Maseru

SWAZILAND

30°S

LESOTHO

Cape Town

10°E   20°E   30°E   40°E   50°E   60°E

CLIMATE
Tropical
■ Tropical rain forest
□ Tropical savanna
Dry
■ Steppe (partly dry)
□ Desert
Mid-Latitude
■ Mediterranean
■ Humid subtropical

Azimuthal Equal-Area projection

0   250   500 mi.
0   250   500 km

Hilly grasslands cover northern Angola. The southern part of the country, however, is a rocky desert. A low strip of land winds along the 900-mile (1,448-km) Atlantic coastline. This lowland area has little natural vegetation, except for a few rain forests in the north.

If you traveled across Angola, you would experience three kinds of climates. The map above shows you that coastal areas have a steppe or desert climate. The inland plateau has a tropical savanna climate of wet and dry seasons. This area receives enough rainfall for farming.

The Economy Angola's major economic activity is agriculture. Nearly 58 percent of Angola's 12 million people live in the countryside. Subsistence farmers there grow corn, cotton, and sugarcane. Coffee is the leading export crop. Some Angolans also herd cattle, sheep, and goats.

Oil and mining, however, provide most of Angola's income. Look at the map on page 558. It shows you that Angola is rich in diamonds, iron ore, manganese, and petroleum. Oil is the country's leading export. Most of the oil deposits are located off the shore of Cabinda.

Influences of the Past Most of Angola's people are Africans who belong to several ethnic groups. They trace their ancestry to the ancient Bantu-speaking peoples. In the 1400s the Kongo kingdom ruled a large part of northern Angola.

From the 1500s until its independence in 1975, Angola was a colony of Portugal. Portuguese influence still remains strong in Angola. Although African languages are spoken, Portuguese is the official language. Almost 70 percent of Angolans practice the Roman Catholic faith brought to Angola by Portuguese missionaries.

Angola's capital and major port is located on the country's long coastline.
LOCATION: What is the name of Angola's exclave?

After independence, civil war among different political and ethnic groups broke out in Angola. The fighting brought great suffering to Angola's people. Today the fighting has diminished, but rebuilding Angola will be a very difficult task.

Place

Namibia

South of Angola lies Namibia, one of Africa's newest countries. Namibia became independent in 1990 after 75 years of South African rule. The map on page 548 shows you the dates of independence for other African countries.

A Plateau and Deserts Namibia has a land area of 317,870 square miles (823,283 sq. km)—about one-half the size of Alaska. If you look at the map on page 546, you will see that a large plateau runs through the center of Namibia. It is the most populated area of the country. The rest of Namibia is made up of deserts. The Namib Desert runs almost the entire length of Namibia's Atlantic coast. The Kalahari Desert stretches across the southeastern part of the country. As you might guess, most of Namibia has a hot, dry climate.

The Economy Namibia's economy depends on the export of minerals. The country has rich deposits of diamonds, manganese, copper, silver, and zinc. Many Namibians work in the country's mines.

Because of the desert environment, Namibia has difficulty feeding its people. Farmers in the plateau region receive barely enough rainfall to grow corn. Most Namibians are herders who raise cattle, goats, and sheep.

Land of Few People Only 1.6 million people live in Namibia—one of the most sparsely populated countries in Africa. Most Namibians belong to African ethnic groups. A small number are of European ancestry. English and Afrikaans are the official languages, but most Namibians speak African languages.

SECTION 2 ASSESSMENT

REVIEWING TERMS AND FACTS

1. **Define the following:** exclave.

2. **HUMAN/ENVIRONMENT INTERACTION** What is Angola's most important mineral export?

3. **PLACE** What European country once ruled Angola?

4. **LOCATION** Where do most Namibians live?

MAP STUDY ACTIVITIES

5. Turn to the climate map on page 553. What climate regions are found in Angola and Namibia?

MAKING CONNECTIONS

| MATH | SCIENCE | **HISTORY** | LITERATURE | TECHNOLOGY |

Long before Europeans arrived on Africa's shores, inland African kingdoms were wealthy from the gold trade. Between A.D. 1000 and 1500, a Bantu-speaking people—the Shona—built nearly 300 stone-walled fortresses throughout southern Africa. The largest fortress was called Great Zimbabwe (zihm•BAH•bwee)—meaning "house of stone." It was the religious and trading center of the Shona kingdom.

A GIGANTIC CAPITAL Great Zimbabwe included fortress walls, temples, marketplaces, and homes spread out on a fertile, gold-rich plateau south of the Zambezi River. The city was the capital of the Shona kingdom and displayed the wealth and power of its kings. Its shape was oval, and its buildings were made of stone. This made the city stand out from the traditional circular mud-and-thatch villages.

The most impressive part of the city was an area called the Great Enclosure. The Great Enclosure's outer wall was 16.5 feet (5 m) thick and 32 feet (9.8 m) high. It was built of 900,000 granite stones fitted together without mortar. Enclosed within this wall, a maze of interior walls and hidden passages pro-

tected the circular house of the king, the Great Temple, and other religious buildings.

🌐 **Remains of Zimbabwe Fortress**

INTERNATIONAL TRADE During the A.D.1400s, Shona kings controlled the trade routes between inland goldfields and seaports on the Indian Ocean coast. Civil wars and attacks from European invaders brought an end to the empire. Archaeologists in the late 1800s concluded that Great Zimbabwe, however, enjoyed a profitable trade with overseas lands. In Great Zimbabwe's ruins, they uncovered articles from India, China, and Asia.

🌐 **Making the Connection** ▽ △

1. Why did Great Zimbabwe become so powerful?
2. With whom did Great Zimbabwe trade?

SECTION 3 ·
Inland Southern Africa

PREVIEW

Words to Know
- copper belt
- sorghum

Places to Locate
- Zambia
- Malawi
- Zimbabwe
- Botswana
- Zambezi River
- Harare

Read to Learn . . .
1. what Zambia's most important natural resource is.

2. how Zimbabwe's recent history is similar to South Africa's.
3. why most of Botswana's people live in the eastern part of the country.

So much mist and noise rise from this waterfall that people in the area call it the "smoke that thunders." You are looking at the waters of the Zambezi River as they plunge 350 feet (107 m) to the bottom of a deep gorge. They form Victoria Falls, one of Africa's great natural wonders.

The Zambezi (zam•BEE•zee) River is one of southern Africa's longest rivers. For part of its journey, the Zambezi forms the border between two inland countries in southern Africa—Zambia and Zimbabwe. Two other countries—Malawi (muh•LAH•wee) and Botswana (baht•SWAH•nuh)—are also located in this region.

Place
Zambia

Zambia is a landlocked country of 287,020 square miles (743,382 sq. km). Zambia—slightly larger than the state of Texas—lies near the Equator. Zambia's high elevation gives most of the country a mild climate.

The map on page 546 shows you that a high plateau covers much of Zambia. Several rivers, including the Zambezi, cross the country. Kariba Dam—one of Africa's largest hydroelectric projects—spans the Zambezi River.

The Economy Zambia is one of the world's largest producers of copper. This resource provides more than 80 percent of Zambia's income. You will find a

copper belt, a large area of copper mines, in northern Zambia near the border with the Democratic Republic of the Congo.

Because copper is so valuable, Zambians have paid less attention to developing agriculture. As a result, Zambia must import much of its food. Railroads link landlocked Zambia to ports in neighboring countries, such as Mozambique, Angola, and Tanzania.

The People Zambia's 9.5 million people belong to more than 70 different ethnic groups. They speak 8 major African languages. Once a British colony, Zambia became independent in 1964. Because of the British influence, many Zambians today also speak English.

About 40 percent of Zambia's population live in urban areas. Lusaka is the capital and largest city. Many urban dwellers work in mining and service industries. The rest of Zambia's people live in rural villages. They grow corn and other subsistence food crops for their families.

Place
Malawi

East of Zambia lies Malawi, a small, narrow country of only 36,320 square miles (94,069 sq. km). In some places, Malawi is no more than 50 miles (81 km) wide! Green plains and savanna grasslands cover western Malawi. Part of the Great Rift Valley runs through the country from north to south. In the middle of the valley lies beautiful Lake Malawi. High plateaus and tall mountains rise on both sides of the lake. Malawi lies in the tropics, but its plateaus and mountains give it a mild climate.

The Economy Because of its mountains, only one-third of the land in Malawi is suitable for farming. Agriculture, however, is the country's major economic activity. Some of Malawi's farmers work on large commercial farms that grow tea for export. Others cultivate sugarcane, corn, cotton, peanuts, and **sorghum** for local use on small family plots. Sorghum is a tall grass with seeds like corn. People in Malawi also fish and herd livestock.

The People Look at the map on page 562. It shows you that Malawi is one of the most densely populated countries in Africa. Malawi has about 269 people per square mile (104 people per sq. km). Most of its 9.8 million people belong to many different African ethnic groups and live in small villages.

All in a Day's Work

Fish from Lake Malawi are an important part of the diet of local residents.
REGION: Why does the tropical country of Malawi have a mild climate?

Place
Zimbabwe

South of Zambia lies the country of Zimbabwe. With 149,363 square miles (386,850 sq. km), Zimbabwe is a little larger than the state of Montana. Like other countries in southern Africa, most of Zimbabwe occupies a high plateau. North and south of the plateau are lowlands. The Zambezi River crosses the northern lowlands. The Limpopo River winds through the southern lowlands.

Most of Zimbabwe has a tropical savanna climate of wet and dry seasons. High elevations, however, keep temperatures cool and pleasant in many parts of the country.

The Economy Gold, asbestos, and copper are three of the most important minerals mined in Zimbabwe. Although mining provides most of Zimbabwe's income, a majority of its people are farmers. About half of Zimbabwe's land is fertile enough for farming. Farmers grow corn and herd cattle and livestock. Some people also work on large commercial farms that grow coffee and tobacco for export.

The People Zimbabwe has about 11 million people. Most of the people belong to two African ethnic groups—the Shona and the Ndebele (ehn•duh•BEH•leh). The country takes its name from the famous trading center—Great Zimbabwe—built by the Shona in the A.D. 1400s.

In the 1890s the British began to rule the area that is now Zimbabwe. As in South Africa, Europeans ran the government and owned all the best farmland.

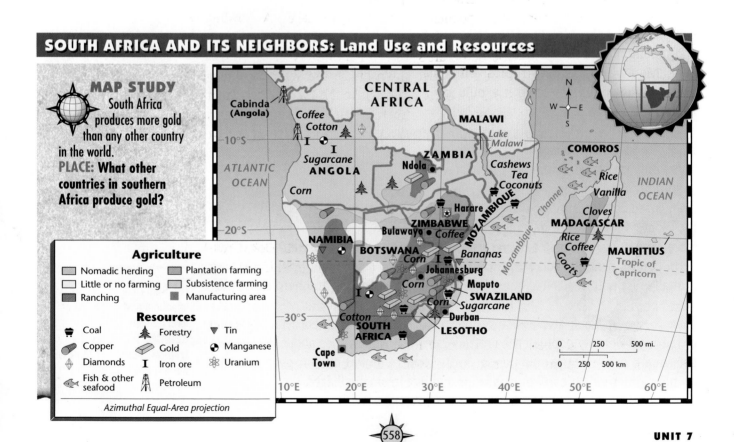

SOUTH AFRICA AND ITS NEIGHBORS: Land Use and Resources

MAP STUDY South Africa produces more gold than any other country in the world.
PLACE: What other countries in southern Africa produce gold?

Agriculture
- ▢ Nomadic herding
- ▢ Little or no farming
- ▢ Ranching
- ▢ Plantation farming
- ▢ Subsistence farming
- ▢ Manufacturing area

Resources
- Coal
- Copper
- Diamonds
- Fish & other seafood
- Forestry
- Gold
- I Iron ore
- ▼ Tin
- Manganese
- Uranium
- Petroleum

Azimuthal Equal-Area projection

The Kariba Dam harnesses the power of the mighty Zambezi River.
LOCATION: The Zambezi forms the border between Zambia and what other country?

They refused to share power with the Africans. The Africans, in turn, resisted European rule. Peace talks finally led to a settlement. In 1980 free elections brought an independent African government to power.

Today about 70 percent of the population live in rural villages. A growing number, however, are moving to cities to find factory jobs. The largest city is Harare, the national capital. One of Harare's residents is fifteen-year-old Moses Katiyo. He and his family live in a crowded township in the suburbs. When school is not in session, Moses and his older brother take the train to Bulawayo (BOO•luh•WAY•oh), a city in the south. From there they often visit the Khami Ruins. These are the remains of stone buildings built hundreds of years ago by Moses' Bantu-speaking ancestors.

Place
Botswana

Botswana is the most isolated country in inland southern Africa. Desert covers much of its land area of 218,810 square miles (566,718 sq. km). If you look at the map on page 546, you will see that the vast Kalahari Desert spreads over southwestern Botswana. The Kalahari Desert is hot and dry, but parts of it have some vegetation. In eastern Botswana the land rises to a savanna region of grasses, bushes, and trees. The Okavango River flows through northwestern Botswana. There it forms one of the largest swamp areas in the world.

On National Day in Zimbabwe, a flame is lit to mark the country's independence from the United Kingdom.
PLACE: What industry generates the most income for Zimbabwe?

The Economy Like its neighbors, Botswana is rich in mineral resources. Among its treasures are diamonds, copper, and uranium. Mining, however, provides jobs for only a small number of Botswana's people. Most raise livestock or farm the land, but growing crops is difficult in Botswana. During the 1980s a severe drought brought hardships to many of Botswana's farmers. To earn a living, many young people work in the Republic of South Africa for several months a year.

The People The map on page 562 shows you that Botswana has few people for its large size. Most of the 1.4 million people live in eastern Botswana, where the land is most fertile. About 75 percent of the people live in rural areas, but thousands of them move to the cities each year. Gaborone is the capital and largest city.

SECTION 3 ASSESSMENT

REVIEWING TERMS AND FACTS

1. Define the following: copper belt, sorghum.

2. HUMAN/ENVIRONMENT INTERACTION What is Zambia's most abundant natural resource?

3. PLACE What lake lies in Malawi's Great Rift Valley?

4. REGION Why do many young people from Botswana go to the Republic of South Africa?

MAP STUDY ACTIVITIES

5. Look at the land use map on page 558. What are the major resources of Zambia?

PREVIEW

Words to Know
- slash-and-burn farming

Places to Locate
- Mozambique
- Madagascar
- Comoros
- Mauritius
- Indian Ocean

Read to Learn . . .
1. why Mozambique's ports are so valuable.
2. why Madagascar has unusual plants and animals.
3. how the islands of Comoros and Mauritius were formed.

Modern buildings rise along the tree-lined streets of Maputo, capital of the country of Mozambique (MOH•zuhm•BEEK). Maputo is a port city near the Indian Ocean. It provides a link to the outside world for the landlocked countries of southern Africa.

Mozambique is a large country in southern Africa that borders the Indian Ocean. Three island countries—Madagascar (MA•duh•GAS•kuhr), Comoros (KAH•muh•ROHS), and Mauritius (maw•RIH•shuhs)—also form part of southern Africa's Indian Ocean region.

Place
Mozambique

The map on page 544 shows you that Y-shaped Mozambique stretches along the Indian Ocean. Its landscapes include sandy lowlands, high plateaus, and tall mountains. Sand dunes, swamps, and fine natural harbors line Mozambique's Indian Ocean coast. In the center of the country lies a flat plain covered with grasses and tropical forests. Beyond the plain, the land slowly rises to form high plateaus and mountains along Mozambique's border with Malawi.

A number of rivers cross the northern and central parts of Mozambique. The most important river is the Zambezi. The Cabora Bassa Dam on the Zambezi provides electric power for much of the country. The map on page 553 shows you that most of Mozambique has a tropical savanna climate with wet and dry seasons. Rainfall is heaviest in the north. Much of the southwest is dry.

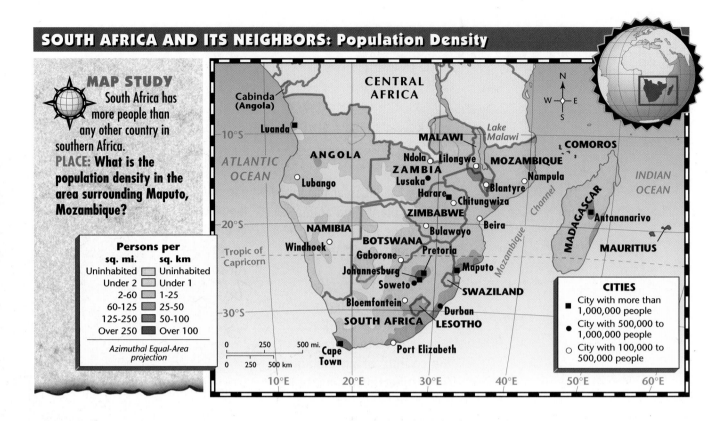

MAP STUDY
South Africa has more people than any other country in southern Africa.
PLACE: What is the population density in the area surrounding Maputo, Mozambique?

Persons per

sq. mi.	sq. km
Uninhabited	Uninhabited
Under 2	Under 1
2-60	1-25
60-125	25-50
125-250	50-100
Over 250	Over 100

Azimuthal Equal-Area projection

CITIES
■ City with more than 1,000,000 people
● City with 500,000 to 1,000,000 people
○ City with 100,000 to 500,000 people

The Economy Most people in Mozambique are farmers. Some of them practice **slash-and-burn farming.** This means they cut and burn forest trees to clear areas for planting. Mozambique's major crops are cashews, coconuts, tea, sugarcane, and bananas. Fishing provides an income for people living along the Indian Ocean.

Mozambique's major source of income, however, comes from its seaports. South Africa, Zimbabwe, Swaziland, and Malawi pay Mozambique to use docks at Maputo and other ports.

During the 1980s and early 1990s, a fierce civil war slowed industrial growth in Mozambique. Although the fighting has ended, the country's economy still cannot provide enough jobs. In order to earn their livings, many people from Mozambique work in the Republic of South Africa.

The People About 18.6 million people live in Mozambique. Nearly all of them belong to African ethnic groups. Most live in the southern part of the country around Maputo, the capital. Portugal governed Mozambique from the early 1500s until 1975, when Mozambique became independent. Portuguese is still the official language. Most of the people, however, speak African languages.

Place
Madagascar

Madagascar is an island nation in the Indian Ocean. Look at the map on page 546. It shows you that Madagascar lies about 250 miles (400 km) southeast of the African mainland. Cool highland areas cross the middle of

Madagascar. Most coastal areas have warm, humid plains and fertile river valleys. Partly dry grasslands cover southern Madagascar.

In the past, Madagascar's island location kept it isolated from other parts of the world. Today Madagascar has many plants and animals that are not found elsewhere on the earth. One of these unusual animals is the lemur, which looks like a monkey. Lemurs have large eyes that appear to glow at night as they leap from tree to tree in the forests.

Madagascar's Wildlife

The Economy Agriculture is the chief economic activity in Madagascar. Farmers grow food crops—mainly rice—and herd cattle. Coffee is the leading export. If you like vanilla ice cream, thank Madagascar! It produces most of the world's vanilla beans.

The People About 14 million people live in Madagascar. Most trace their ancestry to Southeast Asian and African groups that settled the island centuries ago. Malagasy, the major language, is similar to languages spoken in Southeast Asia. French is also spoken in Madagascar's cities. Madagascar was a colony of France from the 1890s until it gained its independence in 1960.

The ring-tailed lemur is one of the many unusual species of animals that live in Madagascar.
LOCATION: Why does Madagascar have so many unusual plants and animals?

Place
Small Island Countries

Far from Africa in the Indian Ocean are two other island republics—Comoros and Mauritius. The people of both countries have many different backgrounds.

Comoros Comoros is a group of four mountainous islands that lie between mainland Africa and Madagascar. Volcanoes formed all four islands thousands of years ago. Thick tropical forests cover Comoros today. At one time a French territory, Comoros became independent in 1975.

Most of Comoros's 500,000 people are farmers. They grow rice, cassava, corn, vanilla, sugarcane, and coffee. The people of Comoros are a mixture of Arabs, Africans, and people from Madagascar. They speak Arabic and Swahili, and most practice the religion of Islam.

Mauritius The island republic of Mauritius lies about 500 miles (800 km) east of Madagascar in the Indian Ocean. Mauritius includes one major island and a number of smaller islands, which are territories. Mauritius gained its freedom from the United Kingdom in 1968.

Horse-drawn wagons share the streets with cars in Madagascar's capital.
PLACE: What flavoring is Madagascar known for?

Like Comoros, the islands of Mauritius were formed by volcanoes. Bare, black peaks rise sharply above green fields, and palm-dotted white beaches line the coasts. The climate is subtropical. Because Mauritius has few natural resources, its economy depends on agriculture. Sugar and sugar products are the island country's major exports.

Mauritius has about 1.2 million people who come from many different backgrounds. About 70 percent are descendants of settlers from India. The rest are of African, European, or Chinese ancestry. The population of Mauritius is increasing rapidly. Providing food, jobs, and housing for more people will be a major challenge for Mauritius in the years ahead.

SECTION 4 ASSESSMENT

REVIEWING TERMS AND FACTS

1. Define the following: slash-and-burn farming.

2. HUMAN/ENVIRONMENT INTERACTION What is Mozambique's major source of income?

3. LOCATION Where is Madagascar located?

4. MOVEMENT What major groups settled in Madagascar?

MAP STUDY ACTIVITIES

5. According to the population density map on page 562, where are Mozambique's most densely populated areas?

Chapter 21 Highlights

Important Things to Know About South Africa and Its Neighbors

SECTION 1 REPUBLIC OF SOUTH AFRICA

- Most of inland South Africa is a high plateau.
- Because of its mineral resources, South Africa has the most developed economy in Africa.
- In recent years, South Africa has given equality to citizens of all ethnic groups and is struggling to become a more complete democracy guaranteeing the rights of all its citizens.

SECTION 2 ATLANTIC COUNTRIES

- Angola has large deposits of oil, diamonds, and iron ore.
- Two major deserts cross Namibia. This harsh landscape limits agriculture in Namibia.

SECTION 3 INLAND SOUTHERN AFRICA

- Zambia is one of the world's largest producers of copper.
- Tiny, mountainous Malawi has a large lake running almost the entire length of its eastern border.
- Zimbabwe has many mineral resources and good farmland.
- Desert covers a large part of Botswana.

SECTION 4 INDIAN OCEAN COUNTRIES

- Mozambique's ports on the Indian Ocean are used by landlocked countries in southern Africa.
- Madagascar's people trace their ancestry to African and Southeast Asian settlers.

Celebrating South Africa National Independence Day ▶

REVIEWING KEY TERMS

Match the numbered terms in Set A with their lettered definitions in Set B.

A

1. exclave
2. high veld
3. sorghum
4. enclave
5. township
6. apartheid

B

A. flat, grass-covered plains in South Africa

B. official system of racial separation in South Africa

C. crowded African neighborhood outside a city

D. part of a country separated from the main part

E. tall grass with seeds like corn

F. small country surrounded by another country

Mental Mapping Activity

Draw a freehand map of South Africa and its neighbors. Label the following:
- South Africa
- Indian Ocean
- Angola
- Zambia
- Zimbabwe
- Johannesburg

REVIEWING THE MAIN IDEAS

Section 1

1. PLACE What oceans border South Africa?
2. MOVEMENT How did the coming of Europeans affect life for Africans in South Africa?

Section 2

3. LOCATION Describe Namibia's coast.
4. HUMAN/ENVIRONMENT INTERACTION What is Angola's main agricultural export?

Section 3

5. PLACE How does the population density of Malawi compare with populations in other African countries?
6. MOVEMENT What two African ethnic groups are found in Zimbabwe?

Section 4

7. LOCATION Why has industrial development been slow in Mozambique?
8. REGION How were Comoros and Mauritius formed?

CRITICAL THINKING ACTIVITIES

1. **Formulating Questions** Write three questions you would ask a South African about his or her way of life and recent changes there.

2. **Identifying Central Issues** What are some of the challenges faced by countries in southern Africa as they try to develop their economies?

CONNECTIONS TO GOVERNMENT

Cooperative Learning Activity Work in small groups to learn more about democracy in southern Africa today. Each group will choose one country and research the following topics: (a) form of government before independence; (b) present form of government; or (c) current political challenges faced by the government. As a group, create a skit or news broadcast to share with the class.

GeoJournal Writing Activity

Choose a country in southern Africa and imagine that you are a travel writer visiting the place. Write a short article about the country. You may want to describe its major natural attractions or its cities, or give interesting information about its people.

TECHNOLOGY ACTIVITY

Developing Multimedia Presentations

Use the Internet, as well as other reference works, to research the kingdom of the Shona and its capital city of Great Zimbabwe. Then, imagine that you are a tour guide. Create a short multimedia presentation about the political and religious center of the Shona kingdom. Incorporate photographs, drawings, maps, music, and video, if possible, into your presentation.

PLACE LOCATION ACTIVITY: SOUTH AFRICA AND ITS NEIGHBORS

Match the letters on the map with the places of South Africa and its neighbors. Write your answers on a separate sheet of paper.

1. Mauritius
2. Lake Malawi
3. Drakensberg Mountains
4. Botswana
5. Cape Town
6. Luanda
7. Kalahari Desert
8. Zambezi River
9. Madagascar
10. Mozambique

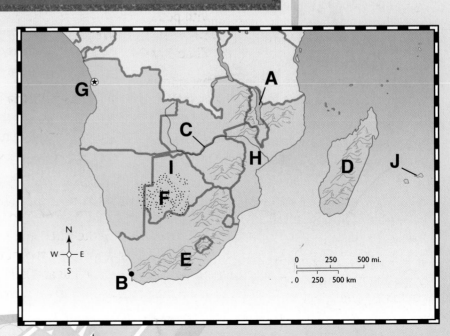

EYE ON THE ENVIRONMENT

Africa South of the Sahara

DESERTIFICATION

PROBLEM

Desertification, or the "making of deserts," is occurring in parts of Africa south of the Sahara. Poor conservation methods lead to soil erosion and dried-up wells and water holes. Trees and shrubs are cut down and used as fuel. Vegetation dies out as herds of goats and cattle overgraze fragile lands. Dry winds and fierce sandstorms blow away the naked topsoil and cover once-productive fields with sheets of sand. Marching dunes bury fields, roads, and entire villages. And so the desert advances.

SOLUTIONS

● Farmers in some areas confine grazing animals to fenced areas and bring food to the animals.

● To help save rapidly dwindling forests, Mali, Niger, and Burkina Faso have started national campaigns to place more efficient clay or metal cookstoves in every household.

● In Niger, more than 435 miles of windbreaks have been planted to slow advancing sand and protect fragile seedlings.

● In Ethiopia, communities are planting hundreds of trees, and in Kenya, people are terracing farmland to control erosion.

Camels drink at a well in Niger. Ironically, use of wells and water holes can contribute to desertification as animals trample and kill plants when they drink.

DESERTIFICATION FACT BANK

🍃 Thirty-four percent of Africa's land is at risk from desertification.

🍃 The Sahel, a fragile belt of grasses and forests that stretches 3,000 miles across Africa, has lost 30 percent of its trees in the last 20 years.

🍃 Once 40 percent of Ethiopia was forested; in 1990 less than 4 percent of the forests remained.

🍃 In Mali, the Sahara has spread more than 400 miles in 20 years.

🍃 During the last 20 years, many villages in the Sahel have lost as much as one-half of their farmland to desertification.

Dwellings nestle in a sandy area of trees and shrubs in Kenya. Drought and overuse could easily turn a semiarid region into desert.

TEEN TRIBUTE

Eleven-year-old George Otieno lives on a farm in Kenya. George's family could not grow enough food to feed themselves because the soil on their farm was so poor. Then at school George learned that planting trees can help reverse desertification. George planted several hundred seedlings he received from CARE, an environmental relief group. The trees restored nutrients to the soil—and improved the family's harvest and their standard of living.

environmental activities

BE NICE TO TREES

TAKE A CLOSER LOOK

1 What is desertification?

2 What are the main causes of desertification in Africa south of the Sahara?

WHAT CAN YOU DO?

🍃 Adopt a piece of land in your own neighborhood. Clean it up, plant flowers, and put up a No Littering sign.

🍃 Conserve water—cut your shower water use by half by a using a low-flow shower head.

🍃 Volunteer at your local zoo. Help with programs to save endangered species.

🍃 Support agencies that help African countries to reforest empty land.

Men on camels herd cattle in Chad. As livestock strip away vegetation, soil blows away, and bare patches become desert.

569

Unit 8 Asia

What Makes Asia a Region?

The people share . . .

- high population densities.
- deadly natural hazards—typhoons, floods, volcanic activity.
- expanding economies.
- rich natural resources.
- ancient religions and cultures.

To find out more about Asia, see the Unit Atlas on pages 572–585.

Bali Island, Indonesia

EXPLORING THE INTERNET

To learn more about Asia, visit the Glencoe Social Studies Web site at **www.glencoe.com** for information, activities, and links to other sites.

GeoJournal Activity
As you read about Asia, follow news reports about the region either on television or in newspapers and magazines. Organize your information in your journal under the following categories: Environment, Culture and Daily Life, and Economy.

UNIT 8 ATLAS

NATIONAL GEOGRAPHIC SOCIETY

IMAGES *of the* WORLD

1. Horse breeder, Mongolia
2. Traffic in the heart of the city, Hong Kong
3. Harvesting coffee, Indonesia
4. Mosque in a rainstorm, Indonesia
5. Girl holding the Quran, Indonesia
6. Father and son readying their boat, Indonesia
7. Pilgrims bathing in the Ganges River, India

All photos viewed against snow-covered peaks in the Karakoram Range, Pakistan.

Regional Focus

Asia is both a continent and a region. The *continent* of Asia covers the eastern part of the large landmass called Eurasia. The *region* of Asia is a smaller part of the Asian continent. As a region, Asia reaches from the rugged mountains of Pakistan in the west to the volcanic islands of the Philippines in the east. From north to south, the region of Asia stretches from the cool highlands of northeastern China to the tropical islands of Indonesia.

Region
The Land

The Gobi

The Gobi stretches across southern Mongolia and northern China. It is colder and farther north than any other desert in the world.
PLACE: What other desert lies in central Asia?

The region of Asia covers about 7.8 million square miles (20.2 million sq. km). Look at the physical map on page 578. Asia's winding coastlines touch the Indian and Pacific oceans as well as many seas. Within Asia's vast land area are mountains, deserts, plains, and great rivers.

Mountains Rugged mountain ranges sweep through central Asia. The best known are the Himalayas. Their highest peak—Mount Everest—is the world's tallest mountain. Mountains also cross northeastern China, the Korean Peninsula, and Southeast Asia. Off Asia's coasts you will find mountainous groups of islands, such as Japan, the Philippines, and Indonesia. They lie in the Ring of Fire, a Pacific area known for its earthquakes and active volcanoes.

Deserts and Plains Large deserts—the Taklimakan (tah•kluh•muh•KAHN) and the Gobi—stretch across inland parts of central Asia. Fertile plains lie in northern India, eastern China, and mainland Southeast Asia. Many of Asia's rivers flow through these plains to the ocean or sea. The most important rivers are the Indus in Pakistan; the Ganges and Brahmaputra in India; the Mekong in Southeast Asia; and the Chang Jiang and Huang He in China.

Region
Climate and Vegetation

Because of its vast size, Asia has diverse climates. They include cold highlands and hot deserts in north and central Asia, mild climates in the east, and tropical climates in the south.

Monsoons Winds called monsoons affect climate in much of Asia. In winter, monsoons from the north bring cool, dry weather. In summer, the winds reverse direction and blow from the seas that lie south of Asia. The summer monsoons bear hot, humid weather. Wet monsoons often bring heavy rains and floods.

Vegetation Dry areas in central Asia have little plant life, except for grasses. These grasses provide food for livestock. You will find many trees and plants in eastern Asia, which gets plenty of rain. The warm, wet climate of Southeast Asia favors tropical vegetation. Many Southeast Asian exports are hardwoods from rain forests.

Place
The Economy

Asia is made up of 23 countries. Economies differ from country to country, although agriculture is the major economic activity. Important crops include rice, wheat, cotton, rubber, and tea. Because of Asia's mountainous areas and large populations, farmers try to produce as much food as they can on small areas of land, including terraced mountain slopes.

Mining and Industry Asia is rich in mineral resources. Coal, iron ore, manganese, and mica come from India and China. Southeast Asia is a major supplier of tin, oil, and bauxite. Most of Asia's manufacturing takes place in Japan, South Korea, Taiwan, China, and India. Factories in these countries produce cars, electronic equipment, ships, and textiles. In other parts of Asia, industry is less developed. Governments in these areas, however, are trying to create new industries and improve old ones.

Transportation Industrial growth in Asia has led to better means of transportation. Roads and railroads link major urban areas. In Asian cities you can see modern buses, trains, and cars. In rural areas people often travel by foot or use carts pulled by animals. People living along rivers rely on small boats for traveling and transporting goods.

Mohenjo-Daro, Pakistan

Ruins of one of the world's first civilizations were discovered in the Indus River Valley in Pakistan.
PLACE: What forms of transportation might have been used by these early people?

Movement

The People

About 3.3 billion people live in Asia. China, India, Indonesia, Bangladesh, and Japan are among the world's most heavily populated countries. Most Asians make their homes in river or mountain valleys or near seacoasts. Among the most crowded areas of Asia are Bangladesh, eastern China, northern India, southern Japan, and the island of Java in Indonesia.

Population People are not evenly distributed throughout Asia. Few Asians live in the mountains or desert areas. Only about 34 percent of them live in urban areas. However, Asian cities such as Calcutta (India), Shanghai (China), Tokyo (Japan), and Seoul (South Korea) are rapidly growing. Many are over-crowded, and some show great contrasts of wealth and poverty.

Ethnic Groups and Religions Japan and the Koreas each have one main ethnic group. China also has one, but many small non-Chinese groups live on the edges of the country. India and Southeast Asia, however, have many ethnic groups with different languages and cultures.

Anuradhapura, Sri Lanka

The graceful domes of Buddhist temples are a common sight in the island country of Sri Lanka.
PLACE: What religion is practiced by more Asians than any other religion?

Religion has strongly influenced the arts and cultures of Asia. Hinduism is the primary religion of the region. Hindus generally live in India, where the religion began thousands of years ago. Islam—Asia's most widespread faith—began in Southwest Asia and later reached India, western China, and Southeast Asia. Buddhism, which started in India, is now a major religion of Southeast Asia, China, Korea, and Japan. Smaller numbers of Asians practice Christianity or local religions.

History Some of the world's oldest civilizations arose in Asia. Until the 1500s, Asia was more advanced than Europe in culture and technology. Early Asians founded cities, set up states, and carved out trade routes.

Europeans arrived in Asia around 1500. By the early 1800s many Asian lands had fallen under European control. The Europeans brought Western ideas to Asia. In the early 1900s, Japan became the world's leading Asian power. World War II led to Japan's defeat, but it also ended Europe's hold on Asia. Nearly all of the Asian lands ruled by foreigners became independent in the mid-1900s.

From the late 1940s to the 1970s, Asia was caught in the global struggle between communist and noncommunist countries. Today communist governments still rule in China, Vietnam, and North Korea. Other Asian nations, however, are governed by monarchs (Nepal), military leaders (Myanmar), or democratic systems (India, Japan, and the Philippines).

Multimedia Unit 8 Activities

NATIONAL GEOGRAPHIC SOCIETY

Picture Atlas of the World

Asian Ethnic Groups, Religions, and Languages

Using the CD-ROM Asia is a land of many languages, ethnic groups, and religions. Japan and the Koreas each have one main ethnic group, as does the populous country of China. In addition, many small non-Chinese groups live throughout China. India and the countries of Southeast Asia claim a variety of ethnic groups, which practice different religions and speak various languages. Develop a cultural distribution map of Asia, which marks the locations of the major religions, languages, and ethnic groups. Follow these steps to create your map:

1. Access the *World Globe* and click on the continent of Asia.
2. Then click on the individual countries of Asia.
3. Note the major religions and languages of each country. Use an encyclopedia or other reference work to find the major ethnic groups of each country.
4. On a blank map of Asia, indicate the major religions, languages, and ethnic groups of each Asian country. Be sure to include a map legend.

Surfing the "Net"

The Past, Present, and Future of Asia

The last several years have brought great change to Asia. Japan, once admired for its strong economy, experienced a slowed economic growth rate in the 1990s. At the same time, several other Asian countries were experiencing rapid industrialization. In China, Communists held tightly to their control of the government despite a movement for democracy. To find out more about the recent history of Asia, search the Internet.

Getting There

Follow these steps to gather information on Asia's past and present:
1. Use a search engine. Type in *Asia*.
2. Enter the names of individual Asian countries and words like the following to focus your search: *history, economy, news,* and *culture.*
3. The search engine will provide you with a number of links to follow.

What To Do When You Are There

Organize the class into groups of three to four students. Each group should focus on one Asian country and work to compile information on the country's past and present, and also make predictions for the future. Groups should present their information to the rest of the class. As a class, determine whether each group's "predictions" for their country's future seem likely to occur.

Physical Geography

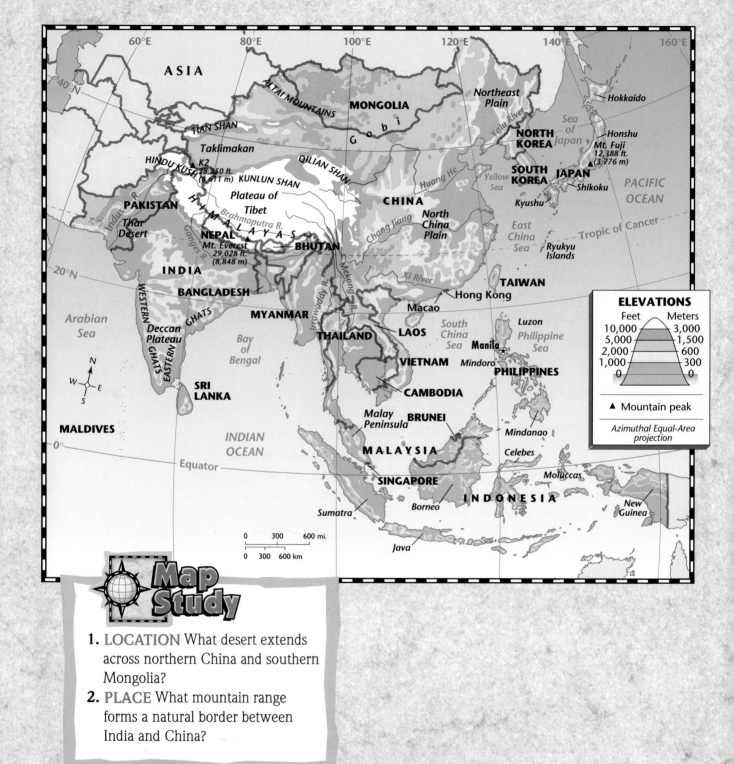

ASIA

60°E — 40°N — ALTAI MOUNTAINS — MONGOLIA — Gobi — Northeast Plain — Hokkaido

80°E — TIAN SHAN — Taklimakan — Yalu River — NORTH KOREA — Sea of Japan — Honshu — Mt. Fuji 12,388 ft. (3,776 m)

HINDU KUSH — K2 28,250 ft. (8,611 m) — KUNLUN SHAN — QILIAN SHAN — Huang He — Yellow Sea — SOUTH KOREA — JAPAN — PACIFIC OCEAN — Shikoku

PAKISTAN — Plateau of Tibet — CHINA — North China Plain — Kyushu — Tropic of Cancer

Indus R. — Thar Desert — Brahmaputra R. — HIMALAYAS — Chang Jiang — East China Sea — Ryukyu Islands

Ganges R. — NEPAL — Mt. Everest 29,028 ft. (8,848 m) — BHUTAN — 20°N

INDIA — Xi River — TAIWAN — Hong Kong

BANGLADESH — WESTERN GHATS — Macao

Arabian Sea — Deccan Plateau — EASTERN GHATS — MYANMAR — Irrawaddy R. — Mekong R. — THAILAND — LAOS — South China Sea — Manila — Luzon — Philippine Sea

Bay of Bengal — VIETNAM — Mindoro — PHILIPPINES

SRI LANKA — CAMBODIA

MALDIVES — INDIAN OCEAN — Malay Peninsula — BRUNEI — Mindanao

Equator — MALAYSIA — Celebes

0° — SINGAPORE — Borneo — INDONESIA — Moluccas — New Guinea

Sumatra — Java

0   300   600 mi.
0   300   600 km

ELEVATIONS

Feet	Meters
10,000	3,000
5,000	1,500
2,000	600
1,000	300
0	0

▲ Mountain peak

Azimuthal Equal-Area projection

Map Study

1. **LOCATION** What desert extends across northern China and southern Mongolia?

2. **PLACE** What mountain range forms a natural border between India and China?

ELEVATION PROFILE: South Asia

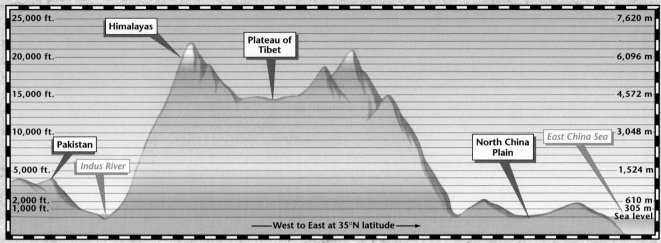

25,000 ft.	7,620 m
	Himalayas
	Plateau of Tibet
20,000 ft.	6,096 m
15,000 ft.	4,572 m
10,000 ft.	3,048 m
	North China Plain
	East China Sea
	Pakistan
	Indus River
5,000 ft.	1,524 m
2,000 ft.	610 m
1,000 ft.	305 m
	Sea level

◄—— West to East at 35°N latitude ——►

Source: *Goode's World Atlas,* 19th edition

Highest point: Mount Everest (Nepal-Tibet) 29,028 ft. (8,848 m) high

Longest river: Chang Jiang (China) 3,964 mi. (6,378 km) long

Largest desert: Gobi (Mongolia and China) 500,000 sq. mi. (1,300,000 sq. km)

Highest waterfall: Jog (Gersoppa) Falls (India) 830 ft. (253 m) high

ASIA AND THE UNITED STATES: Land Comparison

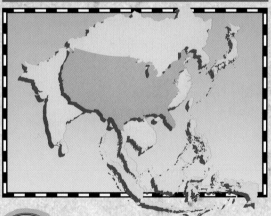

1. **PLACE** What is the average elevation of the Himalayas?

2. **PLACE** What is the difference in elevation between the Plateau of Tibet and the North China Plain?

UNIT 8 ATLAS

Cultural Geography

ASIA

60°E · 80°E · 100°E · 120°E · 140°E · 160°E

40°N

Ulan Bator
MONGOLIA

NORTH KOREA
Pyongyang
Sea of Japan

Beijing
Seoul
SOUTH KOREA
Tokyo

Islamabad
CHINA
JAPAN
Yellow Sea

PAKISTAN
East China Sea
Tropic of Cancer

New Delhi
NEPAL
Kathmandu
BHUTAN
Thimphu
Taipei

20°N
INDIA
Dhaka
TAIWAN

Arabian Sea
BANGLADESH
Hanoi
Hong Kong
PACIFIC OCEAN

MYANMAR
LAOS
Vientiane
Macao
South China Sea
Philippine Sea

Bay of Bengal
Yangon
THAILAND
Bangkok
VIETNAM
Manila

CAMBODIA
Phnom Penh
PHILIPPINES

MALDIVES
Colombo
SRI LANKA

Male
INDIAN OCEAN
BRUNEI
Bandar Seri Begawan

0°
Kuala Lumpur
MALAYSIA

Equator
Singapore
SINGAPORE
INDONESIA

Jakarta

0 · 300 · 600 mi.
0 · 300 · 600 km

— National boundary
⊛ National capital

Azimuthal Equal-Area projection

N
W E
S

Map Study

1. **LOCATION** What country lies at the southern tip of mainland Malaysia?
2. **PLACE** What country borders northwestern India?

COMPARING POPULATION: Asia and the United States

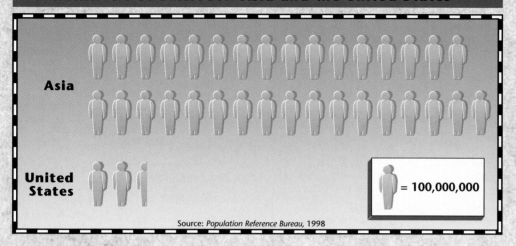

Asia

United States

= 100,000,000

Source: *Population Reference Bureau, 1998*

GeoFacts

Biggest country (land area): China 3,600,930 sq. mi. (9,326,409 sq. km)

Smallest country (land area): Maldives 116 sq. mi. (300 sq. km)

Largest urban area (population): Tokyo (1995) 28,447,000; (2015 projected) 30,000,000

Highest population density: Singapore 16,415 people per sq. mi. (6,338 people per sq. km)

Lowest population density: Mongolia 4 people per sq. mi. (2 people per sq. km)

Asian Population

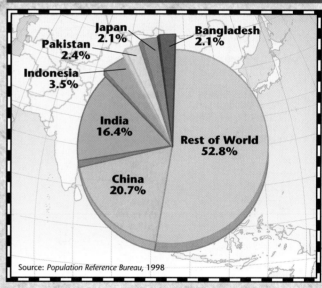

Japan 2.1%
Pakistan 2.4%
Indonesia 3.5%
Bangladesh 2.1%
India 16.4%
Rest of World 52.8%
China 20.7%

Source: *Population Reference Bureau, 1998*

Graphic Study

1. **PLACE** About how many times more people live in Asia than live in the United States?
2. **REGION** What two Asian countries have the largest populations?

Countries at a Glance

Bangladesh

Dhaka

LANDMASS:
50,260 sq. mi./
130,173 sq. km
MONEY:
Taka
MAJOR EXPORT:
Garments
MAJOR IMPORT:
Textile Yarn

CAPITAL:
Dhaka
MAJOR LANGUAGE(S):
Bengali, Chakma, Magh
POPULATION:
123,400,000

Bhutan

Thimphu

LANDMASS:
18,150 sq. mi./
47,009 sq. km
MONEY:
Ngultrum
MAJOR EXPORT:
Electricity
MAJOR IMPORT:
Petroleum

CAPITAL:
Thimphu
MAJOR LANGUAGE(S):
Dzongkha, Nepali and
Tibetan dialects
POPULATION:
800,000

Brunei

Bandar Seri
Begawan

LANDMASS:
2,035 sq. mi./
5,271 sq. km
MONEY:
Brunei Dollar
MAJOR EXPORT:
Petroleum
MAJOR IMPORT:
Machinery

CAPITAL:
Bandar Seri Begawan
MAJOR LANGUAGE(S):
Malay, English, Chinese
POPULATION:
300,000

Cambodia

Phnom
Penh

LANDMASS:
68,154 sq. mi./
176,519 sq. km
MONEY:
Riel
MAJOR EXPORT:
Rubber
MAJOR IMPORT:
Machinery

CAPITAL:
Phnom Penh
MAJOR LANGUAGE(S):
Khmer, French
POPULATION:
10,800,000

China

Beijing

LANDMASS:
3,600,930 sq. mi./
9,326,409 sq. km
MONEY:
Yuan
MAJOR EXPORT:
Textile Products
MAJOR IMPORT:
Machinery

CAPITAL:
Beijing
MAJOR LANGUAGE(S):
Mandarin, Yue,
Wu Hakka, Xiang
POPULATION:
1,242,500,000

India

New Delhi

LANDMASS:
1,147,950 sq. mi./
2,973,191 sq. km
MONEY:
Rupee
MAJOR EXPORT:
Diamonds
MAJOR IMPORT:
Fuels

CAPITAL:
New Delhi
MAJOR LANGUAGE(S):
Hindi, English
POPULATION:
988,700,000

Indonesia

Jakarta

LANDMASS:
705,190 sq. mi./
1,826,442 sq. km
MONEY:
Rupiah
MAJOR EXPORT:
Petroleum
MAJOR IMPORT:
Machinery

CAPITAL:
Jakarta
MAJOR LANGUAGE(S):
Bahasa Indonesia
(Malay), Javanese
POPULATION:
207,400,000

Japan

Tokyo

LANDMASS:
145,370 sq. mi./
376,508 sq. km
MONEY:
Yen
MAJOR EXPORT:
Motor Vehicles
MAJOR IMPORT:
Fuels

CAPITAL:
Tokyo
MAJOR LANGUAGE(S):
Japanese
POPULATION:
126,400,000

Laos

Vientiane

LANDMASS:
89,110 sq. mi./
230,795 sq. km
MONEY:
New Kip
MAJOR EXPORT:
Wood
MAJOR IMPORT:
Food Products

CAPITAL:
Vientiane
MAJOR LANGUAGE(S):
Lao
POPULATION:
5,300,000

Malaysia

Kuala
Lumpur

LANDMASS:
126,850 sq. mi./
328,542 sq. km
MONEY:
Ringgit
MAJOR EXPORT:
Machinery
MAJOR IMPORT:
Machinery

CAPITAL:
Kuala Lumpur
MAJOR LANGUAGE(S):
Malay, English, Chinese,
Indian languages
POPULATION:
22,200,000

Countries not drawn to scale.

Maldives

CAPITAL:
Male
MAJOR LANGUAGE(S):
Divehi (Sinhalese dialect)
POPULATION:
300,000

LANDMASS:
116 sq. mi./
300 sq. km
MONEY:
Rufiyaa
MAJOR EXPORT:
Clothing
MAJOR IMPORT:
Food and Beverages

Mongolia

CAPITAL:
Ulan Bator
MAJOR LANGUAGE(S):
Khalkha, Mongolian
POPULATION:
2,400,000

LANDMASS:
604,825 sq. mi./
1,566,500 sq. km
MONEY:
Tugrik
MAJOR EXPORT:
Minerals and Metals
MAJOR IMPORT:
Machinery

Myanmar

CAPITAL:
Yangon
MAJOR LANGUAGE(S):
Burmese, Karen, Shan
POPULATION:
47,100,000

LANDMASS:
253,880 sq. mi./
657,549 sq. km
MONEY:
Kyat
MAJOR EXPORT:
Agricultural Products
MAJOR IMPORT:
Machinery

Nepal

CAPITAL:
Kathmandu
MAJOR LANGUAGE(S):
Nepali
POPULATION:
23,700,000

LANDMASS:
52,820 sq. mi./
136,804 sq. km
MONEY:
Rupee
MAJOR EXPORT:
Food and Live Animals
MAJOR IMPORT:
Machinery

North Korea

CAPITAL:
Pyongyang
MAJOR LANGUAGE(S):
Korean
POPULATION:
22,200,000

LANDMASS:
46,490 sq. mi./
120,409 sq. km
MONEY:
Won
MAJOR EXPORT:
Minerals
MAJOR IMPORT:
Petroleum

Pakistan

CAPITAL:
Islamabad
MAJOR LANGUAGE(S):
Urdu, Punjabi, Sindhi,
Pushtu
POPULATION:
141,900,000

LANDMASS:
297,640 sq. mi./
770,888 sq. km
MONEY:
Rupee
MAJOR EXPORT:
Textile Fabrics
MAJOR IMPORT:
Petroleum

Philippines

CAPITAL:
Manila
MAJOR LANGUAGE(S):
Pilipino, English,
Cebuano, Bicol
POPULATION:
75,300,000

LANDMASS:
115,120 sq. mi./
298,161 sq. km
MONEY:
Peso
MAJOR EXPORT:
Food and Live Animals
MAJOR IMPORT:
Machinery

Singapore

CAPITAL:
Singapore
MAJOR LANGUAGE(S):
Chinese, Malay,
Tamil, English
POPULATION:
3,900,000

LANDMASS:
236 sq. mi./
611 sq. km
MONEY:
Dollar
MAJOR EXPORT:
Office Machines
MAJOR IMPORT:
Petroleum

South Korea

CAPITAL:
Seoul
MAJOR LANGUAGE(S):
Korean
POPULATION:
46,400,000

LANDMASS:
38,120 sq. mi./
98,731 sq. km
MONEY:
Won
MAJOR EXPORT:
Machinery
MAJOR IMPORT:
Machinery

Sri Lanka

CAPITAL:
Colombo
MAJOR LANGUAGE(S):
Sinhalese, Tamil
POPULATION:
18,900,000

LANDMASS:
24,950 sq. mi./
64,621 sq. km
MONEY:
Rupee
MAJOR EXPORT:
Food and Live Animals
MAJOR IMPORT:
Machinery

UNIT 8 ATLAS

Countries at a Glance

Taiwan

Taipei

LANDMASS:
13,970 sq. mi./
36,182 sq. km
MONEY:
New Taiwan Dollar
MAJOR EXPORT:
Machinery
MAJOR IMPORT:
Machinery

CAPITAL:
Taipei
MAJOR LANGUAGE(S):
Mandarin Chinese,
Taiwanese, Hakka
POPULATION:
21,700,000

Thailand

Bangkok

LANDMASS:
197,250 sq. mi./
510,878 sq. km
MONEY:
Baht
MAJOR EXPORT:
Machinery
MAJOR IMPORT:
Machinery

CAPITAL:
Bangkok
MAJOR LANGUAGE(S):
Thai, Chinese, Malay,
regional dialects
POPULATION:
61,100,000

Vietnam

Hanoi

LANDMASS:
125,670 sq. mi./
325,485 sq. km
MONEY:
Dong
MAJOR EXPORT:
Fuels
MAJOR IMPORT:
Machinery

CAPITAL:
Hanoi
MAJOR LANGUAGE(S):
Vietnamese, Chinese
POPULATION:
78,500,000

Thimphu, Bhutan

Bhutan's king rules from this
fortified hillside monastery in
Bhutan's capital.

Guangxi Province, China

These unusual cone-shaped hills
in southern China were formed
by erosion and weathering.

Countries not drawn to scale.

UNIT 8

Angkor Wat, Cambodia

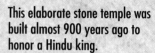

This elaborate stone temple was built almost 900 years ago to honor a Hindu king.

World's Tallest Mountains

GRAPHIC STUDY

Kilimanjaro is the highest mountain peak in Africa.

PLACE: What is the highest mountain peak in the world?

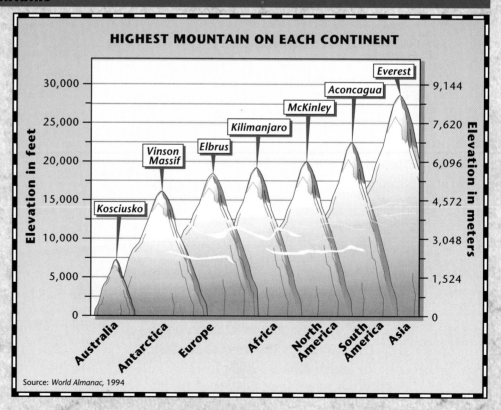

HIGHEST MOUNTAIN ON EACH CONTINENT

Elevation in feet

| 30,000 |
| 25,000 |
| 20,000 |
| 15,000 |
| 10,000 |
| 5,000 |
| 0 |

Kosciusko — Vinson Massif — Elbrus — Kilimanjaro — McKinley — Aconcagua — Everest

Elevation in meters

| 9,144 |
| 7,620 |
| 6,096 |
| 4,572 |
| 3,048 |
| 1,524 |
| 0 |

Australia · Antarctica · Europe · Africa · North America · South America · Asia

Source: *World Almanac, 1994*

Chapter 22 South Asia

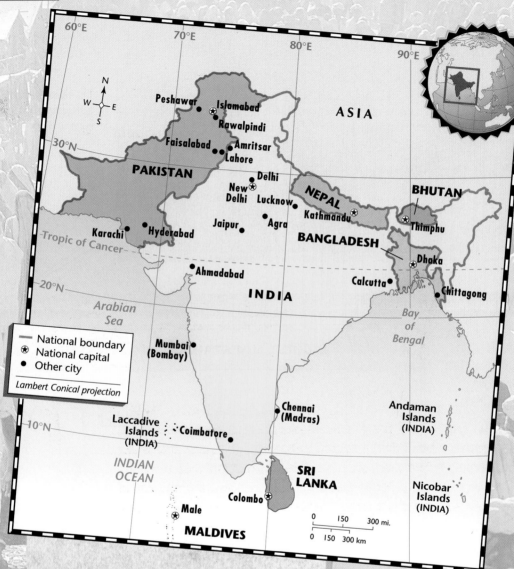

National boundary
National capital
Other city

Lambert Conical projection

60°E
70°E
80°E
90°E

Peshawar
Islamabad
Rawalpindi
Amritsar
Faisalabad
Lahore

PAKISTAN

Delhi
New
Delhi
Lucknow
Jaipur
Agra

Karachi
Hyderabad

ASIA

NEPAL
Kathmandu

BHUTAN
Thimphu

BANGLADESH
Dhaka

Calcutta
Chittagong

Tropic of Cancer

30°N

Ahmadabad

INDIA

20°N

Arabian
Sea

Mumbai
(Bombay)

Bay
of
Bengal

Andaman
Islands
(INDIA)

Chennai
(Madras)

Laccadive
Islands
(INDIA)

Coimbatore

10°N

INDIAN
OCEAN

SRI
LANKA

Colombo

Nicobar
Islands
(INDIA)

Male

MALDIVES

0   150   300 mi.
0   150   300 km

MAP STUDY ACTIVITY

As you read Chapter 22, you will
learn about India and the
other countries in South Asia.

**1. What is the largest country
in South Asia?**

**2. What two island countries lie off
the coast of southern India?**

3. What is the capital of India?

SECTION 1

India

PREVIEW

Words to Know

- subcontinent
- monsoon
- jute
- cottage industry

Places to Locate

- India
- Himalayas
- Ganges River
- Deccan Plateau
- Calcutta
- Mumbai (Bombay)
- New Delhi

Read to Learn . . .

1. what a subcontinent is.
2. how seasonal winds affect India's climate.
3. what religions India's people follow.

This is the Ganges River in the South Asian country of India.

Hindus—the largest religious group in India—consider the Ganges a holy river. About 1 million Hindus from all areas of India pray and bathe in this river each year.

India and several other countries—Pakistan, Bangladesh (BAHN•gluh•DEHSH), Nepal, Bhutan, Sri Lanka, and the Maldives—make up the area known as South Asia. South Asia is often called a **subcontinent,** or a landmass that is like a continent, only smaller. The map on page 588 shows you that towering mountains in the north separate South Asia from the rest of the continent of Asia. Bordering South Asia are three large bodies of water: the Indian Ocean, the Arabian Sea, and the Bay of Bengal.

Location

The Land

India takes up about 75 percent of the land area of South Asia. India covers 1,147,950 square miles (2,973,191 sq. km)—about one-third the area of the United States. The map on page 588 shows you that mountains run along three sides and through the middle of the country. Plateaus and desert plains run through the rest of the country.

Mountains Two huge walls of mountains—the Karakoram (KAR•uh•KOHR•uhm) Range and the Himalayas (HIH•muh•LAY•uhs)—form India's northern border. The Himalayas—made up of several ranges—stretch more than 1,500 miles (2,414 km) across northern South Asia. They are the tallest mountains in the world. Their snowcapped peaks average more than 5 miles (8 km) in height!

In the center of India lies another mountain range—the Vindhya (VIHN•dyuh) Mountains. These low mountains divide India in half. They have helped create two very different cultures in northern India and southern India.

Two chains of hills and mountains called the Eastern Ghats (GAHTS) and Western Ghats edge the southern coasts of India. These mountain chains lie just inland from the Bay of Bengal and the Arabian Sea.

Plains and Plateaus Sweeping through northern India between the Himalaya and Vindhya ranges is the Ganges Plain. It boasts some of the most fertile soil in the country. Look at the map below. You see that the Ganges River begins in the Himalayas and flows through the Ganges Plain to the Bay of Bengal. At the western edge of the Ganges Plain lies the Thar (TAHR) Desert. It is about 500 miles (804 km) long and 275 miles (442 km) wide.

The Deccan Plateau is located south of the Vindhya Mountains. This triangle-shaped landmass makes up the southern two-thirds of India. Forests, fertile farmland, and rich deposits of minerals make the Deccan Plateau a valuable region.

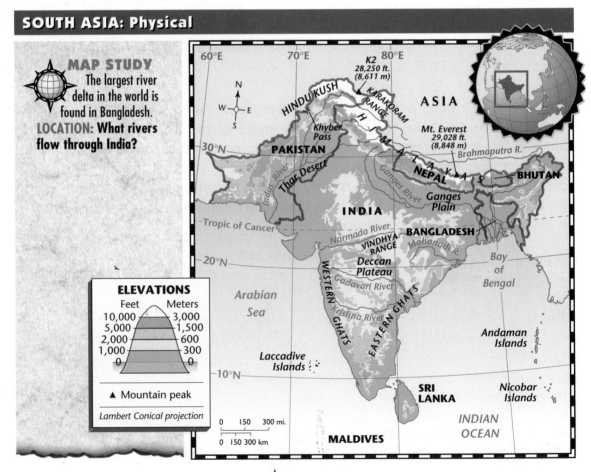

SOUTH ASIA: Physical

MAP STUDY The largest river delta in the world is found in Bangladesh.
LOCATION: What rivers flow through India?

ELEVATIONS
Feet / Meters
10,000 — 3,000
5,000 — 1,500
2,000 — 600
1,000 — 300
0 — 0
▲ Mountain peak
Lambert Conical projection

The Thar Desert is one of the most sparsely populated areas of India *(left)*. In contrast, Calcutta is one of the most crowded cities in the world *(right)*.
LOCATION: In what part of India is Calcutta located?

Climate Most places in India are warm or hot most of the year. The people of India can thank the Himalayas for their warm climate. These mountains block cold northern air from entering India. The map on page 595 shows you the climate regions of India.

The **monsoons,** or seasonal winds, are another important influence on India's climate. Most of India has three seasons—cool, hot, and rainy. During the cool season—November through February—and the hot season—March through April—monsoon winds from the north bring dry air. During the rainy season—May through October—monsoon winds reverse direction and bring moist air from the Indian Ocean.

Human/Environment Interaction
The Economy

Agriculture and industry are equally important to India's economy. The government of India has set up plans to increase the production of farm and industrial goods. The goal is to develop India's resources and to improve the standard of living for India's people.

Agriculture Most of India's best farmland lies in the Ganges Plain and the Deccan Plateau. In both places you will see many farmers work small plots of land. India is the world's second-largest producer of rice. Other important crops are tea, sugarcane, wheat, barley, cotton, and jute. **Jute** is a plant fiber used in making rope, twine, and burlap bags.

Industry and Mining Huge factories in India turn out cotton textiles and produce iron and steel. Oil and sugar refineries loom over the industrial landscape

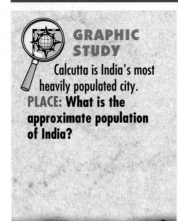

GRAPHIC STUDY

Calcutta is India's most heavily populated city.
PLACE: What is the approximate population of India?

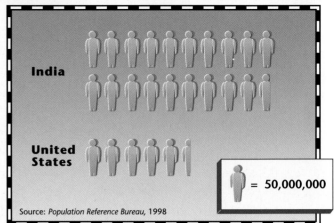

India

United States

= 50,000,000

Source: *Population Reference Bureau, 1998*

of many cities. Other factories produce locomotives, cars, cement, and chemical products. India's mica mines produce much of the world's supply of this mineral.

Cottage industries are also the source of many Indian products. A **cottage industry** is based in a rural village where family members use their own equipment. Goods produced in cottage industries include cotton cloth, silk cloth, rugs, leather products, and metalware.

Human/Environment Interaction

The People

India has more than 988 million people. It is the world's second-largest country in population. Only China is larger. The graph above shows you how India's population compares with that of the United States. The people of India speak 14 major languages and more than 1,000 other languages and dialects. Hindi is India's official language, but English is commonly spoken in government and business.

Influences of the Past About 4,000 years ago, the first Indian civilization built well-planned cities in present-day Pakistan. About 1500 B.C. warriors known as Aryans (AR•ee•uhnz) entered the subcontinent from central Asia. They set up kingdoms in northern India, forcing many of the earlier Indian peoples to move southward. The Aryans brought the religious teachings of Hinduism and the traditional system of social groups to India's culture.

Many other groups came later to the area that is now India. Beginning in the A.D. 700s, Muslims from Southwest Asia brought Islam to India. In the 1500s they founded the Mogul (MOH•guhl) Empire, which lasted more than 200 years.

The British ruled large areas of India from the 1700s to the mid-1900s. They built roads, railroads, and seaports. They also made large profits from the plantations, mines, and factories they set up in India. An Indian leader named Mohandas Gandhi (moh•HAHN•das GAHN•dee) led a movement that brought

India independence from the United Kingdom in 1947. Since independence India has been a democracy.

Religion About 80 percent of India's people are Hindus. Hindus believe that all living things have souls that belong to one eternal spirit. After the body dies, the soul is reborn and returns to the earth. This process is repeated many times until the soul reaches perfection in a higher state of existence.

Islam also claims many followers in India. India's 140 million Muslims form one of the largest Muslim populations in the world. Other religions in India are Buddhism, Sikhism (SEE•KIH•zuhm), Jainism (JY•NIH•zuhm), and Christianity.

The Arts Religion has influenced the arts of India. Ancient Hindu builders constructed temples with tall, elaborately carved towers. Hindu writers left stories, poems, and legends. Hindu artists developed dances and music that are still performed today.

Muslims also have added to India's artistic heritage. One of the finest Muslim buildings in India is the Taj Mahal. A ruler in the mid-1600s built the Taj Mahal as a monument to a beloved wife. By the 1800s European influences affected the arts of India. Today Indian arts reflect a blend of both East and West, old and new.

Movie-Mania in New Delhi, India

A large billboard advertises a popular movie—one of hundreds produced every year in India.
PLACE: What other forms of entertainment are popular in India?

A Typical Morning in New Delhi

Rani Singh lives in a two-room apartment in New Delhi with her parents, grandparents, and uncles. Before school each day Rani and her grandmother walk to the bazaar in their neighborhood to buy food for that day's meals. Rani has chosen her favorite *sari* to wear today. A *sari* is a long piece of fabric that is draped over the shoulders and head. The round dot that Rani wears in the middle of her forehead is called a *kumkum*. It is worn as a decoration—much like lipstick or jewelry.

Daily Life About 74 percent of India's people live in rural villages. Some villagers have brick homes. Others live in mud-and-straw shelters. Both men and women work in nearby fields. The Indian government tries to provide electricity, drinking water, better schools, and paved roads for many villages. Life is still difficult, however, for many Indian villagers. In recent years large numbers of people from the countryside have moved to urban areas.

India's cities today are very crowded. Delhi, Calcutta, and Mumbai are among the cities in India that have more than 5 million people each. Modern high-rise apartment and office buildings rise above the skyline. Next to these modern buildings stand slum areas of poverty. Bicycles, carts, animals, and people fill the narrow streets lined by small family-owned shops.

Fifteen-year-old Rajinder Swarup lives in Mumbai. Like many Indians, he wears a mix of traditional and modern clothing. He favors light, loose clothes because of India's hot climate. Like you, Rajinder loves going to the movies. Filled with action, adventure, and romance, movies are the most popular form of entertainment in India today. India makes more movies than any other country in the world. Rajinder also enjoys sports such as rugby and soccer. His favorite holiday is Diwali—the festival of lights. It is a Hindu festival marking the coming of winter and the victory of good over evil.

SECTION 1 ASSESSMENT

REVIEWING TERMS AND FACTS

1. Define the following: subcontinent, monsoon, jute, cottage industry.

2. LOCATION What two mountain ranges lie along India's northern border?

3. PLACE What kinds of landforms are found in India?

4. HUMAN/ENVIRONMENT INTERACTION What is India's leading food crop?

MAP STUDY ACTIVITIES

5. Look at the physical map on page 588. What three major bodies of water border India?

6. Turn to the political map on page 586. What is the capital of India?

BUILDING GEOGRAPHY SKILLS

Reading a Circle Graph

A circle graph is like a sliced pie. Often it is even called a pie chart. In a circle graph, the complete circle represents a whole group—or 100 percent. The circle is divided into "slices," or wedge-shaped sections representing parts of the whole.

Suppose, for example, that 25 percent of your friends watch five hours of television per day. On a circle graph representing hours of TV watching, 25 percent of the circle would represent friends who watch five hours. To read a circle graph, follow these steps:

- Read the title to find out what the subject is.
- Study the labels or key to determine what the parts or "slices" represent.
- Compare the parts to draw conclusions about the subject.

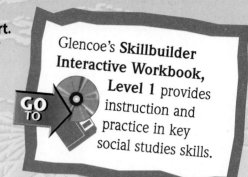

Glencoe's **Skillbuilder Interactive Workbook, Level 1** provides instruction and practice in key social studies skills.

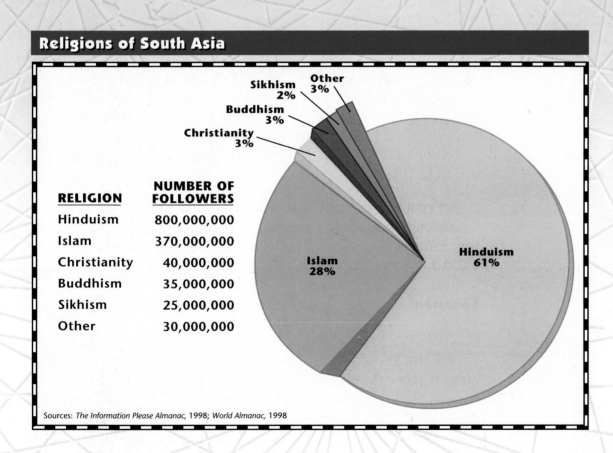

Religions of South Asia

RELIGION	NUMBER OF FOLLOWERS
Hinduism	800,000,000
Islam	370,000,000
Christianity	40,000,000
Buddhism	35,000,000
Sikhism	25,000,000
Other	30,000,000

Hinduism 61%
Islam 28%
Christianity 3%
Buddhism 3%
Sikhism 2%
Other 3%

Sources: *The Information Please Almanac, 1998; World Almanac, 1998*

Geography Skills Practice

1. Which religion in Asia has the most followers?
2. What percentage of South Asians are Muslim?
3. What is the combined percentage of Buddhist and Christian followers?

SECTION 2 Muslim South Asia

Words to Know
- tributary
- delta
- cyclone
- teak

Places to Locate
- Pakistan
- Bangladesh
- Hindu Kush
- Brahmaputra River
- Indus River
- Karachi
- Islamabad
- Dhaka

Read to Learn . . .
1. where Pakistan and Bangladesh are located.
2. how people in Pakistan and Bangladesh earn their livings.
3. how Islam influences the cultures of Pakistan and Bangladesh.

This beautiful domed mosque soars above the city skyline of Lahore, Pakistan. About 1,000 years old, Lahore boasts many buildings that reflect the religion of Islam.

Two countries in South Asia—Pakistan and Bangladesh—are largely Muslim. Although they share a religion, the peoples of these two countries have very different cultures. They also are separated from each other by more than 1,000 miles (1,610 km).

Location
Pakistan

Pakistan lies northwest of India. Its 297,640 square miles (770,888 sq. km) make it about the size of Texas. Pakistan is an independent country today because of religion. Its Muslim population did not want to be part of largely Hindu India. When British rule of India ended in 1947, the western and eastern parts of India became Pakistan. In 1971 the eastern section became Bangladesh.

Land and Climate Snow-topped mountains, high plateaus, fertile plains, and sandy deserts make up Pakistan. In the north you will find the Hindu Kush. Mountain passes cut through these rugged peaks. Centuries ago, the Khyber Pass was an important passage for people entering into South Asia from the north.

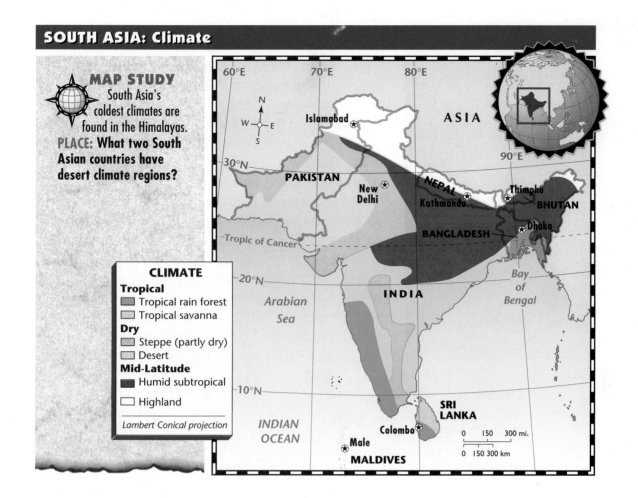

MAP STUDY
South Asia's coldest climates are found in the Himalayas.
PLACE: What two South Asian countries have desert climate regions?

CLIMATE
Tropical
◼ Tropical rain forest
◻ Tropical savanna
Dry
◼ Steppe (partly dry)
◻ Desert
Mid-Latitude
◼ Humid subtropical

◻ Highland

Lambert Conical projection

Plains cover most of eastern Pakistan. They are rich in fertile soil deposited by rivers. The major river system running through Pakistan's plains is the Indus River and its tributaries. A **tributary** is a small river that flows into a larger river or body of water. West of the Indus River valley, the land rises to form a plateau. A large part of this plateau is dry and rocky with little vegetation. Another vast barren area—the Thar Desert—sweeps east of the Indus River valley.

Pakistan has desert and steppe climates, with hot summers and cool winters. Rainfall in most regions is less than 10 inches (25 cm) a year. As in India, mountains in northern Pakistan block cold air from central Asia.

The Economy Many Pakistanis earn their livings by farming. Farmers use irrigation to grow wheat, cotton, and corn. They also raise cattle, goats, and sheep. Pakistan's factories manufacture cotton textiles for export. Other manufactured goods include cement, chemicals, fertilizer, and steel. Many craftworkers participate in cottage industries making metalware, pottery, and carpets.

The People Although 97 percent of Pakistanis are Muslim, the country's population includes many different ethnic and language groups. Among the major languages of Pakistan are Punjabi and Urdu. English is widely spoken in government and business.

Pakistani women prepare the evening meal while one of the men spins wool.
PLACE: What is the major religion of Pakistan?

About 72 percent of Pakistan's people live in rural villages. Most follow traditional customs and live in small homes of clay or sun-dried mud. In spite of its largely rural population, Pakistan has many large cities. Karachi, a seaport on the Arabian Sea, has about 9.3 million people. To the far north lies Islamabad, the national capital. It is a well-planned, modern city of 210,000 citizens. The government of Pakistan built Islamabad to draw people inland from crowded coastal areas.

Most people in Pakistan's cities are factory workers, shopkeepers, and craftworkers. They live in crowded neighborhoods. Many come from the countryside and keep close ties to their villages. Wealthier city dwellers live in modern homes and follow modern ways of life. In recent years, women in Pakistan have become active in politics. Benazir Bhutto served as prime minister of Pakistan during the late 1980s and the mid-1990s. She was the first woman to head a modern Muslim country.

Location

Bangladesh

The South Asian republic of Bangladesh has an area of 50,260 square miles (130,173 sq. km)—about the size of the state of Wisconsin. The map on page 586 shows that Bangladesh is nearly surrounded by India. Although Bangladesh is an Islamic nation, it shares many cultural features with eastern India.

From 1947 to 1971, Bangladesh was part of Pakistan, but separated from the rest of Pakistan by the nation of India. In 1971 Bangladesh gained its independence from Pakistan after a nine-month civil war.

The Land Flat, low plains cover most of Bangladesh. Two major rivers—the Brahmaputra (BRAH•muh•POO•truh) and the Ganges—flow through these plains. Find these rivers on the map on page 588. The people of Bangladesh depend on the rivers for transportation and for farming. The rivers often overflow their banks, leaving fertile soil. These deposits have formed a **delta,** or land formed by mud and sand at the mouth of a river. The largest river delta in the world has formed in Bangladesh where the Brahmaputra and Ganges rivers flow into the Indian Ocean.

Bangladesh has tropical and humid subtropical climates. As in India, the monsoons affect Bangladesh. During the summer monsoons, **cyclones** may cause flooding. A cyclone is an intense storm with high winds and heavy rains. Cyclones kill many people and animals and ruin crops.

The Economy Bangladesh's economy depends on farming. Farmers there raise rice, sugarcane, and jute in the wet, fertile soil. Bangladesh produces the world's largest supply of jute. Its thick forests provide **teak,** a wood used for

Almost 80 percent of the land in Bangladesh is part of a large delta.
HUMAN/ENVIRONMENT INTERACTION: What is the advantage and one disadvantage of living in a delta area?

shipbuilding and fine furniture. Although it has natural riches, Bangladesh is a struggling country. Its farmers have few modern tools and use outdated farming methods. The country often suffers from disastrous floods and food shortages.

The People Bangladesh—with about 123 million people—is one of the world's most densely populated countries. It has about 2,454 people per square mile (948 people per sq. km). About 86 percent of Bangladesh's people live in rural areas. Their homes are made of bamboo with thatched roofs. In recent years, many people have moved to crowded neighborhoods in urban areas to find work in factories. Their most common choice is Dhaka (DA•kuh), Bangladesh's capital and major port.

Most of Bangladesh's people speak the Bengali language, which is also spoken in neighboring India. About 87 percent of the people in Bangladesh are Muslim. Muslim influences are strong in the country's art, literature, and music.

SECTION 2 ASSESSMENT

REVIEWING TERMS AND FACTS
1. **Define the following:** tributary, delta, cyclone, teak.
2. **LOCATION** Describe the landforms of Pakistan.
3. **PLACE** How have rivers affected life in Bangladesh?
4. **REGION** What is the major religion of Pakistan and Bangladesh?

MAP STUDY ACTIVITIES

5. Turn to the climate map on page 595. What are the three major climate regions of Pakistan?

MAKING CONNECTIONS

Have you ever stopped to think where the number system you use came from? Centuries ago, Hindu scholars in India made great discoveries in mathematics.

NUMBER SYSTEM As early as 300 B.C., Hindu scholars came up with a number system based on 10. This number system used just 9 symbols, or numerals: 1, 2, 3, 4, 5, 6, 7, 8, 9. The strength of their number system lay in the idea of *place value.* The same symbol could represent different amounts, depending on its place in a number. For example, the symbol "3" represents 3 ones in the number 13, and 30 ones in the number 31. The Hindu scholars discovered that an infinitely large number could be represented using only 9 symbols.

ZERO About A.D. 600, Hindu thinkers invented a tenth symbol, *sunya,* which means "empty." This symbol, which became known as zero, was used to fill an empty place in a number. Until the invention of *sunya,* Hindus wrote a word to show the place value of each numeral above 99. For example, the number we would write as 306 was written "3 *sata* 6."

Zero ended the need for written place names, so "3 *sata* 6" became "306."

From Zero to Microchips

SPREAD OF NUMBERS For several centuries, only well-educated Hindus used the number system. Later, Arabs from Southwest Asia began using the Hindu system and found it much easier than their own. They adopted the system and eventually brought it to Europe. There it became known as the Arabic-Hindu number system. It replaced the long, clumsy system of numerals that had been used by the Romans.

Making the Connection

1. Hindu scholars developed 9 symbols and what special new symbol as part of their number system?

2. What people later used the Hindu number system?

The Himalayan Countries

PREVIEW

Words to Know
- dzong

Places to Locate
- Nepal
- Bhutan
- Mount Everest
- Kathmandu
- Thimphu

Read to Learn . . .
1. why Nepal and Bhutan were historically isolated.
2. how the people of Nepal and Bhutan earn their livings.
3. what influence religion has on the Himalayan countries.

In the Himalayas, north and east of India, you will find two small kingdoms—Nepal and Bhutan. The people of these

Himalayan countries visit street markets like this one every day to shop for food and visit with their neighbors.

Nepal and Bhutan both lie hidden among the towering peaks of the Himalayas. For centuries their mountain location kept them isolated. Today, airplanes and roads open up Nepal and Bhutan to the rest of the world.

Place
Nepal

A well-known myth of the Himalayas tells about a gigantic, apelike beast that roams the mountainous wilderness of Nepal. This "abominable snowman" seems almost too large for the country of Nepal, a small kingdom of only 52,820 square miles (136,804 sq. km)—an area about the size of North Carolina.

The Land The map on page 588 shows you that the Himalayas cover about 80 percent of Nepal's land area. The Himalayas are actually three mountain chains running side by side. Steep river valleys cut through the ice and snow of these mountains. You will find 8 of the 10 highest mountains in the world in the Himalayas of Nepal. Mount Everest is the highest peak at 29,028 feet (8,848 m).

Hills and valleys lie south of the Himalayas. Thick forests of trees and bamboo grasses grow in this region. A flat fertile river plain runs along Nepal's southern border with India. The plain includes farmland, rain forests, and swamps. Animals such as tigers and elephants roam these forests.

Differences in elevation affect Nepal's climate. Mountainous areas have long, harsh winters and short, cool summers. Valley areas have a cool climate with heavy summer rains. The plains in the south have a humid subtropical climate.

The Economy Nepal's economy depends almost entirely on farming. The farmers of Nepal grow enough wheat, barley, rice, and potatoes to feed themselves and their families. At markets farmers trade surplus crops for other items.

With few railroads or roads, Nepal carries on limited trade with the outside world. Herbs, jute, rice, spices, and wheat are exported to India. In return Nepal imports gasoline, machinery, and textiles. The country's rugged mountains, which attract thousands of climbers and hikers each year, have helped build a growing tourist industry.

The People Nepal has a population of 23.7 million. Most of its people are related to peoples in northern India and in Tibet to the north. One group in Nepal—the Sherpa—are known for their skills as mountain guides. About 90 percent of Nepal's people live in rural villages. A growing number of Nepal's people live in Kathmandu, the capital and largest city. Most Nepalese follow a form of Hinduism mixed with Buddhist practices.

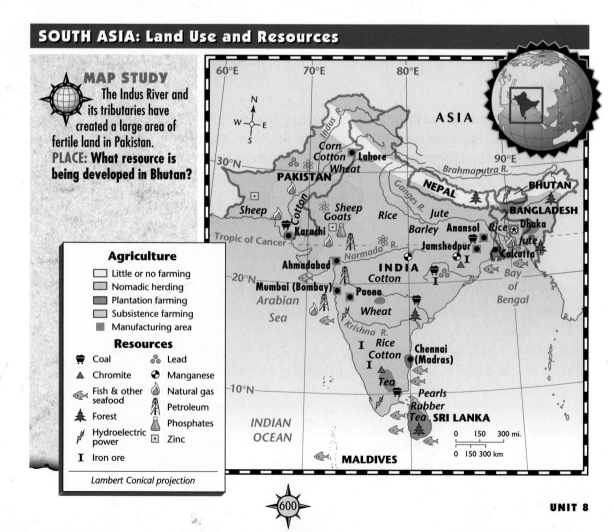

SOUTH ASIA: Land Use and Resources

MAP STUDY
The Indus River and its tributaries have created a large area of fertile land in Pakistan.
PLACE: What resource is being developed in Bhutan?

Agriculture
- ☐ Little or no farming
- ☐ Nomadic herding
- ■ Plantation farming
- ☐ Subsistence farming
- ■ Manufacturing area

Resources
- Coal
- Chromite
- Fish & other seafood
- Forest
- Hydroelectric power
- Iron ore
- Lead
- Manganese
- Natural gas
- Petroleum
- Phosphates
- Zinc

Lambert Conical projection

Place
Bhutan

East of Nepal lies the even smaller kingdom of Bhutan. The map on page 586 shows you that a small part of India separates Bhutan from Nepal. Bhutan has a land area of about 18,150 square miles (47,009 sq. km)—about the size of New Hampshire and Vermont put together.

The Land and Economy As in Nepal, the Himalayas are the major landform of Bhutan. They rise more than 24,000 feet (7,320 m) in many places. Mountainous areas of the country have snow and ice all year. In the foothills of the Himalayas the climate is mild. Thick forests cover much of this area. To the south along Bhutan's border with India lies an area of subtropical plains and river valleys.

Most of Bhutan's people are subsistence farmers. They live in the fertile mountain valleys and grow barley, rice, and wheat. In the severe climate of the mountain areas people herd cattle and yaks, or longhaired oxen. With India's help, Bhutan is developing its economy. It has set up commercial farms and built hydroelectric plants.

The People Bhutan has about 800,000 people. Most of them speak the Dzongkha dialect. Many live in rural villages that dot valleys and plains in the southern part of the country. Thimphu, the national capital, is in this area.

Bhutan once was called the "hidden holy land." Bhutan's people are not as isolated as they used to be. They are deeply loyal to the Buddhist religion, which teaches that people can find peace from life's troubles by living simply, doing good deeds, and praying. An Indian holy man named Siddhartha Gautama first preached Buddhism in India in the 500s B.C. Buddhism later spread eastward and northward to other parts of Asia. In Bhutan, Buddhist centers of prayer and study called **dzongs** have shaped the country's art and culture.

The Sherpas of Nepal

The Sherpas of northern Nepal serve as guides in the Himalayas.
LOCATION: What is the highest peak in the Himalayas?

SECTION 3 ASSESSMENT

REVIEWING TERMS AND FACTS
1. Define the following: dzong.
2. **PLACE** What climate areas are found in Nepal?
3. **LOCATION** Where do most people in Nepal and Bhutan live?

MAP STUDY ACTIVITIES
4. Turn to the land use map on page 600. What natural resource is found in eastern Nepal?

PREVIEW

Words to Know
- atoll
- lagoon

Places to Locate
- Sri Lanka
- Maldives
- Colombo

Read to Learn . . .
1. where Sri Lanka and the Maldives are located.
2. what major products are grown in Sri Lanka.
3. what formed the islands of the Maldives.

Two countries of South Asia—Sri Lanka and the Maldives—are islands in the Indian Ocean. Sri Lanka's capital city, Colombo, is a major seaport. Most of its people are Buddhists.

Sri Lanka and the Maldives lie south of India in the Indian Ocean. The sea affects the ways in which the peoples of both island countries live and earn their livings.

Place
Sri Lanka

Pear-shaped Sri Lanka lies about 20 miles (32 km) off the southeastern coast of India. Covering 24,950 square miles (64,621 sq. km), Sri Lanka is about the size of West Virginia. Sri Lanka, located on an important ocean route between Africa and Asia, became a natural stopping place for traders. Beginning in the 1500s, Sri Lanka—then known as Ceylon—came under the control of European countries. The British ruled the island from 1802 to 1948, when it became independent once again. In 1972 Ceylon took the name Sri Lanka, an ancient term meaning "brilliant land."

Sri Lanka is a land of white beaches and thick forests. Much of the country along the coast is rolling lowlands. Highlands cover the center. Rivers flow down the low mountains, providing irrigation for crops.

The Economy Farming is Sri Lanka's major economic activity. Many farmers grow rice and other food crops in lowland areas. Water buffaloes help to plow their fields. In higher elevations, tea, rubber, and coconuts grow on large plantations. Sri Lanka is one of the world's leading producers of tea and rubber.

Sri Lanka, rich in natural resources, is famous for its sapphires, rubies, and other gemstones. Sri Lanka's forests contain many valuable woods such as ebony and satinwood, and a variety of birds and animals. To protect wildlife, the government of Sri Lanka has set aside land for national parks.

Sri Lanka is developing its industrial economy. Factories in Sri Lanka's cities produce textiles, fertilizers, cement, leather products, and wood products for export. Colombo, the capital city, is a bustling port on the country's west coast.

The People About 19 million people live in Sri Lanka. They belong to two major ethnic groups, the Sinhalese (SIHN•guh•LEEZ) and the Tamils (TA•muhlz). Forming about 74 percent of the population, the Sinhalese live in the southern and western parts of the island. They speak Sinhalese and are Buddhists. The Tamils make up about 18 percent of the population. They live in the north and east, speak Tamil, and are Hindus. Some Tamils have fought to set up a separate country in northern Sri Lanka.

Whatever their ethnic background, people in Sri Lanka value their history and the arts—still visible in the remains of ancient cities. These remains hold

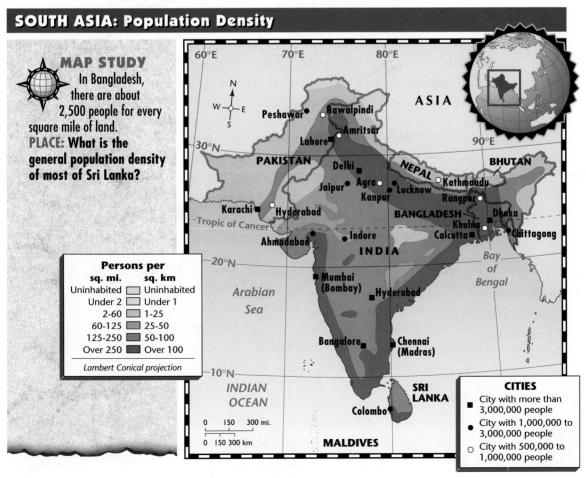

SOUTH ASIA: Population Density

MAP STUDY
In Bangladesh, there are about 2,500 people for every square mile of land.
PLACE: What is the general population density of most of Sri Lanka?

Persons per

sq. mi.	sq. km
Uninhabited	Uninhabited
Under 2	Under 1
2-60	1-25
60-125	25-50
125-250	50-100
Over 250	Over 100

Lambert Conical projection

0   150   300 mi.
0 150 300 km

CITIES
■ City with more than 3,000,000 people
● City with 1,000,000 to 3,000,000 people
○ City with 500,000 to 1,000,000 people

Workers harvest tea leaves on large plantations in Sri Lanka's highlands *(left)*. Fish are the most important food and export in the Maldives *(right)*. **LOCATION: Which of these two island countries is closer to India?**

many old palaces, temples, and statues of Buddha. Sri Lankan craftspeople also take pride in their carved wooden masks, brass work, and handmade cloth.

Region
Maldives

About 370 miles (595 km) south of India lie the Maldives, the smallest country of South Asia. The Maldives are made up of about 1,200 coral islands. Many of the islands are **atolls,** or low-lying islands that circle pools of water called **lagoons.** Only 200 islands have people living on them.

The climate of the Maldives is warm and humid throughout the year. Monsoons bring plenty of rain. Fish and tortoises fill the tropical waters around the islands. The people of the Maldives are skilled sailors, and fishing is their major economic activity. Fish and rice are the major foods. In recent years, the islands' palm-lined sandy beaches and coral formations have attracted tourists.

About 300,000 people live in the Maldives. Some 30,000 make their homes in Male, the capital. Most are Muslims. The islands, which came under British rule during the late 1800s, became independent in 1965.

SECTION 4 ASSESSMENT

REVIEWING TERMS AND FACTS
1. Define the following: atoll, lagoon.
2. PLACE What two major ethnic groups are found in Sri Lanka?
3. REGION What is the major economic activity of the Maldives?

MAP STUDY ACTIVITIES
4. Look at the population density map on page 603. What part of Sri Lanka has the highest population density?

Chapter 22 Highlights

Important Things to Know About South Asia

SECTION 1 INDIA

- India is the largest country of South Asia in size and population.
- The Himalayas and the monsoons, or seasonal winds, affect India's climate.
- India's economy is based on farming and industry.
- Although the peoples of India belong to many religious and language groups, most are Hindus.

SECTION 2 MUSLIM SOUTH ASIA

- Islam has shaped the cultures of Pakistan and Bangladesh.
- Pakistan is made up of mountains, deserts, and fertile river valleys.
- The Ganges and Brahmaputra rivers form a fertile delta area in Bangladesh.
- Monsoons bring plenty of rain to Bangladesh's low, fertile plains.

SECTION 3 THE HIMALAYAN COUNTRIES

- The Himalayas are the major landform of Nepal and Bhutan.
- Most people in Nepal and Bhutan are farmers or herders. They live in the southern parts of their countries.
- The Hindu and Buddhist religions have shaped the cultures of Nepal and Bhutan.

SECTION 4 ISLAND COUNTRIES

- Sri Lanka and the Maldives are island republics in the Indian Ocean.
- Sri Lanka is one of the world's major tea-producing countries.
- The Maldives are made up of about 1,200 coral islands. Only 200 are inhabited.

Festival market in Rajasthan, India ▶

Chapter 22 Assessment and Activities

REVIEWING KEY TERMS

Match the numbered terms in Set A with their lettered definitions in Set B.

A

1. cottage industry
2. tributary
3. jute
4. atoll
5. subcontinent
6. lagoon
7. teak
8. cyclone

B

A. landmass like a continent, only smaller
B. low-lying island circling a pool of water
C. small river that flows into a larger river
D. wood used for shipbuilding and fine furniture
E. where family members use their own equipment to create a product
F. seasonal wind
G. plant fiber used in making burlap bags
H. pool of water circled by an island

Mental Mapping Activity

Draw a freehand map of South Asia. Label the following:
- Calcutta
- Arabian Sea
- Hindu Kush
- Bay of Bengal
- Bhutan
- Deccan Plateau
- Indian Ocean

REVIEWING THE MAIN IDEAS

Section 1
1. **LOCATION** Where is the Deccan Plateau located?
2. **REGION** What major river flows through northern India?

Section 2
3. **HUMAN/ENVIRONMENT INTERACTION** What has helped increase food production in Pakistan?
4. **MOVEMENT** What is the major means of transportation in Bangladesh?

Section 3
5. **PLACE** What climates are found in Nepal?
6. **PLACE** Why was Bhutan once called the "hidden holy land"?

Section 4
7. **HUMAN/ENVIRONMENT INTERACTION** What agricultural products are grown on plantations in Sri Lanka?
8. **PLACE** What is the major religion of the Maldives?

CRITICAL THINKING ACTIVITIES

1. **Drawing Conclusions** Why do you think the description "brilliant island" applies to Sri Lanka?
2. **Cause and Effect** How have the environments of Nepal and Bhutan affected the way the countries' people live?

CONNECTIONS TO WORLD CULTURES

Cooperative Learning Activity Work in groups of three to learn more about the major religions of South Asia: Hinduism, Islam, and Buddhism. Choose one of the religions and research how it has affected one of the following areas: (a) the arts; (b) diet; and (c) family life. After your research is complete, share your information with your group. As a group, prepare a chart and a poster that present the group's findings.

GeoJournal Writing Activity

Imagine what it would be like to visit the Himalayas. Write about your experiences to a friend. Describe the types of transportation that you might use and the scenes that you might see on your journey.

TECHNOLOGY ACTIVITY

Using a Database Conduct research about the following religions of India: Buddhism, Hinduism, Jainism, Sikhism, Christianity, and Islam. Organize your research into a database with the following fields: number of followers, basic beliefs, and major leaders. Share your database with the rest of the class.

PLACE LOCATION ACTIVITY: SOUTH ASIA

Match the letters on the map with the places and physical features of South Asia. Write your answers on a separate sheet of paper.

1. Karachi
2. Himalayas
3. New Delhi
4. Ganges River
5. Mumbai (Bombay)
6. Bangladesh
7. Thar Desert
8. Sri Lanka
9. Indus River
10. Western Ghats

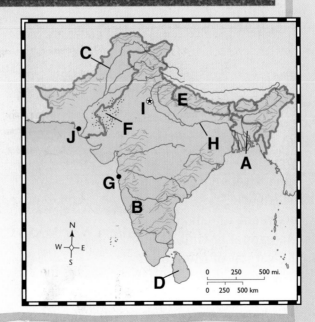

Chapter 23 China

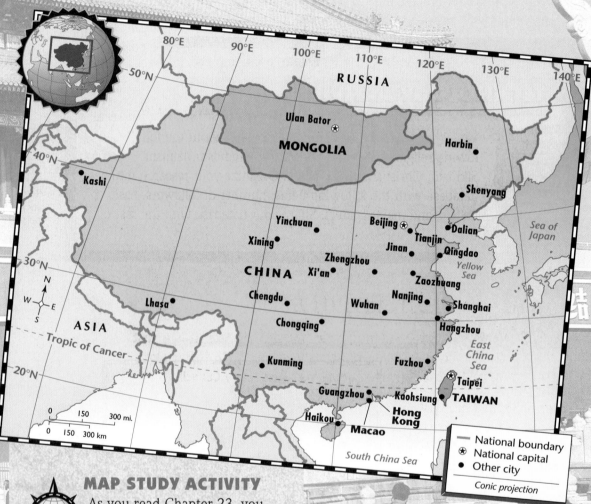

National boundary
⊛ National capital
● Other city

Conic projection

MAP STUDY ACTIVITY

As you read Chapter 23, you will learn about China and its neighbors.

1. What area is an island?
2. What is the capital of China?
3. What is the capital of Mongolia?

The Land

PREVIEW

Words to Know
- loess
- dike
- typhoon

Places to Locate
- China
- Plateau of Tibet
- Kunlun Shan
- Taklimakan
- Gobi
- Huang He
- Chang Jiang
- Xi

Read to Learn . . .
1. why the Plateau of Tibet is called the "Roof of the World."
2. where most of China's people live.
3. what climates are found in China.

The Great Wall of China is one of the wonders of the world. The longest structure ever built, the wall stretches about 4,000 miles (6,437 km). The Chinese built the wall centuries ago to keep out invaders.

The People's Republic of China lies in the eastern part of Asia. It is a vast country that covers more than 3,600,930 square miles (9,326,409 sq. km). China is the world's third-largest country in area, after Russia and Canada.

Region
Landscapes

If you travel by train across China, you will be amazed by the many different landscapes and climates. Look at the map on page 610. You see that China has tall mountains, vast deserts, fertile plains, and mighty rivers.

Mountains Rugged mountains cover about one-third of China. They lie in the western part of the country. The Himalayan range sweeps across the south on the border between China and Nepal. The center of China's mountainous area holds the largest plateau in the world—the Plateau of Tibet. It is called the "Roof of the World" because its average height is about 13,000 feet (4,000 m) above sea level. Scattered shrubs and grasses cover the countryside. The eastern edge of the plateau is home to pandas, golden monkeys, and other rare animals.

Deserts In the north of China, you notice that the mountain ranges circle desert basins. One of these deserts is the Taklimakan. It is an isolated region with some of the world's highest temperatures. Sandstorms there may last for days and create huge, shifting dunes.

Farther east lies another desert, called the Gobi. The Gobi is made up of rock and gravel instead of sand. Temperatures there can rise to 110°F (43°C) during summer days. Because the Gobi is far north and at a high altitude, its temperatures during winter nights can drop as low as −30°F (−34°C)!

Plains and Highlands Plains cover the eastern part of China. Look at the map below. You see that these plains run along the coasts of the South China and East China seas. This area has fertile land and is home to almost 90 percent of China's people.

The Northeast Plain lies in the center of Manchuria, a region in northeast China. To the east of the plain lies a heavily forested, hilly area near China's border with Korea. To the south is the wide, flat North China Plain. Other plains areas extend farther south near the coast.

Inland in the southeast, the land changes from plains to green highlands. This region is one of the most scenic areas in China. Many tourists come to see its many scattered limestone hills that rise 100 to 600 feet (30 to 182 m).

Rivers Three of China's major rivers—the Huang He (HWAHNG HUH), the Chang Jiang (CHAHNG jee•AHNG), and the Xi (SHEE)—flow through the

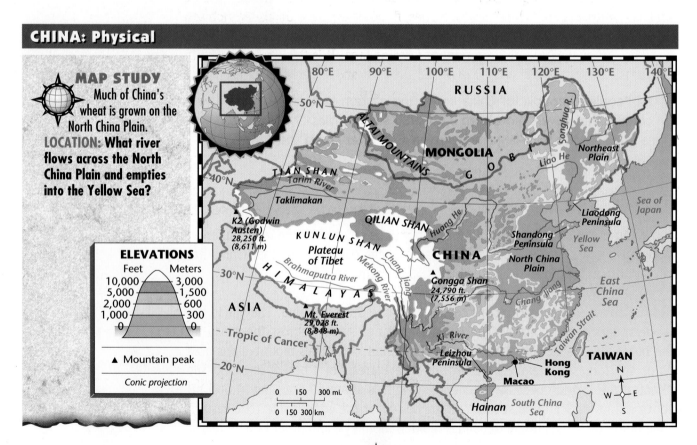

CHINA: Physical

MAP STUDY
Much of China's wheat is grown on the North China Plain.
LOCATION: What river flows across the North China Plain and empties into the Yellow Sea?

ELEVATIONS

Feet	Meters
10,000	3,000
5,000	1,500
2,000	600
1,000	300
0	0

▲ Mountain peak

Conic projection

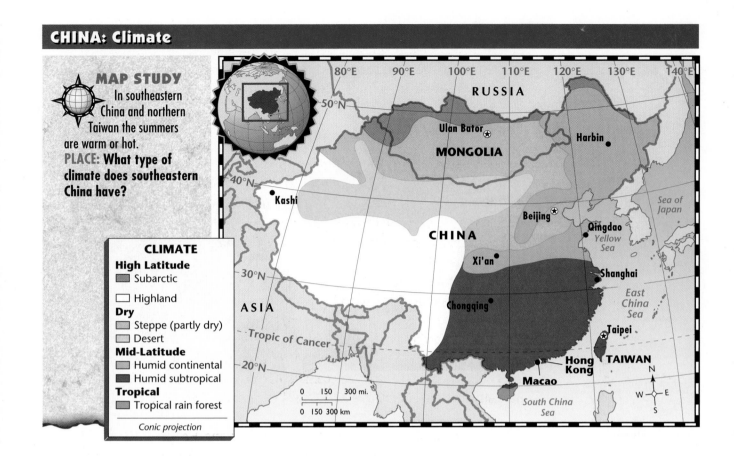

In southeastern China and northern Taiwan the summers are warm or hot.
PLACE: What type of climate does southeastern China have?

CLIMATE

High Latitude
- Subarctic

- Highland

Dry
- Steppe (partly dry)
- Desert

Mid-Latitude
- Humid continental
- Humid subtropical

Tropical
- Tropical rain forest

Conic projection

plains and southern highlands. They serve as important transportation routes and as a source of soil. How? For centuries, floodwaters have deposited rich soil to form flat river basins.

The Huang He basin is an important farming area. It is thickly covered with **loess,** a fertile, yellow-gray soil deposited by wind and water. The name *Huang He* means "yellow river," referring to the loess carried by the river.

The rivers of China have brought disasters as well as blessings. The Chinese, for example, call the Huang He "China's sorrow." In the past, its flooding cost hundreds of thousands of lives and caused much damage. To control floods, the Chinese have built dams and **dikes,** or high banks of soil, along the sides of this river. In the late 1990s work began on the Three Gorges Dam on the Chang Jiang. The dam—the world's largest construction project—will prevent serious flooding and provide electricity for industries.

Region
Climate

Like the United States, China has many different climate regions. The map above shows you the seven climate regions of China. Location, elevation, and wind currents affect the type of climate found in any particular area of the country. Southeastern China has a humid subtropical climate with hot, humid summers. The northeast has a humid continental climate with cold winters. In desert areas to the northwest, summers are hot and dry and winters are cold

Rice fields are a common sight in warm, humid southeastern China *(left)*. Beijing has a humid continental climate with cold, snowy winters *(right)*.
LOCATION: What are hurricanes in the Pacific Ocean called?

and windy. In the southwestern part of China, the high Plateau of Tibet has bitterly cold winters and cool summers.

Monsoons greatly affect China's climates. In winter cold, dry air flows from central Asia across China. In summer the monsoons blow in from the sea, bringing warm, moist air. The summer monsoons often bring typhoons to coastal areas in the south. **Typhoons**—called hurricanes in the Atlantic Ocean—are tropical storms with strong winds and heavy rains.

SECTION 1 ASSESSMENT

REVIEWING TERMS AND FACTS

1. Define the following: loess, dike, typhoon.

2. LOCATION Where is the Plateau of Tibet located?

3. PLACE How would you describe the landscape of southeastern China?

4. REGION How do monsoons affect China's climates?

MAP STUDY ACTIVITIES

5. Look at the physical map on page 610. What three major rivers flow through China?

6. Study the climate map on page 611. What are the three major climate regions of western China?

UNIT 8

BUILDING GEOGRAPHY SKILLS

Comparing Two or More Graphs

Comparing two or more graphs gives you a better understanding of a topic. When comparing graphs, first see how they are related. Are they on the same topic? Are they the same kind of graph? Are the same units of measurement used? Then look for similarities and differences in the information on the graphs. To compare two or more graphs, apply these steps:

- Read the title and labels on each graph to see how they are related.

- Look for similarities and differences in the information given on each graph. Draw conclusions based on the comparison.

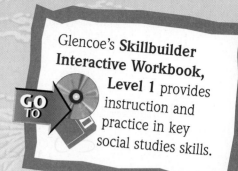

Glencoe's **Skillbuilder Interactive Workbook, Level 1** provides instruction and practice in key social studies skills.

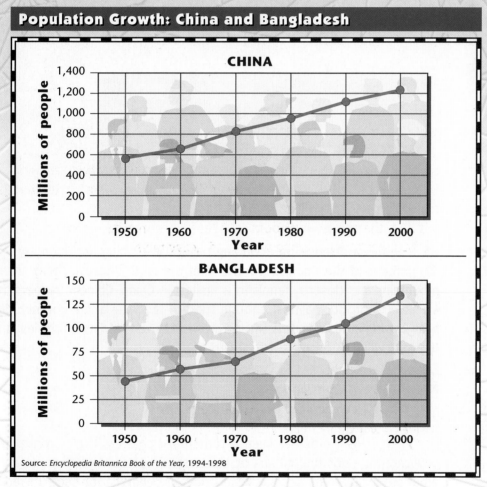

Population Growth: China and Bangladesh

Source: *Encyclopedia Britannica Book of the Year, 1994-1998*

Geography Skills Practice

1. What is the subject of the graphs?

2. What was Bangladesh's population in 1960?

3. By how much did China's population increase from 1980 to 1990?

4. What conclusions can you draw from the information on these graphs?

SECTION 2 · The Economy

Words to Know
- invest
- consumer goods
- tungsten
- terraced field

Places to Locate
- Shanghai
- Beijing
- Wuhan
- Guangzhou

Read to Learn . . .
1. how most Chinese earn their livings.
2. what the three economic regions of China are.
3. how China's economy has changed in recent years.

Shanghai is China's largest city and its major port. The name *Shanghai* means "on the water." Shanghai is a leader in trade and business. Because Shanghai is a port city, it has many contacts with the rest of the world.

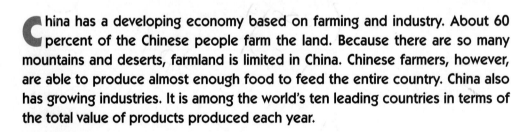

China has a developing economy based on farming and industry. About 60 percent of the Chinese people farm the land. Because there are so many mountains and deserts, farmland is limited in China. Chinese farmers, however, are able to produce almost enough food to feed the entire country. China also has growing industries. It is among the world's ten leading countries in terms of the total value of products produced each year.

Region
A New Economy

Since 1949 China has been a communist country. Under communism, the government—not individuals or businesses—decides what crops are grown, what products are made, and what prices are charged for both. In recent years China's government has made many changes in the economy. Why? It hopes to make China a modern, industrialized nation. It also wants to increase trade with the United States and other Western countries. Without completely giving up communism, the government has allowed many features of free enterprise to develop in China.

Making a Living Since the 1980s, China has increased its ties with foreign countries. Eager to learn about new business methods, it has asked other countries to **invest,** or put money, in Chinese businesses. Many businesses in China are now jointly owned by Chinese and foreign businesspeople.

Instead of telling people what jobs they should have, the Chinese government now wants individuals to choose where to work and even to start their own businesses. Shop owners, dentists, and barbers are among the people competing for customers in China's major cities.

Changes have also come to China's countryside. Farmers once spent most of their time working on government farms. Now they grow extra crops on private plots of land. Farmers can keep the money they make from selling these crops.

Because of all these changes, farming, manufacturing, and trade are booming in China. Some Chinese now enjoy a good standard of living. They can afford **consumer goods,** or products such as television sets, cars, and motorcycles. Not everyone, however, finds it easy to adjust to the new economy. Many Chinese are finding that the prices of food and goods are rising faster than their incomes. Some Chinese are getting very rich, while others remain poor.

As China industrializes, it faces environmental problems. Many factories dump poisonous chemicals into rivers. Others, powered by coal, send out air-polluting smoke. The building of Three Gorges Dam also has raised concerns. About 2 million people will be moved before the dam's lake covers up farms, villages, and canyons. Critics claim that the project is destroying one of the world's most scenic areas, as well as centuries-old temples.

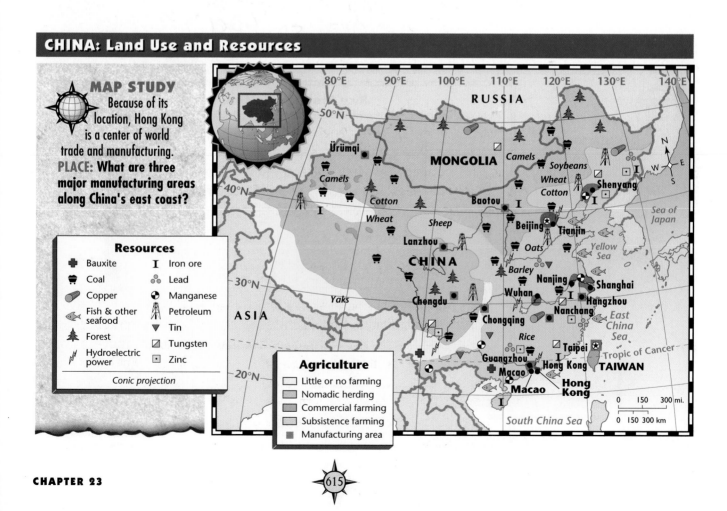

CHINA: Land Use and Resources

MAP STUDY
Because of its location, Hong Kong is a center of world trade and manufacturing.
PLACE: What are three major manufacturing areas along China's east coast?

Resources

✚ Bauxite	I Iron ore
🛒 Coal	♣ Lead
Copper	◓ Manganese
🐟 Fish & other seafood	⌁ Petroleum
🌲 Forest	▼ Tin
⚡ Hydroelectric power	▧ Tungsten
	⊡ Zinc

Conic projection

Agriculture
Little or no farming
Nomadic herding
Commercial farming
Subsistence farming
■ Manufacturing area

Region
Economic Regions

The physical geography of China influences its economy. Many geographers divide China into three economic regions: the north, the south, and the west.

The North The north region includes the plains and highlands of northeastern China. It has a variety of economic activities, including manufacturing, mining, farming, and commercial fishing. Because of the partly dry, often cold climate, farmers in the region grow hearty grains such as wheat, cotton, and soybeans. In isolated grassland areas, nomads herd livestock.

The map on page 615 shows you that the north is rich in natural resources such as coal, petroleum, iron ore, and tungsten. **Tungsten** is a metal used in electrical equipment. China is a world leader in the mining of coal and iron ore. A number of large urban areas in the north manufacture many goods. Workers in these industrial centers produce textiles, chemicals, electronic equipment, farm machinery, airplanes, and other metal goods. Beijing (BAY•JIHNG), China's capital and major industrial city, is located in the north.

Fourteen-year-old Lai Sau Chun has lived in Beijing all of her life. She explains that, like many Chinese cities, Beijing is a blend of old and new. Many factories have been built near the city, but Sau Chun prefers the old Beijing. She takes visitors to the Imperial Palace in the heart of the city. Chinese rulers built it in the early 1400s. The palace was called "the forbidden city" because only the ruler's family and government officials were allowed to enter.

World's Leading Rice-Producing Countries

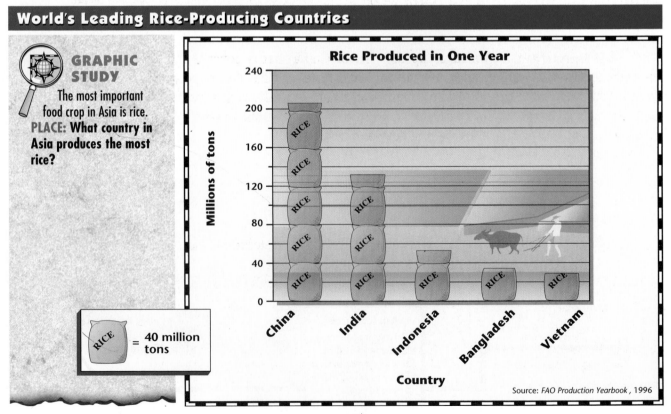

GRAPHIC STUDY
The most important food crop in Asia is rice.
PLACE: What country in Asia produces the most rice?

RICE = 40 million tons

Rice Produced in One Year

Source: FAO Production Yearbook, 1996

A worker puts the finishing touches on a vase in a Beijing factory *(left)*. A sheepherder tends his flock in northwestern China *(right)*.
PLACE: What natural resources are found in northern China?

The South The south region includes the southern half of eastern China. It has fertile soil, a humid climate, and a long growing season. If you travel through hilly areas, you will see farmers growing crops on terraced fields. A **terraced field** has strips of land cut out of a hillside like stair steps. Rice is the south's major crop. Farmers also grow tea, jute, silk, fruits, and vegetables.

In addition to fertile soil, the south is rich in bauxite, iron ore, tin, and other minerals. It also has many urban manufacturing areas. Some cities, such as Shanghai, are located on or near the coast. Others, such as Wuhan (WOO•HAHN) or Guangzhou (GWAHNG•JOH), are on rivers. Workers in these industrial cities make ships, machinery, textiles, and electrical equipment.

The West The west region takes up the part of China that is covered by mountains, deserts, and grasslands. The dry and cold Plateau of Tibet is not very good for farming. It provides only limited grazing for hardy animals such as yaks. In low-lying fertile areas, farmers are able to grow cotton, wheat, and other food crops. The west is also rich in petroleum, coal, and iron ore.

SECTION 2 ASSESSMENT

REVIEWING TERMS AND FACTS
1. **Define the following:** invest, consumer goods, tungsten, terraced field.
2. **PLACE** How do most Chinese earn their livings?
3. **HUMAN/ENVIRONMENT INTERACTION** How has China's economy been changing?
4. **REGION** What region is the major rice-producing area of China?

MAP STUDY ACTIVITIES
5. Look at the map on page 615. What minerals are found in northeastern China?
6. Where do you think oranges and other tropical crops grow in China?

MAKING CONNECTIONS

MATH **SCIENCE** **HISTORY** **LITERATURE** **TECHNOLOGY**

Chinese poets aimed to create a certain mood and atmosphere in their works. Many of their poems describe brief moments of deep feelings, such as those expressed in these poems by Wang Wei.

NORTH HILL

North Hill stands out above the lake
Against thick evergreens gleams startlingly
    a vermilion [red-orange] gate.
Below, South River zig-zags toward the horizon,
Glistening, here and there, beyond the treetops of the
    blue forest.

SAILING ON THE RIVER TO CH'ING-HO

Sailing on the great river
The gathered waters reach to the very rim of heaven.
Suddenly the sky-high waves part,
Disclosing country of a thousand and cities of ten
    thousand roofs.
And sailing further, still other villages and
    market-towns appear
Mid mulberry trees and growing flax.
Looking back towards my native village . . .
The vast waters have joined the clouds.

Chinese Painting

From *Poems by Wang Wei,* translated by Chang Yin-nan and Lewis C. Walmsley. Copyright © 1958 by Charles E. Tuttle Co., Inc., Tokyo, Japan. Reprinted by permission.

Making the Connection

1. What geographic features are mentioned in these Chinese nature poems?
2. How does the poet describe the environment?

The People

PREVIEW

Words to Know
- dynasty
- calligraphy
- pagoda

Places to Locate
- Taiwan
- Beijing

Read to Learn . . .
1. what groups influenced the culture of China.
2. how city life differs from country life in China.
3. what arts and recreation the Chinese enjoy.

Along city streets in China, you will see many young people dressed as you are—in jeans. Western-style clothing is very popular. People are prepared to pay high prices for this clothing.

Crowded city streets are common in China. The country has more people than any other place in the world. About 1.2 billion people—one-fourth of the world's population—make China their home. The graph on page 620 shows you how China's population compares with that of the United States.

About 94 percent of China's people are Han Chinese. They have a unique culture, although they speak different dialects of the same Chinese language. The remaining 6 percent belong to 55 other ethnic groups. Most of these groups, such as the Tibetans, live in the western part of the country. For years the Tibetans have struggled to keep their culture and win their independence from China.

Place
Influences of the Past

Chinese civilization is more than 4,000 years old. For centuries **dynasties,** or ruling families, governed China. Each dynasty was made up of a line of rulers from the same family. Under the dynasties, China set up a strong government and enlarged its borders.

COMPARING POPULATION: China and the United States

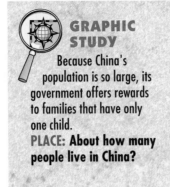

GRAPHIC STUDY

Because China's population is so large, its government offers rewards to families that have only one child.

PLACE: About how many people live in China?

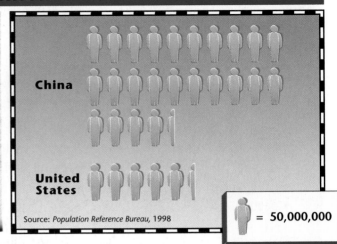

China

United States

Source: *Population Reference Bureau, 1998*

 = 50,000,000

Ideas and Inventions The early Chinese developed a rich civilization. They believed that learning was a key to good behavior. A philosopher named Konfuzi, or Confucius, taught that people should be polite, honest, brave, and wise. Children were to obey their parents, and every person was to respect the elderly and obey the country's rulers. For more than 2,000 years, the Chinese followed the teachings of Confucius. They blended these teachings with those of Buddhism, which reached China from India about A.D. 100.

Did you know that the early Chinese were using paper and ink before people in other parts of the world? Early Chinese developed a number of outstanding inventions. Besides paper and ink, these inventions include the clock, the compass, the first printed book, and fireworks.

Modern China The Chinese tried to avoid contact with other parts of the world. In the 1700s and 1800s, however, European powers forced the Chinese to trade and tried to make them accept Western ways. This angered the Chinese.

In 1912 a national uprising overthrew the last dynasty, and China became a republic. Soon civil war divided the country. Nationalist forces, led by General Chiang Kai-shek (jee•AHNG KY•SHEHK), fought the Chinese Communists, led by Mao Zedong (MOW ZUH•DUNG). In 1949 the Communists won the civil war, driving Chiang's government to the offshore island of Taiwan.

Mainland China became the communist People's Republic of China. The Communists wanted to make China a modern, industrialized nation, but they denied freedoms to the Chinese people. In 1989 Chinese students in Beijing's Tiananmen Square called for democracy. The government answered their protest with gunfire, killing many and jailing others.

Meanwhile, the people of Tibet, a mountainous region of southwestern China, demanded more freedom and independence. Once a separate Buddhist kingdom, Tibet came under direct Chinese rule in the 1950s. The Dalai Lama, Tibet's spiritual leader, leads a worldwide movement for Tibetan rights from his place of exile in India.

Tiananmen Square, Beijing

In 1989 hundreds of students were killed here when they protested against the Chinese government.
PLACE: Why do you think the Chinese government would kill protesting students?

UNIT 8

Rural Life About 70 percent of China's people live in rural areas. The map below shows you that most people are crowded into the fertile river valleys of eastern China. Families in the countryside work very hard in their fields. They often use simple hand tools because farming equipment is too expensive.

Village life, though, has improved in recent years. Most rural families live in three- or four-room houses. They have enough food and some modern appliances. Many villages have community centers where people can watch movies, read books, and play sports such as table tennis and basketball.

City Life More than 370 million Chinese—about 30 percent of the population—live in cities. China's cities are growing rapidly as people from the countryside move to cities hoping to find work. Living conditions are crowded, but most homes and apartments have heat, electricity, and running water. Many people now earn enough money to buy extra clothes and television sets. They also have more leisure time to attend concerts or Chinese operas, walk in parks, visit zoos, or talk with friends.

Region
The Arts

China is famous for its traditional arts. Chinese craftworkers make bronze bowls, jade jewelry, glazed pottery, and fine porcelain. The Chinese are also known for their painting, sculpture, and architecture.

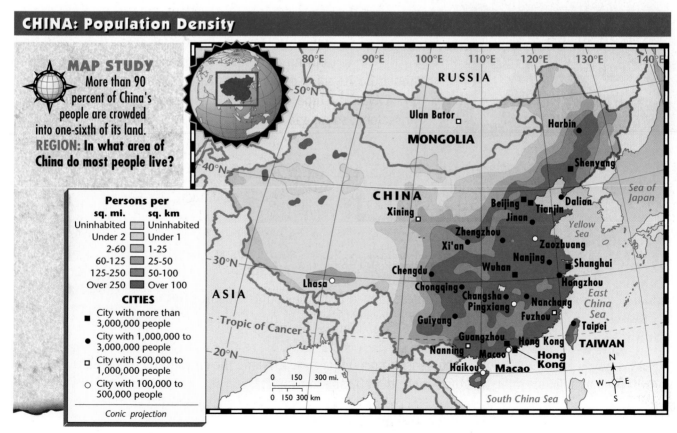

CHINA: Population Density

MAP STUDY More than 90 percent of China's people are crowded into one-sixth of its land.
REGION: In what area of China do most people live?

Persons per

sq. mi.	sq. km
Uninhabited	Uninhabited
Under 2	Under 1
2-60	1-25
60-125	25-50
125-250	50-100
Over 250	Over 100

CITIES
■ City with more than 3,000,000 people
● City with 1,000,000 to 3,000,000 people
□ City with 500,000 to 1,000,000 people
○ City with 100,000 to 500,000 people

Conic projection

Art The Chinese love of nature has influenced painting and poetry in China. Chinese artists paint on long panels of paper or silk. Artwork often shows scenes of mountains, rivers, and forests. Artists attempt to portray the harmony between nature and the human spirit.

Many Chinese paintings include a verse or poem written in **calligraphy,** the art of beautiful writing. Chinese writing is different from the print you are reading right now. It uses characters that represent words or ideas instead of sounds. There are more than 50,000 Chinese characters, but the average person recognizes only about 8,000.

Chinese pottery is famous for its painted decoration. The Chinese developed the world's first porcelain centuries ago. Porcelain is made from coal dust and a fine, white clay. Porcelain vases from early Chinese history are considered priceless today.

Architecture Most buildings in Chinese cities are modern in style, but a few traditional buildings stand. Some have large tiled roofs with edges that curve gracefully upward. Others are Buddhist temples with many-storied towers called **pagodas.** Large statues of the Buddha are found in these buildings.

Foods Cooking differs greatly from region to region in China. In coastal areas fish, crab, and shrimp dishes are common, while central China is famous for its spicy dishes. Most Chinese, however, eat very simply. A typical Chinese meal includes vegetables with bits of meat or seafood, soup, and rice or noodles.

Teen Scene

At Home in the Air

Acrobat Choi Hong See has been in training since she was three years old. Her family performs in Shanghai's Acrobatics Theater. She can walk on high wires, do triple somersaults, and balance on a human pyramid. Acrobatics have been popular in China for more than 2,000 years. Many acrobatic groups are supported by the government and are often invited to perform in Europe or the United States.

SECTION 3 ASSESSMENT

REVIEWING TERMS AND FACTS

1. Define the following: dynasty, calligraphy, pagoda.

2. PLACE How was early China governed?

3. MOVEMENT Why have many Chinese moved from the countryside to cities?

4. REGION How is life in urban China different from life in rural China?

MAP STUDY ACTIVITIES

5. Look at the population density map on page 621. Name the cities in the People's Republic of China that have more than 3 million people.

622

SECTION 4

China's Neighbors

PREVIEW

Words to Know
- empire
- yurt
- high-technology industry

Places to Locate
- Mongolia
- Taiwan
- Hong Kong
- Macao

Read to Learn . . .
1. how people live and earn their livings in Mongolia.
2. what products are made in Taiwan.
3. why Hong Kong is one of the world's busiest ports.

North of China lies Mongolia. If you flew to Mongolia, you would arrive in Ulan Bator (OO•LAHN BAH•TAWR), the capital and largest city. Ulan Bator began as a Buddhist community in the early 1600s. Today it is a modern cultural and industrial center.

The large inland territory of Mongolia and the island of Taiwan are two countries that border or lie close to the People's Republic of China. Throughout their history, the people of these lands have had close ties to China. Mongolia once had a communist government similar to China's. Many of Taiwan's people are Chinese who left the mainland and set up a noncommunist Chinese government on Taiwan. On China's southern coast are two other Chinese territories—Hong Kong and Macao. After years of European rule, Hong Kong and Macao in the late 1990s became part of the People's Republic of China.

Place
Mongolia

Mongolia lies in east central Asia between China and Russia. The map on page 608 shows you that Mongolia is a landlocked country. Its land area of 604,825 square miles (1,566,500 sq. km) is more than twice the size of Texas.

Centuries ago, Mongolia was the home of an Asian people called the Mongols. During the 1200s powerful Mongol armies carved out the largest land empire in history. An **empire** is a collection of territories under one ruler. The

Workers assemble electronic circuit boards in a Taiwan factory.
MOVEMENT: Why do the Republic of China and mainland China have different governments?

Mongol Empire reached from China to eastern Europe. Today Mongolia is smaller in size, but its people still honor their Mongol past.

The Land Rugged mountains and high plateaus cover most of Mongolia. In the southeast, river basins dot the mountainous landscape. The bleak desert landscape of the Gobi spreads through central and southeastern Mongolia. Most of the country receives only small amounts of rain, and temperatures are usually very cold or very hot.

The Economy Mongolia's economy relies on the raising of livestock. Many Mongolians herd sheep, goats, cattle, or camels on grasslands that cover large areas of the country. For centuries the people of Mongolia have been famous for their skills in raising and riding horses. In limited areas, farmers grow wheat, barley, and oats.

The People Nearly all of Mongolia's 2.4 million people are Mongols who speak the Mongol language. About 57 percent of them live in urban areas such as Ulan Bator. Mongolians in the countryside live on farms. A few still follow the nomadic way of their ancestors. These herder-nomads live in **yurts,** or large circle-shaped tents made of animal skins.

Mongolians traditionally followed a form of Buddhism. When Mongolia was under a rigid communist government from 1924 to 1990, however, the Communists discouraged religious practice. Now that Mongolia is independent, people are free to observe their Buddhist beliefs again.

Place
Taiwan

About 100 miles (161 km) off the southeast coast of China lies the island country of Taiwan. In 1949 the Chinese Communists conquered mainland China. The defeated Chinese government of Chiang Kai-shek moved to Taiwan and established the Republic of China. Today Taiwan is still governed separately from the communist-ruled mainland.

The Land Thickly forested mountains run through the center of Taiwan. A flat plain along the western coast is home to most of the island's people. Like southeastern China, Taiwan has a humid subtropical climate of mild winters and hot, rainy summers.

The Economy Taiwan has one of the world's most prosperous economies. Taiwan's wealth rests largely on high-technology industries, manufacturing, and trade with foreign countries. **High-technology industries** produce computers and other kinds of electronic equipment. Taiwanese workers produce many different products, including computers, clothing, furniture, ships, radios, and televisions. You have probably seen goods from Taiwan sold in stores in your community.

Taiwan's mountains limit available farmland. Farmers often terrace hills to provide more land for growing rice. Other crops grown in Taiwan include bananas, citrus fruits, corn, peanuts, sugar, tea, and vegetables.

The People About 75 percent of Taiwan's 21.7 million people live in urban areas. Taipei is the capital and largest city. Skyscrapers and busy streets are major features of Taipei, which has about 2.6 million people.

Taiwan is a modern country, but many of its people still follow Chinese traditions. About 95 percent of the people are Chinese and speak Chinese dialects. Buddhism is the leading religion of Taiwan.

WHAT IN THE WORLD?

Growing Land?
You've heard of growing crops, but have you heard of growing land? Hong Kong has many people, and there is no place to put them all. Therefore, workers are blasting the mountains on Hong Kong's Kowloon Peninsula. They dump the rock and dirt from the blastings onto the shore of the South China Sea—growing new land.

Place
Hong Kong and Macao

Big changes recently occurred in Hong Kong and Macao, which lie on the southeastern coast of China. Both are made up of hilly peninsulas and groups of islands. The United Kingdom ruled Hong Kong until 1997, when it returned to Chinese rule. Another European country—Portugal—governed Macao before the territory became part of China in 1999.

Busy Port Hong Kong has a land area of about 382 square miles (990 sq. km). Natural harbors make Hong Kong one of the world's busiest ports and a major center of trade and business. It is also one of

Hong Kong Souvenirs

the world's most crowded places. About 6.7 million people live in Hong Kong. This means that the population density is about 16,383 people per square mile (6,325 per sq. km).

Nearly all the people of Hong Kong are Chinese. They live mostly in urban areas in crowded, high-rise apartments. Others make their homes on boats in harbors. Many of Hong Kong's people work in manufacturing, banking, government, foreign trade, and tourism.

A gift shop in Hong Kong sells souvenirs commemorating the territory's return to China.
PLACE: What country ruled Hong Kong before 1997?

Hindu Wedding in Hong Kong

At a Hindu wedding, the bride and groom dress in traditional clothing and sit in front of a sacred fire.
PLACE: How is this wedding different from or similar to the weddings you have attended?

Once A Bit of Portugal About 40 miles (64 km) west of Hong Kong lies the territory of Macao. The map on page 608 shows you that Macao covers about 8 square miles (20 sq. km). It has about 500,000 people, most of whom are Chinese. Most of the territory's food is imported from China. Macao's economy is based on industry, trade, and tourism.

Macao has an interesting mix of Chinese and Portuguese cultures. In the older sections of Macao, you will find Roman Catholic churches and old European-looking houses next to Chinese temples and Chinese-owned shops. Across town high-rise apartments and modern office buildings crowd the skyline. There is much new construction here in response to the political change in 1999.

SECTION 4 ASSESSMENT

REVIEWING TERMS AND FACTS

1. Define the following: empire, yurt, high-technology industry.

2. PLACE What desert area crosses Mongolia?

3. HUMAN/ENVIRONMENT INTERACTION What is the major economic activity of Mongolia?

4. PLACE Where do most of Taiwan's people live?

MAP STUDY ACTIVITIES

5. Look at the climate map on page 611. What is the major climate region of Mongolia?

6. Turn to the physical map on page 610. What body of water touches Taiwan, Hong Kong, and Macao?

Chapter 23 Highlights

Important Things to Know About China

SECTION 1 THE LAND

- China has more people than any other country in the world. It is the world's third-largest country in area, after Russia and Canada.
- Three important rivers—the Huang He, Chang Jiang, and Xi—flow through fertile river valleys in eastern China.
- The western part of China is largely mountains and deserts.
- China's climate is affected by monsoons.

SECTION 2 THE ECONOMY

- Modern China's communist leaders have moved China's economy toward free enterprise.
- Most of China's people are farmers. Rice is the major food crop in the south.
- China's largest urban areas lie near rivers or on the coast.

SECTION 3 THE PEOPLE

- China has one of the world's oldest civilizations.
- The early Chinese passed on many inventions to the rest of the world, including paper, ink, porcelain, and the first printed book.

SECTION 4 CHINA'S NEIGHBORS

- Mongolia has rugged mountains, plateaus, and desert.
- Taiwan has a prosperous economy based on high-technology industries, manufacturing, and trade.
- Hong Kong, once under British rule, became part of China in 1997. Macao, once a Portuguese colony, returned to Chinese rule in 1999.

Tiananmen Square, Beijing ▶

REVIEWING KEY TERMS

Match the numbered terms in Set A with their lettered definitions in Set B.

A

1. loess
2. typhoon
3. dike
4. terraced field
5. tungsten
6. yurt
7. high-technology industry
8. calligraphy

B

A. level strip of land cut out of a hillside
B. the art of beautiful writing
C. tent made of animal skins
D. production of computers and electronic equipment
E. yellow-gray soil deposited by the wind
F. tropical storm with strong winds and heavy rains
G. bank of earth built to prevent floods
H. metal used in electrical equipment

Mental Mapping Activity

Draw a freehand map of China and its neighbors. Label the following:
- Beijing
- Chang Jiang
- Gobi
- Taiwan
- Ulan Bator

REVIEWING THE MAIN IDEAS

Section 1
1. REGION How does China compare in size to other countries?
2. PLACE Why was the Huang He called "China's sorrow"?

Section 2
3. MOVEMENT Why has China made changes in its economy?
4. HUMAN/ENVIRONMENT INTERACTION How has the growth of industry affected China's economy?

Section 3
5. REGION What percent of China's people live in rural areas?
6. PLACE On what ideas did the ancient Chinese base their way of living?

Section 4
7. PLACE What religion is practiced in Mongolia?
8. PLACE Why does Taiwan have limited farmland?
9. PLACE What happened to Macao in 1999?

CRITICAL THINKING ACTIVITIES

1. **Determining Cause and Effect** How have recent changes in China's economy affected the Chinese people?
2. **Drawing Conclusions** How have landscape and climate affected where the Chinese people live?

CONNECTIONS TO WORLD HISTORY

Cooperative Learning Activity Work in a group of three to learn about the flags of China, Mongolia, and Taiwan. Each group will choose the flag of one country, and research the following: (a) flag symbols and their meanings; (b) flag colors and their meanings; or (c) the historical events related to the creation of the flags. Then prepare a group report describing the symbols, colors, and history of the flag.

GeoJournal Writing Activity
Imagine that you are a reporter traveling on one of the major rivers of China. Write a news story about the river and the people who live along it.

TECHNOLOGY ACTIVITY

Using a Computerized Card Catalog Use a computerized card catalog to locate sources of information about the Three Gorges Dam Project on the Chang Jiang in China. Write a report describing its size, how much electricity it generates, and how many villages and people it will ultimately displace. Summarize your report by explaining your opinion about the dam's development.

PLACE LOCATION ACTIVITY: CHINA

Match the letters on the map with the places and physical features of China and its neighbors. Write your answers on a separate sheet of paper.

1. Huang He
2. Mongolia
3. Taklimakan
4. Beijing
5. Hong Kong
6. Taipei
7. Shanghai
8. South China Sea
9. Tian Shan
10. Himalayas

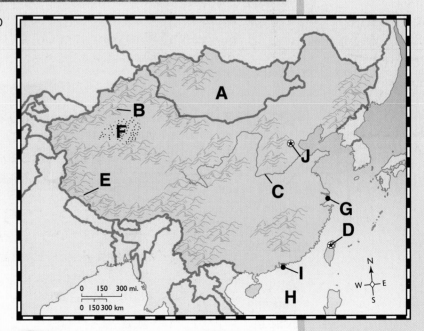

GeoLab
ACTIVITY

From the classroom of
Don Mendenhall,
Coleman Junior High,
Van Buren, Arkansas

EROSION: Saving the Soil

Background

Because of the enormous amount of rainfall in many Asian countries, growing crops has become very difficult. Too much rain erodes topsoil and results in the destruction of crops. In this activity, you will try various ways to solve the problem of erosion.

Terraced Fields

Monsoon rains drench Asia every year, bringing much-needed water. But they also cause serious flooding. In 1887 the Huang He overflowed its banks because of heavy rains and the spring thaw. About 1 million people died as a result—the greatest flood disaster in history.

Materials

- 5 aluminum 8- or 9-inch pans
- leaves or grass clippings
- 500-ml beaker
- water
- ruler
- newspaper
- pebbles
- potting soil
- watering can
- dishpan

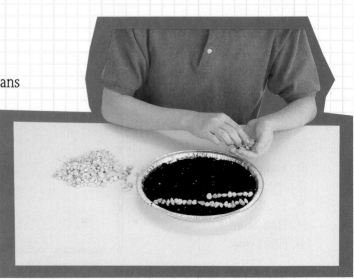

What To Do

A. Fill each pan almost to the rim with soil. Pat the soil until the surface is firm and flat. Soak the soil by pouring 100 ml of water into each pan.

B. In one pan, cover the surface with a layer of grass clippings or leaves.

C. In the second pan, lay newspaper strips across the surface.

D. In the third pan, build a terrace by forming two small walls of pebbles.

E. In the fourth pan, use your finger to make curved grooves across the surface. This represents contour plowing.

F. Set the fifth pan aside.

G. Now follow this procedure with all five pans:

H. Measure 200 ml of water into the watering can.

I. Hold the pan so that one side dips slightly into the dishpan.

J. Slowly pour the water onto the soil at the top of the pan. Wait until the excess water runs across the soil and into the dishpan. **Note:** Be sure to hold the terracing and contour plowing pans so the water runs across the pebble walls and the grooves.

K. After repeating step J for each pan, pour the water and soil that ends up in the dishpan into the beaker. Measure the amount and record the results. When the soil settles to the bottom, measure and record its height in the beaker.

Lab Activity Report

1. Which pan had the least erosion?

2. Which pan had the most erosion?

3. **Drawing Conclusions** What did you learn about controlling erosion?

Go A Step Further!

Activity

Experiment with different types of soils to see if soil composition—as well as soil covering—affects erosion. Mix potting soil with sand, clay, or flour. Which mixture best prevents erosion?

Chapter 24 Japan and the Koreas

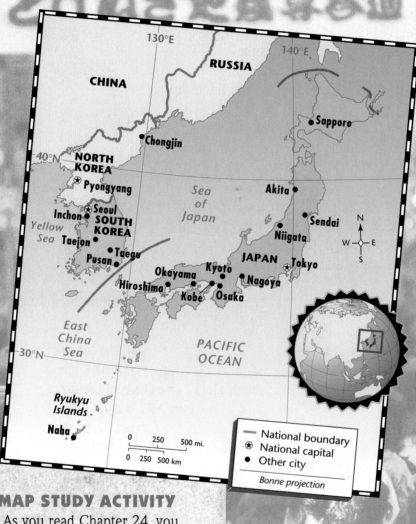

Map labels

130°E · 140°E

RUSSIA

CHINA

Sapporo

Chongjin

40°N **NORTH KOREA**
✪ Pyongyang

Akita

Inchon ✪ Seoul
SOUTH KOREA

Yellow Sea

Taejon

Taegu

Pusan

Sea of Japan

Sendai

Niigata

JAPAN

Okayama · Kyoto
Hiroshima · Kobe · Osaka · Nagoya

✪ Tokyo

N W E S

East China Sea

PACIFIC OCEAN

30°N

Ryukyu Islands

Naha

0 250 500 mi.
0 250 500 km

── National boundary
✪ National capital
● Other city

Bonne projection

MAP STUDY ACTIVITY

As you read Chapter 24, you will learn about Japan and the Koreas.

1. What is the capital of North Korea?
2. What sea lies between Japan and North Korea?
3. What is Japan's capital?

SECTION 1

Japan

PREVIEW

Words to Know
- archipelago
- tsunami
- intensive cultivation
- clan
- samurai
- shogun
- megalopolis

Places to Locate
- Japan
- Tokyo
- Honshu
- Hokkaido
- Kyushu
- Shikoku

Read to Learn . . .
1. where Japan is located.
2. why Japan has a strong economy.
3. what religions influenced the culture of Japan.

On a clear day in Tokyo, Japan, you can see beautiful Mount Fuji. Mount Fuji, Japan's highest peak, is an inactive volcano. During the summer thousands of Japanese make their way to the top of the mountain.

Love of nature is a major part of the Japanese way of life. The sun, the mountains, and the sea have always been important to the people of Japan. The map on page 632 shows you that Japan is an **archipelago,** or a chain of islands. The Japanese islands form a curve off the coast of eastern Asia between the Sea of Japan and the Pacific Ocean. Four main islands make up Japan. They are Hokkaido (hah•KY•doh), Honshu, Shikoku (shih•KOH•koo), and Kyushu (kee•OO•shoo). Thousands of smaller islands are also part of Japan.

Location
The Land

Japan has a land area of 145,370 square miles (376,508 sq. km)—slightly smaller than California. The Japanese islands are the peaks of a great mountain range rising 20,000 to 30,000 feet (6,000 to 9,000 m) from the floor of the Pacific Ocean. They lie in the Ring of Fire—an area surrounding the Pacific Ocean where the earth's crust often shifts. As a result, earthquakes and volcanic eruptions are common in Japan and other parts of the Ring of Fire.

MAP STUDY Japan is made up of four main islands and thousands of smaller islands.
PLACE: What are the names of Japan's four main islands?

CHINA
RUSSIA
Hokkaido
Yalu R.
Sea
of
Japan
40°N
NORTH
KOREA
Taedong
R.
Korean
Peninsula
Shinano-
Gawa R.
Honshu
Taebaek
Sanmaek
Mountains
Yellow
Sea
Han R.
JAPAN
Naktong R.
SOUTH
KOREA
Kanto
Plain
Cheju
Korea Strait
Shikoku
Mt. Fuji
12,388 ft.
(3,776 m)
N
Kyushu
Inland
Sea
W   E
East
China
Sea
PACIFIC
OCEAN
140°E
S
30°N
Ryukyu
Islands
0   150   300 mi.
0   150   300 km
130°E

ELEVATIONS

Feet	Meters
10,000	3,000
5,000	1,500
2,000	600
1,000	300
0	0

▲ Mountain peak

Bonne projection

Earthquakes Japan experiences about 5,000 earthquakes a year. Most are minor and cause little damage. But every few years, severe earthquakes shake the land. One of the worst earthquakes took place in 1995, near the port city of Kobe (KOH•bee). Thousands of people died, and damage was estimated at $50 billion to $70 billion. People in Japan and other areas of the Ring of Fire also have to face **tsunamis.** These huge sea waves caused by undersea earthquakes are very destructive along Japan's Pacific coast.

Mountains The physical map above shows you that rugged mountains and steep hills cover most of Japan. Rocky gorges and narrow valleys cut through the mountains. Thick forests grow on the lower slopes. Swift, short rivers flow through the mountains in thundering waterfalls before running to the sea.

Sea and Plains No part of Japan is more than 70 miles (113 km) from the sea. Coastlines are rough and jagged. In bay areas along the coasts, you will find many fine harbors and ports. One of Japan's most important seacoast areas lies along a body of water called the Inland Sea. This sea winds its way among the islands of Honshu, Shikoku, and Kyushu.

Narrowly squeezed between the seacoast and the mountains are the plains areas of Japan. The largest is the Kanto Plain, on the east coast of Honshu where Tokyo and Yokohama are located. You will find most of Japan's cities, farms, and industries in the coastal plains.

The Climate Ocean and wind currents affect Japan's climate. Look at the map below. You see that climate differs in northern and southern Japan. Hokkaido and northern Honshu have a humid continental climate. Cold ocean and wind currents from the Arctic area and Siberia result in cold winters here. Southern Honshu and the other two major islands have a humid subtropical climate. This climate is caused by a warm current flowing north from the Pacific Ocean.

Human/Environment Interaction
The Economy

Although Japan's land area is relatively small, the country is an economic giant. Japan is one of the world's major industrial powers. It makes use of every resource—people, technology, land, and sea. The Japanese people have one of the highest standards of living in the world.

Farming Farmland in Japan is very limited. Most of Japan's farms are small—about three acres (one ha). Japanese farmers, however, raise about 60 percent of their country's food on privately owned plots of land. How do they do this? Japanese farmers use fertilizers and modern machinery to produce high crop yields. They also practice **intensive cultivation** in which farmers grow crops on

Inland Sea, Japan

The Inland Sea is an important waterway linking the Pacific Ocean and the Sea of Japan.
PLACE: What is a tsunami?

JAPAN AND THE KOREAS: Climate

MAP STUDY
In northern Japan, the winters are cold and snowy.
REGION: What type of climate is found in northern Japan and all of North Korea?

CHINA

RUSSIA

Sapporo

Chongjin

NORTH KOREA

Sea of Japan

40°N

Pyongyang

Yellow Sea

Seoul

SOUTH KOREA

JAPAN

Tokyo

Kobe

Osaka

East China Sea

PACIFIC OCEAN

140°E

30°N

Ryukyu Islands

130°E

N
W   E
S

CLIMATE
Mid-Latitude
☐ Humid continental
■ Humid subtropical

Bonne projection

0   150   300 mi.
0   150   300 km

every available piece of land. If you visit Japan, you will see farmers growing rice on terraces cut into the hillsides. You will also see rows of crops growing between buildings and highways. Using these methods, Japanese farmers are able to plant and harvest three crops a year in the warmer regions of the country. They grow rice, wheat, fruit, vegetables, and tea.

The Japanese also rely on the sea for food. People who fish in Japan catch salmon, tuna, and snapper. Huge catches make Japan one of the world's leading fishing countries.

Cars Made in Japan

GRAPHIC STUDY

Japan produces about as many passenger cars as the United States does.

MOVEMENT: By how much did Japanese car production increase between 1965 and 1990?

Source: *The World Almanac, 1998*

Manufacturing Japan is known around the world for the variety and quality of its manufactured goods. It is a leader in manufacturing because of its use of high technology, including robots. Japan produces and exports steel, cars, ships, computers, cameras, televisions, and textiles. The graph above shows you car production in Japan since 1965. You probably own a piece of electronic equipment—TV, camera, or VCR—that was made in Japan or has Japanese parts.

In addition to the latest technology, Japan's economy has benefited from its highly skilled workers. The Japanese people value education, hard work, and cooperation. In recent years, however, Japan's economy has been hard hit by global economic troubles. Some Japanese banks have failed, industrial production has dropped, and an increasing number of Japanese workers have lost their jobs.

Trade Japan has limited mineral resources and must import oil, coal, iron ore, and other materials for manufacturing. In return, Japan exports manufactured goods to other countries. The chart on page 648 compares Japan's trade with that of other Asian countries in the Pacific Rim. The term "Pacific Rim" often refers to the economically important Asian countries that border the Pacific Ocean.

Environment Japan's industrial growth has created many environmental challenges. By the 1970s, waste products from factories had polluted the air and water. The Japanese government urged industries to prevent pollution. Today Japan's pollution-control laws are among the strictest in the world. Japan also leads the world in the development of advanced public transportation systems. You can read about Japan's famous bullet trains on page 641.

Movement
Influences of the Past

The heritage of the Japanese people is long and rich. The Japanese trace their ancestry to various **clans,** or groups of related families, that lived and ruled in Japan as early as the late A.D. 400s.

Early Japan The Japanese developed close ties with the Chinese on the Asian mainland. Ruled by emperors, Japan modeled its society on the Chinese way of life. The Japanese also borrowed the Chinese system of writing and accepted the Buddhist religion brought by Chinese missionaries.

Beginning in the 790s, the Japanese emperors' power declined. From the late 1100s to the 1860s, Japan was ruled by the **samurai**—the warrior class—and their leaders—the **shoguns.** Like China, Japan worked to keep out foreign countries that wanted to trade with them. In 1853 the United States government sent a fleet headed by Commodore Matthew Perry to Japan to demand trading privileges. On threat of war, the Japanese opened their country to the rest of the world.

Modern Japan In the late 1800s, Japan began to change into a modern country. Japanese leaders used European and American ideas to improve education and to set up industries. By the early 1900s, Japan had become the world's leading Asian power.

In the 1930s Japan needed more resources for its growing population. It began taking over land in China. In 1941 the Japanese attacked the American naval base at Pearl Harbor in Hawaii and began the Pacific conflict in World War II. The fighting between Japan and the United States and other countries ended in 1945 with Japan's defeat. After the war, many of Japan's cities lay in ruins.

Exercising at the Office

Japanese office workers in Kyoto stand for their morning exercises.
MOVEMENT: How can Japan produce so many goods when it has so few natural resources?

The Japanese economy was near collapse. With help from the United States, Japan became a democracy and rebuilt its economy.

Movement

The People

Although only the size of California, Japan has about one-half the population of the entire United States. Look at the graph below to compare Japan's population of 126.4 million people with that of the United States. About 78 percent of Japan's people are crowded into urban areas on the coastal plains. More than one-third of these city dwellers live in four large cities located on the plains of Honshu. These four cities are Tokyo, Yokohama, Osaka, and Nagoya. They form a **megalopolis,** a huge supercity made up of several large cities and the smaller cities near them.

City Life Japanese cities are very modern. They are crowded with tall office buildings, busy streets, and freeways. Homes are small and close to one another. In the suburbs, you will find small apartment buildings and wooden houses with tiny gardens.

In Japan's cities, the past remains very much a part of urban life. Along the narrow side streets of most cities, small shops sell traditional items. Parks, gardens, and religious shrines are found close to modern neighborhoods. In Tokyo and other cities, it is common to see a person dressed in a traditional garment known as a kimono walking with another person wearing a T-shirt and jeans!

Country Life Only 22 percent of Japan's people live in rural areas. These people generally earn less than city dwellers, but their standard of living has improved in recent years. Most have a small house, a TV, and modern kitchen

Paper Art

People in Japan make an art—called origami—out of folding paper. From paper they create decorative objects such as animals, fish, or flowers. There are about 100 traditional origami patterns. Each year new patterns are developed.

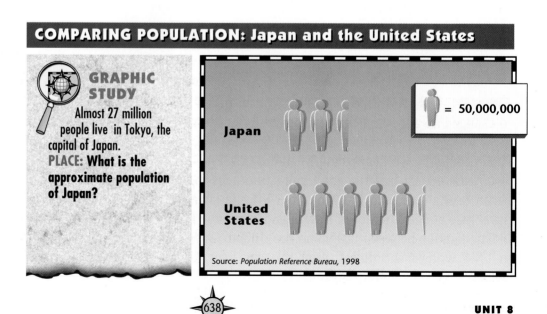

COMPARING POPULATION: Japan and the United States

GRAPHIC STUDY

Almost 27 million people live in Tokyo, the capital of Japan.
PLACE: What is the approximate population of Japan?

Japan

United States

= 50,000,000

Source: *Population Reference Bureau, 1998*

appliances. In some rural areas, a small farmer might have three jobs. For example, a person might farm one day, fish or cut timber another day, and work occasionally in a factory.

Family Life In traditional Japan the family was the center of one's life. Each family member had to obey certain rules. Grandparents, parents, and children all lived in one house. In recent years much has changed. Family ties remain strong, but each family member is allowed more freedom. Today many Japanese also live in small family groups that consist only of parents and children.

Fifteen-year-old Michiko Wakita lives with her parents and older brother Naohiro in Tokyo. Michiko, like most Japanese teenagers, studies hard six days a week in school. After graduation from high school, she plans to study marine biology at a local university.

Despite Japan's strong emphasis on education, life is not all work for Michiko and other Japanese young people. She enjoys rock music, modern fashions, television, and movies. Her brother Naohiro likes motorcycles and comic books. When Michiko isn't watching game shows on television, Naohiro watches baseball and sumo, a Japanese form of wrestling. Judo is also popular. It is a form of martial arts that developed from ancient Asian fighting skills. Today martial arts are practiced for self-defense and for exercise. They are also very popular in the United States.

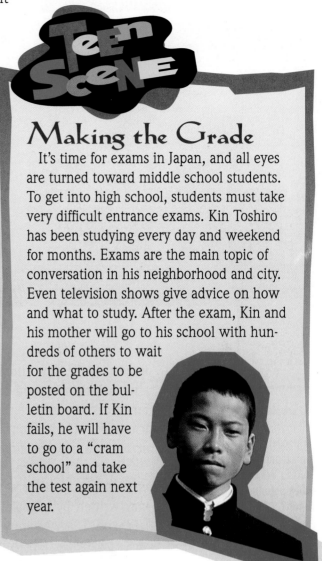

Making the Grade

It's time for exams in Japan, and all eyes are turned toward middle school students. To get into high school, students must take very difficult entrance exams. Kin Toshiro has been studying every day and weekend for months. Exams are the main topic of conversation in his neighborhood and city. Even television shows give advice on how and what to study. After the exam, Kin and his mother will go to his school with hundreds of others to wait for the grades to be posted on the bulletin board. If Kin fails, he will have to go to a "cram school" and take the test again next year.

Place
Culture

Over the centuries the Japanese have developed many customs and art forms to express their delight in nature. They have passed on some of these traditions to the rest of the world. At the same time, the Japanese have accepted cultural practices from foreign countries. Many of Japan's people, for example, enjoy Western movies, hamburgers, and rock music.

Religion Japan has two major religions—Shinto and Buddhism. Shinto began in Japan centuries ago. It teaches a respect for nature, a love of simple things, and a concern for cleanliness and good manners. Buddhism came to Japan from China. It teaches respect for nature and the need for inner peace. Many Japanese practice both religions.

The people of Japan love baseball *(left)*. A colorful float used for puppet shows is part of a festival parade in Japan *(right)*. **PLACE: What are some other popular pastimes in Japan?**

The Arts You may sense a peaceful feeling of nature in Japanese painting, music, dance, and literature. Many paintings show scenes of the countryside and include verses in calligraphy. The Japanese enjoy traditional and modern forms of drama. Since the 1600s Japanese theatergoers have attended the historical plays of the Kabuki theater. Kabuki plays are set on colorful stages with actors wearing brilliantly colored costumes.

Japanese literature includes poetry and novels. Many scholars believe that the world's first novel came from Japan. This novel is called *The Tale of Genji* and was written by a noblewoman of the emperor's court about A.D. 1000. Haiku (HY•koo) is a well-known type of Japanese poetry that is written according to a very specific formula. A haiku has only 3 lines and 17 syllables.

SECTION 1 ASSESSMENT

REVIEWING TERMS AND FACTS

1. **Define the following:** archipelago, tsunami, intensive cultivation, clan, samurai, shogun, megalopolis.

2. **PLACE** What is the major landform of Japan?

3. **HUMAN/ENVIRONMENT INTERACTION** What are the main economic activities that take place in Japan?

4. **MOVEMENT** Where do most Japanese live?

MAP STUDY ACTIVITIES

5. Look at the physical map on page 634. About how many miles is Japan from north to south?

6. Study the climate map on page 635. What are Japan's two climate regions?

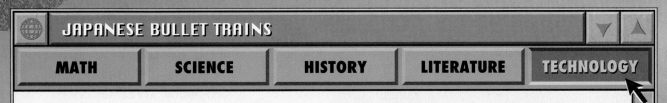

MAKING CONNECTIONS

MATH | **SCIENCE** | **HISTORY** | **LITERATURE** | **TECHNOLOGY**

What travels almost as fast as a speeding bullet? Japan's famous bullet trains first whizzed across the Japanese countryside in 1964. The trains completely changed ground travel in Japan. They also became the model for public transport vehicles in other densely populated countries.

Speeding Along

FROM TOKYO TO OSAKA The first blue-and-ivory bullet-nosed train raced at 125 miles per hour (242 km per hour) between Tokyo and Osaka. These two cities lie 320 miles (515 km) apart. The new train gained instant popularity. The number of people riding the train daily grew by 300 percent during its first five years.

THE BULLETS INCREASE The bullet train became the first Japanese train to run at a profit. It was so successful that other trains were built in different parts of Japan. About 260 bullet trains run daily now, covering a distance equal to circling the earth three times. Lines stretch from Tokyo to cities north and south. Eventually a bullet train will run under the sea through a tunnel to Hokkaido, the northernmost Japanese island.

MEETING CHALLENGES The bullet trains carry 400,000 passengers every day—135 million each year—at speeds up to 170 miles per hour (274 km per hour). They have a 99 percent on-time record, and have never had a fatality.

The trains help the Japanese successfully overcome several challenges. They streamline transportation in tiny, crowded Japan. Most Japanese now prefer the comfort and speed of the trains to cars. As a result, the popular trains help to keep Japan's overseas oil purchases under control.

Making the Connection

1. How fast and how far do bullet trains travel daily?
2. What problems in Japan do the bullet trains help solve?

The Two Koreas

Words to Know

- anthracite

Places to Locate

- North Korea
- South Korea
- Seoul
- Pyongyang

Read to Learn . . .

1. where the Korean Peninsula is located.
2. why the Koreas are divided.
3. how life in South Korea differs from life in North Korea.

Forests—often shrouded in mist—cover much of Korea. You feel a sense of peace and calm as you look at the mountainous Korean landscape. Because of its geography, Korea was once known as "land of the morning calm."

In modern times Korea has been anything but calm. Today it is divided into two nations—communist North Korea and noncommunist South Korea. The two governments of Korea are bitter enemies, although efforts are being made to bring them together.

Region

A Divided Country

Although Korea is now divided, it has a long history as a united country. The land of Korea lies on the Korean Peninsula. This peninsula juts out from northern China, between the Sea of Japan and the Yellow Sea.

The Korean Past The Koreans trace their ancestry to people who settled the Korean Peninsula thousands of years ago. From A.D. 668 to 935, a Korean kingdom called Silla united much of the peninsula for the first time under one government. Under Silla rulers, Korea enjoyed a period of cultural and scientific advances. For example, the Koreans built one of the world's earliest astronomical observatories in the A.D. 600s.

For centuries the Korean Peninsula was a bridge between Japan and the mainland of Asia. Trade and ideas went back and forth. China and Japan, however, wanted to control Korea. China ruled part of Korea from the 100s B.C. until the early A.D. 300s. Japan later governed Korea from 1910 until the end of World War II in 1945.

Division of Korea After Japan's defeat in World War II, troops from the communist Soviet Union took over the northern half of Korea. American troops occupied the southern half. Korea eventually divided along the 38th parallel. A noncommunist government ruled in what became South Korea, while a communist government governed what became North Korea.

The Korean War In 1950 North Korean armies attacked South Korea. Their plan was to bring all of Korea under communist rule—a plan that led to the Korean War. Noncommunist United Nations countries, led by the United States, rushed to aid South Korea. China's communist leaders eventually sent troops to help North Korea. The fighting finally ended in 1953—without a peace treaty or a victory for either side. By the 1960s, two separate countries with their own ways of life had developed in the Korean Peninsula. Today the United States continues to encourage peaceful relations between the two Koreas.

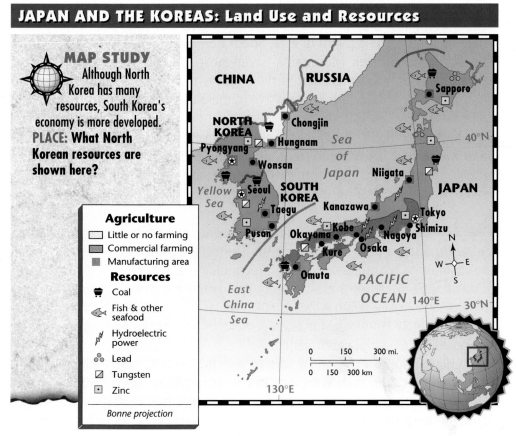

JAPAN AND THE KOREAS: Land Use and Resources

MAP STUDY Although North Korea has many resources, South Korea's economy is more developed. **PLACE: What North Korean resources are shown here?**

Agriculture
- ☐ Little or no farming
- ▨ Commercial farming
- ■ Manufacturing area

Resources
- 🛒 Coal
- 🐟 Fish & other seafood
- ⚡ Hydroelectric power
- ⚬⚬ Lead
- ▨ Tungsten
- ⊡ Zinc

Bonne projection

South Korea's modern capital city is home to almost one-fourth of its people.
PLACE: What are some of South Korea's most important exports?

Place
South Korea

South Korea lies at the southern end of the Korean Peninsula. About 38,120 square miles (98,731 sq. km) in area, South Korea is slightly larger than Indiana. The Taebaek Sanmaek Mountains cover most of central and eastern South Korea. Plains with hills and fertile river valleys spread along the southern and western coasts. Most South Koreans live in these coastal areas.

The Climate South Korea has a climate affected by monsoons. As a result, fierce winds and heavy rains are common to South Koreans. During the summer, a monsoon from the south brings hot, humid weather. In winter, a monsoon blows in from the north, bringing cold weather.

The Economy South Korea's mineral resources include iron ore, tungsten, and **anthracite**, a hard coal. During the 1980s, South Korea had a strong industrial economy. Manufacturing, high-technology, and service industries grew tremendously. Many of these privately run companies made products for overseas markets. Today South Korea is still a major exporter of ships, cars, textiles, and computers. However, mismanagement in key banks and businesses has weakened its industrial performance.

Agriculture also plays an important role in South Korea's economy. South Korean farmers own their land. Most of their farms are very small—about 2.7 acres (1.1 ha) in size. The major crops are rice, apples, barley, Chinese cabbage, onions, and potatoes. Many farmers add to their income by fishing. The map on page 643 shows you that South Korea, like Japan, is a major fishing country.

The People The people of the two Koreas belong to the same Korean ethnic group. South Korea has a population of 47 million people. About 79 percent of South Koreans live in cities and towns in the coastal plains. Since the mid-1950s, South Korea's capital of Seoul has grown into a busy, modern city of more than 11 million people.

Most people in South Korea's cities live in tall apartment buildings and modern homes. Many own cars, but they also rely on buses, subways, and trains to get from one place to another. In rural areas many people live in small, one-story houses made of bricks or concrete blocks. Some rural dwellers still follow traditional ways. A large number of South Koreans have also emigrated to the United States since the end of the Korean War.

Religion and the Arts South Koreans enjoy freedom of religion. Although most practice Buddhism or the teachings of Confucius, some are Christians. The traditional arts of Korea were influenced by Chinese religion and culture. The Koreans, however, have developed their own culture. The most widely practiced art forms are music, poetry, pottery, sculpture, and painting.

If you visit South Korea, you can see examples of traditional Korean art and architecture. In Seoul stand ancient palaces that were modeled after the Imperial Palace in Beijing, China. Ancient Buddhist temples dot the hills and valleys of the Korean countryside. Within these temples are beautifully sculpted figures of the Buddha in stone, iron, and gold. One of the great achievements of early Koreans was well-crafted pottery. Even today, Korean potters still make bowls and dishes that are admired throughout the world.

Towering Buddha

Statues of the Buddha are common sights at temples in South Korea.
MOVEMENT: What two neighboring countries have had a major influence on the history and culture of Korea?

Place
North Korea

Communist North Korea lies at the northern end of the Korean Peninsula. Slightly larger than South Korea, North Korea covers 46,490 square miles (120,409 sq. km). Rugged mountains run through the center of North Korea. A group of narrow valleys separate these ranges. Plains and lowlands run along the western and eastern coasts.

The Climate Monsoons also affect climate in North Korea. Most of North Korea has hot summers and cold, snowy winters. Because the central mountains block the full effects of the winter monsoon, the east coast generally has warmer winters than the rest of the country.

The Economy Like South Korea, North Korea has an industrial economy. With more minerals than the south, the northern part of the peninsula was the major industrial area for many years. The map on page 643 shows you that North Korea has minerals such as anthracite, tungsten, and zinc. After the 1950s, however, South Korea surpassed North Korea in industrial growth because of its dynamic, free enterprise economy.

In North Korea the communist government owns and runs factories, businesses, and farms. Under the government's direction, North Korean industrial workers produce chemicals, iron and steel, machinery, and textiles. Most farmers work on government-run farms, where they raise livestock and grow rice and vegetables. Rice is the basic food item. A popular Korean side dish is *kimchi*, a spicy blend of Chinese cabbage, onions, and other vegetables.

The People North Korea has about 22.2 million people. About 59 percent of the population live in urban areas along the coasts and river valleys. Pyongyang is North Korea's capital and largest city. Largely rebuilt since the Korean War, Pyongyang has many modern buildings and monuments to communist leaders.

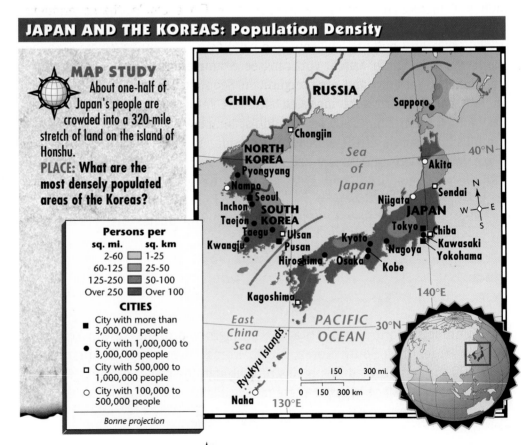

JAPAN AND THE KOREAS: Population Density

MAP STUDY
About one-half of Japan's people are crowded into a 320-mile stretch of land on the island of Honshu.
PLACE: What are the most densely populated areas of the Koreas?

Persons per

sq. mi.		sq. km
2-60		1-25
60-125		25-50
125-250		50-100
Over 250		Over 100

CITIES
■ City with more than 3,000,000 people
● City with 1,000,000 to 3,000,000 people
□ City with 500,000 to 1,000,000 people
○ City with 100,000 to 500,000 people

Bonne projection

CHINA RUSSIA Sapporo Chongjin NORTH KOREA Sea of Japan 40°N Akita Pyongyang Nampo Seoul Sendai Inchon Niigata JAPAN Taejon SOUTH KOREA Tokyo Chiba Taegu Ulsan Kyoto Kawasaki Kwangju Pusan Nagoya Yokohama Hiroshima Osaka Kobe 140°E Kagoshima East China Sea PACIFIC OCEAN 30°N Ryukyu Islands 0 150 300 mi. 0 150 300 km Naha 130°E

A modern hotel is surrounded by office buildings and apartments in North Korea's capital. **PLACE: Before Korea was divided, why was most industry located in North Korea?**

Under communist rule North Koreans have had to put aside much of their traditional culture in favor of communist ways. For example, the government discourages the practice of religion, although many people still hold on to their beliefs. The government also places the needs of the communist system over the needs of individuals or families. During the 1990s North Korea's government and economy became unstable. Longtime leader Kim Il Sung died in 1994. His son Kim Jong Il succeeded him after a power struggle. By the late 1990s, crop failures resulted in severe food shortages for many North Koreans.

SECTION 2 ASSESSMENT

REVIEWING TERMS AND FACTS

1. Define the following: anthracite.

2. LOCATION What body of water borders Korea on the east?

3. MOVEMENT How did Korea's location on a peninsula affect its history?

4. HUMAN/ENVIRONMENT INTERACTION What farm products are grown in South Korea?

MAP STUDY ACTIVITIES

5. Look at the land use map on page 643. What are three of South Korea's large manufacturing areas?

6. Turn to the population density map on page 646. What are the two most densely populated South Korean cities shown?

BUILDING GEOGRAPHY SKILLS

Using a Map and a Graph

A map and a graph both show you a picture. They can show geographic information but in different ways. Maps show locations, while graphs display information—usually data in numbers—about those locations. Comparing maps and graphs can help you understand more about a location. Comparing also helps show relationships between locations. In using a map and a graph, apply these steps:

- Read the map/graph title to see how the two are related.

- On the map, locate places included in the graph.

- Use information on the graph to draw conclusions about the places on the map. Use the map to help explain the information on the graph.

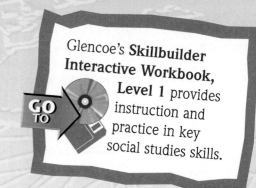

Glencoe's **Skillbuilder Interactive Workbook, Level 1** provides instruction and practice in key social studies skills.

ASIA'S PACIFIC RIM

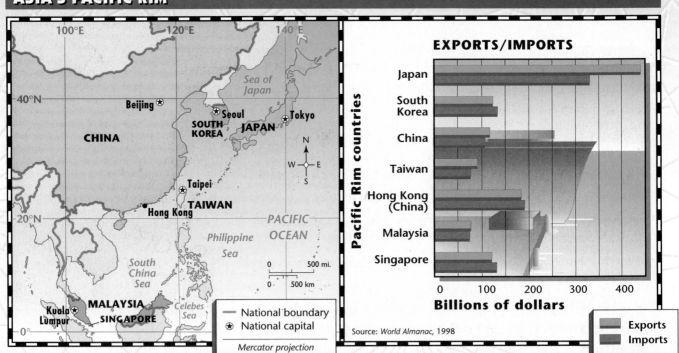

EXPORTS/IMPORTS

Pacific Rim countries: Japan, South Korea, China, Taiwan, Hong Kong (China), Malaysia, Singapore

Billions of dollars: 0, 100, 200, 300, 400

Source: *World Almanac*, 1998

- National boundary
- ★ National capital

Mercator projection

Exports
Imports

Geography Skills Practice

1. How are the map and the graph related?

2. Which country shown on the map and listed on the graph has the largest land area?

3. Which country has the most exports and imports?

4. How does the physical geography of this region explain the trading activity shown on the graph?

Chapter 24 Highlights

Important Things to Know About Japan and the Koreas

SECTION 1 JAPAN

- Japan is made up of four main islands and thousands of smaller islands.
- Japan lies on the Pacific Ring of Fire and experiences many earthquakes.
- Japan's strong economy is based on manufacturing and trade.
- Japan—a densely crowded country—holds a population almost half the size of that of the United States.
- The Japanese follow modern ways of life but keep many of their traditions.
- The religions of Shinto and Buddhism have influenced the Japanese arts.

SECTION 2 THE TWO KOREAS

- The Korean Peninsula is divided between communist North Korea and noncommunist South Korea.
- After World War II, communist troops from the Soviet Union took over the northern half of Korea. American troops occupied the southern half. Korea eventually divided along the 38th parallel.
- Both Koreas have inland mountains and coastal plains.
- Monsoon winds affect the climate of the Koreas, bringing hot weather in the summer and cold weather in winter.
- South Korea has a free enterprise economy that exports many industrial goods to other countries.
- A communist government controls North Korea's economy.
- South Korea has about 46.4 million people; North Korea, about 22.2 million.
- Most South Koreans practice Buddhism or the teachings of Confucius. Under communism, North Koreans are discouraged from practicing religion.

Tokyo, Japan ▶

REVIEWING KEY TERMS

Match the numbered terms in Set A with their lettered definitions in Set B.

A

1. megalopolis
2. anthracite
3. shogun
4. samurai
5. intensive cultivation
6. archipelago
7. tsunami

B

A. an urban area made up of several large cities
B. the warrior class of early Japan
C. huge sea wave created by an undersea earthquake
D. farming nearly every area of land
E. a group of islands
F. hard coal
G. Japanese military leader

REVIEWING THE MAIN IDEAS

Mental Mapping Activity

Draw a freehand map of Japan and the two Koreas. Label the following:

- Honshu
- North Korea
- Pacific Ocean
- Kyushu
- Kobe
- Tokyo

Section 1

1. **REGION** What is the Ring of Fire?
2. **PLACE** Why does Hokkaido have cold winters?
3. **HUMAN/ENVIRONMENT INTERACTION** Why are Japanese farmers able to produce high crop yields?
4. **HUMAN/ENVIRONMENT INTERACTION** How has Japan tried to improve its environment?
5. **PLACE** What do the religions of Shinto and Buddhism teach?

Section 2

6. **REGION** Why was Korea divided after World War II?
7. **REGION** How do monsoons affect climate in the two Koreas?
8. **PLACE** What is South Korea's capital and largest city?
9. **LOCATION** Where is most mineral wealth on the Korean Peninsula found?
10. **PLACE** How many North Koreans live in urban areas?

CRITICAL THINKING ACTIVITIES

1. **Drawing Conclusions** Do you think other countries would want to model their economies after Japan's economy? Explain.
2. **Making Comparisons** How do ways of life differ in North Korea and South Korea?

CONNECTIONS TO ECONOMICS

Cooperative Learning Activity The United States is an important trading partner of Japan and South Korea. Work in groups to find out what products from Japan and South Korea the members of your group use. Look at home and school for products made in these two countries. Make a list of the products and where they were made. As a group, prepare a list of your findings and share the list with the class.

GeoJournal Writing Activity

A haiku is a three-line poem, usually about nature and human emotions. The traditional haiku requires 17 syllables—5 in the first line, 7 in the second line, and 5 in the third line. In your journal, write your own haiku in which you poetically sketch a scene from nature, such as a snowfall or a sunrise.

TECHNOLOGY ACTIVITY

Building a Database Collect information about North Korea and South Korea. Organize your findings in a database comparing the two countries. Include information about population, major imports and exports, literacy rate, birthrate, death rate, and life expectancy. Use your database to write a paragraph that summarizes the differences and similarities between the two Koreas.

PLACE LOCATION ACTIVITY: JAPAN AND THE TWO KOREAS

Match the letters on the map with the places and physical features of Japan and the two Koreas. Write your answers on a separate sheet of paper.

1. Seoul
2. Honshu
3. Taebaek Sanmaek Mountains
4. Hokkaido
5. East China Sea
6. Inland Sea
7. Sapporo
8. Tokyo
9. Pyongyang
10. Sea of Japan

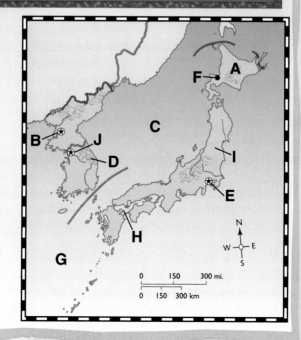

Cultural HERITAGE: ASIA

MUSIC ▶ ▶ ▶ ▶

The music of India is played mostly on guitarlike stringed instruments and drums. Ancient melodies, called *ragas,* sound very different from Western music because they use a different scale.

ARTIFACTS ▲ ▲ ▲ ▲

This tiny stone carving was one of many discovered at Mohenjo-Daro in the Indus River valley. At least 4,000 years old, it was probably used by traders as a stamp or seal to mark their merchandise.

ARCHITECTURE ▶ ▶ ▶ ▶

Buddhist pagodas can be found throughout southern Asia. Most pagodas are eight-sided with an odd number of stories. Each story has an umbrella-shaped roof made of tiles.

◀ ◀ ◀ ◀ THEATER

Actors with wigs, painted faces, and elaborate costumes perform in Japan's popular Kabuki theaters. Kabuki performers act out historic events or real-life experiences in a lively, dramatic way. All Kabuki actors are men who begin their training at a very young age.

ISLAMIC ARCHITECTURE ▶ ▶ ▶ ▶

The magnificent Taj Mahal was built by an Indian ruler in the 1600s as a tomb and monument to his beloved wife. The white marble dome and minarets reflect the Islamic influence in India at the time. It took 20,000 laborers almost 20 years to complete this elaborate structure.

◀ ◀ ◀ ◀ DECORATIVE ART

These colorful, hand-painted umbrellas are an art form in Thailand. The umbrellas are made from paper and bamboo and decorated with colorful flowers, dragons, and birds.

APPRECIATING CULTURE

1. Compare and contrast the architecture of the Taj Mahal and the Buddhist pagoda.

2. The animal seal from Mohenjo-Daro was probably used as a stamp or trademark to indicate ownership of goods. Design your own personal trademark.

Chapter 25 Southeast Asia

SOUTHEAST ASIA: Political

100°E · 110°E · 120°E · 130°E

ASIA

Tropic of Cancer

Mandalay

MYANMAR

Chiang Mai · Hanoi

LAOS

Yangon · Vientiane

South China Sea

20°N

THAILAND · Da Nang

Bangkok · VIETNAM · Quezon City · Philippine Sea

CAMBODIA · Manila

Phnom Penh · PHILIPPINES

Ho Chi Minh City · Iloilo · Cebu

PACIFIC OCEAN

Pinang · 10°N

Zamboanga · Davao

Medan · BRUNEI

Kuala Lumpur · Bandar Seri Begawan

Singapore · MALAYSIA · Celebes Sea

Equator · SINGAPORE

Pontianak · 0°

INDIAN OCEAN · Palembang · INDONESIA

Java Sea

Jakarta

Surabaya

| National boundary |
| National capital |
| Other city |

Mercator projection

0 250 500 mi.
0 250 500 km

MAP STUDY ACTIVITY

As you read Chapter 25, you will learn about Southeast Asia.

1. What countries in Southeast Asia border China?
2. What Southeast Asian countries are made up of islands?
3. What three countries share one island?

Mainland Southeast Asia

PREVIEW

Words to Know
- alluvial plain
- deforestation
- delta

Places to Locate
- Myanmar
- Thailand
- Laos
- Cambodia
- Vietnam
- Bangkok
- Mekong River

Read to Learn . . .
1. what countries make up mainland Southeast Asia.
2. what minerals are found in mainland Southeast Asia.
3. why the economies of Laos, Cambodia, and Vietnam are not fully developed.

Imagine visiting a village deep in the forests of Myanmar, a country in Southeast Asia. You would see that many village homes are built on poles. Homes are built this way to protect the occupants from floods and wild animals.

South of China and east of India lies a region known as Southeast Asia. The peninsulas and thousands of islands that make up this region cover an area of about 1,750,000 square miles (4,536,000 sq. km). Including its seas, Southeast Asia spreads over an area as big as the contiguous United States. On the peninsulas of mainland Southeast Asia, you will find the countries of Myanmar, Thailand, Laos, Cambodia, Vietnam, and part of Malaysia.

Location
Myanmar

Southeast Asia's northernmost country is Myanmar. Once called Burma, Myanmar for many years was part of British India. It became an independent republic in 1948. Since independence, military leaders have turned Myanmar into a socialist country. The government runs the economy and forbids any criticism of its policies. It also limits Myanmar's contacts with the outside world.

The Land and Economy Myanmar's 253,880 square miles (657,549 sq. km) make it about the size of Texas. Rugged mountains run through the eastern

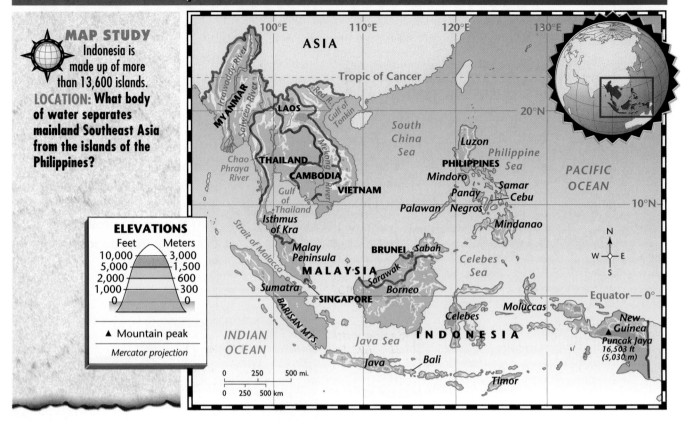

MAP STUDY
Indonesia is made up of more than 13,600 islands.
LOCATION: What body of water separates mainland Southeast Asia from the islands of the Philippines?

ELEVATIONS

Feet	Meters
10,000	3,000
5,000	1,500
2,000	600
1,000	300
0	0

▲ Mountain peak

Mercator projection

and western parts of the country. Between the mountain ranges flow two major rivers—the Irrawaddy (IHR•uh•WAH•dee) and the Salween. The land along these rivers contains Myanmar's most fertile soil.

Southern Myanmar has tropical climates. Monsoon winds from the Indian Ocean bring rain during Myanmar's summers. Dry monsoons from the north occur during winter. As is typical in high elevations, temperatures are cooler in Myanmar's northern mountains.

Myanmar's economy is based on farming. Rice is the major food crop. Some farmers work their fields with tractors, but most rely on plows drawn by water buffalo. Forests cover large areas of Myanmar and provide about 80 percent of the world's teakwood. Mines yield minerals such as tin, copper, and tungsten. Gemstones—rubies, sapphires, and jade—also come from Myanmar.

The People About 75 percent of Myanmar's 47.1 million people live in small villages. They build their homes on poles above the ground to protect themselves from floods and wild animals. The most densely populated area of the country is the fertile Irrawaddy River valley.

Many rural people in Myanmar follow traditional ways. In cities you will notice a blend of old and new. The capital and largest city, Yangon, is famous for both its modern university and its gold-covered Buddhist temples. Buddhism is the major religion of Myanmar.

Thailand

Southeast of Myanmar lies Thailand. The map on page 654 shows you that Thailand looks like a flower on a stem. The "flower" is the northern part of Thailand, located on the mainland. The "stem" is a very narrow strip of Thailand that runs southward to the Malay Peninsula.

Once called Siam, Thailand means "land of the free." It is the only Southeast Asian country that has never been a European colony. The people of Thailand have skillfully blended modern ways with traditional practices and a loyalty to Thailand's royal family.

The Land and Economy Mountains, plateaus, and river valleys spread through northern Thailand. In the central part of the country, the Chao Phraya (chow•PRY•uh) River has formed a large alluvial plain. An **alluvial plain** is a land area built up by a river's soil deposits.

Because of its fertile soil, central Thailand has most of the country's people, farms, and industries. Farther south the thin strip of Thailand along the Malay Peninsula boasts thick forests, rubber trees, and mineral wealth.

The map below shows you that Thailand has tropical and subtropical climates. Monsoon winds bring dry and wet seasons. Rainfall is heaviest in the south and southeast. Thailand is an agricultural country. Thai farmers grow rice

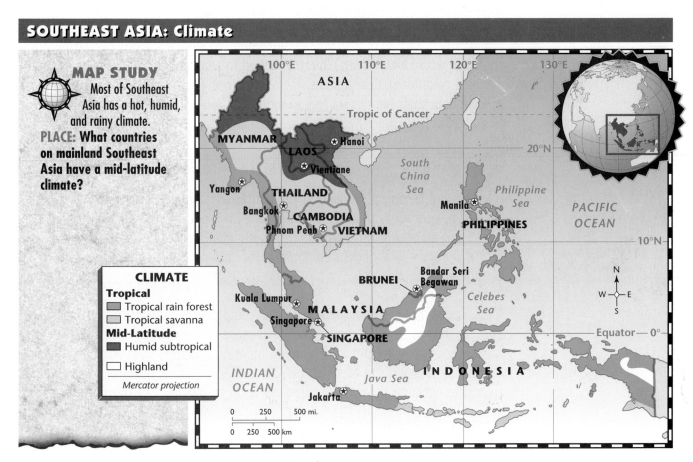

SOUTHEAST ASIA: Climate

MAP STUDY
Most of Southeast Asia has a hot, humid, and rainy climate.
PLACE: What countries on mainland Southeast Asia have a mid-latitude climate?

CLIMATE
Tropical
- Tropical rain forest
- Tropical savanna
Mid-Latitude
- Humid subtropical
- Highland

Mercator projection

Traffic on the crowded streets of Thailand's capital has created an air pollution problem. **HUMAN/ENVIRONMENT INTERACTION: What percentage of Thailand's people live in cities?**

as their major crop. Plantations in the south provide rubber for export. Teak and other wood products come from Thailand's forests. In recent years developers have cut down so many forests that many Thai people fear the complete loss of one of their country's most valuable resources. The government now tries to limit **deforestation,** or the widespread cutting of trees.

Thailand is rich in mineral resources. It is one of the world's leading exporters of tin and gemstones. Most of Thailand's manufacturing is located around Bangkok, the capital and largest city. Thai workers there produce textiles, cement, and paper products.

The People Most of Thailand's 61.1 million people belong to the Thai ethnic group and follow the Buddhist religion. Hundreds of Buddhist temples called *wats* dot cities and the countryside. Bangkok alone has more than 300 Buddhist temples! Every morning Buddhist monks, or holy men, carrying small bowls and wearing yellow-orange robes walk among the people to receive food offerings. Buddhism inspires many Thai festivals celebrated with parades and colorful flowers.

About 70 percent of Thailand's people live in rural villages. Bangkok, however, is growing rapidly as more and more people move into the city from rural areas. Today Bangkok has about 6.5 million people. Thirteen-year-old Chuan Soonsiri notices the contrasts of life in Bangkok. He points out beautiful Buddhist temples and royal palaces next to modern skyscrapers and crowded streets filled with cars and people. Although he worries about the air pollution that gets worse every year, Chuan is proud to see dancers in elaborate costumes perform traditional Thai dances. He also enjoys Thai boxing, in which the boxers use both their hands and feet to fight.

Place
Laos

Laos lies north of Thailand in the center of mainland Southeast Asia. Once a French colony, Laos became independent in 1953. A civil war soon tore Laos apart, and a communist government finally came to power in 1975. In the late 1980s the government relaxed some of its rigid policies and opened Laos to the outside world.

The Land and Economy The only landlocked country in Southeast Asia, Laos is covered by rugged mountains. Because of the cooling effect of the mountains, Laos has a humid subtropical climate. Southern Laos includes a fertile farming area along the Mekong (MAY•KAWNG) River. Southeast Asia's longest river, the Mekong provides landlocked Laos with access to the ocean. It flows about 2,600 miles (4,180 km) before emptying into the South China Sea.

Laos's communist government runs the country's economy, which is mostly agricultural. Farmers grow fruits, vegetables, and rice along the fertile banks of the Mekong River. Industry is still largely undeveloped because of isolation and years of civil war.

The People About 80 percent of Laos's 5.3 million people live in villages and towns along the Mekong River. You find Vientiane (vyehn•TYAHN), the largest city and capital, near the river. The communist government discourages religion, but most of Laos's people remain Buddhists.

Place
Cambodia

South of Thailand and Laos is Cambodia. About 1,000 years ago, Cambodia was the center of a large empire that ruled much of Southeast Asia. Today you can still see the remains of Angkor, the ancient empire's capital. In modern times Cambodia was under French rule, finally becoming independent in 1953.

Cambodians have faced almost constant warfare since the 1960s. A communist government took control of the country in the mid-1970s, and the people suffered great hardships. More than 1 million Cambodians died. Many fled overseas to the United States and other countries. Since the mid-1990s political disputes have divided the country.

The Land and Economy Low mountains, lakes, and forested plains stretch across Cambodia. The Mekong River and a number of smaller rivers crisscross the country, providing water, fertile soil, and waterways for transportation. The map on page 657 shows you that Cambodia has tropical climates.

For many years Cambodia was a rich farming country that exported rice and rubber. By the 1980s its economy was in ruins because of war and harsh communist rule. Today the government is trying to rebuild the economy. Rice and soybeans are the major crops.

The People Most of Cambodia's 10.8 million people belong to the Khmer (kuh•MEHR) ethnic group. They live in rural villages and farm the land. Only about 14 percent live in cities such as the capital, Phnom Penh (NAHM PEHN). Buddhism is the major religion of Cambodia.

Teen Scene

Happy New Year!

In Thailand the New Year is celebrated in April, one of the hottest months of the year. Vina Charas and her sister Uma look forward to the celebration. This year they will be in their village parade. After the parade, the fun begins. On New Year's Day, people drench anyone and everyone with buckets of water. The tradition started many years ago by people hoping for rain for the crops. Young people splash anyone who comes near. "No one seems to mind because it's so hot!" Vina says.

Rush-hour traffic in Vietnam's largest city means curb-to-curb mopeds.
PLACE: What is the capital of Vietnam?

Place
Vietnam

Vietnam lies along the east coast of mainland Southeast Asia. A few decades ago, Americans talked about Vietnam every day. United States forces were fighting a war there. Today Vietnam is a communist country, but its leaders are opening Vietnam to Western ideas and practices.

The Land and Economy Vietnam has a long coastline bordering the South China Sea and the Gulf of Thailand. Rugged mountains and high plateaus cover most of Vietnam. The rest of the country is made up of coastal lowlands and river deltas. A **delta** is an area of land formed by soil deposits at the mouth of a river. Two important river deltas lie at opposite ends of Vietnam. One is the Red River delta in the north; the other is the Mekong River delta in the south. Vietnam has tropical and subtropical climates. Monsoons bring wet and dry seasons.

Vietnam's communist government runs the economy. But it has reduced controls since the late 1980s. Coal, zinc, and other mineral resources are plentiful in north and central Vietnam. Most of Vietnam's industries are located there. Years of warfare have kept Vietnam's industries from fully developing. The country is rich in fertile soil, and agriculture is the major economic activity.

The People About 85 percent of Vietnam's 78.5 million people belong to the Vietnamese ethnic group. The rest are Chinese, Cambodians, and other Asian ethnic groups. Most Vietnamese are crowded along the coasts and river deltas. The largest cities are Ho Chi Minh (HOH chee MIHN) City in the Mekong delta, and Hanoi, the capital, in the north. The most widespread religion is Buddhism.

SECTION 1 ASSESSMENT

REVIEWING TERMS AND FACTS
1. **Define the following:** alluvial plain, deforestation, delta.
2. **LOCATION** Where is Myanmar located?
3. **REGION** How do monsoons affect climate in mainland Southeast Asia?
4. **PLACE** What major food crop is grown in mainland Southeast Asia?

MAP STUDY ACTIVITIES
5. Look at the physical map on page 656. What river flows through Cambodia?
6. Study the map on page 657. What are the major climates of mainland Southeast Asia?

BUILDING GEOGRAPHY SKILLS

Reading a Contour Map

Hills and valleys have highs and lows. How do you show this on a flat map? One way is with a contour map. Contour maps use lines, called *isolines,* to outline the shape—or contour—of the landscape. Each isoline connects all points that are at the same elevation. If you walked along one isoline, you would always be at the same height above sea level.

Where isolines are far apart, the land rises gradually. Where they are close together, the land rises steeply. To read a contour map, follow these steps:

- Name the area shown on the map.
- Read the numbers on the isolines to find out how much the elevation increases with each line.
- Identify areas of steep and level terrain.

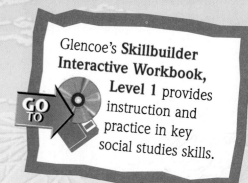

Glencoe's **Skillbuilder Interactive Workbook, Level 1** provides instruction and practice in key social studies skills.

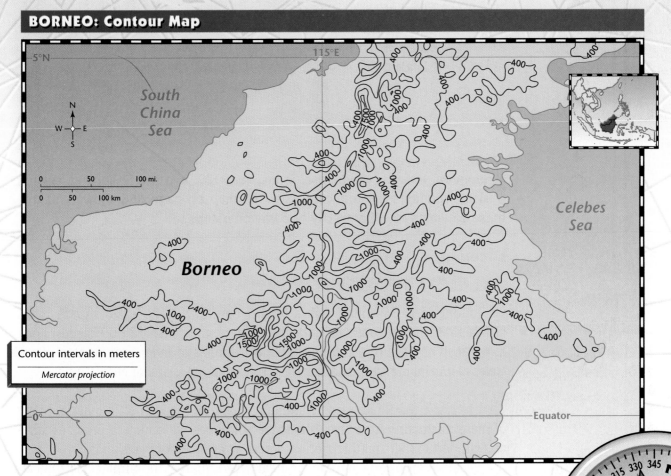

BORNEO: Contour Map

Contour intervals in meters

Mercator projection

Geography Skills Practice

1. What area of Southeast Asia is shown on the map?

2. What elevations are shown by the isolines?

Island Southeast Asia

PREVIEW

Words to Know
- cassava
- free port
- abaca

Places to Locate
- Indonesia
- Malaysia
- Singapore
- Brunei
- Philippines
- Jakarta
- Manila
- Java

Read to Learn . . .
1. what groups of people live in island Southeast Asia.
2. why people in Malaysia, Singapore, and Brunei have high standards of living.

This volcano is part of a chain on the island of Java, one of thousands of islands that make up the country of Indonesia. The islands of Southeast Asia lie along the Pacific Ocean's Ring of Fire. People living here constantly face the threat of volcanic activity and earthquakes.

Indonesia, Malaysia, Singapore, Brunei (bru•NY), and the Philippines are the island countries of Southeast Asia. The map on page 656 shows you that these countries lie either wholly or partly on islands.

Location
Indonesia

Indonesia is Southeast Asia's largest country. Its 705,190 square miles (1,826,443 sq. km) consist of an archipelago of more than 13,600 islands. Look at the map on page 656 and find the major islands of Indonesia: Sumatra, Java, Celebes (SEH•luh•BEEZ), and the western part of the island of New Guinea.

The Land and Economy Mountains tower over much of Indonesia. Many of these mountains are active or inactive volcanoes. Volcanic ash provides farmers with fertile soil, and thick forests spread over much of the islands.

Because Indonesia lies on or near the Equator, its climate is tropical. Monsoons blow across the islands, bringing a wet season and a dry season. Indonesia's major economic activities are farming, mining, and manufacturing. Indonesian farmers grow rice, coffee, spices, and **cassava,** a plant root used in making flour. Forests supply teak and other hardwoods. Indonesia also mines tin, petroleum, nickel, and other minerals. The graph below shows that Indonesia is one of the world's leading producers of tin.

Low labor costs attract manufacturing and processing industries to Indonesia from all over the world. You find most of Indonesia's industries on the island of Java. Factories there churn out steel products, chemicals, textiles, and cement.

The People Indonesia has about 208 million people—the largest population of any country in Southeast Asia. Its rapidly growing population has made it one of the most densely populated countries in the world. About 60 percent of Indonesia's people live on the island of Java. There you will find Jakarta (juh•KAHR•tuh), Indonesia's capital and largest city. Jakarta has modern buildings and streets crowded with cars and bicycles.

Most Indonesians belong to the Malay ethnic group. They are divided into about 250 to 300 smaller groups with their own languages and dialects. The official language, Bahasa Indonesia, is taught in schools. Some urban Indonesians also understand the Dutch language. Why? The Dutch ruled Indonesia from the 1600s to the mid-1900s. In 1945 Indonesia declared itself an independent republic.

Islam—brought to the islands some 500 years ago by Arab traders—is the major religion of the country. Some Indonesians are Hindus or Christians, or they practice traditional religions. In past centuries, cultural influences from China and India affected Indonesian ways of life. Today European and American influences are also widespread.

Waxing Cloth?
Indonesia and Malaysia are known for a striking art form called *batik*—a Javanese word meaning "drop." To create a *batik* pattern, wax is first dripped onto fabric. When the fabric is dyed, the dye reaches only the fabric not covered with wax. After the dye has dried, the fabric is boiled to remove the wax. A beautiful pattern is revealed.

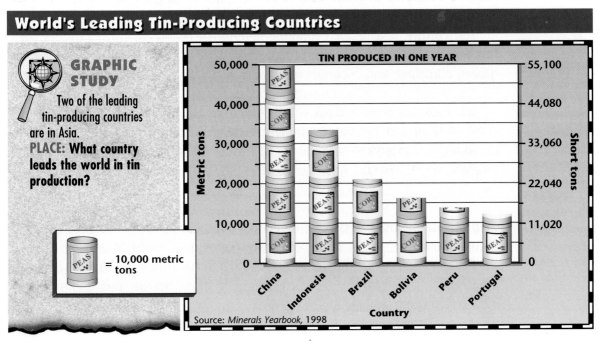

World's Leading Tin-Producing Countries

GRAPHIC STUDY

Two of the leading tin-producing countries are in Asia.

PLACE: What country leads the world in tin production?

= 10,000 metric tons

TIN PRODUCED IN ONE YEAR

Metric tons: 50,000 / 40,000 / 30,000 / 20,000 / 10,000 / 0

Short tons: 55,100 / 44,080 / 33,060 / 22,040 / 11,020 / 0

Country: China, Indonesia, Brazil, Bolivia, Peru, Portugal

Source: *Minerals Yearbook,* 1998

Young boys join in a dance at a Bali religious festival *(left)*. A Malaysian rubber worker taps a rubber tree *(right)*.
PLACE: What is the most important economic activity in Malaysia?

Place
Malaysia

North of Indonesia is Malaysia, a Southeast Asian country with an abundance of natural resources. Malaysia has two separate areas. One is the mainland part, called Malaya, on the Malay Peninsula. The other is made up of two territories—Sarawak (suh•RAH•wahk) and Sabah (SAH•buh)—on the northern edge of the island of Borneo.

The Land and Economy Mountains cover the center of Malaya. Low-lying plains are found east and west of these mountains. In Sarawak and Sabah, coastal swamps, rain forests, and rugged mountains make up most of the landscape. The map on page 657 shows you that both parts of Malaysia have a tropical rain forest climate. They have high temperatures and plenty of rain.

Agriculture is a major economic activity in Malaysia. Most farmers own small plots of land and raise rice, fruits, and vegetables. Plantation owners grow rubber, oil palms, and coconuts for export. Malaysia leads the world in the production of natural rubber and palm oil.

Malaysia is rich in tin, bauxite, petroleum, and iron ore. It is rapidly developing its oil industry. The government uses money from tin and oil exports to develop new industries that produce consumer goods. Malaysians perform manufacturing tasks for industries across the globe. For example, many computer parts are produced in Malaysia and then shipped to the United States.

The People About 22.2 million people live in Malaysia. Most belong to the Malay ethnic group, but a large Chinese population lives in the cities. People from India, Pakistan, Bangladesh, and Sri Lanka also make their homes in Malaysia. In the marketplaces you can hear Malay, Chinese, Tamil, and English spoken. Most Malaysians are Muslims, but there are large numbers of Hindus, Buddhists, Christians, and followers of traditional Chinese religions.

UNIT 8

About 57 percent of Malaysia's people live in urban centers; the rest live in rural areas. In rural villages many people live in thatched-roof homes built on posts a few feet off the ground. Most people in the cities live in high-rise apartments. Kuala Lumpur (KWAH•luh LUM•PUR), located on the Malay Peninsula, is the capital and largest city.

Movement

Singapore

Singapore is made up of 58 islands off the southern shore of the Malay Peninsula. With only 236 square miles (611 sq. km), Singapore is the smallest country in Southeast Asia. The city of Singapore takes up much of the largest island. In the early 1800s, Singapore was covered with rain forests and swamps. Today it is one of the world's leading commercial and trading centers.

The Land and Economy The central part of Singapore is hilly; coastal areas are flat. Urban centers with modern buildings, streets, and highways have replaced much of the rain forest. Like Malaysia, Singapore has a hot, humid climate.

Singapore's economy relies on trade and manufacturing. The city of Singapore has one of the world's busiest harbors. It is a **free port,** a place where goods can be loaded, stored, and shipped again without payment of import taxes. Singapore's workers add to the country's wealth by making industrial goods for export.

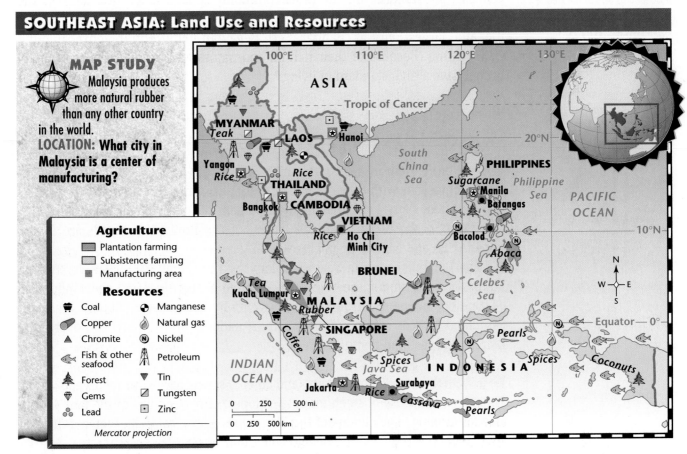

SOUTHEAST ASIA: Land Use and Resources

MAP STUDY
Malaysia produces more natural rubber than any other country in the world.
LOCATION: What city in Malaysia is a center of manufacturing?

Agriculture
- Plantation farming
- Subsistence farming
- Manufacturing area

Resources
- Coal
- Copper
- Chromite
- Fish & other seafood
- Forest
- Gems
- Lead
- Manganese
- Natural gas
- Nickel
- Petroleum
- Tin
- Tungsten
- Zinc

Mercator projection

Singapore, the largest port in Southeast Asia, has been described as one of the cleanest and safest cities in the world. **MOVEMENT: Why is Singapore called a *free port*?**

The People Most of Singapore's nearly 4 million people are Chinese, but Malays, Indians, and Europeans also live there. Singapore's people enjoy a high standard of living. More than 90 percent of them live in the capital city of Singapore.

Place
Brunei

On the northern coast of Borneo lies another small nation—Brunei. Brunei's 2,035 square miles (5,271 sq. km) make it about the size of Delaware. The map on page 654 shows you that Brunei borders the South China Sea. Malaysia divides Brunei into two parts. The western part has low plains and coastal swamps. The eastern part has forested hills and mountains. All of Brunei experiences a hot and humid tropical climate.

The Economy and People Oil and natural gas are Brunei's main exports and sources of income. With money from these resources, the government of Brunei is building new industries. Farming and fishing are other important economic activities. Brunei's farmers grow rice, pepper, coconuts, and fruit.

Most of Brunei's 300,000 people are either Malays or Chinese. Malay, Chinese, and English are the most commonly spoken languages. Islam is the leading religion. About 67 percent of the people of Brunei live in cities. The capital and largest city is Bandar Seri Begawan (BUHN•duhr sehr•EE buh•GAH•wuhn). Because of their rich oil and natural gas resources, the people of Brunei have a high standard of living.

Place
The Philippines

East of Vietnam in the South China Sea lies the Philippines. The Philippines is an archipelago with more than 7,000 islands. Only about 900 of the islands are inhabited. Only 11 islands hold 95 percent of the Filipino people. The map on page 656 shows you that the largest of the Philippine islands are Luzon (loo•ZAHN) and Mindanao (MIHN•duh•NAH•oh).

The Philippines was a Spanish colony for more than 300 years. In the late 1800s, it came under the rule of the United States. In 1946 the Philippines became an independent democracy. Twenty-five years later, Filipino President Ferdinand Marcos set up a dictatorship. A national uprising in 1986 overthrew Marcos and restored democracy.

The Land Volcanic mountains and forests spread over the landscape of the Philippines. Because of steep mountain slopes, only one-third of the land is suitable for farming. The climate of the Philippines is tropical, with wet and dry seasons. Typhoons often cause flooding and much loss of life and property.

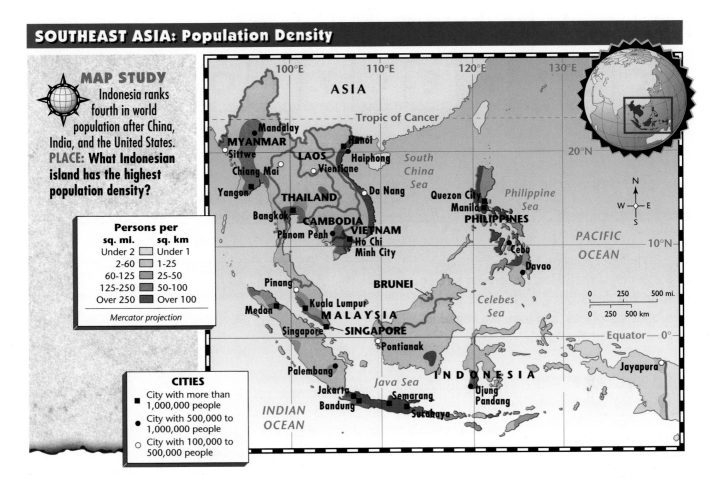

MAP STUDY
Indonesia ranks fourth in world population after China, India, and the United States. **PLACE:** What Indonesian island has the highest population density?

Persons per

sq. mi.	sq. km
Under 2	Under 1
2-60	1-25
60-125	25-50
125-250	50-100
Over 250	Over 100

Mercator projection

CITIES
■ City with more than 1,000,000 people
● City with 500,000 to 1,000,000 people
○ City with 100,000 to 500,000 people

The Economy and People Agriculture, mining, and industry are the major economic activities in the Philippines. Filipino farmers harvest rice, sugarcane, pineapples, and **abaca,** a plant fiber used to make rope. Filipino workers produce textiles, chemicals, electronic products, and other manufactured goods.

About 75.3 million people live in the Philippines. Most Filipinos live on the island of Luzon. Manila, the capital and largest city, is located on Luzon and is one of the great commercial centers of Asia.

The Filipinos are Malay in ethnic background. Their culture, however, blends Malay, Spanish, and American influences. About 90 percent of Filipinos practice the Roman Catholic faith, brought to the islands by Spanish missionaries. American influences show up in the products sold along busy city streets.

SECTION 2 ASSESSMENT

REVIEWING TERMS AND FACTS

1. Define the following: cassava, free port, abaca.

2. PLACE What are the four major islands of Indonesia?

3. REGION What are Singapore's two important economic activities?

MAP STUDY ACTIVITIES

4. Look at the population density map above. What two cities in the Philippines have more than 1 million people?

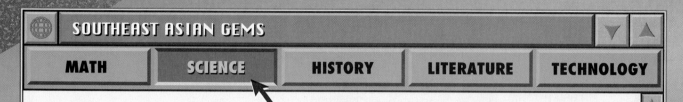

MAKING CONNECTIONS

MATH   **SCIENCE**   **HISTORY**   **LITERATURE**   **TECHNOLOGY**

Buried deep in the soils of Southeast Asia are minerals that take the form of rare and sparkling gemstones. Most of the world's rubies, sapphires, and jade are mined in Myanmar and Thailand.

RUBIES AND SAPPHIRES Rubies and sapphires come from the same mineral—corundum. Only one other pure mineral—the diamond—has a hardness greater than corundum. Impurities in corundum determine whether a gem is a ruby or a sapphire.

Corundum that has a trace of chromium oxide produces rubies—the rarest of gemstones. A ruby's color may be pink or deep bluish-red. Myanmar produces the finest rubies. A perfect Myanmar ruby is large and red and worth twice as much as a diamond of the same size. Thailand's rubies are yellow-red.

Corundum containing iron and titanium produces blue sapphires. Star sapphires—the most precious of all the blues—reflect light as white, glowing, starlike rays. Sapphires come in a rainbow of colors, but blue sapphires are the most prized.

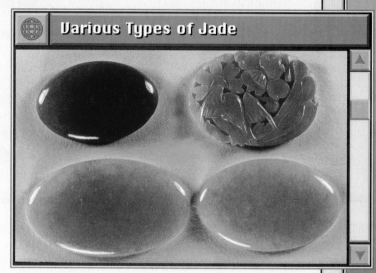

Various Types of Jade

JADE Jade is a hard, tough gemstone with a soft, shiny finish. Its unusual inner structure of interlocking needles makes jade strong and therefore good for delicate detailing. It is used in jewelry and carvings. The finest jade, found mainly in Myanmar but also in California and Mexico, comes from the rare mineral jadeite. The most beautiful colors of this gemstone include rich shades of light and dark greens, as well as softer shades of lilac and pink. The rarest and best-quality jadeite appears as a clear, bright green.

Making the Connection

1. What precious gemstones come from Southeast Asia?
2. Which Southeast Asian countries produce the gems?

Important Things to Know About Southeast Asia

SECTION 1 MAINLAND SOUTHEAST ASIA

- Mainland Southeast Asia includes the countries of Myanmar, Thailand, Laos, Cambodia, and Vietnam.
- Plentiful water supplies and fertile soil enable farmers to grow rice as a major food crop.
- Most of this area of Southeast Asia has tropical and subtropical climates.
- Most mainland Southeast Asians live in rural villages, but cities are growing in population.
- Most people in mainland Southeast Asia practice Buddhism.

SECTION 2 ISLAND SOUTHEAST ASIA

- The islands of Southeast Asia lie along the Pacific Ocean's Ring of Fire. Their people face the threat of volcanic activity and earthquakes.
- Indonesia is the largest country in Southeast Asia in area and population.
- The city of Singapore is one of the world's busiest trading and commercial centers.
- The people of Malaysia come from many different ethnic backgrounds.
- The culture of the Philippines shows Malay, Spanish, and American influences.

Floating market in Bangkok, Thailand ▶

Chapter 25 Assessment and Activities

REVIEWING KEY TERMS

Match the numbered terms in Set A with their lettered definitions in Set B.

A

1. cassava
2. deforestation
3. abaca
4. delta
5. alluvial plain
6. free port

B

A. land area formed by a river's soil deposits
B. plant fiber used in making rope
C. place where goods are unloaded, stored, and reshipped without import taxes
D. land formed by soil deposits at a river's mouth
E. widespread cutting of trees
F. plant root used in making flour

Mental Mapping Activity

Draw a freehand map of Southeast Asia. Label the following:
- Java
- Kuala Lumpur
- South China Sea
- Thailand
- Bangkok
- Vietnam

REVIEWING THE MAIN IDEAS

Section 1

1. PLACE Which Southeast Asian country was never a colony?
2. HUMAN/ENVIRONMENT INTERACTION What two major agricultural products are grown in Thailand?
3. HUMAN/ENVIRONMENT INTERACTION Where do most people live in Myanmar and Thailand?
4. MOVEMENT Why did many people flee Cambodia during the 1970s?

Section 2

5. PLACE What is the major religion of Indonesia and Malaysia?
6. HUMAN/ENVIRONMENT INTERACTION Why is only one-third of the Philippines suitable for farming?
7. MOVEMENT Why is Singapore important to the economies of Southeast Asia?
8. REGION How are Indonesia and the Philippines alike geographically?

CRITICAL THINKING ACTIVITIES

1. **Drawing Conclusions** Why is the Mekong River important to the people of Southeast Asia?
2. **Predicting Outcomes** Experts believe that Brunei has enough oil reserves to last until 2018. What might happen to the country's economy and standard of living at that time?

CONNECTIONS TO WORLD CULTURES

Cooperative Learning Activity Work in a group of three to learn about one of the countries of Southeast Asia. Each group should select one country. Then each member should choose one of the following topics to research: (a) the art; (b) holidays and festivals; or (c) the music and literature. Share your information with your group and then as a group, prepare a written report with illustrations or photos that illustrate the group's findings.

GeoJournal Writing Activity

Imagine you are a news reporter on assignment in Southeast Asia to report on an active volcano. Write about the volcanic activity as if you were an eyewitness.

TECHNOLOGY ACTIVITY

Using E-Mail Use your local library or access the Internet to locate an E-mail address for the United States Embassy in Manila, the Philippines. Compose a letter requesting information about the eruption of Mount Pinatubo in 1991. Using this information, create a bulletin board showing the chronology of the volcanic eruption. Provide photographs and captions with your display.

PLACE LOCATION ACTIVITY: SOUTHEAST ASIA

Match the letters on the map with the places and physical features of Southeast Asia. Write your answers on a separate sheet of paper.

1. Myanmar
2. Mekong River
3. Singapore
4. Jakarta
5. Sumatra
6. Java Sea
7. Vietnam
8. Brunei
9. Manila
10. Indian Ocean

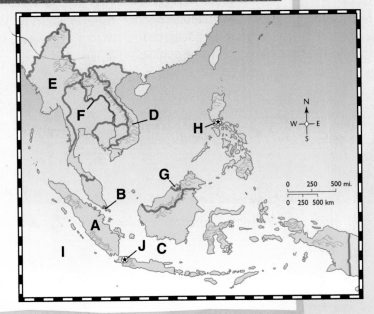

EYE ON THE ENVIRONMENT

Asia

HABITAT LOSS

PROBLEM

Over the last 100 years, Asia has suffered greater destruction of habitats than any other continent. Asian countries have drained wetlands, felled forests, and plowed under grasslands to feed growing populations and boost their economies. As the demand for resources increases, species disappear at an alarming rate. Scientists estimate that every few hours one plant or animal species in Asia becomes extinct.

SOLUTIONS

● India has created more than 370 national parks and preserves since 1960 to protect endangered wildlife such as the tiger. Since 1970 the population of Indian tigers in the wild has doubled.

● Countries in Asia have passed tough laws protecting wildlife. In China, poachers are executed.

● China has established 12 reserves for its giant pandas over the past few decades and plans to create 14 additional protected areas.

● In Indonesia, villagers now learn how to use the forests without destroying them. Some locals are hired as park rangers and tourist guides; others make money raising and selling butterflies or growing forest crops of nuts and palm leaves.

This tiger in India's Ranthambhore National Park is part of the success story of Project Tiger. Since the effort to protect tigers was launched in 1973, the number of tigers living in the park's forest has grown from around 14 to more than 40.

HABITAT LOSS FACT BANK

🌿 In 1900, 40,000 tigers lived in the wild in India alone; today about 6,000 survive, mostly in Asia.

🌿 In 1900, 40 percent of India was forested; today, only 15 percent is forested—and the number of endangered species has increased tenfold.

🌿 Indonesia has 210 endangered plant and animal species—more than any other nation in the world.

🌿 Javan and Sumatran rhinos are among the world's most endangered species. Fewer than 800 animals survive.

🌿 An estimated 1,200 giant pandas survive in bamboo forests in China. Poachers kill 40 or more pandas a year.

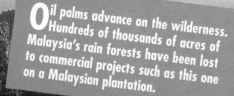

TEEN TRIBUTE

Maria Martinez, a student at Boston English High School in Boston, Massachusetts, wanted to do something to save endangered Asian wildlife. So she got a job. Maria works two to three hours a day at the World Society for the Protection of Animals. She sends out petitions that call for the end of inhumane treatment of endangered Asiatic black bears in China. Maria also answers letters from other young people who want to help.

environmental activities

These men in Foshan, China, carry bamboo poles for sale in local markets. Pandas and villagers often compete for bamboo.

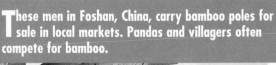

TAKE A CLOSER LOOK

1 What has happened to Asia's habitats in the last 100 years?

2 What growing demand has caused the most destruction of Asia's plants and animals?

WHAT CAN YOU DO?

🍃 Talk to your parents. Ask them to vote to keep our national parks and nature preserves.

🍃 Save a tree—carry your own shopping bag. The average American uses up seven trees a year.

🍃 Stuff envelopes or help a group raise money to protect an endangered animal.

🍃 Make a bird feeder. Spread a pinecone with peanut butter and hang it from a tree.

Save the Ancient Forests

Unit 9
Australia, Oceania, and Antarctica

What Makes Australia, Oceania, and Antarctica a Region?

Most people share . . .

- a location mostly south of the Equator.
- a location near or in the Pacific Ocean.
- unique animals and vegetation.
- specialized economies.

To find out more about Australia, Oceania, and Antarctica, see the Unit Atlas on pages 676–687.

The Ross Sea, Antarctica

EXPLORING THE INTERNET

To learn more about Australia, Oceana, and Antarctica, visit the Glencoe Social Studies Web site at **www.glencoe.com** for information, activities, and links to other sites.

GeoJournal Activity

Many unique and inter-esting animals live in Australia, Oceania, and Antarctica. As you read this unit, choose one of the animals mentioned and write a story about it that would appeal to younger students. Illustrate your story with colorful drawings and a sketch map of where this animal lives.

UNIT 9 ATLAS

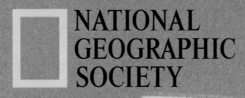

NATIONAL
GEOGRAPHIC
SOCIETY

IMAGES
of the
WORLD

1. **Kangaroo and joey, Australia**
2. **Three cowboys, Australia**
3. **Chinstrap penguins, Antarctica**
4. **Stone faces, Easter Island**
5. **Sutherland Falls, New Zealand**
6. **Boab tree, Western Australia**
7. **Traditional fishing, Vanuatu**

All photos viewed against the soaring roofline of Sydney's Opera House, Australia.

Regional Focus

The region of Australia, Oceania, and Antarctica lies in the Pacific Ocean southeast of Asia. It includes two continents—Australia and Antarctica. Australia is the only place on the earth that is both a continent *and* a country. Antarctica is the world's coldest and iciest continent. This region also takes in Oceania—an area of thousands of islands in the Pacific Ocean.

Region
The Land

The region of Australia, Oceania, and Antarctica lies mostly south of the Equator. It spreads out across millions of square miles of the Pacific Ocean. Long distances and rugged landscapes have kept many parts of the region isolated from each other and the rest of the world.

Kangaroo Crossing?

NEXT 8 km

In some parts of Australia, kangaroo populations are so large that signs are posted to warn motorists. **HUMAN/ENVIRONMENT INTERACTION: In what ways do you think humans could be a threat to the kangaroo?**

Australia A vast spread of dry, flat land makes up most of Australia. Through the central part of the country, however, runs a thick ribbon of pastureland. Underground pools of water allow farming and ranching to take place in this area. Hills and mountains stretch along Australia's east coast. A thin coastal strip faces the Pacific Ocean.

Oceania Oceania is made up of different kinds of Pacific islands. Some of the islands are mountainous. They are known for volcanic and earthquake activity. Others are massive formations of rock that rise from the ocean floor. Still other islands are low-lying, ring-shaped coral islands.

Antarctica Antarctica spreads over the southern end of the globe. It lies under an enormous sheet of ice. Mountains run through Antarctica and divide it into two parts. To the east is a high, flat plateau. A landmass largely below sea level lies to the west.

Place
Climate and Vegetation

Location and wind and ocean currents account for the extremes of temperature in Australia, Oceania, and Antarctica. Because of centuries of isolation, the region has animals and vegetation found nowhere else in the world.

Australia's eastern coast receives Pacific Ocean rains and enjoys mild climates year-round. Mountains block this moisture from reaching inland areas, which have dry or partly dry climates.

Oceania has mostly tropical climates, with wet and dry seasons. Warm days follow each other in an almost unbroken chain. Tropical plants, such as coconut palms, cover many of the islands.

Antarctica is one of the coldest places on earth. It is also one of the windiest, highest, and driest of the continents. The icy winds that sweep down from the plateau are unable to carry much moisture. As a result, only an average of 2 inches (50 mm) of precipitation, in the form of snow, falls each year.

Emperor Penguins

Penguins thrive in the Southern Hemisphere where icy Antarctic currents flow.
REGION: What do you think makes up the main portion of a penguin's diet?

Human/Environment Interaction
The Economy

Agriculture is the major economic activity of Oceania. Because of great distances and high development costs, manufacturing is limited to Australia and the island country of New Zealand. The mining of minerals, such as uranium, bauxite, iron ore, copper, nickel, and gold, is important to Australia. Modern means of transportation have increased trade between the region of Australia, Oceania, and Antarctica and the rest of the world.

Australia Only a small area of Australia is good for farming. Australian farmers make good use of their country's limited land and water resources. Australia has many cattle and sheep ranches. It is a major exporter of beef, lamb, and wool. Australians also mine uranium, bauxite, iron ore, copper, nickel, and gold. Factories produce cars, machinery, food products, and textiles.

UNIT 9 ATLAS

Oceania The soil in much of Oceania limits the growth of crops for export. Most people raise only enough food to feed themselves. Some larger islands, however, have rich volcanic soil and abundant rain that allows for cash crops, such as tropical fruits, sugar, coffee, and coconut products. New Zealand, located between Australia and Oceania, relies on both agriculture and manufacturing. It exports mainly meat, wool, and dairy products. Fishing is an important industry throughout Oceania.

Docking at Suva, Fiji

Fiji's capital and largest city is a busy seaport.
PLACE: What physical features do you think are necessary for a city to be a seaport?

Antarctica Because there are no permanent settlements and because of its harsh climate, Antarctica produces no farm crops. Its icy coastal waters, however, are a rich source of seafood. The continent is believed to be rich in minerals. To preserve Antarctica as the world's last wilderness, many nations have agreed—for now—not to mine this mineral wealth.

Movement
People

Australia and Oceania are a blend of European, traditional Pacific, and Asian cultures. In Australia, the original settlers were the Aborigines; in New Zealand, they were the Maori. In the late 1700s and early 1800s, the British settled and ruled Australia and New Zealand. These lands became independent in the early 1900s. Other South Pacific areas won their freedom after World War II. Today most countries in the region have democratic forms of government.

Population In spite of its vast size the region today has only about 30 million people. Nearly 18.7 million are in Australia, and most of them live in coastal cities, such as Sydney and Melbourne. Another 3.8 million live in New Zealand, which also has large urban populations. The rest of the South Pacific islands are less urbanized. Many Pacific islanders, especially young people, leave their villages and head to the region's major cities to find work. Although Antarctica has no permanent population, multinational groups of scientists live there for brief periods to carry out research.

The Arts Traditional Pacific cultures created many art forms still practiced today. For example, Aborigines paint and Maori carve beautiful wooden masks and figures. European settlers brought traditions from their homelands. In recent years traditional Pacific and European ways have blended to form a new culture unique to the region.

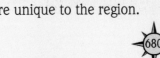

Multimedia Unit 9 Activities

NATIONAL GEOGRAPHIC SOCIETY

Picture Atlas of the World

Comparing and Contrasting Places

Using the CD-ROM The lands of Oceania, Australia, and Antarctica are unique. Oceania is an area of thousands of islands. Australia is the only place on earth that is both a continent and a country. Antarctica is the world's coldest and iciest continent. Explore these places using the *Picture Atlas of the World* CD-ROM. Imagine you are an anthropologist and create a file that uses photographs, statistics, or essays to compare and contrast these places with the United States. Then use the file as the basis of a class presentation. Explain the differences and similarities you've discovered, and why you think these differences and similarities exist.

Surfing the "Net"

The Antarctic Ozone Hole

The ozone layer is a thin layer of atmosphere that protects us from the sun. The ozone blocks the sun's most dangerous ultraviolet, or UV, rays from reaching the earth. However, scientists recently have recorded a steady loss of ozone in the earth's upper atmosphere. After discovery of an "ozone hole," scientific teams raced to research the problem and environmental agencies have pushed national governments to do something about it. To find out more about the ozone hole, search the Internet.

Getting There

Follow these steps to gather information on the ozone hole.

1. Use a search engine. Type in the words *ozone layer*.

2. Enter words like the following to focus your search: *South Pole, ozone depletion, ozone hole, CFC, Antarctica*.

3. The search engine will provide you with a number of links to follow.

What To Do When You Are There

Click on the links to navigate through the pages of information and gather your findings. Use your findings to create a pamphlet on ozone depletion. Include facts and figures to help explain why ozone depletion affects our lives and how we can do our part to reverse the problem. Be sure to explain the problem and resolution using terms and illustrations that a general audience could easily understand. Share your pamphlet with the class.

UNIT 9 ATLAS

Physical Geography

AUSTRALIA, OCEANIA, AND ANTARCTICA: Physical

ELEVATIONS

Feet	Meters
10,000	3,000
5,000	1,500
2,000	600
1,000	300
0	0

▲ Mountain peak
☐ Ice cap
☐ Ice shelf

Mercator projection

PACIFIC OCEAN

MICRONESIA

MELANESIA

POLYNESIA

Equator

Tropic of Capricorn

INDIAN OCEAN

AUSTRALIA

Great Sandy Desert
Gibson Desert
Great Victoria Desert
Macdonnell Ranges
Lake Eyre
Great Artesian Basin
Darling R.
Murray R.
Great Dividing Range
Great Australian Bight
▲ Mt. Kosciusko 7,310 ft. (2,228 m)

Coral Sea
Great Barrier Reef

PACIFIC OCEAN

Tasmania
Tasman Sea

North Island
Mt. Cook 12,349 ft. (3,764 m)
NEW ZEALAND
South Island
Southern Alps

0   250   500 mi.
0   250   500 km

South Pole
Vinson Massif 16,066 ft. (4,897 m)
Antarctic Circle

0   250 mi.
0   250 km

Map Study

1. **LOCATION** What reef lies off the northeastern coast of Australia?
2. **PLACE** About how much of Antarctica is covered by ice cap?

ELEVATION PROFILE: Australia

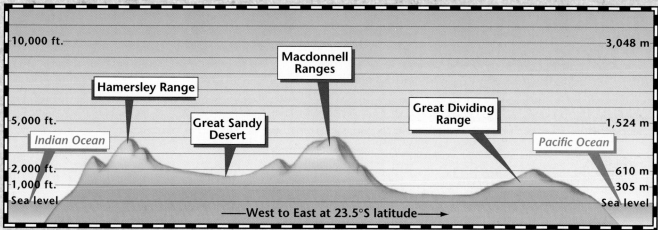

10,000 ft.   3,048 m

Macdonnell Ranges

Hamersley Range

5,000 ft.   1,524 m

Great Sandy Desert

Great Dividing Range

Indian Ocean

Pacific Ocean

2,000 ft.   610 m
1,000 ft.   305 m
Sea level   Sea level

←——West to East at 23.5°S latitude——→

Source: *Goode's World Atlas*, 19th edition

Highest point: Vinson Massif (Antarctica) 16,066 ft. (4,897 m) high

Lowest point: Lake Eyre (Australia) 52 ft. (16 m) below sea level

Longest river: Murray-Darling (Australia) 2,310 mi. (3,717 km) long

Largest lake: Lake Eyre (Australia) 3,600 sq. mi. (9,324 sq. km)

Largest waterfall: Sutherland Falls (New Zealand) 1,904 ft. (580 m)

Largest desert: Gibson Desert (Australia) 120,000 sq. mi. (310,800 sq. km)

AUSTRALIA/OCEANIA/ANTARCTICA AND THE UNITED STATES: Land Comparison

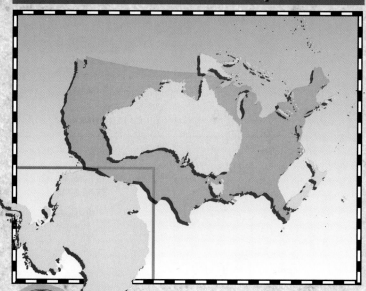

1. **PLACE** What is the highest point in the Australia/Oceania/Antarctica region?
2. **PLACE** How would you describe the size of the United States compared with the combined sizes of Australia, Oceania, and Antarctica?

Cultural Geography

AUSTRALIA AND OCEANIA: Political

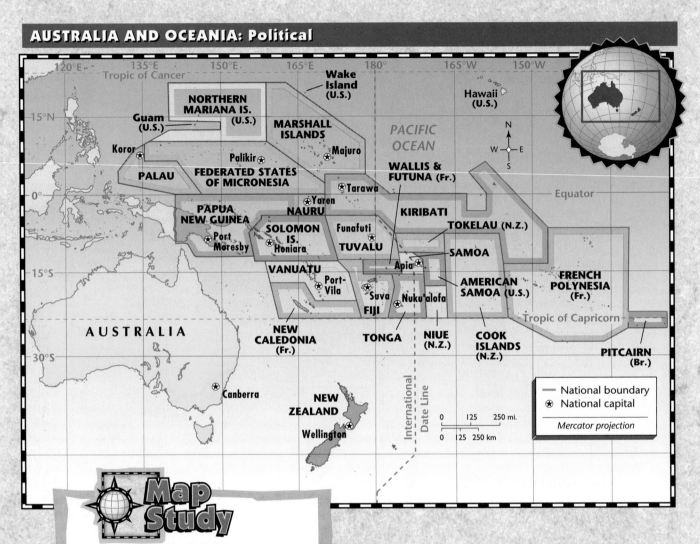

120°E 135°E 150°E 165°E 180° 165°W 150°W

Tropic of Cancer

Wake Island (U.S.)

Hawaii (U.S.)

15°N

Guam (U.S.)

NORTHERN MARIANA IS. (U.S.)

MARSHALL ISLANDS

PACIFIC OCEAN

N
W E
S

Koror

Majuro

PALAU

Palikir

FEDERATED STATES OF MICRONESIA

WALLIS & FUTUNA (Fr.)

0°

Tarawa

Equator

PAPUA NEW GUINEA

Yaren
NAURU

KIRIBATI

TOKELAU (N.Z.)

Port Moresby

SOLOMON IS.
Honiara

Funafuti

TUVALU

SAMOA

15°S

VANUATU

Port-Vila

Apia

Suva

AMERICAN SAMOA (U.S.)

FRENCH POLYNESIA (Fr.)

Nuku'alofa

FIJI

Tropic of Capricorn

AUSTRALIA

NEW CALEDONIA (Fr.)

TONGA

NIUE (N.Z.)

COOK ISLANDS (N.Z.)

PITCAIRN (Br.)

30°S

Canberra

NEW ZEALAND

Wellington

International Date Line

— National boundary
✪ National capital

Mercator projection

0 125 250 mi.
0 125 250 km

Map Study

1. **PLACE** What city is Australia's national capital?
2. **LOCATION** Which Pacific island capital lies closest to both the Equator and the International Date Line?

COMPARING POPULATION:
Australia/Oceania and the United States

Australia/
Oceania

= 25,000,000

United
States

Source: *Population Reference Bureau, 1998*

GeoFacts

Biggest country (land area):
Australia 2,951,521 sq. mi.
(7,644,439 sq. km)

Smallest country (land area): Tuvalu
9 sq. mi. (23 sq. km)

Largest city (population): Sydney
(1997) 3,800,000; (2015 projected) 3,900,000

Highest population density:
Nauru 1,267 people per sq. mi.
(489 people per sq. km)

Lowest population density: Australia
6 people per sq. mi. (2 people
per sq. km)

AUSTRALIA: Population Growth

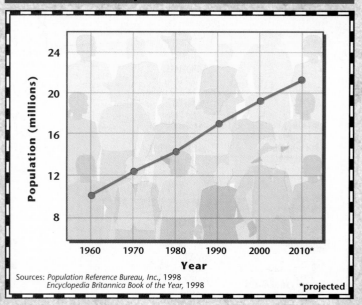

Sources: *Population Reference Bureau, Inc., 1998*
Encyclopedia Britannica Book of the Year, 1998

*projected

Graphic Study

1. **REGION** What is the combined population of Australia and Oceania?
2. **PLACE** How much did Australia's population grow between 1960 and 1990?

Countries at a Glance

Australia

Canberra ⭐

CAPITAL:
Canberra
MAJOR LANGUAGE(S):
English, Aboriginal languages
POPULATION:
18,700,000

LANDMASS:
2,951,521 sq. mi./
7,644,439 sq. km
MONEY:
Australian Dollar
MAJOR EXPORT:
Crude Oil
MAJOR IMPORT:
Machinery

FEDERATED STATES OF Micronesia

Palikir ⭐

CAPITAL:
Palikir
MAJOR LANGUAGE(S):
English
POPULATION:
100,000

LANDMASS:
270 sq. mi./
699 sq. km
MONEY:
U.S. Dollar
MAJOR EXPORT:
Copra
MAJOR IMPORT:
Food and Beverages

Fiji

Suva ⭐

CAPITAL:
Suva
MAJOR LANGUAGE(S):
English, Fijian, Hindi
POPULATION:
800,000

LANDMASS:
7,054 sq. mi./
18,270 sq. km
MONEY:
Fiji Dollar
MAJOR EXPORT:
Sugar
MAJOR IMPORT:
Machinery

Kiribati

Tarawa ⭐

CAPITAL:
Tarawa
MAJOR LANGUAGE(S):
English, Gilbertese
POPULATION:
82,449

LANDMASS:
313 sq. mi./
811 sq. km
MONEY:
Australian Dollar
MAJOR EXPORT:
Copra
MAJOR IMPORT:
Machinery

Marshall Islands

Majuro ⭐

CAPITAL:
Majuro
MAJOR LANGUAGE(S):
English, Marshallese, Japanese
POPULATION:
100,000

LANDMASS:
70 sq. mi./
181 sq. km
MONEY:
U.S. Dollar
MAJOR EXPORT:
Crude Coconut Oil
MAJOR IMPORT:
Food and Live Animals

Nauru

Yaren ⭐

CAPITAL:
Yaren
MAJOR LANGUAGE(S):
Nauruan, English
POPULATION:
10,390

LANDMASS:
21 sq. mi./
54 sq. km
MONEY:
Australian Dollar
MAJOR EXPORT:
Phosphates
MAJOR IMPORT:
Food

New Zealand

Wellington ⭐

CAPITAL:
Wellington
MAJOR LANGUAGE(S):
English, Maori
POPULATION:
3,800,000

LANDMASS:
103,470 sq. mi./
267,987 sq. km
MONEY:
New Zealand Dollar
MAJOR EXPORT:
Food and Live Animals
MAJOR IMPORT:
Machinery

Palau

Koror ⭐

CAPITAL:
Koror
MAJOR LANGUAGE(S):
English, Palauan
POPULATION:
20,000

LANDMASS:
190 sq. mi./
492 sq. km
MONEY:
U.S. Dollar
MAJOR EXPORT:
Tuna
MAJOR IMPORT:
Food

Papua New Guinea

Port Moresby ⭐

CAPITAL:
Port Moresby
MAJOR LANGUAGE(S):
English, Melanesian, Papuan languages
POPULATION:
4,300,000

LANDMASS:
174,850 sq. mi./
452,862 sq. km
MONEY:
Kina
MAJOR EXPORT:
Gold
MAJOR IMPORT:
Machinery

Samoa

Apia ⭐

CAPITAL:
Apia
MAJOR LANGUAGE(S):
Samoan, English
POPULATION:
200,000

LANDMASS:
1,090 sq. mi./
2,823 sq. km
MONEY:
Tala
MAJOR EXPORT:
Coconut Oil
MAJOR IMPORT:
Food

Countries not drawn to scale.

Delivering Supplies to Antarctica

Solomon Islands

Honiara

CAPITAL:
Honiara
MAJOR LANGUAGE(S):
English, Papuan, Melanesian, Polynesian languages
POPULATION:
400,000

LANDMASS:
10,810 sq. mi./
27,998 sq. km
MONEY:
Solomon Islands Dollar
MAJOR EXPORT:
Fish Products
MAJOR IMPORT:
Machinery

Tonga

Nuku'alofa

CAPITAL:
Nuku'alofa
MAJOR LANGUAGE(S):
Tongan, English
POPULATION:
107,335

LANDMASS:
290 sq. mi./
751 sq. km
MONEY:
Pa'anga
MAJOR EXPORT:
Squash
MAJOR IMPORT:
Food and Live Animals

Tuvalu

Funafuti

CAPITAL:
Funafuti
MAJOR LANGUAGE(S):
Tuvaluan, English
POPULATION:
10,297

LANDMASS:
9 sq. mi./
23 sq. km
MONEY:
Australian Dollar
MAJOR EXPORT:
Copra
MAJOR IMPORT:
Food

Vanuatu

Port-Vila

CAPITAL:
Port-Vila
MAJOR LANGUAGE(S):
Bislama, French, English
POPULATION:
200,000

LANDMASS:
4,710 sq. mi./
12,199 sq. km
MONEY:
Vatu
MAJOR EXPORT:
Copra
MAJOR IMPORT:
Machinery, Food

A Coast Guard icebreaker cuts through the ice to deliver passengers and supplies to a science station in Antarctica.

Overlooking Sydney, Australia

A bridge over Sydney Harbor carries traffic from the main business district to the suburbs.

Chapter 26 Australia and New Zealand

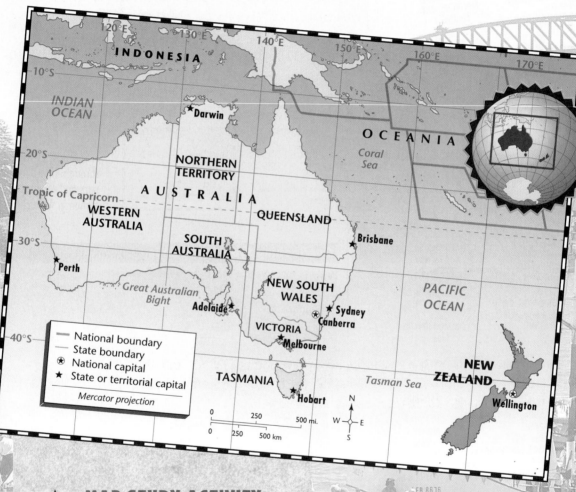

- National boundary
- State boundary
- ⊛ National capital
- ★ State or territorial capital

Mercator projection

0 250 500 mi.

0 250 500 km

MAP STUDY ACTIVITY

In this chapter you will read about Australia and New Zealand, which are completely surrounded by water.

1. **What is the national capital of Australia?**
2. **What sea separates Australia and New Zealand?**

SECTION 1

Australia

PREVIEW

Words to Know
- coral
- outback
- station
- marsupial
- bush

Places to Locate
- Australia
- Great Barrier Reef
- Great Dividing Range
- Western Plateau
- Canberra
- Sydney
- Melbourne

Read to Learn . . .
1. why most Australians live in coastal areas.
2. why Australia has a strong economy.
3. what groups have influenced Australia's culture.

"**I** started the morning routine—boil the tea, pack the gear, saddle the camels—and head south once more. . . . At Areyonga I filled my drinking-water bag with rainwater and set off for Ayers Rock. . . ." Explorer Robyn Davidson describes her recent trek across the desert of Australia. Ayers Rock—a huge red island of stone rising 1,000 feet (305 m)—is a landmark to all who travel through central Australia.

Australia lies far south on the globe between the Indian Ocean and the Pacific Ocean. People often call Australia the Land Down Under because it is located entirely within the Southern Hemisphere.

Location
The Land

Surrounded by water, Australia is a country that is a continent all by itself! Why isn't it called an island? It is too large. Instead geographers describe it as a continent—the smallest one in the world. The world's largest island, Greenland, has only about one-third of Australia's land area. With a land area of 2,951,521 square miles (7,644,439 sq. km), Australia is the world's sixth-largest country—about the size of the continental United States.

Not only is Australia the smallest continent—it is also the oldest. Australia does not have a wide variety of physical features. It is made up of mostly deserts and dry grasslands, with a few low mountains and coastal plains. Winds and time have worn the land into the flattest and lowest continent in the world.

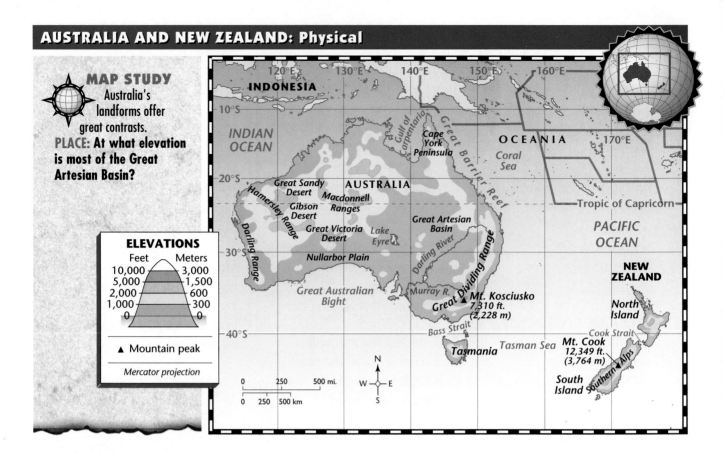

MAP STUDY
Australia's landforms offer great contrasts.
PLACE: At what elevation is most of the Great Artesian Basin?

ELEVATIONS

Feet	Meters
10,000	3,000
5,000	1,500
2,000	600
1,000	300
0	0

▲ Mountain peak

Mercator projection

Off Australia's northeastern coast lies one of the world's great natural wonders—the Great Barrier Reef. It is a 1,250-mile (2,011 km) chain of colorful **coral** formations and islands rising out of blue-green Pacific waters. Coral is a hard, rocklike material made of the skeletons of small sea animals. To Australia's southeast lies the small island of Tasmania.

The Great Dividing Range The Great Dividing Range stretches along Australia's Pacific coast. It divides the flow of Australia's rivers—eastward toward the Pacific Ocean or westward to inland areas. Two of Australia's major rivers—the Murray and the Darling—begin in the Great Dividing Range and flow westward into central Australia.

In Australia's southeastern corner, you find a narrow coastal plain. This fertile flatland boasts Australia's best farmland. You can see on the map on page 700 that it also includes some of Australia's most densely populated areas. The southern end of the Great Dividing Range hugs this narrow plain. In this area rise the peaks of the Australian Alps, Australia's highest mountains. At 7,310 feet (2,228 m), Mount Kosciusko (KAH•zee•UHS•koh) is Australia's highest peak.

The Central Lowlands As you move inland, the Great Dividing Range levels off into pastureland known as the Central Lowlands. This and other inland parts of Australia are called the **outback**. Partly dry grasslands cover much of the Central Lowlands. Here in the outback, mining towns and huge **stations**—Australian sheep and livestock ranches—are scattered over a vast area.

UNIT 9

A resource prized above all others, water is scarce or lacking across 75 percent of the country. An unusual feature of the Central Lowlands is the Great Artesian Basin, where water is found deep underground. The basin is like a huge pail that catches water from eastern mountain rivers. Ranchers drill wells to make the underground water available for irrigating grazing land.

The Western Plateau Picture a carpet of sand twice as large as Alaska, Texas, California, and New Mexico combined. That's about the size of Australia's Western Plateau. Travel across the plateau, which covers about 75 percent of Australia, is mostly by airplane.

The map on page 690 shows you the huge deserts that spread across much of the Western Plateau. This part of the outback, however, is not a wasteland—it is rich in mineral wealth. To the south, a narrow coastal plain is home to almost all of the Western Plateau's farms and people.

The Climate Look at Australia's climate regions on the map on page 692. You can see that most of the country is dry with desert or partly desert grassland. But there are several other climate regions. Northern Australia—the part of the country closest to the Equator—has a tropical climate with rainy and dry seasons and year-round hot or warm temperatures. Coastal areas in the south and southeast have warm summers and mild winters. Only the Australian Alps and the southern island of Tasmania have winter temperatures that fall below freezing for more than a day.

Animals and Plants Unless you visit a zoo, you probably don't see kangaroos or koalas near your home. Kangaroos and koalas are native only to Australia. These animals and about 120 similar kinds are called **marsupials**, or mammals that carry their young in a pouch. Why do they live only in Australia?

For millions of years, Australia was isolated from other continents. This separation led to a very special wildlife population. Dingoes, wombats, kookaburras, emus, bandicoots, and dugongs are just some of the animals found in Australia. One particularly unusual Australian animal—the platypus—lays eggs but has a furry body, a large bill, and webbed feet. Rare plants such as the eucalyptus are also native to Australia. The long, leathery leaves of the eucalyptus are the only food a koala will eat.

A Busy Station in the Outback

Sprawling cattle and sheep ranches, called stations, dot the landscape of Australia's outback.
LOCATION: Where in Australia do you find the outback?

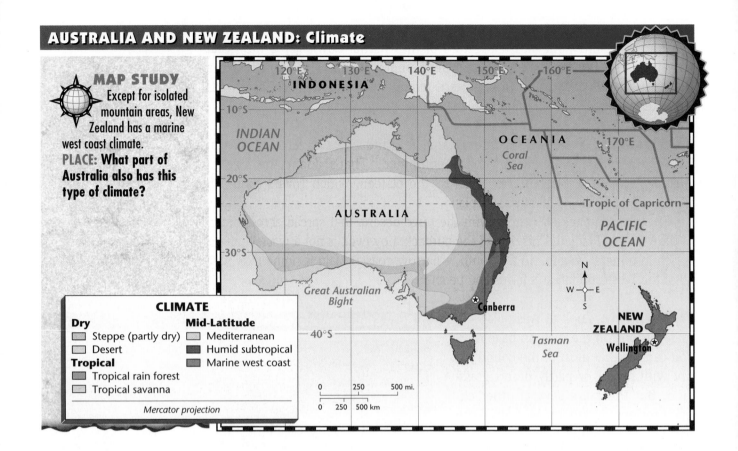

MAP STUDY
Except for isolated mountain areas, New Zealand has a marine west coast climate.
PLACE: What part of Australia also has this type of climate?

CLIMATE

Dry
Steppe (partly dry)
Desert

Tropical
Tropical rain forest
Tropical savanna

Mid-Latitude
Mediterranean
Humid subtropical
Marine west coast

Mercator projection

Movement
The Economy

A treasure chest overflowing with mineral resources—that's Australia. The continent's strong economy relies on exporting these resources. In the past the United Kingdom ranked as Australia's most important trading partner. Today Australia trades more with Japan and the United States.

Many of the products that are manufactured in the country meet local needs. Australian factories produce cars, trains, computers, and consumer goods. Tourism is also an important industry. Many tourists flock to Australia each year, and the country's service industries thrive on the tourist trade.

Agriculture Australia's dry climate limits farming. Only about 10 percent of Australia's land is good for growing crops. Farmers rely on machines and irrigation to improve crop yields. Wheat is the main crop. Other grains, sugarcane, fruits, and vegetables are also grown. Most of Australia's land is set aside for grazing. Ranchers raise cattle for beef and sheep for meat and wool. The graph on page 693 shows that Australia is the world's leading wool-producing country.

Mining When gold was discovered in Australia in the mid-1800s, thousands flocked to the outback. But gold is only one of the country's many mineral resources. Australia is the world's leading producer of bauxite and lead. The map on page 698 shows you the location of Australia's coal, iron ore, uranium, silver, opal, and zinc mines.

692

Place
The People

For its huge area, Australia has very few people—only 18.7 million. Most Australians live in scattered urban areas along the coasts. Australia has long needed more skilled workers to develop its open land and boost its economy. To meet this goal, the Australian government has encouraged people from other countries to settle in Australia.

Past and Present A small part of the continent's current population were the first Australians—the Aborigines (A•buh•RIHJ•neez). The Aborigines came to Australia from Asia about 30,000 to 40,000 years ago. They moved throughout the continent hunting animals and gathering plants. The Aborigines developed a unique culture. Ancient cave paintings show the work of their artists.

The voyages of Captain James Cook provided the British with a claim to Australia in the late 1700s. At first the United Kingdom used Australia as a colony for prisoners. Then other British emigrants settled in Australia. These settlers took land from the Aborigines and forced them far into the outback. Many Aborigines died of European diseases.

In 1901 the British colonies became states and united to form the independent Commonwealth of Australia. Australia is divided into seven states as shown on the map on page 688. Australians created a British-style parliamentary

WHAT IN THE WORLD?

On Target

Hunters around the world could take a lesson from Australia's Aborigines. They invented a hunting tool called the boomerang. Shaped like a bent wing, the wooden boomerang— thrown by a hunter—sails out toward its target. If the boomerang misses, it returns to land near the person who threw it!

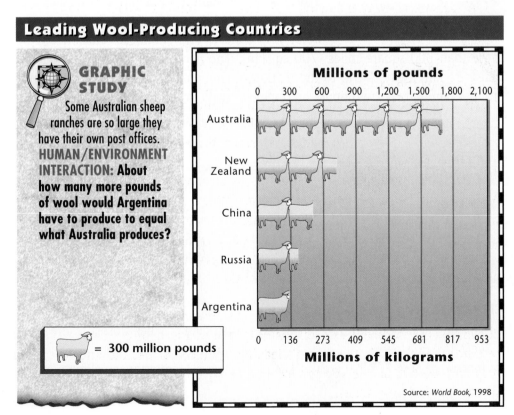

Leading Wool-Producing Countries

GRAPHIC STUDY

Some Australian sheep ranches are so large they have their own post offices. **HUMAN/ENVIRONMENT INTERACTION: About how many more pounds of wool would Argentina have to produce to equal what Australia produces?**

= 300 million pounds

Millions of pounds

	0	300	600	900	1,200	1,500	1,800	2,100
Australia								
New Zealand								
China								
Russia								
Argentina								

Millions of kilograms
0   136   273   409   545   681   817   953

Source: *World Book,* 1998

democracy and preserved much of their British heritage. Australia has developed its own national character, however. The "new" Australia comes from the diversity of its people. Many present-day Australians have Italian, Greek, Slavic, Aboriginal, or Chinese backgrounds.

City Life About 85 percent of Australia's population live in cities. Sydney and Melbourne—on the southeastern coast—are the largest cities. With populations of more than 3 million, both are lively cultural and commercial centers. Australia's next largest cities—Adelaide, Brisbane, and Perth—also are located on the coast. Canberra, the national capital, lies a relatively short distance inland from Sydney and Melbourne. With about 300,000 people, Canberra was developed through a government plan designed to draw people away from the coasts and into the outback.

Rural Life Only about 15 percent of Australians live in rural areas known as the **bush**. Many rural people live on stations that raise sheep or cattle. Thirteen-year-old Alice Adams lives on a sheep station in the Central Lowlands. She and her family live in a large, comfortable farmhouse. Working on the station is a family effort. Alice and other rural Australians sometimes find life in the bush very lonely. The Adamses' station covers more than 300 square miles (700 sq. km). The nearest settlement is about 50 miles (81 km) away. A trip to the market or shopping center is not quick. Alice and her family travel two hours on unpaved roads to reach a distant rural town.

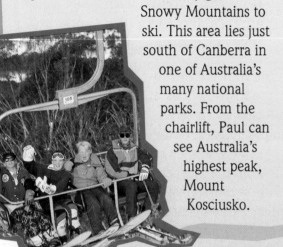

Teen Scene

Winter in June

It's June in Australia—time to get out the skis! Because the seasons in Australia are opposite of those in the Northern Hemisphere, Paul Noone and his friends plan their winter activities from June to September. Paul and his family go to Snowy Mountains to ski. This area lies just south of Canberra in one of Australia's many national parks. From the chairlift, Paul can see Australia's highest peak, Mount Kosciusko.

SECTION 1 ASSESSMENT

REVIEWING TERMS AND FACTS

1. **Define the following:** coral, outback, station, marsupial, bush.

2. **PLACE** What is the landscape of the central three-fourths of the country like?

3. **PLACE** What important agricultural products does Australia produce?

4. **MOVEMENT** Who were the first people to settle Australia?

MAP STUDY ACTIVITIES

5. Compare the maps on pages 688 and 690. What large Australian city lies closest to the Great Barrier Reef?

6. Turn to the climate map on page 692. What area of Australia enjoys a Mediterranean climate?

BUILDING GEOGRAPHY SKILLS

Using Technology

Analyzing a LANDSAT Image

Look carefully at the photograph below. What does it show? A LANDSAT satellite carrying cameras took this picture of southern Florida from space. The cameras record millions of energy waves invisible to the human eye. LANDSAT satellites relay these signals to receiving stations here on the earth. Computers then change this information into pictures of the earth's surface.

LANDSAT carries scanning instruments that pick up sunlight reflected off of water, rocks, plants, and other objects. Geographers use the different colors of reflected light to analyze the area. In LANDSAT images, the colors red or pink mean healthy vegetation. Deep blue shows areas of clear water, while light blue shows unclear water. These images help to pinpoint polluted areas.

To interpret a LANDSAT image, apply these steps:

- Name the area shown in the image.
- Point out the natural and human-made features.
- Look for areas of healthy vegetation and clean water, and areas showing signs of pollution.
- Look for connections between the areas of pollution and the human-made features.

Southern Florida

Technology Skills Practice

1. What geographic area is shown in this image?

2. Where do you see the healthiest vegetation?

3. Where in the picture do you see water that might be polluted or muddy?

New Zealand

PREVIEW

Words to Know
- *manuka*
- fjord
- geothermal power

Places to Locate
- New Zealand
- North Island
- South Island
- Wellington
- Auckland
- Southern Alps
- Canterbury Plains

Read to Learn . . .
1. what two large islands make up most of New Zealand.
2. what products New Zealand exports to other countries.
3. how New Zealanders spend their leisure time.

What country has more than 15 times as many sheep as people? New Zealand is the answer. Rolling green hills provide excellent grazing land for some 55 million sheep. New Zealand is one of the world's leading producers of lamb and wool.

New Zealand is an island country in the southwest Pacific Ocean. It lies about 1,200 miles (1,930 km) southeast of its nearest neighbor, Australia. For centuries, distance and isolation separated New Zealand from other parts of the world. Today it is a modern Pacific nation with one of the highest standards of living in the world.

Human/Environment Interaction

The Land

If you were shown two photographs—one of a desert and one of a lush green landscape—you could probably guess which one was Australia and which New Zealand. Although New Zealand lies relatively close to Australia, its landscape is very different from that of its larger neighbor. With 103,470 square miles (267,987 sq. km), New Zealand is about the size of the state of Colorado. The map on page 690 shows you that New Zealand is a long, narrow country. Anywhere you stand in New Zealand, you are not more than 80 miles (129 km) from the sea. New Zealand includes two main islands—North Island and South Island—as well as many smaller islands.

North Island A large volcanic plateau forms the center of North Island. Three active volcanoes and the perfectly shaped cone of inactive Mount Egmont rise above the island's fertile plains. Small shrubs called *manuka* grow well in the plateau's volcanic soil.

Farther north on the plateau you see geysers spewing hot water as high as 100 feet (30 m) in the air. Nearby flow icy cold streams. Engineers use natural steam from the geysers and other volcanic activity to produce electricity for New Zealand's industries, farms, and urban areas.

Fertile lowlands, forested hills, and sandy beaches surround the North Island's volcanic plateau. On the plateau's slopes, sheep and cattle graze. Fruits and vegetables are grown on the coastal lowlands. Seaside resorts offer tourists deep-sea fishing for marlin, shark, and tuna.

South Island A massive mountain chain called the Southern Alps runs along South Island's west coast. Snowcapped Mount Cook, the highest peak in New Zealand, soars 12,349 feet (3,764 m) in the Southern Alps.

This mountainous region reveals some of New Zealand's most beautiful scenery. Glaciers lie on mountain slopes high above thick, green forests and sparkling lakes. Along the southwestern coast, long narrow fingers of the sea called **fjords** cut into the land. New Zealand's fjords are similar to those in Norway.

From the Southern Alps to the east coast stretch the Canterbury Plains. They form New Zealand's largest area of flat or nearly flat land. Farmers grow grains and ranchers raise sheep here. The map on page 698 shows you the kinds of products New Zealand produces.

Animals and Plants As in Australia, animals and plants in New Zealand are found nowhere else. New Zealanders take pride in this fact. They chose the unusual kiwi (KEE•wee)—a long-beaked bird that cannot fly—as their national symbol. You may have eaten a fruit called a kiwi. This fruit—really a Chinese gooseberry—originated in New Zealand and was named for the kiwi bird.

Giant Australian kauri trees once covered all of North Island. European settlers cut down much of the kauri forest to build homes and ships. Today the kauri trees are protected. One of them is more than 2,000 years old!

The Climate New Zealand has a mild, wet climate. Compare the map on page 692 to the map on page 94. You see that New Zealand lies in a marine west coast region similar to the Pacific Northwest coast of the United States.

Thermal Springs in Rotorua, New Zealand

"Big Splash" is one of the many spectacular geysers created by volcanic activity in northern New Zealand.
HUMAN/ENVIRONMENT INTERACTION: How are New Zealand's engineers using geysers as a source of energy?

Ocean breezes from the west warm the land in winter and cool it in summer. As in the Pacific Northwest, rain falls throughout the year in New Zealand.

Because New Zealand lies south of the Equator, its seasons are the opposite of those in the Northern Hemisphere. July is New Zealand's coolest month, while January and February are the warmest months. Temperatures get cooler as you move from the north to the south.

Movement
The Economy

About 55 percent of New Zealand's land—its greatest resource—is cropland or pastureland. With its mild climate and fertile soil, New Zealand supports a thriving farm economy and exports many farm products. In recent years, manufacturing and tourism also have become important sources of income for New Zealand.

Farming A return to the use of natural fibers in clothing has boosted New Zealand's economy. A large percentage of the world's wool comes from New Zealand ranches. Its farmers and ranchers also produce dairy products and meat. The major agricultural exports of New Zealand are wool, lamb, butter, and cheese. Wheat, barley, oats, apples, and kiwifruit are also grown there.

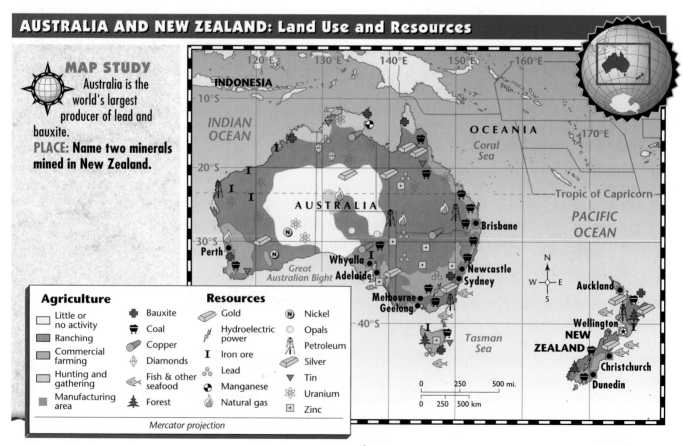

AUSTRALIA AND NEW ZEALAND: Land Use and Resources

MAP STUDY Australia is the world's largest producer of lead and bauxite.
PLACE: Name two minerals mined in New Zealand.

Agriculture
- Little or no activity
- Ranching
- Commercial farming
- Hunting and gathering
- Manufacturing area

Resources
- ✚ Bauxite
- 🚃 Coal
- Copper
- ✦ Diamonds
- 🐟 Fish & other seafood
- 🌲 Forest
- Gold
- Hydroelectric power
- I Iron ore
- Lead
- Manganese
- Natural gas
- N Nickel
- Opals
- Petroleum
- Silver
- Tin
- Uranium
- Zinc

Mercator projection

A Maori wood-carver practices his skill *(left)*. The British sport of lawn bowling is popular with New Zealanders of European descent *(right)*.
PLACE: What percentage of New Zealand's population is Maori?

Mining and Manufacturing New Zealand has only a few minerals but is rich in **geothermal power**, or electric power produced from natural steam. The major source of power, however, is hydroelectricity.

In recent years, New Zealand has developed its industrial economy. Its factories now produce cars, furniture and other wood products, clothing, and electronic equipment. Most manufacturing, however, involves making food products such as butter, milk, and cheese.

Place
The People

Picture this scene: The year is about A.D. 1300. Small canoes paddled by a Polynesian people called the Maori reach the New Zealand shore. The lush, green landscape is a welcome sight. A quick hunt produces a large mao bird and a lizard for dinner. The Maori decide to stay.

The arrival of the first people in New Zealand could have happened just this way. Today about 9 percent of the population are Maori (MOWR•ee). Most of the rest of New Zealand's 3.8 million people trace their ancestry to British settlers who came to the islands in the 1800s. Immigrants still come to New Zealand from the United Kingdom and other countries.

Influences of the Past The Maori are believed to have arrived in New Zealand about 700 years ago from Pacific islands far to the northeast. They developed skills in farming, weaving, fishing, and bird hunting and became well-known for their artistic wood carvings.

A few European explorers first came to New Zealand in the mid-1600s. Almost 200 years passed before European settlers—most of them British—arrived in large numbers. Then the British government decided to add New Zealand to its empire. In 1840 British officials signed a treaty with Maori leaders. The Maori agreed to accept British rule in return for the right to keep their lands. British settlers, however, eventually moved into Maori territory. Although the Maori fought to keep their lands, the British eventually defeated them.

New Zealand is a country of "firsts." In 1893 it became the first country in the world to give women the right to vote. New Zealand was also among the first countries to provide government help to the elderly, the sick, and those without jobs.

In 1907 New Zealand became independent of the United Kingdom. A parliamentary democracy, the New Zealand government works to improve housing, job opportunities, and education for all of its citizens, including Maori. Many Maori also want the government to return lands taken from them under British rule. In 1994 the New Zealand government paid one Maori group millions of dollars for land taken by the British.

Way of Life About 85 percent of New Zealanders live in urban areas. You can see where these cities are located on the map on page 698. The largest cities are Auckland (AW•kluhnd), a major port, and Wellington, the national capital. Both lie on North Island where three-fourths of New Zealanders live. Two smaller cities—Christchurch and Dunedin—are manufacturing centers on South Island.

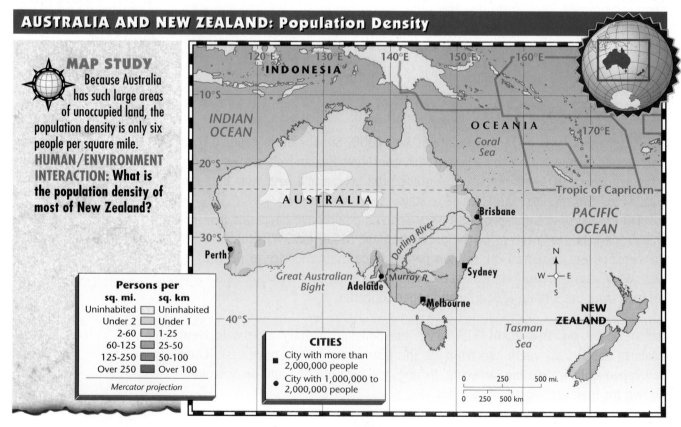

AUSTRALIA AND NEW ZEALAND: Population Density

MAP STUDY
Because Australia has such large areas of unoccupied land, the population density is only six people per square mile.
HUMAN/ENVIRONMENT INTERACTION: What is the population density of most of New Zealand?

Persons per
sq. mi.	sq. km
Uninhabited	Uninhabited
Under 2	Under 1
2-60	1-25
60-125	25-50
125-250	50-100
Over 250	Over 100

Mercator projection

CITIES
■ City with more than 2,000,000 people
● City with 1,000,000 to 2,000,000 people

New Zealanders love outdoor activities. In New Zealand's mild climate, they enjoy camping, hiking, hunting, boating, and mountain climbing in any season. The people of New Zealand also enjoy sports such as cricket, a British game played with a bat and ball, and rugby, a British form of football.

Many New Zealanders still observe British customs. Yet immigrants from Asia, the Pacific, and various European nations have brought cultural diversity to New Zealand. Today, the growing Maori population also receive more recognition than they did in the past.

New Zealand is rapidly becoming a modern Pacific nation that blends European and Maori ways. This mix is evident in the life of 15-year-old Ruth Erutoe, a Maori teenager who lives in a suburb of Auckland. Ruth and her family listen to Italian opera and admire Kiri Te Kanawa (KEE•ree teh KAN• ah•wah), the world-famous opera star from New Zealand. The Erutoes also remain close to their Maori heritage. For weddings and other special occasions, they go to a *marae*, or a Maori meetinghouse. Some meetinghouses are simple buildings. Others are decorated with elaborate wood carvings. Turn to page 708 to see an example of a Maori wood carving.

Bustling Auckland, New Zealand

Auckland is New Zealand's largest city, chief seaport, and industrial center.
PLACE: What is the capital of New Zealand?

SECTION 2 ASSESSMENT

REVIEWING TERMS AND FACTS
1. Define the following: *manuka*, fjord, geothermal power.
2. REGION What physical feature lies in the center of North Island?
3. MOVEMENT From where did the Maori come?
4. LOCATION Where is the city of Wellington in relation to Auckland?

MAP STUDY ACTIVITIES
5. Look at the land use map on page 698. What mineral resources does New Zealand have?
6. Turn to the population density map on page 700. Where do the fewest people of New Zealand live?

MAKING CONNECTIONS

Australia is home to more than 120 kinds of marsupials—mammals whose young grow and mature inside a pouch on the mother's belly. Wallabies, wombats, and other marsupials are unfamiliar to people outside Australia. Kangaroos and koala bears, however, are known throughout the world.

KANGAROOS No matter what their size or where they live, all kangaroos have one thing in common: big hind feet. Either red or gray, kangaroos bound along at about 20 miles (32 km) per hour. Some may race up to 40 miles (65 km) per hour. In a single jump a kangaroo can hop 10 feet (3 m) high and cover a distance of 45 feet (14 m)!

Kangaroos also have big ears. Sensitive instruments, a kangaroo's ears are never still. One ear points forward, the other backward. Female kangaroos have front pouches that open at the top to hold their joeys, or infants. These big pockets provide nourishment and protection for several months.

KOALAS Despite their name, koalas are not bears. Like their kangaroo cousins, female koalas protect and nourish their young in pouches on their bellies. A female koala's pouch, however, opens at the bottom. Strong

muscles keep the pouch shut and the young koalas safe inside.

Koala Mom and Joey

Koalas make their homes in the eucalyptus trees of southeastern Australia. Quiet, calm, and sleepy, the cuddly-looking animals spend at least 20 hours a day dozing high in a favorite tree. Koalas are fussy eaters. They eat nothing but eucalyptus leaves—about 2.5 pounds (1.2 kg) a day. The leaves also provide the animals with all the moisture they need.

Making the Connection

1. About how many kinds of marsupials live in Australia?
2. How are kangaroos and koalas alike? How are they different?

Chapter 26 Highlights

Important Things to Know About Australia and New Zealand

SECTION 1 AUSTRALIA

- Australia is both a country and the world's smallest continent.
- Deserts and grasslands cover most of the country west of the Great Dividing Range.
- Australia's economy relies mostly on farming and mining.
- Australia is the world's leading wool-producing country. It also leads in the production of bauxite and lead.
- The first Australians were the Aborigines. Others are of European and Asian descent.
- Most Australians live in cities. Sydney, Melbourne, Adelaide, Brisbane, and Perth are the largest.

SECTION 2 NEW ZEALAND

- New Zealand includes two major islands—North Island and South Island—and many smaller islands.
- New Zealand has a mild, wet climate.
- New Zealand's agricultural exports—especially wool—are the mainstays of its economy.
- New Zealand gets most of its electricity from geothermal energy.
- The Maori, a Polynesian ethnic group from the South Pacific, were the first people to settle New Zealand.
- New Zealand's culture at one time was mainly influenced by British traditions. Now it is a mix of many cultures, including those of the Maori and other immigrants.

Sydney, Australia ▶

Chapter 26 Assessment and Activities

REVIEWING KEY TERMS

Match the numbered terms in Set A with their lettered definitions in Set B.

A
1. *manuka*
2. bush
3. geothermal power
4. outback
5. marsupial
6. fjord
7. coral
8. station

A. inlet of the sea that cuts into the land
B. hard material made of the skeletons of small sea animals
C. electricity produced from natural steam
D. Australian sheep or cattle ranch

B
E. vast inland areas of Australia
F. small shrubs that grow on New Zealand's volcanic soil
G. Australian word for rural areas
H. mammal that carries its young in a pouch

REVIEWING THE MAIN IDEAS

Mental Mapping Activity

Draw a freehand map of Australia and New Zealand. Label the following:
- Sydney
- Melbourne
- Auckland
- North Island
- South Island
- Wellington

Section 1
1. **LOCATION** Why is Australia called the Land Down Under?
2. **MOVEMENT** What two countries are Australia's major trading partners?
3. **PLACE** Why was the city of Canberra built?

Section 2
4. **PLACE** What large mountain range runs down the western coast of South Island?
5. **MOVEMENT** What are New Zealand's major agricultural exports?
6. **HUMAN/ENVIRONMENT INTERACTION** What leisure activities do New Zealanders enjoy that are made possible by the country's climate?

CRITICAL THINKING ACTIVITIES

1. **Making Comparisons** How does the climate of Australia compare with that of New Zealand?
2. **Making Inferences** Why do you think most Australians live in coastal cities?

704

CONNECTIONS TO WORLD CULTURES

Cooperative Learning Activity Work in a group of three to learn more about the people and cultures of Australia and New Zealand. Each group will choose a country. Then each member will select *one* of the following topics to research: (a) leisure activities; (b) arts and crafts; or (c) music and dance. Share your information with the rest of your group. Prepare a written report, a poster, or a travel brochure that presents your group's findings.

GeoJournal Writing Activity

Imagine that you are visiting Australia. Using the map on page 688, prepare a list of places to see—one place for each state of the country. Write an itinerary describing the places you want to visit, what you hope to see there, and how many miles (km) you will travel from place to place.

TECHNOLOGY ACTIVITY

Using a Computerized Card Catalog

Use a computerized card catalog in your school or local library to locate sources about Australia's Great Barrier Reef. Write a report describing the formation of the reef, its animal life, and environmental threats to it. Be sure to include a map showing the location of the reef. Share your report with the rest of the class.

PLACE LOCATION ACTIVITY: AUSTRALIA AND NEW ZEALAND

Match the letters on the map with the places and physical features of Australia and New Zealand. Write your answers on a separate sheet of paper.

1. Brisbane
2. Great Barrier Reef
3. Wellington
4. Darling River
5. Canberra
6. Great Dividing Range
7. Auckland
8. Southern Alps
9. Perth
10. Tasmania

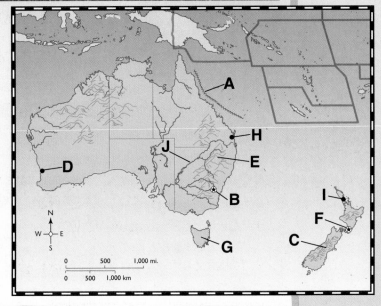

GeoLab
ACTIVITY

From the classroom of
Rebecca Revis, Sylvan
Hills Junior High School,
Sherwood, Arkansas

GLACIERS: Earth's Scouring Pads

Background

Antarctica, the world's coldest and iciest region, has mountains, valleys, and lowlands underneath its ice cap. Like the Arctic Circle, the Antarctic Circle also has many glaciers. Glaciers and rivers erode the land and change it a great deal. In this activity, you'll observe how glaciers and rivers change the earth's surface.

Glacier-created valley, New Zealand

Believe it or NOT!

Glaciers cover about one-tenth of the earth's land and hold about 85 percent of all freshwater. The power of glaciers is enormous. Consider that the Great Lakes were formed by a glacier scouring out a river valley.

Materials

- water
- sand, gravel, and clay
- stream table with sand
- lamp with reflector
- ruler

(NOTE: You will be using electrical equipment near water in this GeoLab. Please keep these items away from one another.)

What To Do

A. Set up the stream table and lamp as shown.

B. Make an ice block by mixing water with sand, gravel, and clay in a container that measures about 2 inches (5 cm) by 8 inches (20 cm) by 1 inch (2 cm). Then freeze this container.

C. Make a V-shaped river channel. Measure and record its width and depth. Draw a sketch that includes these measurements.

D. Place the ice block, to act as a moving glacier, at the upper end of the stream table.

E. Gently push the glacier along the river channel until it is under the light, halfway between the top and bottom of the stream table.

F. Turn on the light and allow the ice to melt. Observe and record what happens.

G. Measure and record the width and depth of the glacial channel. Draw a sketch of the channel and include these measurements in your journal.

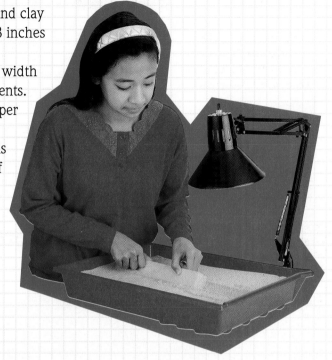

Lab Activity Report

1. How can you figure out the direction from which a glacier traveled?

2. How can you tell how far down the valley the glacier traveled?

3. How do glaciers affect the surface over which they move?

4. **Drawing Conclusions** How can you identify land that was once covered by a glacier?

Go A Step Further

+2   +2

Before your "glacier" melts completely, take one small part of it and rub it over a piece of wood. What happens in the wood's surface? How do glaciers make similar patterns in rocks?

Cultural HERITAGE:
AUSTRALIA AND OCEANIA

ARCHITECTURE ▶ ▶ ▶ ▶

The opera house in Sydney's harbor is a landmark recognized throughout the world. The overlapping white shells of its roofline were designed to imitate the shapes of boat sails in the harbor.

◀ ◀ ◀ ◀ WOOD CARVING

The Maori of New Zealand are expert wood-carvers. Many of their intricately textured carvings portray ancestors. The carvings decorate homes and community meetinghouses.

PAINTING ▶ ▶ ▶ ▶

Nature plays an important role in the lives of the Australian Aborigines, and it is reflected in their art. This painting, done on bark, shows many animal, plants, and landscape features.

This mask, with its large eyes and bright colors, represents the face of an owl. The people of Papua, New Guinea, make animal masks to use in traditional dances.

THATCHED BUILDINGS ► ► ► ►

New Guinea builders use the trunks of palm trees to build the frames for their beautiful spirit houses. The walls and roof are woven from palm leaves. Only men are allowed in these buildings, which hold each village's religious relics.

BOOMERANGS ▲ ▲ ▲ ▲

Australian boomerangs come in many shapes and sizes. The Aborigines use boomerangs mostly for hunting. Some are carved or painted for use in religious ceremonies. The small boomerangs are toys for children.

APPRECIATING CULTURE

1. What natural materials do the Aborigines, Maori, and people of New Guinea need to create their artwork and buildings?

2. How would you describe the Sydney Opera House to someone who has not seen it?

CHAPTER 26

Tropic of Cancer

150°E   165°E   180°   165°W   150°W

15°N

Guam (U.S.)

NORTHERN MARIANA IS. (U.S.)

Wake Island (U.S.)

MARSHALL ISLANDS

Hawaii (U.S.)

O C E A N I A

PACIFIC OCEAN

N
W   E
S

Koror

PALAU

Palikir ⊛

FEDERATED STATES OF MICRONESIA

Majuro

0°   Tarawa ⊛

WALLIS & FUTUNA (Fr.)

Equator

PAPUA NEW GUINEA

Yaren ⊛ NAURU

KIRIBATI

Port Moresby ⊛

SOLOMON IS.
⊛ Honiara

Funafuti ⊛
TUVALU

TOKELAU (N.Z.)

SAMOA
Apia ⊛

15°S

VANUATU

Port-Vila ⊛

⊛ Suva

AMERICAN SAMOA (U.S.)

COOK ISLANDS (N.Z.)

FRENCH POLYNESIA (Fr.)

AUSTRALIA

NEW CALEDONIA (Fr.)

FIJI   Nuku'alofa ⊛

PITCAIRN (Br.)

30°S

TONGA

NIUE (N.Z.)

Tropic of Capricorn

— National boundary
⊛ National capital

Mercator projection

0   150   300 mi.
0   150   300 km

45°S

NEW ZEALAND

International Date Line

ANTARCTICA (inset map)

Russia

U.K. Argentina

ATLANTIC OCEAN   0°

South Africa   Russia

Argentina

India   Japan   40°E

Chile   U.S.   U.K.   Japan   Russia

U.K.   U.K.   Argentina   Argentina   Australia

U.S.   Russia   80°E

A N T A R C T I C A

South Pole

U.S.   Australia

100°W   Russia   Russia

PACIFIC OCEAN

Russia   New Zealand   U.S.   Australia

140°W   Russia   France

0   250 mi.
0   250 km

Antarctic Circle

180°

INDIAN OCEAN

■ Major scientific research station

National Claims
- Argentina
- Australia
- Chile
- France
- United Kingdom
- New Zealand
- Norway

Azimuthal Equal-Area projection

MAP STUDY ACTIVITY

You are ending your geographic "tour" of the world by learning about Oceania and Antarctica.

1. What is the capital of Papua New Guinea?
2. How many research stations does the United States have in Antarctica?

SECTION 1

Oceania

PREVIEW

Words to Know
- continental island
- copra
- high island
- low island
- atoll
- phosphate
- trust territory

Places to Locate
- Melanesia
- Micronesia
- Polynesia

Read to Learn . . .
1. what geographic areas make up Oceania.
2. how land and climate affect the way people in Oceania earn their livings.
3. what groups have settled in Oceania.

What do you picture when you hear the word "paradise"? You might see one of the more than 30,000 tropical islands that dot the blue waters of the Pacific Ocean. Some of the islands, like New Guinea shown here, are mountainous.

The Pacific island region is called Oceania. Great distances separate this region from other parts of the world. The islands of Oceania—about 25,000 of them—are spread out across 70 million square miles (181.3 million sq. km). Because of Oceania's vast distances, geographers group the islands into three main regions: Melanesia, Micronesia, and Polynesia.

Location

Melanesia

North and east of Australia lie the islands of Melanesia (MEH•luh•NEE•zhuh). Because of their dense vegetation, Melanesia's islands are often called the "black islands." Melanesia's land and ocean area combined almost match the land area of the continental United States. The largest country in size and population is Papua New Guinea (PA•pyuh•wuh noo GIH•nee). It lies on the eastern half of the island of New Guinea.

Southeast of Papua New Guinea are three other independent island countries: the Solomon Islands, Fiji, and Vanuatu (VAN•WAH•TOO). Near these countries is New Caledonia, a group of islands ruled by France.

The Climate and Economy Melanesia consists mostly of continental islands. A **continental island** is formed by chunks of land split off from a larger continent or through the erosion of a link of land that once connected the island to the mainland. Rugged mountains and thick rain forests cover the continental islands of Melanesia. Strips of fertile plains hug island coastlines.

Thermometers record little change in temperatures in Melanesia year-round. The map on page 720 shows you that most of Melanesia has a tropical climate. The temperature along island coasts seldom falls below 70°F (21°C) or rises above 80°F (27°C).

Most Melanesians are subsistence farmers, growing only enough to feed their families. Papua New Guinea and Fiji, however, cultivate agricultural products for export. Farmers in Papua New Guinea grow cacao, tea, and coffee. Those in Fiji raise sugarcane. A major money-making product of Melanesia—and the rest of Oceania—is **copra**, or dried coconut meat. Countries around the world use coconut oil from copra to make margarine, soap, and other products.

Some Melanesian islands, such as Papua New Guinea and New Caledonia, hold rich mineral resources. Rugged mountains and vast shipping distances, however, make it costly to develop mining and other industries.

The Past and Present Melanesians share many different cultures and languages. More than 700 languages are spoken in Papua New Guinea alone! To ease communication, English is the language of business and government throughout most of Melanesia. From the 1800s to the mid-1900s, Western countries ruled most of Melanesia. Many island countries today are a blend of Western and Melanesian cultures.

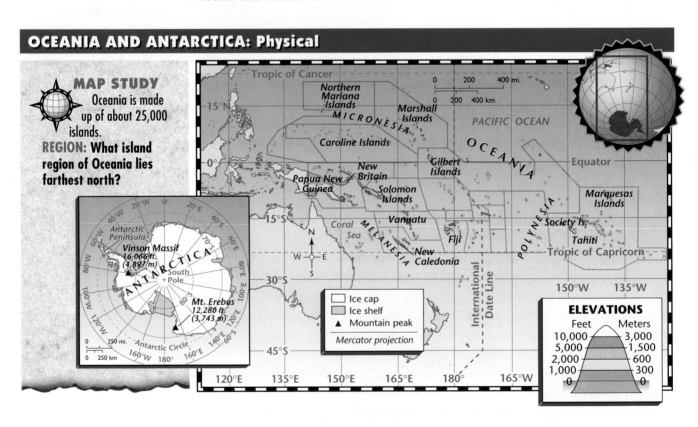

OCEANIA AND ANTARCTICA: Physical

MAP STUDY
Oceania is made up of about 25,000 islands.
REGION: What island region of Oceania lies farthest north?

UNIT 9

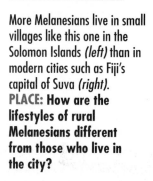

Many Melanesians still live in small villages in houses made of grass and other natural materials. They keep strong ties to their local group and hold on to traditional customs. Only a few Melanesians live in cities. Port Moresby in Papua New Guinea and Suva in Fiji are among the largest urban areas. Many people in these cities live in modern houses and have jobs in business and government.

More Melanesians live in small villages like this one in the Solomon Islands *(left)* than in modern cities such as Fiji's capital of Suva *(right)*.
PLACE: How are the lifestyles of rural Melanesians different from those who live in the city?

Region
Micronesia

North of Melanesia lies another island region—Micronesia. The name refers to the "tiny islands" in this region. Look at the map on page 712 to see the contrasts between the two regions. The map shows you that the islands of Micronesia are scattered across a vast ocean area. The independent countries of Micronesia are the Federated States of Micronesia, the Marshall Islands, Palau (puh•LOW), Nauru (nah•OO•roo), and Kiribati (KIHR•uh•BAS). The Northern Mariana Islands and the island of Guam are territories linked to the United States.

High and Low Islands Rich soil, ribbonlike waterfalls, and thick green vegetation blanket Micronesia's **high islands**. Volcanic activity formed most of these rugged islands many centuries ago. The rest of Micronesia includes **low islands** usually made of coral, and have little vegetation. Skeletons of millions of tiny sea animals formed the low islands over thousands of years. Most of the low islands are **atolls**, or ring-shaped coral reefs that surround small lagoons. You often will find atolls built up on the rims of underwater volcanoes that no longer erupt. The chart on page 714 shows you how a typical atoll is formed.

The Climate and Economy Micronesia has a tropical climate. Daily temperatures average about 80°F (27°C) year-round, and rain is plentiful across the region. Tropical breezes flowing over these islands are not always gentle, however. From July to October, fierce storms called typhoons sometimes strike the islands, causing loss of life and much destruction.

Subsistence farming is the major economic activity on Micronesia's high islands. Farmers there grow food crops in the fertile soil and raise cattle, sheep, and goats. On some high islands, copra is a major export. People on the low coral islands rely on the sea for food.

Most Micronesian island countries depend on aid from the United States, Australia, and Europe. With this aid they are building roads, ports, and airfields so that industry can develop in the future. The only exception is the tiny island republic of Nauru. For years the 10,400 people of Nauru have prospered from the mining of **phosphate,** a mineral used in making fertilizers. With Nauru's phosphate deposits running out, however, the people are looking for other sources of income.

The Past and Present Southeast Asians first settled Micronesia about 4,000 years ago. European explorers, traders, and missionaries settled on the Micronesian islands during the 1700s and early 1800s. By the early 1900s, many European countries, the United States, and Japan owned colonies in this region.

During World War II, the United States and Japan fought a number of bloody battles on Micronesian islands. Why? Look at the map on page A30 to see the islands' location in relation to Japan and the United States. Following World War II, most of Micronesia was turned over to the United States as **trust territories**. These territories were under temporary United States control. Since the 1970s, many of the Micronesian islands have become independent.

Formation of an Atoll/Coral Reef

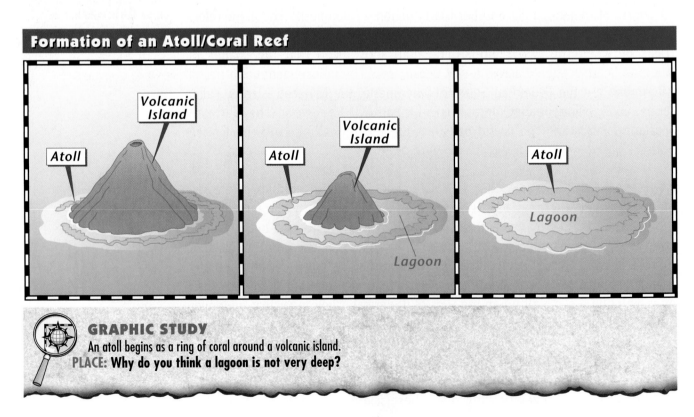

GRAPHIC STUDY
An atoll begins as a ring of coral around a volcanic island.
PLACE: Why do you think a lagoon is not very deep?

Today Micronesia's 400,000 people speak 11 languages and several dialects. Most of them live in villages headed by local chiefs. In recent years many young people have left the villages to find jobs in towns.

Place
Polynesia

East of Melanesia and Micronesia lie the tropical islands often pictured on travel brochures—the islands that make up Polynesia. The name refers to the "many islands" in the region. The map on page 712 shows you that Polynesia is the largest of the three Pacific island regions. Three independent countries—Samoa, Tonga, and Tuvalu—are found in Polynesia. Other island groups are under French rule and are known as French Polynesia. Tahiti, Polynesia's largest island, is part of this French territory.

The Climate and Economy Most Polynesian islands are high volcanic islands, some with tall, rugged mountains. Thick rain forests cover mountain valleys and coastal plains. Other Polynesian islands are of the low coral type. With little soil, the only vegetation is scattered coconut palms. Because most of Polynesia lies in the tropics, the climate is hot and humid.

Most people in Polynesia raise crops or fish for their food. Some farmers raise coconuts and tropical fruits for export. Tourism is the fastest-growing industry of Polynesia. Each year thousands of visitors come by air or sea to its emerald-green mountains and white, palm-lined beaches. New roads, hotels, shops, and restaurants serve these tourists.

The Past and Present Settlers came later to Polynesia than to the other two Pacific regions. The first Polynesians were probably Melanesians or Micronesians who sailed to the islands in canoes.

During the late 1800s, the stronger European powers divided up Polynesia among themselves. Military bases were built on these well-located islands in the 1900s. They were perfect refueling stops for airplanes crossing the Pacific. Beginning in the 1960s, some Polynesian territories chose independence while others decided to remain under Western rule.

Today about 580,000 people live in Polynesia. Modern influences have affected the islands, yet some Polynesians live traditionally in villages. Women in Samoa and Tonga still make tapa cloth. They first strip the inner bark from paper

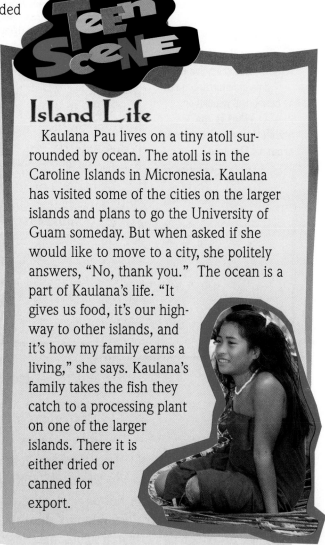

Teen Scene

Island Life

Kaulana Pau lives on a tiny atoll surrounded by ocean. The atoll is in the Caroline Islands in Micronesia. Kaulana has visited some of the cities on the larger islands and plans to go the University of Guam someday. But when asked if she would like to move to a city, she politely answers, "No, thank you." The ocean is a part of Kaulana's life. "It gives us food, it's our highway to other islands, and it's how my family earns a living," she says. Kaulana's family takes the fish they catch to a processing plant on one of the larger islands. There it is either dried or canned for export.

OCEANIA AND ANTARCTICA: Population Density

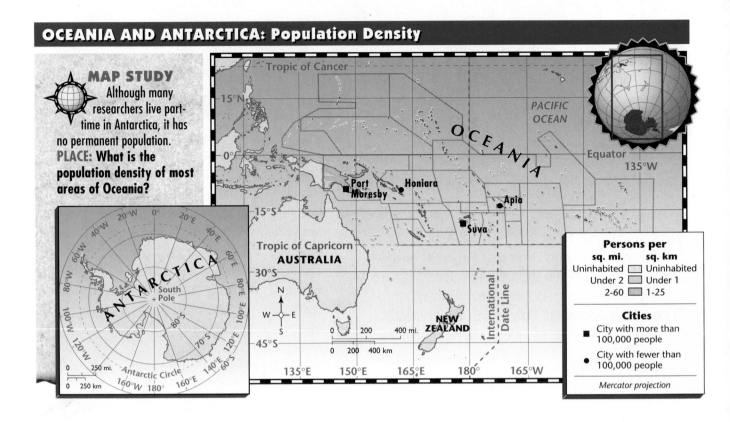

MAP STUDY
Although many researchers live part-time in Antarctica, it has no permanent population.
PLACE: What is the population density of most areas of Oceania?

mulberry trees. Then they soak the bark and beat it with wooden clubs to form the cloth which is used to make clothing.

An increasing number of Polynesians live in cities or towns. Thirteen-year-old Henri Moeino lives in Papeete (PAH•pee•AY•tee), the largest city and port on the island of Tahiti. Henri points out that Papeete and its surrounding area are home to more than 100,000 people. He and other Tahitians belong to an interesting mix of peoples. Henri has Polynesian and French ancestry. Like most urban Polynesians, Henri follows both European and American ways.

SECTION 1 ASSESSMENT

REVIEWING TERMS AND FACTS

1. Define: continental island, copra, high island, low island, atoll, phosphate, trust territory.

2. REGION What are the three island regions of Oceania?

3. PLACE How do the high and low islands in Micronesia differ in appearance?

4. HUMAN/ENVIRONMENT INTERACTION How do most people in Oceania make a living?

MAP STUDY ACTIVITIES

5. Turn to the political map on page 710. What is the capital of Fiji?

6. Look at the physical map on page 712. What is the largest island in Oceania?

UNIT 9

BUILDING GEOGRAPHY SKILLS

Analyzing a Photograph

You can learn much about the geography and culture of a place from a photograph. The key is to examine clues in the photo. Because photographs show only a small "slice" of reality, however, they can be misleading. For example, a picture of the wealthiest part of a city does not accurately show the living standards of all of the city's people. To accurately interpret a photograph, follow these steps:

- Read the caption.
- Name the landforms, climate, vegetation, cultural features, and human activities in the photograph.
- Think about what may be excluded from this view.

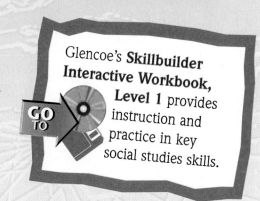

Glencoe's **Skillbuilder Interactive Workbook, Level 1** provides instruction and practice in key social studies skills.

Drying Copra on the Solomon Islands

A woman collects dried coconuts, called copra, which will be pressed to produce coconut oil. About 30 coconuts are needed to make 1 gallon of oil. The oil is then used to make soaps, margarine, detergents, cosmetics, and many other products.

Geography Skills Practice

1. What is the main topic of this photograph?
2. What natural features are included in the photograph?
3. What information about culture is shown?
4. What information do you think is left out of the photograph?

SECTION 2

Antarctica

Words to Know
- rookery
- ice shelf
- crevasse
- krill
- ozone

Places to Locate
- Antarctica
- Transantarctic Mountains
- South Pole

Read to Learn . . .
1. where Antarctica is located.
2. why Antarctica is called a polar desert.
3. how Antarctica is like a scientific laboratory.

Who doesn't love penguins? Their shiny "tuxedos" and waddling walk have fascinated people for years. Scientists study the habits of penguins, the largest group of marine animals in Antarctica. For example, studying where they build their nests—or **rookeries**—gives scientists more clues to understanding the vast continent of Antarctica.

Antarctica—surrounded by icy ocean water—sits on the southern end of the earth. It covers about 5,400,000 square miles (14,000,000 sq. km). It is larger in size than either Europe or Australia. Mysterious Antarctica is the least explored of all the continents.

Location

Landforms

Picture Antarctica—a rich, green land covered by forests and lush plants. Does this description match your mental picture of the continent? Fossils discovered there tell scientists that millions of years ago Antarctica's landscape was inhabited by dinosaurs and small mammals.

Today a huge ice cap buries nearly 98 percent of Antarctica's land area. In many places, the ice cap is 1.5 to 2 miles (2.4 to 3.2 km) thick—about the height of 10 tall skyscrapers. This sea of ice holds about 70 percent of the world's freshwater. Where the ice cap spreads beyond the land and over the ocean, it forms an **ice shelf.**

Icy Coasts The ice cap is heavy. In some areas, it forms **crevasses**, or cracks, more than 100 feet (30 m) deep. As glaciers meet the sea, large chunks of ice break off to form icebergs. Antarctica borders the Atlantic, Pacific, and Indian oceans. As these oceans reach Antarctica, they become cooler and less salty, which allows them to freeze. Some of the fish in these waters have a special antifreeze—a protein in their blood that keeps it from freezing.

Landforms Beneath the packed layers of the ice cap lie highlands, valleys, and lowlands—the same features you would find on other continents. A long mountain range called the Transantarctic Mountains crosses Antarctica and splits it in two. The highest peak on the continent—Vinson Massif—rises here to 16,066 feet (4,897 m). The Transantarctic Mountains sweep out into the Antarctic Peninsula, which reaches within 600 miles (965 km) of South America's Cape Horn.

A high, flat plateau covers the area east of the Transantarctic Mountains. The earth's southernmost point, the South Pole, lies on the plateau at the center of Antarctica. West of the mountains is a group of low islands buried under layers and layers of ice. On Ross Island rises the peak of Mount Erebus. At 12,220 feet (3,743 m), it is Antarctica's most active volcano.

Place
Climate

Now that you have a mental picture of Antarctica's ice cap, think about this: Antarctica receives so little precipitation that it is the world's largest, coldest desert! Inland Antarctica receives no rain and hardly any new snow each year.

A scientist lowers into a crevasse near his research station.
HUMAN/ENVIRONMENT INTERACTION: What may happen to Antarctica's environment if too many people work there?

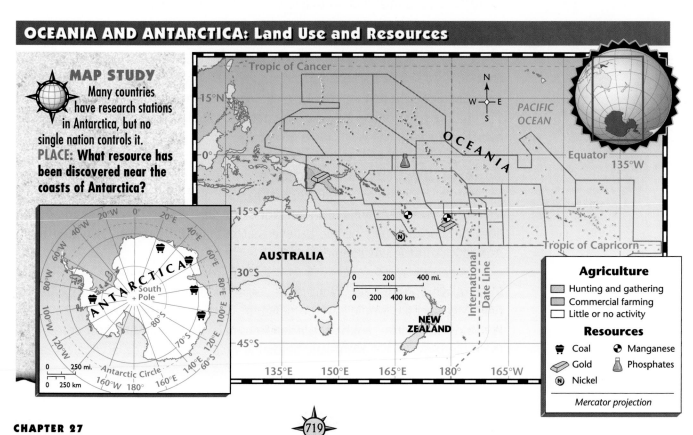

OCEANIA AND ANTARCTICA: Land Use and Resources

MAP STUDY Many countries have research stations in Antarctica, but no single nation controls it.
PLACE: What resource has been discovered near the coasts of Antarctica?

Tropic of Cancer

15°N

0°

PACIFIC OCEAN

OCEANIA

Equator
135°W

15°S

AUSTRALIA

30°S

0   200   400 mi.
0   200   400 km

International Date Line

Tropic of Capricorn

NEW ZEALAND

45°S

ANTARCTICA
South Pole

0   250 mi.
0   250 km
Antarctic Circle

135°E   150°E   165°E   180°   165°W

Agriculture
- Hunting and gathering
- Commercial farming
- Little or no activity

Resources
- Coal
- Gold
- Nickel
- Manganese
- Phosphates

Mercator projection

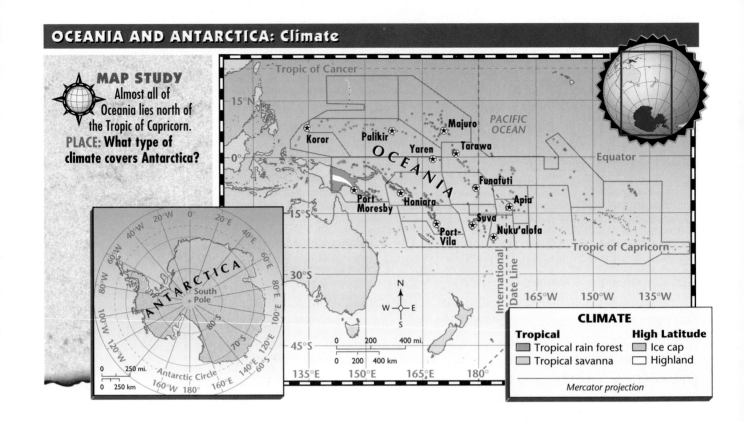

MAP STUDY
Almost all of Oceania lies north of the Tropic of Capricorn.
PLACE: What type of climate covers Antarctica?

The map above shows you that Antarctica has a polar ice cap climate. Imagine summer in a place where temperatures may fall as low as −30°F (−34.5°C) and climb only to 32°F (0°C)! Antarctic summers last from December through February. Winter temperatures along the coasts fall to −40°F (−40°C), and in inland areas to a low of −100°F (−73°C).

Region

Resources

Scientists believe that the Antarctic ice hides a treasure chest of minerals. The mining of resources on this continent is a source of international disagreement, however.

Mineral Resources Antarctica has major deposits of coal, and lesser amounts of copper, gold, iron ore, manganese, and zinc hidden underneath its ice cap. Petroleum may lie offshore. None of Antarctica's mineral resources have been developed. Many people believe that mining these resources would harm Antarctica's fragile environment.

Plants and Animals What is the largest permanent land animal in Antarctica? Most people would say the penguin. They would be incorrect, however. Penguins are marine animals, not land animals. Today the life-form considered to be the largest permanent land animal is *Beligica antarctica*, a wingless insect about one-tenth of an inch long.

Only a few small plants and insects can survive in Antarctica's cold, dry, inland areas. Most plants on the continent are algae and mosses that grow along the coast on rocky surfaces. The waters and coasts around Antarctica, however, are home to penguins, fish, whales, and many kinds of flying birds. A small shrimplike animal called **krill** is a key source of food for other animals in Antarctica.

Human/Environment Interaction

Exploration and Research

Because of its harsh climate, Antarctica is the only continent that has no permanent human settlement. Its largely unspoiled environment has made it a favorable place for scientific research.

WHAT IN THE WORLD?

Operation Highjump

In 1946 the United States carried out the largest exploration of Antarctica up to that time. Captain Richard E. Byrd led 4,700 people in Operation Highjump. With 13 ships and 23 airplanes, Byrd and his group found more than 25 new islands and photographed 1,400 miles (2,300 km) of uncharted coastline.

Explorers and Scientists Europeans first sighted Antarctica during the early 1800s. In 1911 explorers competed with one another to reach the South Pole. You can read about this exciting race on page 722. The discovery of the South Pole opened the rest of Antarctica for exploration.

By the 1960s scientists from 12 countries had set up research centers on Antarctica. The map on page 710 shows you the major research centers. To preserve Antarctica as a peaceful scientific research lab, 12 countries signed the Antarctic Treaty in 1959. In the early 1990s, a number of countries agreed to keep Antarctica free of mining for 50 years.

Much research in Antarctica deals with **ozone,** a form of oxygen in the atmosphere. You learned in Chapter 3 that ozone protects all living things on the earth from certain harmful rays of the sun. In the 1980s scientists discovered a weakening or "hole" in the ozone layer above Antarctica. Studying this layer will help scientists learn more about possible changes in the world's climates.

SECTION 2 ASSESSMENT

REVIEWING TERMS AND FACTS

1. Define the following: rookery, ice shelf, crevasse, ozone.

2. LOCATION What oceans surround Antarctica?

3. PLACE What kind of climate does Antarctica have?

4. HUMAN/ENVIRONMENT INTERACTION Why was the 1959 Antarctic Treaty important?

MAP STUDY ACTIVITIES

5. Look at the map on page 719. What major mineral resource lies in Antarctica?

6. Turn to the climate map on page 720. How do the climates of Oceania and Antarctica differ?

MAKING CONNECTIONS

THE RACE TO THE SOUTH POLE

MATH | SCIENCE | **HISTORY** | LITERATURE | TECHNOLOGY

The big news in the fall of 1911 was the race between Norway's Roald Amundsen and Britain's Robert Scott. The two seasoned explorers took on the challenge of finding the exact location of the South Pole.

READY, SET . . . Arriving at Ross Island, Scott and his crew set up headquarters at Cape Evans. Their equipment included 21 ponies and 3 tractor-like, motorized sleds.

Amundsen and his team pitched camp at the extreme western edge of the Ross Ice Shelf. To make the journey easier, Amundsen brought trained sled dogs and special skis.

. . . GO!! Amundsen's crew and dogsleds pulled out on October 19. Their expedition cut a new route over the Great Ice Barrier and through Antarctica's great central plateau. During the 1,400-mile (2,253-km) trip, the men left supplies in seven well-marked places. They reached the South Pole on December 14. There they put up a small tent on the Pole's exact location and left a letter inside for Scott. The team arrived back at their original campsite on January 25, 1912, in good health and high spirits.

On November 1, Scott and his 15-member team headed for the central plateau on a well-known route across Beardmore Glacier. The trip was a disaster. Their motorized sleds bogged down in soft snow, and supplies ran low. Eleven men and all the ponies died. Only Scott and 4 others made it to the South Pole on January 17, 1912. On the return trip to Cape Evans, however, the men were unable to find their poorly marked supply sites in the vast windy Antarctic wastes. All died from hunger and cold.

Scott's Expedition

Making the Connection

1. What transportation equipment did each team bring?

2. How did Antarctica's geography and climate defeat Scott?

722

Chapter 27 Highlights

Important Things to Know About Oceania and Antarctica

SECTION 1 OCEANIA

- Oceania is made up of three regions—Melanesia, Micronesia, and Polynesia.
- Oceania has three kinds of islands: continental islands separated from continents, high islands formed by volcanoes, and low islands made from coral reefs.
- Most islands of Polynesia were trust territories at one time. Today many are independent.
- The cultures of these island groups are a mixture of local traditions, the cultures of European settlers, and modern influences.
- Farming, fishing, and tourism are the major economic activities of Oceania.
- The largest island region in Oceania by size is Polynesia.

SECTION 2 ANTARCTICA

- A huge ice cap covers most of Antarctica.
- The earth's southernmost point, the South Pole, lies near the center of Antarctica.
- Potential uses of Antarctica's resources include mining of minerals and transforming ice into freshwater that could be used by the rest of the world.
- Animals such as penguins, fish, whales, birds, and krill make their homes in the icy waters and on the coasts of Antarctica.
- Antarctica is the only continent with no permanent human settlement.
- Antarctica is a major center of international scientific research.

Port Ucla, Vanuatu ▶

REVIEWING KEY TERMS

Match the numbered terms in Set A with their lettered definitions in Set B.

A
1. krill
2. atoll
3. crevasse
4. high island
5. copra
6. low island
7. continental island
8. phosphate

B
A. large island thrust up from the ocean
B. large break in the Antarctic ice
C. mineral used to make fertilizer
D. shrimplike animal found in Antarctic waters
E. island formed from coral reefs
F. coral reef surrounding a bay
G. island made from volcanoes
H. dried coconut meat

Mental Mapping Activity

Draw a freehand map of Antarctica. Label the following:
• Atlantic Ocean
• Pacific Ocean
• Antarctic Peninsula
• South Pole
• Antarctic Circle

REVIEWING THE MAIN IDEAS

Section 1
1. LOCATION Where is Oceania located?
2. REGION What is the major climate found in Melanesia, Micronesia, and Polynesia?
3. PLACE What mineral is mined in the island republic of Nauru?
4. MOVEMENT Why do people in Polynesia leave their islands?

Section 2
5. LOCATION What lies under the ice cap of Antarctica?
6. PLACE What is summer like in Antarctica?
7. PLACE What kinds of plants grow in Antarctica?
8. REGION Why is Antarctica considered a good place for scientific research?

CRITICAL THINKING ACTIVITIES

1. **Drawing Conclusions** Why is tourism a growing industry in Oceania?
2. **Predicting Consequences** What might have happened to Antarctica if the Antarctic treaties had not been signed?

CONNECTIONS TO WORLD CULTURES

Cooperative Learning Activity Work in groups to learn more about one of the island countries of Oceania. Each group will choose one country from Melanesia, Micronesia, or Polynesia. Then each group member will select one topic to research for a group essay entitled: "One Day of My Life in _____." The essay should describe what a person's daily activities in that location might be like. Put your group's information together to write an essay. Add illustrations if possible. Share your essays with the rest of the class.

GeoJournal Writing Activity

Imagine that you are a scientist working in Antarctica. Write a letter home describing your work and the work of other scientists at your research station.

TECHNOLOGY ACTIVITY

Using a Database Research the wildlife of Australia, Oceania, and Antarctica. Create a database of your information with separate fields for the following: animal, natural habitat, diet, natural predators, introduced predators, and population status. Use the database to create a map showing the locations of each animal group in the region.

PLACE LOCATION ACTIVITY: OCEANIA AND ANTARCTICA

Match the letters on the map with the places and features of Oceania and Antarctica. Write your answers on a separate sheet of paper.

1. Port Moresby
2. Solomon Islands
3. French Polynesia
4. South Pole
5. Fiji
6. Federated States of Micronesia
7. Antarctic Peninsula
8. Pacific Ocean

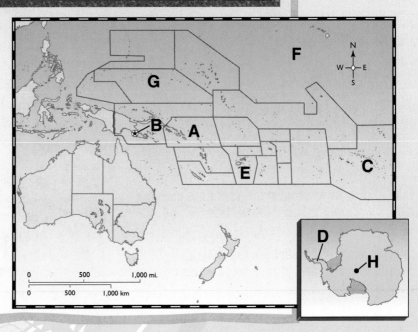

Great Barrier Reef
TROUBLE DOWN UNDER

PROBLEM

Australia's Great Barrier Reef, the world's largest coral reef, is threatened by human activity. Pollution and commercial fishing endanger the reef. The possibility of a major oil spill—hundreds of oil tankers pass near the reef each year—looms as a constant worry. But tourism poses the most serious threat to the reef's health. And every year more than *2 million* people visit the reef!

SOLUTIONS

● To protect the reef and its treasures, the government of Australia works with environmental organizations around the world.

● New laws restrict where and how people may fish, dive, sail, snorkel, or build near the reef.

● Oil tankers, other large ships, and ships carrying hazardous cargo must carry a specially trained pilot to help navigate the reef.

● Specific areas of the reef are set aside for certain tourist activities. Other areas are off-limits except to scientists.

A leopard moray eel pokes its head out of coral. The Great Barrier Reef supports fish, whales, turtles, and countless other sea creatures.

CORAL REEF FACT BANK

🍃 Coral reefs, formed by the accumulated skeletons of coral, are "fish factories" crucial to the world's food chain.

🍃 The Great Barrier Reef covers about 135,000 square miles (350,000 sq. km). It is 1,250 miles (2,011 km) long and contains 2,600 separate reefs and about 300 islands.

🍃 It is home to more than 1,500 species of fish, 400 types of coral, 4,000 species of mollusks, and 22 types of whales.

🍃 The Great Barrier Reef—the largest structure built by living creatures anywhere in the world—is clearly visible from space.

An island—among more than 300 in the Great Barrier Reef—rises from the crystal-clear Coral Sea. Agricultural runoff and other pollution endanger the sea's delicate chemistry and the reef's colonies of coral.

NEW ZEALAND

environmental activities

TEEN TRIBUTE

Five students from Pioneer State High School, in Mackay, Australia, set up a self-guided snorkeling tour of the Great Barrier Reef that keeps tourists off the coral. The students studied an area and then laid out an underwater trail. By carefully choosing the path and placing information markers on the trail, the students help protect the reef—and educate sightseers.

WHAT CAN YOU DO?

TAKE A CLOSER LOOK

1 Where is the Great Barrier Reef located?

2 What human activities have threatened the reef?

🍃 Write to oil companies, encouraging them to be watchful when shipping oil.

🍃 Leave nature the way you find it whenever hiking or camping in state and national parks.

🍃 Find out about a natural area in your community or state that people are trying to protect. Join in!

🍃 Support the International Green Cross—an environmental protection group organized at the recent Earth Summit.

Equal Rights for All Species

727

A diver explores the reef, eyeing table coral and other wonders in this enchanted undersea garden that stretches for more than a thousand miles.

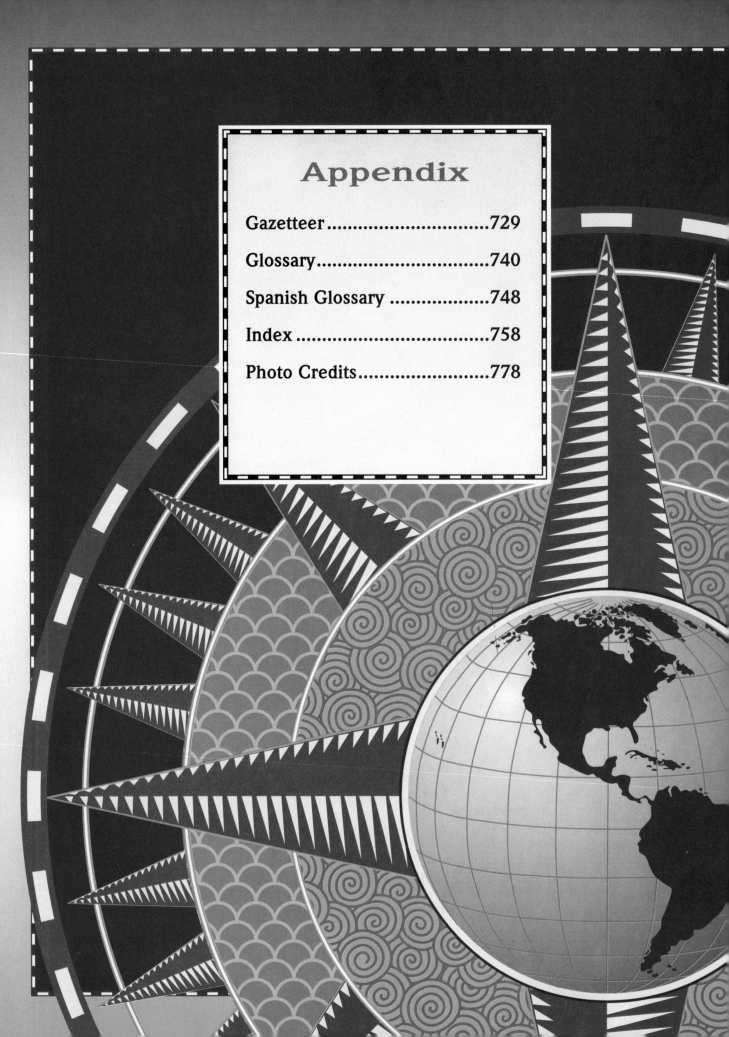

Appendix

Gazetteer

A Gazetteer (GAZ•uh•TIR) is a geographic index or dictionary. It shows latitude and longitude for cities and certain other places. Latitude and longitude are shown in this way: 48°N 2°E, or 48 degrees north latitude and two degrees east longitude. This Gazetteer lists most of the world's largest independent countries, their capitals, and several important geographic features. The page numbers tell where each entry can be found on a map in this book. As an aid to pronunciation, most entries are spelled phonetically.

A

Abidjan [A•bih•JAHN] Capital and port of Côte d'Ivoire, Africa. 5°N 4°W (p. 476)

Abu Dhabi [AH•boo DAH•bee] Capital of the United Arab Emirates, on the Persian Gulf. 24°N 54°E (p. 414)

Abuja [ah•BOO•jah] Capital of Nigeria. 8°N 9°E (p. 476)

Accra [uh•KRAH] Port city and capital of Ghana. 6°N 0° longitude (p. 476)

Addis Ababa [A•duhs A•buh•buh] Capital of Ethiopia. 9°N 39°E (p. 476)

Adriatic [AY•dree•A•tihk] **Sea** Arm of the Mediterranean Sea between the Balkan Peninsula and Italy. 44°N 14°E (p. 246)

Afghanistan [af•GA•nuh•STAN] Country in central Asia, west of Pakistan. 33°N 63°E (p. 414)

Albania [al•BAY•nee•uh] Country on the east coast of the Adriatic Sea, south of Serbia and Montenegro. 42°N 20°E (p. 248)

Algeria [al•JIHR•ee•uh] Country in North Africa. 29°N 1°E (p. 414)

Algiers [al•JIHRZ] Capital of Algeria. 37°N 3°E (p. 414)

Alps [ALPS] Mountain system extending east to west through central Europe. 46°N 9°E (p. 246)

Amazon [A•muh•ZAHN] **River** Largest river in the world by volume and second-largest in length. 2°S 53°W (p. 144)

Amman [ah•MAHN] Capital of Jordan. 32°N 36°E (p. 414)

Amsterdam [AMP•stuhr•DAM] Capital of the Netherlands. 52°N 5°E (p. 248)

Andes [AN•DEEZ] Mountain system extending north to south along the western side of South America. 13°S 75°W (p. 144)

Andorra [an•DAWR•uh] Small country in southern Europe, between France and Spain. 43°N 2°E (p. 248)

Angola [ang•GOH•luh] Country in Africa, west of Zambia. 14°S 16°E (p. 476)

Ankara [ANG•kuh•ruh] Capital of Turkey. 40°N 33°E (p. 414)

Antananarivo [AN•tuh•NA•nuh•REE•VOH] Capital of Madagascar. 19°S 48°E (p. 476)

Arabian [uh•RAY•bee•uhn] **Peninsula** Large peninsula of southwest Asia, extending north to south into the Arabian Sea. 28°N 40°E (p. 412)

Argentina [AHR•juhn•TEE•nuh] Country in the southern part of South America, east of Chile on the Atlantic Ocean. 36°S 67°W (p. 146)

Armenia [ahr•MEE•nee•uh] Southeastern European country between the Black Sea and the Caspian Sea. 40°N 45°E (p. 354)

Ashkhabad [ASH•kuh•BAD] Capital of Turkmenistan. 38°N 58°E (p. 354)

Asmara [az•MAHR•uh] Capital of Eritrea. 16°N 39°E (p. 476)

Astana [ah•STAH•nah] Capital of Kazakhstan. 52°N 72°E (p. 354)

Asunción [uh•SOONT•see•OHN] Capital of Paraguay. 25°S 58°W (p. 146)

Athens [A•thuhnz] Capital and largest city in Greece. 38°N 24°E (p. 248)

Australia [aw•STRAYL•yuh] Country and continent southeast of Asia. 23°S 135°E (p. 682)

Austria [AWS•tree•uh] Country in central Europe, east of Switzerland and south of Germany and the Czech Republic. 47°N 12°E (p. 248)

Azerbaijan [A•zuhr•BY•JAHN] European-Asian country on the Caspian Sea. 40°N 47°E (p. 354)

B

Baghdad [BAG•DAD] Capital of Iraq. 33°N 44°E (p. 414)

Bahama [buh•HAH•muh] **Islands** Country made up of many islands, between Cuba and the United States. 23°N 74°W (p. 145)

Baku [bah•KOO] Port city and capital of Azerbaijan. 40°N 50°E (p. 354)

Balkan [BAWL•kuhn] **Peninsula** Peninsula in southeastern Europe, mostly covered by the country of Greece. 42°N 20°E (p. 246)

Baltic [BAWL•tihk] **Sea** Sea in northern Europe that is an arm of the Atlantic Ocean connected to the North Sea. 55°N 17°E (p. 246)

Bamako [BAH•muh•KOH] Capital of Mali. 13°N 8°W (p. 476)

Bangkok [BANG•KAHK] Capital of Thailand. 14°N 100°E (p. 580)

Bangladesh [BAHNG•gluh•DEHSH] Country in South Asia, bordered by India and Myanmar. 24°N 90°E (p. 580)

Bangui [BAHNG•GEE] Capital of the Central African Republic. 4°N 19°E (p. 476)

Banjul [BAHN•JOOL] Port city and capital of Gambia. 13°N 17°W (p. 476)

Barbados [bahr•BAY•DOHS] Island country between the Atlantic Ocean and the Caribbean Sea. 14°N 59°W (p. 146)

Beijing [BAY•JIHNG] Cultural and industrial center and capital of China. 40°N 116°E (p. 580)

Beirut [bay•ROOT] Capital of Lebanon. 34°N 36°E (p. 414)

Belarus [BEE•luh•ROOS] Eastern European country west of Russia and east of Poland. 54°N 28°E (p. 354)

Belgium [BEHL•juhm] Country in northwestern Europe, south of the Netherlands. 51°N 3°E (p. 248)

Belgrade [BEHL•GRAYD] Capital of Yugoslavia (Serbia). 45°N 21°E (p. 248)

Belize [buh•LEEZ] Country in Central America, east of Guatemala. 18°N 89°W (p. 146)

Belmopan [BEHL•moh•PAN] Capital of Belize. 17°N 89°W (p. 146)

Benin [buh•NIHN] Country in western Africa. 8°N 2°E (p. 476)

Berlin [BUHR•LIHN] Capital of Germany. 53°N 13°E (p. 248)

Bern [BUHRN] Capital of Switzerland. 47°N 7°E (p. 248)

Bhutan [boo•TAHN] Country in the eastern Himalayas, northeast of India. 27°N 91°E (p. 580)

Bishkek [bihsh•KEHK] Largest city and capital of Kyrgyzstan. 43°N 75°E (p. 354)

Bissau [bih•SOW] Capital of Guinea-Bissau. 12°N 16°W (p. 476)

Black Sea Large sea between Europe and Asia. 43°N 32°E (p. 352)

Bogotá [BOH•guh•TAH] Capital of Colombia. 5°N 74°W (p. 146)

Bolivia [buh•LIH•vee•uh] Country in the central part of South America, north of Argentina. 17°S 64°W (p. 146)

Bosnia-Herzegovina [BAHZ•nee•uh HEHRT•suh•goh•VEE•nuh] Southeastern European country between Yugoslavia and Croatia. 44°N 18°E (p. 248)

Botswana [baht•SWAH•nuh] Country in Africa, north of the Republic of South Africa. 22°S 23°E (p. 476)

Brasília [bruh•ZIHL•yuh] Capital of Brazil. 16°S 48°W (p. 146)

Bratislava [BRA•tuh•SLAH•vuh] Capital and largest city of Slovakia. 48°N 17°E (p. 248)

Brazil [bruh•ZIHL] Largest country in South America. 9°S 53°W (p. 146)

Brazzaville [BRA•zuh•VIHL] Port city and capital of the Congo Republic in west central Africa. 4°S 15°E (p. 476)

Brunei [bru•NY) Country in Southwest Asia, on northern coast of the island of Borneo. 5°N 114°E (p. 578)

Brussels [BRUH•suhlz] Capital of Belgium. 51°N 4°E (p. 248)

Bucharest [BOO•kuh•REHST] Capital of Romania. 44°N 26°E (p. 248)

Budapest [BOO•duh•PEHST] Capital of Hungary. 48°N 19°E (p. 248)

Buenos Aires [BWAY•nuhs AR•eez] Capital of Argentina, on the Río de la Plata. 34°S 58°W (p. 146)

Bujumbura [BOO•juhm•BUR•uh] Capital of Burundi in eastern Africa near Lake Tanganyika. 3°S 29°E (p. 476)

Bulgaria [BUHL•GAR•ee•uh] Country in southeastern Europe, south of Romania. 42°N 24°E (p. 248)

Burkina Faso [bur•KEE•nuh FAH•soh] Country in western Africa, south of Mali. 12°N 3°E (p. 476)

Burundi [bu•ROON•dee] Country in central Africa at the northern end of Lake Tanganyika. 3°S 30°E (p. 476)

C

Cairo [KY•ROH] Capital of Egypt, on the Nile River. 31°N 32°E (p. 414)

Cambodia [kam•BOH•dee•uh] Country in Southeast Asia, south of Thailand. 12°N 104°E (p. 580)

Cameroon [KA•muh•ROON] Country in west Africa, on the northeast shore of the Gulf of Guinea. 6°N 11°E (p. 476)

Canada [KA•nuh•duh] Northernmost country in North America. 50°N 100°W (p. 84)

Canberra [KAN•buh•ruh] Capital of Australia. 35°S 149°E (p. 684)

Cape Town Port city and legislative capital of the Republic of South Africa. 34°S 18°E (p. 544)

Caracas [kuh•RAH•kuhs] Capital of Venezuela. 11°N 67°W (p. 146)

Caribbean [KAR•uh•BEE•uhn] **Sea** Part of the Atlantic Ocean, bordered by the West Indies, South America, and Central America. 15°N 76°W (p. 144)

Caspian [KAS•pee•uhn] **Sea** Salt lake between Europe and Asia that is the world's largest inland body of water. 40°N 52°E (p. 352)

Caucasus [KAW•kuh•suhs] **Mountains** Mountain range in the southern part of the former Soviet Union. 43°N 42°E (p. 352)

Central African Republic Country in central Africa, south of Chad. 8°N 21°E (p. 476)

Chad [CHAD] Country in central Africa, west of Sudan. 18°N 19°E (p. 476)

Chang [CHANG] **River** Principal river of China that rises in Tibet and flows into the East China Sea near Shanghai. 31°N 117°E (p. 578)

Chile [CHIH•lee] South American country, west of Argentina. 35°S 72°W (p. 146)

China [CHY•nuh] Country in eastern and central Asia, known officially as the People's Republic of China. 37°N 93°E (p. 580)

Chisinau [KEE•shih•NOW] Largest city and capital of Moldova. 47°N 29°E (p. 354)

Colombia [kuh•LUHM•bee•uh] Country in South America, west of Venezuela. 4°N 73°W (p. 146)

Colombo [kuh•LUHM•BOH] Capital of Sri Lanka. 7°N 80°E (p. 580)

Comoros [KAH•muhr•ROHS] **Islands** Small island country in Indian Ocean between the island of

Madagascar and the southeast African mainland. 13°S 43°E (p. 476)

Conakry [KAH•nuh•kree] Capital of Guinea. 10°N 14°W (p. 476)

Congo [KAHNG•GOH] **Republic** Country in equatorial Africa, south of the Central African Republic. 3°S 14°E (p. 476)

Congo [KAHNG•GOH] **Democratic Republic of the** African country on the Equator, north of Zambia and Angola. 1°S 22°E (p. 476)

Copenhagen [KOH•puhn•HAY•guhn] Capital of Denmark. 56°N 12°E (p. 248)

Costa Rica [KAHS•tuh REE•kuh] Central American country, south of Nicaragua. 11°N 85°W (p. 146)

Côte d'Ivoire [KOHT dee•VWAHR] West African country, south of Mali. 8°N 7°W (p. 476)

Croatia [kroh•AY•shuh] Southeastern European country on the Adriatic Sea. 46°N 16°E (p. 248)

Cuba [KYOO•buh] Island in the West Indies. 22°N 79°W (p. 146)

Cyprus [SY•pruhs] Island country in the Mediterranean Sea, south of Turkey. 35°N 31°E (p. 248)

Czech [CHEHK] **Republic** European country south of Germany and Poland. 50°N 15°E (p. 248)

D

Dakar [DA•KAHR] Port city and capital of Senegal. 15°N 17°W (p. 476)

Damascus [duh•MAS•kuhs] Capital of Syria. 34°N 36°E (p. 414)

Dar es Salaam [DAHR•ehs suh•LAHM] Capital of Tanzania. 7°S 39°E (p. 520)

Denmark [DEHN•MAHRK] Country in northwestern Europe, between the Baltic and North Seas. 56°N 9°E (p. 248)

Dhaka [DA•kuh] Capital of Bangladesh. 24°N 90°E (p. 580)

Djibouti [juh•BOO•tee] Country in east Africa, on the Gulf of Aden. 12°N 43°E (p. 476)

Dodoma [DOH•doh•mah] Future capital of Tanzania. 7°S 36°E (p. 476)

Doha [DOH•HAH] Capital of Qatar. 25°N 51°E (p. 414)

Dominican [duh•MIH•nih•kuhn] **Republic** Country in the West Indies on the eastern part of Hispaniola Island. 19°N 71°W (p. 174)

Dublin [DUH•bluhn] Port city and capital of Ireland. 53°N 6°W (p. 248)

Dushanbe [doo•SHAM•buh] Largest city and capital of Tajikistan north of Pakistan. 39°N 69°E (p. 354)

E

Ecuador [EH•kwuh•DAWR] Country in South America, south of Colombia. 0° latitude 79°W (p. 146)

Egypt [EE•juhpt] Country in northern Africa on the Mediterranean Sea. 27°N 27°E (p. 414)

El Salvador [el SAL•vuh•DAWR] Country in Central America, southwest of Honduras. 14°N 89°W (p. 146)

Equatorial Guinea [EE•kwuh•TOHR•ee•uhl GIH•nee] Country in western Africa, south of Cameroon. 2°N 8°E (p. 476)

Eritrea [EHR•uh•TREE•uh] Country in eastern Africa on the Red Sea, north of Ethiopia. 17°N 39°E (p. 476)

Estonia [eh•STOH•nee•uh] Northern European country on the Baltic Sea, north of Latvia. 59°N 25°E (p. 248)

Ethiopia [EE•thee•OH•pee•uh] Country in eastern Africa, north of Somalia and Kenya. 8°N 38°E (p. 476)

Euphrates [yu•FRAY•TEEZ] **River** River in southwestern Asia that flows through Syria and Iraq and joins the Tigris River. 36°N 40°E (p. 412)

F

Fiji [FEE•JEE] Country comprised of an island group in the southwest Pacific Ocean. 19°S 175°E (p. 684)

Finland [FIHN•luhnd] Country in northern Europe, east of Sweden. 63°N 26°E (p. 248)

France [FRANTS] Country in western Europe, south of the English Channel. 47°N 1°E (p. 248)

Freetown [FREE•TOWN] Port city and capital of Sierra Leone, in western Africa. 9°N 13°W (p. 476)

French Guiana [gee•AH•nah] French-owned territory in northern South America. 5°N 53°W (p. 146)

G

Gabon [ga•BOHN] Country in western Africa on the Atlantic Ocean, west of Congo. 0° latitude 12°E (p. 476)

Gaborone [GAH•buh•ROH•NAY] Capital of Botswana, in southern Africa. 24°S 26°E (p. 476)

Gambia [GAM•bee•uh] Country in western Africa, along the Gambia River. 13°N 16°W (p. 476)

Georgetown [JAWRJ•TOWN] Capital of Guyana. 8°N 58°W (p. 146)

Georgia [JAWR•juh] Asian-European country bordering the Black Sea, south of Russia. 42°N 43°E (p. 354)

Germany [JUHR•muh•nee] Country in north central Europe, officially called the Federal Republic of Germany. 52°N 10°E (p. 248)

Ghana [GAH•nuh] Country in western Africa on the Gulf of Guinea, east of Côte d'Ivoire. 8°N 2°W (p. 476)

Great Plains The continental slope extending through the United States and Canada. 45°N 104°W (p. 82)

Greece [GREES] Country in southern Europe, mostly on the Balkan Peninsula. 39°N 22°E (p. 246)

Greenland [GREEN•luhnd] Island in northwestern Atlantic Ocean and the largest island in the world. 74°N 40°W (p. 84)

Guatemala [GWAH•tuh•MAH•luh] Country in Central America, south of Mexico. 16°N 92°W (p. 146)

Guatemala City Capital of Guatemala and the largest city in Central America. 15°N 91°W (p. 146)

Guinea [GIH•nee] West African country on the Atlantic coast north of Sierra Leone. 11°N 12°W (p. 476)

Guinea-Bissau [GIH•nee bih•SOW] West African country on the Atlantic coast. 12°N 20°W (p. 476)

Gulf of Mexico Gulf on the southeast coast of North America. 25°N 94°W (p. 82)

Guyana [gy•A•nuh] South American country on the Atlantic coast, between Venezuela and Suriname. 8°N 59°W (p. 146)

H

Haiti [HAY•tee] Country on Hispaniola Island in the West Indies. 19°N 72°W (p. 174)

Hanoi [ha•NOY] Capital of Vietnam. 21°N 106°E (p. 580)

Harare [huh•RAH•RAY] Capital of Zimbabwe. 18°S 31°E (p. 476)

Havana [huh•VA•nuh] Seaport, capital city of Cuba, and largest city of the West Indies. 23°N 82°W (p. 146)

Helsinki [HEHL•SIHNG•kee] Capital of Finland. 60°N 24°E (p. 248)

Himalayas [HI•muh•LAY•uhs] Mountain range in South Asia, bordering the Indian subcontinent on the north. 30°N 85°E (p. 578)

Honduras [hahn•DUR•uhs] Central American country, on the Caribbean Sea. 15°N 88°W (p. 146)

Hong Kong [HAHNG KAHNG] Administrative district and port in southern China. 22°N 115°E (p. 580)

Huang He [HWAHNG•HE] River in north central and eastern China, also known as the Yellow River. 35°N 114°E (p. 578)

Hungary [HUHNG•guh•ree] Central European country, south of Slovakia. 47°N 18°E (p. 248)

I

Iberian [y•BIHR•ee•uhn] **Peninsula** Peninsula in southwest Europe, occupied by Spain and Portugal. 41°N 1°W (p. 246)

Iceland [YS•luhnd] Island country between the North Atlantic and the Arctic oceans. 65°N 20°W (p. 248)

India [IHN•dee•uh] South Asian country, south of China and Nepal. 23°N 78°E (p. 580)

Indonesia [IHN•duh•NEE•zhuh] Group of islands that forms the Southeast Asian country known as the Republic of Indonesia. 5°S 119°E (p. 580)

Indus [IHN•duhs] **River** River in Asia that rises in Tibet and flows through Pakistan to the Arabian Sea. 27°N 68°E (p. 578)

Iran [ih•RAHN] Southwest Asian country that was formerly named Persia, east of Iraq. 31°N 54°E (p. 414)

Iraq [ih•RAHK] Southwest Asian country, south of Turkey. 32°N 43°E (p. 414)

Ireland [YR•luhnd] Island west of England, occupied by the Republic of Ireland and by Northern Ireland. 54°N 8°W (p. 248)

Islamabad [ihs•LAH•muh•BAHD] Capital of Pakistan. 34°N 73°E (p. 580)

Israel [IHZ•ree•uhl] Country in southwest Asia, south of Lebanon. 33°N 34°E (p. 414)

Italy [IH•tuhl•ee] Southern European country, south of Switzerland and east of France. 44°N 11°E (p. 248)

J

Jakarta [juh•KAHR•tuh] Capital of Indonesia. 6°S 107°E (p. 580)

Jamaica [juh•MAY•kuh] Island country in the West Indies. 18°N 78°W (p. 146)

Japan [juh•PAN] Country in east Asia, consisting of the four large islands of Hokkaido, Honshu, Shikoku, and Kyushu, plus thousands of small islands. 37°N 134°E (p. 580)

Jerusalem [juh•ROO•suh•luhm] Capital of Israel and a holy city for Christians, Jews, and Muslims. 32°N 35°E (p. 414)

Jordan [JAWR•duhn] Country in southwest Asia, south of Syria. 30°N 38°E (p. 414)

K

Kabul [KAH•buhl] Capital of Afghanistan. 35°N 69°E (p. 414)

Kampala [kahm•PAH•luh] Capital of Uganda. 0° latitude 32°E (p. 476)

Kathmandu [KAT•MAN•DOO] Capital of Nepal. 28°N 85°E (p. 580)

Kazakhstan [KA•ZAK•STAN] Large Asian country south of Russia, bordering the Caspian Sea. 48°N 59°E (p. 354)

Kenya [KEH•nyuh] Country in eastern Africa, south of Ethiopia. 1°N 37°E (p. 476)

Khartoum [kahr•TOOM] Capital of Sudan. 16°N 33°E (p. 476)

Kiev [KEE•ehf] Capital of Ukraine. 50°N 31°E (p. 354)

Kigali [kih•GAH•lee] Capital of Rwanda, in central Africa. 2°S 30°E (p. 476)

Kingston [KIHNG•stuhn] Capital of Jamaica. 18°N 77°W (p. 146)

Kinshasa [kihn•SHAH•suh] Capital of the Democratic Republic of the Congo. 4°S 15°E (p. 476)

Kuala Lumpur [KWAH•luh LUM•PUR] Capital of Malaysia. 3°N 102°E (p. 580)

Kuwait [ku•WAYT] Country between Saudi Arabia and Iraq, on the Persian Gulf. 29°N 48°E (p. 414)

Kyrgyzstan [KIHR•gih•STAN] Small central Asian country on China's western border. 41°N 75°E (p. 354)

L

Lagos [LAY•GAHS] Port city of Nigeria. 6°N 3°E (p. 495)

Laos [LOWS] Southeast Asian country, south of China and west of Vietnam. 20°N 102°E (p. 580)

La Paz [luh PAHZ] The administrative capital of Bolivia, and the highest capital in the world. 17°S 68°W (p. 146)

Latvia [LAT•vee•uh] Northeastern European country on the Baltic Sea, west of Russia. 57°N 25°E (p. 248)

Lebanon [LEH•buh•nuhn] Country on the Mediterranean Sea, south of Syria. 34°N 34°E (p. 414)

Lesotho [luh•SOH•TOH] Country in south Africa, within the borders of the Republic of South Africa. 30°S 28°E (p. 476)

Liberia [ly•BIHR•ee•uh] West African country, south of Guinea. 7°N 10°W (p. 476)

Libreville [LEE•bruh•VIHL] Port city and capital of Gabon. 1°N 9°E (p. 476)

Libya [LIH•bee•uh] North African country on the Mediterranean Sea, west of Egypt. 28°N 15°E (p. 414)

Liechtenstein [LIHK•tuhn•SHTYN] Small country in central Europe, between Switzerland and Austria. 47°N 10°E (p. 248)

Lilongwe [lih•LAWNG•way] Capital of Malawi, in southeastern Africa. 14°S 34°E (p. 476)

Lima [LEE•muh] Capital of Peru. 12°S 77°W (p. 146)

Lisbon [LIHZ•buhn] Port city and capital of Portugal. 39°N 9°W (p. 248)

Lithuania [LIH•thuh•WAY•nee•uh] Northeastern European country on the Baltic Sea, west of Belarus. 56°N 24°E (p. 248)

Ljubljana [lee•OO•blee•AH•nuh] Largest city and capital of Slovenia. 46°N 14°E (p. 248)

Lomé [loh•MAY] Port city and capital of Togo in Africa. 6°N 1°E (p. 476)

London [LUHN•duhn] Capital of the United Kingdom, on the Thames River. 52°N 0° longitude (p. 248)

Luanda [lu•AN•duh] Port city and capital of Angola. 9°S 13°E (p. 476)

Lusaka [loo•SAH•kuh] Capital of Zambia. 15°S 28°E (p. 476)

Luxembourg [LUHK•suhm•BUHRG] Small European country between France, Belgium, and Germany. 50°N 7°E (p. 248)

M

Macao [muh•KOW] Administrative district and port in southern China. 22°N 113°E (p. 580)

Macedonia [MA•suh•DOH•nee•uh] **Former Yugoslav Republic of** Southeastern European country north of Greece. 42°N 22°E (p. 248). Macedonia also refers to a geographic region in the Balkan Peninsula.

Madagascar [MA•duh•GAS•kuhr] Island in the Indian Ocean off the southeast coast of Africa. 18°S 43°E (p. 476)

Madrid [muh•DRIHD] Capital of Spain. 40°N 4°W (p. 248)

Malabo [mah•LAH•BOH] Capital of Equatorial Guinea. 4°N 9°E (p. 476)

Malawi [muh•LAH•wee] Southeastern African country, south of Tanzania and east of Zambia. 11°S 34°E (p. 476)

Malaysia [muh•LAY•zhuh] Federation of states in Southeast Asia on the Malay Peninsula and the island of Borneo. 4°N 101°E (p. 578)

Maldive [MAWL•DEEV] **Islands** Island country in the Indian Ocean near South Asia. 5°N 42°E (p. 580)

Mali [MAH•lee] Country in western Africa, east of Mauritania and south of Algeria. 16°N 0° longitude (p. 476)

Managua [muh•NAH•gwuh] Capital of Nicaragua. 12°N 86°W (p. 146)

Manila [muh•NIH•luh] Port city and capital of the Republic of the Philippines. 15°N 121°E (p. 580)

Maseru [MA•suh•ROO] Capital of Lesotho, in southern Africa. 29°S 27°E (p. 476)

Mauritania [MAWR•uh•TAY•nee•uh] Western African country, north of Senegal. 20°N 14°W (p. 476)

Mauritius [maw•RIH•shuhs] Small island country in the Indian Ocean east of Madagascar. 21°S 58°E (p. 476)

Mbabane [EHM•bah•BAH•nay] Capital of Swaziland, in southeastern Africa. 26°S 31°E (p. 476)

Mediterranean [MEH•duh•tuh•RAY•nee•uhn] **Sea** Large inland sea surrounded by Europe, Asia, and Africa. 36°N 13°E (p. 246)

Mekong [MAY•KAWNG] **River** River in Southeast Asia that rises in Tibet and empties into the South China Sea. 18°N 104°E (p. 578)

Mexico [MEHK•sih•KOH] Country in North America, south of the United States. 24°N 104°W (p. 146)

Mexico City Capital and most populous city of Mexico. 19°N 99°W (p. 146)

Minsk [MIHNTSK] Capital of Belarus. 54°N 28°E (p. 354)

Mississippi [MIH•suh•SIH•pee] **River** Large river system in the central United States that flows southward into the Gulf of Mexico. 32°N 92°W (p. 82)

Mogadishu [MAH•guh•DIH•SHOO] Major seaport and capital of Somalia, in eastern Africa. 2°N 45°E (p. 476)

Moldova [mahl•DOH•vuh] Small European country between Ukraine and Romania. 48°N 28°E (p. 354)

Monaco [MAH•nuh•KOH] Small country in southern Europe, on the French Mediterranean coast. 44°N 8°E (p. 248)

Mongolia [mahn•GOH•lee•uh] Country in Asia between Russia and China 46°N 100°E (p. 580)

Monrovia [MUHN•ROH•vee•uh] Major seaport and capital of Liberia, in western Africa. 6°N 11°W (p. 476)

Montevideo [MAHN•tuh•vuh•DAY•OH] Seaport and capital of Uruguay in South America. 35°S 56°W (p. 146)

Morocco [muh•RAH•KOH] Country in northwestern Africa on the Mediterranean Sea and the Atlantic Ocean. 32°N 7°W (p. 414)

Moscow [MAHS•KOH] Capital and largest city of Russia. 56°N 38°E (p. 354)

Mount Everest [EHV•ruhst] Highest mountain in the world, in the Himalayas between Nepal and Tibet. 28°N 87°E (p. 578)

Mozambique [MOH•zuhm•BEEK] Country in southeastern Africa, south of Tanzania. 20°S 34°E (p. 476)

Muscat [MUHS•KAT] Seaport and capital of Oman. 23°N 59°E (p. 414)

Myanmar [MYAHN•MAHR] Country in Southeast Asia, south of China and India, formerly called Burma. 21°N 95°E (p. 580)

N

Nairobi [ny•ROH•bee] Capital of Kenya. 1°S 37°E (p. 476)

Namibia [nuh•MIH•bee•uh] Country in southwestern Africa, south of Angola on the Atlantic Ocean. 20°S 16°E (p. 476)

Nassau [NA•SAW] Capital of the Bahamas. 25°N 77°W (p. 146)

N'Djamena [EN•juh•MAY•nuh] Capital of Chad. 12°N 15°E (p. 476)

Nepal [nuh•PAWL] Mountain country between India and China. 29°N 83°E (p. 580)

Netherlands [NEH•thuhr•luhndz] Western European country on the North Sea, north of Belgium. 53°N 4°E (p. 248)

New Delhi [NOO DEH•lee] Capital of India. 29°N 77°E (p. 580)

New Zealand [NOO ZEE•luhnd] Major island country in the South Pacific, southeast of Australia. 42°S 175°E (p. 684)

Niamey [nee•AH•may] Capital and commercial center of Niger, in western Africa. 14°N 2°E (p. 476)

Nicaragua [NI•kuh•RAH•gwuh] Country in Central America, south of Honduras. 13°N 86°W (p. 146)

Nicosia [NIH•kuh•SEE•uh] Capital of Cyprus. 35°N 33°E (p. 248)

Niger [NY•juhr] Landlocked country in western Africa, north of Nigeria. 18°N 9°E (p. 476)

Nigeria [ny•JIHR•ee•uh] Country in western Africa, south of Niger. 9°N 7°E (p. 476)

Nile [NYL] **River** Longest river in the world, flowing north and east through eastern Africa. 19°N 33°E (p. 412)

North Korea [kuh•REE•uh] Asian country in the northernmost part of the Korean Peninsula. 40°N 127°E (p. 580)

Norway [NAWR•WAY] Country on the Scandinavian Peninsula. 64°N 11°E (p. 248)

Nouakchott [noo•AHK•SHAHT] Capital of Mauritania. 18°N 16°W (p. 476)

O

Oman [oh•MAHN] Country on the Arabian Sea and the Gulf of Oman, east of Saudi Arabia. 20°N 58°E (p. 414)

Oslo [AHZ•loh] Capital and largest city of Norway. 60°N 11°E (p. 248)

Ottawa [AH•tuh•wuh] Capital of Canada. 45°N 76°W (p. 84)

Ouagadougou [WAH•guh•DOO•goo] Capital of Burkina Faso, in western Africa. 12°N 2°W (p. 476)

P

Pakistan [PA•kih•STAN] South Asian country on the Arabian Sea, northwest of India. 28°N 68°E (p. 580)

Palau [pah•LOW] Island country in the Pacific Ocean. 7°N 135°E (p. 684)

Panama [PA•nuh•MAH] Country in Central America, on the Isthmus of Panama. 9°N 81°W (p. 146)

Panamá Capital city of Panama. 9°N 79°W (p. 146)

Papua New Guinea [PA•pyuh•wuh NOO GIH•nee] Independent island country in the Pacific Ocean north of Australia. 7°S 142°E (p. 684)

Paraguay [PAR•uh•GWY] Country in South America, north of Argentina. 24°S 57°W (p. 146)

Paramaribo [PAIR•uh•MAIR•uh•BOH] Port city and capital of Suriname. 6°N 55°W (p. 146)

Paris [PAIR•uhs] River port and capital of France. 49°N 2°E (p. 248)

Persian [PUHR•zhuhn] **Gulf** Arm of the Arabian Sea between Iran and Saudi Arabia. 28°N 51°E (p. 412)

Peru [puh•ROO] Country in South America, south of Ecuador and Colombia. 10°S 75°W (p. 146)

Philippines [FIH•luh•PEENZ] Country in the Pacific Ocean, southeast of Asia. 14°N 125°E (p. 580)

Phnom Penh [NAWM•PEN] Capital of Cambodia. 12°N 106°E (p. 580)

Poland [POH•luhnd] Country on the Baltic Sea in eastern Europe. 52°N 18°E (p. 248)

Port-au-Prince [POHRT•oh•PRINTS] Port city and capital of Haiti. 19°N 72°W (p. 146)

Port Moresby [MOHRZ•bee] Port city and capital of Papua New Guinea. 10°S 147°E (p. 684)

Port of Spain [SPAYN] Capital of Trinidad and Tobago in the West Indies. 11°N 62°W (p. 146)

Porto-Novo [POHR•tuh•NOH•voh] Port city and capital of Benin, in western Africa. 7°N 3°E (p. 476)

Portugal [POHR•chih•guhl] Country on the Iberian Peninsula, south and west of Spain. 39°N 8°W (p. 246)

Prague [PRAHG] Capital of the Czech Republic. 51°N 15°E (p. 248)

Pretoria [prih•TOHR•ee•uh] Capital of South Africa. 26°S 28°E (p. 476)

Puerto Rico [PWEHR•tuh•REE•koh] Island in the Caribbean Sea; U.S. Commonwealth. 19°N 67°W (p. 146)

Pyongyang [pee•AWNG•YAHNG] Capital of North Korea. 39°N 126°E (p. 580)

Q

Qatar [KAH•tuhr] Country on the southwestern shore of the Persian Gulf. 25°N 53°E (p. 414)

Quito [KEE•toh] Capital of Ecuador. 0° latitude 79°W (p. 146)

R

Rabat [ruh•BAHT] Capital of Morocco. 34°N 7°W (p. 414)

GEOGRAPHY: The World and Its People

Reykjavík [RAY•kyuh•VIK] Capital of Iceland. 64°N 22°W (p. 248)

Rhine [RYN[**River** River in western Europe that flows into the North Sea. 51°N 7°E (p. 246)

Riga [REE•guh] Capital and largest city of Latvia. 57°N 24°E (p. 248)

Rio Grande [REE•OH•GRAND] River that forms the boundary between the United States and Mexico. 30°N 103°W (p. 82)

Riyadh [ree•YAHD] Capital of Saudi Arabia. 25°N 47°E (p. 414)

Rocky Mountains Mountain system in western North America. 50°N 114°W (p. 82)

Romania [roo•MAY•nee•uh] Country in eastern Europe, east of Hungary. 46°N 23°E (p. 248)

Rome [ROHM] Capital of Italy. 42°N 13°E (p. 248)

Russia [RUH•shuh] Largest country in the world, covering parts of Europe and Asia. 60°N 90°E (p. 354)

Rwanda [roo•AHN•duh] Country in Africa, south of Uganda. 2°S 30°E (p. 476)

S

Sahara [suh•HAIR•uh] Desert region in northern Africa that is the largest desert in the world. 24°N 2°W (p. 412)

Saint Lawrence [LAWR•uhns] **River** River that flows from Lake Ontario to the Atlantic Ocean and forms part of the boundary between the United States and Canada. 48°N 70°W (p. 82)

San'a [sa•NAH] Capital of Yemen. 15°N 44°E (p. 414)

San José [SAN•uh•ZAY] Capital of Costa Rica. 10°N 84°W (p. 146)

San Marino [SAN•mah•REE•noh] Small European country, located in the Italian Peninsula. 44°N 13°E (p. 248)

San Salvador [san•SAL•vuh•DAWR] Capital and industrial center of El Salvador. 14°N 89°W (p. 146)

Santiago [SAN•tee•AH•goh] Capital and major industrial center of Chile. 33°S 71°W (p. 146)

Santo Domingo [SAN•tuh•duh•MIN•goh] Capital of the Dominican Republic. 19°N 70°W (p. 146)

São Tomé and Príncipe [SOWN•tuh•MAY PRIN• suh•puh] Small island country in Gulf of Guinea off the coast of Central Africa. 1°N 7°E (p. 474)

Sarajevo [SAR•uh•YAY•voh] Capital of Bosnia-Herzegovina. 43°N 18°E (p. 248)

Saudi Arabia [SOW•dee uh•RAY•bee•uh] Country on the Arabian Peninsula. 23°N 46°E (p. 412)

Senegal [SEH•nih•GAWL] Country on the coast of western Africa, on the Atlantic Ocean. 15°N 14°W (p. 476)

Seoul [SOHL] Capital of South Korea. 38°N 127°E (p. 580)

Serbia [SUHR•bee•uh] European country south of Hungary. 44°N 21°E (p. 248)

Seychelles [say•SHELZ] Small island country in the Indian Ocean near East Africa. 6°S 56°E (p. 476)

Sierra Leone [see•EHR•uh•lee•OHN] Country in western Africa, south of Guinea. 8°N 12°W (p. 476)

Singapore [SIHNG•uh•POHR] Multi-island country in Southeast Asia near tip of Malay Peninsula. 2°N 104°E (p. 578)

Skopje [SKAW•PYAY] Capital of the Former Yugoslav Republic of Macedonia, in southeastern Europe. 42°N 21°E (p. 248)

Slovakia [sloh•VAH•kee•uh] Central European country south of Poland. 49°N 19°E (p. 248)

Slovenia [sloh•VEE•nee•uh] Small central European country on the Adriatic Sea, south of Austria. 46°N 15°E (p. 248)

Sofia [SOH•fee•uh] Capital of Bulgaria. 43°N 23°E (p. 248)

Solomon [SAW•lah•mahn] **Islands** Island country in the Pacific Ocean, northeast of Australia. 7°S 160°E (p. 684)

Somalia [soh•MAH•lee•uh] Country in Africa, on the Gulf of Aden and the Indian Ocean. 3°N 45°E (p. 476)

South Africa [A•frih•kuh] Country at the southern tip of Africa. 28°S 25°E (p. 476)

South Korea [kuh•REE•uh] Country in Asia on the Korean Peninsula between the Yellow Sea and the Sea of Japan. 36°N 128°E (p. 580)

Spain [SPAYN] Country on the Iberian Peninsula. 40°N 4°W (p. 246)

Sri Lanka [sree•LAHNG•kuh] Country in the Indian Ocean south of India, formerly called Ceylon. 9°N 83°E (p. 580)

Stockholm [STAHK•HOHLM] Capital of Sweden. 59°N 18°E (p. 248)

Sucre [SOO•kray] Constitutional capital of Bolivia. 19°S 65°W (p. 146)

Sudan [soo•DAN] Northeast African country on the Red Sea. 14°N 28°E (p. 476)

Suriname [SUHR•uh•NAH•muh] South American country on the Atlantic Ocean between Guyana and French Guiana. 4°N 56°W (p. 146)

Suva [SOO•vuh] Port city and capital of Fiji. 18°S 177°E (p. 684)

Swaziland [SWAH•zee•LAND] South African country west of Mozambique, almost entirely within the Republic of South Africa. 27°S 32°E (p. 476)

Sweden [SWEE•duhn] Northern European country on the eastern side of the Scandinavian Peninsula. 60°N 14°E (p. 246)

Switzerland [SWIT•suhr•luhnd] European country in the Alps, south of Germany. 47°N 8°E (p. 246)

Syria [SIHR•ee•uh] Country in Asia on the east side of the Mediterranean Sea. 35°N 37°E (p. 414)

T

Taipei [TY•PAY] Capital of Taiwan. 25°N 122°E (p. 580)

Taiwan [TY•WAHN] Island country off the southeast coast of China, and the seat of the Chinese Nationalist government. 24°N 122°E (p. 580)

Tajikistan [tah•JIH•kih•STAN] Central Asian country east of Turkmenistan. 39°N 70°E (p. 354)

Tallinn [TA•luhn] Largest city and capital of Estonia. 59°N 25°E (p. 248)

Tanzania [TAN•zuh•NEE•uh] East African country on the coast of the Indian Ocean, south of Uganda and Kenya. 7°S 34°E (p. 476)

Tashkent [tash•KENT] Capital of Uzbekistan and a major industrial center. 41°N 69°E (p. 354)

Tbilisi [tuh•BEE•luh•see] Capital of the Republic of Georgia. 42°N 45°E (p. 354)

Tegucigalpa [tuh•GOO•suh•GAL•puh] Capital of Honduras. 14°N 87°W (p. 146)

Tehran [TAY•RAN] Capital of Iran. 36°N 52°E (p. 414)

Thailand [TY•LAND] Southeast Asian country south of Myanmar. 17°N 101°E (p. 580)

Thimphu [THIHM•boo] Capital of Bhutan. 28°N 90°E (p. 580)

Tigris [TY•gruhs] **River** River in southeast Turkey and Iraq that merges with the Euphrates River. 35°N 44°E (p. 412)

Tiranë [tih•RAH•nuh] Capital of Albania. 42°N 20°E (p. 248)

Togo [TOH•goh] West African country between Benin and Ghana, on the Gulf of Guinea. 8°N 1°E (p. 474)

Tokyo [TOH•kee•OH] Capital of Japan. 36°N 140°E (p. 580)

Trinidad and Tobago [TRIH•nih•DAD tuh•BAY•goh] Island country between the Atlantic Ocean and the Caribbean Sea, near Venezuela. 11°N 61°W (p. 146)

Tripoli [TRIH•puh•lee] Capital of Libya. 33°N 13°E (p. 414)

Tunis [TOO•nuhs] Port city and capital of Tunisia. 37°N 10°E (p. 414)

Tunisia [too•NEE•zhuh] North African country on the Mediterranean Sea between Libya and Algeria. 35°N 10°E (p. 414)

Turkey [TUHR•kee] Country in southeastern Europe and western Asia. 39°N 32°E (p. 414)

Turkmenistan [TUHRK•MEH•nuh•STAN] Central Asian country on the Caspian Sea. 41°N 56°E (p. 354)

U

Uganda [oo•GAN•duh] East African country south of Sudan. 2°N 32°E (p. 476)

Ukraine [yoo•KRAYN] Large eastern European country west of Russia, on the Black Sea. 49°N 30°E (p. 354)

Ulan Bator [OO•LAHN•BAH•TAWR] Capital of Mongolia. 48°N 107°E (p. 580)

United Arab Emirates [ih•MIR•uhts] Country made up of seven states on the eastern side of the Arabian Peninsula. 24°N 54°E (p. 414)

United Kingdom Country in western Europe made up of England, Scotland, Wales, and Northern Ireland. 57°N 2°W (p. 248)

United States of America Country in North America made up of 50 states, mostly between Canada and Mexico. 38°N 110°W (p. 84)

Uruguay [UR•uh•GWY] South American country, south of Brazil on the Atlantic Ocean. 33°S 56°W (p. 146)

GEOGRAPHY: The World and Its People

Uzbekistan [OOZ•BEH•kih•STAN] Central Asian country south of Kazakhstan, on the Caspian Sea. 42°N 60°E (p. 354)

V

Vanuatu [VAN•WAH•TOO] Country made up of islands in the Pacific Ocean, east of Australia. 17°S 170°W (p. 684)

Vatican [VA•tih•kuhn] **City** Headquarters of the Roman Catholic Church, located in the city of Rome in Italy. 42°N 13°E (p. 253)

Venezuela [VEH•nuh•ZWAY•luh] South American country on the Caribbean Sea, between Colombia and Guyana. 8°N 65°W (p. 146)

Vienna [vee•EH•nuh] Capital of Austria. 48°N 16°E (p. 248)

Vientiane [VYEHN•TYAHN] Capital of Laos. 18°N 103°E (p. 580)

Vietnam [vee•EHT•NAHM] Southeast Asian country, east of Laos and Cambodia. 18°N 107°E (p. 580)

Virgin Islands Island territory of the United States, east of Puerto Rico in the Caribbean Sea. 18°N 65°W (p. 146)

W

Warsaw [WAWR•SAW] Capital of Poland. 52°N 21°E (p. 248)

Washington, D.C. Capital of the United States, in the District of Columbia. 39°N 77°W (p. 84)

Wellington [WEH•lihng•tuhn] Capital of New Zealand. 41°S 175°E (p. 684)

West Indies [IHN•deez] Islands in the Caribbean Sea, between North America and South America. 19°N 79°W (p. 144)

Windhoek [VIHNT•HOOK] Capital of Namibia, in southwestern Africa. 22°S 17°E (p. 476)

Y

Yamoussoukro [YAH•muh•SOO•kroh] Second capital of Côte d'Ivoire, in western Africa. 7°N 6°W (p. 496)

Yangon [YAHN•GOHN] Capital of Myanmar. 17°N 96°E (p. 580)

Yaoundé [yown•DAY] Capital of Cameroon, in western Africa. 4°N 12°E (p. 476)

Yemen [YEH•muhn] Country on the Arabian Peninsula, south of Saudi Arabia on the Gulf of Aden and the Red Sea. 15°N 46°E (p. 414)

Yerevan [YEHR•uh•VAHN] Largest city and capital of Armenia. 40°N 44°E (p. 354)

Z

Zagreb [ZAH•GREHB] Largest city and capital of Croatia. 46°N 16°E (p. 248)

Zambia [ZAM•bee•uh] Country in south central Africa, south of the Democratic Republic of the Congo and Tanzania. 14°S 24°E (p. 476)

Zimbabwe [zim•BAH•bwee] Country in south central Africa. 18°S 30°E (p. 476)

Glossary

A

abaca [a•buh•KAH] plant fiber used to make rope (p. 671)

absolute location the exact position of a place on the earth's surface (pp. 5, 21)

acid rain precipitation in which water carries large amounts of chemicals, especially sulfuric acid (pp. 68, 100, 283)

adobe [uh•DOH•bee] sun-dried clay brick (p. 166)

alluvial [uh•LOO•vee•uhl] **plain** plain built up from soil deposited by a river (pp. 435, 657)

altiplano [AL•tih•PLAH•NOH] high plateau region of the Andes (p. 222)

altitude [AL•tuh•TOOD] height above sea level (pp. 156, 202)

anthracite [AN•thruh•SYT] hard coal (p. 644)

apartheid [uh•PAHR•TAYT] South African legal restrictions and practices separating racial and ethnic groups, word means "apartness" (p. 549)

aquifer [AH•kwuh•fuhr] underground rock layer that stores large amounts of water (p. 39)

archipelago [AHR•kuh•PEH•luh•GOH] large group or chain of islands (pp. 183, 633)

atmosphere thick cushion of gases surrounding the earth, made up mainly of the gases nitrogen and oxygen (p. 32)

atoll [A•TAWL] ring-shaped coral reef or island surrounding a small bay (pp. 604, 713)

autobahn [AW•toh•BAHN] a superhighway in Germany (p. 284)

autonomy [aw•TAH•nuh•mee] self-government (p. 532)

axis (plural, **axes**) the horizontal (bottom) or vertical (side) line of measurement on a graph (p. 13)

B

bar graph graph in which vertical or horizontal bars represent quantities (p. 13)

basin broad flat lowland area surrounded by higher land (pp. 155, 195, 511)

bauxite [BAWK•SYT] mineral ore from which aluminum is taken (pp. 178, 497)

bazaar a marketplace (p. 451)

Bedouin [BEH•duh•wuhn] member of the nomadic desert peoples of North Africa and Southwest Asia (p. 429)

bilingual having or speaking two languages (p. 129)

birthrate number of children born each year for every 1,000 people (p. 62)

bog low-lying marshy land (pp. 261, 323)

buffer state small country located between larger, often hostile, states (p. 208)

bush Australian term for remote rural areas (p. 694)

C

cacao [kuh•KOW] tropical tree whose seeds are used to make cocoa and chocolate (p. 487)

calligraphy [kuh•LIH•gruh•fee] the art of beautiful handwriting (p. 622)

campesino [KAM•puh•SEE•noh] farmer in Latin America (p. 220)

canopy topmost layer of a rain forest, which shades the forest floor (p. 505)

cardinal directions the basic directions on the earth: north, south, east, west (p. 7)

casbah [KAZ•BAH] old area of cities with narrow streets and small shops in North Africa (p. 458)

cash crop crop grown to be sold, often for export (p. 220)

cassava [kuh•SAH•vuh] plant whose roots are ground into flour and eaten; in tapioca pudding, for example (pp. 210, 495, 663)

caudillo [kaw•DEE•yoh] in Latin American history, a military dictator (p. 203)

chart graphic way of presenting information clearly (p. 17)

chicle juice of the sapodilla tree, used in making chewing gum (p. 177)

circle graph round or pie-shaped graph showing how a whole is divided (p. 14)

city-state in history, an independent city and the lands around it (p. 306)

civil war conflict involving different groups within one country (pp. 430, 534)

civilization highly developed culture, usually with organized religion and laws (p. 56)

clan group of people or families related to one another (pp. 538, 637)

climate usual pattern of weather events in an area over a long period of time (pp. 15, 41)

climate region a broad area of earth with the same climate (p. 12)

climograph combination bar and line graph giving information about temperature and precipitation (p. 15)

cloves spice made from flower buds of clove trees (p. 529)

cold war era from late 1940s to early 1990s in which the United States and Soviet Union competed for world influence (p. 371)

colony overseas settlement made by a parent country (pp. 103, 187)

command economy economic system in which the government owns land, resources, and means of production and makes all economic decisions (p. 364)

commonwealth term for a republic or state governed by the people; the official title of Puerto Rico (p. 189)

communism authoritarian political system in which the central government controls the economy and society (pp. 187, 282, 319, 364)

compass rose device drawn on maps to show the directions (p. 7)

compound in rural Africa, a group of houses surrounded by a wall (p. 488)

condensation process in which water vapor changes into a liquid form (p. 38)

constitutional monarchy form of government headed by a monarch but with most power given to an elected legislature (pp. 259, 430)

consumer goods products for personal use, such as clothing and household goods (pp. 336, 366, 449, 615)

contiguous referring to areas that touch or share a boundary (p. 89)

continent one of the major land areas of the earth (p. 30)

continental island land originally part of a continent, separated by rising water or by geological activity (p. 712)

contour line on a contour map, line connecting all points at the same elevation (p. 9)

cooperative in Cuba, a farm owned and operated by the government (p. 188)

copper belt in Zambia, region of rich copper mines (p. 557)

copra dried inner meat of coconuts (p. 712)

coral rocklike material formed of the skeletons of small sea animals (pp. 521, 690)

coral reef low-lying ocean ridge made of coral (p. 93)

cordillera [KAW•duhl•YEHR•uh] group of mountain chains that run side by side (pp. 118, 217)

core central layer of the earth, probably composed of hot iron and nickel, solid in the inner core and molten in the outer (p. 29)

cottage industry home- or village-based industry in which family members supply their own equipment to make such goods as cloth and metalware (p. 590)

crevasse [krih•VAS] deep crack in an ice cap or glacier (p. 718)

crust outer layer of the earth (p. 29)

cultural diffusion process by which knowledge and skills spread from one area to another (p. 56)

culture way of life of a group of people who share similar beliefs and customs, including language and religion (p. 55)

currents moving streams of water, warm or cold, in the oceans; also streams of air (p. 42)

cyclone intense storm system with heavy rain and high winds blowing in a circular pattern (p. 596)

czar [ZAHR] title of the former emperors of Russia (p. 370)

D

death rate number of deaths each year for every 1,000 people (p. 62)

deforestation loss of forests due to widespread cutting of trees (p. 658)

delta triangle-shaped area at a river's mouth, formed of mud and sand deposited by water (pp. 448, 596, 660)

demographer scientist who studies population (p. 62)

desertification process in which grasslands become drier and desert areas expand (p. 491)

developed country country that is industrialized rather than agricultural (p. 62)

developing country country in the process of becoming industrialized (p. 62)

diagram drawing that shows steps in a process or parts of an object (p. 16)

dialect local form of a language (pp. 188, 284, 300)

dictatorship government under the control of a single all-powerful leader (p. 455)

dike high bank of soil or concrete built to hold back the water of a river or ocean (p. 611)

drought [DROWT] extended period of extreme dryness and water shortages (pp. 490, 537)

dry farming method of wheat farming in dry areas that conserves moisture by deep plowing (p. 98)

dynasty a family of rulers (p. 619)

dzong Buddhist center for prayer and study (p. 601)

E

earthquake violent jolting or shaking of the earth caused by movement of rocks along a fault (p. 30)

elevation height above sea level (pp. 9, 48, 310)

elevation profile cutaway diagram showing changes in elevation of land (p. 9)

emigrate to move from one's native country to another country (pp. 60, 312)

empire group of lands or states under the rule of one nation or ruler (pp. 223, 623)

enclave small nation or distinct region located inside a larger country (p. 545)

environment natural surroundings (p. 23)

equinox day in March and September when the sun's rays are directly overhead at the Equator, making day and night of equal length (p. 27)

erg desert region of shifting sand dunes (p. 458)

erosion wearing away of the earth's surface, mainly by water, wind, or ice (p. 31)

escarpment steep cliff separating two fairly flat land surfaces, one higher than the other (pp. 196, 522, 546)

estancia [ehs•TAHN•syah] a large ranch in Argentina (p. 230)

ethnic group people who share a common cultural background, including ancestry and language (pp. 104, 371, 438, 487)

evaporation process in which water from oceans, lakes, and streams is turned into a gas or vapor by the sun's heat (p. 38)

exclave small part of a nation separated from the main part of the country (p. 552)

F

famine lack of food, affecting a large number of people (p. 63)

farm belt region of the midwestern United States with flat land and fertile soil (p. 98)

fault crack in the rocks of the earth's crust along which movement occurs (pp. 30, 391, 522)

favela [fuh•VEH•luh] in Brazil, term used for urban slum areas (p. 199)

fellahin [fehl•uh•HEEN] Egyptian farmers and farm workers (p. 450)

fjord [FYOHRD] long, narrow, steep-sided inlet from the sea (pp. 266, 697)

flow chart type of diagram that shows direction or movement (p. 16)

food processing industry in which foods are prepared and packaged for sale (p. 393)

fossil fuels group of nonrenewable mineral resources—coal, oil, natural gas—formed in the earth's crust from plant and animal remains (pp. 66, 124)

free enterprise basis of a market economy in which people start and run businesses to make a profit with little government intervention (p. 58)

free enterprise system economic system based on free enterprise (see above) (pp. 96, 365)

free port area where goods can be loaded, stored, and reshipped without payment of import-export taxes (p. 666)

G

galaxy huge system in the universe including millions of stars (p. 25)

gaucho in southern South America, a cowhand on horseback (pp. 208, 230)

geography the study of the earth and its land, water, and plant and animal life (p. 21)

geothermal power electric power produced by natural underground sources of steam (p. 699)

geyser hot spring that spouts steam and hot water through a crack in the earth (p. 269)

glasnost [GLAZ•nohst] Russian term for "openness," used for President Gorbachev's policy in late 1980s (p. 371)

great circle route ship or airplane route following a great circle, the shortest distance between two points on the earth (p. 6)

grid system network of imaginary lines on the earth's surface, formed by the crisscrossing patterns of the lines of latitude and longitude (pp. 5, 22)

groundwater water stored in rock layers beneath the earth's surface (p. 39)

H

hajj religious journey made by Muslims to Makkah (p. 432)

harmattan [HAHR•muh•TAN] dry wind that blows southward from the Sahara in winter (p. 486)

heavy industry production of industrial goods such as machinery and military equipment (p. 365)

hemisphere one-half of the globe; the Equator divides the earth into Northern and Southern hemispheres; the Prime Meridian divides it into Eastern and Western hemispheres (pp. 4, 22)

hieroglyphs ancient form of Egyptian writing using pictographs (p. 451)

high island in Micronesia, a mountainous volcanic island (p. 713)

high-technology industry factory that produces computers or other electronic equipment (p. 624)

high veld flat, grassy plateau region of southern Africa (p. 546)

Holocaust [HOH•luh•KAWST] term for the mass imprisonment and slaughter of European Jews and others by German Nazis during World War II (pp. 284, 426)

humid continental climate climate in continental interiors, with cold winters and short, hot summers (p. 47)

humid subtropical climate warm, mild, rainy mid-latitude climate (p. 47)

hurricane fierce tropical storm system with high-speed winds, formed over warm oceans (pp. 43, 177)

hydroelectricity/hydroelectric power electric power generated by falling water (pp. 68, 202, 449, 507)

I

ice shelf part of an ice cap that extends from the land over the ocean (p. 718)

immigrant person who moves from one place to make a permanent home in another (p. 102)

industrialized describing a country in which industry has replaced farming as the main economic activity (p. 161)

intensive cultivation farming method using all available land, producing several crops a year (p. 635)

interdependent referring to countries or people that rely on one another (p. 100)

intermediate direction any direction between the cardinal directions, such as southeast or northwest (p. 7)

invest to put money into a company in return for a share of its profits (pp. 329, 615)

isthmus [IHS•muhs] narrow piece of land connecting two larger pieces of land (pp. 32, 181)

J-K

jute [JOOT] plant fiber used in making rope and burlap bags (p. 589)

key on a map, an explanation of the symbols used (p. 7)

krill small shrimplike shellfish, source of food for many sea animals (p. 721)

L

ladino in Guatemala, a person who speaks Spanish and follows a Spanish-American, not Native American, lifestyle (p. 180)

lagoon pool of water surrounded by reefs or sandbars (p. 604)

land bridge narrow strip of land that joins two larger landmasses (p. 153)

landlocked describing a country that has no land on a sea or ocean (p. 207)

language family group of languages that comes from a common ancestor (p. 57)

latitude location north or south of the Equator, measured by imaginary lines (parallels) numbered in degrees north or south (pp. 5, 22, 155)

leap year year with 366 days to account for the extra one-fourth day in Earth's revolution around the sun (p. 26)

light industry production of consumer goods, such as food products and household goods (pp. 188, 365)

line graph graph in which one or more lines connect dots representing changing quantities (p. 14)

literacy rate percentage of adults in a society who can read and write (pp. 58, 181)

llanos [LAH•nohs] large, grassy plains regions of Latin America (pp. 202, 218)

loch [LAHK] long, narrow bay with mountains on either side; also, a lake (p. 256)

loess [LEHS] fine-grained fertile soil deposited by the wind (pp. 283, 611)

longitude location east or west of the Prime Meridian, measured by imaginary lines (meridians) numbered in degrees east or west (pp. 5, 22)

low island in Micronesia, a low-lying island formed of coral (p. 713)

M

magma [MAG•muh] melted rock within the earth's mantle (p. 30)

mainland the main landmass of a country, as contrasted with nearby islands (p. 309)

mangrove tropical tree that grows in swampy land, it has roots both above and below the water (p. 485)

mantle middle layer of the earth, composed of thick hot rock (p. 29)

manuka shrub typical of volcanic soil in New Zealand (p. 697)

maquiladoras [mah•KEE•luh•DOH•rahz] factories that assemble parts shipped in from other countries, producing automobiles, stereos, etc. (p. 161)

marine west coast climate coastal climate with mild winters and cool summers (p. 47)

marsupial type of mammal that carries its young in a pouch as the infants mature (p. 691)

Mediterranean climate mild mid-latitude coastal climate with rainy winters and hot, dry summers (p. 47)

megalopolis [MEH•guh•LAH•puh•luhs] area in which neighboring urban areas blend into one "super city" (pp. 91, 638)

mestizo [meh•STEE•zoh] person of mixed Native American and European ancestry (p. 165)

migrate to move to another place (p. 421)

mobile moving from place to place (p. 105)

monotheism belief in one God (p. 425)

monsoon seasonal winds that bring rain to parts of Asia (pp. 43, 589)

moor in the United Kingdom, a treeless, windy highland area (p. 256)

mosque [MAHSK] place of worship for followers of Islam (pp. 338, 421)

movement one of the geographic themes, describing how people from different places interact (p. 23)

multicultural referring to something that includes many different cultures (p. 104)

multinational firm a company that has offices in several countries and does business in several countries (p. 289)

N

national park public lands set aside for recreation and wilderness protection (p. 106)

natural resource anything from the natural environment that people use to meet their needs (p. 65)

nature preserve land that is set aside by a government to protect plants and wildlife (pp. 332, 387)

navigable [NA•vih•guh•buhl] describing a body of water wide and deep enough for ships to pass (pp. 222, 278)

neutrality policy of refusing to take sides in international disputes and wars (p. 290)

newsprint type of paper used for printing newspapers (p. 124)

nomads people who move from place to place, often with herds of sheep or other animals (p. 396)

nonrenewable resources metals and other minerals that cannot be replaced once they are used up (p. 66)

nutrients minerals in soil that supply growing plants with food (p. 177)

O

oasis (plural, **oases**) place with water and green vegetation surrounded by desert (pp. 398, 432)

oil shale layered rock that contains oil (p. 320)

orbit elliptical path that a planet follows in revolving around the sun (p. 25)

outback remote inland regions of Australia (p. 690)

ozone form of oxygen (O_3) in the atmosphere (pp. 72, 721)

P

pagoda many-storied tower built as a temple or shrine (p. 622)

parliamentary democracy form of government in which voters elect representatives to a law-making body (parliament), which then chooses a leader, the prime minister (pp. 128, 259)

peat partly decayed plant matter, found in bogs, used for fuel and fertilizer (p. 261)

peninsula piece of land surrounded by water on three sides (pp. 32, 153)

permafrost permanently frozen lower layers of soil in Arctic regions (pp. 47, 362)

pesticide chemicals used to kill insects and other pests (p. 67)

phosphate mineral salt containing phosphorus, used in fertilizers (pp. 424, 714)

pictograph graph in which small symbols represent quantities (p. 15)

place one of the geographic themes, describing the typical characteristics that distinguish one place from another (p. 22)

plantation large farm on which a single cash crop is raised (pp. 161, 177)

plateau flat landform whose surface is raised above the surrounding land, with a steep cliff on one side (pp. 32, 297)

plate tectonics theory in geology that the earth's crust is made up of huge, moving plates of rock (p. 30)

poacher person who hunts and kills animals illegally (p. 523)

polder [POHL•duhr] in the Netherlands, an area of land reclaimed from the sea (p. 289)

pollution putting impure or poisonous substances into land, water, or air (p. 67)

pope title of the head of the Roman Catholic Church (p. 326)

population density average number of people living in a square mile or square kilometer (pp. 12, 61)

potash [PAHT•ASH] mineral salt, often used in fertilizers (p. 424)

prairie rolling, inland grassland area with fertile soil (p. 118)

precipitation water that falls to the earth as rain, snow, or sleet (p. 38)

Prime Meridian Line of 0° longitude from which east and west locations are measured, runs through Greenwich, United Kingdom (p. 22)

prime minister government leader chosen in a parliamentary democracy by members of parliament (p. 128)

projection in mapmaking, a way of drawing the round Earth on a flat surface (p. 2)

Q

quinoa [KEEN•WAH] cereal grain grown in Bolivia (p. 226)

R

rain forest dense forest in tropical regions with heavy, year-round rainfall (p. 45)

rain shadow area on the inland side of mountain ranges, where little rain falls (p. 43)

reef narrow ridge of coral, rock, or sand near the surface of water (p. 522)

refugee person who has had to flee to another country for safety from disaster or danger (pp. 60, 534)

region one of the geographic themes, defining parts of the earth that share common characteristics (p. 23)

relative location the position of a place on the earth's surface in relation to another place (p. 22)

relief differences in height in a landscape; how flat or rugged the surface is (p. 9)

renewable resource resource that is replaced naturally or can be grown quickly (p. 66)

republic government without a monarch, in which people elect important officials (pp. 103, 280)

revolution for a planet, one complete trip around the sun (p. 26); in politics, a sudden radical change in government (p. 103)

rookery nesting place for large numbers of birds or marine animals (p. 718)

rural relating to the countryside, not the city (p. 91)

S

samurai class of warrior-nobles in feudal Japan (p. 637)

savanna tropical grassland with scattered trees, usually with wet and dry seasons (pp. 46, 485)

scale the relationship between distance on a map and actual distance on the earth (p. 8)

scale bar on a map, a divided line showing the map scale, usually in feet, miles, or kilometers (p. 8)

selva thick tropical forests of the Amazon basin in Brazil (p. 195)

serfs farm laborers bound to the land they worked (p. 370)

service industry business that provides services to people rather than producing goods (pp. 67, 97, 159, 310, 333)

shah title held by kings of Iran, formerly Persia (p. 437)

shogun military ruler or dictator of feudal Japan (p. 637)

silt particles of soil carried and deposited by running water (p. 448)

sirocco [shuh•RAH•koh] hot, dry wind that blows across southern Europe from North Africa (p. 304)

sisal [SY•suhl] plant fiber used to make rope (pp. 197, 529)

slash-and-burn farming farming techniques in which areas of forest are cleared for farmland by burning (p. 562)

smog fog mixed with smoke and chemicals (p. 162)

socialism economic system in which government sets economic goals and may own some businesses (p. 58)

sodium nitrate mineral used in making fertilizer and explosives (p. 227)

solar energy power produced by the heat of the sun (p. 68)

solar system group of planets and other bodies that revolve around the sun (p. 25)

solstice day in June and December when the sun is directly overhead at the Tropic of Cancer (23 1/2° N) or Tropic of Capricorn (23 1/2° S), marking the beginning of summer or winter (p. 26)

sorghum tall grass used as grain and to make syrup (p. 557)

standard of living measure of quality of life based on income and material possessions (p. 58)

station Australian term for a sheep or cattle ranch (p. 690)

steppe dry treeless grasslands, often found at the edges of deserts (pp. 48, 384)

strait narrow body of water lying between two pieces of land (pp. 32, 419)

subcontinent large landmass that is part of another continent but distinct from it, such as South Asia (p. 587)

subsistence farm/farming farm that produces only enough to support a family's needs (pp. 67, 161)

suburbs smaller communities surrounding a central city (p. 311)

T

table graphic way of organizing and presenting statistics or facts (p. 17)

taiga [TY•guh] huge, subarctic evergreen forests in northern Europe, Asia, and North America (p. 361)

tannin substance from tree bark used in turning hides into leather (p. 229)

teak tropical wood used in furniture and shipbuilding (p. 597)

terraced field hillside field cut in steplike strips to hold water (p. 617)

textile fabrics and clothing (p. 299)

timberline elevation above which trees cannot grow (p. 48)

township in South Africa, term for some settlements outside cities (p. 550)

tributary small stream or river that flows into a larger river (p. 595)

tropics region of the world located between the Tropics of Cancer and Capricorn (about 30° North and 30° South), with a generally hot climate (p. 41)

trust territory region placed under control of a another country by international agreement (p. 714)

tsetse fly insect whose bite causes sleeping sickness (p. 512)

tsunami [tsoo•NAH•mee] huge sea wave caused by an earthquake on the ocean floor (pp. 30, 634)

tundra broad, dry, treeless plain in the high latitudes (pp. 47, 361)

tungsten hard, grayish metallic element used in electrical equipment (p. 616)

typhoon a hurricane, or tropical storm system, that forms in the Pacific Ocean (pp. 43, 612)

U-V

urban related to a city or densely populated area (p. 91)

urbanization tendency of a country's people to move from rural areas to cities (p. 62)

vegetation plant life in a region (p. 37)

W-X-Y-Z

water cycle process by which the earth's water moves from the oceans to the air to the land and back to the oceans (p. 37)

water vapor water in the form of a gas (p. 37)

watershed a region that drains into a common waterway or body of water (pp. 290, 533)

weather changes in temperature and precipitation over a short period of time (p. 41)

weathering process by which surface rocks are broken down into smaller pieces by water, chemicals, or frost (p. 31)

welfare state country in which government money is used to provide needy people with health care, unemployment benefits, and so on (pp. 208, 268)

yurt large, round tents of animal skin, used by nomads in Mongolia (p. 624)

Spanish Glossary

A

abaca [a•buh•KAH]/**abacá** fibra de una planta la cual es usada para hacer soga (p. 671)

absolute location/localización absoluta la posición exacta de un lugar en la superficie de la Tierra (pp. 5, 21)

acid rain/lluvia ácida precipitación la cual posee grandes cantidades de químicos, especialmente ácido sulfúrico (pp. 68, 100, 283)

adobe [uh•DOH•bee]/**adobe** ladrillo secado al Sol (p. 166)

alluvial [uh•LOO•vee•uhl] **plain/plano aluvial** plano que fue creado por la tierra acumulada de un río (pp. 435, 657)

altiplano [AL•tih•PLAH•NOH]/**altiplano** región altiplana de los Andes (p. 222)

altitude [AL•tuh•TOOD]/**altitud** altura sobre el nivel del mar (pp. 156, 202)

anthracite [AN•thruh•SYT]/**antracita** carbón duro (p. 644)

apartheid [uh•PAHR•TAYT]/**apartheid** restricciones y prácticas legales sudafricanas que segregan a los grupos raciales y étnicos, la palabra significa "aparte" (p. 549)

aquifer [AH•kwuh•fuhr]/**acuífer** capa de tierra subterránea que contiene grandes cantidades de agua (p. 39)

archipelago [AHR•kuh•PEH•luh•GOH]/**archipiélago** grupo grande o cadenas de islas (pp. 183, 633)

atmosphere/atmósfera cojín espeso de gases que rodean a la Tierra, formados principalmente de los gases nitrógeno y oxígeno (p. 32)

atoll [A•TAWL]/**atolón** arrecife o isla coralina de forma anular que rodea a una laguna pequeña (pp. 604, 713)

autobahn [AW•toh•BAHN]/*autobahn* una autopista en Alemania (p. 284)

autonomy [aw•TAH•nuh•mee]/**autonomía** auto gobierno (p. 532)

axis (plural, **axes**)/**axis** la línea o eje horizontal o vertical de medida usado en una gráfica (p. 13)

B

bar graph/gráfica de franjas gráfica en la cual las franjas verticales u horizontales representan cantidades (p. 13)

basin/cuenca extensa área de tierra baja plana, la cual está rodeada por tierra más alta (pp. 155, 195, 511)

bauxite [BAWK•SYT]/**bauxita** mineral metálico del cual se saca el aluminio (pp. 178, 497)

bazaar/bazar un mercado público (p. 451)

Bedouin [BEH•duh•wuhn]/**beduino** miembro de los nómadas del desierto del África del norte y del sudoeste de Asia (p. 429)

bilingual/bilingüe persona que habla dos idiomas (p. 129)

birthrate/índice de natalidad el número de niños que nace cada año por cada 1,000 personas (p. 62)

bog/pantano área de tierra baja y pantanosa (pp. 261, 323)

buffer state/estado intermedio pequeño país que sirve de valla entre dos naciones rivales (p. 208)

bush/área remota término australiano que designa a las áreas rurales remotas (p. 694)

C

cacao [kuh•KOW]/**cacao** árbol tropical cuyas semillas se emplean para hacer cacao y chocolate (p. 487)

calligraphy [kuh•LIH•gruh•fee]/**caligrafía** el arte de escribir con letra bonita (p. 622)

campesino [KAM•puh•SEE•noh]/*campesino* granjero de Latinoamérica (p. 220)

canopy/bóveda capa superior de un bosque tropical, la cual resguarda el suelo del bosque (p. 505)

cardinal directions/puntos cardinales las direcciones básicas: el norte, sur, este y oeste (p. 7)

casbah [KAZ•BAH]/*casbah* la parte vieja de las ciudades en África del norte donde hay calles estrechas y mercados pequeños (p. 458)

cash crop/cosecha al contado cosecha que se cultiva para la venta, a menudo con fines de exportación (p. 220)

cassava [kuh•SAH•vuh]/**yuca** planta cuyas raíces se transforman en harina y se comen, tal como en el pudín de tapioca, por ejemplo (pp. 210, 495, 663)

caudillo [kaw•DEE•yoh]/*caudillo* un dictador militar en la historia de Latinoamérica (p. 203)

GEOGRAPHY: The World and Its People

chart/diagrama forma gráfica de representar datos claramente (p. 17)

chicle/chicle jugo del árbol de zapotillo, el cual se usa en la producción de la goma de mascar (p. 177)

circle graph/gráfica circular gráfica redonda, la cual muestra cómo se divide un conjunto (p. 14)

city-state/ciudad estado en la historia, consiste de una ciudad independiente y las tierras a su alrededor (p. 306)

civil war/guerra civil conflicto que incluye a diferentes grupos de ciudadanos en un mismo país (pp. 430, 534)

civilization/civilización cultura altamente desarrollada, usualmente posee la religión organizada y leyes (p. 56)

clan/clan grupo de personas o familias relacionadas unas a las otras (pp. 538, 637)

cloves/clavos de especia especia que se produce de los capullos de las flores del clavero (p. 529)

climate/clima los patronos normales del clima de un área a través de un largo período de tiempo (pp. 15, 41)

climate region/región climática un área extensa del mundo con el mismo clima (p. 12)

climograph/climograma una combinación de gráfica lineal y de franjas, la cual provee información sobre la temperatura y precipitación (p. 15)

cold war/la guerra fría época desde los finales del 1940 al comienzo de los 1990 en la cual los EE. UU. y la Unión Soviética compitieron por la influencia mundial (p. 371)

colony/colonia colonización extranjera hecha por un país matriarcal (pp. 103, 187)

command economy/economía autoritaria sistema económico en el cual el gobierno toma posesión de las propiedades, los recursos y medios de producción, y hace todas las decisiones económicas (p. 364)

commonwealth/estado libre asociado término usado para designar a una república o un estado gobernado por la gente; el título oficial de Puerto Rico (p. 189)

communism/comunismo sistema político autoritario en el cual el gobierno central controla a la economía y a la sociedad (pp. 187, 282, 319, 364)

compass rose/rosa de los vientos emblema

que aparece dibujado en los mapas para señalar direcciones (p. 7)

compound/campamento grupo de viviendas rodeadas por una muralla en el África rural (p. 488)

condensation/condensación paso del vapor de agua al estado líquido (p. 38)

constitutional monarchy/monarquía constitucional forma de gobierno encabezado por un monarca pero donde el mayor control es otorgado a la legislatura electa (pp. 259, 430)

consumer goods/bienes del consumidor productos para el consumo personal, tales como la ropa y los productos caseros (pp. 336, 366, 449, 615)

contiguous/contiguo se refiere a las áreas adyacentes o que comparten una frontera (p. 89)

continent/continente una de las áreas de población más grandes de la Tierra (p. 30)

continental island/isla continental tierra que fue originalmente parte de un continente, separada por agua creciente o por alguna actividad geológica (p. 712)

contour line/curva de nivel en un mapa topográfico, la línea que une a todos los puntos en la misma elevación (p. 9)

cooperative/cooperativa en Cuba, es una finca controlada y operada por el gobierno (p. 188)

copper belt/región cobreña en Zambia, región de valiosas minas de cobre (p. 557)

copra/copra médula del coco seco (p. 712)

coral/coral material rocoso formado de los esqueletos de animales marinos pequeños (pp. 521, 690)

coral reef/arrecife de coral arrecife que yace en un nivel bajo del océano y está hecho de coral (p. 93)

cordillera [KAW•duhl•YEHR•uh]**/cordillera** grupo de montañas enlazadas entre si (pp. 118, 217)

core/núcleo capa central de la Tierra, probablemente compuesta de hierro caliente y níquel, es sólida en su núcleo interno y fundida en su exterior (p. 29)

cottage industry/industria autosuficiente industria casera o iniciada en un pueblo donde los miembros de la familia suplen su propio equipo de trabajo para producir bienes tal como las telas y efectos de metal (p. 590)

crevasse [krih•VAS]**/grieta** rajadura profunda en una capa de hielo o glaciar (p. 718)

crust/corteza la capa externa de la Tierra (p. 29)

cultural diffusion/difusión cultural el proceso por el cual los conocimientos y las destrezas se diseminan de un área a la otra (p. 56)

culture/cultura forma de vida de un grupo de gente que comparten costumbres y creencias similares, tanto como un mismo idioma y religión (p. 55)

currents/corrientes corrientes de agua, caliente o fría, en los océanos (p. 42)

cyclone/ciclón tormenta intensa que produce fuertes lluvias y vientos veloces que giran en forma circular (p. 596)

czar [ZAHR]**/zar** título de los emperadores antiguos de Rusia (p. 370)

D

death rate/índice de mortalidad número de muertes que ocurren cada año por cada 1,000 personas (p. 62)

deforestation/deforestación la pérdida de los bosques debido a la extensa práctica de cortar los árboles (p. 658)

delta/delta área triangular en la boca de un río, la cual se forma debido al lodo y la arena depositadas por el agua (pp. 448, 596, 660)

demographer/demógrafo científico que estudia la población (p. 62)

desertification/desiertificación proceso por el cual los prados se secan y las áreas desiertas aumentan (p. 491)

developed country/país desarrollado país que es industrial en vez de ser agricultural (p. 62)

developing country/país en vías de desarrollo país en proceso de industrialización (p. 62)

diagram/diagrama dibujo que muestra los pasos de un proceso o las partes de un objeto (p. 16)

dialect/dialecto variante regional de un idioma (pp. 188, 284, 300)

dictatorship/dictadura un gobierno que está bajo el control de un sólo líder que posee todo el poder (p. 455)

dike/dique banco alto hecho de tierra o concreto, el cual es construido para contener el agua de un río u océano (p. 611)

drought [DROWT]**/sequía** un largo período de extrema sequía y escasez de agua (pp. 490, 537)

dry farming/cultivo seco método de cultivar el trigo en áreas secas, el cual conserva humedad por medio del arado profundo (p. 98)

dynasty/dinastía una familia de gobernantes (p. 619)

dzong/dzong centro budista para el rezo y el estudio (p. 601)

E

earthquake/terremoto estremecimiento violento o temblor de la tierra causado por el movimiento de rocas en una falla (p. 30)

elevation/elevación la altura sobre el nivel del mar (pp. 9, 48, 310)

elevation profile/perfil de elevación diagrama transversal que muestra los cambios en la elevación de la Tierra (p. 9)

emigrate/emigrar trasladarse de su país nativo a otro país (pp. 60, 312)

empire/imperio grupo de países o estados bajo el mando de una nación o un gobernante (pp. 223, 623)

enclave/enclave pequeña nación o región específica establecida dentro de un país más grande (p. 545)

environment/medio ambiente el ambiente natural (p. 23)

equinox/equinoccio día en los meses de marzo y septiembre en que los rayos del Sol caen directamente sobre el ecuador, causando que el día y la noche tengan la misma duración (p. 27)

erg/ergio región desértica con dunas de arena movedizas (p. 458)

erosion/erosión desgaste de la superficie de la Tierra, principalmente por el agua, el viento o el hielo (p. 31)

escarpment/escarpadura acantilado alto que separa a dos superficies de terreno plano, una más alta que la otra (pp. 196, 522, 546)

estancia [ehs•TAHN•syah]**/estancia** rancho grande en Argentina (p. 230)

ethnic group/grupo étnico personas que comparten una historia cultural común, tanto como la misma raza e idioma (pp. 104, 371, 438, 487)

evaporation/evaporación el proceso por el cual

el agua de los océanos, lagos y arroyos se transforma en un gas o vapor por el calor del Sol (p. 38)

exclave la parte pequeña de una nación que está separada del área principal del país (p. 552)

F

famine/hambre escasez de alimentos que afecta a un gran número de personas (p. 63)

farm belt/zona de cultivo la región del centro occidental de los EE. UU. donde el terreno es plano y la tierra es fértil (p. 98)

fault/falla fractura de las rocas en la corteza de la Tierra, a lo largo de la cual ocurren desplazamientos (pp. 30, 391, 522)

favela [fuh•VEH•luh]/*favela* en el Brasil, término que se refiere a los barrios urbanos pobres (p. 199)

fellahin [fehl•uh•HEEN]/*fellahin* granjeros egipcios y trabajadores de la granja (p. 450)

fjord (FYOHRD)/**fiordo** entradas que provienen del mar, las cuales son largas, estrechas y de lados profundos (pp. 266, 697)

flow chart/diagrama de progreso tipo de diagrama que muestra dirección o movimiento (p. 16)

food processing/proceso de comidas la industria en la que las comidas son preparadas y empaquetadas para la venta (p. 393)

fossil fuels/combustibles fósiles grupo de recursos minerales no renovables — el carbón, el petróleo, el gas natural — formados en la corteza de la Tierra de los restos de plantas y animales (pp. 66, 124)

free enterprise/empresa libre la base de una economía de mercados en la cual las personas comienzan y manejan sus propios negocios para lograr una ganancia con poca intervención del gobierno (p. 58)

free enterprise system/sistema libre de empresa sistema económico basado en la libre empresa (ver definición anterior) (pp. 96, 365)

free port/puerto libre el área donde los artículos pueden ser cargados, almacenados y reembarcados sin pagar impuestos de importación-exportación (p. 666)

G

galaxy/galaxia sistema enorme en el universo, el cual incluye a millones de estrellas (p. 25)

gaucho/**gaucho** en la parte sur de América Latina, es un ganadero a caballo (pp. 208, 230)

geography/geografía el estudio de la Tierra y sus regiones, su agua y su vida vegetal y animal (p. 21)

geothermal power/energía geotérmica energía eléctrica producida por fuentes de vapor naturales subterráneas (p. 699)

geyser/géiser manantial caliente que arroja vapor y agua caliente a través de una rajadura en la Tierra (p. 269)

glasnost [GLAZ•nohst]/*glasnost* término ruso que significa "política abierta"; lema usado por Gorbachev en su política a finales de los 1980 (p. 371)

great circle route/ruta del gran círculo ruta tomada por un barco o avión en la cual éste hace un círculo grande, la distancia más corta entre dos puntos de la Tierra (p. 6)

grid system/sistema de cuadrícula red de líneas imaginarias sobre la superficie de la Tierra, formadas cuando las líneas de latitud y longitud se cruzan, creando patrones (pp. 5, 22)

groundwater/agua subterránea el agua que está almacenada en las capas de roca bajo la superficie de la Tierra (p. 39)

H

hajj/hajj viaje religioso que los musulmanes hacen al Makkah (p. 432)

harmattan [HAHR•muh•TAN]/**harmatán** viento seco que sopla hacia el sur desde el Sahara en el invierno (p. 486)

heavy industry/industria pesada producción de bienes industriales tales como las maquinarias y equipo militar (p. 365)

hemisphere/hemisferio una mitad del globo terráqueo; el ecuador divide a la Tierra en los hemisferios norte y sur; el primer meridiano lo divide en los hemisferios oriental y occidental (pp. 4, 22)

hieroglyphs/jeroglífico estilo antiguo de escritura egipcia, el cual usa pictografías (p. 451)

high island/isla montañosa es una isla volcánica montañosa en Micronesia (p. 713)

high-technology industry/industria de tecnología avanzada fábrica que produce computadoras u otros equipos electrónicos (p. 624)

high veld/estepa región altiplana y herbosa de Sudáfrica (p. 546)

Holocaust [HOH•luh•KAWST]**/Holocausto** término usado para describir el encarcelamiento y matanza de judíos europeos por los nazi alemanes durante la Segunda Guerra Mundial (pp. 284, 426)

humid continental climate/clima húmedo continental el clima en el interior del continente, con inviernos fríos y veranos cortos y calientes (p. 47)

humid subtropical climate/clima húmedo subtropical el clima cálido, templado y lluvioso de la altitud media (p. 47)

hurricane/huracán tormenta tropical devastadora, con fuertes vientos, la cual se forma en las aguas cálidas del océano (pp. 43, 177)

hydroelectric power (hydroelectricity)/ energía hidroeléctrica (hidroelectricidad) energía eléctrica generada por el agua que cae (pp. 68, 202, 449, 507)

I

ice shelf/capa de hielo parte de una capa espesa de hielo que se extiende desde la tierra hasta el océano (p. 718)

immigrant/inmigrante persona que se muda de un lugar y forma un hogar permanente en otro lugar (p. 102)

industrialized/industrializado término que describe a un país en el que la industria ha reemplazado al cultivo como actividad económica principal (p. 161)

intensive cultivation/cultivo intenso método de cultivo que utiliza toda la tierra disponible para producir varias cosechas al año (p. 635)

interdependent/interdependiente se refiere a los países o a personas que dependen unos de otros (p. 100)

intermediate direction/dirección intermedia cualquier dirección que está entre los puntos cardinales, tal como el sudeste o el noreste (p. 7)

invest/invertir poner dinero en una compañía a cambio de recibir un porcentaje de sus ganancias (pp. 329, 615)

isthmus [IHS•muhs]**/istmo** lengua de tierra que une a dos partes más grandes (pp. 32, 181)

J-K

jute [JOOT]**/yute** fibra de una planta, se usa para hacer cordeles y telas de saco (p. 589)

key/clave la explicación de los símbolos usados en un mapa (p. 7)

krill/crustáceo marisco pequeño parecido al camarón, el cual es fuente de alimento para muchos animales marinos (p. 721)

L

*ladino***/ladino** en Guatemala, se refiere a una persona que habla español y que sigue el estilo de vida español-americano, no el nativo americano (p. 180)

lagoon/laguna cuerpo de agua rodeado de arrecifes o bancos de arena (p. 604)

land bridge/puente terrestre pedazo estrecho de tierra que une a dos áreas mayores (p. 153)

landlocked/rodeado de tierra término que describe a un país cuya tierra no tiene salida al mar o al océano (p. 207)

language family/familia de idiomas grupo de idiomas que proviene de la misma lengua predecesora (p. 57)

latitude/latitud distancia al norte o sur del ecuador, la cual se mide con líneas imaginarias (paralelos) contadas en grados al norte o sur (pp. 5, 22, 155)

leap year/año bisiesto año que tiene 366 días, establecido para compensar por la fracción adicional de un cuarto de día que se produce cuando la Tierra gira alrededor del Sol (p. 26)

light industry/industria liviana producción de bienes al consumidor, tales como productos comestibles y del hogar (pp. 188, 365)

line graph/gráfica lineal gráfica en la cual una o más líneas conectan los puntos que representan cantidades variantes (p. 14)

literacy rate/índice de alfabetización porcentaje de adultos en una sociedad que pueden leer y escribir (pp. 58, 181)

llanos [LAH•nohs]**/llanos** las amplias regiones llanas y verdosas de Latinoamérica (pp. 202, 218)

loch [LAHK]**/ensenada** bahía larga y estrecha con montañas a cada lado; también, un lago (p. 256)

loess [LEHS]**/loes** tierra fértil de grano fino, depositada por el viento (pp. 283, 611)

longitude/longitud distancia al este u oeste del primer meridiano, la cual se mide con líneas imaginarias (meridianas) contadas en grados al este u oeste (pp. 5, 22)

low island/isla baja es una isla en Micronesia situada a nivel bajo y formada de coral (p. 713)

M

magma [MAG•muh]**/magma** roca derretida situada dentro de la capa intermedia de la Tierra (p. 30)

mainland/territorio continental el área principal y de más volumen de tierra que tiene un país, en contraste con las islas adyacentes (p. 309)

mangrove/mangle árbol tropical que crece en tierra pantanosa, tiene raíces tanto sobre como bajo la superficie del agua (p. 485)

mantle/capa capa intermedia de la Tierra, compuesta de roca espesa y caliente (p. 29)

manuka/*manuka* arbusto típico de la tierra volcánica de Nueva Zelandia (p. 697)

maquiladoras [mah•KEE•luh•DOH•rahz]**/maquiladoras** factorías que ensamblan las partes recibidas de otros países para producir automóviles, sistemas de sonido estereofónicos, etc. (p. 161)

marine west coast climate/clima marino de la costa occidental clima costeño, el cual tiene inviernos templados y veranos frescos (p. 47)

marsupial/marsupial tipo de mamífero que lleva a sus crías en una bolsa abdominal donde terminan su desarrollo (p. 691)

Mediterranean climate/clima mediterráneo clima templado costeño a mitad de latitud, el cual tiene inviernos lluviosos y veranos calientes y secos (p. 47)

megalopolis/la megalópolis región en la que las áreas urbanas adyacentes están integradas, formando así una "gran ciudad" (pp. 91, 638)

mestizo [meh•STEE•zoh] persona nacida de padres procedentes de las razas americana nativa y europea (p. 165)

migrate/migrar mudarse a otro lugar (p. 421)

mobile/móvil mudarse de lugar a lugar (p. 105)

monotheism/monoteísmo la creencia en un sólo Dios (p. 425)

monsoon/monzón vientos estacionales que traen la lluvia a ciertas partes de Asia (pp. 43, 589)

moor/páramo un terreno alto y sin árboles pero ventoso en el Reino Unido (p. 256)

mosque [MAHSK]**/mezquita** lugar de devoción de los que siguen la religión del Islam (pp. 338, 421)

movement/movimiento uno de los temas geográficos que describen cómo las personas de diferentes lugares se comunican (p. 23)

multicultural/multicultural se refiere a un asunto que incluye a muchas culturas diferentes (p. 104)

multinational firm/empresa multinacional una compañía que tiene oficinas y comercia en varios países (p. 289)

N

national park/parque nacional terrenos públicos que se reservan para el recreo de los visitantes al área y para la protección de su estado virgen (p. 106)

natural resource/recurso natural cualquier cosa que proviene del medio ambiente natural y que las personas usan para satisfacer sus necesidades (p. 65)

nature preserve/santuario terreno reservado por el gobierno para proteger a las plantas y a la fauna silvestre (pp. 332, 387)

navigable/navegable término que describe a un cuerpo de agua lo suficientemente ancho y

SPANISH GLOSSARY

profundo para permitir el paso a los barcos
(pp. 222, 278)

neutrality/neutralidad la política que rechaza la
preferencia de una u otra posición en lo relativo
a desacuerdos y guerras internacionales (p. 290)

newsprint/papel de periódico tipo de papel
usado para imprimir los periódicos (p. 124)

nomads/nómadas personas que se mudan de un
lugar a otro, a menudo se llevan consigo a sus
rebaños de ovejas u otros animales (p. 396)

**nonrenewable resources/recursos no reno-
vables** metales y otros minerales que no pueden
ser reemplazados después de ser utilizados (p. 66)

nutrients/nutrientes los minerales de la tierra
que suplen alimento a las plantas en crecimiento
(p. 177)

oasis (plural, **oases**)**/oasis** lugar donde el agua y
la vegetación verdosa está rodeada por un
desierto (pp. 398, 432)

oil shale/esquisto roca arcillosa en capas, la cual
contiene aceite (p. 320)

orbit/órbita trayectoria elíptica que sigue un planeta cuando da vueltas alrededor del Sol (p. 25)

outback/llanura desértica regiones distantes del
interior de Australia (p. 690)

ozone/ozono forma de oxígeno (O_3) en la atmós-
fera (pp. 72, 721)

P

pagoda/pagoda una torre de muchos pisos, la
cual fue construida para servir de templo o lugar
de adoración (p. 622)

**parliamentary democracy/democracia parla-
mentaria** estilo de gobierno en el cual los
votantes eligen a representantes para formar
parte de un cuerpo creador de leyes (parlamento), el cual después escoge a un líder, el
primer ministro (pp. 128, 259)

peat/turba materia de las plantas en un estado
parcialmente descompuesto, se encuentra en
pantanos y es usada como combustible y fertil-
izante (p. 261)

peninsula/península pedazo de tierra rodeado
por agua en tres de sus lados (pp. 32, 153)

permafrost/permagel capas inferiores de la
tierra en las regiones árticas, las cuales están per-
manentemente congeladas (pp. 47, 362)

pesticide/insecticida químicos usados para
matar a los insectos y a otros animales dañinos
(p. 67)

phosphate/fosfato sal mineral que contiene fós-
foro, usado en fertilizantes (pp. 424, 714)

pictograph/pictografía gráfica en la que los símbolos pequeños representan cantidades (p. 15)

place/lugar uno de los temas geográficos, el cual
describe las características típicas que distinguen
a un lugar de otro (p. 22)

plantation/hacienda granja de gran tamaño
donde se cultiva sólo una cosecha al contado
(pp. 161, 177)

plateau/meseta forma terrestre plana cuya superficie está por encima de la tierra que le rodea,
con una colina muy alta en uno de sus lados
(pp. 32, 297)

plate tectonics/tectónicas de lámina teoría en
el campo de geología la cual dice que la corteza
de la Tierra está compuesta de enormes láminas
movedizas de roca (p. 30)

poacher/cazador furtivo persona que caza y
mata a los animales ilegalmente (p. 523)

polder [POHL•duhr]**/pólder** área de tierra en los
Países Bajos, la cual fue recuperada del mar
(p. 289)

pollution/contaminación el arrojar substancias
impuras o venenosas en la tierra, agua o aire
(p. 67)

pope/papa título del cabecilla de la Iglesia
Católica Romana (p. 326)

population density/densidad de la población
número promedio de personas que viven en
una milla cuadrada o kilómetro cuadrado
(pp. 12, 61)

potash [PAHT•ASH]**/potasio** sal mineral; usada a
menudo en fertilizantes (p. 424)

prairie/pradera terreno ondulado de pastoreo, se
encuentra en áreas interiores con tierra fértil (p.
118)

precipitation/precipitación agua que cae sobre
la Tierra en forma de lluvia, nieve o aguanieve
(p. 38)

Prime Meridian/primer meridiano la línea de longitud de 0° desde la cual se miden las distancias al este y al oeste, atraviesa a Greenwich, Reino Unido (p. 22)

prime minister/primer ministro líder gubernamental el cual es seleccionado en una democracia parlamentaria por los miembros del parlamento (p. 128)

projection/proyección en la cartografía, es una manera de dibujar la redondez de la Tierra en una superficie plana (p. 2)

Q

quinoa [KEEN•WAH]*/quinoa* grano de cereal que se crece en Bolivia (p. 226)

R

rain forest/bosque tropical bosque espeso en las regiones tropicales, donde llueve mucho todo el año (p. 45)

rain shadow/sombra de lluvia región en la parte interior de las cordilleras de montañas, donde cae muy poca lluvia (p. 43)

reef/arrecife hilera estrecha de coral, roca o arena cerca de la superficie del agua (p. 522)

refugee/refugiado persona que ha tenido que escapar a otro país para protegerse de un desastre o de peligro (pp. 60, 534)

region/región uno de los temas geográficos, el cual define a las partes de la Tierra que comparten características comunes (p. 23)

relative location/ubicación relativa la ubicación de un lugar en la superficie de la Tierra con relación a otro lugar (p. 22)

relief/relieve diferencias en la altura de un terreno; cuán plano o desigual es su superficie (p. 9)

renewable resource/recurso renovable recurso que puede ser reemplazado naturalmente o que se puede desarrollar con rapidez (p. 66)

republic/república un gobierno sin monarca, en el cual las personas eligen a los oficiales importantes (pp. 103, 280)

revolution/revolución al referirse a un planeta, es un recorrido completo alrededor del Sol

(p. 26); en la política, es un cambio radical y repentino de gobierno (p. 103)

rookery/criadero nido para un gran número de pájaros o animales marinos (p. 718)

rural/rural relativo al campo, no a la ciudad (p. 91)

S

samurai/samurai clase de guerreros aristocráticos en el Japón feudal (p. 637)

savanna/sabana terreno de pastoreo tropical con árboles dispersos, generalmente tiene épocas lluviosas y secas (pp. 46, 485)

scale/escala la relación entre la representación de una distancia en un mapa y la distancia verdadera en la Tierra (p. 8)

scale bar/barra de escala la línea dividida en un mapa, la cual muestra la escala del mapa, generalmente en pies, millas o kilómetros (p. 8)

*selva/***selva** bosque tropical espeso en la cuenca del Amazonas de Brasil (p. 195)

serfs/siervos labradores vinculados a la tierra en la que han trabajado (p. 370)

service industry/industria de servicio comercio que provee un servicio a las personas en vez de producir bienes para el consumidor (pp. 67, 97, 159, 310, 333)

shah/sha título de los reyes de Irán, país anteriormente llamado Persia (p. 437)

shogun/shogún mandatario militar o dictador del Japón feudal (p. 637)

silt/sedimento partículas de tierra que son arrastradas y depositadas por el agua (p. 448)

sirocco [shuh•RAH•koh]**/siroco** viento caliente y seco que sopla a través de Europa desde el África del Norte (p. 304)

sisal [SY•suhl]**/sisal** fibra de una planta, la cual se usa para hacer cordeles (pp. 197, 529)

slash-and-burn farming/agricultura de tala y quemado técnicas agrícolas en las cuales ciertas áreas forestales son despejadas para el cultivo por medio de incendios (p. 562)

smog/smog humo mezclado con niebla y productos químicos (p. 162)

socialism/socialismo sistema económico en el que el gobierno establece las metas económicas y puede poseer algunos comercios (p. 58)

SPANISH GLOSSARY

sodium nitrate/nitrato de sodio mineral que se usa al preparar los abonos y en explosivos (p. 227)

solar energy/energía solar la energía producida por el calor del Sol (p. 68)

solar system/sistema solar grupo de planetas y otros cuerpos celestes que dan vuelta alrededor del Sol (p. 25)

solstice/solsticio día en junio y diciembre cuando el Sol está directamente sobre el Trópico de Cáncer (23 1/2° N) o el Trópico de Capricornio (23 1/2° S), lo que marca el comienzo del verano o el invierno (p. 26)

sorghum/melaza hierba alta que se usa como un grano y también para hacer almíbar (p. 557)

standard of living/nivel de vida término que mide la calidad de vida de una persona con base en su ingreso y posesiones materiales (p. 58)

station/estación término australiano que se refiere a un rancho de ovejas o ganado (p. 690)

steppe/estepa tierras secas y sin árboles utilizadas para el pastoreo, a menudo se encuentran a los bordes de los desiertos (pp. 48, 384)

strait/desfiladero estrecho cuerpo de agua que yace entre dos pedazos de tierra (pp. 32, 419)

subcontinent/subcontinente gran área de tierra que forma parte de otro continente, pero es distinto a éste, tal como el Asia del Sur (p. 587)

subsistence farm (farming)/granja de subsistencia/cultivo granja que sólo produce lo suficiente para satisfacer las necesidades de una familia (pp. 67, 161)

suburbs/suburbios comunidades más pequeñas que rodean a una ciudad central (p. 311)

T

table/tabla forma gráfica de organizar y representar datos o estadísticas (p. 17)

taiga [TY•guh]**/taiga** enormes selvas subárticas de árboles siempre verdes en el norte de Europa, en Asia y América del Norte (p. 361)

tannin/tanino substancia que proviene de la corteza de los árboles y que se usa para convertir las pieles de animales en cuero (p. 229)

teak/teca madera tropical que se usa en la construcción de muebles y barcos (p. 597)

terraced field/campo abancalado campo donde la ladera ha sido cortada en franjas parecidas a escaleras para contener el agua (p. 617)

textile/textil telas y ropa (p. 299)

timberline/altura límite límite de elevación sobre el cual los árboles no pueden crecer (p. 48)

township/municipio término usado para describir a varios pueblos a las afueras de las ciudades en el África del Sur (p. 550)

tributary/tributario arroyo pequeño o río que desemboca en un río más grande (p. 595)

tropics/tropical región del mundo ubicada entre los Trópicos de Cáncer y Capricornio (alrededor de 30° al norte y 30° al sur), generalmente es de clima caliente (p. 41)

trust territory/territorio bajo administración fiduciaria región que está bajo el control de otro país debido a un pacto internacional (p. 714)

tsetse fly/mosca tse-tsé insecto cuya picada causa la enfermedad del sueño (p. 512)

tsunami [tsoo•NAH•mee]**/tsunami** ola gigantesca de mar causada por un terremoto en el suelo del océano (pp. 30, 634)

tundra una llanura extensa, seca y sin árboles en las latitudes altas (pp. 47, 361)

tungsten/tungsteno elemento metálico duro y de color gris, usado en los equipos eléctricos (p. 616)

typhoon/tifón un huracán o tormenta tropical, el cual se forma en el Océano Pacífico (pp. 43, 612)

U-V

urban/urbano relacionado a una ciudad o área densamente poblada (p. 91)

urbanization/urbanización la tendencia de la gente de un país a mudarse de áreas rurales a las ciudades (p. 62)

vegetation/vegetación la vida vegetal de una región (p. 37)

W-X-Y-Z

water cycle/ciclo de agua proceso por el cual el agua de la Tierra se mueve de los océanos al aire, a la tierra y vuelve otra vez a los océanos (p. 37)

water vapor/vapor de agua agua en forma de gas (p. 37)

watershed/cuenca arrecife alto del cual los ríos de un área fluyen en direcciones diferentes (pp. 290, 533)

weather/clima cambios de temperatura y de precipitación que ocurren en un período corto de tiempo (p. 41)

weathering/acción corrosiva proceso donde las rocas en la superficie se desmoronan en trozos más pequeños debido a la acción corrosiva del agua, los elementos químicos o la congelación (p. 31)

welfare state/estado benefactor país donde se usa el dinero del gobierno para proveer cuidado de la salud, beneficios de desempleo y otros servicios a las personas necesitadas (pp. 208, 268)

yurt/yurta tiendas de campaña grandes y redondas, las cuales están hechas de piel de animales y son usadas por los nómadas en Mongolia (p. 624)

SPANISH GLOSSARY

Index

GEOGRAPHY: The World and Its People

INDEX

D

GEOGRAPHY: The World and Its People

H

I

GEOGRAPHY: The World and Its People

INDEX

T

U

INDEX

Photo Credits

Cover, F. Mels/Westlight; **vi,** Rich Buzzelli/Tom Stack & Associates; **vii,** James P. Blair/National Geographic Society; **viii,** Jodi Cobb/National Geographic Society; **ix,** Sheila Nardulli/Liaison International; **x,** (t) Paul A. Zahl/National Geographic Society, (b) Sam Abell/National Geographic Society; **xii,** Martin Harvey/The Wildlife Collection; **18-19,** Bob Sacha/used with permission, Christian Science Mapparium, Boston, Massachusetts; **20,** ©Telegraph Colour Library/FPG International; **21,** ©Superstock, Inc.; **22,** (l) ©Ann & Myron Sutton/Superstock, Inc., (r) ©Galen Rowell/Mountain Light; **23,** ©Guy Marche/FPG International; **25,** NASA/National Geographic Society; **28,** James P. Blair/National Geographic Society; **29,** Otis Imboden/National Geographic Society; **30,** (l) Walter Meayers Edwards/National Geographic Society, (r) ©Douglas Faulkner/Photo Researchers, Inc.; **33,** ©Telegraph Colour Library/FPG International; **36,** ©F. Stuart Westmorland/Photo Researchers, Inc.; **37,** Steve Raymer/National Geographic Society; **39,** Robert W. Madden/National Geographic Society; **41,** James P. Blair/National Geographic Society; **43,** (l) ©Michael P. Gadomski/Photo Researchers, Inc., (r) David S. Boyer/National Geographic Society; **45,** Susan Snyder; **48,** Walter Meayers Edwards/National Geographic Society; **49,** ©F. Stuart Westmorland/Photo Researchers, Inc.; **52,** (l) ©Stephen J. Krasemann/Photo Researchers, Inc., (r) Dwight Kuhn; **53,** Matt Meadows; **54,** Dick Durrance II/National Geographic Society; **55,** Ed Kim/National Geographic Society; **58,** (l) James L. Amos/National Geographic Society, (r) ©Superstock, Inc.; **60,** David Alan Harvey/National Geographic Society; **63,** James P. Blair/National Geographic Society; **64,** David R. Frazier Photolibrary; **65,** ©C. Shambroom/Photo Researchers, Inc.; **68,** (l) ©Simon Fraser/Science Photo Library/Photo Researchers, Inc., (r) Bryan Hodgson/National Geographic Society; **69,** Dick Durrance II/National Geographic Society; **72,** Robert W. Madden/National Geographic Society; **72-73,** (background) Aaron Haupt; **73,** (t) NASA/GSFC/National Geographic Society, (br) Bruce Dale/National Geographic Society; **74-75,** Phil Schemeister/National Geographic Society; **76,** (l) George F. Mobley/National Geographic Society, (b) James P. Blair/National Geographic Society; **76-77,** (t,b) James P. Blair/National Geographic Society, (c) James L. Amos/National Geographic Society; **77,** (t,b) James P. Blair/National Geographic Society, (r) George F. Mobley/National Geographic Society; **78,** Walter Meayers Edwards/National Geographic Society; **79,** George F. Mobley/National Geographic Society; **80,** Johnny Johnson; **88,** Mark Burnett; **89,** ©John W. Warden/Superstock, Inc.; **90,** (l) David R. Frazier Photolibrary, (r) Tim Courlas; **91,** ©James Blank/FPG International; **93,** Mark E. Gibson; **96,** Robert W. Madden/National Geographic Society; **98,** (t) Bruce Dale/National Geographic Society, (b) David M. Dennis; **99,** ©Superstock, Inc.; **101,** John Youger; **102,** James L. Amos/National Geographic Society; **103,** Larry Hamill; **104,** ©Mark Rightmire/The Orange County Register; **107,** Mark Burnett; **110,** (l) Rich Buzzelli/Tom Stack & Associates, (r) Matt Meadows; **111,** Matt Meadows; **112,** (t) Craig Lowell/Viesti Associates, Inc., (c) Tim Courlas, (b) Bridgeman Art Library/Art Resource; **113,** (t) Jodi Cobb/National Geographic Society, (bl) George F. Mobley/National Geographic Society, (br) Art Resource; **114,** Rick Stewart/Allsport; **115,** Gordon W. Gahan/National Geographic Society; **116, 119,** David R. Frazier Photolibrary; **121,** George F. Mobley/National Geographic Society; **124,** (l) George F. Mobley/National Geographic Society, (r) David R. Frazier Photolibrary; **126,** Archive Photos; **127,** George F. Mobley/National Geographic Society; **128,** (l) George F. Mobley/National Geographic Society, (r) David S. Boyer/National Geographic Society; **130,** Bruce Dale/National Geographic Society; **131,** Rick Stewart/Allsport; **134,** James L. Stanfield/National Geographic Society; **134-135,** (background) Aaron Haupt; **135,** (t,bl) Joseph H. Bailey/National Geographic Society; **136-137,** William Albert Allard/National Geographic Society; **138,** (l) Otis Imboden/National Geographic Society, (b) Walter Meayers Edwards/National Geographic Society; **138-139,** (c) Bruce Dale/National Geographic Society, (background) James P. Blair/National Geographic Society; **139,** (tl,br) James P. Blair/National Geographic Society, (tr) O. Louis Mazzatenta/National Geographic Society, (cr) Otis Imboden/National Geographic Society; **140,** ©Tom Tracy/FPG International; **141,** ©Telegraph Colour Library/FPG International; **142,** ©Superstock, Inc.; **151,** Jan Butchofsky/Dave G. Houser; **152,** Ted Streshinsky/Photo 20-20; **153,** Buddy Mays/Travel Stock; **155, 159,** Robert Frerck/Odyssey; **163,** Dallas & John Heaton/Westlight; **164,** David Ryan/Photo 20-20; **165,** Robert Frerck/Odyssey; **166,** David R. Frazier Photolibrary; **166,** Robert Frerck/Odyssey; **167,** Ted Streshinsky/Photo 20-20; **168,** Donna Carroll/Travel Stock; **169,** Ted Streshinsky/Photo 20-20; **172,** (l) Reggie David/Pacific Stock, (r) Matt Meadows; **173,** Matt Meadows; **174,** D. Donne Bryant/DDB Stock Photo; **175, 179,** ©Superstock, Inc.; **180,** James C. Simmons/Dave G. Houser; **183,** ©Messerschmidt/FPG International; **185,** (l) Suzanne Murphy-Lamrond/DDB Stock Photo, (r) Winfield Parks/National Geographic Society; **188,** ©Carol Lee/Tony Stone Images; **191,** D. Donne Bryant/DDB Stock Photo; **194,** ©Ary Diesendruck/Tony Stone Images; **195,** ©Tom McHugh/Photo Researchers, Inc.; **198,** Robert Fried/DDB Stock Photo; **201,** Robert W. Madden/National Geographic Society; **203,** ©Marco Corsetti/FPG International; **205,** Buddy Mays/Travel Stock; **206,** Robert Frerck/Odyssey; **207, 208,** ©Superstock, Inc.; **210,** (l) O. Louis Mazzatenta/National Geographic Society, (r) Peter Lang/DDB Stock Photo; **211,** ©Ary Diesendruck/Tony Stone Images; **214,** (t) Janice Sheldon/Photo 20-20, (c) Barry W. Barker/Odyssey, (b) O. Louis Mazzatenta/National Geographic Society; **215,** (t) Scalkwijk/Art Resource, (r) Photo 20-20, (b) Robert Frerck/Odyssey; **216,** David L. Perry; **217,** ©Sheryl McNee/FPG International; **221,** Bates Littlehales/National Geographic Society; **222,** Gilbert M. Grosvenor/National Geographic Society; **224,** David L. Perry; **225,** Gordon W. Gahan/National Geographic Society; **227,** (l) ©Philip & Karen Smith/Tony Stone Images, (r) George F. Mobley/National Geographic Society; **228,** James P. Blair/National Geographic Society; **230,** (l) ©John Warden/Tony Stone Images, (r) ©Neil Beer/Tony Stone Images; **233,** David L. Perry; **236,** Paul A. Zahl/National Geographic Society; **236-237,** (background) Aaron Haupt; **237,** (t & br) James P. Blair/National Geographic Society; **238-239,** James L. Stanfield/National Geographic Society; **240,** Bruce Dale/National Geographic Society; **240-241,** (c,background) Sam Abell/National Geographic Society, (b) O. Louis Mazzatenta/National Geographic Society; **241,** (tl,bl,br) James L. Stanfield/National Geographic Society, (tr) George F. Mobley/National Geographic Society; **242,** ©Bruce Stoddard/FPG International; **243,** TRIP/J. Bartos; **244,** ©Al Michaud/FPG International; **254,** ©Tony Stone Images; **255,** ©David H. Endersbee/Tony Stone Images; **256,** George F. Mobley/National Geographic Society; **257,** ©Marcus Brooke/FPG International; **260,** ©James P. Rowan/Tony Stone Images; **261,** ©Ric Ergenbright Photography; **262,** Winfield Parks/National Geographic Society; **266,** Bryan & Cherry Alexander; **267,** ©Superstock, Inc.; **268,** ©Chad Ehlers/Tony Stone Images; **271,** ©Tony Stone Images; **274,** (t) ©Superstock, Inc., (c) James L. Stanfield/National Geographic Society , (b) Smithsonian Institution; **275,** (t) Department of the Environment, London, (c,b) Scala/Art Resource; **276,** ©Thomas Craig/FPG International; **277,** ©Charlie Waite/Tony Stone Images; **279,** (l) Bruce Dale/National Geographic Society, (r) James L. Stanfield/National Geographic Society; **282,** ©Peter Turnley/Black Star; **283,** ©Josef Beck/FPG International; **285,** Patrick Piel/Liaison International; **286,** Anthony Suau/Liaison International; **287,** ©David Noble/FPG International; **290,** Thomas J. Abercrombie/National Geographic Society; **291,** James L. Stanfield/National Geographic Society; **293,** ©Thomas Craig/FPG International; **295,** Rudi von Briel; **296,** ©Robert Frerck/Tony Stone Images; **297,** ©George Hunter/Tony Stone Images; **299,** (l) ©Guy Marche/FPG International, (r) Ric Ergenbright Photography; **300,** ©Travelpix/FPG International; **303,** ©Telegraph Colour Library/FPG International; **304,** ©Superstock, Inc.; **306,** Albert Moldvay/National Geographic Society; **307,** ©J.E. Stevenson/FPG International; **308,** James P. Blair/National Geographic Society; **310,** (t) Winfield Parks/National Geographic Society, (b) Otis Imboden/National Geographic Society; **313,** Robert Frerck/Tony Stone Images; **316,** (l) ©Ian Bradshaw/Zefa-UK/The Stock Market,

(r) Matt Meadows; **317,** Matt Meadows; **318,** ©B. Bisson/Sygma; **319,** TRIP/W. Jacobs; **321,** TRIP/T. Noorits; **323,** ©Superstock, Inc.; **325, 326,** James P. Blair/National Geographic Society; **327,** The Granger Collection; **328,** ©Superstock, Inc; **329,** TRIP/M. Barlow; **331,** ©Superstock, Inc.; **332,** James L. Stanfield/National Geographic Society; **333,** ©Peter Turnley/Black Star; **335,** TRIP/I. Wellbelove; **338,** James L. Stanfield/National Geographic Society; **339,** ©B. Bisson/Sygma; **342,** James P. Blair/National Geographic Society; **342-343,** (background) Aaron Haupt; **343,** (t) James L. Stanfield/National Geographic Society, (b) Albert Moldvay/National Geographic Society; **344-345,** National Geographic Society; **346,** Dean Conger/National Geographic Society; **346-347,** (c) Bruce Dale/National Geographic Society, (background) Bruce Dale/National Geographic Society; **347,** Steve Raymer/National Geographic Society; **348, 349,** Novosti/Sovfoto/Eastfoto; **350,** Bruce Dale/National Geographic Society; **357,** (t) ©Shinichi Kanno/FPG International, (b) Steve Raymer/National Geographic Society; **358,** Dean Conger/National Geographic Society; **359,** Sovfoto/Eastfoto; **362,** Steve Raymer/National Geographic Society; **364,** ©Bill Swersey/Liaison International; **365,** ©Superstock, Inc.; **369,** ©G. Pinkhassov/Magnum Photos, Inc.; **370,** ©De Keerle/Sygma; **373,** (l) Steve Raymer/National Geographic Society, (r) ©Dave Bartruff/Artistry International; **374,** Steve Raymer/National Geographic Society; **375,** Dean Conger/National Geographic Society; **378,** (l) Wolfgang Kaehler, (r) Matt Meadows; **379,** Matt Meadows; **380,** (t) ©Superstock, Inc., (c) ©TRIP/S. Pozharskij, (b) Tate Gallery, London/Art Resource; **381,** (t) Dean Conger/National Geographic Society, (bl) ©Forbes Magazine Collection/Superstock, Inc., (b) George F. Mobley/National Geographic Society; **382,** Steve Raymer/National Geographic Society; **383,** Sovfoto/Eastfoto; **387,** ©Jeremy Hartley/Panos Pictures; **389,** ©TRIP/V. Kolpakov; **390,** C. Romantsova from Sovfoto/Eastfoto; **391, 392,** George F. Mobley/National Geographic Society; **394,** ©TRIP/V. Slapinia; **395,** Buddy Mays/Travel Stock; **397,** Sovfoto/Eastfoto; **398,** Tass/Sovfoto/Eastfoto; **399,** Steve Raymer/National Geographic Society; **402,** George F. Mobley/National Geographic Society; **402-403,** (background) Aaron Haupt; **403,** (t) William H. Bond/National Geographic Society, (br) Steve Raymer/National Geographic Society; **404-405,** Kenneth Garrett/National Geographic Society; **406,** (l) James L. Stanfield/National Geographic Society, (b) Steve Raymer/National Geographic Society; **406-407,** (t,background) David Alan Harvey/National Geographic Society, (c) James L. Stanfield/National Geographic Society; **407,** (t,br) Jodi Cobb/National Geographic Society, (b) Winfield Parks/National Geographic Society; **408,** ©Nik Wheeler; **409,** Robert Harding Picture Library; **410,** Robert Harding Picture Library; **418, 419,** James L. Stanfield/National Geographic Society; **421,** file photo; **423,** ©Richard T. Nowitz; **425,** (t) Aaron Haupt, (b) Jodi Cobb/National Geographic Society; **426,** Sygma; **427,** ©Katrina Thomas/Photo Researchers, Inc.; **428, 430,** ©Superstock, Inc.; **431,** TRIP Photographic; **432,** ©J. Langevin/Sygma; **433,** ©Nabeel Turner/Tony Stone Images; **434,** ©Christina Dameyer/Photo 0-20; **436,** James P. Blair/National Geographic Society; **437,** ©Benard Sioberstein/FPG International; **439,** James L. Stanfield/National Geographic Society; **442,** (l) ©Stephen Marks/The Image Bank, (r) Matt Meadows; **443,** Matt Meadows; **444,** (t) Victoria & Albert Museum/Art Resource, (c) D.W. Funt/Art Resource, (b) Erh Lessing/Archaeological Museum, Cairo, Egypt/Art Resource; **44,** (t) ©Superstock, Inc., (c) ©Josef Beck/FPG International, (b) ©Richard T. Nowitz; **446,** ©Murray & Associates/Tony Stone Images; **447,** Leo de Wys, Inc./©Fridmar Damm; **449,** ©F. Lazi/FPG International; **452,** Robert Harding Picture Library; **454,** ©Superstock, Inc.; **456,** Thomas J. Abercrombie/National Geographic Society; **457, 458,** ©Dave Bartruff/Artistry International; **460,** The Granger Collection; **461,** ©Murray & Associates/Tony Stone Images; **464,** James L. Stanfield/National Geographic Society; **464-465,** (background) Aaron Haupt; **465,** (t) James L. Stanfield/National Geographic Society, (b) David Alan Harvey/National Geographic Society; **466-467,** James P. Blair/National Geographic Society; **468,** (l) George F. Mobley/National Geographic Society, (br)

Steve Raymer/National Geographic Society; **468-469,** (c) James L. Stanfield/National Geographic Society, (background) Thomas J. Abercrombie/National Geographic Society; **469,** (tl) James L. Stanfield/National Geographic Society, (tr) Steve Raymer/National Geographic Society, (cr,br) George F. Mobley/National Geographic Society; **470,** Jeremy Hartly/Panos Pictures; **471,** ©Gerry Ellis Nature Photography; **472,** N. Durrell McKenna/Panos Pictures; **483,** (tl) Jeremy Hartley/Panos Pictures, (tr) Betty Press/Panos Pictures, (bl) Jason Lauré; **484,** Panos Pictures; **485,** Bruce Paton/Panos Pictures; **487,** Betty Press/Panos Pictures; **490,** Jeremy Hartley/Panos Pictures; **492,** Marcus Rose/Panos Pictures; **493,** Sara Leigh/Panos Pictures; **494, 497,** Dave G. Houser; **498,** Ron Giling/Panos Pictures; **499,** Panos Pictures; **502,** (t,c) Boltin Picture Library, (b) ©Travelpix/FPG International; **503,** (t) ©C.M. Hardt/Liaison International, (bl) Boltin Picture Library, (br) ©Superstock, Inc.; **504,** Nick Robinson/Panos Pictures; **505,** Robert Harding Picture Library; **507,** Jason Lauré; **509,** Dave G. Houser; **511,** ©Marco Corsetti/FPG International; **514,** Martin Adler/Panos Pictures; **517,** Nick Robinson/Panos Pictures; **520,** ©Gerald Cubitt/FPG International; **521,** Emory Kristof/National Geographic Society; **523,** Joseph J. Scherschel/National Geographic Society; **525,** Trygue Bolstad/Panos Pictures; **527,** Bruce Dale/National Geographic Society; **529,** Jason Lauré; **530,** Robert F. Sisson/National Geographic Society; **531,** Sarah Leen/National Geographic Society; **533,** Emory Kristof/National Geographic Society; **534,** Marc Schlossman/Panos Pictures; **535,** TRIP/Helen Rogers; **537,** (l) James P. Blair/National Geographic Society, (r) Liba Taylor/Panos Pictures; **538,** Betty Press/Panos Pictures; **539,** ©Gerald Cubitt/FPG International; **542, 543,** Matt Meadows; **544,** Johann Van Tonder/Images of Africa Photobank; **545,** James P. Blair/National Geographic Society; **547,** (l) Volkmar Wentzel/National Geographic Society, (r) Neil Cooper/Panos Pictures; **549,** (l) Liaison International, (r) Jason Lauré; **550,** David Atchinon-Jones/Robert Harding Picture Library; **552,** Rob Cousins/Panos Pictures; **554,** Bruce Paton/Panos Pictures; **555,** Walter Meayers Edwards/National Geographic Society; **556,** James L. Stanfield/National Geographic Society; **557,** Trygve Bolstad/Panos Pictures; **559,** Gary John Norman/Panos Pictures; **560,** James L. Stanfield/National Geographic Society; **561,** Mark Schlossman/Panos Pictures; **563,** Alan Carey; **564,** Steve Raymer/National Geographic Society; **565,** Johann Van Tonder/Images of Africa Photobank; **568,** James L. Stanfield/National Geographic Society; **568-569,** (background) Aaron Haupt; **569,** (t) Emory Kristof/National Geographic Society, (b) Gordon W. Gahan/National Geographic Society; **570-571,** Paul Chesley/National Geographic Society; **572,** (l) Dean Conger/National Geographic Society, (br) George F. Mobley/National Geographic Society; **572-573,** (c) Jodi Cobb/National Geographic Society, (background) George F. Mobley/National Geographic Society; **573,** (tl) Sam Abell/National Geographic Society, (tr,cl) James L. Stanfield/National Geographic Society, (br) David Alan Harvey/National Geographic Society; **574,** Dean Conger/National Geographic Society; **575,** James L. Stanfield/National Geographic Society; **576,** Dallas & John Heaton/Westlight; **584,** (t) ©Superstock, Inc., (b) James P. Blair/National Geographic Society; **585,** ©Josef Beck/FPG International; **586,** ©N. Shah/Superstock, Inc.; **587,** George F. Mobley/National Geographic Society; **589,** (l) James L. Stanfield/National Geographic Society, (r) George F. Mobley/National Geographic Society; **591,** Steve Raymer/National Geographic Society; **594,** James L. Stanfield/National Geographic Society; **596,** George F. Mobley/National Geographic Society; **597,** Dick Durrance II/National Geographic Society; **598,** Steve Raymer/National Geographic Society; **599,** ©Naomi Duguid/Asia Access; **601,** Paul von Stroheim/Westlight; **602,** ©Travelpix/FPG International, Inc.; **604,** (l) ©Superstock, Inc., (r) James L. Stanfield/National Geographic Society; **605,** ©N. Shah/Superstock, Inc.; **608, 609,** ©Superstock, Inc.; **612,** (l) Bruce Dale/National Geographic Society, (r) ©Antoinette Jong/Superstock, Inc.; **614,** ©Superstock, Inc.; **617,** (l,r) Dean Conger/National Geographic Society; **618,** The Metropolitan Museum of Art, Edward Elliott Family Collection, Purchase, The Dillon Fund Gift, 1982.; **619,** Nik Wheeler/Westlight;